MARINE DESIGN XIII

9781138340763-1
T0376800

PROCEEDINGS OF THE 13TH INTERNATIONAL MARINE DESIGN CONFERENCE (IMDC 2018), 10–14 JUNE 2018, ESPOO, FINLAND

Marine Design XIII

Editors

Pentti Kujala & Liangliang Lu
Marine Technology, Department of Mechanical Engineering, School of Engineering, Aalto University, Finland

VOLUME 1

CRC Press
Taylor & Francis Group
Boca Raton London New York Leiden

CRC Press is an imprint of the
Taylor & Francis Group, an **informa** business

A BALKEMA BOOK

Cover photo: Meyer Turku shipyard

CRC Press/Balkema is an imprint of the Taylor & Francis Group, an informa business

© 2018 Taylor & Francis Group, London, UK

Typeset by V Publishing Solutions Pvt Ltd., Chennai, India

All rights reserved. No part of this publication or the information contained herein may be reproduced, stored in a retrieval system, or transmitted in any form or by any means, electronic, mechanical, by photocopying, recording or otherwise, without written prior permission from the publisher.

Although all care is taken to ensure integrity and the quality of this publication and the information herein, no responsibility is assumed by the publishers nor the author for any damage to the property or persons as a result of operation or use of this publication and/or the information contained herein.

Published by: CRC Press/Balkema
    Schipholweg 107C, 2316 XC Leiden, The Netherlands
    e-mail: Pub.NL@taylorandfrancis.com
    www.crcpress.com – www.taylorandfrancis.com

ISBN: 978-1-138-54187-0 (set of 2 volumes + CD in volume 1)
ISBN: 978-1-138-34069-5 (Vol 1)
ISBN: 978-1-138-34076-3 (Vol 2)
ISBN: 978-1-351-01004-7 (eBook set of 2 volumes)
ISBN: 978-0-429-44053-3 (eBook, Vol 1)
ISBN: 978-0-429-44051-9 (eBook, Vol 2)

Table of contents

Preface	xiii
Committees	xv

VOLUME 1

SoA report

State of the art report on design methodology D. Andrews, A.A. Kana, J.J. Hopman & J. Romanoff	3
State of the art report on cruise vessel design P. Rautaheimo, P. Albrecht & M. Soininen	17

Keynote paper

Disruptive market conditions require new direction for vessel design practices and tools application P.O. Brett, H.M. Gaspar, A. Ebrahimi & J.J. Garcia	31
Towards maritime data economy using digital maritime architecture T. Arola	49
Is a naval architect an atypical designer—or just a hull engineer? D. Andrews	55
New type of condensate tanker for arctic operation M. Kajosaari	77

Education

HYDRA: Multipurpose ship designs in engineering and education R.J. Pawling, R. Bilde & J. Hunt	85
Development and lessons learned of a block-based conceptual submarine design tool for graduate education A.A. Kana & E. Rotteveel	103

Design methodology

Intelligent general arrangement A. Yrjänäinen & M. Florean	113
Vessel.js: An open and collaborative ship design object-oriented library H.M. Gaspar	123
Exploring the blue skies potential of digital twin technology for a polar supply and research vessel A. Bekker	135
Combining design and strategy in offshore shipping M.A. Strøm, C.F. Rehn, S.S. Pettersen, S.O. Erikstad, B.E. Asbjørnslett & P.O. Brett	147

System engineering based design for safety and total cost of ownership 163
P. Corrignan, V. Le Diagon, N. Li, S. Torben, M. de Jongh, K.E. Holmefjord, B. Rafine, R. Le Nena, A. Guegan, L. Sagaspe & X. de Bossoreille

Optimization of ship design for life cycle operation with uncertainties 173
T. Plessas, A. Papanikolaou, S. Liu & N. Adamopoulos

Handling the path from concept to preliminary ship design 181
G. Trincas, F. Mauro, L. Braidotti & V. Bucci

A concept for collaborative and integrative process for cruise ship concept design—from vision to design by using double design spiral 193
M.L. Keiramo, E.K. Heikkilä, M.L. Jokinen & J.M. Romanoff

High-level demonstration of holistic design and optimisation process of offshore support vessel 203
M. de Jongh, K.E. Olsen, B. Berg, J.E. Jansen, S. Torben, C. Abt, G. Dimopoulos, A. Zymaris & V. Hassani

HOLISTIC ship design optimisation 215
J. Marzi, A. Papanikolaou, J. Brunswig, P. Corrignan, L. Lecointre, A. Aubert, G. Zaraphonitis & S. Harries

A methodology for the holistic, simulation driven ship design optimization under uncertainty 227
L. Nikolopoulos & E. Boulougouris

Performance analysis through fuzzy logic in set-based design 245
H. Yuan & D.J. Singer

Managing epistemic uncertainty in multi-disciplinary optimization of a planing craft 255
D. Brefort & D.J. Singer

Quantifying the effects of uncertainty in vessel design performance—a case study on factory stern trawlers 267
J.J. Garcia, P.O. Brett, A. Ebrahimi & A. Keane

The role of aesthetics in engineering design—insights gained from cross-cultural research into traditional fishing vessels in Indonesia 275
R.W. Birmingham & I. Putu Arta Wibawa

When people are the mission of a ship—design and user research in the marine industry 285
M. Ahola, P. Murto & S. Mallam

Human-centered, collaborative, field-driven design—a case study 291
E. Gernez, K. Nordby, Ø. Seim, P.O. Brett & R. Hauge

Seeing arrangements as connections: The use of networks in analysing existing and historical ship designs 307
R.J. Pawling & D.J. Andrews

Process-based analysis of arrangement aspects for configuration-driven ships 327
K. Droste, A.A. Kana & J.J. Hopman

A design space generation approach for advance design science techniques 339
J.D. Strickland, T.E. Devine & J.P. Holbert

An optimization framework for design space reduction in early-stage design under uncertainty 347
L.R. Claus & M.D. Collette

Design for resilience: Using latent capabilities to handle disruptions facing marine systems 355
S.S. Pettersen, B.E. Asbjørnslett, S.O. Erikstad & P.O. Brett

Design for agility: Enabling time-efficient changes for marine systems to enhance operational performance 367
C. Christensen, C.F. Rehn, S.O. Erikstad & B.E. Asbjørnslett

Design for Decommissioning (DfD) of offshore installations *C. Kuo & C. Campbell*	377
Understanding initial design spaces in set-based design using networks and information theory *C. Goodrum, S. Taylordean & D.J. Singer*	385

Structural design

Probabilistic assessment of combined loads for trimarans *H.C. Seyffert, A.W. Troesch, J.T. Knight & D.C. Kring*	397
Trimaran structural design procedure for a large ship *J.C. Daidola*	411
Analysis of calculation method of hull girder residual strength for cruise ship *Y. Pu & G. Shi*	421
Integrated knowledge-based system for containership lashing bridge optimization design *C. Li & D. Wang*	429
Enhanced structural design and operation of search and rescue craft *F. Prini, R.W. Birmingham, S. Benson, R.S. Dow, P.J. Sheppard, H.J. Phillips, M.C. Johnson, J.M. Varas & S. Hirdaris*	439
The anti-shock design of broadside structure based on the stress wave theory *Z.-f. Meng, J.-c. Lang, S.-b. Xu, C. Feng & P.-p. Wang*	453
Multiobjective ship structural optimization using surrogate models of an oil tanker crashworthiness *P. Prebeg, J. Andric, S. Rudan & L. Jambrecic*	459
Improved ultimate strength prediction for plating under lateral pressure *M.V. Smith, C. Szlatenyi, C. Field & J.T. Knight*	471
Experimental reproduction of ship accidents in 1:100 scale *M.A.G. Calle, P. Kujala, R.E. Oshiro & M. Alves*	479

Hydrodynamic design

Experimental validation of numerical drag prediction of novel spray deflector design *C. Wielgosz, M. Fürth, R. Datla, U. Chung, A. Rosén & J. Danielsson*	491
Experimental and numerical study of sloshing and swirling behaviors in partially loaded membrane LNG tanks *M. Arai, T. Yoshida & H. Ando*	499
A numerical trim methodology study for the Kriso container ship with bulbous bow form variation *M. Maasch, E. Shivachev, A.H. Day & O. Turan*	507
Hull form hydrodynamic design using a discrete adjoint optimization method *P. He, G. Filip, J.R.R.A. Martins & K.J. Maki*	517
Potential effect of 2nd generation intact stability criteria on future ship design process *Y. Zhou, Y. Hu & G. Zhang*	527
Operational profile based evaluation method for ship resistance at seas *P.Y. Feng, S.M. Fan, Y.S. Wu & X.Q. Xiong*	535
On the importance of service conditions and safety in ship design *R. Grin, J. Bandas, V. Ferrari, S. Rapuc & B. Abeil*	543
First principle applications to docking sequences *C. Weltzien*	555

Ship mooring design based on flexible multibody dynamics *H.W. Lee, M.I. Roh & S.H. Ham*	563

Ship concept design

Managing complexity in concept design development of cruise-exploration ships *A. Ebrahimi, P.O. Brett & J.J. Garcia*	569
Concept design considerations for the next generation of mega-ships *K.M. Tsitsilonis, F. Stefanidis, C. Mavrelos, A. Gad, M. Timmerman, D. Vassalos & P.D. Kaklis*	579
Optimization attempt of the cargo and passenger spaces onboard a ferry *P. Szymański, G. Mazerski & T. Hinz*	589
Application of a goal based approach for the optimization of contemporary ship designs *O. Lorkowski, K. Wöckner-Kluwe, J. Langheinrich, R. Nagel, H. Billerbeck & S. Krüger*	595
Parametric design and holistic optimisation of post-panamax containerships *A. Priftis, O. Turan & E. Boulougouris*	603
Optimization method for the arrangements of LNG FPSO considering stability, safety, operability, and maintainability *S.H. Lee, M.I. Roh, S.M. Lee & K.S. Kim*	613
Container ship stowage plan using steepest ascent hill climbing, genetic, and simulated annealing algorithms *M.A. Yurtseven, E. Boulougouris & O. Turan*	617
A concept study for a natural gas hydrate propulsion ship with a fresh water supply function *H.J. Kang*	625
Development and initial results of an autonomous sailing drone for oceanic research *U. Dhomé, C. Tretow, J. Kuttenkeuler, F. Wängelin, J. Fraize, M. Fürth & M. Razola*	633
Author index	645

VOLUME 2

Risk and safety

Collision accidents analysis from the viewpoint of stopping ability of ships *M. Ueno*	651
Collision risk factors analysis model for icebreaker assistance in ice-covered waters *M.Y. Zhang, D. Zhang, X.P. Yan, F. Goerlandt & P. Kujala*	659
Collision risk-based preliminary ship design—procedure and case studies *X. Tan, J. Tao, D. Konovessis & H.E. Ang*	669
Using FRAM to evaluate ship designs and regulations *D. Smith, B. Veitch, F. Khan & R. Taylor*	677
Using enterprise risk management to improve ship safety *S. Williams*	685
Using system-theoretic process analysis and event tree analysis for creation of a fault tree of blackout in the Diesel-Electric Propulsion system of a cruise ship *V. Bolbot, G. Theotokatos & D. Vassalos*	691
Safe maneuvering in adverse weather conditions *S. Krüger, H. Billerbeck & A. Lübcke*	701
Design method for efficient cross-flooding arrangements on passenger ships *P. Ruponen & A.-L. Routi*	709

Pro-active damage stability verification framework for passenger ships 719
Y. Bi & D. Vassalos

Weight and buoyancy is the foundation in design: Get it right 727
K.B. Karolius & D. Vassalos

SmartPFD: Towards an actively controlled inflatable life jacket to reduce death at sea 737
M. Fürth, K. Raleigh, T. Duong & D. Zanotto

Arctic design

Numerical simulation of interaction between two-dimensional wave and sea ice 747
W.-j. Hu, B.-y. Ni, D.-f. Han & Y.-z. Xue

Azimuthing propulsion ice clearing in full scale 757
P. Kujala, G.H. Taimuri, J. Kulovesi & P. Määttänen

Removable icebreaker bow with propulsion 769
H.K. Eronen

Azimuthing propulsor rule development for Finnish-Swedish ice class rules 777
I. Perälä, A. Kinnunen & L. Kuuliala

A method for calculating omega angle for the IACS PC rules 783
V. Valtonen

Probabilistic analysis of ice and sloping structure interaction based on ISO standard by using Monte-Carlo simulation 789
C. Sinsabvarodom, W. Chai, B.J. Leira, K.V. Høyland & A. Naess

Research on the calculation of transient torsional vibration due to ice impact on motor propulsion shafting 801
J. Li, R. Zhou & P. Liao

Simulation model of the Finnish winter navigation system 809
M. Lindeberg, P. Kujala, O.-V. Sormunen, M. Karjalainen & J. Toivola

Ice management and design philosophy 819
S. Ruud & R. Skjetne

Towards holistic performance-based conceptual design of Arctic cargo ships 831
M. Bergström, S. Hirdaris, O.A.V. Banda, P. Kujala, G. Thomas, K.-L. Choy, P. Stefenson, K. Nordby, Z. Li, J.W. Ringsberg & M. Lundh

Comparison of vessel theoretical ice speeds against AIS data in the Baltic Sea 841
O.-V. Sormunen, R. Berglund, M. Lensu, L. Kuuliala, F. Li, M. Bergström & P. Kujala

Autonomous ships

The need for systematic and systemic safety management for autonomous vessels 853
O.A.V. Banda, P. Kujala, F. Goerlandt, M. Bergström, M. Ahola, P.H.A.J.M. van Gelder & S. Sonninen

Do we know enough about the concept of unmanned ship? 861
R. Jalonen, E. Heikkilä & M. Wahlström

Towards autonomous shipping: Operational challenges of unmanned short sea cargo vessels 871
C. Kooij, M. Loonstijn, R.G. Hekkenberg & K. Visser

Towards the unmanned ship code 881
M. Bergström, S. Hirdaris, O.A. Valdez Banda, P. Kujala, O.-V. Sormunen & A. Lappalainen

Autonomous ship design method using marine traffic simulator considering autonomy levels 887
K. Hiekata, T. Mitsuyuki & K. Ito

Toward the use of big data in smart ships 897
D.G. Belanger, M. Furth, K. Jansen & L. Reichard

Simulations of autonomous ship collision avoidance system for design and evaluation 909
J. Martio, K. Happonen & H. Karvonen

Energy efficiency

Feedback to design power requirements from statistical methods applied to onboard measurements 917
T. Manderbacka & M. Haranen

Reducing GHG emissions in shipping—measures and options 923
E. Lindstad, T.I. Bø & G.S. Eskeland

Alternative fuels for shipping: A study on the evaluation of interdependent options for mutual stakeholders 931
S. Wanaka, K. Hiekata & T. Mitsuyuki

On the design of plug-in hybrid fuel cell and lithium battery propulsion systems for coastal ships 941
P. Wu & R.W.G. Bucknall

Estimation of fuel consumption using discrete-event simulation—a validation study 953
E. Sandvik, B.E. Asbjørnslett, S. Steen & T.A.V. Johnsen

Voyage performance of ship fitted with Flettner rotor 961
O. Turan, T. Cui, B. Howett & S. Day

Time based ship added resistance prediction model for biofouling 971
D. Uzun, R. Ozyurt, Y.K. Demirel & O. Turan

Hull form design

Utilizing process automation and intelligent design space exploration for simulation driven ship design 983
E.A. Arens, G. Amine-Eddine, C. Abbott, G. Bastide & T.-H. Stachowski

Smart design of hull forms through hybrid evolutionary algorithm and morphing approach 995
J.H. Ang, V.P. Jirafe, C. Goh & Y. Li

Hull form resistance performance optimization based on CFD 1007
B. Feng, H. Chang & X. Cheng

Development of an automatic hull form generation method to design specific wake field 1015
Y. Ichinose & Y. Tahara

Hull form optimization for the roll motion of a high-speed fishing vessel based on NSGA-II algorithm 1019
D. Qiao, N. Ma & X. Gu

Propulsion equipment design

The journey to new tunnel thrusters, the road so far, and what is still to come 1033
N.W.H. Bulten

Study on the hydrodynamic characteristics of an open propeller in regular head waves considering unsteady surge motion effect 1043
W. Zhang, N. Ma, C.-J. Yang & X. Gu

Application of CAESES and STARCCM + for the design of rudder bulb and thrust fins 1057
F. Yang, W. Chen, X. Yin & G. Dong

Design verification of new propulsion devices 1065
X. Shi, J.S. He, Y.H. Zhou & J. Li

Navy ships

An approach for an operational vulnerability assessment for naval ships using a Markov model *A.C. Habben Jansen, A.A. Kana & J.J. Hopman*	1073
Early stage routing of distributed ship service systems for vulnerability reduction *E.A.E. Duchateau, P. de Vos & S. van Leeuwen*	1083

Offshore and wind farms

An innovative method for the installation of offshore wind turbines *P. Bernard & K.H. Halse*	1099
Loads on the brace system of an offshore floating structure *T.P. Mazarakos, D.N. Konispoliatis & S.A. Mavrakos*	1111
Downtime analysis of FPSO *M. Fürth, J. Igbadumhe, Z.Y. Tay & B. Windén*	1121

Production

Prediction of panel distortion in a shipyard using a Bayesian network *C.M. Wincott & M.D. Collette*	1133
Author index	1141

Preface

This book collects the contributions to the 13th International Marine Design Conference, IMDC 2018, held in Espoo, Finland between 10 and 14 June 2018. This is the thirteenth in the IMDC conference series. In spring 1982, the first of the IMDC series of conferences was held in London (United Kingdom). Successive conferences were held every three years, namely 1985 in Lyngby (Denmark), 1988 in Pittsburgh (USA), 1991 in Kobe (Japan), 1994 in Delft (The Netherlands), 1997 in Newcastle (United Kingdom), 2000 in Kyongju (Korea), 2003 in Athens (Greece), 2006 in Ann Arbor-Michigan (USA), 2009 in Trondheim (Norway), 2012 in Glasgow (United Kingdom) and 2015 in Tokyo (Japan).

The aim of IMDC is to promote all aspects of marine design as an engineering discipline. The focus of this year is on the key design challenges and opportunities in the area of current maritime technologies and markets, with special emphasis on:
- Challenges in merging ship design and marine applications of experience-based industrial design
- Digitalisation as technological enabler for stronger link between efficient design, operations and maintenance in future
- Emerging technologies and their impact on future designs
- Cruise ship and icebreaker designs including fleet compositions to meet new market demands

To reflect on the conference focus, the book covers the following research topic series from worldwide academia and industry:
- State of the art ship design principles – education, design methodology, structural design, hydrodynamic design
- Cutting edge ship designs and operations – ship concept design, risk and safety, Arctic design, autonomous ships
- Energy efficiency and propulsions – energy efficiency, hull form design, propulsion equipment design
- Wider marine designs and practices – navy ships, offshore and wind farms and production

In total, the book contains 111 papers, including 2 state of the art reports related to the design methodologies and cruise ships design and 4 keynote papers related to the new direction for vessel design practices and tools, digital maritime traffic, naval ship designs and new tanker design for the Arctic.

The articles in this book were accepted after peer-review process, based on the full text of the papers. Many thanks are sincerely given to the reviewers of IMDC 2018 who helped the authors deliver better papers by providing constructive comments. Meanwhile, we also would like to thank the sponsors of IMDC 2018: ABB Marine, Aker Arctic, Arctech Helsinki shipyard, Elomatic, Meyer Turku shipyard, Royal Caribbean Cruise Ltd.

Hope the proceedings of IMDC 2018 contribute to marine design research and industry.

Pentti Kujala
Local Chairman, IMDC2018
Vice Dean, Professor, Aalto University

Committees

INTERNATIONAL COMMITTEE

David Andrews (Chairman), *Professor, University College London, United Kingdom*
Apostolos Papanikolaou, *Professor, Hamburgische Schiffbau-Versuchsanstalt GmbH, Germany*
Makoto Arai, *Professor, Yokohama National University, Japan*
Richard Birmingham, *Professor, University of Newcastle, United Kingdom*
Stein Ove Erikstad, *Professor, Norwegian University of Science and Technology, Norway*
Sheming Fan, *Professor, Marine Design and Research Institute of China, China*
Stefan Krüger, *Professor, Technical University of Hamburg, Germany*
Patrik Rautaheimo, Dr., *Managing Director, Elomatic Oy, Finland*
Hiroyuki Yamato, *Professor, The University of Tokyo, Japan*
David Singer, *Associate Professor, University of Michigan, United States of America*
Dracos Vassalos, *Professor, University of Strathclyde, United Kingdom*
Hans Hopman, *Professor, Delft University of Technology, The Netherlands*
Per Olaf Brett, Dr., *Deputy Managing Director, Ulstein International AS, Norway*
Kelly Cooper, *Program Manager, Ship Systems and Engineering Research, US Navy, United States of America*
Chris Mckesson, Dr., *University of British Columbia, Canada*

LOCAL ORGANIZING COMMITTEE (FINLAND)

Pentti Kujala (Chairman), *Professor, Aalto University*
Patrik Rautaheimo, *Managing Director, Elomatic*
Mervi Pitkänen, *Head of External Funding, Rolls-Royce*
Reko-Antti Suojanen, *Managing director, Aker Arctic*
Niko Rautiainen, *Senior Vice President, Design, Arctech Helsinki shipyard*
Riku-Pekka Hägg, *Vice-President Ship Design, Wärtsilä*
Mikko Ilus, *Head of Ship Theory, Meyer Turku shipyard*
Tommi Arola, *Head of Unit, Finnish Transport Safety Agency*
Elina Vähäheikkilä, *Secretary General, Finnish Maritime Industries*
Marjo Keiramo, *Senior Program Manager, Royal Caribbean Cruises Ltd*
Andrei Korsstrom, *Product Manager, ABB Marine*
Teemu Manderbacka, *Senior R&D Engineer, NAPA Shipping Solutions*
Jani Romanoff, *Professor, Aalto University*
Otto Sormunen, *Postdoctoral Researcher, Aalto University*
Liangliang Lu, *Doctoral Researcher, Aalto University*
Sophie Cook, *Project Manager, HRG Nordic*

ADDITIONAL AALTO TEAM

Heikki Remes, *Professor, Aalto University*
Kari Tammi, *Professor, Aalto University*
Markus Ahola, *Project Manager, Experience Platform, Aalto University*
Tommi Mikkola, *Lecturer, Aalto University*

Floris Goerlandt, *Lecturer, Aalto University*
Osiris A. Valdez Banda, *Postdoctoral Researcher, Aalto University*
Martin Bergström, *Postdoctoral Researcher, Aalto University*
Mihkel Korgesaar, *Postdoctoral Researcher, Aalto University*
Jakub Montewka, *Postdoctoral Researcher, Aalto University*
Mikko Suominen, *Doctoral Researcher, Aalto University*
Fang Li, *Doctoral Researcher, Aalto University*
Lei Du, *Doctoral Researcher, Aalto University*

SoA report

State of the art report on design methodology

David Andrews
Department of Mechanical Engineering, University College London, London, UK

A.A. Kana & J.J. Hopman
Department of Maritime and Transport Technology, Delft University of Technology, Delft, The Netherlands

Jani Romanoff
Department of Mechanical Engineering, Aalto University, Espoo, Finland

1 INTRODUCTION TO THE DESIGN METHODOLOGY STATE OF ART REPORT

1.1 Overall 2018 SoA reports summary and introduction

In the introduction to the IMDC 2015 Design Methodology State of the Art report (Andrews & Erikstad, 2015) it was remarked that that Design Methodology State of the Art report was the first DM Report since IMDC 2009. There had also been a Design for X Report edited by A. Papanikalaou and covering Safety, Performance, Artic Operations and Producability. There had also been a Design for X in 2012, which largely focused on Design for Layout and recent cooperative work by the University of Michigan, Technical University Delft and University College London. It also included an introduction to Design for X (by A. Papanikalaou) and Design for Production (particularly Operations Research in Ship Production Logistics) by F. Dong & D. Singer. There were two other SoA Reports in 2012 covering D for Safety (by D. Vassalos) and LNG Carriers (by A. Murakami & Y. Takaoko).

However it is now considered worthwhile summarising the introductory remarks made to the IMDC 2015 Design Methodology State of the Art report, where it outlined the overall history of SoA Reports to recent IMDCs, which have now become a unique feature of the IMDC series of conferences. The IMDC International Committee prior to the Sixth IMDC held at Newcastle in May 1997 decided that a new activity at that conference would be the presentation of a series of State of Art reports which would also be discussed in open plenary session and the discussions recorded and published along with the discussion on each of the presented papers in the main sessions of the Conference. The motivation behind this new feature of the 1997 IMDC was the desire to raise the status of IMDC within the marine technology field to be comparable to the long established fora dealing with marine hydrodynamics (the ITTC) and marine structures (the ISSC). It was felt by the proponents for SoA reports that production of such reports by a team of experts in each of the intended topics presented and discussed at the triannual conference, would complement the presentation and discussion of specific international research and practice in marine design provided by the normal medium of the technical conference papers. Such a set of SoA reports could, after the initial conference, where a degree of wider review and scene setting would be appropriate, then constitute a statement on the developments and current issues in the component topics in marine design that have arisen since the previous IMDC.

The 2015 Design Methodology State of the Art report then went on to outline the various SoA reports produced in 1997, 2006 and 2009. These have been on generic design and "ship" design issues as well as the design of specific types of ships/marine structures. It is suggested that the 2015 comprehensive listing of the various SoA reports from 1997 onwards is consulted for further detail on the scope covered in both respects above. The characteristic of the SoA reports since 2009 has been that the current organisation putting on that IMDC has sponsored SoA reports on specific ship design issues of particular interest to that nation. Thus for this IMDC, reflecting Finnish marine design expertise, there one other SoA report, in addition to this Design Methodology report is being presented:

• SoA Report on Cruise Vessel Design.

Coming to the current Design Methodology SoA report, this has three sets of review, rather like the two distinct sets of reviews for the IMDC 2015 SoA Report, which consisted of ten short reviews of very recent key design methodological papers of (by D. Andrews) and a separate essay like review (by S.O. Erikstad) on current design methodological developments. This was a break in format for Design Methodology State of the Art reports and it was suggested at the report's presentation at IMDC 2015 that this was an opportunity to debate how IMDC State of the Art reports should be presented in future IMDCs. With little feedback it has been decided to broadly continue with is approach, although for IMDC 2018 there are now three elements to the DM SoA Report:

- Three long reviews of three recent substantial books on generic design and scientific method from a perspective of their relevance to the design of complex marine vessels (by Professor David Andrews, UCL);
- A review of a larger number of recent ship design research activities (by J.J. Hopman & A. Kana, TU Delft);
- A very specific and more detailed item of a design methodological nature in addressing material and structural selection in early stage ship design (The example is specific to cruise ship design application but is presented as how traditionally downstream naval architectural analysis might in future be introduced into Early Stage Ship Design.) (by J. Romanoff, Aalto University).

A final remark by way of introduction (and a lead into possible review and discussion for the SoA DM Report for IMDC 2021) is to remark that a special edition of the International Journal of Maritime Engineering (and Transactions of RINA) addressing "The Sophistication of Early Stage Design of Complex Vessels" is due to be published as IMDC 2018 takes place. This special edition of IJME will not just include a substantial paper (by David Andrews) addressing this topic, but also several comments directly to the paper by eminent "ship design" practitioners but also leading researchers into marine design philosophy, methods and practice. This should be a very useful State of the art document. It was intended to be produced ahead of IMDC 2018 but is now hoped that it can be considered as a notable input to IMDC 2021, as part of the intent behind IMDC to advance understand and appreciation of the importance of the practice of marine design as a sophisticated example of engineering design.

1.2 Generic design issues

1.2.1 Introductory remarks

The three substantial book reviews below are presented as part of the IMDC Design Methodology SoA report as they follow on from the early pattern of the DM reports to consider wider design practice and recent publications of potential relevance to marine design. In some respect it might be argued by practicing ship designers, that we do not need to be aware of such developments—having enough direct engineering problems in producing something as complex as a modern ship or marine structure? However, the intent behind the inauguration of the IM(S)DC series of conferences in 1982 by Stian Erichsen and his committee of eminent ship designers was to raise the awareness of marine design to the wider maritime and design community. This was also the intent behind the IMDC SoA Reports, as commented in the overall introductory remarks to this report. The motivation to raise awareness and the intellectual rigour of marine design practice stems from a belief that (including many in the ship design community) ship design is simple and it's the applied science in naval architecture that is the intellectual challenge. IMDC denies this and advances in CAD and digital/graphical computation reinforces this belief. Thus IMDC's awareness and discussion of "the sophistication of ship design" can only help to improve ship design practice and the belief of the "marine design community" as to the rigour of our field of endeavour.

The three publications reviewed below are contrasted in that the first takes a very (excessively broad?) view of what constitutes "Design". The second as a directly philosophical work seems to then reduce its consideration to excluding engineering design (and hence marine design). The final book reviewed takes the work of Karl Popper, the founding philosopher of the Philosophy of Science and greatly admired by many practicing scientists, and with its own (philosophical) agenda is reviewed for its engineering design applicability. Thus these three somewhat broad design view are presented to encourage a wider sense of marine design intellectual position as the previous paragraph concludes.

1.2.2 Review of "The Design Way" by Harold G. Nelson and Erik Stolterman (2nd Edition M.I.T. Press 2012)

The authors have a rather expansive view of design: "When we create new things—technologies, organisations, processes, systems, environments, ways of thinking—we engage in design." They see design as having its own culture of enquiry and action. Their revised book provides a formulation of this "design culture's" fundamental core of ideas. However they see this as applicable to "an infinite variety of (so-called) design domains—not just "architecture and graphic design" but also "organisational, educational, interaction and healthcare design". Quite what the third of the last quartet of "design domains" means may be obvious to the authors, as two US senior business/public policy academics—one in public policy and the other in informatics—but seems a far cry from engineering design in general and marine design at its most complex (high) end.

Taking this very wide interpretation of design (see the contrasting distinction in Parson's "Philosophy of Design" also reviewed here), it is surprising the authors do not mention Bruce Archer's (as the first Professor of Design at the Royal College of Art (and Design)) picture of design as the "Third Culture". Archer sees this as distinct from the humanities and the sciences—and with engineering design as close to (but distinct from) the sciences, due to its artistic element, like the even more artistic fellow discipline of architecture.

Nevertheless this substantial publication is worth reading in this second greatly revised edition not just because it "helps develop a way of seeing, thinking, understanding, and acting .... (to) become more client-centred, creative, and adaptive to others' ever-changing environments". In that respect, due in large part to the new tables and diagrams, it can be seen to be providing a good set of overarching principles and check off lists to ensure all those in the design process are aware of the wider sociological aspects involved in "design" in its broadest possible sense. Thus I particularly liked the new chapters "Becoming a designer" and "Being a designer", both of which are considered in more detail below.

DETAILED COMMENTS:
Initial chapters cover both the "wicked problem" (by listing ten characteristics but not the key issue of Rittel & Webber's term, which is that finding out what is really wanted is the real problem, since "after that design is easier") and "wise action or design wisdom", which is probably a better characteristic of design than problem solving. Also Chapter 3 has become "Systemics" (changed from "Systems"), well explained by figures on "systemic stances and standpoints", "systemic categories" and "systems thinking"—covering both systems science and the wider systems approach in a 'softer' manner.

Under what most would call "needs" or less accurately "requirements", N&S call this "Desiderata" and then talk of vision rather that needs, which is consistent with their view of mankind's desired outcomes, but hardly appropriate designing even the most complex building or construction—so it seems questionable as to its relevance to Real design? Still under this heading, desiderata is seen as the desire to "create situations, systems of organisation or (at last!) concrete artefacts"—however it would seem from a Real design point of view only the latter constitutes actual design and that's done to meet a needs—however "wickedly" that is vaguely perceived. In the same chapter N&S talk of the "traditional design process, which first develops a concept and then implementation plans.. .... (with) all improvement occurs during the final redesign process." [My view is that in 50 years of ship design study and practice, I find this hard to recognise as anything like design].

The next chapter is entitled "Metaphysics" but seems to be addressing the ethical aspects (of N&S' broad view of design). Thus they consider the "evil of design" and the "splendour of design", which lead on to a discussion of "value" and "meaning" (i.e, "what is a good design") and to "timelessness" (which is a term used by Alexander (1990) in his rejection of his original hard systems approach (1970) to architectural design). [All this to me seems like skirting around the key design choice of "Style" (Andrews 2017)].

Next N&S adopt the concept of "g.o.d. ("guarantor-of-design") which is based on Churchman (1970) "G.O.D." (Destiny). Thus they see g.o.d. being the legitimacy and certainty of the designer's actions and accountability. Then the designer can avoid responsibility for "design" by choosing a method ("operant") and leaving the client to make decisions ("facilitator") with some "internal inspirations". There then seem to be several options with a conduit as a messenger for internal (designer) inspiration: "slough (ing) off" responsibility, either by "religion or administration", which seems further removed in the former case from Real design; the scientific approach; ecological sustainability; or, finally, chance (or fate – the designer "can only do so much"). So while some of this seems terribly cerebral, every designer of significant products does at times face ethical issues and the above options do reveal some get out clauses some of us have had to make recourse to, in our worse moments of design choice.

N&S go on to say "Design is about creating a new reality", which leads them on to say creativity comes down to the designer's character drawing on the designer's "values beliefs, skills, sensibility, reason, ethics and aesthetics". To me this sounds a bit like Daley's (1980) set of personal schemas (visual, verbal and values) that I have adopted in a visual representation of an integrated approach to ship synthesis (Andrews, 1986) and publicised in previous IMDC Design Methodology reports (Andrews et al, 2009, 2012). However N&S also say "Anyone can become a designer or design connected". Given the latter is a pointless truism (in that their definition of design is so broad, in a connected world of man-made artefacts, of course everyone is connected to design outcomes), their highly questionable all-encompassing view of "design", verges on being anything to anyone. If we think of design rather as actually having an end product, however broad that might be in hard and software terms, then we are back to real design. Left with N&S' too universal a stance they make design synonymous with healthy living/the good life and other social platitudes. This can be seen to be in stark contrast to Bruce Archer's (1980) more sophisticated view of design as a Third Culture alongside sciences and the humanities, and with modelling as its mode of communication (in comparison to the mathematical and linguistic modes of the other two cultures). Such a view also sees design, varying from crafts through graphic art to engineering design, the latter seen as design at its most scientific, as still about professional practice—something N&S as sociologists seem to be fundamentally uneasy with? The final remark they make on this topic is the need to "evaluate the development of design abilities by a reflective utilisation of useful schemas…", which might just enable designers to invoke Daley's schemas, to explain the human element in design synthesis.

In pulling together all the extensive scope of what N&S mean by design, there are some useful remarks on design philosophy, which can be put alongside thoughts by design philosophers that the topic is still at a very formative stage (Galle, 2002) as remarked in previous IMDC DM State of Art reports. Thus they show twelve purposes and thirteen assumptions associated with design philosophy (Figures 14.1 and 14.2). In talking of "meta-design" (i.e. understanding design at the level above direct design practice by involving clients and the environment) they make the useful point that "designers need to engage in meta-design". Given that any significant design practice involves clients and modern concerns for environmental issues is now axiomatic, this may seem another stating of the obvious. I would further argue that the issue of "constraints" (see Andrews, 2017) covers the need to be aware of the design environment not just environmental issues. In fact the whole emphasis in the IMDC SoA reports on design methodology (in the proper sense of methodology) can be seen as meta-design awareness. However it is also worth remarking that engineering designers in general are very wary of such "philosophical musings" and this could be a partial contribution to the historically poor intellectual status of engineering design alongside the engineering sciences. In teaching ship design by starting at the meta-design level (UCL MSc in Naval Architecture), I observed resistance by engineers to taking joint responsibility for requirements elucidation (Andrews 2011) with "clients", preferring to see "ship design" as starting with a specification (produced by the client or requirements owner) and thus abrogating the designer of responsibility in the "top level" design decisions.

This reduces the ship designer (and designers of other large-scale artefacts) to a mere technician or even just a CAD jockey. At least in this regard it would seem N&S generally excessive scope for design is making an important point about real design practice.

Continuing on their philosophical view N&S consider epistemology as a "reflective study of enquiry" with four possible stances:

> "the abandoned centre" leading to ever more specialised disciplines—a clear danger with the esoteric developments in (say) hydrodynamics and marine structures, which general ship designers find hard to keep up with;
> "the soft centre" with an emphasis on universal or generic truths—strongly focused on multidisciplinary issue, which are often the source of design errors;
> "the hard centre", emphasising shared principles or even laws and common curricula, where there are seen to be dangers in the rigidity of professional associations to innovation;
> "the liquid centre", which encourages mixed or enriched (even supersaturated) solutions and seen by N&S as the preferable approach avoiding reductionist, hard systems practice, so that every solution is different requiring multiple perspectives rather than integrative. The latter might be seen as an ideal in the divergent/exploratory stages of the concept phase but clearly as integration is required for synthesising large complex systems, which N&S don't really regard as their main focus.

Finally on "becoming" and "being a designer" the book has some quite insightful diagrams on "design scholarship" (namely, discovery, integration, application and teaching); "design milieu" and "design inquiry" said to lead to design as a third way contrasted to technology/applied science (presumably associated with engineering design) and craft/applied art (which is what most people think of as design?). N&S finally conclude with a figure listing seven "designer qualities" consistent with their very broad vision of design.

1.2.3 Review of "The Philosophy of Design" by Glenn Parsons, Polity Press, Cambridge, 2016

Parsons is Associate Professor of Philosophy at Ryerson University and comes to design philosophy from an aesthetics focus, stating this book "is the first introduction to the philosophy of design". This is a bold statement, given some considerable considerations of design method by both eminent philosophers of the past (such as Pierce) and more recently views on design philosophy by theorists in design journals, that have been highlighted in previous IMDC SoA reports (Andrews et al, 2006). He distinguishes between "Design", which he denotes as the practice or profession, and "design" as a general sort of "cognitive activity", which I take to be somewhat matching the extremely broad term used by Nelson and Stolterman, which the above review argues is stretching "design" way beyond a useful set of boundaries.

Parsons goes on to say he is examining "Design systematically from the perspective of contemporary philosophy", seeing the key areas as being "aesthetics, epistemology, metaphysics and ethics", which is consistent with his existing research interests. At this point, he then (bizarrely) distinguishes Design from engineering. This is done because he views engineering as concerned with wiring and plumbing systems, which shows the problem engineers have with other professions in them understanding what we do. He further compounds the crime by seeing "Design practice" as concerned with the "surface of things", which later leads him to problems with architectural design and reveals (in stark contrast to N&S) the narrowness of his boundaries of "Design", which seem to consist as largely the practice of industrial or product design. His dismissal of engineering (design) in general and ignorance of design engineering on a grand scale, such as is relevant to the design of Physically Large and Complex Systems, whether architectural or engineering (e.g. civil constructions, marine vehicles and structures, chemical plants), means this book also has to be treated circumspectly for any insights it might provide.

He tries to get round his term "surface" by saying this is more than visual with qualities such as shape and colour with "interactive dynamics" as being the way an object is used and the way it responds to use. This would seem to be very mass product focused. He goes on to say "Designer's…. view of user and components/aspects that figure in the user's relationship to the object" are contrasted to engineers who "must often focus on elements that, although vital to an object's functioning, do not figure in the user's interaction with it…" This seems a very simplistic view of design.

[The author's distinction regarding the engineer's lack of focus on "the user's interaction" clearly doesn't apply to sophisticated ship design. While the ship "engineering sciences" (such as strength and stability maybe taken for granted by the operator, seakeeping, speed (and endurance) and manoeuvrability clearly are operator priorities. This also applies to much captured by the term Style (see Andrews 2017 Table 1 for a comprehensive listing of style issues), such as human factors and ILS. However there are some more designer focused topics like margin policy and choice of design processes that are of little interest to most users, despite their clear importance to the end artefact. All this just reveals the rather narrow scope of Parsons' approach to design].

Parsons further implies (page 24) that engineers are only "Designer(s) of structures" – only concerned with "surfaces". This is ludicrous unless his view is (probably) limited to consumer durables (a common mistake if one looks a popular books on "Design", but hardly worthy of a philosophical text). To disregard the design of large-scale structures, such as in civil and maritime engineering, which are partially concerned with synthesising and analysis the performance of complex three-dimensional structures under extreme random loading is bizarre.

The author goes on to talk about the "rise of the Designer" in the early industrial revolution but then uses surface pattern in domestic ceramics to justify the limitation on "Design" to "surface design". Parsons does note the "early precursors (*of the Designer*) in ancient professions, such as architecture and shipbuilding", yet seeing "architecture as anomalous in not typically involving mass production". [Rather, I would argue Parsons is being far too narrow in not recognising the spectrum of architectural and engineering "design on a grand scale" (Fuller, 198?) or the whole nature of the design of Physically Large and Complex (PL&C) systems (Andrews, 2012)].

Interestingly, Parsons defines Design as "Design is the intentional solution of a problem, by the creation of plans for a new sort of thing, where plans would not be immediately seen, by a reasonable person, as an inadequate solution." He does this to rule out encompassing the Designer imagining (say) a "time machine" rather than actually the Designer designing such a thing, which is a sensible thing to exclude but is much broader in scope than his actual very limited bounding of design practice. In tackling the "Design Process" Parsons quotes Christopher Alexander who early on (1964), as an architectural theorist, approached the process in a somewhat reductionist manner, which he later rejected for a more romantic historical crafts approach, seeing the Designer as "bewildered, the form-maker stands alone". Thus the issue of creativity is highlighted as is his question "Are all Design problems ill-defined?" (page 32). However he then goes on to attack Rittel and Webber's widely accepted notion of the "Wicked Problem", which he sees as okay for difficult policy planning but not appropriate for Design, which is not surprising given his limited view as being the design of workspaces and furniture. However his reason doesn't really address the key point of their idea (i.e. determining what is wanted is more important/difficult than the subsequent task of technical design (Andrews 2012)). This just highlights again Parsons restricted focus on furniture, industrial design and graphic design—rather than also addressing architecture (which he acknowledges is difficult—despite so much of design theory having been written by architectural practitioners/theorists) and engineering design (especially of PL&C systems), which he excludes from his "Philosophy". Even software engineers (without the complication of large scale physicality of ships and civil engineering constructions) seem to have a more sophisticated view of design (see Brooks 2008).

Parsons looks at the Designer's Creativity seeing designs as like scientific hypotheses, but in realising Designers do not have the scientists need to test the designs, concludes "Designers do not require that sort of knowledge". While this might be the case for industrial designers (and the rest) concerned with appearance/aesthetics and marketability, engineering designers (and most architects) spend most of their time testing their designs. An engineer (especially designing PL&C systems) having created a conceptual design outline of a new artefact immediately tests its viability with engineering analysis and works it up with a constant concern for safety, balance and economy. Parsons in saying (page 46) "Designers only has an obligation to come up with designs, not justify their efficacy..." again shows he doesn't understand design of complex systems—part of synthesis is to achieve balance etc. [Thus issues which distinguish the design of PL&C systems involve choosing the decision process (to solve the "wicked problems" and work up the design); the nature of constraints (Andrews 1981); and the importance of choosing and exploring a design's "style" (Andrews 2017). All these key aspects are ignored in this very limited philosophical view of design practice].

Parsons reveals his lack of understanding of engineering design in saying (page 48) that the American industrial designer Raymond Loewy (1988) "produced designs for... aircraft, battleships...", which is as much nonsense as that Kaiser Wilhelm II produced "Pre-Dreadnought designs", which could not float. Proper design of complex engineering artefacts have to firstly be compliant with the laws of Physics—aesthetics of such PL&C systems (which was Loewy's primary contribution) is only "surface appearance" and of secondary concern in most complex engineering productions. A similar over focus on appearance results in a chapter on Modernism, with little relevance to engineering design, beyond Parsons perpetuating the form follow function myth rather than recognising selection of an overall form for a new design requires a choice of style—be that a mono or multi hull or a piece of Brutalist architecture or, even (at the product design level) a Modernist kettle. The author has a section of his Chapter 5 on "Objections to an Evolutionary Theory for artefact function", which is inevitably commodity focused. However given a lot of commercial shipping is essentially evolutionary (see Andrews (2012) Table 2), this might have some bearing beyond product industrial design.

On page 104 Parsons says "Designers' progeny leave them (unlike artists), yet the latter is debatable and the former wrong as even in the collective endeavour of ship design it is often quite clear "who was the designer" (see Brown 1983). And on the same page he then says "function always underdetermines form", which is an odd re-phasing of the "functional" belief, that many engineering designers are also mistakenly wedded to—hence approaches such as requirements engineering. This is despite the fact that it is possible to come up with an infinite set of possible forms and therefore designers actually have to exercise choice with regard to form selection. Having excluded engineering Parson then (page 105) argues electronic form is hidden (meaning in regard to his "surface packaging" limitation) and so not determined by function, despite the fact that the functioning of electronic devices, such as phones, radios, are determined by the need to function electronically through physical circuitry. Furthermore, the outward packaging must meet the user functions of holding/carrying/controlling, while still having scope for form/style selection. This doesn't seem to reflect the totality of industrial design practice?

On the next page Parsons sees ship hull design as an exception where form is determined by function, which he thinks is solely due to hydrodynamic efficiency. But of course he fails to realise there is much more to hull form selection, with stability concerns (both intact and damage states) driving the waterline beam and the displacement of that form driven by the demands of payload, structure and outfit, crew needs, fuel economy, operability, etc. This is well shown in the ostensibly hydrodynamic design of the new QUEEN ELIZABETH Class aircraft carriers (Campbell-Roddis 2017).

Parsons discussion in Chapter 6 "Function, Form and Aesthetics" ends up largely on the latter, again revealing the restricted scope of his term Design. It clearly is of importance in product industrial design where surface design (but also form) is crucial. This is what this book on Design is largely about, despite occasional comments and examples from architecture and even bridge design, despite the author stating engineering is not Design. So we still seem to be some way from a useful philosophical outline of the nature of design at its complex engineering end of the spectrum. The book proved a salutary read in showing a very narrow view of design all too common in not tackling the challenge of design at its most challenging and sophisticated—that of designing Physically Large and Complex systems, exemplified by complex vessels (Andrews, 2012).

1.2.4 Review of "Karl Popper, Science and Enlightenment" by Nicholas Maxwell, UCL Press, Sept 2017

Maxwell is Emeritus Reader in the Philosophy of Science at UCL and although his very recent book is more about the wider implication of his development of a Popperian approach to wide academic practice and, indeed, Western post-Enlightenment culture, it is worth reviewing to a marine design audience as engineering as a whole struggles to form a coherent philosophy of engineering design.

As the twentieth century philosopher, who initiated the field of the philosophy of science, Karl Popper has always had a good reception from eminent practicing scientists (e.g. Medawar (19xx) and Popper's co-author Ecceles (19xx)). This is due to his seminal idea that "falsification" (rather than induction) is the best philosophical explanation as to how science proceeds and therefore what makes it distinct from (and better than) other forms of human endeavour. That he extended his approach to the social sciences to particularly critique Plato and Marx political philosophies and Freud's psychological ideas, might lead one to consider whether his approach might also be applicable to engineering practice and particularly engineering design. Recent articles in the design research fraternity still invoke Popper as a key philosophical source ( ). In part this is because design is a much more disparate and sociological endeavour than just "applied science" and Popper's ideas are therefore seen as potentially applicable to design theory is not necessarily engineering design. One engineering design issue Popper's approach to scientific discovery does not seem to help, in comparison with say Pierce's concept of abduction (Magani 2001) is in regard to design synthesis. Popper in addressing the scientific equivalent of conceptually producing a new design, namely a new scientific hypothesis is that such creativity just "arises" in a metaphysical way—beyond scientific methods, quite unlike his belief in rational behaviour downstream in the scientific process when falsification comes into play to test the relative truth of the newly discovered scientific hypothesis.

Turning to Maxwell he is supportive of Popper's seminal contribution, however considers it needs to be extended since it is based on what Maxwell considers profound flaws in the whole Enlightenment (and subsequent academic establishment's false attachment to applying the scientific method (in a pre-Popperian manner) to society through social science (i.e. knowledge alone) rather than applying "generalised progress-achieving methods of science to social life itself" (i.e. advancing society rather than just acquiring knowledge). Much of Maxwell's book spells out both Popper's ideas and the Enlightenment's "false track" before outlining his solution in the key substantial final chapter (Chapter 10) "Karl Popper and the Enlightenment Programme", in ten sections which the rest of this review will briefly summarise before seeing what relevance this might have to design methodology.

Thus Maxwell propounds his extension of Popper's philosophy, given in his first key publication "The Logic of Scientific Discovery" (Popper, 1959) and developed in his three other main works the last of which extends this to the political and social sciences and hence the Enlightenment Programme. Thus there are seen to be three steps to "put the Enlightenment idea into practice correctly":

i. "Progress-achieving methods of science need to be identified";
ii. These need to be made applicable to "human endeavour", not just improving knowledge;
iii. Then exploited to make social progress towards "an enlightened, civilised world".

Maxwell then provides four rules which he considers an improved version of Popper's critical rationalism, which he calls "problem-solving rationality" [which sounds to an engineering designer not that far from engineering intentions]. These rules are:

1. Articulate, and improve, the problem to be solved – [sounds like tackling the "wicked problem"];
2. Propose and critically assess possible solutions – [sounds like design solutions and assessment leading hopefully to requirements elucidation, 'though Maxwell doesn't seem to go that extra step looping back to the problem(s)];
3. When necessary break down the problem further "to work gradually towards a solution to the basic problem" – [this is very like designer's breaking down the design into its component parts or discrete elements of analysis];
4. Interconnect attempts to solve basic and specialised problems (from rule 3) – [this sounds very like design integration—with, in our case of complex systems, needing to ensure balance and coherence is achieved and maintained as the design is worked up].

The rest of the chapter goes on to develop Maxwell's idea of "New Enlightenment" in three steps:

i. from falsification to aim-oriented empiricism;
ii. from critical to aim-oriented rationalism;
iii. from knowledge to wisdom.

These are spelt out in some detail followed by dealing with objections and concluding with the implications for academic enquiry, with some twenty three points to put this into practice. This is rather academia focused which is understandable for a philosopher, since this has been the main mode of that practice since Kant. However from the point of view of design methodology, what is interesting is that this reformulation of Popper's approach to make what Maxwell considers is a more practical application across the breadth of academic involvement in society's concerns, doesn't sound that radical to those involved in not just engineering but more specifically direct engineering applications and even more so those in engineering design. This latter field of endeavour whether in actual engineering practice or the academic areas of research and teaching can be seen, in Maxwell's terms, already engaged in aim-oriented practice, which further methodological considerations might just ensure is also rational.

2 DESIGN RESEARCH AND PRACTICE PERSPECTIVE

2.1 Introduction

This state of the art report was written to complement the David Andrews' Design Methodology State of the Art Report focusing on Generic Design Issues. The focus of this contribution is on design research and practice, with special attention paid to recent developments in both industry practice and academic research. The structure is designed to match the 2015 State of the Art Report (Andrews & Erikstad, 2015), where several important research topics have been identified with several view points and representative papers are discussed within each. The authors have chosen these topics as they believe they represent areas that have received considerable attention in the previous 3 years, or which show a continuation of some of the topics addressed in the previous State of the Art report. The selected topics are listed below and described in detail in their respective sections.

- Advances in complex ship design processes
- Handling uncertainty in future contexts
- Understanding emergent design failure
- Architecture of early stage distributed system design

2.2 Advances in complex ship design processes

There have been several advances in the design process of complex vessels over the last several years. This report will discuss two recent PhD projects and one continuing research area. These topics cover a novel interactive concept exploration method, one focused on controlled innovation of complex objects, and one continuing research theme on improving how designers handle the large amounts of data developed during the ship design process.

Duchateau (2016) developed an interactive evolutionary concept exploration method that assists the designer in the task of balancing customer's desires and elucidating vessel requirements. He aims to address issues related to the combinatorically large problem of generating a large set of solutions, challenges with identifying promising designs, and bridging the gap between the design space and the solution and performance space. His work is a continuation and extension of the work of van Oers (2011), which employs a packing approach to the early stage design of complex vessels. Duchateau (2016) developed a method that allows the designer to interactively adjust criteria while exploring the design space. This enables the designer to explore the space without the need for well-defined objective of "what to look for", as the designer can interactively adjust their search based on new knowledge gained throughout the process.

Van Bruinessen (2016) explored ways to improve the coevolution of various innovative solutions within a design process. He applied CK theory to his problem to help model both the concept space and the knowledge space needed to properly define the creative aspects of various design strategies. He developed a model that accounts for system-of-system interaction and individual system descriptions of Form, Characteristics, Performance, and Function. He then applied his design strategy to the development of two active ship design projects at Ulstein Design and Solutions B.V. with considerable impact.

Data-Driven Documents (D3) has been a new approach that has been pursued by Henrique Gaspar (Gaspar et al., 2014, Calleya et al., 2016) to help address the problem of efficiently understanding the large amounts of data present during ship design. It enables the use of modern visualization and interactive techniques via a JavaScript library to better inform the designer throughout the design process.

They have applied D3 to the Whole Ship Model (Calleya et al., 2016) to explore its benefits for a complicated design problem involving various emission reduction options. They were able to perform new analyses for both a single design and for a range of designs to study trends. New insights were gained through various novel visualization methods that focused on clustering, highly dimensional data, and dependencies of design variables.

2.3 Handling uncertainty in future contexts

This topic of handling uncertainty in future contexts is included in part to continue the discussion from the 2015 State of the Art Report on design methodology, and also because it continues to be an important aspect influencing major ship design projects today. Environmental regulations continue to evolve, economics and fuel prices continue to fluctuate, and technology development both influences these aspects as well as is impacted by these elements. One example of this is the impact of Emission Control Areas, changing fuel prices, and the advancements of LNG as a viable fuel. Two approaches from recently completed PhD theses are discussed which work to include these future uncertainties in ship design and decision making: one using Markov decision process, and one using stochastic optimization.

Kana (2016) employed the use Markov decision processes (MDPs) to analyze design decisions in the face of uncertain future contexts. He used two techniques to study the impact of uncertain environmental policies and economic scenarios on technology selection of shipping vessels. His work moved beyond traditional MDPs, which primarily focus on identifying the optimal decision policy through time which maximizes a reward function accounting for temporal uncertainty. Instead, the focus of his work was two-fold. First, he used Monte Carlo simulations to model the stochastics which better account for the true uncertainty in modeling future contexts. Second, he introduced a new perspective within the MDP model by employing eigenvalue and eigenvector analysis of the system. Eigenvalue analysis enabled the ability to forecast all viable life cycle decision paths without the need to recursively test all initial conditions or simulations. The process to obtain the eigenvalues and eigenvectors is described in detail in Kana & Singer (2016).

Patricksson (2016), at the Norwegian University of Science and Technology (NTNU) has pursued stochastic optimization to model this problem. Part of his work focused on the machinery selection and configuration problem, with a specific focus on regulatory compliance and a minimum cost objective. His method moves beyond traditional deterministic decision support methods to include aspects of flexibility, modularity, and robustness in the design as well as handling the variability in fuel prices. He employs a two stage optimization approach to handle both the *here-and-now* decisions, and the possible recourse actions related to machine reconfiguration.

2.4 Understanding emergent design failure

Understanding ahead of time why designs fail is a key indicator for future design success, especially during early stage design activities. Many times early stage design or team issues emerge as clear problems during detailed design and engineering. This topic has been one of the ship design problems that has been explored at the University of Michigan under Associate Professor David Singer. Two PhDs are discussed here, one on studying the impact of error propagation through the design team, and one focused on a knowledge centric perspective of design using network theory.

Strickland (2015) looked into team aspects of the design activity and explored the impacts of error variability propagation that may stem from communication or cognitive skills errors. He developed the Process Failure Estimation Technique (ProFET) to help evaluate the likelihood of a design process success. This technique has its roots in state space Stream of Variation modeling. His results show that our probabilistic intuition of how error propagates through the process is not always accurate. Higher order effects may impact the results of a team process in unforeseen ways that may challenge our natural intuition and thus impact the final design in unexpected ways.

Shields (2017) investigates emergent design failures using a Knowledge-Action-Decision Framework. He argues that sudden and unexpected cost increases and schedule delays of large acquisition programs of complex ships is not caused by physical product failure, but instead they emerge from the complexity, learning, and decision making throughout the design activity. He proposes a knowledge-centric perspective of design which can be analyzed via a network representation routed in complex systems theory. This representation helps capture the temporal path dependencies present in the knowledge structure. His results provided new insights into how knowledge structures can help identify design conditions that cause increased risk of future design failures and can help identify when decisions made by the designer may influence this risk.

2.5 Architecture of early stage distributed system design

The increasing complexity of distributed systems, especially for naval vessels, has necessitated a drive to address this during early stage design especially as it pertains to ship survivability and vulnerability. Increases in interconnected systems, higher energy requirements, and the push, in some areas, towards all electric ships has created a need to address this topic from a design perspective. Several research groups are actively pursuing research in this field, many of whom have detailed work in these 2018 IMDC proceedings.

There is a Naval International Cooperative Opportunities in Science and Technology Program (NICOP) actively researching this with partners: the University of Michigan, Delft University of Technology, University College London, and Virginia Tech. This cooperation aims to better understand the relationship between the architecture of distributed systems, the vessel layout, and its operations, all within in the context of survivability. They have developed a new framework that is specifically suited for early stage ship design which can be used to describe and analyze distributed naval ship systems. The framework decomposes the system into three separate architectures, as well as their relations. This framework and the specific architectures are described in detail in Brefort et al. (2018). The primary architectures that describe the distributed system are:

- *Physical architecture*: Spatial architecture describing the ship arrangements, and the physical attributes of components and their position in space.
- *Logical architecture*: A description of the connections between system components, from a macroscopic view, by focusing on interactions and flows exchange, and by structuring it into larger-scale modules.
- *Operational architecture*: A description of the tasks, operational elements, and information flows required to accomplish or support a war fighting function in time.

The university partners have contributed several articles to this 2018 IMDC conference which cover their technical and theoretical contributions to this project in greater detail.

There is another research group in the United States working in this area: the Electric Ship Research and Development Consortium (ESRDC). Within ESRDC, there is a smaller team working on developing software tools known as Smart Ship Systems Design (S3D), which is designed to support evaluating the performance of distributed systems during early stage design across a range of mission scenarios. They are working to develop collaborative software tools which incorporate appropriate levels of detail of distributed systems for early stage design. They are also working to incorporate a multi-discipline physics-based performance analysis that is necessary in early stage distributed system design. Their original platform was cloud-based and accessed via a web browser, while their current efforts are towards a more traditional desktop package to be used within navy laboratories. Emphasis is placed on collaboration between engineers during the design process via 3D geometrical views of the ship layout and system. A more detailed explanation of their project, including motivations, challenges, and successes can be found in Dougal & Langland (2016).

Other active work is being carried out in the Netherlands by Peter de Vos (2014) and the Netherlands Defence Material Organization (Duchateau et al., 2018). Their focus is on topology generation, component sizing estimation, and vulnerability assessment during early stage ship design using network theory and first principles.

From an industry perspective, the Society of Naval Architects and Marine Engineers (SNAME) has dedicated two special issues of their quarterly magazine *Marine Technology* on distributed systems and vulnerability respectively (Kelly 2016, 2017). Those issues present the current industry focuses in this area. The focus has been on distributed system layouts and routing, energy efficiency and storage, electrification, and increasing reliability and reducing vulnerability.

3 STATE OF THE ART ON DESIGN METHODS—EXAMPLE OF MATERIAL AND STRUCTURAL SELECTION TOOLS AT EARLY DESIGN STAGES

3.1 *Strength analysis in ship concept design*

As noted already by Brown & Andrews (1980) in the S^5-design ideology strength is still one of the most fundamental aspect of ship design. Today, when materials and production methods develop at accelerating speed, the strength assessment becomes more challenging as material and production method selection will affect the allowed strength values and this way also the concept to be evaluated. The benefits of better materials and production methods can be only utilised, if the design methods are at the same level and interlinked to realistic production quality; see for instance implementation of high strength steels to bulkhead structures and thin-deck structures as presented by Remes et al. (2013) and Lillemäe et al. (2017) for cruise ships and Fig. 1.

Today, ship geometry and topology can be so complex that simple beam theory based assessment of strength is not accurate enough to be used even at the conceptual design level (ISSC, 1997). Therefore, recently the Finite Element Method has been developed recently to the direction that allows modeling of any material and structural configuration using a single FE-mesh with equivalent shell and beam element formulations. In this formulation the homogenisation and orthotropic shell (and beam) theory is used for stiffness as proposed already by Hughes (1983) to ship design, but this is complement with localisation approaches that allow extraction of the strength from the homogenised solution (Romanoff and Varsta, 2007); these concepts are derived from scientific field of multi-scale modeling used nowadays in materials science and engineering (e.g. Miehe et al, 2002; Geers et al., 2010). In order to do this, another important aspect need to be handled that is related to the length scale interac-

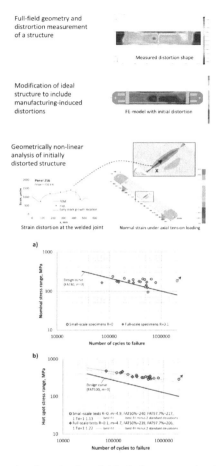

Figure 1. Introduction of thin deck structures (t = 4 mm) to cruise ships by combing production, advanced geometry measurements and Finite Element Analyses to ship design process. Figures from Lillemäe et al. (2017).

alternatives, the design selection becomes enormous challenge; see Fig. 2.

The challenge is increased by the fact that many modern structural layouts are such that the simple beam theory is not valid assumption even for the simplest modeling stages and therefore 3D Finite Element Models are needed (ISSC, 1997). Then, the starting point of the material and structural selection is 3D-model of the ship geometry in which the location of primary (e.g. bulkheads, decks) and secondary structural members (e.g. double bottom and side height, girders and webframes) are already defined by ship functions and general arrangement. This means that the geometrical reference planes of ship panels and reference lines of beam type structures are defined in ship 3D product model; see Fig. 3. Then, the structural behaviour is described based on first principles of continuum and structural mechanics, e.g. local approaches and First-Order Shear Deformation Theory for beams, plates and shells. The benefit of this approach in design is that the time-consuming modeling and post-processing stages are performed only once and those focus on integration of main structural elements and general arrangement instead of modeling that actual material and structural definitions directly to the

tion that violates the classical division of structural analysis between primary, secondary and tertiary responses. During optimisation, the computations might visit regions of design space where the two consecutive length scales are close (i.e. characteristic lengths of displacement or stress). This type of situation leads to violation of the fundamental assumptions of continuum mechanics that in turn question the validity of equivalent beam and shell theories. In next chapters we go through some of the recent developments that extend the design space of Strength in S^5-design ideology.

3.2 Material and structural selection

Today, there are over 100000 materials from which engineer can select. When this spectrum is complemented with various geometrical and topological

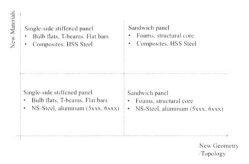

Figure 2. Increase of structural efficiency by use of new materials, geometry and topology.

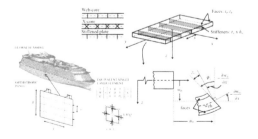

Figure 3. Modeling complex ship geometry by use of equivalent shell and beam elements. Reference planes and lines given as dashed lines.

3D-model. Instead we model only the equivalent descriptions of material and structural designs directly to the FEA input file (text file modifications). This becomes beneficial when for example structural optimisation is performed.

In this kind of approach the challenge is the local determination of strain and stress. In classical continuum mechanics the stress is defined fully by the local values of strain at a point of interest and by the relationship between stress and strain. This assumption is feasible if the two consecutive length-scales are far apart in terms of characteristic lengths defines by deformation, stress or vibration modes, i.e. $l_{higher} \gg l_{smaller}$. In structural optimisation of T-girders and structural core sandwich panels the situation might occur where "the material" scale (i.e. periodic) is visible at higher length scale. Especially at the locations of high strain gradients (e.g. pillars, hard points) this causes violation of the assumptions at the strain and stress tensor assumptions as bending effects are neglected there. Therefore, in recent years so-called non-local theories have been developed and utilised in analysis of ship structures where the stress definition at the point is affected by both strain at the point and its first derivative (e.g. Mindlin, 1963; Eringen, 1972; Reddy, 2011). The accuracy of these type of beam, plate and shell theories are superior when compared to the accuracy of classical theories. This applies to various limit states.

3.3 Limit state analysis

Limit state analysis is fundamental part of ship strength assessment. Often this is handled by splitting the analysis to serviability, fatigue, ultimate and accidental limit states. The methodology described above have been tested for all these over the last 5–10 years.

The *serviability limit state* covers prediction of stresses and deflections, but also the vibratory responses. The papers by Avi et al. (2015), Romanoff et al., (2007), Reinaldo Goncalves et al. (2016) reveal that the stress responses can be predicted very accurately at the levels of hull girder, plate-girder interaction and down to tertiary level by using the equivalent shell and beam element techniques. Also, as shown in Avi et al. (2015), Jelovica et al. (2016) and Reinaldo Goncalves et al. (2017), also the eigenfrequencies can be accurately captured by use of this type of modeling approaches.

Fatigue limit state is somewhat more challenging as this focuses on statistical effects governed by very local geometrical and material effects. It is also known that local deformation can have large effect on the localization of fatigue critical stresses (e.g. Eggert et al. 2012) and as ships contain hundreds of kilometers of welds the actual shape is enormous challenge to be included into design process. As shown by Lillemäe et al. (2017) initial distortions are crucial when local analysis is to be done

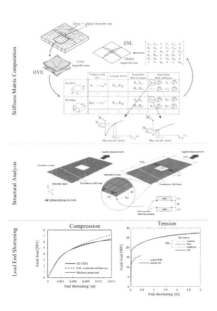

Figure 4. Analysis of different limit states by use of equivalent shell and beam elements.

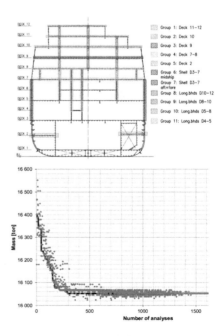

Figure 5. Analysis of different limit states by use of equivalent shell and beam elements. Optimization of a main frame of a passenger ferry. (Raikunen, 2016).

accurately and if this can be properly modelled, then accuracy of present design methods is very good. This means that the steel designers must have knowledge of the production-induced initial geometry. In practice this is done by databases of produced and measured geometries, accurate Finite Element Analyses and increased quality control.

Ultimate limit state refers here to buckling failure under compressive loads and ductile fracture under tensile loads. In case of ductile fracture the key question is the up-scaling of fracture strain and progressive fracture from material to structural scale and futher to the level of hull girder. In Körgesaar (2016) this process is described and excellent agreement between detailed 3D failure analysis and simplified equivalent analysis is demonstrated. In buckling analysis the panel level failure is presented in Reinaldo Goncalves et al. (2017) where the initial deformations due to production are included into the model at two consecutive length scales. These two types of sub-problems could be coupled to handle the *Accidental Limit State*. This extension work is still ongoing. However, as shown by Körgesaar and Ehlers (2010) this work practically can be handled only to feasible design by use of parametric Finite Element Modeling in which the otherwise feasible structure (i.e. serviceability, fatigue and ultimate limit states) is directly meshed for crashworthiness analysis with detailed FE-mesh of the actual 3D-topology.

3.4 *Optimisation*

The benefit of the homogenization is that the pre-processing of large complex structure in terms of time-consuming mesh generation must be done only once; see Fig. 3. The change of scantlings affects only the equivalent stiffness properties and therefore the mesh itself remains unaffected. As the stiffness matrix and localization scheme can be computed using only the geometry and material properties of the structure, this allows the ease to perform structural optimization. Raikunen (2016) performed optimization of a passenger ferry using the ship global 3D FE-model and ESL approach (see Figure 10). This structural analysis approach was coupled with particle swarm optimization (PSO) code capable of searching effectively global optimums with only 291 structural analyses with weight reduction from 16560 tons to 16050 tons (Raikunen, 2016), that is 3% weight reduction. It should be also mentioned, that the computational savings in analysis are enormous especially in the ultimate strength analysis as the solution with ESL requires significantly less iterations than full 3D analysis. This computational saving is due to the fact that local failure defines the required times step in FEA. Due to this the computational speed at the moment of local failure; this relates to characteristic lengths of buckling, i.e. $l_{secondary} \gg l_{tertiary}$, where the effect of tertiary buckling length is included to pre-computed load-end-shortening curve. A case study by Metsälä (2016) on simple tanker type box-beam shows 64 times less memory requirement than 3D FEA and the analysis is carried out in minutes to hours rather than in days (in case of 3D-FEA).

REFERENCES

Andrews, D., & Erikstad, S.O. 2015. The Design Methodology State of the Art Report. 12th International Marine Design Conference 2015. Tokyo, Japan. Vol 1: 90–105.

Andrews, D, Papanikolaou, A, & Singer, D. (2012). Design for X. State of Art Report presented at IMDC12 – The 11th International Marine Design Conference, Glasgow, Scotland.

Andrews, D. (1981), Creative Ship Design, Trans RINA, 1981.

Andrews, D. (1986), An Integrated Approach to Ship Synthesis, Trans RINA, 1986.

Andrews, D. (2011), Marine Requirements Elucidation and the Nature of Preliminary Ship Design, IJME Vol. 153 Part A1 2011.

Andrews, D. (2012), Art and Science in the Design of Physically Large and Complex Systems, Proc Royal Society, April 2012.

Andrews, D. (2017), The Key to Ship Design – Choosing the Style of a New Design, COMPIT, Cardiff, May 2017.

Andrews, D. (2018),The Sophistication of Early Stage Design of Complex Vessels. Special Edition IJME Part A June 2018.

Andrews, D., et al (2009, The Design Methodology State of the Art Report. 10th International Marine Design Conference. Trondheim, June 2009.

Avi, E., Lillemäe, I., Niemelä, A. and Romanoff, J. (2015), Equivalent Shell Element for Ship Structural Design, Ships and Offshore Structures, Vol. 10, No. 3, pp. 239–255.

Brefort, D., Shields, C. Habben Jansen, A., Duchateau, E., Pawling, R., Droste, K., Jaspers, T., Sypniewski, M., Goodrum, C., Parsons, M.A., Kara, M.Y., Roth, J., Singer, D.J., Andrews, D., Hopman, H., Brown, A., Kana, A.A. 2018. An Architectural framework for distributed naval ship systems. Ocean Engineering 147: 375–385.

Brooks, F.P. (2010), The Design of Design – Essays from a Computer Scientist, Addison-Wesley, Upper Saddle River, NJ, 2010.

Brown, D.K. (1983), A Century of Naval Construction, Conway Maritime Press, London, 1983.

Brown, D.K. & Andrews, D.J. (1980), The Design of Cheap Warships, Proc. of International Naval Technology Expo 80, Rotterdam, June 1980. (Reprinted in Journal of Naval Science April 1981).

Callya, J., Pawling, R., Ryan, C., Gaspar, H. 2016. Using Data Driven Document (D3) to explore a whole ship model. Proc. 11th System of Systems Engineering Conference (SoSE 2016). June 12–15 Kongsberg, Norway.

De Vos, P. 2014. On the application of network theory in naval engineering – Generating network topologies. Proc. of 12th International Naval Engineering Conference (INEC2014), Amsterdam, IMarEST, London.

Dougal, R.A., & Langland, B. 2016. Catching it early: modeling and simulating distributed systems in early stage design. Marine Technology (MT). January. SNAME, Alexandria, VA, USA.

Duchateau, E. 2016. Interactive evolutionary concept exploration in preliminary ship design. PhD Thesis, TU Delft.

Duchateau, E.A.E., de Vos, P., van Leeuwen, S. 2018. Early stage routing of distributed ship service systems for vulnerability reduction. 13th International Marine Design Conference 2018. Helsinki, Finland.

Eggert, L., Fricke, W., Paetzhold, H. (2012), Fatigue strength of thin-plated block joints with typical shipbuilding imperfections, Welding in the World. Vol. 56 No. 11-12, pp. 119–128.

Eringen, A.C. (1972), Nonlocal Polar Elastic Continua, International Journal of Engineering Sciences. Vol. 10 No. 1, pp. 1–16.

Gaspar, H.M, Brett, P.O., Ebrahim, A., Keane, A. 2014. Data-Driven Documents (D3) Applied to Conceptual Ship Design Knowledge. 13th Conference on Computer and IT Applications in the Maritime Industries (COMPIT'14). May 12–14, Redworth, UK.

Hughes, O. (1983) Ship structural design: a rationally-based, computer-aided optimization approach. New York: Wiley, 566 p.

ISSC (1997), Committee II.1 – quasi-static response. Proceedings of the 13th International Ship and Offshore Structures Congress; 1997 Aug 18–22; Trondheim, Norway. pp. 158–165. Oxford (England): Elsevier Science.

Jelovica, J., Romanoff, J. and Klein, R., (2016), Eigenfrequency Analyses of Laser-Welded Web-core Sandwich Panels, Thin-Walled Structures, Vol. 101, pp. 120–128.

Kana, A.A. 2016. Enabling decision insight by applying Monte Carlo simulations and eigenvalue spectral analysis to the ship-centric Markov decision process framework. PhD Thesis, University of Michigan.

Kana, A.A., & Singer, D.J. 2016. A ship egress analysis method using spectral Markov decision processes. 13th International Symposium on Practical Design of Ships and Other Floating Structures – PRADS'2016. September 4–8, Copenhagen, Denmark. ISBN: 978-87-7475-473-2.

Kelly, D.R. 2016. Power trip: designing more efficient distributed systems [special issue]. Marine Technology (mt). January. SNAME, Alexandria, VA, USA.

Kelly, D.R. 2017. Vulnerability. Survivability. Recoverability. [special issue]. Marine Technology (mt). July. SNAME, Alexandria, VA, USA.

Körgesaar, M. and Ehlers, S. (2010) An Assessment Procedure of the Crashworthiness of an LNG Tanker Side Structure, Ship Technology Research, Vol. 57, No. 1, pp. 50–64.

Körgesaar, M., Reinaldo Goncalves, B., Romanoff, J. and Remes, H. (2016), Behaviour of orthotropic web-core steel sandwich panels under multi-axial tension, International Journal of Mechanical Sciences. Vol. 115–116, pp. 428–437. DOI: 10.1016/j.ijmecsci.2016.07.021.

Maxwell, N. 2017, Karl Popper, Science and Enlightenment, UCL Press, London, Sept 2017.

Metsälä, M. (2016), Geometrically Nonlinear Bending Response of Steel Sandwich Box Girder Using Equivalent Single Layer Theory, M.Sc. Thesis. Aalto University, School of Engineering. 2016.

Mindlin, R.D. (1963), Micro-structure in linear Elasticity, Department of Mechanical Engineering, University of Brittish Columbia.

Nelson, H.G. & Stolterman, E. 2012 "the design way", 2nd Edition, M.I.T. Press, Cambridge, MA 2012.

Parsons, G. 2016, The Philosophy of Design, Polity Press, Cambridge UP, 2016.

Patricksson, O.S. 2016. Decision support for conceptual ship design with focus on a changing life cycle and future uncertainty. PhD Thesis, Norwegian University of Science and Technology.

Popper, K., (1959), The Logic of Scientific Discovery, Routledge, London, 1959.

Raikunen Joni (2015), Optimization approach for passenger ship structures using Finite Element Method, Aalto University, School of Engineering.

Reddy, JN. (2011), Microstructure-dependent couple stress theories of functionally graded beams, Journal of the Mechanics and Physics of Solids. Vol. 59, pp. 2382–2399. DOI: 10.1016/j.jmps.2011.06.008.

Reinaldo Goncalves, B., Jelovica, J. and Romanoff, J. (2016), A homogenization method for geometric nonlinear analysis of sandwich structures with initial imperfections, International Journal of Solids and Structures. Vol. 87 No. 1 pp. 194–205. DOI: 10.1016/j.ijsolstr.2016.02.009.

Reinaldo Goncalves, B., Karttunen, A., Romanoff, J. and Reddy, JN (2017), Buckling and free vibration of shear-flexible sandwich beams using a couple-stress-based finite element, Composite Structures, Vol. 165, pp. 233–241.

Remes, H., Korhonen, E., Lehto, P., Romanoff, J., Niemelä, A., Hiltunen, P. and Kontkanen, T. (2013), Influence of surface integrity on the fatigue strength of high- strength steels, Journal of Constructional Steel Research Vol. 89, pp. 21–29.

Rittel, H.M.J, & Webber, M.W. (1973), Dilemmas in the general theory of planning policy sciences, Policy sciences, Vol. 4, 1973.

Romanoff, J. and Varsta, P. (2007), Bending response of web-core sandwich plates, Composite Structures. Vol. 81 No. 2, pp. 292–302. DOI: 10.1016/j.compstruct.2006.08.021.

Romanoff, J., Varsta, P. and Remes, H. (2007), Laser-Welded Web-Core Sandwich Plates under Patch-Loading, Marine Structures, Vol. 20, No. 1–2, 2007, pp. 25–48.

Shields, C.P.F. 2017. Investigating emergent design failures using a knowledge-action-decision framework. PhD Thesis, University of Michigan.

Strickland, J. 2015. A design process centric application of state space modeling as a function of communication and cognitive skills assessment. PhD Thesis, University of Michigan.

Van Bruinessen, T. 2016. Towards controlled innovation of complex objects. PhD Thesis, TU Delft.

Van Oers, B.J. 2011. A packing approach for the early stage design of service vessels. PhD thesis, TU Delft.

State of the art report on cruise vessel design

Patrik Rautaheimo, Peter Albrecht & Mikko Soininen
Elomatic Oy, Turku, Finland

1 INTRODUCTION

The development of the cruise vessel fleet and the newbuilding orderbook has been record fast in the past few years including many new designs. The orderbook includes large cruise vessels as well as niche market cruise vessels. The International Maritime Organisation (IMO) has adopted new regulations for safety and environmental issues, which must be implemented both for the newbuilding fleet and partly also for the existing fleet. The designers must understand the requirements of the new regulations, how to implement the requirements into the design, bearing in mind the economical and operational requirements of the shipowner. A good, previous state of arts for cruise ships have been done e.g. by Levander (1991, 2004) and this documents concentrates mainly on the very recent development.

2 CRUISE VESSEL ORDERBOOK

The record order book for cruise vessels comprises more than 90 vessels and includes more new designs than ever, totally 29 prototypes (Fig. 1). The last delivery is in 2026, 8 years from today.

All the big players have very large cruise vessels under construction reaching from 150 000 GT to 250 000 GT. The average size of all large cruise vessels has been growing linearly over the years (Fig. 2).

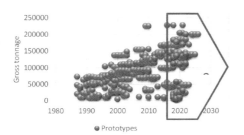

Figure 1. Cruise vessel deliveries, current orderbook and prototypes.

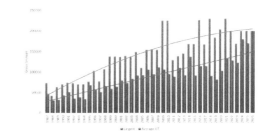

Figure 2. Development of size of cruise vessels.

3 A ROUGH GRADING OF THE CURRENT CRUISE VESSEL MARKET

The cruise market can technically be divided into four segments according to the space ratio (Gross tonnage/Double occupancy = GT/PAX) (Fig. 3) and PAX/Crew ratio (Fig. 4).

3.1 Mass market segment

In Berlitz's star rating system the mass-market cruise ships attain most often 2.5–3.5 stars of five. But some have gained even rating of 4–. The passenger capacity in mass-market ships is today typically >2000 passengers. 20 years ago, average size of the cruise ship was 50–60000 GT with 1500...2000 passengers. At that time the biggest cruise ship, the Voyager of the Seas, reached about 138000 GT with 3480 passengers and the crew of 1181.

In mass-market ships the space ratio of GT/PAX is around 40 (Fig. 3). The PAX/crew ratio varies typically between 2 and 3.5 depending on the shipping company (Fig 4). Today 150000...200000 GT cruise ships are not anymore isolated orders but common in mass-market ships. The largest cruise ships of today are the Oasis-class ships with 227000 GT and 6780 passengers.

Mass-market ships are more prone to be family oriented than luxury/premium or expedition segment ships. Adventurous pass-time features are common in these ships such as large pool areas with water slides, climbing walls, zip-lines etc.

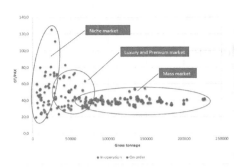

Figure 3. Space ratio.

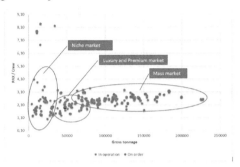

Figure 4. Passenger/crew ratio.

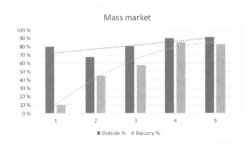

Figure 5. Increase of outside and balcony cabins in the mass-market segment.

Figure 6. Adventure of the Seas—Royal Caribbean (MAS).

Figure 7. Norwegian Star—NCL (MAS).

Figure 8. Allure of the Seas—Royal Caribbean (RCCL).

Figure 9. Artania ex Royal Princess—Phonix Cruises (MAS).

Figure 10. MSC Poesia—MSC Cruises (MAS).

3.2 *Luxury and premium segment*

The Luxury and Premium segments cannot clearly be separated from each other. In Berlitz's rating system premium cruise ships attain better than 4-stars and luxury cruise ships attain better than

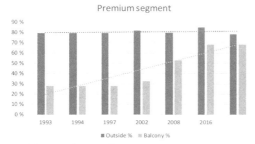

Figure 11. Increase of outside and balcony cabins in the luxury and premium segment.

Figure 12. Seabourn Quest—Seabourn Cruises (MAS).

Figure 13. Crystal Symphony—Crystal Cruises (MAS).

Figure 14. Marina—Oceania (MAS).

Figure 15. Rotterdam—Holland America Line (MAS).

5-stars. One difference between premium and luxury is that in the Luxury segment all staterooms are outside or balcony staterooms. Premium ships may also have some inside rooms.

The outside and the balcony ratio have increased during the years as shown in Fig. 11.

In the Luxury and Premium segments the feeling of space and immaculate service are musts. GT/PAX and PAX/crew ratios are higher in the Luxury segment than in the Premium segment. Typically the GT/PAX ratio is above 40 in premium ships while in luxury ships it can be up to 80 (Fig. 3). A luxury cruise ship with 100 GT/pax is on order. Pax/crew ratio in premium ships is about 2, and in luxury ships it is well below 2, typically between 1.5…1.9 (Fig. 4).

Some of these ships are 'adults-only' ships meaning that no facilities for children or teens are available. However, most of these ships have some spaces and personnel dedicated for children and teens. Spa and fitness areas are relatively larger than in mass-market ships. Phenomena of "deck chair hogs" are practically unknown on these ships.

3.3 *Niche market/expedition segment*

Expedition ships are high valued in Berlitz's star rating system. Variation is, however, large from abt. 3 to 5+. That is why some expedition cruise ships are also in the luxury/premium category. The

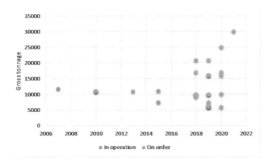

Figure 16. There is a boom in the Niche cruise vessel market.

Figure 17. Le Boreal (Fincantieri).

size of the expedition ships is between 5 000…30 000 GT and generally have a range of 75–250 passengers. PAX/crew ratio can vary in large scale between 1.5 up to 8.

Expedition ships have typically extra storage spaces for zodiacs and their tendering spaces. Also some spaces are needed for sporting and diving equipment, bicycles etc. A space where all passengers can have a seat for briefing of coming shore excursions exists. Otherwise, choices for different dining venues are often limited and only basic features for spa and fitness are available. The larger expedition ships do not have these limitations.

Figure 18. Fram (Fincantieri).

There has been an increase in the number of newbuilding orders for exhibition and polar cruise vessels. These vessels are purpose-built for demanding operations in the Arctic and Antarctic areas. During the last 10 years, only a few purpose built expedition cruise vessels have been constructed, most of the vessels in operation have been converted from other vessels, while the current orderbook comprises more than 20 niche cruise vessels (Fig. 16). The vessels carry a limited group of passengers who want to experience the nature or exotic ports where bigger vessels cannot enter.

So far, only one residence ship exists, The World. In this ship, staterooms are apartments varying from studio staterooms up to three bedroom and three bathroom apartments. Every apartment has its own galley. Apartment shareholders have meetings where they agree on the ship's itinerary. Projects for residence ships exist but none of them has so far materialized. Also cruise ships with apartments have been designed and are entering the market.

Figure 19. The World—Residences at Sea (MAS).

and fill up the ship. When passengers on top of the ticket price spend money onboard for wine, drinks and special dining, spa and health services, shopping, in casinos and shore excursions, significant additional turnover is created.

4 ATTRACTIONS ONBOARD

The passenger spaces are the parts of the ship that create the turnover for the shipowner. In addition to the cabins there must be a variety of passenger amenities and activities to attract the passengers

4.1 *Food and beverage*

Main dining rooms with fixed seating are no more so important today but they still exist. Instead a variety of small restaurants are gaining popularity. Main dining rooms (MDR) are used practically

Figure 20. MDR—Oasis of the Seas (RCCL).

Figure 21. MDR entrance—Crystal Serenity (MAS).

Figure 22. Teppanyaki – (Carnival Cruise).

Figure 23. Sushi Bar—Crystal Serenity (MAS).

Figure 24. Italian restaurant—Crystal Serenity (MAS).

Figure 25. Asian restaurant—Crystal Serenity (MAS).

on every ship as a main dining venue and they are more or less used like restaurants onshore. However, MDRs can have separate sections for fixed seating and open seating. Depending on size of the ship number of seats in one deck MDR can vary between 100…600 seats, and if MDR extends through several decks number of seats can be up to 3000 seats. Total number of seats in MDR are generally about half of the total number of passenger capacity.

In MDR themed food is served during some evenings. These themes can be from general food style to specialities at visited ports.

As special or alternative dining, more Asiatic restaurants, like Teppanyaki, Sushi Bars and restaurants dedicated for Asiatic food have been available on western markets. These restaurants are musts in Asian markets.

Other themed alternative restaurants are like Italian (considered luxurious in Asian countries), French cuisine or beef venues. More display kitchens have been utilized. Most commonly number of seats in alternative restaurants are from 30 upwards. Some special restaurants have also onshore company name and license. Vintage restaurants where number of seats are between 10…30 are coming more common.

Snack bars or trademarked and franchised fast food restaurants. The same trademarks that are familiar on-shore have also taken their root on-board.

4.2 Theatres and casinos

Theatre or Show Lounge is for large production shows and lectures. These lounges are for few hundreds of passengers up to multi-deck theatres for

Figure 26. Bistro—Crystal Symphony (MAS).

Figure 27. Lido Outside—Crystal Symphony (MAS).

Figure 28. Pool area—Crystal Serenity (MAS).

Figure 29. Lifting pod (North Star®).

more than thousand viewers. Sound systems are usually equipped with the latest sound and lighting technics and professional stage managers operate them.

Practically all cruise ships have casinos. Their size are heavily depending on demographic of cruise guests. For instance, casinos are very important and large in Asian mass-market, but in Japan casinos are not very important.

4.3 Different venues and shops

"Automated bartenders", "Bionic Bar", robots as bartenders. More small bars and venues with themed purpose are available like British Pub, Jazz Club etc.

Every cruise ship has a space that could be named as an Observation Lounge. These lounges are on the top of superstructure facing to forward. It is important to have observation deck outside where cruise guests see forward. This is especially important for expedition cruise ships.

4.4 Open decks

Lido comes originally from Italian where it means beach or beach area. On Italian liners pool areas were called Lido or Lido Deck and that naming has been adopted commonly to pool areas on cruise ships. This name has also been adopted to

Figure 30. Aqua Theater—Royal caribbean cruise line.

Figure 31. Genting Dream Sports deck (Meyer Werft).

Figure 32. Inside stateroom—Norwegian Breakaway (NCL).

Figure 33. Outside—Eurodam (Holland America Line).

Figure 34. Balcony—Crystal Serenity (MAS).

restaurant and outside dining venues adjacent to pool areas.

New spectacular outdoor attractions have been introduced especially on the mass-market cruise vessels, such as the Aqua theater on Oasis of the Seas, SkyRide and Waterworks on Carnival Vista, Aqua Park and Outdoor Sports deck on Genting Dream, North Star® lifting pod on Quantum of the Seas.

4.5 Staterooms

Area of the standard staterooms in mass-market—inside, outside or with balcony—vary between 14 m^2 and 19 m^2 without balconies. As a rule of thumb, the older the ship, the smaller the stateroom. In premium and luxury segment area of a standard stateroom varies between 18 m^2 to 30 m^2. Equipment in standard staterooms are very alike, but surfaces and services vary. A larger stateroom allows larger bathrooms with both shower and bathtub, walk-in-closets and overall more spacious living area. Queen size bed is standard but larger staterooms can have king size bed. Standard staterooms are generally between one web frame long which is abt. 2.6...3.0 m.

Two cruise guests occupy staterooms usually. In some staterooms sofa is convertible to an extra bed. In addition, some staterooms can be connected for families or groups. Pullman beds are some- times used in mass-market ships, but in premium and luxury segment they do not exist. If only one cruise guest occupies a stateroom, extra charges that varies from +10% up to +100% depending on stateroom grade or season are applied. That is why

Figure 35. Studio stateroom for solo cruise guest—Norwegian Breakaway (NCL).

Figure 36. Penthouse—Crystal Serenity (MAS).

Figure 37. Haven—Norwegian Bliss (NCL).

Figure 38. Owner's Suite—Riviera (Oceania).

Figure 39. Loft Suite—Oasis of the Seas (RCCL).

Figure 40. Crystal Penthouse—Crystal Serenity (Crystal Cruises).

some operators have special cabins for solo cruise guests and those cabins are about 10…12 m².

Suites have different names: Mini-suite, suite, penthouse, penthouse suite etc. Anyway, these staterooms are larger than standard staterooms. Their depth is usually the same as standard staterooms, but their length are 1.5…2 times longer.

Finally, depending on the ship, there is at least one large suite with an area above 60 m² and up to 150 m². There are also loft suites that extend two decks. Common features for these large suits are one or two separated bedrooms with own bath facilities, very large living space including own dining area.

5 SAFETY AND ENVIRONMENT

5.1 *Safe return to port*

The Safe Return to Port regulation has been in force since 1 July 2010 and has been adopted for all the large cruise vessels constructed lately.

The basic principle with the regulation is that the ship is the safest place in case the casualty does not exceed a defined casualty threshold. Safe areas shall be available for the passengers and crew, and the ship shall be able to proceed to a safe port with own propulsion.

This regulation covers most cruise vessels, only the smallest cruise vessels with a length less than 120 m and less than 3 main vertical fire zones are ruled out. For the smaller vessels for which the Safe Return to Port regulations must be applied, the size is the biggest challenge. The requirements are the same as for the large vessels, but there is less space available. For the Expedition vessels the range and time at sea is even more demanding.

Bearing in mind the huge number of persons onboard, an evacuation procedure is always a challenge and through the Safe Return to Port philosophy and the increased safety policy of the shipowners, the safety of the vessels has improved considerably.

5.2 Polar code

IMO has adopted the International Code for Ships Operating in Polar Waters (Polar Code) and related amendments. The code covers requirements for the design, construction, equipment, operation, training, search and rescue and environmental protection matters (Fig. 42). The Polar Code entered into force on 1 January 2017 for new ships and 1 January 2018 for existing ships.

Cruise vessels concerned by the Polar code are mainly the small expedition cruise ships intended for Polar waters. The Code affects the whole ship design, and the requirements for the structures and safety equipment onboard dramatically increases the size and weight of the vessels compared to expedition vessels without Polar Code requirements. No ship has been certified up to date, so the design of the new ships is a learning process of understanding and interpreting the requirements of the code.

5.3 Environmental footprint and how to reduce it

The rapidly expanding cruise industry puts a lot of effort into reducing the environmental footprint. The industry's total footprint should grow as little as possible. Emissions to the air and sea are main topics. Vessels entering service today will have an expected operational lifetime of 30 to 40 years and it is unlikely that all equipment can be retrofitted when new technologies become available. Therefore, it is important to select and install equipment and

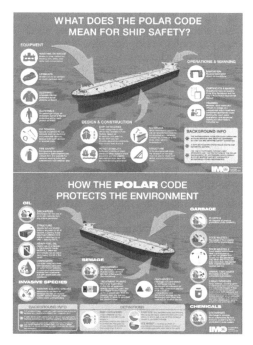

Figure 42. The impact of the Polar code for ship safety and the environment.

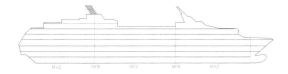

Figure 41. Safe return to port must be applied for ships with L > 120 m or 3 or more main fire zones.

systems that reduce the environmental footprint to meet both todays and expected future requirements.

For the existing fleet, the requirements to reduce exhaust gas emissions and discharge of black and grey water as well as the requirements for ballast water treatment has resulted in extensive retrofit installations, which will continue for some years ahead.

Much effort is invested in reducing energy consumption onboard as it both decreases the fuel bill as well as reduces the emissions. However, despite the reduced absolute amount of emissions, requirements can be met only by using alternative fuels or by cleaning the exhaust gases.

There are many new techniques available and under development which can contribute to the powering of the vessel such as:

– continuous improvement of the hydrodynamics (hull form and appendages)
– air lubrication
– rotor sails
– generation of electricity from waste heat
– fuel cells
– energy storage
– hybrid propulsion, eg battery power

Batteries are used to store power when excess power is available and use it to smooth the peak loads instead of starting an additional diesel engine. Battery power may also come to question for small cruise vessels and especially for cruise vessels operating in sensitive areas such as the Antarcitc and Arctic. With today's technology, batteries can be used for a short period when entering areas where silent and emission free sailing is important e.g. for experiencing the nature.

5.4 *The 2020 global sulphur limit*

IMO has set a global limit for fuel in oil used on board ships of 0.50% m/m (mass by mass) from 1 January 2020. To meet this requirement the ships have various alternatives:

– use low-sulphur compliant fuel (MGO, Low Sulphur fuel oil)
– use alternative liquid fuel such as methanol
– use gas as fuel (LNG)
– use exhaust gas cleaning systems (scrubbers) to clean the emissions before releasing to the atmosphere

5.5 *Low-sulphur compliant fuel*

All cruise vessels can use low-sulphur compliant fuels without considerable modifications. However, this will cause additional fuel costs and the majority of the shipowners look for alternative solutions.

IMO prohibits heavy grade oils to be carried onboard in the Polar areas and thus expedition vessels generally are designed for low-sulphur compliant fuel like MGO.

5.6 *Exhaust gas cleaning systems*

Marine exhaust gas cleaning systems enable the ship operator to run on cost-efficient high Sulphur fuel and still be compliant with the 2020 0.5% SOx cap as well as the 0.1% SOx cap in ECA. However, the scrubbers will cause additional investment costs, take space onboard, increase energy consumption and generate maintenance work. The extra expenditures will with today's fuel prices return with cost savings in a couple of years.

In order to comply with the limit, several cruise companies have installed scrubber technology, which uses seawater to wash the exhausts, on their ships, and new installations are in the pipeline. Most of the cruise vessels on order, which will not use LNG or MGO as main fuel, will have exhaust gas scrubbers installed and are thus able to use cheaper high-sulphur fuel.

6 LNG

In order to reduce the emissions and meet the coming environmental regulations, LNG has been an interesting option and recently selected as the main fuel for several of the cruise vessels on order.

The International Code of Safety for Ships using Gases or other Low-flashpoint Fuels (IGF Code) was adopted on 11th June 2015 and made it easier to design the LNG containment system and machinery onboard.

Space requirements for LNG storage are greater than for conventional fuel, both due to the density and calorific value of the gas as well as due to the safety distances to the ship's hull defined in IGF code. Thus, the capacity of the fuel tanks and the operational range on LNG is less than for the traditional HFO vessels. The bunkering frequency has to be increased. With dual fuel engines installed, the range can be extended using low Sulphur compliant fuel.

The first cruise vessels fueled with LNG is the Mitsubishi built vessels AIDAprima introduced in 2016. AIDAprima and her sister vessel AIDAperla use LNG while in port.

There are totally 16 vessels under construction to be powered with LNG, Eight for Carnival Brands, three for Disney and two for Royal Caribbean at Meyer shipyards in Papenburg and Turku. In addition two MSC vessels at STX France

Figure 43. Installation of the exhaust gas scrubber on Allure of the Seas.

and one Polar Expedition Icebreaker at Vard in Norway will be equipped with LNG machinery.

6.1 Ballast water treatment

The Ballast Water Management Convention (BWM) entered into force 8 September 2017. According to the convention, all ships in international traffic are required to manage their ballast water and sediments to a certain standard, according to a ship-specific ballast water management plan.

There are two ballast water management standards, D-1 and D-2. The D-1 standard requires ships to exchange their ballast water in open seas, 200 nautical miles from land and in at least 200 m deep waters. The D-2 standard specifies the maximum number of viable organisms that may be discharged.

Ballast water is used for trimming, heeling and stability control of the ship. Passenger vessels have low-volume ballast requirements, typically 4000 m^3 in 10–20 tanks for a large cruise vessel, but they are subject to the same installation complexities and compliance. It is a true challenge to design a ballast free vessel and, in practice, all new ships must be equipped with a ballast water management system, which meets the D-2 standard.

Existing vessels must comply with the D-1 ballast water exchange standard until a ballast water management system is installed. The latest installation is in time connected to the next IOPP renewal survey, which is phased in over time so that all vessels will comply on 8 September 2024 (Fig. 44).

When installing a BWM system, and especially as a retrofit, the following things need to be considered:

- Route of the ship
- Footprint of the equipment including e.g. filter and UV reactor
- Power consumption of BWM unit and electricity available onboard
- Pressure losses due to extra piping, impact on ballast water pump capacity
- Redundancy of the BWM system
- 3D laser scanning for design of the BWM system and for pre-fabrication of pipe spools
- Required amount of extra piping and piping connections
- Installation
- Risk assessment, e.g handling hazardous materials
- CAPEX, OPEX and Life Cycle Cost

In general, good knowledge of marine engineering, naval architecture is required to ensure that the ship can operate safely and efficiently.

6.2 Emissions to the ocean

Cruise vessels may according to current MARPOL Annex 4, release untreated sewage into the ocean if they are at least 12 nautical miles from land.

Special areas where disposal of untreated water is prohibited have been have been defined by IMO and comprise polar waters (Polar Code) and the Baltic Sea (HELCOM). The restrictions in polar water has already entered into force and for the Baltic Sea, new ships have to comply with the regulations as from 1 June 2019, while for existing cruise ships the time limit is 1 June 2021. The Alaska cruise ship law also prohibits cruise ships to discharge untreated water into state water areas. In these special areas, water may be discharged if an approved wastewater treatment system is in use.

The great number of persons in a large cruise ship produce an enormous amount of black and grey water and the treatment system will take a considerable space onboard (Fig. 45). In addition, there must be holding tanks for both dirty and clean water.

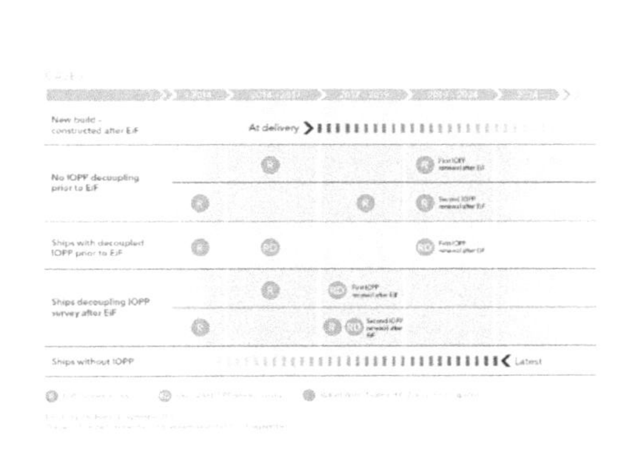

Figure 44. Latest installation date for D-2 compliant ballast water management system (DNV-GL).

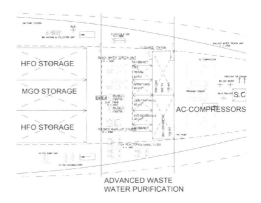

Figure 45. Space required for Advanced Waste water Purification system (AWP).

7 STABILITY

Stability can be divided into two area: 1) Intact stability, that describes ship's ability to float and stay afloat in different intact operational conditions, and 2) damage stability, that describes how the ship stays afloat in damaged conditions.

Intact stability criteria are practically same as decades ago. The wind criteria is the most critical in most cruise ships because of large wind profiles. The location of maximum GZ-value is perhaps the second critical criteria because hulls have become wider and limited operational draught. In IMO's Marine Safety Committee (MSC) have for several years been developed new criteria so-called Second Generation Intact Stability (SGIS). These criteria consider parametric roll, dead ship conditions, surfing or broaching, excessive stability and pure loss of stability. These criteria will not replace existing intact stability requirement but might affect to present constructions when becoming mandatory for new projects in the future.

MSC 98 adopted a draft for amendments to SOLAS2009 that raise the damage stability requirements for passenger vessels in the event of flooding caused by a collision. The amendments raise the 'required index R', the damage stability requirement representing the ship's capability to remain stable and afloat in the event of flooding after a collision. This stricter requirement is supposed to be valid from 2020 onward. Figure 46 will show differences between present and suggested R-index. Green curve is the most probable to be adopted. This would mean that R-index will raise in 3000 pax cruise ship with about 10%. Compared to existing cruise ships that are based mostly on pre SOLAS2009 – rules, superstructures should become lighter and/or hull shape different. Both being uneconomical to cruise operators because of less space and increasing fuel consumption. Also this might lead to increasing number of watertight compartments and then length of individual compartment will become shorter. If SOLAS2009 – ship and SOLAS2020 – ship both have full survivability when whatever two adjacent compartments are flooded, the damage length in the newer SOLAS2020 – ship is shorter.

8 CONCLUSION

The work in creating a successful, safe and environmentally friendly cruise vessel fleet is a challenging work involving cruise vessel operators, naval architects, interior architects, classification societies, national and international maritime organizations.

Bearing in mind the development towards larger vessels and specialized niche vessels, the spectacular exterior and interior design features, the new fuels and systems and last but not least the new regulations entering into force, we have a tremendous work ahead of us. The long orderbook reaching to 2026 may give the designers some more time to develop the prototypes to meet the requirements of the operator for an economical operation and a smaller environmental footprint compared to the existing fleet.

REFERENCES

Levander, K., "System Based Passenger Ship Design," Proceedings/IMDC91, Kobe 1991.

Levander, K., 2004. Chapter 37 "Passenger ships" in Ship design and construction. The Society of Naval Architects and Marine Engineers, SNAME. New York.

IMO Polar Code: International Code for Ships Operating in Polar Waters.

IMO BWM: International Convention for the Control and Management of Ships' Ballast Water and Sediments.

IMO MARPOL Annex IV: Regulations for the Prevention of Pollution by Sewage from Ships.

IMO IGF Code: The International Code of safety for ships using gases or other low-flashpoint fuels.

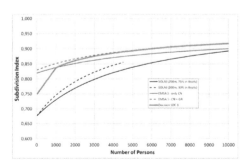

Figure 46. R-index comparison. Black curves as SOLAS 2009, Red and green curves suggested SOLAS 2020.

Keynote paper

Disruptive market conditions require new direction for vessel design practices and tools application

P.O. Brett
Ulstein International AS, Ulsteinvik, Norway
Department of Marine Technology, Norwegian University of Science and Technology (NTNU), Trondheim, Norway

H.M. Gaspar
Department of Ocean Operations and Civil Engineering, Norwegian University of Science and Technology (NTNU), Ålesund, Norway

A. Ebrahimi & J.J. Garcia
Ulstein International AS, Ulsteinvik, Norway
Department of Marine Technology, Norwegian University of Science and Technology (NTNU), Trondheim, Norway

ABSTRACT: Disruptive market conditions challenge existing vessel design practices and tools application. Consequently, this paper argues that design firms must reinvent their business models and work practices. They must also enhance the rationality and logic of consequential and preference-based decision-making processes and revise their solution-making tools to better meet the expectations of a new interrelationship with ship owners, charterers, suppliers, bankers and other relevant stakeholders. These are means to survive and retain future attractiveness and competitiveness. Ulstein Group, as many other maritime industry operations, has recently experienced such a transition and has already made significant changes to its market position and customer orientation, knowledge building, vessel design practices and naval architectural "toolbox" to ensure long-term survivability. This paper discusses why, what and how Ulstein has handled its transition process within the suggested areas of concern—from being an offshore vessel designer and builder to become an exploration-cruise vessel project maker in very short time.

The paper summarizes the challenges of transformation aspects experienced. It describes how market conditions directly have impacted the shift in vessel buyer behavior, knowledge and experience building in the company, and why particularly, the vessel concept design processes and analyses toolbox, have been refurbished and design work practices executed differently than in the past. Examples are review and discussed as to how data-driven methods (accelerated business development (ABD™), fast track vessel concept design analyses (FTCDA™), Ulstein vessel performance analysis (UVPA™) and Ulstein project making (UPM™)) can effectively support the necessary change process and the mitigation of the "wicked problem" effect in vessel design, and ship design firms renewal. The paper concludes with a call for a business centered naval architecture, with the aid of modern data-driven tools.

Keywords: Vessel design, ship design approaches, design tools, market influences, business development

1 INTRODUCTION

Within a short time period, Ulstein has experienced a dramatic change to its previous core market segment—offshore oil & gas related support vessels. In the period 2014 through 2017, Ulstein core market segments worldwide, did reduced from almost 400 newbuilding orders entered into per year, to 15 in 2017, of which Ulstein has earned a market share, like many of our competitors, varying from 2 to 5 percent depending upon the ship type and segment. Under no circumstances, could this setback be compensated by increased market shares and hence, an almost stop in production, in several business areas, was experienced. A dramatic and disruptive change, eliminating in practical terms, the complete basis for our existing operation. In some cases, our traditional customers have stopped ordering new vessels and many of them have consolidated into larger outfits with major fleet overcapacity following and little propensity to realize ship newbuilding. Making the situation worse, is the fact that this disruptiveness has happened pretty much to all ship segments in the world, with a few exceptions. Formerly, such crises were more often confined to their industry segment or geography of origin. This time the situation is different, spreading among all sectors of the maritime industry. Figure 1 shows this market dramatics, when it comes new building contracting activity levels in the period 1996 through 2017.

Such disruptive change situations required Ulstein to react. It was considered that the environmental factors changing so dramatically over a short time period, spurred the need for an adaptive approach to renewal of the firm to a new market situation. Adaptation is not always easy, but when a firm misses the incremental steps of adaptation, then a large, riskier one-shot change lures in the form of a corporate turnaround, bankruptcy or even a close (Reeves et al. 2015). In Ulstein, it was considered unnecessary to initiate a full turnaround operation, instead a transition based on renewal principles was instituted. After all, this was not the first time Ulstein has been involved in such a market shift situation. In the early 70 s the company changed from being a passenger and fishing vessel builder to become a designer and builder of highly specialised offshore vessels. Most recently we have returned to some of our historical vessel segments and expanded the scope to become a complete project maker.

When all-encompassing changes occur, most people focus on the main effects—the spurs in renewed growth of new types of products and the falling demand for others (Reeves et al. 2015). Ulstein have dug deep to understand the forces underlaying the main effects of this dynamism and have developed a point of view about the first and second-order and derivative market changes that have been set in motion recently.

Consequently, and not surprisingly has the performance yield of vessel design firms—integrated ones with yards, larger independent and smaller independent ones, seen their accounts' top-line and bottom-lines drastically shrink. Some have closed due to bankruptcies or lack of adequate financial funding/support—unavailable lending opportunities, some have been bought by larger yards and/or investor constellations. Many have become smaller, fewer remains independent, and overall, the ship design industry ended up in a dire situation.

In Figure 2 the performance of a selection of Norwegian based ship design firms from 2000 and onwards is presented and compared. Such situations have been experienced before and the consequences recorded are unfortunate; loss of talent; loss of expertise; less interest in developing new tools and industry enhancing technologies. Less attractiveness of such declining industries leads to less recruitment to the universities and naval architecture and marine engineering studies—an evil spiral that should be watched and mitigated.

In the midst of this industry implosion, and our own belief in survival from the dire situation, the old way we have approached the market place and customers, performed ship design activities—how we have set focus and priorities, the tools and technologies we have used, the work processes by which vessel design packages are developed, the interphase and how we relate to yards and suppliers, have come under scrutiny and a recent revelation that things have had to change fast has become paramount.

We argue here that such dramatic events and challenges need to be met with creativity, in—and foresight, focus, dedication, openness, and not least persistence (Reeves et al. 2015). Time has, perhaps, run out for specialisation and one-off ship

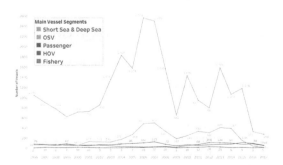

Figure 1. New building activity by numbers of vessels contracted per year for different fleet groups. (Brett & Garcia 2018).

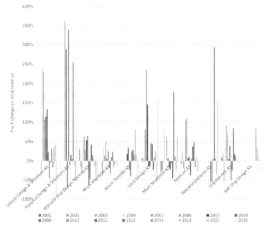

Figure 2. The development of revenue (y-o-y change %) for Norwegian based ships design firms for the period 2000–2016. (Data from Proff 2018).

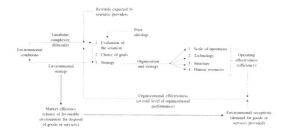

Figure 3. An interdisciplinary complexity model to reveal the full potential of a viable ship design firm renewal (Adopted from Child 1997).

solutions? Flexibility, adaptability and standardisation are contingency measures, which count (Rummelt 2011). Such approaches have been recognised and practised by shipowners like Stena Line or Calmac (Moore 2016). In Ulstein, it was concluded that an interdisciplinary approach was necessary to reveal the full potential of a viable renewal of our operation—market positioning, stakeholder relationship building, organisational development, leadership practice, cost level achievements, products and services adaptations, work processes and data integration systems implementation, vessel design knowledge and expertise and analyses tools development. They all needed to be upgraded, finetuned and applied for an everyday situation, such that future success could be achieved and survival in the short time-frame secured. In some cases, the same approach is recommended when a ship design firm in distress is needing a proper turnaround. However, in such situations the mitigation process would be different and more radical—change management, organizational restructuring, cost cutbacks, asset reduction, and revenue improvement (Tikici et al. 2011).

In the following section, we, attempt to contemplate and build comprehension "bridges" among the mitigation factors leading to a renewal of a vessel design firm, and what has been applied in Ulstein. Figure 3 outlines a generic model showing the interrelationships among factors influencing the viability of a ship design firm renewal and outline typical causal relationships—what influences what? What we don't know are their causal relationship strengths, except for anecdotal material and theoretical elaboration available in the main-stream management literature. Several of these factors constitute what we would call the ship design approach and commercial methodology.

Many alternative ways for improvement can be applied, and there is no *the* way to take us there. Yet, there exist proven prescriptions to be selected and partly or fully applied, gradually bringing us out of the difficult situation.

In the following chapter, we will discuss how we have met the market disruption, set new goals and targets for the renewal process and how we have dealt with the overall transformation process to a more viable operational situation. Special attention is given to the effects, such change processes have on the ship design approach, its influential externality and internality factors.

2 MEETING THE CHALLENGE OF DISRUPTIVE MARKET CHANGE

Let us start looking at the way we plan and prepare for the actual ship design activities—the way we have done it in the past and how we preferably should do it in the future to contribute to and enhance a viable renewal of our operation? Firstly, a major effort had to be done to reduce resistance to change in all aspects of the transformation, by creating awareness of the actual situation. Then, identify what needs to be changed, to be followed up by implementing change initiatives and practice new work behaviour. At the same time, auditing should be carried out closely of what is happening—are we taking our own medicine and feeling better by taking it? Such situations are always full of resistance within and among employees and sufficient understanding of the critical situation might not be grounded in each and one person. Figure 4 applies the classical Kubler-Ross change model a firm and its' employees go through, mentally that is, meeting disruptive market change.

Anticipation of a difficult situation—a disruptive market change, is often looked upon as a process of standard scenario forecasting (Rummelt 2011). Typically, if you do scenario forecasting, you wind up with a graph with three lines labelled "high", "medium" and "low". Everyone looks at

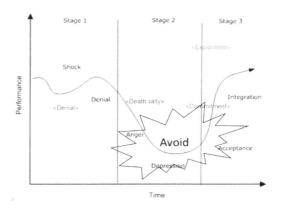

Figure 4. The change denial Kluber-Ross curve applied to firms—what to do about it?

it and believes that they have paid attention to the uncertainty and complexity involved (Ison 2010). Then, of course, they plan on "medium"! But they are missing the risk. The risk is not that the price of oil may be high or may be low. The risk is that it will go high, motivating you into a major investment for god times, and then, suddenly, turn and dive to low, leaving you with a useless asset, as it has happened in the offshore oil & gas industry (Garcia et al. 2016).

More surprisingly, is the situation that many of our customers are not really understanding the situation either. Business drivers and enablers seem to alter faster than the contemplation of the market players – the stakeholders in the market place. Cascading effects, like – lower oil price – leads to less cash flow – leads to reduced E&P spending – leading to less chartering of vessels, lower charter rates, new building prices and finally no need for newbuildings are not well understood in the maritime industry. The authors have yet to recognise such causal chains being considered carefully, when newbuilding orders with their respective expectations are integrated in the macro context of the vessel design approach.

A similar chain of causal relationships is envisaged in the ship design industry: slow economic growth; – bad shipping market; – less new buildings; – less work for ship designers; – lower prices; – less funds for fun; – cost consciousness; – more for less achievements; – product and service obsolescence. The same situation occurs, when in a good market situation, but the causal chain will be different and their consequences and implications quite opposite of the bad times. Thus, these factors all interfere with the design firm, the design approach, the vessel design solutions and not least the interrelations ship among stakeholders and the ship designers. Typically, and paradoxically, in bad times new technology, innovation and the need for enhanced knowledge, new products and services appear and more prominently than in good times. In good times, most firms are too busy with customer projects to explore new grounds, products and services, reap advantages from new technology advances and innovate, because of lack of need and/or man resources. It is, therefore, argued in this paper that these relationships cannot be underestimated or departed from. Their profound effect on the boundary conditions for the transformation of the design firm, ship design process, methodologies and tools and product and services' development are important to understand and to include in the solution space context definition.

A central place to start searching for improvement initiatives is, therefore, in the recent advances of ship design processes and methodology, most of which are, in among other prominent sources, well documented in papers presented in several IMDC Proceedings (Erikstad 2009, Ulstein & Brett, 2009, 2012 and 2015). A very extensive review of the subject was presented in the meta-study of Andrews et al. IMDC *2009 State of art report on design methodology* (Andrews et al. 2009). Several authors, researchers and practitioners in the field, claim the importance of looking at the ship design approach or process in the framework of a holistic view, accounting for an increased number of design variables to be included in the decision-making process compared to the past in many of the cases. However, it is not a clear how wide and deep the scope of the solution space should be. This paper, argues that such a holistic or systemic ship design process must also include the ship design firm's organization – resources, tools and system infrastructure and expertise and market considerations, as a minimum of factors to be included in the analysis. Traditionally, the rather limited approaches have been dealt with in a very static, present time, fragmented and conventional way. Current customer desires, needs and expectations of their fleet of vessels and the individual vessels as a new building project, are typically defined as – what is it that the vessel needs to do now? In what way? And to what cost? Such a traditional approach, always treat the design firm as a static, as well as the customers' needs, and expectations are considered in a very narrow context.

The recent disruptive market situation has accentuated and brought forward a set of additional questions to the vessel designers and the ship design situation, like: In what context and market is the vessel going to operate and function and for how long? What would be the boundary conditions within which the vessel design solution should prove its expected performance yield – commercially, operationally and technically – now, in the immediate future and or in the far future? Ships typically, have a life cycle of 25 years or more, so much can and will happen with respect to the influence of externalities and internalities (factors) over such a long time-period. Furthermore, how vulnerable – or inversely, how robust is the vessel design solution in relation to such a changing environment – and boundary condition settings? If these conditions change, what would then be the fit of the vessel design solution and its performance yield in the years and operational situations to come? How sensitive would the solution be in relation to environmental or solution space boundary condition changes?

What would be the externality – for example, market condition and internality – for example, new technology, factors influencing the vessel design solution and its performance over time, – incremental or radical changes impact on the recommended solution at hand? Over time—short or long, what would be the consequences and implications of such changes—weak or strong, insignificant or profoundly? Are these challenges one problem or several problems to handle, and is it at all possible to expect a rational and elucidated handling of the nature of such a "wicked" problem? (Andrews 2003). He continues: "Identifying what is the nature of the problem is the main problem, and that attempting to do so without recourse to potential material solutions verges on making a difficult operation impossible". The "wicked problem" is, therefore, much more than identifying all the expectations, requirements and needs of the ship owner or close by stakeholder – it is truly about identifying what is the nature of the greater problem, which is the main problem. The nature of the greater problem is an all-encompassing feature, and we are, therefore, of the opinion that also such features as the market situation, the economic situation in general and the involved firms' condition—qualities, capacities, capabilities and robustness, play a significant part in the overall nature and dynamics of the "wicked problem".

It is still a limiting consideration and an interpretation of the original "wicked problem" concept of Rittel & Webber (1973), which prevails in the ship design world among naval architects. In recent discussions of this issue, most of the classic naval architect and marine engineers have fallen back to his or her roots and have in most cases, aborted the real challenge of the "wicked problem". This is demonstrated by the way such custodians of the ship design community, repeatedly rationalising the problem by setting strict and limiting boundaries to the solution space and, thereby, partly avoiding the complexity and uncertainty aspects of the problem at hand. Traditionally, they have defined or restricted the solution space in the past, by bolstering the solution space by a set of very few rational and relevant presumptions and assumptions. We argue that to seriously and sincerely challenging the way we deal with the "wicked problem", it must be addressed and dealt with in a systemic way such that more complexity and uncertainty factors are involved in the solution space definition. It is also important to understand that the solution space will change over time as a consequence of good and bad times. Only in few user-cases known to the authors, has the dynamics of the market situation been explicitly handled as a direct influencing factor in defining the final effective solution space. Newly developed methods and practical approaches to cater for a much broader approach to the "wicked problem" is recommended, even if the task implies higher complexity and more uncertainty when applied to facilitate such a sincere task of mitigating the "wicked problem" challenges (Ebrahimi et al. 2015a, b, Ulstein & Brett 2012 and 2015).

Furthermore, it is argued by the Authors that in only some cases are vessel design experts moving out of the comfort zone w.r.t expectations and alternative perspective exploration in the relation to the "wicked problem". Very rarely, are future possible needs and expectations covering all relevant performance perspectives that may be relevant included in the analyses.

Vessels are traditionally, designed pretty much to service a market place under ideal—predictive and attractive market conditions. Time series-based market segment studies, show clearly that for most of the time do ships operate in real life, only for short periods in good markets and for most of the time, markets are bad, in the sense that they barely can support vessel CAPEX, OPEX and VOYEX costs (Ulstein & Brett 2015, Stopford 2009). In some cases, market rates only cover parts of VOYEX and OPEX costs and no CAPEX support (Stopford 2009). In such situations, we experience over again, that contemporary ship design firms and their design solutions are not well adapted to the real market situation and in many cases a far too complex, expensive and costly business case and vessel design solution are resulting. In the offshore service vessel market segment many such ship designs were conceptualised and built in the period 2015 and onwards, particularly in the Offshore Construction Vessel (OCV) segment.

Thus, it is important to cover as many objectives and constraints including, but not necessarily limiting to commercial, operational and technical related aspects of a ship new building project. Figure 5 shows a set of advised perspectives, which may assist the "wicked problem" elaborations at metalevel.

According to the authors, at least, one additional aspect needs more attention to stimulate the further advancements of ship design methodology. This extra factor is the given boundary condition awareness.

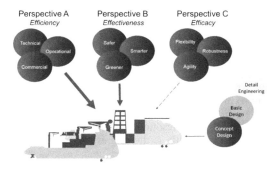

Figure 5. Advised perspectives, which may assist the "wicked problem" elaborations. (Adopted and adapted from Ulstein & Brett 2015).

From a resource-based strategy perspective (Cyert & March 1992), the boundaries of the design firm would seem to derive their importance from the fact that they determine the firm's sourcing of resources (in-house or external sourcing), co-determine the terms at which resources may be acquired and practised, influence the extent to which rents may be appropriated from, for example, valuable knowledge and expertise. Thus, in such a scheme, the boundaries' issue is seen to be directly relevant to the issue of sustained competitive advantage, arguably the key issue in a ship design firms' strategy. This is because knowing something about the design firm's boundaries (externalities and internalities), and therefore, its context and organization, also tell us something about how efficiently strategic resources are organized.

From a scientific and a normative perspective, we wish to know which entities and mechanisms determine observed and felt boundary choices; New and maturing technology, inventions and innovation? Design and production costs? Knowledge accumulation and data retrieval? Dependence – causes and effects considerations? Transaction costs and time? Transparency and trust among stakeholders?

Another important aspect of the holistic or systemic-based ship design approach in a ship design firm context, not so frequently discussed in scientific journals and proceedings, is the balance between the process and solution orientation of the ship design project. It is, therefore, also suggested in this paper that more attention is needed on the actual ship new building project execution in its broadest context, to arrive at improved ship design solutions and new building project effectiveness under conditions of disturbance and high levels of uncertainty.

3 RETAINING FUTURE ATTRACTIVENESS, COMPETITIVENESS AND SURVIVAL

A renewal strategy, renews the vitality and competitiveness of a design firm when it is operating in a harsh or bad market environment. Such a challenge can be caused by protracted mismatch between the firm's approach to strategy and its environment or by an external shock. When the external circumstances are so difficult that our current way of doing business cannot be sustained, changing course to preserve and free up resources, and then later redirect toward new growth, is the only way to not merely survive, but to eventually thrive again (Reeves et al. 2015). A firm must first notice and react to the deteriorating environment as early as possible. Then, the firm needs to economize to decisively address its immediate impediments to financial viability or even its survival. To do so, it focuses the business, cut cost, and preserve capital while also freeing up resources to fund the next part of the renewal journey. This is just what Ulstein did in the years from 2015 and onwards. Finally, the firm needed to pivot to one strategy to ensure long-term growth and competitiveness, by resetting the strategic direction of the company in line with its environment and innovating strategically.

Early in the process, it was decided to look for opportunities, which could strengthen both the accounts top-line and the bottom-line. More critical was the top-line improvement. Hence, both niche markets, represented by typically 5 to 10 new buildings (NBs) in the world per year and big volume markets 30 and more NBs were looked at. The niche markets consisting of: i) Offshore Oil & Gas (OO&G) market segments, which do show some improvements in the next 3 to 5 years. ii) Offshore Wind Energy Generation (OWEG) for Service Operational Vessels (SOVs) and Heavy Lift Vessels (HLV), which will experience a continued moderate growth in European waters, and a high growth in China and later in the USA. iii) Exploration-Cruise vessels, which will experience a flat, reasonable demand in the years to come. iv) RoPax-ferry vessels, which will also experience a flat, reasonable demand. v) RoPax-roro and PCTC/Ls, which will experience a flat, reasonable demand. The potential big volume markets, typically > 30 to several hundreds of NBs in the world per year: vi) Trawlers. vii) General Dry Cargo ships. viii) Tankers. ix) Bulkers, and finally x) Container carriers, eventually had to be considered. Priority was given to: a) Expand the market share in the Cruise-Exploration segment—yet a moderate volume. b) Expand the market share in the SOV segment—yet a moderate volume. c) Expand the market share in the HLV segment including jack-up vessels—yet a small volume. d) Expand market share in the RoPax-ferry segment—yet a moderate volume. e) Prepare for next generation of OSVs sales—moderate volumes—with particular attention to the ERRVs, OCVs, DSVs. f) Prepare for the next generation Factory Stern Trawlers—large volumes. g) Prepare for «web-based» sales of standardized catalog ship designs (concepts and class drawing packages) to third party without project making, packages, site support and commissioning—stock sales of design solutions.

It was decided that focus should be: Selling ship projects rather than ships. Performing Project Making rather than only designing ships and building them. Be proactive through identifying fleet renewal opportunities, retrofits and upgrading of existing ships rather than pushing new ships. Increase our sales through Web Based Vessel Marketing and ship configuration rather than traditional "no cure no pay" developments of one-off ship design solutions. This new strategy is reflected in Figure 6.

The renewal approach is unique both because it is contemporary and because it is a combination of at least two approaches to strategy, each with its own distinct logic (Reeves et al. 2015). The combination is challenging because the two approaches' requirements are in some ways diametrically opposed—work processes, knowledge and expertise, tools used, business proposition and concepts are different. Such a mismatch can come about, either because a firm chose the wrong strategy or, more often, because the environment has changed, and the strategy didn't, leading to chronic under-performance. In our case, our existing strategy concept was considered adequate, by adding the "web-based ship design concept, since our original strategy had not lead to under-performance. It was the disruptive environment—market conditions that made the world tops-turn for us.

The different steps of the Ulstein value creation process, digitalization, and automation of design processes are continually being applied to reduce production time and required resources, while at simultaneously enhancing robustness of the decisions and actions that are being made. Figure 7 illustrates the process within a product lifecycle for ships. This is a new product development process where different business intelligence approaches, tools and techniques are applied to improve overall system effectiveness based on extensive data analysis.

4 RE-INVENTING OUR BUSINESS MODEL

4.1 Business model innovation

A business model innovation is a natural consequence of the renewal strategy discussed in the previous section. It focuses on the changes of the multiple elements of the way we service customers and create value. It can perhaps best be defined as the orchestration of all assets and capabilities of the firm to realize a disruptive value proposition (Rummelt 2011). Hence, the business model innovation requires a quantum leap, rather than incremental and individual changes in service, products, or operations. It might include changing the distribution or revenue model or in our case, our value chain footprint to fully harness the power of new technology, and the reconceptualization of our products or services provided. For this reason, the new or revised business model and proposition of Ulstein still, tend to be nebulously broad affirmations of our existing firms' current business model.

In Figure 8, we have advised a format upon which a renewed business model can be built, which includes a merge between value proposition, including product, segments and revenue, and operations, including value chain processes, cost assessment and internal expertise. It is the format being used by Ulstein to renew its business model.

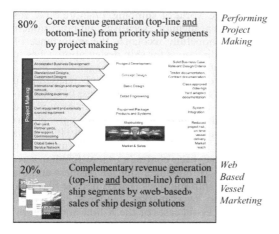

Figure 6. Ulstein multilevel, dual ship design development strategy.

Figure 7. Ulstein value creation process in vessel design. (Adopted from Keane et al. 2017).

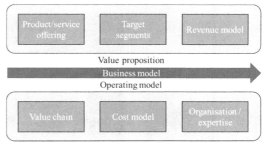

Figure 8. Business model innovation framework. (Adapted from Reeves et al. 2015).

4.2 The revised products and service portfolio

This model allowed us to think about the possibilities of combining a product portfolio based on all the offshore support vessel segments mentioned in Section 3, but expanded upon with the exploration-cruise segment, the complementary RoPax—Ferry segment, the positioning of ourselves in the OWEG market with SOVs—resembling the OSV markets and a new technology challenge—commercially, operationally and technically, the Factory Stern Trawler ship segment. Together with a new operating model to follow that leveraged product and service portfolio, the power of data-driven analytics, digitalisation of production, e-commerce, and a partner centric revenue approach, eventually constituted the renewal concept.

Given that we set a visionary approach involving fixed goals, but flexible means to overcome hurdles to achieve it, our approach is more like a long-distance road map that allows for flexibility along the way. Because we are, by definition, charting unknown territory with these extra segments, we can be sure that some obstacles will force us to adjust course underway. Therefore, our approach is not relying on the kind of elaborate documentation of detailed financial and operational milestones the you might see and prepare for a more classical approach, even if our investors may expect and require them. Instead, we have defined high-level milestones to keep us pointing in the right direction and moving speedily toward our end vision. Figure 9 exemplifies how Ulstein is expanding from its core market segments in the offshore support vessel sector to new important market and ship segments.

Within each new ship type segment, it is important to expand the business opportunities within each and one ship market segment. Figure 10 shows the direction Ulstein has taken, in for example, the OWEG segment. It is important to make the most out of each new segment. They are expensive, risky and knowledge demanding to enter in a successful way.

Figure 9. Ulstein renewed expansion strategy.

Figure 10. Ulstein segment opportunity expansion within the OWEG segment.

Since we are creating new market opportunities, we must be aware that we are communicating to the unconverted. Suddenly, do not do the established brand and reputation, make the future work. At least to lesser extent than in pursuing the existing portfolio of ship type designs.

Proactive professionals throughout the field of ship design are recommended to gradually shift their focus from pure heuristics and static-serial oriented analysis to critical systems thinking, dynamic modelling, and system-based ship design approaches. In the hypercompetitive, globally competitive world of shipping, offshore oil & gas operations academics, executives and practitioners alike will hopefully recognize the importance of such new directions for ship design approaches to achieve lasting, superior performance.

Finally, our vision will not be truly realised until it has been accepted by a critical mass of our existing and particularly new customer groups and related stakeholders' attitudes. Our approach has been naturally met with scepticism since it presents something new that not only may be unfamiliar, but also may even contradict more familiar ways of doing and thinking about the business and the integrated ship solutions. This is not the first time that Ulstein levers market standards, exemplified by the successful cases of the X-bow and X-Stern. Therefore, we have re-fined our solution portfolio in such a way that it can be communicated in a convincing way to customers and investors. Finally, we have concentrated our efforts and resources to demonstrate that our approach has traction and is credible. The use of data-driven methods, digitalisation and analytics have become very instrumental means to achieve these results. Facts-based decision making in such a process is a must.

4.3 Customization or standardization of ships

It is clear to the authors that alternatively, in the future, should simply more ship owners be offered a standard, off-the-shelf catalogue type vessel, designed with pre-defined capacities, capabilities and pre-qualified ship system choices for a broader range of their needs? In this way can complexity be reduced, and certainty of functional performance, quality, price and delivery time be secured. A standard catalogue vessel is, for example, a ship concept where hull optimization, layout, crew comfort, environmental impact, safety and reliability, efficiency at sea and in port, are backed up by pre-defined user expectations and fabrication friendliness, by the designer, yard and pilot owners and benchmarked with all other relevant vessels in the segment. Thus, simplistic standard vessels can be outstanding too, even if they are considerably less fancy and ready-made than the customized, hi-tech versions of the most renowned ship owners. At the end of the day, the true purpose of most ship design approaches is to find better solutions, which, on the other hand, are not too costly. Low building and operational costs of vessels, which are efficient at sea and in port, can typically be met by the offerings of such standard catalogue vessels. As a bonus, the complex, uncertain and ambiguous process of initiating and realizing a ship design new building project process can partly be removed and resources spent on alternatively more important value creating tasks. Such strategies are also relevant for traditionally more "exclusive" market segments like Exploration-cruise, RoPax Ferries and Factory Stern Trawlers.

It is therefore, the opinion of the authors that customization and standardization in ship design will develop side by side depending upon the market situations and owner's needs. Special needs will continue to be serviced by special equipment and not so special needs to be well serviced by more standardized vessel solutions. Figure 11 exemplifies recent vessel portfolio developments in Ulstein.

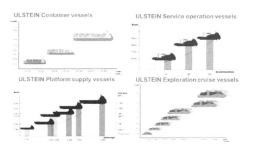

Figure 11. Ulstein's standard container vessel, SOV, PSV and exploration-cruise fleet of product families.

Figure 12. Strategic value chain positioning for ship design firms.

4.4 The value chain expansion

Our focus—is to exploit a wave of change—disruptive markets are largely exogenous—they are mostly beyond the control of anyone firm, but they can be favourably reacted to. This is exactly what Ulstein is trying to do among all phases of the value chain. Figure 12 shows different strategies with regards to the different value chain positions (A, B, C, D, and E) a ship design firm can take. Marked by green, are those activities typically being practiced by design firms. This paper has documented that the wealth creation will vary with position in the value chain, but also other factors count.

By prioritizing certain phases of the process according to the business concept, Ulstein can act efficiently within the cost/commitment boundaries of its projects. We are able to follow the ship within all the phases, from project to delivery (concept E, Figure 12), but this may not be the most profitable strategy for all NBs. Therefore, selecting the right business concept in the initial stage of the process is paramount. Ulstein is now practicing business concepts D and E, depending upon whether we follow up new buildings at own or in a 3rd party yard.

4.5 Utilizing business intelligence and market analysis

The "Ulstein Business Intelligence (BI) Methodology" is intended to create, map and organize the necessary resources and tasks that streamline the process from an initial BI need to final BI product. The preconceived idea of an Ulstein methodology was originally sketched as shown in Figure 13. The ensuing process of refining this model was performed together with experienced domain experts from external companies. The methodology is designed purposefully to be iterated upon, such that during or after each case specific feedback or experiences should be continuously integrated, adjusted and updated. In Ulstein, marketing research and analytics support executive, marketing, sales, and product development managers with

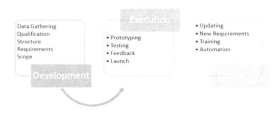

Figure 13. Sketch of the BI methodology at Ulstein. (Adopted from Keane et al. 2017).

more effective every day direction-setting and strategic decision-making.

Ulstein is today a proprietor of vast market information, and as such actively subscribes to various sources consisting of news subscriptions, PDF reports, and databases to name a few. The databases constitute a driving portion of the performed market analysis, and traditionally would have to be individually cleaned, modified, and processed before any analysis would occur—a task that could extend up to several hours without difficulty (Keane et al. 2017).

BI advantages can be summarized as the ability to efficiently store and present non-trivial data, such as large processes databases, long time series and hierarchical information, in an easy to share setting. Its high level of customization allows intensive multi-platform interactivity, as well as association with proprietary simulations tools and design databases (Gaspar et al. 2014).

Moreover, the proper interpretations of exorbitant amounts of information, whether obtained via experience or simulation database, is not just relevant in order to increase our limited ability to gain value through cognitively processing raw data, but also to uncover the value decreasing aspects.

4.6 *The benefits of data-driven methods*

Following the emergence of Industry 4.0, the prevalence of digitization, and the ensuing deluge of information and knowledge that has surfaced because of it, the topic of data-driven methods has been at the centre of many discussions in terms of defining what it is, and how it can be done.

Data-driven methods are an efficient way to map, access, handle, visualize, and interact with a large amount of model-based systems information. It presents a collaborative foundation, able to combine powerful visualization components and a data-driven approach to designs' manipulation. Collecting and handling vast information during all the phases of the ship design value chain (Figure 12) is being leveraged in many ways to gain insights into phenomena and create predictive models. Advances in statistical analysis and machine learning can be combined with such data-driven methods (e.g. big data), shifting perception from the delivery of goods by a ship with a size X and power Y to providing service A and B within safety, economic, and environmental constraints.

When considering methods and techniques of approach, implementing, and governing Big Data initiatives, there are many options to consider. Prominent techniques used in the context of analysis and prediction include data mining, clustering, regression, classification, association analysis, anomaly detection, neural networks, generic algorithms, multivariate statistical analysis, optimisation, heuristic search (Chen et al. 2012). Ulstein has paid much attention to the potential opportunities of these techniques and most of it is already in place and being tested out.

As learning outcome, connecting and storing data in extensive repositories, which can be accessible for simulations and knowledge building in different phases of the ship life cycle is a significant achievement of this initiative. Ulstein Big Data model in test is already able to connect 2 main phases of the ship life cycle and weather data in a general data ware-house (Abbasian et al., 2018).

4.7 *Moving towards intelligent decision-making and Machine Learning (ML) in ship design practices*

Artificial intelligence (AI), at its most fundamental, is the practice of using algorithms to parse data, learn from it, and then make a determination or pre-diction about something in the world (Copeland, 2016, Bravo et. al. 2014). So rather than hand-coding software routines with a specific set of instructions to accomplish a particular task, for instance updating an excel spreadsheet to calculate hull coefficients or re-doing a CFD analysis to evaluate wave resistance, the machine is "trained" using large amounts of data and algorithms that give it the ability to learn how to perform the tasks.

Within a maritime domain, the use of AI and machine learning (ML) yields little results in terms of academic research literature. Identified examples have shown application regarding image processing for ship detection (Tang et al. 2015), and the response or load prediction of offshore floating structures (Mazaheri 2006, Uddin et al. 2012, Maslin 2017).

From Ulstein's perspective, ML must be used primary for intelligent decision-making in the conceptual phase, tackling it from different perspectives: i) To differentiate among different solutions; ii) to have better understanding of consequence for any small input change; iii) to measure goodness of fit between final product and requirements; iv) to have more meaningful benchmarking with market competitors; v) to make better and more robust decision-making in vessel design; and vi) to support the development of effective sales arguments of—what is a better vessel. Furthermore, Ulstein appreciates that proper and effective decision-making should be based on the fact that ship design is a multi-variable-based decision-making process, and big data oriented to secure proper balancing of new vessel designs with appropriate trad-off among requirements to resolve the inherent complexity of ship design. It is essential that the decision-making model can demonstrate, separate and distinguish among the effects of design parameters (main dimensions, power, mission attributes, machinery, etc.) on final vessel design performance yield (Ebrahimi et al. 2015b). The Ulstein approach typically, integrates both multi criteria evolutionary problems with multiple objective optimization problems to come up with the better solution. Ulstein applies complementary methods for benchmarking of ship designs: i) Ranking based vessel design including indices developed based on vessel missions, scoring by indices, and ranking by statistics. Ulstein also apply Hierarchical multivariate based vessel design benchmarking according to smarter, safer, and greener performance perspectives by i) hierarchical factor categorization, ii) metric attribution of design factor causal map matrices and iii) hierarchical comparative based ranking. Including these tools in consistent ML framework is the challenge for the next years.

4.8 *PLM in ship design*

Product lifecycle management (PLM) systems promise, already for a couple of decades, to keep control of products' digital data structuring, using dedicated (and expensive) software for improving the management and collaboration of the team throughout the product development process. Product lifecycle management (PLM) was introduced in the 1990s to better manage information and expand the scope of computer aided tools throughout all phases of a product's lifecycle, which from a manufacturer's standpoint is comprised of imagination, definition, realisation, support, and retirement (Lee et al. 2008). Such an approach enhances the overall effectiveness of vessel design (Ulstein & Brett 2015). If modern PLM systems deliver on the promise of handling the challenges of multi-taxonomy/disciplinarily issues, it is reasonable to assume that we can start to use PLM as a foundation towards handling and storing diverse data, and furthermore develop other emergent technologies.

The drawbacks of a decision to go completely via PLM are, however, well known. It is vital to understand how complicated and time consuming the implementation of a PLM project might become depending on the company requirements. Often maritime companies consider PLM system as too time and resource consuming before bringing benefits and tend to avoid it or postpone its implementation. Another drawback is the failures of previous implementation attempts (Levisauskaite et al. 2016). PLM systems have been promising these implementation *miracles* for years. Many ship design companies faced difficulties in the past, including Ulstein, when managing the large scale of data using 10–20 year old PLM systems, and such a poor experience is clearly reflected when introduced to more modern and agile software. Thus, Ulstein is about to develop and implement a relevant PLM system.

5 VESSEL DESIGN WORK PROCESSES AND PRACTICE

State-of-the-art ship design methodology and vessel design approaches are all taking place within the realms of an industry setting and a firm's—a design outfit, a shipyard and or an industrial complex providing integrated vessel design and ship building and equipment provision services in each market segment and condition—a bad or a good one.

Evidently, the issue of market conditions and strategic factors such as diversification, outsourcing, partnering, alliances, ship design firm product and services, integration with the customer, etc. have a significant influence on how the designer plan, prepare and execute his or her design work processes and what complementary analyses tools are used to meet the challenges of a highly volatile market place.

Ulstein Group is searching for performance yield and improved competitiveness not only through Vision projects like a next new *X-Bow-type concept*, but also looking for innovation in our existing products and processes. At Ulstein, to date, no common framework is able to integrate all the ship modules into a common library of parts, neither to efficiently communicate the evaluation of the modules to owners and suppliers, incorporating feedback into the process.

We suggested that more attention is needed on the actual ship new building project execution in its broadest context, to arrive at improved ship design solutions and new building project effectiveness under conditions of disturbance and high levels of uncertainty. Therefore, that one such systemic approach including critical systems thinking that identifies and relates to all relevant aspects of commercial, operational and technical expectations and priorities, objectives and purposes, and business strategies, constraints and concerns, can reduce or eliminate deficiencies and the negative implications of existing and past ship design approaches

Another important aspect is the incorporation of the downstream value chain when creating, analyzing and evaluating the modules during the early stages, once that the design of highly demanding vessels, such as offshore support vessels, cannot ignore the large range of operations. Environmental concerns are also connected to the part of the value chain, such as air emissions and scrapping and how to make more robust decision making in choosing the better solution among peers. Figure 14 shows that while conceptualisation of new ships consumes typically less than 5% of total project's design and construction time, almost 90% of the innovation potential resides at this step. It also shows the importance of spending more time and resources early in the design process to explore the solution space. Further, it also indicates the asymmetry of time spent and where is it important to spend it.

Different activities have been carried out in Ulstein to explore the opportunities for upgrade design work processes and practices: i) Develop a cost-effective framework for design and engineering of vessels based on a modularized and standardized approach (based on three different principles: Customize to order; Configure to order; and Standardize to order) through the whole value-chain, from the conceptual design of the vessel until scrapping. ii) Develop a system theory-based prototype design tool able to concurrently integrate the framework with the current module work at the value-chain, testing the efficiency gain of the proposed solution against the current status quo. iii) Test and implement the framework within the value chain elements. A cost-driven study of the most time-consuming modeling tasks, using this new knowledge as input for the effectiveness of the framework: a) Improved design protocol, with the documentation of what decision are made → why is it made → what are the consequences. b) Incorporate client expectations in the framework, with a collaborative systems communication tool able to manage feedback and feed forward about the design among designers, suppliers and owner (stakeholders). c) Modules compatibility through the whole value chain, preferably using a common parametric 3D library of modules, systems and interfaces. d) Quality control of detailed engineering, documentation and administration to 3rd parties. e) Integration of analytical tools (e.g. CFD, FEA, Stability) into the framework, easing the model creation and conversion. f) Partial expert-systems capability (inference-machine thinking).

Typically, vessel families are developed along three main dimensions: Type of ship; Capacity (size); Capability (functionality) and vessel system configurations (diesel-mechanical or diesel electric, etc.). A good example are the four families of vessels presented in Figure 11. This problem is tackled mainly via efficient usage of computational tools and libraries. During the last 30 years, where personal computers have been available, each engineering discipline has developed their own specialized computer tools and software. The performance of each tool is recognized, but when designing complex systems as offshore vessels, cruise and trawlers with thousands of input variables, the specialized tools often lead to a large amount of information to be handle during early stages, consuming a large amount of time for integrating these models and its evaluation into the value-chain activities, as well as difficult in documenting these results towards stakeholders. Nowadays a significant part of high-qualified employees' job is to take data from one set of systems, manipulating and feeding it into other systems, consuming a large part of the time.

Exploring alternative dominant new building project development logic necessitates first unlearning the existing dominant logic. When it comes to the adaptation to change and adoption of new business concepts and work procedures, incremental advances typically prevail. Radical change among maritime firms is not so frequently experienced. In several cases where the authors have been involved in the facilitation of systemic-based ship design and holistic oriented new building development projects, it has been demonstrated how important it is to unlearn old habits as it is to open up for new advances in systems-based ship design approaches (Brett et al. 2006).

The continuation of improvement activities will encompass, but not be restricted to such expansions such as online web-based vessel configuration, algorithmic intelligence supported naval architecture and vessel engineering work, and artificial intelligence-based tools' expansions. Continual improvement work along these strategic avenues and steadily inclusion of new internally and externally developed analytics tools are bound to happen, thus accelerating the effectiveness of Ulstein doings, discussed in the next section.

6 VESSEL DESIGN SOLUTION DEVELOPMENT TOOLBOX

6.1 Digitalization, simulation and virtual prototyping

The ship design environment has been the one using the most simulation and virtual prototyping within the maritime industry (Keane et al. 2017). The lack of information at the early concept design stages, together with the influence on final perform-

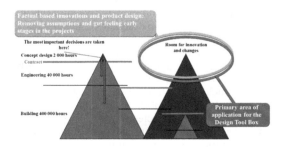

Figure 14. Conceptual model of Ulstein—where to increase attention in ship design.

ance of decisions taken at this stage-up to 80% of the costs are fixed in the concept design (Erikstad 2007) – spurred the need of understanding the consequences and implications of decisions in technical, operational, and commercial performances (Ulstein & Brett 2015). Specific, single-attribute simulation tools could not solve this complexity problem, therefore holistic, multi-attribute simulation tools have been the core focus of recent applications. Concept design workbenches developed both by universities and industry, pursue the acceleration of the concept design development process and to increase the potential number of alternatives being evaluated during consideration of changing contextual factors. These workbenches approach the concept design development from an alternative perspective. Rather than focusing purely on design parameters their approach embraces the selection of functional requirements, and which mission the vessel is intended for, as a premise to design a better vessel: "it is only when having the correct set of requirements that we can decide upon the correct vessel" (Gaspar et al. 2016).

Ulstein has undertaken a substantial digitalisation effort in recent years and developed and implemented a set of new internally manufactured digital data-driven tools and knowledge repositories to enhance its ship design activity. It is envisaged that this effort is only a good start in exploring the benefits and potential of digitalisation and virtualization of business and work processes. Further, Ulstein is now working with the implementation of tools and work procedures simulating future scenarios in conceptual design phase and virtual prototyping.

6.2 *Accelerated Business Development (ABD™)*

Ulstein has over the years introduced and implemented an Accelerated Business Development methodology (ABD) to enhance and strengthen our capability to effectively solicit relevant stakeholders' expectations and desires when it comes to the realization of ship designs and new building projects (Ulstein & Brett 2009).

The core elements of the ABD approach (Figure 15) have been extensively discussed in previous IMDC editions (Ulstein & Brett 2009, 2012), which aims to better guide ship designers, yards, cargo, and ship owners in realizing a business opportunity within intermodal transport or offshore field development work whereby ship design is utilized to achieve a competitive advantage. The approach advocates that a new or improved solution system, where the ship plays a significant role, shall fulfil the needs and expectations of all the involved stakeholders in the best possible way through the multi-attribute decision making ABD-

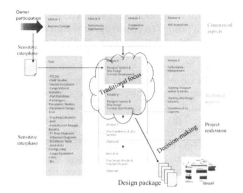

Figure 15. The Ulstein's ABD "wicked problem" search methodology. (Adopted and adapted from Brett et al. 2006).

approach. This approach makes it possible to follow the complex and normally fragmented processes of business development related to maritime transport, cruise-exploration, trawler industry, offshore oil & gas field, and the pertinent ship design in a systemic and explicit way.

Historically, separate documents like outline, contract and/or building specifications and drawings have constituted the communicational instrument among the players in the overall decision-making process. Owners' specifications are typically formulated based mainly on their experience in ship operations. Expanding on what is or has been the experiences of the past is more typical than what it is we really need. Yards or designers, on the other hand, typically optimize a vessel with respect to preferred engineering criteria, such as installed engine power, speed, or lane meters and frequently their own production facilities. The ABD approach counteracts these discrepancies and inefficiencies and secures a holistic management of complex data such as metric, film/video, sensor signals and the like.

Ulstein has carried out more than 30 such ABD processes on own development projects and with customers. Comprehensive data analytics processes have been carried as complimentary fact finding following such ABD approaches.

6.3 *Fast Track Vessel Concept Design Analyses (FTCDA™) and Ulstein Vessel Performance Analytics (UVPA™)*

Companies operating in the design of maritime units are challenged by the need of incorporating flexibility, innovation, speed, and agility to their business model (Ulstein & Brett 2009). The conventional concept design development process, based on work processes relating to the traditional design spiral for vessels has proven to be non-effective when it comes to ensuring very short customer response time and robustness of the results. It is too time consuming and resource demanding, and drastically limits the number of alternatives to potentially be evaluated for goodness of fit.

In response to this, Ulstein has developed a Fast-Track Concept Design Analysis tool (FTCDA). This simulation tool combines multivariate statistics, network resources and design knowledge/expertise to accelerate effective decision making in vessel concept design. The FTCDA is an integration tool which gathers different modules of the conceptual design process in a unified digital platform. A holistic approach, combining technical, commercial, and operational perspectives, among others, ensures a more balanced and robust design solution. The overall concept design development is benchmarked with peer vessel alternatives, including existing vessels. Hence, the concept design is validated, and potential points of improvement can be identified and rectified to improve the overall performance of vessel design solutions proposed.

This comprehensive approach requires a multidisciplinary design platform, combining the different aspects of maritime systems. Technical analyses such as stability, structural strength, and calm water and waves propulsion resistance. Hydrodynamic aspects such as seakeeping and operability, combined with the evaluation of capacities and capabilities give the operational perspective. The feasibility analysis of the configured solution, is assessed simultaneously in the tool, including the commercial perspective. Newbuilding price and operational expenses are then contrasted with the potential revenue capability and costs of the design solution. This fast-track evaluation of design performance enables designers and decision makers to better perceive the implications and consequences of individual design changes such as: main dimensions, mission equipment, operational environment, crew nationality, material, or build country.

Figure 16 shows a collage of result communication templates from the FTCDA resulting from real life project applications.

The implementation of FTCDA in early design phases has demonstrated three principal advantages: more robust decisions, higher quality of vessel design solutions—due to the availability of additional information at an early stage of the concept design process of the problem at hand. Other achievements include a significant reduction of response time and committed resources, and the capability of evaluating (visually and analytically) multiple design solutions. In addition, it brings the possibility of performing sensitivity analyses of cost, capacities, and capabilities towards

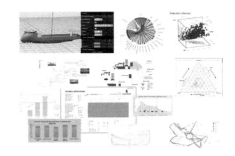

Figure 16. Collage of example results from Ulstein's vessel design solution toolbox.

specific design parameters. Complementary use of the FTCDA, ABD and other data-driven analytics tools allow us to validate and verify promising solutions very quickly. This again, has dramatically reduced the response time with customers.

6.4 *Ulstein Project Making (UPM™)*

Among practitioners and business executives there are obvious and concerning consequences and implications of not advancing the design approaches faster and with a broader scope: – The concept design and detailed engineering time for completing a vessel solution takes too long of a time and the overall process is too costly. – Sometimes the quality control and assurance are also suffering. – The communication and decision-making process is too often ineffective and partly high risk-based. – The fabrication phase also quite often is experiencing significant quality challenges and realized vessel solutions typically become more expensive than they had to! – Sometimes they are not even really "fit for purpose", because the purpose shift so dramatically and fast with fluctuating market conditions and business-related opportunities. Sometimes they also become severely delayed.

When a typical ship design-new building-project-process takes say, 24 to 36 months or more to complete, what seemed to be a perfect timing of an initiative can easily turn into a disaster business wise, because the peak and trough periods did not appear as expected. Exploring alternative dominant new building project development logic necessitates first unlearning the existing dominant logic. Historically, traditions and existing business and work procedures dominate the maritime industry and its members (Ulstein & Brett 2009).

Ulstein has responded to these challenges by building up a service portfolio consisting of the elements: Accelerated Business Development; catalogue and customised designs; systems engineering and integration; systems procurement; site support; commissioning assistance; and delivery follow up. Such a service concept is normally offered to 3rd party yards to secure that vessels are built to budget, delivery time and quality. These services have been offered to tenths of yards around the world.

6.5 *Field studies and big data analytics*

As a slight divergence from the most common interpretation of Big Data, field studies are a well-known method for the acquisition of operational data and the facilitation of detailed, holistic, and accurate information that typically is contained as tacit knowledge. They play an important role in acquiring contextual, systems-oriented, and human-centered knowledge from on-site operations and during execution generate an extensive amount of data. Sources of information include video, audio, pictures, interviews, physiological monitoring, notes, diagrams, and models in addition to the plethora of both on-board systems and provided third party sources such as the integrated automation system (IAS), automation systems for winches and cranes, dynamic positioning system, accelerometers, cargo load calculator, route planner, weather forecasts, radar imagery, task plans, operation logs, and so on (Abbasian et al. 2018).

The goal of field studies in a ship design process is to enable the ship designer to personally and physically experience the context for which he or she is designing, as well as to interact with the users he or she is designing for in their living and working context. This experience is expected to enrich the designer's judgment capacity (Lurås & Nordby 2014), which is an important foundation for the designer's ability to deliver creative and innovative solutions. Workshops are used in the field study as a process to work out, collect data, validate initial thoughts, and expand the field findings in a collaborative way.

Ulstein has introduced the human-centered, collaborative, field-driven design processes related to its ship design development work. It comes with the need for developing catalogue vessels with no given taker and shipowner present in the process. Figure 17 relates to the mapping between operation and design resulting from performing field studies.

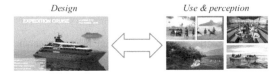

Figure 17. Examples of targets for field study investigation.

7 CONCLUSIONS

The way disruptive market conditions challenge existing vessel design firms, practices and tools application is discussed in this paper. Ulstein Group is gradually applying, getting recognition for, and sees the benefits of the initiatives here mentioned in enhancing and introducing new ways of handling the challenges of the dynamics in the boundary condition for operation. As stated, when the financial crises hit the world markets in 2008 and an economic downturn following from 2009 up until 2018, defaults on making profits and serving dept deteriorated quickly. International seaborne trade and offshore exploration and production activities have plummeted in this period (Brett & Jose 2018). Ships, particularly in the offshore oil and gas industry are laid up. In 2017 more than 1000 out of approximately 6000 offshore vessels was in layup. Only in Norway more than 150 vessels found its home, moored somewhere along the coastline. In less than 2 years (2015 and 2016) it became clear that such circumstances called for a drastic response.

We saw quickly the importance on how response time to customers and development of fast track concept design solutions could be used to save costs and resources. It is also concluded how renewal processes and means for change could be prescribed and implemented when ship design firms are subject to disruptive markets.

Our understanding is not that ship design firms have not seen the disruptive change coming. In most cases they have. Nor did they initially, lack resources to confront them. Most of them did have talented managers and specialists, strong product portfolios, first rate technological knowhow, and partly deep pockets, but typically, very specialised, few ships and narrow scope of their product and service portfolio. Of the approximately 600 recognised ship design firms in the world, of a certain size and recognition (Ulstein & Brett 2012), most of them, over time has developed some specialties and ship segment preferences. Only a few of them, but perhaps the largest, have a broad set of ship types included in their portfolio. This made the many ship design firms more vulnerable, than otherwise would be the case.

The paper continues describing how market conditions directly have impacted the shift in vessel buyer behavior, knowledge and experience building in the company, and why particularly, the vessel concept design process and analyses toolbox, needed to be refurbished and design work processes be executed differently than in the past.

We argue that elucidation and anticipation—the insight into, which predictability aspects of market behavior can be turned into own advantage, if the "wicked problem" is expanded upon in scope rather than being narrowed. It is also advised as to what interdisciplinary management tools from the decision-making paradigm can be applied to control better the uncertainties and complexity issues related to the renewal of ship design firms.

Based on the analysis and reasoning in this paper, it is concluded that ship design firms will have to grow larger to become viable. Healthy growth, however, is not engineered. It is the outcome of growing demand for special capabilities or of expanded or extended capabilities. It is the outcome of a firm having superior products and skills. It is the reward for successful innovation, cleverness, efficiency, effectiveness, efficacy and creativity. No one has an advantage at everything. Firms have advantages in certain kinds of rivalry under certain conditions and they are not lasting forever. We should press where we think we have an advantage and side step situations in which we don't. They have to improve their skills and expertise in a broader range of ship types. They have to broaden their portfolio of products and services and expand their engagement in customer relationships and knowledge development and adoption. Moving from being a design drawing package factory to become a partner-centered project maker stand out as vital factor for future renewal and success. More emphasis on playing the integrator role between the yard, suppliers and the ship owners stand-out as a must. A continued and strengthened interest and engagement in the pre-qualification of relevant and promising technology for ships design and related systems are paramount. More proactive participation with a broader range of stakeholders in the maritime industry is necessary. Time sitting and waiting for the ship-owner to call in for project work is over. If you cannot prove the extra value creation of your vessel design solutions, it is hard to believe that the new building projects involving your organization will take place.

The days of the "no cure no pay" practice in ships design industry is up for revision and new and better contractual relationships representing less asymmetry with respect to risk taking and cash flow requirements are overdue—ship designer cannot any longer bare the risk of delayed cash flow. Hence, new setups with a dominant Front-End Engineering Design (FEED)-oriented approach to ship design solution development are required. The customer has to pay for the development work. Finally, but not least, more ships owners must in more cases accept to choose off-the-shelf catalog vessels to retain competitiveness from a cost level standpoint. Series building of one and the same ship type is desirable—the days of the one-off solution is not over, but too much lesser extent will there be financial room and justification for special designs.

As final encouragement, we believe that the future naval architect must be able to develop both, the business case and ship design solution as one integrated delivery. It can be mentioned that at NTNU, where Ulstein is an effective industrial partner, and the authors also affiliated to, several courses have been and will be adjusted in such a way that students can get trained in such business-centered and modern data-driven methods in the naval architecture approaches. It is not so much the design of the vessel, as it is the design of the integrated business and vessel design solution that counts.

ACKNOWLEDGEMENTS

This work reports on the advances made in the Ulstein Group ASA and its affiliated companies to expand their future attractiveness and competitiveness. Much of the theoretical foundation and the exploration of improved methodology are based on previous developments being presented in IMDC 2006, 2009, 2012 and 2015 IMDC Proceedings. The authors acknowledge their sincere appreciation of contributions from colleagues in the Ulstein Group, as well as collaborating experts and researchers in industry and in academia, nationally and internationally.

Yet, this paper reflects and presents the authors' viewpoints and the respective companies mentioned are not liable or guaranties for any statements made or used, that may be made of the information contained in this paper.

REFERENCES

Abbasian, N.S., Salajegheha, A., Gaspar, H.M. & Brett, P.O. 2018. Improving early OSV design robustness by applying 'Multivariate Big Data Analytics' on a ship's life cycle. *Journal of Industrial Information Integration.*

Andrews, D.J. 2003. A Creative Approach to Ship Architecture. Discussion and Author's response *International Journal of Maritime engineering*, Trans RINA Vol. 145, 146.

Andrews, D., Papanikolaou, A., Erichsen, S. & Vasudevan, S. 2009. State of the art report on design methodology. *Proc. 10th International Marine Design Conference.* Vol II. Trondheim. Norway.

Brett, P.O., Carneiro, G., Horgen, R., Konovesis, D., Oestvik, I., & Tellkamp, J. 2006. LOGBASED: Logistics-based ship design. *Proc. 9th International Marine Design Conference.* Vol II. Ann Arbor: Michigan. USA.

Brett, P.O. & Garcia, J.J. 2018. *UIN Monthly Report.* January 2018. Ulstein International AS. Ulsteinvik. Norway.

Bravo, C.E. et al. 2014. State of the Art of Artificial Intelligence and Predictive Analytics in the E&P Industry: A Technology Survey. *SPE Journal*, 19(4), pp. 547–563.

Chen, H., Chiang, R.H. & Storey, V.C. 2012. Business intelligence and analytics: From big data to big impact. *MIS quarterly*, 36(4), pp. 1165–1188.

Child, J. 1997. Strategic Choice in the Analysis of Action, Structure, Organizations and Environment: Retrospect and Prospect. *Organizational Studies*, 18, pp.43–76.

Copeland, M. 2016. *What's the Difference Between Artificial Intelligence, Machine Learning, and Deep Learning.* NVIDIA.

Cyert, R.B & March J.G. 1992. A Behavioural Theory of the Firm. Prentice Hall, Englewood Cliffs, New Jersey. USA. ISBN 0-6311-7451-6.

De Mauro, A., Greco, M. & Grimaldi, M. 2014. What is big data? A consensual definition and a review of key research topics. Proc. 4th International Conference on Integrated Information. Madrid. Spain. pp. 97–104.

Ebrahimi, A., Brett, P.O., Gaspar, H.M., Garcia, J.J. & Kamsvåg, Ø. 2015a. Parametric OSV Design Studies – precision and quality assurance via updated statistics. Proc. 12th *International Marine Design Conference.* Tokyo. Japan.

Ebrahimi, A., Brett, P.O., Garcia, J.J., Gaspar, H.M. & Kamsvåg, Ø. 2015b. Better decision making to improve robustness of OCV designs. Proc. 12th *International Marine Design Conference.* Tokyo. Japan.

Erikstad, S.O. 2007. Efficient Exploitation of Existing Corporate Knowledge in Conceptual Ship Design. *Ship Technology Research.* 54(4), pp. 184–193.

Erikstad S.O. 2009.Proceedings of the 10th International Marine Design Conference, *Proc. 10th International Marine Design Conference,* Vol I and II. Trondheim. Norway.

Garcia, J.J., Brandt, U.B. & Brett, P.O., Unintentional consequences of the golden era of the Offshore Oil & Gas Industry. *Proc. 1st International Conference on Ships and Offshore Structures.* Hamburg. Germany.

Gaspar, H.M., Brett, P.O., Ebrahimi, A. & Keane, A. 2014. Data-Driven Documents (D3) Applied to Conceptual Ship Design Knowledge, *Proc. 13th International Conference on Computer Applications and Information Technology in the Maritime Industries.* Redworth. UK.

Gaspar, H.M., Hagen, A. & Erikstad, S.O. 2016. On designing a ship for complex value robustness. *Ship Technology Research.* 63(1). pp. 14–25.

Glen, I. 2001. Ship Evacuation Simulation: Challenges and Solutions. *SNAME Transactions*, Volume 109. pp. 121–139.

Ison, R. 2010. Systems Practice: *How to Act – In situations of uncertainty and complexity in a climate-change world.* The Open Univeristy. Springer-Verlag London Ltd. London. UK. ISBN 978-1-4471-7350-2.

Keane, A., Brett, P.O., Ebrahimi, A., Gaspar, H.M. & Garcia, J.J. 2017. Proc. 16th International Conference on Computer Applications and Information Technology in the Maritime Industries. Cardiff. UK.

Lee, S.G., Ma, Y.S., Thimm, G.L. & Verstraeten, J. 2008. Product lifecycle management in aviation maintenance, repair and overhaul. *Computers in industry.* 59(2). pp. 296–303.

Levisauskaite, G., Gaspar, H.M. & Ulstein, B. 2017. 4GD Framework in ship design. Proc. 16th International Conference on Computer Applications and Information Technology in the Maritime Industries. Cardiff. UK.

Li, G., Skogeng, P.B., Deng, Y., Hatledal, L.I. & Zhang, H. 2016. Towards a Virtual Prototyping Framework for Ship Maneuvering in Offshore Operations. *Proc. IEEE Oceans.* Shanghai. China.

Ludvigsen, K.B., Jamt, L.K., Husteli, N. & Smogeli, Ø. 2016. Digital Twins for Design, Testing and Verification Throughout a Vessel's Life Cycle. *Proc. 15th International Conference on Computer and IT Applications in the Maritime Industries.* Lecce. Italy, pp. 448–456.

Lurås, S. & Nordby, K. 2014. Field studies informing ship's bridge design at the ocean industries concept lab. *Proc. International Conference on Human Factors in Ship Design & Operation.* London. UK.

Maslin, E. 2017. Neural networking by design. *Offshore Engineer Magazine.* pp. 26–27.

Mazaheri, S. 2006. The Usage of Artificial Neural Networks in Hydrodynamic Analysis of Floating Offshore Platforms. *International Journal of Maritime Technology.* 3(4). pp. 48–60.

Meyer, G.D. & Heppard, K.A. 2000. *Entrepreneurship as strategy: Competing on the Entrepreneurial Edge.* SAGE Publications, Inc. London. UK. ISBN 0-7619-1580.

Moore, R. 2016. Calmac and Stena aim for standardized fleet, *Passenger Ship Technology*, 28.04.2016.

Proff. 2018. Proff.no. [Online] Accessed: 24.01.2018.

Reeves, M, Haanes, K & Sinha, J, (2015) Your Strategy needs a Strategy. *Harvard Business Review Press.* Boston: USA. ISBN 978-1-62527-586-8

Rittel, H.W.J & Webber, M.M. 1973. Dilemmas in the general theory of planning policy sciences. *Policy sciences.* 4(2). Pp. 155–169.

Rummelt, R. 2011. *Good Strategy – Bad Strategy: The Difference and Why It Matters.* Profile Books Ltd. London. UK. ISBN 978-1-84668-4807.

Stopford, M. 2009. Maritime Economics. 3rd ed. Unwin Hyman Ltd. London. UK. ISBN 0-415-27558.

Tang, J., Deng, C., Huang, G.B. & Zhao, B. 2015. Compressed-domain ship detection on spaceborne optical image using deep neural network and extreme learning machine. *IEEE Transactions on Geoscience and Remote Sensing.* 53(3). pp. 1174–1185.

Tikici, M., Omey, E., Derin, N., Seckin, S.N. & Cereoglu, M. 2011. Operating turnaround strategies during crises periods: a research on manufacturing firms. *Procedia Social and Behavioral Sciences.* 24. Pp. 49–60.

Uddin, M., Jameel, M., Razak, H.A. & Islam, A.B.M. 2012. Response prediction of offshore floating structure using artificial neural network. *Advanced Science Letters.* 14(1). pp. 186–189.

Ulstein, T. & Brett, P.O. 2009. Seeing what's next in design solutions: Developing the capability to build a disruptive commercial growth. *Proc. 10th International Marine Design Conference.* Vol I. Trondheim. Norway.

Ulstein, T. & Brett, P.O. 2012. Seeing what's next in design solutions: Developing the capability to build a disruptive commercial growth. *Proc. 11th International Marine Design Conference.* Vol I. Glasgow. UK.

Ulstein, T. & Brett, P.O. 2015. What is a better ship? – It all depends... *Proc. 12th International Marine Design Conference.* Vol I. Tokyo. Japan.

Towards maritime data economy using digital maritime architecture

Tommi Arola
Head of Unit, Mobility Innovations and R&D, Finnish Transport Safety Agency, Finland

1 INTRODUCTION

The robotics and intelligent automation is becoming our daily business in transport sector. Today digitalization and platform economy is challenging our economic structures and our way of doing business. Tomorrow robotics may also challenge our business areas and traditional step-by-step processes. This has happened or happening in many sectors such as banking. The Economist journal (2017) referred data as a fuel of the future: "Data are to this century what oil was to the last one: a driver of growth and change".

Shipping is a traditional business area but including digital capabilities. Maritime in overall includes collection step-by-step processes empowered by wide variety organisations with specified tasks. There are signs that also data and platform economy entering to the shipping business as well. The data is seen as a crucial player to make the change happen and to adapt to change and make added value to the customer. Maritime is traditional in many aspects but the level of automation in ship operations is high together with the vast usage of electronic information in decision making. The level of automated processes and operations varies on-board, onshore and harbours. In general the level of automation varies in different parts of maritime domain and the maritime voyage. Still a lot of computer aided routine work is done by human such as port operations, cargo optimization or coordinating the supportive task.

In political context the value of EU data economy is said to be worth of €285 billion according to European Union (2014) definition of digital single market. Additionally the data should be accessible and reusable by most stakeholders in an optimal way.

In 9th November 2017 Ms. Violeta Bulc, the EU Transport Commissoner (Bulc, 2017), had a speech at Digital Days 2017 Tallinn. She pointed out that it is important that we build a digital architecture and an efficient, open but secure digital ecosystem. She continued that it is important that we have open and common standards and interfaces where interoperability happens. Interoperability is the keyword which also sets the great challenge because the legacy systems are in place.

In maritime the interoperability is quite interesting because the maritime is intensive to processes, various stakeholders and value making. In maritime small changes can open a great opportunities makes the maritime be a motivation.

2 PROBLEM SETTING AND AIM OF THE PAPER

In maritime we see an oncoming pressure to increase interaction between stakeholders. It happens due to make maritime traffic more efficient, the processes more flexible and digitally connected. Digital connectivity increases in all organization due to the pressure outside by customers and other organizations. Looking from the multimodality point of view it's crucial to understand that freight should move seamlessly from a traffic mode to another. One big unsolved question is how to deal with interoperability and whose role is to take care of it. In maritime the organization boundaries are changing and the roles seems to be mixing. At the same time the transport data should flow from a transport mode to another and enable new business to grow around all modes of transport.

Today many organizations should think their digital connectivity around the system they are and the new added value to be created. Apparently there's no general framework or architecture for digital connectivity or data description. Actually, such a digital connectivity framework is missing in other transport modes as well which delimits the understanding of digital ecosystem and interoperability. In data economy the question is not anymore who has the monopoly for the data. The question is how to make the company interoperable and build new capabilities to exchange the data with the domain around your businesses.

This is a background discussion paper for a keynote presentation held in 13th International Marine Design Conference. This paper is not a scientific but opens new discussion and further research needs in maritime digitalization. In this paper we discuss the need of maritime domain specific digital architecture to increase interoperability. We present a case example of domain specific digital architecture and an application of it in maritime specific domain focusing on data and information interoperability. We discuss the elements of data economy and how these fit to maritime domain. We draw also attention to research needs in maritime domain specific digital architecture and the maritime data.

3 REFERENCES AND LITERATURE

3.1 *The digital architecture*

Wikipedia defines enterprise architecture as principles and practices to guide organizations through the business, information, process, and technology changes necessary to execute their strategies. There is no clear definition available for the digital architecture but digital architecture is defined as digital business requirements for enterprise architecture. Bossert et al (2014) states various functionality that the digital enterprise requires based on enterprise architecture:

1. Two-speed architecture for fast customer centric front end and slow speed transaction focused backend
2. Instant cross-channel deployment of functionality. New microservices defining only a small amount of functionality should be deployable in an hour rather than in several weeks.
3. Zero downtime. In digital global operations, days-long maintenance windows are no longer an option.
4. Real-time data analytics. Customers generate data with every move they make within an app.
5. Easy process configuration. Business users themselves should be able to change automated processes.
6. Product factory. Industries that provide digital products, such as banking and telecommunications, need to decouple the products from the processes.
7. Automated scaling of IT platforms. In a digital business, workloads expand and become harder to predict.
8. Secure architecture. In a digital business model, cybersecurity must be an integral part of the overall application.

3.2 *Digital business*

Digital business has been defined by Andrew (2015) that it is specifically focused on the peer exchange and communication between business (including process and information), people and physical things as equal entities. Digital business is blurring the physical and digital worlds. Andrew (2015) raises that a digital business must include the connection or integration with assets (business, people and things) beyond IT and beyond the control of any one company.

3.3 *Maritime domain*

Maritime domain is defined by IMO and US Navy. White (2014) defines the maritime domain as follows: "the Maritime Domain is all areas and things of, on, under, relating to, adjacent to, or bordering on a sea, ocean, or other navigable waterway, including all maritime-related activities, infrastructure, people, cargo, vessels, and other conveyances."

3.4 *The data economy*

EU Commission (2014) presented a prominent feature of a data-driven economy will be an ecosystem of different types of players interacting in a Digital Single Market, leading to more business opportunities and an increased availability of knowledge and capital, in particular for SMEs, as well as more effectively stimulating relevant research and innovation. The data economy in EU level consists of the three issues:

- Availability of good quality, reliable and interoperable datasets and enabling
- Improved framework conditions that facilitate value generation from datasets
- A range of application areas where improved big data handling can make a difference

EU Commission (2014) presented also a vision for the data-driven economy as an ecosystem with different types of players (e.g. data providers, data analytics companies, cloud service providers, companies from the user industries, venture capitalists, entrepreneurs, public services, research institutes and universities), leading to more business opportunities, in particular for SMEs. The availability of good quality, reliable and interoperable datasets was specifically highlighted as an important enabler for new data products.

McKinsey (2014) showed that cross-border data flows generated $2.8 trillion in economic value exceeding the value of global trade in goods. This shows the dynamism of the technology industry, but also the digitization of the economy as a whole. EU Commission (2014) pointed also out that the digital trade is crucial for nearly all firms, from large multinationals to small businesses that rely on online platforms to connect and trade with customers around the world.

Today we are in very beginning establishing data economy but it is emerging continuously. In 2017 Finnish Minister of Transport and Communications Ms. Anne Berner (2017) proposed to define overarching principles for data use that will then be applied through sectorial work. Ms. Berner continued that we should achieve interoperability and the free flow of data within and between different sectors, platforms and services, and we need to acknowledge the rights of individual users, citizens and businesses.

3.5 *Case example A domain-specific architecture framework for the maritime domain*

Weinert et al. 2016 represented an application of maritime domain specific architecture framework (MAF). This was done to reach common methodology to align and integrate existing system architectures in the maritime domain. It's a standardized methodology to assemble existing architectures in a meaningful way to identify interoperability issues, interfaces and links to other (upcoming) systems.

The maritime domain specific architecture framework was researched from eNavigation point of view which aims to enhance ship-shore communication in digital means. Weinert at al reflected that from eNavigation point of view current systems are barely integrated and combined with each other. Currently applied solutions also do not support a domain-wide information exchange. In addition, they do not follow the e-Navigation strategy and thus they are less future-oriented but well embedded in the maritime domain infrastructure.

The paper found out that it's a challenge to harmonize existing systems which are stand-alone solutions for each use case. The challenge is also to integrate new approaches and technologies into existing technical and organizational structures for sustainable, reliable and safe maritime transportation. Currently applied solutions also do not support a domain-wide information exchange.

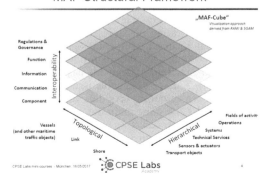

Figure 1. MAF-cube (Weinert et al (2016) and Brinkmann (2017)).

Weinert et al (2016) presented a first known maritime architecture framework (MAF). It has been derived from the energy sector and established on the basis of Smart Grid Architecture Model (SGAM). SGAM is enterprise architecture featured by European Committee for Electrotechnical Standardization (CEN-ELEC). Smart Grid Architecture Model (SGAM) is a successfully established implementation of an Enterprise Architecture Framework to address domain specific issues in the electric utilities domain. SGAM was developed to handle the complexity of the Smart Grid system-of-systems approach with focus on interoperability and standardization aspects for business and governance as well as for technical issues.

Weinert at al. (2016) adopted the MAF from the SGAM model by a community process. The paper reflects the needs to establish clear relationships between technical systems, user and related governance aspects including:

- existing business objectives, that explain the benefits of the systems,
- regulation and governance aspects, which regulates the maritime domain,
- technical functions, that are required to realize the business objectives,
- information exchange between those technical functions including the related information types and/or data models,
- communication protocols to allow the aspired information exchange and
- components, which are required to implement the technical hardware in the system.

The MAF is illustrated as cube (MAF-cube). To support the governance aspects the MAF-cube consist of three different focus (axes) to assess the framework:

1. The topological axis represents the logical location where a technology component is located.
2. The interoperability axis addresses communication, data and information, usage and context of a maritime system.
3. The hierarchical axis substructures management and control systems of the maritime domain interoperability aiming.

Topological axis is a break down from IMO's maritime domain entities for eNavigation and cover the structure of the maritime domain in a logical location: vessel, link and shore. Ships are the maritime traffic objects, links are the physical entities interacting with maritime traffic and shore is side infrastructure, activities and systems on shore including interfaces to logistical movements in/out of maritime domain.

The interoperability axis covers organizational, informational and technical aspects and includes the different levels of interaction: regulation & Governance, function, information (data and information that is being used and exchanged), communication (mostly protocols for information exchange) and component (like systems, actors, applications).

4 ENHANCING THE INFORMATION INTEROPERABILITY

To fully adopt data economy objectives it requires good capabilities and understanding of the data that facilitates the transport operations. The data should also flow from a transport mode to another and enable new business to grow around all modes of transport. In this context there's a need to put effort on the MAF-cube information layer. Weinert et al (2016) defines the MAF-cube information layer from the eNavigation point of view. The scope seems to need more widening in terms of maritime digitalization and data economy needs.

According to White (2014) the maritime domain includes a lot of maritime related functions that relates to sea. There are several data intensive processes which share the same maritime information in information layer: all port operations, pilotage, shipping agents, certifications, reporting formalities, fairway maintenance, hydrography, vessel traffic service, freight forwarding, ship inspection, ship design, ship new building and coast guard functions. These processes represent entities in MAF-cube in a layer called "function". All these processes uses, produce and enrich the maritime domain specific information and data from the layer called "information". The processes use technologies, they communicate in some manner and do some operations.

Next we are going to make a logical data model for the MAF-cube "information"-layer in order to stick on data availability in maritime. We'd like to also create a basis for the interoperability of these maritime processes from a data perspective.

4.1 *Maritime data main data grouping*

The common understanding of information or data needs a main data grouping. The main data grouping is a practical decision of groups how the data is organized between the relevant stakeholders. The idea is each stakeholder can map their own data or relevant parts of the enterprise architecture under the same groups. It isn't exactly a technical description but a logical description of data structure. The main data grouping can be seen as a lowest common denominator in establishing a data or information architecture between several stakeholders who are willing to share their data. It's a part of enterprise architecture modeling in

information architecture but done from a multi-organization point of view.

It's important because the process development face of the very common questions: *"what is the data you have"*, *"what data there is"* and *"how to use it"*. Generally speaking about the data architectures, there are only highly technical IT-descriptions and very simple data lists but nothing between including a logical relationships or commonalities.

The comprehensive grouping of maritime related data or information is very hard to find. A report from Finnish Transport Agency (2010) mapped an overview of maritime related data and information from the cross-sector. The report is limited to authorities, a couple of companies and other stakeholders. The report ended up to categorize maritime data in the following categories (derived from Finnish text):

- Cargo information
- Dynamic traffic information
- Route and timetable information
- Weather conditions
- Traffic limitation information
- Information about maritime authorities and services
- Ship information
- Traffic exemptions
- Hydrography information

White (2014) domain definition leaves a gap in data groups. There are some major data and information which relates to maritime but are missing:

- Information or data about organizations
- Information or data about authorized persons
- Company performance data
- Status data
- Ship design data
- Reporting formalities data
- Certificates

All these main data groups includes a additional entities e.g attributes like x,y co-ordinates or ship's estimated time of arrival (ETA).

Figure 2 is an example of main data classification concentrating on ship. All the main data groups that relates to ship are logically bounding. The headlines for the main data groups can be decided in group and in this case this is just an example. The most valuable issue is to do the mapping each organization's data or information entities under the same main data groups. Ideally each organization can map their data into main data groups and link these into organization internal enterprise architecture descriptions. However, during the modeling it is important to agree about level of description and policy because the data or information might cause conflict of interest in some partners.

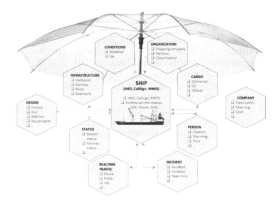

Figure 2. Example of main data grouping for maritime.

5 DISCUSSION: MARITIME DOMAIN SPECIFIC DIGITAL ARCHITECTURE

There is a lot of maritime data but it is scattered. The data is everywhere and you have to see it. There's no general maritime data model available and it makes the maritime process interoperability complex at this time. In maritime there is a lot of data oriented processes but they are happening by a human. According to DNV (2015) these processes are a target for maritime digitalization and automation in the future. The big change is happening in the backend processes such as reporting formalities and information gathering and sharing manually. There are old legacy systems without interoperability and a lot of repetitive work and a lot of manual information searching. The maritime digitalization requires definitely a general data model and a common interoperability framework to fully adopt the all benefits of digitalization.

From Weinert et al (2016) the MAF-cube is a first application to approach this interoperability from a various perspectives. MAF-cube turned out to be a good basis for the digital domain specific architecture and it supports well the data economy principles to improve framework conditions that facilitate value generation from datasets.

We did one example of main data grouping for information layer in MAF-cube. This main data grouping for information layer can be used to establish a discussion with other maritime stakeholders and renewing the processes. The benefit in approaching maritime domain specific architecture is for new business process or a technology adoption. Various stakeholders can logically map themselves for the layers and use that as a communicative tool.

From the research point of view there quite many open questions such as:

- What should be the structure of maritime digital architecture and who maintains it?
- How to model the other layers and the relationships?
- How to make the interoperability and data sharing happen in concrete?
- What standardization needs to be done?
- How different digitalization focus areas are tied together?
- What technologies there are now and in the future.

The research around the digital architecture is very minimal. There is hardly any information about the digital architecture characteristics mentioned in chapter 3.1. It also turned out also that there are hardly any maritime related data, architecture or interoperability research done which is a concrete future need.

6 CONCLUSION

Maritime seems to be a very good example of a domain for a data economy development: large data volume, big international companies and traditional business but rather new technology. The biggest near future opportunities can be seen in back-process automation and digitalization. Digital domain specific architecture seems to be good approach but needs a step-by-step doing, a lot of more definitive work in a concrete focus area like data interoperability. Especially the Bossert et al (2014) requirements for enterprise architecture are good to remember in adopting digital business capabilities for companies.

There were also two political commitments in this paper: data economy and digital architecture. It turned out to be possible to find a context and a solution for these and use maritime in these.

REFERENCES

Berner, A., 2017, From Well Begun to Data-Driven by Default Permanent representation of Finland to the European Union news 5/22/2017.

Bossert Oliver, Chris Ip, Laartz Jürgen, A two-speed IT architecture for the digital enterprise, McKinsey Digital, 2014.

Brikmann, M., 2017, Presentation "CPS Engineering Labs Mini-Courses Testing Maritime Cyber-physical Systems", cited 3.3.2018, url: http://www.cpse-labs.eu/downloads/MunichMay17/2_3.pdf.

Bulc, Violeta, 2017. Speech at Digital days Tallinn 9.11.2017, URL: https://ec.europa.eu/commission/commissioners/2014–2019/bulc/announcements/speech-commissioner-bulc-tallinn-digital-transport-days_en.

CEN-CENELEC-ETSI Smart Grid Coordination Group, Smart Grid Reference Architecture, 2012.

DNV, 2015, Ship Connectivity, DNV GL STRATEGIC RESEARCH & INNOVATION POSITION PAPER 04–2015.

European Commission, Communication on "Towards a thriving data-driven economy", 2014, URL: https://ec.europa.eu/digital-single-market/en/towards-thriving-data-driven-economy).

The Economist journal, May 6th 2017, Data is giving rise to a new economy, url: https://www.economist.com/news/briefing/21721634-how-it-shaping-up-data-giving-rise-new-economy.

Finnish Transport Agency, 2010, Meriliikenteen tietoaineistojen arkkitehtuurin nykytila, in finnish, url: https://julkaisut.liikennevirasto.fi/pdf3/lv_2010-01_meriliikenteen_tietoaineistojen_web.pdf.

McKinsey Global Institute, "Digital Globalization: the New Era of Global Flows", 2016.

Spender, Andrew, 2015, Digital Business 101, Gartner, cited 3.3.2018, url: https://www.gartner.com/smarterwithgartner/digital-business-101-2/.

Weinert, B., Hahn, A., Norkus, O., 2016, A domain-specific architecture framework for the maritime domain, Informatik.

White, Jonathan W., 2014, Advancing Maritime Domain Awareness For The Fleet And The Nation, cited 3.3.2018, url: http://www.doncio.navy.mil/chips/ArticleDetails.aspx?ID=4911.

Is a naval architect an atypical designer—or just a hull engineer?

David Andrews
University College London, London, UK

ABSTRACT: As the demands for future ships become ever greater, due to economic pressures to achieve "value for money" and due to assumptions of more precision in potential ship solutions, then the question to be addressed is whether the naval architectural profession is still best placed to lead in designing complex ships. Other disciplines might be seen to be more relevant in meeting specific ship demands, such as the marine engineer in achieving better fuel efficiency and greener solutions or the combat systems engineer for future naval vessels. Beyond these two disciplines the complexity of particularly naval ship design has led to the generic project management discipline of systems engineering being promoted as more appropriate than naval architecture as the lead discipline. Thus the naval architect becomes a mere "hull engineer" practicing the specific "naval architectural" sub-disciplines, instead of being "primes inter pares" in managing ship design and acquisition. Such a proposal arises both from a belief that the whole ship safety issues need the senior most naval architect's main attention and that skills in systems engineering rather than the naval architect's design skills are best for the overall management of design and acquisition, due to its agnosticism with regards to the cross disciplinary conflicts that arise in such a highly interactive multi-disciplinary exercise.

This issue is explored by considering what are the essential engineering skills employed by a naval architect as the ship equivalent, for large constructional projects, of a terrestrial civil engineer and whether this is just "hull engineering" or something more like the ship equivalent of an architect for major constructions, such as airport termini. This leads on to consideration of the whole ship designer being both the creator of an initial design synthesis as well as maintaining downstream the overall design coherence through exercising design authority for the design's existence. A series of pertinent views on ship design and relevant case examples are considered in order to address these issues beyond broad generalities. These examples include historic cases of "good" and "bad" ship designs and what might have contributed to such subsequent conclusion as to those designs' veracity. Beyond actual built ship designs, case studies produced at UCL both by MSc student and by the author's Design Research centre are presented to provide the basis for refuting the view that disciplines, other than naval architecture, can effectively lead future ship designs. However such a conclusion is only seen to be defensible if the naval architectural profession gives as much emphasis to its understanding and practice of ship design as it gives to its traditional "hull engineering" responsibilities.

> "He ends, of course, by satisfying neither the Commander who is responsible for the men's living conditions nor the Gunnery Officer who is responsible for the guns, but that is the natural fate of the designers of ships – the speed enthusiasts, the gunnery experts and the advocates of armour protection, the men who have to keep the ships at sea and the men who have to handle them in action all combine to curse the designer.
>
> Then comes the day of battle and the mass of compromises, which is a ship of war, encounters another ship of war, which is a mass of different compromises, and then, ten to one, the fighting men on the winning side will take all the credit to themselves and the losers – such of them that survive – will blame the designer all over again."
>
> C.S. Forrester "The Ship" (1942)

Keywords: Naval architect, ship designer, hull engineer, Complex ship design

1 INTRODUCTION – WHAT IS A NAVAL ARCHITECT

It seems appropriate in a regular forum that is focused on marine design to question whether the engineering discipline—that of naval architecture—which has to date dominated ship design practice, still remains best placed to continue in that role. While it is always good to ask such fundamental questions regarding professional practice, it seems we are at a particularly appropriate and pertinent time in the maritime domain to investigate this issue. Not only is the global political, environmental and economic situation evermore highly integrated and interdependent, we are also in an era of very rapid technological change. This is particularly the case with regard to communications, access to data and increasing recourse to expanding application of "artificial intelligence" in the manner in which any sophisticated process, such as marine design, is being and is to be conducted. As this trend is likely to accelerate, the issue as to how we conduct "marine design" is well worth exploring.

The naval architect is that essentially engineering professional concerned, as stated by the Royal Institution of Naval Architects, as the founding professional institution concerned with the maritime sector, from a, mainly, engineering stance that the profession is: "to promote and facilitate the exchange and discussion of scientific and technical developments ... and thereby to improve the design of ships" (Blakeley, 2010).

Interestingly, the name of the engineering discipline "concerned with design etc. of ships etc.", does not contain the term "engineering" in its title (unlike the larger more general engineering institutions in the UK (as the nation that founded such learned societies) and elsewhere). Thus in the early years of the Industrial Revolution, builders of large scale buildings (master masons) became one of two separate professions, civil engineers and architects, while master shipwrights became naval architects (German, 1978 and Brown, 1983). Such developments in professionalization from medieval craft guilds were led by the UK due to its formative role in large-scale industrialisation. The term naval architecture was adopted from the Latin for ship (i.e. not limited to naval ship design) together with the title already adopted in the slightly earlier professionalization of the design of buildings on a grand scale. Thus architects became responsible for the design and the building management, superceding master masons. The origins of the term architect being from two Greek words, that for leader (*archi*) and that for builder (*tecton*), which might also seem to bookend the issue this paper sets out to address, however whether that is also appropriate to the future practice of ship design is open to discussion?

There is also a view, at least with regard to the primary professional education of naval architects, that the nature of naval architecture is essentially "applied mechanical engineering" (Rydill, 1986) rather than warranting a wholly separate first degree. A similar argument could also be made for aeronautical engineering, which also has first degrees in its discipline, and for automotive engineering, which tends to stick with general mechanical engineering for first degree education. Having a broad engineering first degree would seem consistent with the primary distinction between civil and mechanical engineering given the latter focus on machinery, especially steam and internal combustion engines, rather than the civil engineer's focus on structural design. This is somewhat ironic since many naval architects expend a lot of ship design effort on the structural design of their new ships (Rawson & Tupper, 1976), albeit alongside applying hydrostatics and hydrodynamics. This raises the question as to why we aren't "applied civil engineers" as the maritime equivalent of the engineers of buildings and large (civil) engineering complexes, such as bridges, dams and rail, dock and air termini. The alignment with mechanical engineering is much more obvious for the other traditional maritime engineering discipline, namely the marine engineer, who is clearly a "specialised" mechanical engineer. All this would seem to justify the term architect as still being appropriate for the ship (and other maritime structures) equivalent of the architect of the (land based) built environment, rather than just equivalent to the civil engineer. Architect particularly captures the holist and creative design role that is still retained by the urban architect, even if modern "celebrity" architects, like Gehry and Hadid, rely on the structural engineer to practically realise their creations.

It is clear that aside from much of what is seen to constitute naval architecture as "applied (mechanical) engineering", it is the ship design role that gives the naval architect his/her unique role. This is recognised in naval architecture textbooks, thus "ship design is the *raison d'etre* of naval architecture" (Rawson & Tupper, 1976), even though most of such books are taken up with the applied engineering sciences that lead to the discipline being seen as essentially an engineering, rather than a design, discipline. This then leads to the dichotomy posed in the title of this paper, namely, "hull engineer" or designer of ships, which then leads to the corollary that if the naval architect is the latter, then can he or she also exercise the lead discipline in ship design?

The next section of this paper considers the essential engineering skills employed by a naval architect as the ship equivalent of a terrestrial civil engineer, for large constructional projects, and whether this is just "hull engineering" or something more like the ship equivalent of an architect for major construction, such as airport termini. This leads on to reviewing the task of whole ship design, seen as being both that of initial design synthesis as well as maintenance downstream of the overall design coherence and exercising design authority through the design's existence. A series of ship case examples are explored in order to address this task beyond broad generalities. These examples include historic cases of "good" and "bad" ship designs and what might have contributed to such subsequent conclusions as to those designs' veracity. Beyond actual built ship designs, case studies produced at UCL, both by MSc student and by the author's design research team, are presented to provide the basis for refuting the view that disciplines, other than naval architecture, can effectively lead future ship designs. However such a conclusion as to the essence of the role of the naval architect, discussed in the penultimate section, is only seen to be defensible if the naval architectural profession gives as much emphasis to its understanding and practice of ship design as it gives to its traditional "hull engineering" responsibilities.

2 WHAT IS A HULL ENGINEER?

Given that not all engineers qualified as naval architects are directly involved in ship design, the above section's title poses a question that might well need clarifying with regard to both the necessary qualifications to practice and what actually constitutes ship design. Nevertheless, most practicing naval architects are involved in much of what is taken to constitute naval architecture by undertaking tasks, which may or may not be of a direct ship design nature (see the following section) but are applying some of the engineering sub-disciplines taken to constitute naval architecture. These can be conveniently listed under the "hull engineering" banner, to avoid the term naval architecture, given Rawson and Tupper's over-riding emphasis that naval architecture is directly the practice of ship design.

A convenient taxonomy regarding "hull engineering" is provided by the term that the author and a colleague "invented" to cover the naval architect's concerns in designing a ship – "S^5" (see Figure 1 from Brown & Andrews, 1980). The first four of these "S" terms could be said to cover the sub-disciplines of "hull engineering" and the last (that of Style) covers wider design concerns, and so is more appropriately addressed in the next section. The first four terms were identified as: Speed, Stability, Strength and Seakeeping, and at least those other than "Speed" are very much about ship safety, which remains a primary design responsibility, where the naval architect is both "hull engineer" and "whole ship designer". Each of the "S^4 "considerations are discussed below:

2.1 Speed

Speed really encompasses resistance and propulsion plus the need for endurance. Given in most merchant ships this is a significant economic

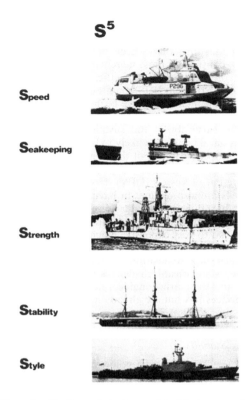

Figure 1. S^5 – Brown & Andrews original design examples.

(and now environmental) driver, the marine engineer and naval architect have to work closely on the design choices. Unlike most merchant vessels, service vessels and particularly naval ships are characterised by the need to operate across a spectrum of speeds. Thus a high top speed (and very good manoeuvrability at top speed) is required for action to pursue enemy units or to take avoiding action. This need for high speed in extremis then governs the choice of main machinery (strictly the marine engineer's "part of ship") and the hull form (very much the naval architect's responsibility), which for reasonably fast monohulls results in a long, slender (L/B > 8 or 9) and a finely shaped underwater form. However such high speeds (typically around 30 knots) are expensive in fuel consumption if sustained for long distances, so for most ocean going naval vessels endurance is usually defined as (say) 6,000 nm at an endurance speed of 18 or 20 knots. This then requires installation of other engines than those providing the power for full speed, which then can deliver considerably less power for these very much less resistful speeds. This then ensures the size of the fuel tanks is kept as small as possible. Such vessels operating at variable speeds tend to operate for most of their careers at relatively slow speeds and so are way off the hydrodynamically efficient top speed for which their hull forms have been optimised. This is in stark contrast to most merchant vessels, which as part of an economically efficient transportation system (Erichsen, 1978) maintain a constant speed for which their hulls have been hydrodynamically optimised.

For survivability and sustainability considerations naval vessels are likely to have two shafts and two propellers in all but the slowest and less capable combatants, such as some corvettes and Offshore Patrol Vessels (OPV). However, many naval vessels are also distinct in having a primary need to reduce their underwater noise signature to minimise their detection by submarines. Not only can this lead to very expensive machinery and propulsion arrangements, which further increase the cost of such ships acquisition and through life support, it poses significant challenges to the naval architect and the marine engineer at their many design interfaces in what are already highly sophisticated vessels.

2.2 *Stability*

The need for all ocean going vessels to resist the most extreme seas and survive a reasonable level of hull damage is more extreme for naval ships (as usually these have to meet the design limit of three major watertight compartments being breached), which has meant recent combatants have proportionately increased their maximum waterline beam. This has had the consequence that trying to maintain high top speeds is usually harder to achieve and results in a clear conflict between these first two "S" aspects, something not seen as starkly in merchant ship design. The need for modern naval vessels to have ever larger radar antennas at the top of high masts, to increase the range of detection in-coming missile (see Figure 2 of the RN Type 45 Destroyer), coupled with light-weight machinery has further exacerbated this trend. A particular safety concern for the naval architect, where large graving docks are not used for assembling and "launching" a new ship, is that of dynamic launching, where the scope for major catastrophe is significant (as occurred with HMS OCEAN (Johnstone-Bryden, 2018)).

One combat driven innovation, the introduction of flare to the hull above the waterline, adopted for radar cross section (RCS) reduction, has led to modern warships being less likely to degrade in stability as their weight increases through life. This increase occurs from unplanned accretions and new (largely combat system related) equipment being installed later in the ship's life. The need to upgrade antenna fits is usually due to emerging

Figure 2. RN Type 45 DARING Class Destroyer (US Navy Credit).

new threats or new technological developments, with the latter arising from electronic equipment, in particular, having shelf lives much shorter than that of most ships. A ship's life is typically 30 years, which means some 50 years design life for a new naval ship class from initiation to last of the class going out of service), unless the ship has a major life extension. (The US Navy is contemplating, for their colossal (100,000 tonne) aircraft carriers, a life of 50 years plus 50 more, after a major rebuild).

The need to survive and continue to function after extensive damage means for naval vessels that particular attention has to be paid to watertight integrity. Thus there are no "closeable" doors in the numerous watertight bulkheads (WTB) below the main access deck. Each watertight section can only be accessed by through deck hatches, closed in higher action states and with scuttles providing a secondary means of escape from compartments below. All penetrations through WTBs are paid special attention with glands and rapid closures of pipework and ventilation trunking on both sides of each WTB. All this contributes to the complexity of naval ship design, construction and maintenance with yet further significant cost implications.

2.3 *Strength*

The structural design of naval vessels is similarly complex and costly, when compared to most merchant ship practice. Investment is made into light and structurally efficient scantlings with closely spaced longitudinal framing and extruded or fabricated "Tee bars". The latter are much more structurally efficient than asymmetric sections adopted for cost reasons in merchantships, which usually have proportionately fewer larger flat or bulb sectioned stiffeners. This naval practice is adopted both to keep the structural weight fraction as low as possible (as the largest single component of such ships' displacement) and ensure the structure better resists explosions, in particular those underwater that impose extreme hull girder longitudinal "whipping" (See Figure 1 example 3). The latter is an extreme form of slam induced whipping that can occur to any ship in heavy seas leading to bow emergence (See Figure 1 example 4) which can, potentially, lead to breaking the ship in two. Also, for both structural efficiency and shock resistance, naval ship structural design practice is to incorporate sophisticated structural joints where transverse frames and deck beams meet and where orthogonal stiffeners cross (Faulkner, 1965). Such good connections ensure strength integrity is maintained.

Three simple guidelines help minimise the likelihood of structural collapse of naval structures under action damage:

i. keeping design stresses low;
ii. having as deep a hull as possible, so less of the hull is likely to be destroyed (This is also good for seakeeping but raises the ship's centre of gravity, with further consequences on increasing waterline beam (see Stability above));
iii. avoiding structural discontinuities—such as break of fo'cstle (especially amidships, as was the practice in WWII destroyers) or not ending deckhouses on WTBs. Thus WTBs also have a major structural function—as so often in ship design, features or components often provide more than one function.

So achieving a robust navalised structural design will cost money in design effort, fabrication and TL support. For this reason navies are increasingly using classification societies to see if some degree of commercial practice (as already has been adopted for some naval auxiliary vessels, such as fleet replenishment tankers) can be adopted in naval combatants to keep cost down, but at some debatable risk.

2.4 Seakeeping

While good seakeeping is a virtue in all ship design, naval vessels are characterised as not being able to adopt weather routing, increasingly reliable for transportation merchant ships, since naval ships must meet operational needs, so requiring their immediate deployment regardless of weather conditions. Thus even before the adoption of computer simulation of ship motions in a (real) random seaway, naval vessels were designed for good seakeeping. This has meant adopting features such as: raised fo'cstles; high freeboards; bridges positioned someway from the bow; roll stabilisation for both gunnery performance and helicopter flight deck operations; and positioning seaboat launch and recovery arrangements in the ship's waists. The waists are also where some replenishment at sea (RAS) takes place; a difficult evolution necessary to keep naval ships refuelled and rearmed/stored for extended ocean deployments.

Flight decks are ubiquitous, even on relatively small frigates (say < 3000 tonnes displacement and some 100 m waterline length) and are increasingly fitted also to non-naval offshore support vessels. Despite the probabilistic nature of sea conditions, it is now possible to model a ship's seakeeping response to assess whether, say, the helicopter or unmanned air vehicle (UAV) can operate off a given ship's flight deck in high sea states and for a range of ship's speeds and headings. The design choices can still remain problematic, since for a combatant the "optimisation for seakeeping" typically balances fo'c'stle wetness, flight deck movement, motions on the bridge and slamming of a bow sonar dome, which all together require conflicting whole ship features to resolved (Lloyd 1992). This is yet another example of naval ship design being a mass of compromises (Purvis 1974). For larger naval vessels, seakeeping is less likely to concern these issues and be more one of needing to ensure in high sea states, for example, that large side openings are positioned to avoid water ingress from high waves (Honnor & Andrews 1982).

2.5 Other disciplines involved in ship design

Part of the issue with the whole problem of what naval architects do both for those designing ships directly and the many more involved either less directly in ship design or "just" in a specific aspect of hull engineering, is that like so much of complex engineering design they do not work in isolation but part of a wider team. Thus we need to consider the roles of the other principal disciplines, not all of whom are engineers.

While Figure 3 shows the various topics that might be involved in the design of a new ship, many are encompassed by those disciplines assigned to the naval architect (e.g. hydrodynam-

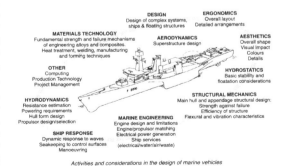

Figure 3. An indication of the topics relevant to ship design (Andrews 1996).

ics, hydrostatics, ship response and structural mechanics—matching the "S^4" categories already outlined). However the obviously separate engineering field (as opposed to design related topics in Figure 3 such as design itself and aesthetics and ergonomics) is that of marine engineering, which can be seen as directly applying main stream mechanical engineering to ship propulsion machinery installed in ships. Thus the marine engineer is responsible for a host of vital equipment and its integration on-board the ship. This then introduces an important set of interfaces with the naval architect, not least being the propeller, which is the design responsibility of the naval architect due to its interaction with the underwater hull design. Despite the marine engineer having a major role in operating the ship (unlike the naval architect who rarely goes to sea) and often having operational responsibility on-board for the naval architect's "part of ship" (e.g. operational actions to deal with damage stability occurrences), he or she is primarily limited in the ship design process to a focus on the machinery spaces.

However, one area where the marine engineer has a growing involvement in modern ship design is with regard to environmental concerns and in particular ship machinery emissions into the atmosphere. The merchantship domain has largely taken the lead due to the extent of global commercial shipping and the greenhouse effect of marine diesel fuel consumed by the world's substantial merchant fleet. However, as with the highly integrated system that is a ship, the challenge to reduce ship emissions is a whole ship design task, one that also heavily involves the naval architect if the issue is to be effectively dealt with (Calea et al., 2015).

While the marine engineer has a core mechanical engineering commonality with the naval architect, in their basic engineering education, this does not apply to the electronics based combat system engineer, the third key player in naval ship design. Despite guns, missiles and torpedoes at the "sharp end" of their business having considerable mechanical functions in their handling and launching from ships, it is the sophistication of the electronic control of them that gives the combat systems their performance edge. Furthermore it is both sensors (radars, sonars and communications) and the combat system management wherein the primary focus of integrating their combat effectiveness lies in a modern naval combatant (Baker, 1990). This is therefore where the tools and practice of systems engineering dovetails closely with the main design responsibility for the overall combat system in a new naval vessel. The interface of the combat systems with the main ship design is thus complex and difficult, as the latter is much more physically grounded (due to whole ship implications, i.e. "S^5") than the data management abstraction governing combat system design. Clearly in merchant ship design, while electronics and automation this enables is growing, there is not the same conflict as there is between the naval ship designer and the combat system designer.

There are also other engineering disciplines involved in ship design, many of which deal with discrete and often distributed systems, such as HVAC which can often be under the ship designer's overall responsibility. Other specialist skills and expertise may be very specific to the particular ship and often arise from operational needs, be they specific cargo handling or even for specialist service vessels, offshore functions such as rig support or autonomous vehicle launch and recovery systems. A very specific and longstanding example of a demanding design interface is that necessary to integrate aircraft operations at sea. Leaving aside the whole issue of large scale naval operations off aircraft carriers (Andrews 2005), many ships, such as offshore support vessels and naval combatants, have helicopter facilities, which can dominate that ship's design. A classic example of this was the Canadian DDH-280 Tribal Class (Farrell et al., 1972) where the design manager considered not just the ship's visible upper works but arrangements deep in the hull in the after half of the ship were dominated by the need to fully support two large helicopters (i.e. Sea Kings). This was because the ship was not just the platform for the aircraft but had to provide the equivalent of a small airfield's facilities. This requires the ship designer's special attention, given the highly sensitive vehicle that has to operate off the ship for extended periods in the demanding maritime environment.

(An aside on the above, which needs flagging, is with regard to the usage of the term "platform" to describe a (naval) ship. The Canadian Tribal Class was providing a platform for those helicopters, however it is wrong to use the term "Platform" to describe a naval vessel as such. It is a whole system of systems and the further split into "payload and platform" (i.e. Combat system and the "rest of the ship") is a very bad design mind-set. It implies the former is good and must be maximised at the expense of the latter. Given the so-called "platform half" of a ship provides flotation and mobility, both of essential military worth, as well as the infrastructure for the personnel (who are "fighting" the ship) and the supporting services without which the combat system elements could not function at all, this shows the utter nonsense of the previous sentence, in not seeing the system as a whole. Thus all the interdependent functions contribute to the vessel's military capability, which means there is no unnecessary "overhead" to be minimised).

One domain where there has been an interesting challenge to the naval architecture discipline has been in offshore extraction of petro-chemicals. Initially the extraction was achieved using fixed concrete structures, which meant that the civil engineer led on such designs, but once deep fields required floating structures with extensive personnel and processing facilities, then these structures became more ship-like. This has now led to FPSOs based on VLCC ship configurations (albeit with extensive processing plant in addition to that provided on a typical oil tanker).

The final and obviously key discipline in ship design is the non-engineering ship operator, distinct from the marine and electronic engineers yet clearly also on-board. As the primary user, if not the owner paying for the design to be realised, the "sailor's" input to the design is critical. Historically the link between the designer and seafarer was the main one until the shipbuilder then executes the design intent. In the merchantship sector major shipowners used to retain their own design teams (Meek, 1982) and this ensured the shipping company's practice was reflected in the design intent, subsequently worked up by the winning shipbuilding team. The naval equivalent to this practice was the substantial in-house design and acquisition organisations maintained by major navies (Brown 1983, Tibbitts, et al., 1993), which until very recently acted as their own classification society drawing on extensive in-house research facilities. (Only the UK, among major navies, has privatised its original towing tank capability.) The relationship between the naval ship operator and the naval ship designer has been key to producing appropriate ship designs but that can be seen to have declined in recent years, just as the social science of ergonomics or human factors has become more scientifically based and sophisticated (Nautical Institute, 2015). The change in merchantship design practice can be attributed to the purchasing of more standard ships driven by fierce competition between shipyards, particularly in the Far East. For naval ship design the change has been more subtle and largely due to the ascendancy of the combat systems engineers spanning both requirements and solution senior management. There has also been an observed decline in the naval architect's role as lead designer/project manager, where often the latter has been undertaken by a generic systems engineer, who may be a non-naval combat engineer that doesn't even see the need to draw upon any maritime experience. This can be seen as a consequence of the mind-set that sees these immensely complex vessels as just another "military platform" and so just requiring project management skill to bring them to physical realisation, rather that needing an intimate knowledge of not just ship engineering but also wider design and operational insights, whereas such insights typify what is acquired in a naval architect's development into a design manager and project director (Andrews, 1993, 2016).

3 WHOLE SHIP DESIGNING

3.1 *The naval architect's design role*

The previous section having outlined the primary elements of "hull engineering" and then considered those involved in ship design other than the naval architect, concluded that the naval architect's hold on their traditional role of ship design lead was increasingly being challenged. This is despite ship design being seen as the *raison d'etre* for naval architecture (Rawson & Tupper 1978). It is therefore now necessary to consider what designing ships, as an engineering practice, is actually about.

The Royal Institution of Naval Architects (RINA) as the premier international professional institution for naval architects lays down what—having obtained the appropriate academic qualifications (see the penultimate section) – a naval architect needs to accomplish beyond those qualifications before they are deemed fully qualified to practice (www.rina.org.uk). The first two years in employment post-graduation include on-the-job training and require three broad areas of post-academic practice to be addressed:

1. Design, which consists of analytical skills being applied, doing design itself, learning communication skills and how to acquire technical information (an obvious burgeoning aspect) plus materials awareness (seen as necessary given the limits in hands off practice at university);
2. Engineering Practice, which is seen to cover shipbuilding production processes, safety and legislature issues, quality control, production management, commissioning new ships and procurement issues;
3. Management Services, which could be said to distinguish wider engineering practice from even applied science, with topics such as accounting & finance, human resources and quality assurance, company structure & organisation, marketing & communications and finally (and somewhat appropriate given the conclusion of previous section on design) and project leadership/management skills.

Thus design is rightly highlighted as key, however the above listing on design also covers the application of the analytical skills acquired in an engineering degree and this then distinguishes engineering design from other (likely to be more artistic) forms of design (Andrews 2012a). Also included in the broad area of design above is communications, which links to management skills and information and recognises the seed change in large scale design practice, where most engineering designers work. Materials seems an odd topic to pull out (being the least analytical of those covered in an engineering degree), but of course it is a vital topic in designing for the extreme environments in which ships have to survive. The other two broad areas above are considered further later in this keynote.

The nature and form of engineering design that designers of the most complex vessels practice has been the subject of considerable study, not least in IMDC papers and State of Art reports (Andrews and Erikstad, 2015). Furthermore the author has characterised the high end of ship design as the design of physically large and complex (PL&C) systems. The design of such systems is seen to also include that of large scale civil engineering and architectural constructions, such as dams, transport termini and public buildings, as well complexes, such as major chemical processing plant. Given all these are one-offs without prototypes, this distinguishes their design and construction from prototyping and production line manufacturing that applies to other vehicles, much smaller than ships and submarines. Again there is a link to the structural design challenges akin to those of the civil engineer and the conflict between that profession and the architect, one that largely does not exist for the naval architect at a design level. However, increasingly for very large and expensive maritime projects, such as major naval vessels and offshore constructions, the coordination and project management role of the naval architect is being questioned. To some degree the demands of "hull engineering" in an understandably more safety obsessed world have led to this, in contrast to the historic view that design leadership and acquisition coordination were the primary function of the naval architect in ship design.

As part of emphasising that the naval architect's role as "the ship designer", rather than just the hull engineer, this author has focused his research and publications particularly on the initial stages of the design of complex vessels. This is because it is acknowledged that the first design phase for such vessels, the concept phase, is the most crucial in that it is then that the major design decisions are made. This is despite the fact that much greater design resources (and hence the generality of engineers and naval architects efforts) are employed to progressively work up a selected design solution—the devil being in the detail (Andrews 2013). For this reason not only has the author pioneered a more comprehensive approach to early stage ship design but done so by insisting that modern ship synthesis be an integration of an architectural ("inside-out") as well as a largely numerical balance of gross weight and space (Andrews, 1986, 2003). This has led to an emphasis in the final "S^5" component that of Style, summarised in the next part of this section.

3.2 *Style in addressing the design transversals and categories of style*

For complex vessels the style to be adopted in a specific ship or submarine design option is seen to be the key design decision for that option and so is the first design decision (beyond deciding that a certain range of solution options is to be investigated). This is indicated in the overall ship design process representation shown in Figure 4 where each step or decision selection is explained more fully in the appendix to Andrews (2013). Thus

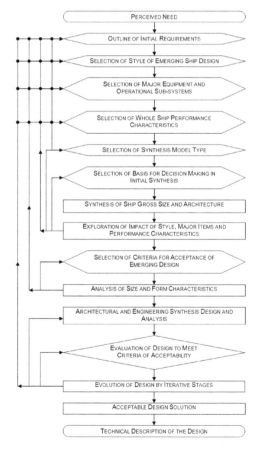

Figure 4. A representation of the overall ship design process emphasising key decisions with style as a critical initial choice.

Selection of the Style of the Emergent Ship Design is the first design choice and can be seen to impact at the macro, major and micro levels of the emergent ship design. Macro level denotes the overall style of a design or preferably a design option, whether it is, for a naval example: a conventional monohull; a more utility or austere design; or a radical configuration, such as a trimaran or SWATH. Below the macro level there can be seen to be some major style choices, such as adopting commercial design standards for a utility helicopter carrier (e.g. HMS OCEAN) or achieving a very low underwater signature for an ASW frigate (e.g. the R.N. Type 23). This level can also cover generic style choices, such as being robust or highly adaptable, or having (say) high sustainability or low manning. It can be seen that these major "Style" choices are largely made by the naval architect, provided they have the requisite skills (both in knowledge and experience) to carry out a new design synthesis. Clearly such design decisions should not be made without involving the key stakeholders and the extent of this involvement is once more the naval architect's call. When such key decisions are not made or choices achieved by default then the naval architect is failing in their role as the primary ship designer. A clear statement of this is due to Baker and is outlined in the next section.

While adopting such style issues is inherent in commencing any design study or a specific option in a series of more exploratory studies, it is important that this is done consciously. This is good design practice since each choice has implications for the eventual design outcome and therefore ought to be investigated before that style aspect is incorporated or rejected. Beyond major style choices are a host of minor style decisions often predicated by the first two levels. These in some sense can be seen as reflecting Ferguson's (1992) observation on engineering design practice, that "Design layout and calculations require dozens of small decisions and hundreds of tiny ones". However, lack of coherence regarding style can mean that at all three levels these decisions are not always made with consistency and so can introduce, at best, inefficiency into the eventual design solution.

The term design style was originally proposed to distinguish a host of disparate issues distinct from the classical engineering sciences applied to ship design (i.e. "S^4" for the naval architect). Many of those issues could be seen to be on the "softer" end of the scientific spectrum drawing on the arts and humanities (Broadbent, 1988), whereas the first four terms under the "S^5" umbrella are, historically, the principal naval architectural (engineering sciences) sub-disciplines associated with a ship's technical behaviour. Thus Style was devised to summarise those other design concerns, which in Table 1 are listed for both a naval combatant and a commercial service vessel (OSV). This very disparate range of issues, have been categorised under some six headings that (ship) designers understand. Thus, for example, concurrent engineering concerns, such as Producibility and Adaptability, are encompassed by the heading Design Style in Table 1.

Importantly these style issues can make a substantial difference to the final outcome of a design, so their relative impact ought, in the case of a complex ship, to emerge from a proper dialogue

Table 1. Comparison of style topics relevant to a naval combatant (all) and a commercial service vessel (OPV) (underlined).

Stealth	Protection	Human factors	Sustainability	Margins	Design style
Acoustic signature	Collision	Accommodation	Mission duration	Space	Robustness
Radar cross section	Fire	Access	Watches	Weight	Commercial
Infra-red	Above water weapon effect	Maintenance levels	Stores	Vertical centre of gravity	Modularity
Magnetic	Underwater weapon effect	Operation automation	Maintenance cycles	Power	Operational serviceability
Visual	NBC contamination	Ergonomics	Refit philosophy	Services	Producibility
	Shock		Upkeep by exchange	Design point (growth)	Adaptability
	Corrosion			Board Margin (future upgrades)	
	Damage control				

63

between designer and client (or in the naval ship design case, the operational requirements owner). Furthermore, most of these issues have been difficult to take into account early in the design process because, usually, initial design exploration has been undertaken with very simple and, largely, numeric models. These can only summarise the likely eventual design definition and give a feel for the cost to acquire the fabric of the ship, which is therefore often dubious (Andrews, 1994). That dialogue can now be informed by also concurrently producing a graphical architectural representation of the ship's configuration and internal architecture, as is reflected in the process summarised by Figure 4. This process reflects the architecturally (rather than solely numerically) driven synthesis and has been demonstrated through applying the author's Design Building Block (DBB) approach (Andrews, 2003). At the critical early design stages, such a computer graphics based approach can then enable the ship designer to take account of likely significant issues, such as those listed in Table 1. Given these are diverse and not readily consistent or comparatively quantifiable, means that designers need to exercise judgement. Preferably, those judgements are informed by dialogues with stakeholders and the dialogues are best achieved using an architecturally driven representation, which are more informative to non-designers and ship operators than tabular data and technical outlines.

The categories adopted in Table 1 reveal the heterogeneous nature of the specific individual style issues, for a complex vessel. Thus, there are the specific naval concerns with the vulnerability of ships to modern weapons addressed by the various items under Stealth through the many different signatures, which a ship has and then these need to be reduced to avoid detection. The Protection items are largely aspects only worth incorporating in the naval ship to mitigate the results of weapon effects, should the Stealth (and any "hard kill" self-defence) fail to be totally effective. However, some of the Protection items are required for any vessel, namely, corrosion control, or measures to mitigate the effects of collisions and fire onboard. The Human Factors aspects are little less coherent (and it might be argued rather more solution oriented than those comprehensively considered by the urban architectural theorist Broadbent (1988)). Broadbent addresses some 21 "human sciences" that he considers are relevant to human habitation—and hence most are also likely to be appropriate to addressing HF in all ships. Some HF issues lead on to the important growth area of automation, which along with micro-ergonomics (e.g. console design) has a strong input to the Protection category, specifically in regard to modern bridge design.

Sustainability is a major consideration in naval ship design and could be said to be a major driver. Hence the relevant measures to be adopted constitute a key hidden decision in the style of such ships from the beginning of any ship design study. For most merchant ships the style of maintenance to be adopted is often fully mandated by the owner's technical team, leaving the ship designer little scope in this regard. The list of Margins just makes the point that there are many features and considerations beyond ensuring there are direct margins on the weight/VCG estimates, where the latter ensure the ship's stability is adequate beyond the day it is accepted into service. Table 1 also distinguishes from Design Margins those margins required for unplanned (but in naval vessels consistently observed) growth in weight and rise in VCG, once in-service. The latter are more rightly categorised as a Design Style issue in Table 1, given they address many likely measures of uncertainty in early design estimates. These margins, across all the weight/space groups, are intended to be absorbed but not exceeded as the design and build process progresses to completion. This has been a major focus for naval ship design due to the potential significance of any of those features being exceeded on a sustained deployment. One reason for resorting to new naval ship designs has been the expense of retrofitting new capabilities, which has also led to the questionable measure of stretched (repeat) designs (e.g. R.N. Type 22 and Type 42 combatants) and to avoid expensive mid-life upgrades by mandating short life designs (e.g. R.N. Type 23, which succeeding UK governments have then failed to abide by). This issue of building in adaptability to anticipate changes in a ship's role and in the market has only recently been seen as an issue in commercial service vessel design (see Gaspar (2013) who applied the M.I.T. Epoch-Era approach to OSV design).

The last category in Table 1 is clearly the most broad and heterogeneous. Also, generally making such choices on these topics can have the biggest impact on the final ship design. But this means they need to be recognised as choices and then properly considered with the owner/requirements team from the beginning of studying any design option. Some of these have been the objects of particular investigations by the author's research group at UCL. They are discussed further in Section 5 as examples of the impact on the ship design of separately considering some of these particular issues, where each could be seen as the specific driver of a design from its initiation. It is noticeable that certain of these issues can only be adequately investigated in the Concept Phase if the architectural synthesis, assumed in Figure 4, is adopted. The other feature of most of the Design Issues listed is that they have a qualitative or fuzzy nature. Thus, say, Robustness implies a greater degree of that quality than the "norm" for that type of vessel. This then raises the point that such a "norm" for a given new design option ought itself be defined but is often just accepted (or inferred) as being "current practice" or by the adoption of existing standards. There are also exceptions in the listing of the Design Style issues, such as Aesthetics, which for most ships, other than mega yachts and some cruise ships, is seen to be "a luxury". However, even this can be a simplification, as in the Cold War there was considerable debate in the US naval ship community as to whether the physical appearance of such a ship was part of its political "armament" (Roach and Meier, 1979). Furthermore, it could be argued that this is another design area where naval architects might have "lost their way" as it is the case that a hansom vessel remains a source of pride to its sailors, despite being hard to precisely quantify ship appearance against the bottom line but nevertheless this ought to remain a significant part of design creativity (and even relevant in marketing a design's acceptability to stakeholders).

3.3 *Recognising design novelty*

The nature of the design of complex ships, such as cruise ships and naval combatants, is such that the need to emphasise the importance and difficulty of early representation of style issues is seen to be a further complication in the practice of designing such vessels. This is due to there being, additionally, a wide range in the practice of such design. This arises from the degree of design novelty adopted in a specific design option, as is indicated by Table 2. This shows a set of examples, across the field of ship design, where the sophistication in the design undertaken ranges from a simple modification of an existing ship, through ever more extensive variations in design practice, to designs adopting, firstly, radical configurations and, beyond that, radical technologies.

The first of the last two categories of Table 2, a radical configuration yet still using current ship technology, is often explored yet such options are still rarely built. This is due to the risk of unknowns usually exacerbated by the lack of investment in a real prototype. Furthermore, radical technology solutions are even more rarely pursued. In part this rarity arises because such radical technology solutions require recourse to design and, indeed, manufacturing practice much more akin to that appropriate to the aerospace industry. Thus new major aircraft projects, typically, require massive development costs (including several full scale physical prototypes, some tested to destruction) and additionally need tooling and manufacturing facilities to be specifically designed and then built, before extensive series production of each new aircraft design can commence. This is of course quite unlike most ship design and manufacture, be it the ubiquitous bulker or the most sophisticated naval vessel. Such distinctions as those of Table 2 for the design of complex ships suggest that, at least, there is a need to consciously recognise that there is a spectrum of possible design approaches consequent on the novelty of each specific design option being pursued. Such choice on design novelty is key to the initial style choice for a given design study or a variant option in a properly conducted concept exploration as the basis for undertaking a properly conducted Requirements Elucidation process for a new vessel (Andrews, 2013).

Both this issue of understanding the novelty of a design option (and thus the use of appropriate design methods and tools) along with the key style issues to be investigated, are the design team's main concerns early in the ship design process. However, it can be seen that all those concerns listed are those relevant to the naval architect, as the ship designer. Furthermore, many are exclusive to that ship design centred discipline and this means that

Table 2. Types of ship design in terms of design novelty.

Type	Example
Second (stretched) batch	RN Batch 2 Type 22 frigate and Batch 3 Type 42 destroyer
Simple type ship	Most commercial vessels and many naval auxiliary vessels
Evolutionary design	A family of designs, such as VT corvettes[1] or OCL container ships[2]
Simple (numerical) synthesis	UCL student designs
Architectural synthesis	UCL (DRC) design studies (see Section 6.2)
Radical configuration	SWATH, Trimaran
Radical technology	US Navy Surface Effect Ship of 1970s[3]

[1](Usher & Dorey, 1982).
[2](Meek, 1970, 1972).
[3](Lavis et al., 1990).

the naval architect is uniquely placed. The inherent sophistication of the earliest stages of the design of complex vessels has often not been recognised, even by many practitioners. This sophistication has been argued by the author from his earliest publications and in a recent series of papers, dealing with elements of the process, which has culminated in comprehensive article encompassing the topics presented in the individual papers and is being published this year (Andrews, 2018). This comprehensive statement and the scope of responsibilities ought to be sufficient to make the argument that this paper is addressing but it is considered that some further issues need to be addressed before a few explanatory case studies are outlined to reinforce the argument.

4 THE NATURE OF DESIGNING A COMPLEX SHIP—WHY THE NAVAL ARCHITECT IS OF THE RIGHT DISCIPLINE

It is now appropriate to draw on the thoughts of three late 20th Century ship designers for their very pertinent views on this subject. Their views seem to further justify the assertion in the title above, making it clear that naval architecture IS the ship design discipline. This then leads on to whether the naval architect should also be the ship design manager, furthermore if a major ship acquisition project should have a project director, who will have more than just design responsibility, should they also be a ship designer? This is probably less of a given, if not just because we have exceptional counter examples in Brunel and Rickover (Rockwell, 2002).

4.1 Stian Erichsen

Erichsen was the founding father of the IMDC, originally called the International Marine Systems Design Conference, in recognition that he saw ship design from a systems perspective. His paper "Some Elements of Ship Design" (Erichsen, 1997) considers the topic from a commercial shipping focus, which is a useful contrast to the current author's largely naval vessel design background, which also applies to the other two significant thinkers on ship design who are considered below.

Erichsen starts by considering design, which he sees as: "developing a description of what is to be". This seems to imply that for a new ship the design constitutes the specification, however this doesn't really address the crucial start of the process, which the previous section has highlighted. However, he goes on to state the need for a "platform of starts", which is taken to mean the prerequisites, which he lists:

- Market demand (wholly appropriate to Erichsen's transportation system);
- Having a concept of the customer, which is seen to be operators and owners but also shipbuilders and the wider public, which continues the transportation vessel pattern but with a systems perspective beyond the ship itself or even its wider transportation system, to reflect a modern perspective of the design's potential environmental impact. There is less straight forward customer perspective for the naval vessel designer in that there are many stakeholders within government and once the ship is in-service the Design Authority (who may or may not be the actual detailed designer) has the responsibility as the technical owner on behalf of government (Andrews 2015);
- Observe ships, which is always a salutary message in keeping a designer's feet on the ground. The fact that most naval architects rarely go to sea once qualified can be felt to be another aspect in which the lead designer might feel unable to exercise a commanding design role, against experienced ship operators. The author was privileged to be part of a post qualification training scheme that gave him over six months at sea in a variety of ships and submarines, gaving him experience into not just the aspects of ship's technical operations but also into the mind-set of the mariner, for whom the ship is their home.
- Design for the future—this inevitably makes the designer conscious of uncertainty and encourages ship designers to rise to the challenge of being creative.

With Erichsen's definition of design he then goes on to make three clear statements about (ship) design:

1. Design is deciding—which concurs with Figure 4 and which he amplifies:
    i. Address overall questions first;
    ii. Design the interfaces before designing "the interfacing elements". (This could be seen as presaging Set-Based Design (Doerry et al., 2009);
    iii. Have a sequence of rational decisions, which Erichsen was suggested well ahead of approaches like DSM (Eppinger and Browning, 2012) now being applied to the ship design process. He suggests such a sequence could be applied to linking functional requirements to decision variables, which sounds a little too like functionalism, which has been critiqued in the case of complex ship design by Andrews (2012a)).
2. Design is teamwork, where he highlights:
    i. The issue of specialists. This is echoed by the extent of specialists involved in a naval ship

design. Andrews (1993) identified over 100 direct specialist authorities dealing with his new project in its Feasibility Phase;
ii. the problem of dealing with contractors. Notably, should the prime contractor's commercial focus lead to them passing on risks down the chain, adding costs and losing design cohesion;
iii. The need for compromise, which is true but the naval architect cannot compromise on safety, nor achieving a sufficiently tight design balance in the overall design;
vi. Agreeing a Measure of Merit (MoM), which is possible in Erichsen's transportation system examples, less so in the author's field of PL&C systems, with the "wicked problem" of requirements and their necessary elucidation (see the previous section).

3. Design aids (meaning methods and tools:
   i. Jones' (1980) work on brain-storming is recommended by Erichsen but has been to some extent been over taken by CAD tools, which produce many variants. However, this introduces the further issue of the limited range of variants likely to be produced and actually then reducing the scope for a wide exploration and inhibiting creativity?
   ii. Different design models can be produced, which can bias the design in detail;
   iii. "Look at other designs." Erichsen says experts benefit from learning what other experts do, which could be a way of enhancing the extent of the concept solution space exploration (Andrews 2013).

4.2 Louis Rydill

Rydill was both the first Professor of Naval Architecture at UCL, setting up the MSc Course, with its strong emphasis on ship design, and an eminent naval vessel designer. Drawing on the latter role he co-authored the definitive textbook on submarine design (Burcher & Rydill, 1994). In a paper presented to the submarine builders at Barrow-in-Furness, entitled "An Idiosyncratic View of Warship Design" (Rydill, 1986) he concluded with a view of ship design practice, which was based on his extensive career in naval construction. The first part of the paper was a summary of his career, which showed his main design achievements were built on deep ship and submarine knowledge at the cutting edge of the technology post WWII: early research work on submarine control; design work including new structural theory on the first post-War UK submarines (PORPOISE Class); practical submarine refit management and even teaching naval architecture. All this was good preparation for being the lead designer on the first British nuclear submarine (HMS/m DREADNOUGHT) from 1957 to 1962, where his skills came to fruition. He was then the design manager for the new aircraft carrier (CVA-01), which was subsequently cancelled. He described that as a relief as it ended "the daunting challenge of getting into a 55,000 ton design the capability of the much larger USS Forrestal Class".

Rydill made some interesting design comments on this set of experiences, which occurred prior to arriving at UCL and that was followed by him holding several very senior positions in UK Defence Procurement. From this he concluded:

- He had been trained as a technical generalist—with benefits and disadvantages;
- He agreed with Rickover's adage that "all knowledge has a half life";
- All naval architects should be taught mechanical engineering to degree level before "specialising" in naval architecture;
- CVA-01's problem was the balance between risk and innovation.

Rydill concluded:

i. The design of naval vessels should be conducted in-house for the critical preliminary design stages as this is necessary to "resist technically bad solutions";
ii. Thus the transition to the shipbuilder for such complex vessels will always be difficult and often arbitrary. (This is interesting to compare with both what has happened in recent years in the UK (Andrews 2012b), and earlier view by the US Navy's then Technical Director (Leopold, 1975), which seems to have been ignored subsequently until the US Secretary of the Navy in 2007 (Winter, 2007) stated the US Navy should take back ship design control (if not undertake its full execution));
iii. He concluded on some UK ship designs that the Type 42 Destroyers showed that "small was not (technically) beautiful", the Type 21 Frig-

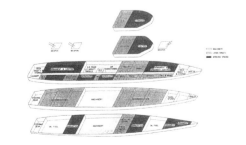

Figure 5. The "Stylised" Layout of Baker's St LAURENT Class Frigates (Baker 1956b).

ates were "pretty but without adequate military value to the Royal Navy" and that the INVINCIBLE CVS Class (Honnor & Andrews, 1980) designed by the Navy and the RCNC was the best post-War ship design;
iv. Submarines were much more free of "bad technical decisions under political pressure" due to their extreme design complexity (Andrews, 2017) but the opposite of that was the 1970's "short fat ship" episode (Andrews, 2012b) on which he stated that that controversy was "the *reductio ad absurdum* of the need to stand up and be counted";
v. Finally he discussed how shipbuilders might take on design responsibility for naval vessels (see references at item ii above) and stated that this should not be by default (or political edict). (He mentions a 1970s report that he commissioned, concluding that adequate intellectual resources need to be applied). This intent had been challenged by Leopold (1975), who had been both sides of the US government/shipbuilder divide. There is the related issue of how design authority, as the bottom line on design responsibility, is exercised. Both Betts (2010) and Andrews (2010) commented on the case of the Type 45 Destroyer class design as to whether Design Authority had been truly exercised by its prime contractor before being restored belatedly to government (Gates, 2009).

4.3 *Rowland Baker*

Sir Rowland Baker was the most eminent naval constructor post-WWII, as argued by his biographer and eminent historian on ship design (Brown, 1986). However what I want to do for the purposes of this paper's argument is to focus on a paper written by Baker when seconded as a Constructor Commodore to the Royal Canadian Navy to lead the project for its first indigenous ship design, that of the St Laurent Class Frigates (a picture is given as the example of Style in Figure 1). The paper was entitled "How to build a ship" (Baker 1956a) and though non-technical and for general naval consumption, wonderfully captures Baker's philosophy, which reached final fruition in his project direction of both DREADNOUGHT and the Resolution Class nuclear submarine projects.

Baker starts by stating:

"'So the chicken comes before the egg and the ship comes before the staff requirement' The naval staff who must decide what the ship must be capable of doing, can only state a requirement in known terms and this means the staff requirements for a new ship must refer explicitly or implicitly to some existing ship ... shipbuilder or designer depends on the staff to say the sort of ship they want, the staff whether they realise it or not, depend on the shipbuilder and designer to indicate what sort of ship is possible".

Thus Baker is consistent with the view of the author in the nature of Requirements Elucidation as key to the Concept Phase for complex vessels (Andrews 2013). He then goes on to list six issues to be necessary to finally produce a new ship:

1. You must have an industry to build a ship.
2. You must have a designer, which Baker interestingly identifies as having:
    i. knowledge, which is to "know what he himself can do";
    ii. be an artist, which Baker sees as not just aesthetics but almost having "feel", since: "if it doesn't look right, it cannot be right (regardless of meeting the staff requirement)";
    iii. achieve a "suitable" arrangement of the compartments (he says more on this below).
3. Someone must require a ship. This Baker sees as the naval staff (for the ships he was procuring), who must:
    i. understand how to fight (It was salutary for the UK Ministry of Defence to re-learn some obvious lessons from its war over the Falkland Islands in 1982 (Meek, 2003));
    ii. sound knowledge of ships, sea and weapons, which has become harder with fewer ships in the fleet;
    iii. what both the crew and materials can do. (Probably the second is the harder for sailors to acquire?)
    vi. have judgement, which Baker considers to be: the power of seeking advice; deciding on its quality; and maintaining decisions once made.
    v. Thus "a ship when completed cannot be a total success, she can well be a total failure", which Baker considers to be the chain like consequence of one significant failure in something as complex and interlinked as a ship. He lists failure as:
        a. The staff requirement can be wrong (see Preston's assertion "There are more bad staff requirements than bad ship designs" (Preston, 2002);
        b. The design can fail to meet the staff requirement, which sounds straight forward but Honnor's adage "the best ship design is that which just fails to meet its requirement" reveals the subtlety of the issue (Honnor, 2006);
        c. Construction can fail to meet the design requirement, which contractually means

the specification (which can never be perfectly watertight);
d. All too common minor failures. (This can in part be down to having no prototypes, with expectations of airplane/car production line delivery, rather than being more like that of the construction industry's practice with inevitable "snagging lists". The obvious solution to (often) poor ergonomics is more money for detailed design and quality of outfitting, which is (sadly) has not traditionally been part of the highly competitive maritime culture).
4. Approving the arrangement provokes two further Baker-isms: "This is probably where the real skill of the designer comes in (of course he has to be a fully qualified naval architect)." "Once everyone is allowed to explore the (n!) "alternatives" the process of design could go on forever. That is NOT the way to build a ship." Baker's solution to the immense difficulty of preventing senior (naval) personnel from insisting that their personal design views on arrangements are incorporated into a new ship design was spelt out in his 1955 paper on the St LAURENT Class (Baker, 1956b). Figure 5 shows how his "Stylised" layout restricts allocation of single functions to specific areas of the layout. This was specifically adopted to ensure that "interested parties" beyond the designer, were constrained from "interfering" in the design. This was devised as a design management approach by Baker, revealing his insight into design being not just "hull engineering" but genuine ship design, that meant the naval architect was the design (and project) leader.

Two further overviews were made by Baker, in his article, one on the traditional design control of the whole design as it went into the Detailed Design phase and the second on the need to judiciously order materials for build assembly. In discussing both aspects he very much reflected practice of sixty years ago, with detailed design approval and close involvement in the build by the navy/government. One comment there is very pertinent to this paper's argument "If the hull is poor, the whole thing fails." This again puts the naval architect in the prime professional position in the ship design process, 'though it could be argued against this that greater safety concerns nowadays for the "S⁴" hull engineering aspects mean the naval architect can't now lead on the overall project too? Baker's and my response to this would be that only the naval architect can be the overall designer and splitting the wider project aspects, so well integrated into Baker's very successful ship and submarine programmes, from the grasp of the person responsible for the vessel's design issues (such as those listed in Table 1) too often courts disaster.

5 SOME CASE STUDY EVIDENCE

5.1 *Historical ship design issues*

It is worth going back to the famous Scandinavian example of the "Vasa" to see what happens when the design and the project authority is excessively exercised by a non-naval architect. (In that case King Gustav Vasa over-ruled the Master Shipwright. The latter's predecessor had been an experienced Dutchman, who might have stood up better to his client than a less experienced Swede in (understandable) awe of his powerful monarch). Such over-bearing was not that uncommon and it might be argued knowledge of the effect of excessive top-weight (due to adding more guns high on the ship) lacked scientific understanding. However if it was observed that a newly loaded ship lolled when alongside then a strong enough designer (the Master Shipwright) would have known not to let the ship sail? To say such things could not happen in the modern era is to ignore the nature of political power—see the last example below.

Preston in his popular book on the world's worst warships (Preston, 2002) looks over a series of examples, not just some very odd nineteenth century designs, but also cases of ships popularly thought to be good designs, often to boost a competing navy's build programme. Some on closer inspection turn out to be far from exemplary (e.g. Hitler's BISMARCK Battleship, US view of Soviet naval design). In the excellent introduction to the specific design examples, Preston considers six factors, which influence warship design:

i. cost, which he rightly points out is rarely due to "gold plating";
ii. perceived threats, which Preston suggests are often over blown by intelligence staff with little understanding of ship design (Preston quotes the UK Director of Naval Intelligence's view on a 1930s Japanese cruiser's speed, to which the Director of Naval Construction retorted that the figures were either wrong (the case) or the ships were built of cardboard!);
iii. industrial capacity, which explained the Royal Navy's former pre-eminence and, now, that of the US Navy and hence why both the German and Soviet navies failed;
iv. technical competence, which was an issue in the rapid technological change of the nineteenth century but unlikely nowadays, if not impossible should politicians (or senior non-designers) ignore designers with regard to clear technical

issues (see again last case in this section). Preston also points out someone has to take final design (and procurement) responsibility, namely "You can devolve as much of the process as you like to industry, but somebody must carry the can if the bloody thing sinks." (Preston, 2002) (See the whole issue of design authority raised at the end of the sub-section on Louis Rydill's views);

v. the operating environment, which item 2 in Figure 1 well exemplifies and highlights that ship designers are designing PL&C systems of systems to operate under uniquely demanding conditions;

vi. incorrect post-battle analysis, which the quote from C.S. Forrester's book well encapsulates. One might add that modern recourse to operational analysis, which is actually fraught with dubious modelling of immensely complex and un-measurable reality, presents the ship designer (and the naval staff, who are over reliant on OA) with the need to vigorously challenge any simplistic conclusions that are used to draw inappropriate design decisions (Wood, 1982).

A good case Preston refers to is the synergy between the pre First War DNC (Sir Philip Watts) and the dynamic and forceful Admiral Jackie Fisher (Brown, 1983). Fisher exhibited many of Vasa's overbearing personality traits and had considerable political influence, however he well understood the competence of Watts and his team. Thus Preston states the radical game changer that was "the battleship DREADNOUGHT was not Fisher's 'creation', but a logical progression from the previous design, as proposed to him by an experienced constructor..." which was then coherently designed and very competently built by Watts' dockyard colleagues—all facilitated by Fisher's drive.

Rowland Baker features in the second (mid twentieth century) example of synergy between a very eminent naval officer and a ship designer— Lord Mountbatten and Baker (Brown, 1986). Having recognised each other's talents before the war in discussing destroyer design, when Mountbatten became Chief of Combined Operations to develop amphibious warfare to liberate Europe, he immediately got Baker on his staff and got him to develop a whole new fleet of invasion ships and craft. These were so successful that the two of them took the designs to Washington and these became the design basis for the US Navy's enormous amphibious fleets for the Pacific theatre, as well as enabling the invasions of occupied Europe (Baker, 1946). This seems a classic example of ship designer being given the scope to design with clarity of "customer direction" and upwards support not downwards (pseudo technical) interference.

The last case is a salutary example that politics is clearly significant in such complex design and acquisition and can (almost) be disastrous if the government process ignores its own design knowledge and experience. The author was asked to address the 1980s controversy of the "Short Fat Ship" that provoked a considerable debate in the UK press following the Falklands War with Argentina. This was done in an article to the UK Naval Review and published in its centenary book (Hore, 2012) as Chapter 23 (Andrews, 2012b). That article commented on a private proposal to produce a "short fat frigate" to meet the Type 23 requirement, which was rejected by the UK MoD, because unlike any new hull configuration it was not presented to the naval architecture profession by the normal means of a scientific paper to the learned society (see debate in published discussion to Bryson (1984) paper on the Type 23). As the Naval Review article says, the hullform of this proposal was an extant small craft (UK NPL Series) planing form, inappropriate for frigate size displacements and speeds, and likely to be too stiff in roll motion due to excessive GM. So the issue was not scientifically considered but initially grasped by politicians as a "miracle cure", tempting when under fiscal pressure. The article concluded the lack of acceptance by the Thatcher government of its own ship designers' advice says "more than a little about the UK's decline from its former position of being the leading industrial power". That such an issue is raised in a technical paper, strongly suggests that a narrow "hull engineering" stance is inadequate if naval architects are to properly exercise their major role as The ship design profession. Design and project leadership, as Rydill said in his comment in the midst of this controversy, requires in such instances as this (and indeed Ro-ro ferry damage stability (Rawson, 1990)) that ship designers "stand up and be counted" (Rydill, 1986).

5.2 *Some ship design examples from academia*

The ship design exercise conducted in the MSc in Naval Architecture at UCL was set up by Rydill 50 years ago as a form of heuristically teaching ship design and differs from previous undergraduate or post-graduate design exercises. This UCL exercise was informed by Rydill's considerable ship and submarine design experience and has the following innovations:

i. the main technical exercise started after the (S^4) lectures, assignments and examinations in the first two terms and was then undertaken full time until late June (Most ship design exercises in universities take place alongside the "more academic" courses typically one day a week.);

ii. each design is allocated to, at most, two naval architects (and a maximum of two marine engineers), who produce a final naval architecture report, which they jointly own and defend, although in the latter development part of the exercise they will divide the tasks between them;
iii. each design is new each year and has only a general outline of need (i.e. a broad statement of need rather than a set of specific ship characteristics—although the hull type (monohull or multihull) will be specified to shorten the exploration of options (unlike a real ship concept exploration (Andrews, 2013));
iv. prior to starting the design full time each team is required to explore the broad requirement (acting as its own naval staff or owner if a merchant vessel), informed by a series of lectures by the MoD appointed Professor of Naval Architecture at UCL (PNA) and invited experts from MoD and industry (this exploration enables the design team to get inside the mind of the requirements owner);
v. One aspect predicated is the procurement cost of the eventual design. (This is so the design team, once they have sized the design meeting their initial specification (of speed, endurance, combat fit, etc.) and then costed it, have to perform trade-off studies to bring the design down to the price ticket—without this the reality of most cost constrained ship design would be missing);
vi. The naval architect(s) (and separately the marine engineers for the final report and examination) present their progress to mandated milestones every two or three weeks, with the final examination presentation and report to a Feasibility first iteration level (testing the students' "S^5" knowledge and understanding) is examined by the UCL MSc course director (the PNA) and their assistant, plus two very eminent ship designers from MoD/industry. Further detail is given in Betts (1986).

While this above outline emphasises the wider design issues beyond simple ship sizing, parametric survey and working up a preliminary design, the exercise is also conducted in a heuristic manner with the aim to give the students exposure to whole ship design and acquisition. This will then prepare them not to be mere hull engineers but total ship designers and, by inference, project managers, leaders and directors of major maritime programmes.

An interesting set of lessons on the nature of the ship design process have been seen from certain of the many ship design exercises the author has been involve with in his several positions at UCL. Some years ago there were marine engineers with electronic (rather than power electrics) background in the ship design groups and thus able to look at the combat system choices rather than just propulsion and power distribution. For naval ships the combat system can be seen to be a major driver and therefore this produced a more intense dialogue as to who could make the significant design decisions than usually occur between the marine engineers and naval architects. This in a way mirrored some of the conflicts in modern warship design where the cultural conflict seemed to lie between what can be considered as physical engineers (mechanical and civil) and electronic or systems engineers, who are less concerned with practical physics and more with data flow and information process.

Having said there is usually more cohesion between marine engineers and naval architects, partly because the former tend to focus on their geographically limited part of ship, there can be tension. Thus for a nominally high speed multihull (in that instance a novel large HYSWAS) there was a conflict between the (sea-going experienced) marine engineers wanting to maximise the propulsion fit and the young inexperienced naval architect. The former couldn't seem to adopt a whole ship perspective—and the bright naval architect soon realised he had to exert whole ship authority. An excellent message that ship design is a lot more than just good hull engineering. Both this and the previous example well emphasise the naval architect as ship designer needs to understand and "control" the other disciplines in the ship design, if the design as a whole is not to lose cohesion and balance

5.3 Some ship research studies exploring discipline clashes

5.3.1 IFEP ship and machinery study

This was a UCL internal study of both naval architecture and marine engineering issues to test out the

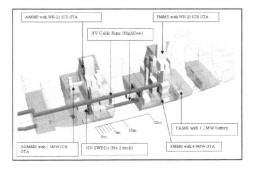

Figure 6. An example of an IFEP investigation for a nominal AAD destroyer (Andrews et al., 2004).

degree to which adoption of IFEP could open up naval combatant layout choices. The DBB approach produced balanced designs for an AAD destroyer with progressively more novel electrical powering features to identify the whole ship impact. Thus without the "tyranny of the shaft line" this could open up the internal compartment disposition, but only if the machinery plant was sufficiently unitised (see Figure 6) (Andrews et al., 2004).

5.3.2 Future weapon design study

This ship and combat system design study explored, using data in the public domain, an appropriate configuration of a large combatant able to deploy future directed energy weapons (DEW). Given the ship fit challenges posed by such prospective large-scale weapon systems, this required a DBB based synthesis approach to produce solutions for both mono-hulled and trimaran variants. It proved essential to comprehend the ship design implications of a novel weapon fit with significant implications for power generation, so spanning all three key engineering disciplines concerns (see Figure 7 from Andrews, et al., 2010).

5.3.3 Various "UXV" design studies

A series of ship concept designs have been produced by the UCL marine design research team (DRC). These have looked at novel ships to host UAVs (Pawling & Andrews, 2009), UUVs (Pawling & Andrews, 2011a) and USVs (Pawling & Andrews, 2013). Given the driver for such dedicated "UXV carriers" is the launch and recovery system (LARS) and the stowage of these autonomous vehicles, an architecturally driven design synthesis is clearly necessary, as is revealed by the SURFCON models of naval architecturally balanced design studies presented in Figure 8.

Figure 7. A mono-hulled variant of a future DEW armed combatant (Andrews, et al., 2010).

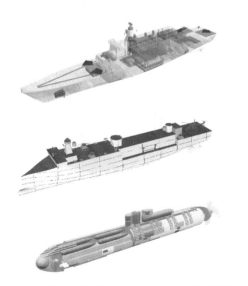

Figure 8. Three UXV carrier vessels—Air Vehicles (Pawling & Andrews, 2009), Surface Vehicles (Pawling & Andrews, 2013) and Submarine "carriers" (Pawling & Andrews, 2011a).

6 WHAT MAKES A NAVAL ARCHITECT?

Having considered various views on ship design, it is now sensible to come back to addressing the question as to what it is that makes a naval architect. The second section outlined the post-graduate training and on the job knowledge and experience a professional awarding body (RINA) considers appropriate on top of the academic qualifications, and the discussion on "hull engineering" identified the main sub-disciplines constituting naval architecture. But it is also worth addressing the nature of the discipline as taught and practiced.

If the fundamental of naval architecture are considered beyond the immediate sub-disciplines encapsulated by at least the first four of the "S⁵"aspects of Figure 1, one comes up against Rydill's belief this is best done by firstly addressing the fundamentals of basic engineering (or even more specifically mechanical engineering). Beyond these engineering topics one then questions whether, in this time of computer based naval architecture, how much what might be called "Basic Naval Architecture" needs to be acquired by 21st Century naval architects. Given there are many basic CAD packages, is the ability to do or even understand tasks like producing displacement sheets, lines plans, and midship sections still necessary when they can all be produced at the push of a button? What about inclining calculations and launch and docking calculations when simple hand held access to programs to "check" these evolutions are readily to hand? Does the naval architect of the future still need to understand basic principles reinforced by some basic calculations only ever preformed when at college? If not what confidence can the owner, or worse the mariner, have in the senior naval architect involved in a new design when they are signing off the safety certificate as the person taking ultimate responsibility for the design efficacy?

When one turns to the advances there have been in recent years to the practice of naval architecture due largely to the ever greater computational power of digital machines and software, then the issue becomes one of how much insight into the fundamentals of finite element methods (FEM) and computational fluid dynamics (CFD) does a design engineer, such as a naval architect involved in ship design, need to acquire. Given the awesome responsibility for the lives of the operators and potential impacts on the wider environment, it would seem that reliance on (separately approved) software in the design analysis of a new ship design is professionally questionable. If anything it is the acquisition of the understanding of the limitations and fundamental assumptions, in any underlying engineering science behind advanced analytical tools, that should be key to any qualification at "professional or chartered engineer/ingenieur" (i.e. the highest professional) level. This is despite the advances in such topics as FEM and CFD applications and the need for the "generalist" ship designer to acquire an ever more widely scoped range of heterogeneous knowledge—not least in the field of human factors, seen as ever more important and inseparable from designing facilities that are the working (and even the living) environments for human beings.

So then turning directly to ship design, this has also seen a great growth in both understanding and methods and (digitally based) tools that continues to expand in scope. Proof of this is shown by the research and practice presented to regular conferences, exemplified by the IMDC series since its precursor conference organised in Trondheim in 1979 by Stain Erichsen (1979). A very comprehensive overview of the IMDC's critical papers and reports was presented to the 2012 IMDC (Andrews, 2012c), yet even since then there have been further advances in the use of new approaches and analyses. Briefly, there are the following areas of ongoing research: data driven documentation (D3) exploiting the big data revolution (Gaspar et al., 2014); use of network analysis to better understand complex ships (Rigterink et al., 2014 and Collins et al., 2015); the use of Operational Analysis to obtain MoE to select new submarine design options (Nordin, 2014); Set Based Design to delay decisions to accommodate evolving subsystems (McKenney, 2013); the DBB approach combined with requirement optimisation (Burger

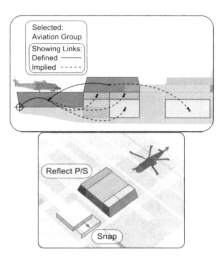

Figure 9. Two suggested techniques to assist in designer led "sketching" in ESSD (Pawling and Andrews, 2011b).

and Horner, 2011); Epoch-Era Analysis to build in through life adaptability (Gasper, 2013); and even research of a methodological and philosophical nature into early stage ship design with the aim of improving design understanding and enhancing its intellectual basis (Andrews, 2012a).

Beyond research into improving the practice of ship design, especially in the earliest formative stages, ship design will become increasingly responsive to wider developments. To a degree the D3 initiative above is of this nature but perhaps more direct is the potential offered by increasingly available tools, such as 3-D printing (The Naval Architect, 2018) and Virtual Reality (Bradbeer, 2016). Both of these are seen as tools to assist designers in better understanding their design choices and being able to better discuss with stakeholders emergent design options. This is likely to significantly alter the way ship design is both taught and practiced. A further area, which the author's group at UCL suggested several years ago, ought to be readily exploitable in making initial ship synthesis more open and creative, is that of Design Sketching. This can be seen as the essential design technique already employed by both designers of Physically Large and Complex (PL&C) systems, and even more so by architects of major buildings and urban structures, prior to recourse to the ever growing capabilities of CAD based design. Pawling and Andrews (2011b) have suggested how a more responsive and innovative sketching like approach to initial ship design might interface directly with an architecturally based technique such as the DBB approach (see Figure 9).

7 CONCLUSIONS

This paper has argued that despite the increasingly demanding safety regime in ship design, that emphasises the naval architect's role as "the hull engineer", the naval architect's primary role remains that of being the overall ship designer. There are seen to be three main reasons for this assertion:

1. "Everyone's problem is the naval architect's problem." This is because should any other part of the ship design get into difficulty then this will impact on the ship's overall weight, space and centroid. It may have other effects but once the budgets allocated by the naval architect are exceeded the knock on effects are likely to require the ship designer to re-balance the design. This shows clearly that architecture drives size, which drives overall form (not just hydrodynamic form). Furthermore it is only the naval architect—not some non-ship design focused manager—who can comprehend the form the style and solution space issues, with direct implications for whole ship cost (see Figure 4 and the Style choices summarised in Table 1).
2. There is a need to have a whole ship perspective to ensure design balance is achieved from the initial synthesis and maintained throughout design development and through life. (Balance in a design is not just floating upright but in obtaining and maintaining total design balance across all the aspects shown in Figure 3). The applicability to complex ship design of Systems Architecture, originating in complex software systems practice, and the achievement of such an approach through a nominated Design Authority was spelt out in Andrews (2015).
3. Architecture is seen to be the key to both initial ship design synthesis and to achieving and maintaining design balance. It is also seen to give the perspective necessary to making coherent style choices and, unlike terrestrial constructions with the lead synthesis role of the architect, this can only be accomplished for marine vessels by the naval architect with the necessary mix of skills in architectural and engineering design and analysis.

The conclusion from this is that if naval architects become just hull engineers then however good the detailed engineering, the overall ship concept will be fundamentally flawed and its development likely to be, at best, incoherent. However this dual role then means that the naval architect, as the primary safety engineer as well as the design lead, has to accept the impossible and thankless burden encapsulated by Forrester's comment on the ship designer at the head of this paper.

ACKNOWLEDGEMENTS

The author would like to acknowledge the insights and pleasure he obtained from the insights of three eminent ship designers (particularly his mentor for many years, the late Louis Rydill), whose views on ship design are at the heart of this paper. He would also like to thank his colleague Dr Rachel Pawling, several of whose design studies are shown here, has demonstrated the value of the author's Design Building Block approach to ship synthesis. The views expressed in the paper, however, are solely those of the author.

REFERENCES

Andrews, D.J. (1986), An Integrated Approach to Ship Synthesis, RINA Transactions, Vol. 128, 1986.
Andrews, D.J. (1993), Management of Warship Design, Trans. RINA Vol. 135, 1993.
Andrews, D.J. (1994), Preliminary Warship Design, Trans. RINA Vol. 136, 1994.
Andrews, D.J. (1996), Trireme to Trimaran – The Fascination of Ship Design, Sept 1994, Inaugural Lecture, UCL, (Reprinted in Journal of Naval Engineering 1996).
Andrews, D.J. (2003), A Creative Approach to Ship Architecture, RINA IJME, Sept 2003, Discussion and Author's response IJME Sept 2004, Trans RINA Vol. 145, 146, 2003, 2004.
Andrews, D.J. (2005), Architectural Considerations in Carrier Design, International Conference 'Warship 2004: Littoral Warfare & the Expeditionary Force', RINA, London, June 2004. RINA International Journal of Maritime Engineering, Sept 2005, Discussion and Author's response IJME Sept 2006, Transactions RINA 2006.
Andrews, D.J. (2010), Discussion to Design Authority of the DARING Class Destroyers by: Gates, PJ, Trans RINA/IJME Vol. 147.
Andrews, D.J. (2012a), Art and science in the design of physically large and complex systems, Proc. Roy. Soc. Lond. A (2012) Vol. 468.
ANDREWS, D.J. (2012b), The View from Bath – a Naval Constructor's Perspective, Chapter 23 in From Dreadnought to Daring – Centenary of the Naval Review, Edited by Captain Peter Hore, January 2013.
Andrews, D.J. (2012c), Is Marine Design now a Mature Discipline? Keynote Paper, Proceedings 11th IMDC, Strathclyde University, Glasgow, June 2012.
Andrews, D.J. (2013a), The True Nature of Ship Concept Design – And what it means for the Future Development of CASD, COMPIT 2013, Cortona, Italy, May 2013.
Andrews, D.J. (2013b), The View from Bath – a Naval Constructor's Perspective, Chapter 23 in From Dreadnought to Daring – Centenary of the Naval Review, Edited by Captain Peter Hore, Seaforth Publishing, Barnsley, UK, January 2013.
Andrews, D.J. (2015), Systems Architecture is Systems Practice in Early Stage Ship Design, IMDC 2015, Tokyo Univ, May 2015.

Andrews, D.J. (2016), Ship Project Managers Need to be Systems Architects not Systems Engineers, RINA Conference on Maritime Project Management, London, Feb 2016.

Andrews, D.J. & Erikstad, S-O. (2015), State of the Art Report on Design Methodology, in IMDC 2015 Proceedings, Edited by Yamoto, H., Tokyo, June 2015.

Andrews, D.J. (2017), Submarine Design is Not Ship Design, Warship 2017: Submarines and UUVs, RINA, Bath, June 2017.

Andrews, D.J. (2018), The Sophistication of Early Stage Design for Complex Vessels, IJME Special Edition to be published 2018.

Andrews, D.J, Bucknall, R & Pawling, R (2010), The impact of integrated electric weapons on future warship design, INEC 2010, HM Naval Base, Portsmouth, May 2010.

Andrews, D.J., Greig, A.R. & Pawling, R., (2004), The Implications of an All Electric Ship Approach on the Configuration of a Warship, INEC 2004, Amsterdam, March 2004. Reprinted in Journal of Naval Engineering June 2005.

Baker, R. (1956b), Habitability in Ships of the Royal Canadian Navy, Trans SNAME 1956.

Baker, L.H. (1990), Combat System Design Developments, Journal of Naval Engineering, Dec 1990.

Baker, R. (1946), Ships of the Invasion Fleet, Trans INA, 1947.

Baker, R. (1956a), How to build a ship, Royal Canadian Navy publication, 1956.

Betts, C.V. (1986), The MSc ship design exercise at UCL, Engineering Design Education, Autumn, 1986.

Betts, C.V. (2010), Discussion to Design Authority of the DARING Class Destroyers by: Gates, PJ, Trans RINA/IJME Vol. 147. Blakeley, T. (2010), Forward, The Royal Institution of Naval Architects 1860–2010, RINA, London, 2010.

Bradbeer, N. (2016), Initial Lessons from the application of Virtual Reality to Warship Design, RINA Warship Conference on Advanced Technologies, Bath, UK, June 2016.

Broadbent, G. (1988), Design in Architecture: Architecture and the Human Sciences, London: David Fulton Publishers, Revised edition, 1988.

Brown, D.K. (1983), A Century of Naval Construction, Conway Maritime Press, London, 1983.

Brown, D.K. (1986), Sir Rowland Baker, RCNC, Warship, Volume IX, Conway Maritime Press, London, 1986.

Brown, D.K. & Andrews, D.J. (1980), The Design of Cheap Warships, Proc. of International Naval Technology.

Bryson, L.S. (1984), The Procurement of a Warship, Trans RINA, Vol. 127, 1984.

Burcher, R. & Rydill, L.J. (1994), Concepts in Submarine Design, Cambridge UP, 1994.

Burger, D. & Horner, D. (2011), The use of Paramarine and mode FRONTIER for ship design space exploration, COMPIT, Berlin, May 2011.

Calea, J., Pawling, R. and Greig, A. (2015), A Data Driven Holistic Early Stage design process to Design Profitable Low Emissions Cargo Ships, Proceedings IMDC 2015, Tokyo, June 2015.

Collins, L. et al (2015), A new approach for the Incorporation of Radical Technologies: Rim Drive for Large Submarines, IMDC 2015, Tokyo Univ., May 2015.

Doerry, N., et al., (2009), What is Set Based Design?, US Naval Engineers Journal, Vol. 121, No. 4, 2009n.

Eppinger, S.D. & Browning, T.R. (2012). Design Structure Matrix Methods and Applications, Cambridge, MA: M.I.T. Press, 2012.

Erichsen, S. (1997), Some Elements of Ship Design, Proceedings IMDC 1997, Newcastle, UK, June 1997.

Erichsen, S. (1998), Management of Marine Design, Butterworths, London, 1998.

Expo 80, Rotterdam, June 1980. (Reprinted in Journal of Naval Science April 1981).

Farrell, K.P. et al., (1972), The DDH 280 Class Design, Marine Technology, Jan 1972.

Faulkner, D. (1965), Welded connections used in warship structures, Trans RINA, Vol. 107.

Ferguson, E.S. (1992), Engineering and the Mind's Eye, Gaspar, H.M., (2013), Handling Aspects of Complexity in Conceptual Ship Design, PhD Thesis, NTNU Trondheim.

Forrester, C.S. (1942), The Ship, Michael Joseph, London, 1942.

Gaspar, H. et al., (2014), Data Driven Documents (D3) Applied to Conceptual Ship Design Knowledge, COMPIT, Darlington, UK, May 2014.

Gates, P.J. (2009), Design Authority of the DARING Class Destroyers, Trans RINA/IJME Vol. 146.

German, W.H. (1978), The Profession of Naval Architecture, Hovering Craft & Hydrofoil, Vol. 17, No.7, Apr 1978.

Honnor, A.F. & Andrews, D.J. (1982), HMS INVINCIBLE - The First of a New Genus of Aircraft Carrying Ships, Trans RINA Vol. 124, 1982.

Honnor, A.F. (2006), quoted in Andrews, D.J. et al (Ed) (2006), State of the Art Report on Design Methodology, in IMDC 2006 Proceedings, Edited by M. Parsons, U of Michigan, Ann Arbor, MN, June 2006.

Hore, P. (2012), in From Dreadnought to Daring – Centenary of the Naval Review, Edited by Captain Peter Hore, Seaforth Publishing, Barnsley, UK, January 2013.

Johnstone-Bryden, R. (2018), H.M.S. Ocean, To be published 2018.

Jones, J.C. (1980), Design Methods, 2nd Edition, Wiley Interscience, London, 1980.

Lavis, D., Rogalski, W. & Spaulding, D. (1990), The Promise of Advanced Naval Vehicles for NATO, Marine Technology, Vol. 27, No. 2. Loyd, A.J. (1992), The Seakeeping Design Package (SDP), Trans RINA, Vol.134.

Leopold, R. (1975), Should the Navy design its own Ships?, US Naval Institute Proceedings, Naval Review, 1975.

Mckenney, T.A. (2013), An Early-Stage Set-Based Design Reduction Decision Support Framework Utilising Design Space Mapping and a Graph Theoretic Markov Decision Process Formulation, PhD Thesis, Univ. of Michigan, 2013. Meek, M., (1970), Encounter Bays – First OCL Container Ships, Trans RINA, Vol. 112.

Meek, M. (1972), Structural Design of the OCL Container Ships, Trans RINA, Vol. 114.

Meek, M. (1982), The Effect of Operational Experience on Ship Design, Proceedings of First IMSDC, RINA, London, June 1982.

Meek, M. (2003), There Go the Ships, The Memoir Club, Spennymore, UK, 2003.

Mit Press, Cambridge, Mass, 1992.

Nautical Institute (2015), Improving Ship Operational Design, The Nautical Institute, London, 2015.

Nordin, M. (2014), A Novel Submarine Design Method, PhD Thesis, Chalmers Univ., Gothenburg, 2014.

Pawling, R. & Andrews, D.J. (2009), The Ship Design Challenge of Naval Unmanned Aerial Vehicles, International Conference 'Warship 2009: Airpower at Sea', RINA, London, June 2009.

Pawling, R. & Andrews, D. (2011a), A Submarine Concept design – The Submarine as an UXV Mothership, Warship 2011: Submarines and UUVs', RINA, Bath, June 2011.

Pawling, R. & Andrews, D.J. (2011b), Design Sketching – The Next Advance in Computer Aided Preliminary Ship Design, COMPIT 2011, Berlin, May 2011.

Pawling, R. & Andrews, D. (2013), Large Unmanned Vehicles and the Minor War Vessel, Warship 2013: Minor Warships, RINA, Bath, June 2013.

Preston, A. (2002), The World's Worst Warships, Conway Maritime Press, London, 2002.

Purvis, M.K. (1974), Post War RN Frigate and Guided Missile Destroyer Design 1944–1969, Trans RINA, Vol. 116.

Rawson, K.J. (1990), Ethics and Fashion in Design, Trans RINA, Vol.133, 1990.

Rawson, K.J. & Tupper, E.C. (1976), Basic Ship Theory, (2nd Ed), Longmans, London, 1976.

Rigterink, D.; Piks, R. & Singer, D.J. (2014), The use of network theory to model disparate ship design information, Int. J. Naval Architecture and Ocean Engineering 6/2.

Roach, J.C. & Meier, H.A. (1979), Warships Should Look Warlike, US Navy Institute Proceedings, June 1979.

Rockwell, T. (2002), The Rickover Effect: How one man made a difference, J. Wiley, New York, NY, 2002.

Rydill, L.J. (1986), An Idiosyncratic View of Warship Design, A Talk to the VSEL Engineering Society, Barrow-in-Furness, UK, Oct 1986.

The Naval Architect, (2018), Testing the waters of 3D printing, Supplement on Innovation in shipbuilding, 2018.

Tibbitts, B. et al., (1993), Naval Ship Design in the 21st Century, Trans SNAME, 1993.

Usher, P. & Dorey, A. (1982), A Family of Warships, Trans RINA, Vol. 124.

Winter, D.C. (2007), Gettig shipbuilding right, US Naval Institute Proceedings, June 2007.

Wood, J.P. (1982), Very Grave Suspicion, RUSI Journal, March 1982.

New type of condensate tanker for arctic operation

Markku Kajosaari
Arctech Helsinki Shipyard Oy, Helsinki, Finland

ABSTRACT: Arctech Helsinki Shipyard Oy is building a new condensate carrier for operation under Arctic conditions in the Barents Sea and the Kara Sea area and along the Northern Sea Route. The vessel features advanced technology and has been designed especially for the demanding operations in the Arctic. The vessel will be delivered by fall 2018.

The condensate carrier will be designed and built for unrestricted open water operation and ice navigation along the Northern Sea Route, including year-round operation on the route from North West Europe to Yamal peninsula through the south western part of Kara Sea and summer navigation from Asia to the port of Sabetta.

The vessel will be designed for energy-efficient operation both in open water as well as in heavy ice conditions. The vessel will be able to move continuously through about 1.9 m thick level ice stern first, and to break about 1.5 m thick ice bow first. In open water the economical service speed will be 13 knots.

The hull form and propulsion arrangement have been developed and tested to minimize the ice resistance and maximize the icebreaking capacity without scarifying the open water performance and economy. The power station consists of a combination of main engines with differing capacity, which enables optimization of the engine load in variable operation conditions. The two azimuthing propulsion units will enhance the manoeuverability and icebreaking performance, especially in heavy ridged and rafted ice.

The vessel represents a new generation of Arctic tankers capable for operation in the extreme cold weather conditions and shallow waters of the river estuaries along the Siberian coast line. The vessel complies with all the relevant rules, regulations and recommendations concerning the safety of navigation and protection of the sensitive environment in the Arctic waters.

The design and construction of the vessel takes place at the facilities of Arctech Helsinki Shipyard Oy. The condensate carrier is a new type of product for the company. The newbuilding project is carried out in close co-operation with a well-established network of suppliers and subcontractors, mainly from various parts of Europe.

Keywords: Icebreaking; Winterization; Northern Sea Route; Yamal; Arctech

1 INTRODUCTION

Arctech Helsinki Shipyard Oy is building a condensate carrier for operation in the Barents Sea and the Kara Sea and along the Northern Sea Route. The vessel features the most advanced technology and has been designed especially for the demanding and cold environmental conditions of the Arctic sea areas. The vessel has the hull number NB-515 and it will be delivered by the fall of 2018.

This paper describes the principal features of the newbuilding project and some important features related to the design and construction of such a special vessel.

2 DESCRIPTION OF THE NEW BUILDING PROJECT

The design of the new type of condensate carrier is the result of a long development project combining the latest features of icebreakers and other heavy icebreaking vessels with the characteristics of a shallow draft tanker. This involves a special hull form with heavy ice strengthening and a diesel-electric power with multiple propulsion units as well as complete cargo and water ballast systems capable for handling several different grades of cargo with simultaneous loading and de-ballasting or discharging and ballasting.

The vessel has been designed for year-round navigation in the Arctic seas. The vessel will primarily operate from the port of Sabetta, located in the Yamal peninsula at the Gulf of Ob, either to North West European ports or directly to Asia along the Northern Sea Route.

The hull form of the vessel features a modern double hulled tanker cargo compartment combined with an icebreaking bow designed for operation in open water and in mild ice conditions. The stern of the vessel has been designed for transit in heavy ice conditions stern first, utilizing the double acting ship concept. With this special arrangement the vessel can operate without icebreaker assistance in heavy ice conditions, but still have good open water and seakeeping performance.

The maximum draft of the vessels visiting the port of Sabetta is limited to 12 m. The main dimensions and hull form design of all the bigger vessels visiting the area have to be adjusted according to this limitation. The water in the estuary of the river Ob is brackish water with density close to that of fresh water. This feature has an additional influence on the scantlings and main dimensions as well as on the ice breaking performance, as the flexural strength of ice in this kind of conditions may be twice the strength of normal sea ice.

The operation conditions in the Arctic are extremely demanding. The lowest ambient temperature for the vessel, its machinery and equipment, their components and related systems, and the accommodation areas, has been set as low as −50°C. Special consideration has been given in the design for the safety and comfort of personnel onboard, particularly as regards the operations and work conditions in low temperatures and the potential for ice accumulation on external working surfaces.

The requirements for the icebreaking performance have been selected to enable independent and efficient operation in ice conditions typically encountered in the Arctic waters of the Northern Sea Route. The vessel can break more than 1.9 m thick level ice when moving stern first at the speed of 2 knots, and when breaking ice bow first, the maximum ice thickness can be 1.5 m. Special consideration has also been given to the vessel's capability to penetrate through ridges and manoeuvre in various ice conditions.

The machinery principle of the vessel is based on the well-proven diesel electric power plant configuration with dual Azipod-type propulsion units. The rudder propellers are arranged to ensure reliable propulsion capability in even the most difficult ice conditions. The plant for the electric power generation consists of four main diesel generator sets and one smaller auxiliary diesel generator set for harbour use. The total installed power of the main diesel generators is 31 MW.

Figure 1. Arctech NB-515—Arctic Condensate Tanker 50000 DWT.

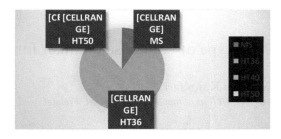

Figure 2. Use of different steel grades in the construction of the steel hull.

The principal concept design for the new vessel has been prepared by Arctech Helsinki Shipyard Oy. The contract for the construction was signed in early 2016, and Arctech, will be responsible for the entire process of construction from the basic design to construction, commissioning and delivery of the vessel, including the responsibility over the performance values.

3 ICE STRENGTHENING

The ice strengthening and ice class for the vessel have been selected according to the rules of the Russian Maritime Register of Shipping (RMRS) for the intended operation profile and area. The ice class is Arc7, which—according to the Rule definitions—enables winter and spring navigation in thick first year ice up to 1,8 m and summer and autumn navigation in second year ice.

The RMRS rules define also the sea areas and seasons of operation for ships with arctic ice class in the Russian arctic seas (RMRS 2017). The selected ice class Arc7 allows independent winter navigation without icebreaker assistance in the Barents Sea area and in the Kara Sea in medium and easy conditions of ice navigation on a year round basis. The ice class Arc7 allows also independent navigation along the Northern Sea Route in all conditions of ice navigation during summer and autumn seasons. For winter navigation along the Norther Sea Route in hard ice conditions ice breaker escort will be needed.

The vessel has been designed for the maximum ice draught of 12,0 m. Due to the double acting ship-principle used in the design of the hull and propulsion systems, the aft part of the hull/stern of the condensate tanker has been strengthened in the same manner as the forward/bow region. This means also some additional steel and weight for the hull in comparison with more conventional designs. In order to balance the extra weight a lot of high tensile steel material has been used in the construction of the hull. About 80% of the steel weight is of high strength steel with yield strength of 355 MPa, about 10% of the steel weight is of high strength steel weight yield strength of 390 MPa or even 500 MPa. Only some 10% of the steel weight is of mild steel used in conventional shipbuilding.

4 ICE BREAKING PERFORMANCE

The hull form and propulsion arrangement of the vessel have been specially developed for the challenging ice conditions of the shallow river estuaries of the Russian Arctic waters. The maximum icebreaking capacity can be obtained when operating utilizing the double acting ship-principle and running the stern first, but the icebreaking capacity with bow first is also sufficient for most ice conditions.

The hull form has been made slightly more narrow than that of other icebreaking tankers operating at the same area. The intention has been to optimize the hull performance in heavy ice conditions. The hull form was developed by Arctech Helsinki Shipyard Oy, first with numerical calculations for optimization of the hull form and preliminary evaluation of the ice resistance. The hull form was later tested in open water conditions at the model basin of HSVA in Hamburg. The final ice model tests were made at Krylov State Research Institute in St. Petersburg.

The icebreaking capacity of NB-515 is extremely good for this size and type of tanker. The maximum capacity, when breaking ice stern first, is higher that of comparable vessels designed according to the dual acting operation principle. And even when breaking ice bow first, the icebreaking capacity of NB-515 is close to the maximum ice breaking capacity of the other vessels.

5 PROPULSION AND MACHINERY

The new vessel will have two azimuthing propulsion units located in the stern. This arrangement provides excellent icebreaking capacity and good performance, especially when breaking ice stern first and when penetrating through pressurized ice fields and ridges. In the astern operation mode the propeller flows of the propulsion units against the hull reduce the friction between the ship´s hull and ice. And when penetrating through ridges, the azimuthing units can be used to break down the ridge and open the route without ramming. The azimuthing propulsion units also enhance remarkably the manoeuverability of the vessel.

The diesel electric machinery configuration is based on the power plant principle. The vessel will have four main diesel engines running AC generators. The total power of the main diesel generator engines is 31 MW, consisting of two smaller and two bigger diesel generator units for flexibility and optimum powering performance and fuel economy. The diesel engines are of type Wärtsilä 12V32 (6720 kW/720 rpm) and Wärtsilä 16V32 (8960 kW/720 rpm). The power of the harbour diesel generator set is 740 kW.

The power transmission is controlled by variable speed converter drives supplying the electrical motors of the propulsion units. The total shaft power for propulsion is 22 MW, divided into two 11 MW Azipod-type units equipped with four bladed propellers made of stainless steel. The propeller diameter is 5600 mm.

6 ENERGY EFFICIENCY

The requirements concerning the energy efficiency index in accordance with MARPOL Annex VI are not applicable for vessels having ice-breaking capacity. The EEDI regulations are also excluded on the basis of the installation of the non-conventional propulsion system that is why no official EEDI calculations or baseline comparisons have been made for the new vessel.

The requirements for the ice class and ice breaking performance bring remarkable addition to the installed power of the vessel in comparison with tankers or condensate carriers of comparable size without ice class and without ice-breaking capacity. The installed propulsion power of a typical 46 000–50 000 DWT handymax tanker is in the range of 9 to 10 MW (MAN Diesel & Turbo), which is

Table 1. Main dimensions and ice breaking performance of modern ice strengthened tankers.

	Mikhail ulyanov	Vasily dinkov	Shturman abanov	NB-515
Length OA	257 m	258 m	249 m	230 m
Beam	34 m	34 m	34 m	32,5 m
Draught dwl	13,6 m	14 m	9 m	12 m
DWT	70000 t	70000 t	42000 t	50000 t
Ice class	LU6 (Arc 6)	LU6 (Arc 6)	Arc 7	Arc 7
Propulsion power	17 MW	20 MW	22 MW	22 MW
Propulsion type	2x Azipod	2x Azipod	2x Azipod	2x Azipod
Speed in ice	3 kn	3 kn	3,5 kn	2 kn
Ice thickness	1,2 m	1,5 m	1,4 m	1,9 m

about 30% of the power needed for an Arc7 tanker with the same cargo capacity. The guidelines for the EEDI calculation recognize both the correction factor for the additional installed power and the capacity correction due to the ice strengthening, but both these correction factors have been defined only for vessels with light to moderate ice class, up to 1ASuper in the Finnish-Swedish ice class.

Arctech has prepared an estimation of the attained EEDI for the arctic tanker, for information purposes, utilizing rough interpretations of the calculation methods for non-conventional propulsion and high ice class. The results indicate the attained EEDI for a 50000 DWT tanker with ice class Arc7 exceeds the current baseline by ca. 100%. If, on the other hand, the calculation will be made for the installed power needed for the normal open water service speed, the current baseline level can be met. With this lower power rating, however, the vessel will hardly comply with the minimum required power for ice class Arc4. Vessels with ice class Arc4 are not be capable nor allowed for year round operation in the Kara Sea or along the Northern Sea Route, and even during favourable seasons and ice conditions ice breaker escort is needed most of the time.

7 THE POLAR CODE AND WINTERIZATION

The vessel has been designed for unrestricted trade in the Arctic waters. The lowest extreme ambient temperature for all design criteria and the relevant components and equipment has been defined as low as −50°C. The design is also in compliance with the RMRS class notation Winterization (−50).

All the external arrangements and selection of equipment, systems, controls, walkways, safety rails, lifesaving appliances etc. have been specially designed and selected to permit ease and safe operation by personnel wearing protective thermal clothing. The design of the vessel contains several features to protect the personnel and equipment from the harsh and cold environmental conditions. E.g. the forward mooring deck has been covered to protect the working area and prevent icing of the equipment. The covered forecastle reduces also ice accretion due to freezing sea spray in some areas of operation.

The requirements for the class notation Winterization (−50) bring also a lot of outfitting for heating and protection of equipment and components. There will standstill heaters for exposed electric motors, the essential walkways and stairs will be provided with heated elements, firefighting systems will be fitted with trace heating, hatches and doorways will have heating to prevent icing, all the wheelhouse windows will also have electrical heating etc. Special outfitting inventories will be provided for personnel thermal protection, ice removal and protection of life saving equipment.

The vessel has been constructed prior to the entry into force of the Polar Code. The design contains, however, all the relevant requirements of the Polar Code applicable for vessels constructed before 1 January 2017. In addition, the recommendations related e.g. to the use of environmentally friendly lubricants will be complied with in order to avoid any accidental discharge directly to sea. The capacities of holding tanks, treated water tanks etc. have been defined on the basis of zero discharge principle during 30 days to avoid any discharge when sailing in the Arctic waters.

The vessel has been designed and built in accordance with applicable rules and regulations but also a full compliance with Bureau Veritas CLEANSHIP C class notation and its RMRS equivalent ECO. In addition, for sailing along the Norther Sea Route and other areas of the environmentally sensitive Arctic waters, the vessel has been designed for zero discharge during 30 days. The capacities of various holding tanks, bilge holding tank and treated water tank have been defined accordingly.

TECHNICAL SPECIFICATIONS

Length	230 m
Breadth maximum	32,5 m
Design draught	11,7 m
Max draught in ice	12,0 m
Depth	18,0 m
Deadweight	50 000 t
Installed power	32 MW
Propulsion power	22 MW
Speed	16 kn
Speed at 1.9 m ice	2 knots
Crew	26 + 4
Ice class	RS Arc 7
Classification	Bureau veritas and russian maritime register of shipping

The vessel will have accommodation capacity for 30 persons. The operating crew will be 26 persons, and all of these will have single cabins. In addition there will be two spare cabins, each for two persons.. All cabins and other accommodation spaces for the crew have been located in the superstructure for easy access and comfort.

The T-shaped and totally enclosed navigation bridge on top of the superstructure provides easy and safe operation both ahead and astern. The double acting ship concept brings some additional features to the arrangement of navigation systems and equipment. There are two control stations on CL of the wheelhouse, one for forward operations at the front of the wheelhouse and another looking stern for afterward operations. On both of the bridge wings there are steering stands, too. The navigation lights will be provided with double installation to provide required signals for both of the primary directions of operation.

8 DESIGN AND CONSTRUCTION

8.1 *Arctech Helsinki Shipyard Oy*

Arctech Helsinki Shipyard Oy is located in the downtown area of the city of Helsinki. There has been a shipyard at the same physical location since 1865. During the time the shipyard area has seen several development phases and several different shipbuilding companies. Today the shipyard operated by Arctech Helsinki Shipyard Oy is a modern facility with covered production spaces. Most of the activities related to the shipbuilding process prior to the launching of the vessel can be done indoors. At the shipyard there are close to 2 billion cubic meters of halls and other enclosed facilities for block assembly, block outfitting painting and hull assembly in addition to the normal workshop spaces, stores and office spaces. These facilities enable effective shipbuilding throughout the year, independently of the weather conditions in the Nordic climate (www.arctech.fi.).

The most important facility at the shipyard is the covered building dock, which measures 280,5 m × 34 m × 8 m and it is suitable for building e.g. cruise vessels up to about 90 000 GT. The maximum lifting capacity of the bridge cranes in the covered building dock hall is 500 t. There are also three large hall for block outfitting and four painting chambers at the shipyard area. The shipyard employs directly about 600 persons, of which close to 100 are working in design activities. In addition, there are some 600–800 persons employed by the subcontractors working within the perimeters of the shipyard area.

Arctech Helsinki Shipyard Oy utilizes so called assembly yard -principle. All the hull blocks are purchased from other shipyards or workshops, typically these suppliers are located at the Baltic Sea area. The transportation of the steel blocks takes place by sea, either with barges or with special vessels available for this kind of transportation. The production facilities at Arctech Helsinki Shipyard Oy do not include steel cutting machines, panel lines or other equipment or premises needed for hull block production. With the assembly yard -principle the shipyard can utilize its capacity for those functions and phases of production for which its facilities are best suited—block outfitting, painting, hull assembly and final outfitting and commissioning. This provides more flexibility in the capacity and production schedules and makes the adjustment of operations to variations in the work load and product types easier.

Arctech Helsinki Shipyard Oy, and its predecessor operating at the same location, have delivered more than 500 newbuilding vessels during the 150 years of operation. This high number of deliveries includes various types of ships, from small tugboats to nuclear powered icebreakers, and from small launches to oceangoing cruise vessels. But NB-515 is the first tanker ever built at Helsinki shipyard.

8.2 *The design and construction process*

The concept design and principal technical features for the new buildings constructed by Arctech Helsinki Shipyard Oy are typically developed by the Naval Architecture – design department of the shipyard. The technical specifications, arrangements and system descriptions created at this phase form the basis for the cost estimation and vessel performance criteria. The specifications are used as a part of the shipbuilding contract documentation, too.

The basic design stage—initiated after the signing of the shipbuilding contract—is primarily executed by the design department of the shipyard, with the assistance of some subcontracted design companies. The detail workshop design is done by the shipyard´s own design personnel, supported by subcontracted design offices, and to some extend by the subcontractors responsible for the supply and installation of specific systems or other installations. The hull workshop design is typically the responsibility of the block supplier(s).

Block outfitting and painting is done either by the hull block suppliers or by Arctech Helsinki Shipyard Oy after the blocks have been transported to yard. The final block painting, lifting to the hull, hull assembly and part of the area outfitting are done in the building dock before launching of the vessel. Final area outfitting and system commissioning are completed at the outfitting quay after the floating out of the hull. All tests and

Figure 3. Arctech Helsinki Shipyard Oy—aerial view.

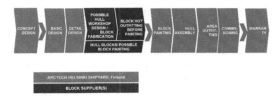

Figure 4. Shipbuilding process at Arctech Helsinki Shipyard Oy.

trials, including also the sea trials are done at this phase of the construction process.

After the delivery of the vessel the shipyard is still responsible for handling and rectifying the possible warranty items during the entire warranty period. In some cases, the final icebreaking trials or similar tests related to the ship operation, are conducted after the delivery of the vessel from the shipyard.

In the case of the condensate carrier NB-515 the hull block purchase and supply was made in two completely different methods, due to the special construction principle selected for the vessel. The aft part of the vessel, containing the stern part with the machinery compartment and accommodation areas, has been build using the typical method of purchasing blocks from subcontractors and taking care of the hull assembly in the building dock of Arctech Helsinki Shipyard Oy. The majority of these blocks has been purchased from the shipyard of Western Baltija Shipbuilding in Lithuania, but three other block suppliers have also supplied some parts for the vessel. The blocks for the stern part have been transported to Helsinki in several dispatches by sea.

The entire fore part of the vessel, containing all the cargo tanks and related systems, as well as the heavily strengthened bow section, has been purchased as a maxi-block from Brodotrogir Shipyard (HBT) in Croatia. Brodotrogir will build and outfit the fore part of the vessel as a complete unit, ready to be connected to the stern part. The bow section will be transported from Croatia to Finland by sea. The final hull assembly will take place in the building dock of Arctech Helsinki Shipyard Oy, where the bow section will be floated and docked in front of the stern part waiting on the keel blocks.

The completion of the ship outfitting and all commission activities will be done by Arctech Helsinki Shipyard Oy after the bow and stern parts of the vessel have been welded together and all the necessary system connections between these have been completed. The testing and commissioning of the vessel will be done after the complete vessel has been launched and floated out from the building dock.

The normal tests and sea trials will be conducted by Arctech Helsinki Shipyard Oy prior to the delivery of the vessel. The sea trials will be done at the Baltic Sea. The icebreaking tests needed to proof the contractual icebreaking capacity can, however, be completed after the delivery of the vessel. These tests will be made while the vessel is in operation and sailing in the Arctic waters, where the thickness of sea ice is sufficient for the testing of the icebreaking capacity.

9 CONCLUSIONS

The new icebreaking condensate carrier will set new standards for cost-efficient, independent and environmentally friendly operation in the Russian Arctic waters on the year round basis. The vessel will comply with all the latest requirements and recommendations related to the ship safety and protection of environment in the harsh and cold Arctic waters. The new NB-515 from Arctech Helsinki Shipyard Inc. will provide an advanced addition to the fleet of vessels capable for regular operation along the Norther Sea Route.

REFERENCES

Russian Maritime Register of Shipping, 2017, Rules for the Classification and Construction of Sea-Going Ships, 2017, Part I Classification, Table 2.2.3.4-1. Saint-Petersburg.

MAN Diesel & Turbo. Propulsion of 46,000–50,000 dwt Handymax Tanker. Available at http://marine.man.eu/docs/librariesprovider6/technical-papers/propulsion-of-46000–50000-dwt-handymax-tanker.pdf.

Education

HYDRA: Multipurpose ship designs in engineering and education

R.J. Pawling
University College London, UK

R. Bilde
MSc Naval Architecture, UCL, UK

J. Hunt
Royal Canadian Navy, Canada

ABSTRACT: The cornerstone of post-graduate naval architecture and marine engineering education at UCL is the Ship Design Exercise. This three-month full-time project sees students placed in small, multi-disciplinary teams and challenged with the concept design of a new vessel based on broad outline requirements provided by the academic staff. This exercise exemplifies the use of design as an integrative teaching method, allowing engineering students to place their academic understanding of technical subjects in a whole-ship concept. This paper describes an innovative design – HYDRA – featuring a single core vessel capable of adaption during build to take on several military or civilian roles. This paper not only describes the technical aspects of the design solution itself, but also discusses the educational implications of setting students the challenge of designing ships to meet multiple, sometimes contradictory requirements. In addition to aligning well with some modern trends in ship design and construction, this type of problem is seen to offer potential benefits in engineering education. These benefits are discussed, in addition to the potential complications they bring to various aspects of the design exercise.

1 INTRODUCTION

Design is widely recognised as being an activity central to engineering education (McLaren, 2008), and many, if not all, naval architecture degrees will feature some ship design activity. Technological and social developments have an impact on education, however, and in discussing the use of new computer-aided approaches in preliminary ship design, Pawling et al (2017) raised the question of "what might be the new key fundamentals of engineering teaching", noting that:

"From 2019, most first-year undergraduates will be fully "21st century students", however it could be argued that some universities are still teaching them using 20th century tools and 19th century methods".

There has been some quantified research, such as that by Collette (2015) investigating the impacts of modern tools (in that case, 3D models) in teaching ship design, but it is still an area for development.

It is proposed that design exercises that cannot be reduced to simple mechanistic analysis or iteration, and which oblige students to make decisions, become more important as the sophistication of modelling and analysis tools available in ship design (and to students) increases. This paper describes the ways in which decision making, as an activity in ship design, are included in the various ship design exercises carried out by undergraduate and postgraduate students at UCL.

2 SHIP DESIGN TEACHING AT UCL

Historically, the Naval Architecture and Marine Engineering (NAME) group, part of the Department of Mechanical Engineering (UCL, 2018) taught ship design at two levels; one and two-year MSc and at the end of three and four-year undergraduate courses. The last undergraduate cohort graduated in 2015, leaving only the postgraduate course, but this is changing with the recent introduction of the "Integrated Engineering Programme" (IEP, 2018), a modular course using the major/minor structure familiar in other countries such as the US. A Maritime Design module, developed by the first author, is available to students in the third year of this course and more detail about this is given later in the paper.

The MSc Naval Architecture and MSc Marine Engineering courses at UCL last for 12 months (with a 12 month foundation year available for students without suitable previous qualifications), and have three main elements; six months of academic teaching and exams; group Ship Design eXercise (SDX); and individual project. Two introductory ship design exercises have recently been added to the MSc timetable, specifically during the academic portion of the course. These are intended to encourage students to consider how their technical tuition integrates with ship design, to illustrate the importance of exercising engineering judgement and introduce the general iterative procedure of initial design. The Introductory Ship Design Exercise, ISDX, takes place in the first month of the MSc, and the Ocean Patrol Vessel Design Exercise (OPV DX) occurs early in the second term.

2.1 *Introductory Ship Design Exercise (ISDX)*

The ISDX is a short exercise, usually taking three hours. The exercise begins with a lecture on the use (and limits of) historical data in ship design. The students rapidly estimate the overall dimensions of a container ship and generate a simple profile view using a highly simplified Excel tool, shown in Figure 1. The primary purpose of the ISDX is to introduce students to the need for engineers to make design decisions and justify them. The sizing relationships are all presented as ranges based on historical data, rather than single lines or algorithms, and the students have to choose where in the range their design is likely to lie, based on the broad implications of "special" requirements, such as icebreaking, gas fuel or high speed.

Additional teaching objectives for the ISDX include providing students with some understanding of the nature, utility and limitations of historical and "type-ship" data in engineering design, to address some of the issues regarding historical lessons raised by Tuttle (1997). The ISDX is deliberately kept more casual than a conventional lecture, to encourage the students to explore the various design options and introduce them to decision making in a "risk-free" environment (as the full SDX involves design reviews with senior staff). After sizing their container ships, the students each present their design to the rest of the group. They are required to state why they chose each parameter in the sizing.

2.2 *Ocean Patrol Vessel Design Exercise (OPV DX)*

The main objective of the OPV DX is to introduce the students to the iterative and interactive nature of ship design, and to the specifics of the procedure and tools used in the main SDX. As with the ISDX it makes use of a greatly simplified dataset and constrained problem, but as the exercise spans two days the students are expected to go into more detail. The students are provided with a requirement that describes the required payload (combat systems), propulsion package and endurance (for fuel and stores), with each requirement having a "special study", a specific key capability which could be; a large flight deck; limited air defence capability; high speed etc.

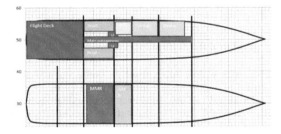

Figure 2. Simplified arrangements in OPV DX Excel tool.

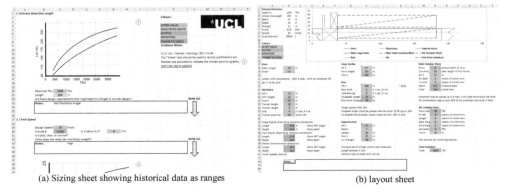

(a) Sizing sheet showing historical data as ranges   (b) layout sheet

Figure 1. The UCL Excel-based ship sizing tool used for the ISDX.

During the OPV DX, the students use a simplified design databook to calculate the overall size of the vessel, then carry out a simplified parametric survey to determine the ratio of volume distribution in the hull and superstructure, to fit within various specified constraints. The design is then worked up with a block layout modelled in an Excel tool (shown in Figure 2) and analysis of stability and powering in Paramarine. The latter makes use of pre-defined template design files representing a typical OPV, so that the students need only enter the dimensions, weights and centres of their designs. The students then present their OPV designs to the group, with questions on the technical aspects of the design.

2.3 Ship Design Exercise (SDX)

The SDX runs for three months full-time between April and June and sees the students split into small groups of 2–4, with a mix of naval architects and marine engineers. The small size of the groups is important to ensure that all students have visibility over all parts of the ship design process, to avoid a student specializing in only a single aspect of the ship design. Each group has a different set of design requirements, mostly for warships and service vessels (due to the students background), and the requirements are characterized by being challenging and relatively open. Table 1 provides some examples of recent design requirements. Although the majority of MSc Naval Architecture students at UCL examine naval vessels, increasingly the course covers other complex service vessels. Unlike "type ship" based approaches—such as the now discontinued undergraduate design exercise for frigates and container ships—the academic staff do not know that the requirements they are setting can even be met.

Another feature of the SDX is the use of several design reviews and consultancy days with academic staff and external subject matter experts. The students must present the progress of their design and answer questions, with a particular emphasis being on the development of their ability to make design decisions and justify them, rather than the precise technical nature of the solution. The UCL MSc SDX groups consist of both naval architects and marine engineers, and each design review covers both domains, and the interactions between them. The whole-ship design decision making is expected to involve inputs from both domains, with more detailed technical analysis being specific to each MSc.

Figure 3 shows an overview of the MSc SDX and its' key stages. The students are provided with a User Requirements Document (URD) developed

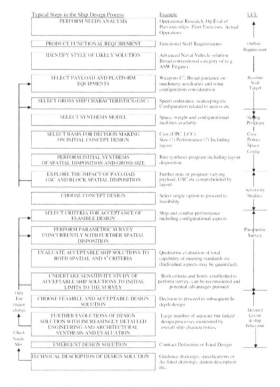

Figure 3. An overview of the MSc SDX process compared with a generic ship design process (Andrews 1986).

Table 1. Examples of UCL MSc SDX design requirements.

Year	Title	Summary
2010	Low Carbon Export Frigate	Adaptable to different requirements, with low emissions cruise mode via fuel cells
2013	Offshore Wind & Marine Current Turbine Support Vessel	To carry staff and parts, with survey and tourism tertiary roles, 30 year life and access capabilities entirely determined by students
2013	Anti-Air Warfare Destroyer	With Anti-Ballistic Missile capability, optional electric weapons and a Lighter-Than-Air sensor system
2016	Mega containership for 2035	Icebreaking container ship carrying 40,000 TEUs at half the EEDI of MSC OSCAR

by the teaching staff, in consultation with industry. The URD sets very high level requirements which can be met in a variety of ways. Few numerical parameters or equipment items are specified, and there are usually several "special studies" or constraints, which may include technologies not yet in operational use. The MSc SDX is notable for a wide variation in design requirements—challenges include unconventional hullforms, technologies or operating restrictions.

The SDX requires the students to integrate the subject specific technical knowledge gained during in the MSc into a coherent design. The design procedure used is a variation on the conventional design spiral. Although this model has been subject to some criticism (Pawling, Andrews & Percival, 2017) it is a relatively simple introduction to design as it represents a linear model of design development, assessing each technical aspect in turn. The particular advantage of using this model of the design process in education is that each type of analysis can be clearly delineated. Emphasis is placed on decision making and justification, and the understanding of influences and interactions in the design. Conceptually the design spiral as implemented in the UCL SDX is closer to the 3D helical model, proposed by Andrews (shown in Figure 4), with its highlighting of the progression of the overall design concept through time (the vertical dimension) and radial constraints and interactions. Importantly this later representation includes the exogenous nature of the constraints, which is not always clear in a simple spiral or linear model.

In addition to numerical sizing and analysis of the design, students are required to consider the configuration of the design as early as possible. As discussed by Pawling & Andrews, (2011), sketching of design options is promoted to aid in exploration and help the students understand the wider interactions in the developing ship design. The format of and tools used for these sketches is not specified; some groups produce hand-drawn sketches, others work directly in CAD tools, whilst other students with a professional or hobbyist background may use computer graphics tools. The importance of developing special modelling in assisting students in developing an understanding of their design has been discussed by Collette (2015), and the sketches in the MSc SDX serve a similar role. Technical aspects are of course examined, as the final reports upon which the students are assessed must contain details of not only the decisions and rationale but also the Naval Architectural and Marine Engineering modelling and assessment for the design.

3 PROJECT HYDRA: AN EXAMPLE UCL MSC SHIP DESIGN

3.1 Introduction

"HYDRA" is a 2017student design developed by the second and third authors to meet a requirement for a "Mediterranean Multirole Coastguard Ship". The development of the design will be summarized in this section, with a particular emphasis on the decision making methods, to illustrate the approach used in the UCL MSc SDX described above. The key feature in this project was that a common design was required able to be completed as either; a coastguard vessel; a research vessel; or a submarine rescue vessel; the three roles giving rise to the name Hydra, the many-headed serpent in Greek mythology. Table 2 summarises the User Requirements Document provided by the academic staff to the design group.

In addition to setting the requirement for multiple, potentially conflicting, roles to be supported by a single, adaptable design, it is notable that the students are encouraged to challenge constraints placed upon them—in this case the displacement cap—with the provision that such challenges must be justified. For this design, no cost cap was specified. It is more typical for MSc URDs to have cost cap, but in this case it was omitted as finding accurate cost data for such vessels can be difficult (UCL's costing database being primarily for naval vessels), and there were seen to be sufficient technical challenges in the URD to occupy the students.

3.2 Decision making processes

Several major phases of decision making and options comparison occur within the MSc SDX

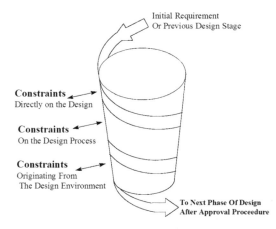

Figure 4. 3-D representation of the ship design spiral (Andrews 1981).

and these were further complicated in the HYDRA design, due to the multifunction requirements. These can be summarized as; operational analysis (OA); payload selection; hullform type; modularity; and parametric survey.

3.2.1 *Operational analysis*

Operational Analysis is required to determine how the proposed vessel will accomplish the very broad requirements given in the URD. The wide variety of ship types examined in the MSc SDX mean that the exact nature of the operational analysis varies, for example a cargo ship may focus on possible routes and economic modelling, whereas an air defence destroyer may compare missile options and magazine capacity. The objective in the HYDRA project was to design a common hull capable of supporting one of three roles, so the operational analysis focused on determining what each of those roles required of the ship.

For each of the three main roles for HYDRA, a survey of existing vessels used for each role was conducted, to determine typical ranges of dimensions, performance and main equipment or ship features. For each of the sub-tasks listed in Table 2 a technical solution with equipment options was proposed. This process was largely conducted by literature review (including the general arrangements of previous vessels), with different methods used to define different technical solutions. For

Table 2. Summary of the HYDRA user requirements document.

Role	A common hull which can be completed as a coastguard vessel, research vessel or submarine rescue vessel. Once completed, there is no requirement to change role during the ship's life.

Primary tasks	Conduct one of the following sets of tasks, chosen at build:
	a. Coastguard
	  i. Very broad area maritime surveillance (visual coverage of 4,000 square nm per hour).
	  ii. Undertake Boarding operations at as long a range as practicable (up to 100 nm.).
	b. Oceanographic & Scientific Research
	  i. Deploy large USV/semi-submersibles for hydrographic survey.
	  ii. Support ocean science research work with laboratory spaces and a working deck.
	c. Submarine Rescue
	  i. Embark and deploy the NATO Submarine Rescue System
	  ii. Provide medical care to the rescued submarine crew, at hyperbaric conditions if required.

Secondary tasks	a. Provide firefighting capability to other vessels and installations. b. Provide humanitarian rescue & evacuation on a large scale, including provision of medical aid. c. Tow disabled ships, up to the largest container ships and bulk carriers.
Area	a. The Mediterranean Sea
Ship life and ISD	a. Ship Life – 40 years b. ISD – 2025
Constraints and special studies	a. Cost – No cost cap b. Speed – Extended low speed loitering, capable of high speed dash for emergency response. c. Displacement – Deep displacement at start of life (in any Role configuration) may not exceed 5,000 tonnes. If required roles cannot be met on this displacement cap, a strong justification should be provided. d. Future Fuels and Emissions – Should design assuming that carbon-containing fuels will increase by an unknown amount between 200% and 500% by 2085, assuming linear increase per year. Strict future emissions regulations should be considered likely during the life of the ship; design to minimise the risk of having to scrap ships early. e. Vulnerability Reduction – Assume collision is likely during ship's life and design to continue operations. f. Must have sufficient gas protection to provide rescue/firefighting services to gas carrier vessels. g. Unoccupied Vehicles (UXVs) – Consider unoccupied vehicles for contributing to surveillance task. Should be able to operate four 11 m RIB/semi-submersible USVs in the survey task. h. Technology Insertion/Mid-Life Update – Scientific facilities should be designed for ease of modification through life. Submarine rescue vessel should be designed to easily accept NSRS successor vehicle. i. Hullform – Multihulls should be considered, and the choice of hullform justified.

example, primary task a (i)—broad area maritime surveillance—led to the proposal for a system of Uninhabited Aerial Vehicles (Figure 5), whilst initial estimates of lab space for the research role could be made from reference papers (Figure 6).

The output of this process was the specification of target equipment fits and design features for each of the three roles. Most notably these targets also contained a justification; e.g. from previous ships, references, or specific capability requirements.

3.2.2 *Payload selection*

Payload selection followed from the operational analysis, performing a cost-capability trade-off using the Equity software (Catalyze 2018). A specific challenge introduced by the multifunctional HYDRA design was that, ideally, each of the three role-specific equipment packages should have the same weight and space requirements, so that no one role was dominating the design. Individual items of equipment were sized based on a UCL database, published datasheets and calculation, and, for each primary role, combined into three functional packages (e.g. UXV equipment, sensors, lab area), each with three levels of capability (e.g. minimum, medium and maximum. Figures 7, 8 and 9 illustrate the equity analysis for each of the three main roles, showing the selected points.

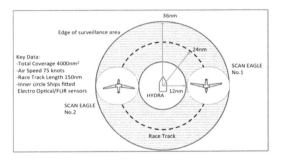

Figure 5. Concept of employment of scan eagle UAVs for maritime surveillance (Hunt 2017).

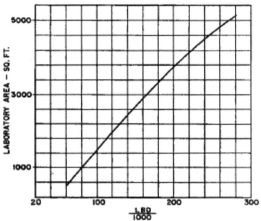

Figure 6. Reference data used for initial estimates of lab area (Rosenblatt, 1960).

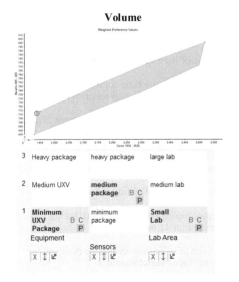
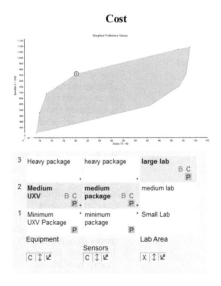

Figure 7. Cost capability trad-off output for HYDRA research (Bilde, 2017).

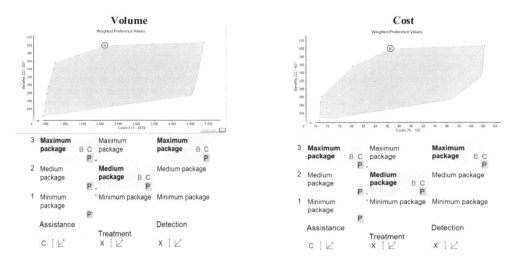

Figure 8. Cost capability trade-off for HYDRA SUBSAR (Bilde, 2017).

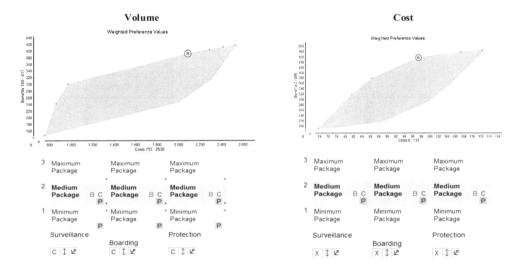

Figure 9. Cost capability trade-off for HYDRA coastguard (Bilde, 2017).

The objective of this analysis was to determine the option providing the highest capability (y-axis) without excessive increases in cost (x-axis) on the right-hand plots. However, due to the HYDRA multi-role requirement, this was complicated by the need to align volume requirements of the ship (left-hand plots). This analysis showed that although the coastguard option was preferable from a cost/capability perspective, it had high volume demands that would drive the design. This allowed the students to challenge the requirements set by the customer (the academic staff), who agreed to a reduction in the minimum number of 11 m boats from 4 to 2.

3.2.3 Hullform type and material

Hullform type and material were selected using a Strengths-Weaknesses-Opportunities-Threats (SWOT) and Weight-Score Method (WSM). Four hullform types were considered: Catamaran (SWATH); Trimaran; conventional catamaran; and conventional monohull. Table 3 gives an example analysis for the catamaran-SWATH option and Table 4 summarises the WSM analysis for hullform type, with the monohull option being preferred. The WSM analysis for structural material compared steel, aluminum and carbon fibre composite, with steel being selected.

Table 3. Example SWOT analysis for the catamaran-SWATH hullform option.

Strengths	Weaknesses
Large righting moment Significant roll damping due to wide beam (in SWATH mode) Beneficial for high speed vessel Shallow draught (in catamaran mode) Operational flexibility in conversion between being a highspeed catamaran or stable SWATH	Very difficult tank layout Difficult to implement podded propulsion High freeboard (in catamaran mode) Wave interference resistance between hulls Constrained layout options RIBs deployment must be at sides ▶ suggest narrow hull
Opportunities	Threats
Large deck area Possibly better work flow at main deck due to a more square shape Innovative solution Requires two shaftlines/azipods	Roll damping results in large relative motions in beam seas Structural issues in beam seas Sensitive to load changes in SWATH mode Requires high level of structural investigation Higher UPC due to complex design

Table 4. WSM decision making matrix for hullform type.

Criteria	Weight [1–5]	Hull form [1–5]							
		Cat-Swath		Trimaran		Conventional Catamaran		Conventional monohull	
Longitudinal strength	3	2	6	3	9	3	9	4	12
Transverse Strength	4	2	8	3	12	1	4	4	16
Resistance at high speeds	4	2	8	3	12	3	12	4	16
Resistance at low speeds	4	2	8	3	12	3	12	4	16
Deck area	2	4	8	4	8	4	8	2	4
Roll damping	4	4	13	3	12	2	8	2	8
Pitch damping	4	5	20	2	8	2	8	4	16
Propulsion arrangement	4	1	4	2	8	3	12	5	20
Sensitivity to load changes	3	1	3	3	9	3	9	4	12
Design security	3	2	6	4	12	3	9	4	12
Project risk	3	4	12	3	9	3	9	5	15
Innovation	3	4	12	4	12	3	9	2	6
Total			106		114		102		140

Modularity was used to allow adaption of the basic vessel design to different options, the decision was taken to concentrate the variant-specific features into a single "Variant Dependent" area, thus encapsulating much of the variation between options. Table 5 summarises the four main approaches to design modularization considered.

The primary consideration for selecting between the options was the interaction of subsequent equipment and functional spaces with operations and capability. Option 2 was selected for further development, as it aligned with a "factory floor" mode of operations; the main deck is a factory floor producing a "product", where operations are managed in offices above and workers are accommodated elsewhere. The functional workflows for each design variant are outlined in Figures 10, 11 and 12 below.

Research Vessel Workflow (Figure 10): The research vessels product is scientific data. Samples are gathered by deployed equipment ▶ The equipment is recovered at the open deck ▶ Equipment goes into storage ▶ Samples are taken out of equipment and transferred to laboratories ▶ Samples are processed and data is gathered.

Submarine Rescue Vessel Workflow (Figure 11): NSRS is deployed and recovered via A-frame at open deck ▶ NSRS interlocks with hyperbaric

Table 5. Modular configuration topologies considered.

Option	Indicative profile
Option 1: 1 deck and 01 deck 1 deck and 01 deck constitutes the variant. Below deck is all accommodation. This requires a low superstructure	
Option 2: Superstructure excl. bridge The variant dependent area is the main deck which becomes a factory floor, where offices and higher rank accommodation are placed above and all lower rank personnel is placed within the hull.	
Option 3: Parallel midbody Variant dependent section is part of the parallel midbody. Allows for length extension if needed	
Option 4: Modular blocks within ship Various blocks within the ship are reserved for the variants. Provides flexibility in allocation of spaces.	

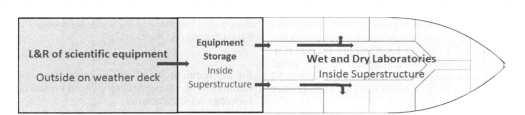

Figure 10. Research vessel workflow (Bilde, 2017).

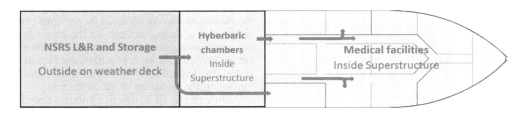

Figure 11. Submarine rescue vessel workflow (Bilde, 2017).

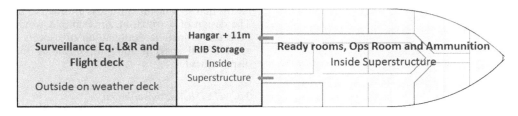

Figure 12. Coastguard vessel workflow (Bilde, 2017).

chambers ▶ Intoxicated submariners are transferred at correct pressure into hyperbaric chambers
▶ Hyperbaric chambers are slowly depressurized
▶ Submariners are transferred to medical facilities. Non-intoxicated submariners are transferred directly to medical facilities.

Coastguard Vessel Workflow (Figure 12): Boarding operations: Boarding team is prepared on ready rooms ▶ Boarding team board transport vehicle ▶ Transport vehicle is deployed.

Surveillance: Surveillance gear is launched and recovered at open deck ▶ Surveillance gear is transferred to storage area.

3.2.4 Parametric survey

The parametric survey is a standard part of the UCL MSc SDX, taking place after the initial sizing process, in which the students construct a parametric model allowing the calculation of ship size, resistance etc. The parametric survey has two broad phases; major and minor. These surveys are based on the ship concept sizing methods described by van Griethuysen (1992). The major parametric survey allows the determination of the overall ship dimensions and ratio of superstructure to hull volume, whilst the minor parametric survey focusses on hullform parameters and their optimization for minimum resistance (or energy consumption). Figure 13 summarises the major parametric survey carried out on the HYDRA design. The variables were; number of internal decks (and thus deckhead height); proportion of internal volume in the superstructure (Vs) and length/depth ratio (L/D). Various constraints were applied to this process, such as a recommended range of Circular M (7–9), consideration of block coefficient suitability

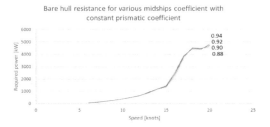
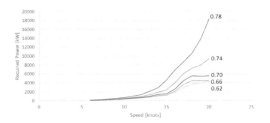

Figure 14. Summary of the resistance-focussed minor parametric survey (Bilde, 2017).

for the required speed and the minimum volume required in the superstructure based on the layout topology selected above.

With the overall dimensions fixed, the minor parametric survey was conducted on the midships coefficient (Cm) and Prismatic coefficient (Cp), as summarized in Figure 14. Just as the major parametric survey was constrained by previous considerations of arrangement etc. so the minor survey was itself constrained by the outputs of the major survey the reductions in prismatic coefficient suggested by Figure 14 were constrained by their impact on the preferred block coefficient.

Although the UCL design guidance does specify the variables to be investigated in the major and minor parametric survey, the wide range of ship and hullform types investigated by the students means that significant latitude is afforded to them. The design of a trimaran, for example, would be expected to include examination of aspects such as side hull spacing and proportion of overall displacement.

3.3 HYDRA design solution

Figure 15 illustrates the HYDRA coastguard option, showcasing the large flight/working deck, 360-degree bridge and twin azipull propulsors, selected for use in dynamic positioning (DP) and including ducted propellers for greater bollard pull during towing operations. The primary external differences between the three options were

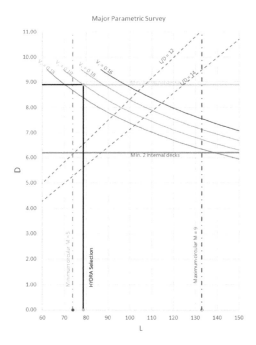

Figure 13. Graphical summary of the major parametric survey, showing constraints on the solution space (Bilde, 2017).

the upperdeck equipment (weapons or cranes) and the replacement of the aft boat bays with increased accommodation in the research and submarine rescue variants. Table 6 summarises the principal particulars for the core design, and Table 7 outlines the primary features unique to each variant.

Figure 16 expands on the topological arrangements shown in Figures 10 to 12, showing the location of the main and auxiliary machinery spaces. HYDRA uses an Integrated Full Electric Propulsion (IFEP) arrangement, with two 4.16MW diesel generators, two 1.4MW fuel cells and two 2.125MWhr battery systems. Given the in-service date of 2025 and the requirement for emissions reduction, the fuel cells were selected for cruising speeds up to 14 knots (also providing a reduction in acoustic noise), with the diesels providing boost power to make the maximum speed of 20 knots. The battery system serves several purposes; a load-levelling and boost system during towing operations (allowing the diesels to operate at constant load); a completely silent mode of operation; and an emergency power source.

It should be noted that the MSc SDX requires the development of the design to some technical detail. Table 8 summarises the naval architectural modelling and analysis activities carried out in the MSc design development. As with the parametric survey, it is expected that all designs in the MSc SDX will cover each of these aspects, but that there may be a focus on specific aspects if the design warrants it (e.g. hullform for a high speed vessel). Figure 17 provides an example of the level of detail in the general arrangement drawing (of the research option in this case).

Figure 15.   External views of the HYDRA coastguard option (Bilde, 2017).

Table 6.   Principal particulars of the core design, common to all variants.

Main Dimensions		**Machinery and Propulsion**	
Length over all	89.5 m	Integrated Fully Electric Propulsion	
Length on waterline	83.5 m	Diesel Generator	2 × 4.2 MW
Beam	15.7 m	Fuel Cells	2 × 4.4 MW
Depth	8.9 m	Steerable Azipull Thrusters	2 × 3.2 MW
Draught (SLL)	4.42 m	Fwd. Tunnel Thruster	1 × 0.8 MW
Design Displacement	3577 tons	Fwd. Retractable Thrusters	2 × 0.8 MW
Cb	0.61	**Capacity**	
Cw	0.8	Total DFO capacity	531 m³
Cp	0.66	Convertible to Future Fuel tank	370 m³
Cm	0.92	Total Freshwater	74 m³
		Deck Area	400 m²
Speed			
Cruise speed	14 knots		
Top speed	20 knots	**Costs**	
		Core vessel	£66.93 M
		Coastguard module	£28.56 M
Platform Features		Rescue module	£20.02 M
Class 1 Lloyds Fire Fighting		Research module	£20.68 M
Towing Bollard Pull	110 tons	**Accommodation**	
Towing Winch Force	300 tons	Single Standard	24
Humanitarian Aid	200 survivors	Single Deluxe	22
Platform TEU capacity (NEO)	18TEU	Double Standard	6
		Officer Suite	2
		Captain Suite	1

Table 7. Features and equipment unique to each of the three variants.

Research	Coast Guard	Rescue
Accommodation for 30 scientist	4 × Scan Eagles, 1 × Helicopter	Organic Decompression facitlites
180 m² Drylab	Large hangar	NSRS compatible
120 m² Wetlab	2 × 11 m RIBs	8 resuscitaiton wards
80 m² equipment bay	2 × 7 m RIBs	20 IC wards
Internal L&R system	1 × State-of-the-art Ops Room	Bloodbank
A-frame	Accommodation for maintenance and boarding teams	

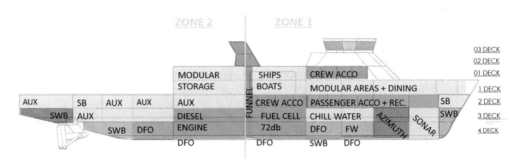

Figure 16. General profile of HYDRA, showing main functional areas and machinery locations (Bilde, 2017).

Table 8. A summary of the naval architectural technical aspects examined in the HYDRA design.

Technical area	Outline
Resistance	Resistance estimation using regression methods; comparison of methods & evaluation of consistency
Propulsion	Maintenance strategy & removal routes, working with the marine engineer
Hullform design	Hullform design in Paramarine or Maxsurf; lines plan generation
Complementing	Complement estimation and rank/task structure
Architecture & arrangement	Layout functional concept; zoning; arrangement detailed to individual space & major equipment level
Structures	Midships bending moment & shear force evaluation; midships section design (including grillage selection); identification of risks & uncertainties
Stability	Intact stability in all loading conditions, beginning & end of life; damaged stability to selected design standards; calculations in Paramarine
Manoeuvring	Evaluation using Paramarine of circle & zig-zag manoeuvres
Seakeeping	Evaluation using Paramarine, motions typically assessed at; bridge; flight deck; boat davits
Costing	Unit Production Cost (UPC) for core vessel & options; operating costs/ required freight rate for commercial vessels; market comparison
Compliance with URD	Statement of compliance/non-compliance with each entry in the URD; mitigation strategies for non-compliance

Table 9 summarises the marine engineering technical modelling and analysis activities carried out in the MSc Marine Engineering component of the SDX. As with the analyses in Table 8, several are carried out independently by the marine engineer, but there are some notable aspects, such as the propulsion system selection, where the two disciplines are expected to work together and to make the decision based on the whole-ship implications. Figure 18 provides a line diagram of the IFEP electrical propulsion and distribution system for HYDRA.

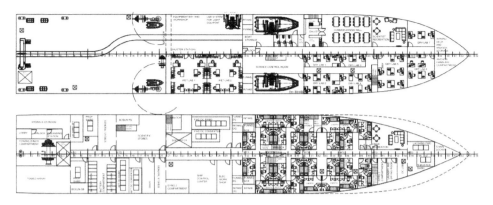

Figure 17. No 1 and No 2 deck of the HYDRA research option (Bilde, 2017).

Table 9. A summary of the naval architectural technical aspects examined in the Marine Engineering MSc.

Technical area	Outline
Complement estimation	Historical data is available, but a detailed breakdown is expected
Environmental legislation	Consideration of appropriate environmental legislation
Fuel choice & consumption	Quantitative & qualitative (technical risk) comparison of fuel options & sizing of tanks
Propulsion system architecture	Consideration of multiple architectures & justification for selection, including working with the naval architect to capture whole-ship impact
Prime mover sizing	Sizing of prime movers, including part-load considerations
Electrical load analysis	Development of electrical load chart
Fault current analysis	Calculation of fault current on each electrical bus & switchboard
Electrical harmonic analysis	Analysis of harmonic distortion in rotating electrical machines
Removal routes & maintenance	Maintenance strategy & removal routes, working with the naval architect
Hotel systems design	Definition of system architecture & sizing of main components for: Chilled water systems, HVAC systems, Fresh water systems, Black & grey water systems, Exhaust treatment systems, High & low pressure sea water systems

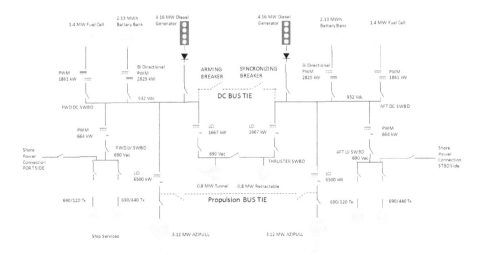

Figure 18. HYDRA power and propulsion system line diagram (Hunt, 2017).

4 SHIP DESIGN IN THE OCEAN ENGINEERING MINOR

4.1 *The IEP*

The Integrated Engineering Programme (IEP) is a multidisciplinary modular teaching framework with eight core engineering disciplines and a range of specialisms, similar to the major/minor degree structure used in some countries. There is a strong emphasis on group and individual design activities as a core part of the teaching. This design education both contextualizes the detailed technical education and also integrates input from industry on "real world" engineering problems for the students to examine.

The IEP Ocean Engineering minor consists of one three-month module in each of the first, second and third years of the course. The first year "Ocean Engineering Fundamentals" provides an introduction to ships and maritime industries, the basic analytical methods of naval architecture, and a practical design-build-test exercise for a small ROV. The second year "Offshore and Coastal Engineering" module turns to the ocean environment, with a port facilities design exercise. The third year "Maritime Design" module was developed by the first author, based in part on the legacy UCL undergraduate ship design exercise.

4.2 *The maritime design module*

In keeping with the design-oriented nature of the IEP programme, the Maritime Design module revolves around a Maritime Design Exercise (MDX) featuring the development of a concept design for a container ship (with additional ship types to be added in future iterations of the module) although as with the MSc course, the OPV DX is used to introduce the students to the design process in a highly simplified form with a constrained problem space. In the MDX, the students are provided with a requirement set that is more open than the OPV DX, but less so than the MSc SDX. Table 10 summarises the design requirements.

The students produce arrangement drawings, preliminary estimates of weights, stability, powering, emissions and cost, along with a 3D printed model of the design. This module has a strong emphasis on individual work, with a limited number of technical lectures and several free-form workshop sessions aligned with various activities in the design process.

4.3 *Challenges in the maritime design module*

One of the primary challenges of the modular IEP course is that, particularly in the later years of the course, students may opt for a specialist minor module without having previously studied that topic. A chemical engineer may decide to only do the third year Maritime Design course without having the first and second year courses which introduce stability, for instance. Another challenge is that specialist CAD software and workstations may not always be available to support the high level of individual work—and the compact three-month timetable highlights the fact that time spent learning design tools is time not spent learning design.

These challenges have impacted on the teaching of the maritime design module in two main ways. Firstly, it limits the depth of technical detail that can be taught and assessed—the students receive summary lectures on stability, powering etc. but they cannot be expected to analyse these to the same level of detail as a dedicated course with dedicated facilities and software. The module is instead focused on the process of design; of integrating different technical assessments in an iterative decision making process under conditions of uncertainty and imperfect knowledge (i.e. Figure 3). Secondly, technical modeling and analysis tasks have to either be simplified enough that the students can construct their own models using Microsoft Excel; or alternatively make use of simplified or constrained software tools, again written in Excel.

Table 11 summarises the key modelling and analysis activities and tools used in the MDX. Notable is the requirement to detail the general arrangement drawing by hand. This serves two purposes; to reduce dependence on specialist CAD software and as a (proposed) means for students to develop the discipline needed to produce clear drawings in future.

4.4 *Decision making and uncertainty in the maritime design module*

As has been noted throughout this paper, a key aspect of ship design education—independent of

Table 10. The initial requirements provided to students in the Maritime Design module.

Requirement	Notes
Role	Feeder, trans-oceanic, general purpose etc.
Capacity	A wide range of capacities of interest.
Speed	A wide range of speeds of interest.
Route	Operating area of interest.
Special study	Special studies to differentiate designs, such as: icebreaking, gas fuel, ConRo, etc.

Table 11. Modelling and analysis tools in the MDX.

Task	Tool	Description
Initial sketches	Excel ISDX tool & hand sketches	The ISDX historical-data based tool is used for initial sketches of possible design solutions.
Numerical sizing	Spreadsheet model	A process & data document is provided to the students and they must construct the iterative sizing model themselves. Individual research is required to complete the dataset.
Costing & economics	Spreadsheet model	As above. Individual research is required to complete the dataset
Emissions analysis	Spreadsheet model	As above. Individual research is required to complete the dataset.
Hullform design	Type-ship based Excel tool	A developed version of the hyperbolic waterlines based approach described by Calleya et al (2015)
Resistance estimation	Type-ship based Excel tool	Spreadsheet implementations of Holtrop & Mennen method.
Intact small angle stability	Hand/Excel calculation	Using hydrostatic outputs from the hullform tool.
Intact large angle stability	Type-ship based Excel tool	A regression-based tool using a database of ship hulls, developed by Ali (2003).
Damage stability	Hand/Excel calculation	A single-hold midships damage case using the added mass or lost buoyancy methods.
Layout (block level)	Excel based tool	A developed version of the OPV DX tool shown in Figure 2.
Layout (detail)	Hand drawing	Hand drawing over printed block drawings; a scaled grid is provided in the printed drawings to assist in area calculations.

the level of technical/analytical detail that may be taught—is in making and justifying decisions, particularly under conditions of incomplete data, changing requirements and uncertain futures. This has led to some specific features of the Maritime Design exercise, including; sketching; options exploration; data provision; and design margin exploration.

The ISDX used in the UCL Naval Architecture and Marine Engineering MSc courses is incorporated into the MDX as part of a market survey/literature survey activity. After receiving their outline ship specifications, the students are tasked with researching the routes, ship sizes and technological or emissions requirements and generating initial "sketch" designs based on historical data and a simplified layout. In addition to the use of historical data (with later comparison to the developed design), this introduces students to the need to sketching (in the methodological sense) as a crucial ship design activity, as described by Pawling and Andrews (2011).

In the legacy UCL undergraduate container ship design exercise, the students were required to develop a simplified parametric sizing model, then conduct an economic design exploration to determine the "optimum" combination of ship speed and capacity for the lowest freight rate, before developing the chosen option. This is retained in the MDX, but with the addition of through-life considerations in a (pseudo) risk based approach influenced by the real options analysis described by Puisa (2015) and scenario planning such as that carried out by Shell (Shell, 2017). Students must propose possible future ranges for fuel prices and technological availability and compare design options for their adaptability (and subsequent financial risk) across these multiple scenarios. This activity also encourages students to investigate efficiency and emissions reduction technologies and their impact on the design, such as those investigated in the Shipping in Changing Climates (SCC) project (Calleya et al, 2016). This wider-ranging study effectively replaces the parametric survey used in the MSc SDX.

The approach to design data provision in the MDX is different to that previously used in the MSc OPV DX and SDX; the students will be provided with a partial dataset, and expected to conduct individual research to obtain additional data. The data they are provided with will include items such as weight of distributed systems (which is difficult to find), but they will be expected to obtain data for major items of equipment by consulting manufacturers webpages, reference books etc. One objective of this aspect is to encourage students to approach sources of data critically, rather than prohibiting them outright (which may be unrealistic).

Design margins have traditionally been handled in student ship design exercises by the suggestion of certain percentages (and locations if appropriate), along with some narrative on the historical and engineering reasons for them to be used in ship design. However, this does not require students to actively engage with the rationale behind design margins, as they become simply a small modification to other numerical data. In the main container ship design exercise for the maritime design module, students are instead required to conduct design explorations, using their parametric models, to assess the impact on the design of using different levels of design margins, and particularly to assess the impact of estimates and assumptions, such as VCG, being incorrect.

5 CONCLUSIONS

This paper has described the ways in which ship design as a subject is taught through practice in the UCL MSc Naval Architecture and Marine Engineering courses, and the Integrated Engineering Programme Maritime Design module. A notable contrast between the four examples (ISDX, OPVDX, SDX and MMDX) is the degree of technical analysis that is expected from the students; the ISDX is very simple, and focusses specifically on the assessment of the use of historical data and identification of design drivers; in contrast the MSc SDX requires the students to both complete a high standard of technical modelling and analysis, and also explore design options and justify decisions to the customer (represented by academic staff). The undergraduate MMDX has the unique requirement that the students may not have completed previous naval architecture modules, so may have a very limited understanding of technical aspects such as stability. The greater technical knowledge of MSc students allows a design process with very broad URDs, which can be interpreted in a range of ways. For the undergraduate course, the design requirements must be more straightforward.

A common feature of all these design exercise, however, is the need for the students to compare options and make decisions, rather than simply follow numerical sizing methods. Both the main exercises—the MSc SDX and undergraduate MMDX—have explicit requirements for options exploration and downselection (e.g. the use of Equity; parametric survey; economic risk analysis). The decision making processes in these cases are supported by analytical methods and tools, but the students are operating under conditions of incomplete knowledge (and an incomplete requirement), so simplistic mechanistic approaches are insufficient, and the students are obliged to intellectually engage with the decision making. The SWOT and WSM approaches used in the HYDRA example are useful in that they can accommodate quantitative and qualitative approaches in a structured way—the key feature being that the decision making is rational and defensible.

A key feature of the HYDRA design is that it is multi-role, and it is proposed that this created a problem ideal for a holistic, integrative approach to teaching ship design. A design for a simple single role vessel may be generated through an effectively linear, mechanistic process, where the student is arguably simply a mechanism to transfer values from a databook to a spreadsheet. Setting design problems where students are obliged to consider multiple, preferably somewhat antagonistic requirements can be an approach to develop decision making approaches and skills in students.

It is important to consider what design tools are available to the students, both from the perspective of time and resources available for tutorials and technical support, and significantly with regards to the impact on their learning. Although it is desirable to introduce students to industry-standard software, the high fidelity and expansive capabilities of these packages can lead to students diving into great design detail, at the expense of visibility over the overall design. The undergraduate MMDX makes use of single-purpose Excel-based software tools, each with limited applicability, in an attempt to address these issues. However, this must be contrasted with the wider range of design types that may be investigated with the more sophisticated tools, so this is highlighted as an area of ongoing discussion.

DISCLAIMER

Although this paper describes current UCL ship design education, the wider conclusions for design education are the opinions of the authors.

ACKNOWLEDGMENTS

The assistance provided by Dr. Nick Bradbeer and Prof. David Andrews in preparing this paper is gratefully acknowledged.

REFERENCES

Ali, H. 2003. GZ Curves of Warships from Form Parameters. MSc Dissertation. London: UCL.
Andrews, D.J. 1981. Creative Ship Design. *Trans. RINA*. Vol. 123. London: RINA.
Andrews, D.J. 1986. An Integrated Approach to Ship Synthesis. *Trans. RINA*. Vol. 128, London: RINA.
Bilde, R. 2017. Project Hydra, Mediterranean Multirole Coastguard, MSc Naval Architecture Report. London: UCL.
Calleya, J. Pawling, R.J. & Greig, A. 2015. A Data Driven Holistic Early Stage Design Process to Design Profitable Low Emission Cargo Ships. *12th International Marine Design Conference (IMDC)*. May 2015. Tokyo, Japan.

Calleya, J. Suárez de la Fuente, S. Pawling, R.J. & Smith, T. 2016. Designing Future Ships for Significantly Lower Energy Consumption. *10th Symposium on High-Performance Marine Vehicles (HIPER)*. 17th–19th October. Cortona, Italy.

Catalyze Consulting homepage. 2018. http://www.catalyzeconsulting.com/software/.

Collette, M. 2015. Studying Student's Experience of the Marine Design Synthesis Problem. *12th International Marine Design Conference (IMDC)*. May 2015. Tokyo, Japan.

Hunt, J. LCdr(RCN). 2017. Project Hydra, Mediterranean Multirole Coastguard, MSc Marine Engineering Report. London: UCL.

McLaren, A. 2008. Approaches to the Teaching of Design. *The Higher Education Academy, Engineering Subject Centre*. ISBN 978-1-904804-802.

Pawling, R.J. & Andrews, D.J. 2011. Design Sketching for Computer Aided Preliminary Ship Design, *Ship Technology Research / Schiffstechnik*. Vol. 58. No. 3. September 2011. Institute of Ship Technology and Ocean Engineering. ISSN 0937-7255.

Pawling, R.J. Piperakis, A.S. & Andrews, D.J. 2015. Developing Architecturally Oriented Concept Ship Design Tools for Research and Education, *12th International Marine Design Conference (IMDC)*. May 2015. Tokyo, Japan.

Pawling, R.J., Percival, V., & Andrews, D.J. 2017. A Study Into the Validity of the Ship Design Spiral in Early Stage Ship Design. *Journal of Ship Production and Design*. Vol. 33. No. 2. May 2017. SNAME. ISSN 2158-2866.

Pawling, R.J. Kouriampalis, N. Esbati, S. Bradbeer, N. & Andrews, D.J. 2017. Expanding the Scope of Early Stage Computer Aided Ship Design. *International Conference on Computer and IT Applications in the Maritime Industries (COMPIT)*. Cardiff, UK. May 2017.

Puisa, R. 2015. Integration of Market Uncertainty in Ship's Design Specification. *International Conference on Computer Applications in Shipbuilding (ICCAS)*, Bremen, Germany: RINA.

Rosenblatt, L. 1960. The Design of Modern Oceanographic Research Ships. *SNAME Proceedings 1960*. 26 May, pp. 193–264.

Shell. 2017. https://www.shell.com/energy-and-innovation/the-energy-future/scenarios.html.

Tuttle, J. 1997. Historical Lessons and Teaching Design. *1997 Annual ASEE Conference*. 15–18 June. Milwaukee, WI, USA.

UCL IEP homepage. 2018. http://www.engineering.ucl.ac.uk/integrated-engineering/.

UCL Mechanical Engineering homepage. 2018. http://mecheng.ucl.ac.uk/.

Van Griethuysen, W.J. 1992. On the Variety of Monohull Warship Geometry. *Trans. RINA*. Vol. 134. London: RINA.

Development and lessons learned of a block-based conceptual submarine design tool for graduate education

A.A. Kana
Department of Maritime and Transport Technology, Delft University of Technology, The Netherlands

E. Rotteveel*
Maritime Research Institution Netherlands, The Netherlands

ABSTRACT: This paper covers the development, challenges, and initial lessons learned from the creation of a new block-based conceptual submarine design tool to support maritime graduate education at Delft University of Technology. The authors developed a design tool to assist students with the weight balancing, general arrangements creation, and automatic visualization of a military submarine. The tool is based in MS Excel and Rhinoceros 3D (Rhino). As an educational tool available to all students, there were certain constraints on the software platform that existed to ensure fair, open access to all students. This paper aims to describe the rationale behind how the current submarine design tool works, and some of the challenges the authors faced when implementing this tool in a 10 week complex marine design course. Lessons learned from both an educational and tool development perspective will be presented as well as areas for future improvements of the tool.

1 INTRODUCTION

This paper covers the development, challenges, and initial lessons learned from the creation of a new block-based conceptual submarine design tool to support maritime graduate education at Delft University of Technology. The course for which this design tool was developed is named "Design of Complex Specials" and focuses on the design of complex and specialized marine vessels. It is a 10 week design course focused on advanced marine design. During the course the students participate in a team design project where they develop a conceptual design of a complex marine vessel. For 2017, the design project was for a military submarine. This paper discusses a background of the course and the design project first, followed by details on the design tool and ends with discussing lessons learned.

1.1 *Background of design of complex specials*

During the 2016–2017 academic year, there was a significant overhaul of the Design of Complex Specials course. The previous version of this course involved a team student design project of a complex surface ship, combined with course

---
*Work performed at Delft University of Technology, Delft, The Netherlands.

work covering high speed marine vehicles. The last design project of the old course involved a team design project focusing on the concept design of a naval surface combatant vessel, where the students used an earlier version of the design tool that is the focus of this paper. At the end of 2015–2016 academic year, this course was updated to change the content and focus of the design project.

Several significant changes to the course have happened for the 2016–2017 academic year. First, the course is now obligatory for all students pursuing an MSc in Marine Technology. There were 77 students enrolled this past year. With project teams of at most 4 students, this meant there were 20 design teams in total. Second, the portion of the course previously devoted to high speed marine vessels has been replaced with course work covering modern advanced marine design techniques. Finally, and most important for this paper, the design project now involves a military submarine. These three changes (more students, different course material, and different design topic) meant that significant effort had to be put in to ensure a successful course. The rest of this paper covers how the authors transitioned this course, with particular focus on the design tool that was given to the students to assist with their design. The objective of the course was to teach students modern techniques of complex marine design, and allow them apply these new techniques to the design of a defined complex vessel for the course.

A military submarine was chosen as the representative "complex vessel" for the 2016–2017 academic year. As such, this was not a submarine design course, but instead this was a course focused on complex marine design in which a submarine was used as a representative complex marine vessel. This subtle distinction was important when developing the design tool to support the student's education. This difference guided the instructor's decisions over which aspects of the submarine to cover, and which aspects of the tool to focus on. For example, while submarine structures is arguably one of the most important aspects of submarine design, this aspect was not specifically listed as a learning objective of the course due to the short time frame and inability to give proper coverage to the topic. Instead, the emphasis of the learning objectives was placed on identifying and rationalizing design tradeoffs for a highly constrained complex vessel.

1.2 Background of design project

The design project involved the concept design of a 2000 ton diesel electric military submarine with a 6 week endurance. Students were provided a list of design requirements that was both tightly constrained in some areas, and allowed design flexibility in others. They were asked to provide a design that balanced capability, cost, and feasibility. Regarding the design specifically, the students were asked to synthesize the submarine's:

- Weight and stability
- Hull form selection and general arrangements
- Resistance and propulsion (for submerged condition)
- Selection of which onboard systems to include
- Marine engineering
- One additional design feature chosen by the design team

The "additional design feature" was included to give students flexibility in their design. Here students were asked to do one of the following:

- A significantly more detailed analysis on one of the required design aspects,
- Perform an analysis on an design aspect not specifically listed (such as structures, maneuvering, manning, etc.), or
- A novel design feature for future submarines (such as modern battery configurations or novel rudder configurations).

It was expected that students would research the additional design feature on their own and provide sufficient justification (both qualitatively and quantitatively) for their design. For example, if the students wanted to use a battery technology that is still in development, they were asked to quantify the benefits regarding potential energy storage and endurance as well as discuss potential risks associated with a yet-to-be proven technology. As an obligatory course, this additional design feature allowed students with specific engineering interests to explore those areas within their design.

2 CHALLENGES OF SUBMARINE DESIGN

2.1 Differences between surface ships and submarines

Some of the challenges faced with developing this tool came from the inherent differences between surface ship and submarine design. The following discussion is not meant to be all inclusive, but instead intended to highlight specific areas of difference that the authors needed to address when developing this tool. Andrews (2017) provides a nice description of these differences. First, the submarine design space is highly constrained compared to surface ships. Submarine concept design requires a higher level of detail compared to concept surface ship design which is typically a more exploratory process searching a broader design space. Nearly all design aspects, including on-board systems, are highly interrelated, leading to a highly dense and complex submarine design. General arrangements of submarines are not as flexible as surface ships leading to only a few high level logical configurations for each concept submarine. The stability of submarines is unique both longitudinally and transversely. While submerged, the lack of a water plane forces the submarine designer to take great care to ensure longitudinal stability via ballast tank design, and transversely by ensuring the center of gravity (COG) is below the center of buoyancy (COB).

Specific engineering differences also make submarine design a unique venture. For example, submarine structures is quite unique due to its need to handle high pressures and its impact on overall layout (in terms of pressure hull configuration). Submarines require unique aspects for marine engineering as well. The limited availability of oxygen for power generators and the need to operate silently while underwater are major design considerations driving systems selection. From a hydrodynamic perspective, submarines operate almost exclusively in one medium (fully submerged in water) as opposed to two (interface between air and water) as is the case for surface ships. This impacts the dominant forms of resistance and thus overall hull shape.

These aspects were considered when developing this tool, while simultaneously considering the overarching objective of the course; to facilitate design education of complex and highly specialized maritime vessels.

2.2 Educational challenges

Regarding these differences, how does one introduce submarines in a 10-week course focused on advanced marine design without spending the entire course on submarine specific topics? Specific decisions had to be made as to which material to cover and how best to do so. One of these decisions is to focus on what is reasonable for students to accomplish within the limited time frame, while still lifting the veil on submarine topics and challenges that students had not seen before. For instance, the students had to decide which missions the submarine should be able to do and which equipment is required for these missions. Additionally, they were provided a tool to assist them in generating general arrangements and investigating weight and stability related problems coming with submarine design.

Submarine specific engineering challenges were handled in each their own way. Submarine structures was one area that was not overly stressed in the course. The importance of structures was presented in a conceptual way, with the introduction of some of the mathematical formulae. Marine engineering was one area where students were expected to place significant effort, including proper system selection and its influence on other design aspects. Hydrodynamics was addressed at a parametric level and only the submerged condition was examined. As previously explained, the students had the option to pursue any of these engineering aspects in greater detail as part of their additional design feature.

3 BACKGROUND OF DESIGN TOOL

A design tool was developed and supplied to the students to assist with weight and sizing calculations, as well as visualization. The original tool was developed several years ago and was designed for surface ships only. This tool was modified to work with submarines in the 2016–2017 academic year by co-author Erik Rotteveel. In future years, maintenance of the tool will be done by co-author and responsible instructor for the course Austin Kana with support from PhD students in the maritime department.

3.1 Objective of submarine design tool

The objective of this submarine design tool was to facilitate weight balancing, layout creation, and visualization of a submarine concept design to help enable fast manual iterations between design iterations. While weight balancing and layout are clearly important aspects of complex marine design (and submarine design specifically), the intention was to facilitate these aspects so that the students had more time to focus on the important design tradeoffs as opposed to focusing on visualization or brute force calculation of submarine geometries. For instance, in previous years, some student teams would spend significant effort developing the hull form, which would leave limited time to address internal layout concerns for the concept design. As modeling specific hull forms was not a learning objective of this course, this aspect was modeled in the tool to allow students to focus on other aspects of the design.

The tool was also explicitly designed not to automatically enable design iterations nor to develop a fully synthesized design. Understanding how to iterate and what design aspects to adjust during the design process in order to develop a fully synthesized design was intentionally left for the students to struggle with themselves.

3.2 Tool requirements for education

When developing this submarine tool, several considerations and requirements had to be accounted for to ensure the focus was on design *education*, and not specifically on detailed submarine design.

The first set of requirements involved focusing it towards education, while the second group involved the software platform. For educational reasons, the design tool cannot be a "black box". Any calculation details had to be open and visible to the students. In addition to this, the tool should not develop a fully synthesized design, but instead leave some room for the students to explore and struggle with the design process on their own.

The tool also needed a short learning curve. The student only had 10 weeks for the entire course, thus creating a need to develop this tool in programs that were already familiar to students. In regards to the software platform, the tool had to work on virtual machines so that each student had open and fair access. This also meant that the tool had to be streamlined enough to run over a network. These requirements lead to a tool that is based in both Excel and Rhino.

3.3 Challenges when converting from surface ship tool

Several challenges were faced when adapting the original tool to one for submarines. The largest differences between the surface ship version and the submarine version were related to the hull form, structural weight estimation and the use of a trim polygon.

The hull form used for surface ships was a baseline frigate-like hull form that could be scaled in three directions. For submarines, the possibilities for a double hull and a semi-double hull had to be added. The use of a double hull added a challenge for object placement; for a surface ship, all objects had to be placed inside the hull. For a double-hull submarine, however, objects can be inside the inner hull, they can be located between the inner and outer hull, or may be penetrating both hulls, which is the case for a divers locker, for example. Ensuring robust and adequate cutting of objects by Rhino and accurate calculation of volume was a main challenge for this part.

Next, structural weights had to be addressed in a different way. For the surface ship, all structural weight was included in the objects to be placed inside the hull. For a submarine, however, the pressure hull is a major weight group. The same applies to ballast tanks, for example. On the other hand, objects related to accommodation are much lighter. Given limitations in submarine specific weight data, adequate estimates gleaned from literature had to be used for different weight groups.

The trim polygon is a feature typical to submarine design. As the trim polygon should envelope all possible load cases, variable filling of ballast tanks, fuel tanks and weight compensation tanks had to be made possible. For the surface ship tool, the only two load cases taken in consideration were full, empty stores/tanks and lightship, whereas the trim polygon should also fit a load case where only a single torpedo has been fired.

Finally, the weight breakdown had to be adapted as well. For the surface ship, the buoyant volume is determined by the hull only. For a submarine, however, the pressure hull as well as objects outside the pressure must be accounted for. Additionally, the COGs of multiple weight groups had to be made clear in order to aid students in obtaining the correct weight and zero trim moment.

4 SUBMARINE DESIGN TOOL DETAILS

4.1 *Tool description and rationale*

The idea of the tool is that students define a set of spatial blocks which combine to form the vessel. Each block is assigned a type, location, and supplemental information that the student can use to define the spatial arrangement within the hull. These blocks are defined with Excel and then exported to Rhino. A script was written to automatically read the block data into a 3D model. This 3D model is then be used to properly trim the blocks within the cylindrical hull to determine area weights, volumes and COGs within the hull, as well as for visualization. Students then export this updated space and weight information back into Excel for more accurate weight balancing estimates. This process is given in Figure 1.

An example visualization from Rhino is presented in Figure 2. This is the parent hull submarine concept the students were provided. The outer hull, semi-double inner hull, the ×-rudder configuration, maneuvering planes, some internal blocks, and the reference planes are visible. Further descriptions of this model are provided in the following sections.

4.1.1 *Block and object definitions*

The spatial blocks combine to make up the physical design. These spatial blocks can vary from specific rooms, to defining the hull form, to the appendages. Each block is defined uniquely by the following characteristics.

- Name of specific block
- Is this block defined by area or volume?
- Type of block (ex. accommodation, engine room, auxiliary room, etc.)
- Mass, COG and orientation

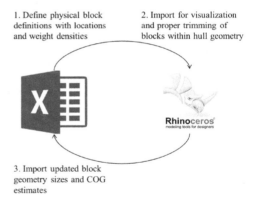

Figure 1. Steps and software platforms for submarine design tool.

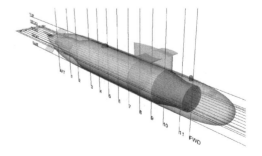

Figure 2. Visualization of parent hull submarine model in Rhino. Note presence of references planes and both sail and bow planes. Students were expected to select which maneuvering planes they choose for their design.

- Spatial relation to X,Y, and Z reference planes
- Optional additional block shaping information
- Ballast fill factor (used with ballast tanks for load cases)
- Is the object located within the inner hull or outer hull?

The specific position of the blocks are defined using reference planes, as opposed to global positioning. These reference planes are further described in the following subsection. In addition to block types provided in the design tool, students were allowed to add additional block types, should they choose.

For each block, three different specific masses can be defined: lightship value, variable mass, and fluid volume. The lightship value accounts for the base structure of what is included in the physical block, including equipment, furniture, and outfitting. The variable mass accounts for changes in mass during operations, or weight change of consumables. The fluid volume accounts for ballast and fuel tanks.

4.1.2 Reference planes

Reference planes are used to define specific locations of each block. Each block location is uniquely defined by a set of X, Y, and Z reference planes, which are also adjustable. There is no requirement that these reference planes are equally spaced, and in most cases were not. The students could decide how many reference planes to add.

The X-reference planes define the X-locations of inner blocks as well as the local dimensions of the pressure hull. The X-reference planes are different than ship stations in classical naval architecture, as reference planes may not be equally spaced. Unequal spacing allowed students to experiment with various shapes of pressure hull, such as single, double, or semi-double. This can be seen in Figure 3 where X-reference planes 0–3 are unequally spaced, but are nevertheless used to define the shape of the semi-double hull.

Similarly, the Y- and Z-reference planes served to define locations in their respective directions. The Z-reference plane was used by many design teams to define decks, or even 0.5 decks if desired.

4.1.3 Secondary objects

Secondary objects were included in the model and they defined objects like engines, motors, sonar arrays, etc. These objects were included for students to perform tradeoffs between functionality and cost of their designs. From a tool perspective, they contributed to ensuring objects were not overlapping, and that there was proper dimensioning within the block for that particular object. Students were provided with a short list of secondary objects to choose, and also instructions on how to generate their own secondary objects not included in the list.

4.2 Defining the hull

Defining the hull consisted of defining the outer hull, appendages, and pressure hull. The addition of a pressure hull had to be included in this version of the tool, as well as the ability for students to choose which type of pressure hull they preferred: single, double, or semi-double.

4.2.1 Outer hull

Defining the outer hull was one of the unique changes that had to be addressed for this updated tool. For simplicity of submarine geometry, the outer hull was defined parametrically as a body of revolution with an entrance, parallel mid-body, and an after end. The parameters defined were done exactly as is presented in Jackson (1992).

4.2.2 Appendage design

Students were able to define their own outer appendages for their design. Students were able to define the size and location of the main sail, the rudder configuration, and the sail or bow planes. Students were then expected to justify the decision for each of these appendage decisions, as each has hydrodynamic, acoustic, and maneuvering influences. As can be seen in Figure 2, the parent hull includes both sail and bow planes. It was left to the students to define which they will use for their specific concept design.

4.2.3 Pressure hull

The pressure hull was defined with a cylindrical shape with the option of defining a single, double,

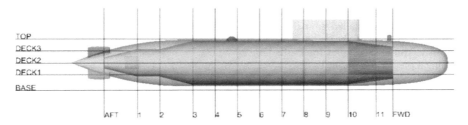

Figure 3. Side view of Submarine parent hull visualized in Rhino. Note the unequal spacing of X-reference planes that help define semi-double hull geometry.

or semi-double configuration. The decision of which pressure hull configuration to use was left to the students. Students were expected to make suitable tradeoffs according to cost, general arrangement implications, and which systems to place outside of the pressure hull. The tool helped with this process by allowing students to define whether systems or secondary objects were placed within the pressure hull or between the pressure hull and outer hull.

Within the Excel spreadsheet, the students could also define the diameter at various portions of the pressure hull defined by the X-reference planes. This can be seen in Figure 3 at stations 1, 2, and 3 where the diameter of the pressure hull clearly changes at those particular locations. It is here where students could define a shape that resembles a single, double, or semi-double hull. Most student concept designs used the semi-double hull. The authors believe this may be because the parent hull form provided to the students was a semi-double hull.

4.3 Weight estimate

The submarine weight estimate had to be simplified because this project did not involve a detailed structural design. The simplified weight breakdown of the submarine consisted of four parts: weight of primary blocks, weight of secondary equipment, outer hull steel weight, and pressure hull weight.

4.3.1 Weight of primary blocks

As a significant portion of the structural weight is included in the estimate for the pressure hull, the weight per block only includes the equipment that belongs to these blocks. The weight estimate per block is based on the volume per block multiplied by the block density. The total weight of blocks is done by summing all blocks, according to Equation 1, where ρ_{block} and ∇_{block} are the density and volume of the individual blocks respectively.

$$W_{blocks} = \sum_{blocks} \rho_{block} * \nabla_{block} \qquad (1)$$

The estimates for these densities are determined from educated guessing, and testing whether the used densities conform to existing data on submarine weight distributions. This approach may not lead to a most accurate estimate, but is almost the only option due to the limitations over detailed submarine weight data. Fortunately, the goal of the design assignment is for students to run into complex trade-offs rather than obtaining a precise submarine design.

4.3.2 Weight of secondary equipment

Secondary equipment, such as torpedo's, sonar arrays, engines, etc., was also included in the weight estimate. These are main weight items in the total weight of the submarine and are therefore separated from the total weight.

Students were provided with a list of some possible secondary equipment, such as weapons and sensor equipment. The data for these items was done using a best estimation from freely available information in the literature (such as Van der Nat (1999) and Rietveld (2017)). Other items, such as engines, the students were expected to research themselves and determine acceptable weight and sizing values for these items, using existing literature or first principles (such as Stapersma & de Vos (2015)).

4.3.3 Outer hull steel weight

The outer hull shape is defined according to the parametric relations from the paper by Jackson (1992). Usually, the steel thickness of the outer hull is relatively small since it does not have to withstand significant pressure. The Rhino design tool returns the steel area in square meters, which can be used to estimate the weight of the outer hull. This is done using Equation 2, where S is the surface area of the steel, t_{steel} is the steel thickness, and ρ_{steel} is the steel density.

$$W_{outerHull} = S * t_{steel} * \rho_{steel} \qquad (2)$$

4.3.4 Pressure hull weight

The weight of the pressure hull depends on the required diving depth, the volume and the type of steel used. For this project, the diving depth is set to a reasonable figure of 300 m. Students were provided with Figure 4 to estimate the pressure hull weight according to their diving depth, pressure hull material, and displacement.

4.4 Weight balancing and trim polygon

The design project only focused on the submerged condition, accounting for the following three weight balances:

- Vertical weight balance
- Longitudinal weight balance
- Roll stability

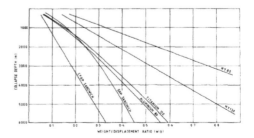

Figure 4. Curves provided to students to estimate pressure hull weight (adopted from Smith (1991)).

The first two are calculated with help from the trim polygon. The trim polygon is a visualization of the capability of the variable ballast system to cope with onboard weight changes or changes of the sea water density. The design tool specifically did not automatically generate a trim polygon for the students, as this was considered an aspect of the course they had to calculate themselves. However, the design tool did facilitate this effort by allowing students to adjust the filling rate of each individual ballast tank, which the students would then use to generate the different cases used to create the trim polygon.

4.5 *3D visualization*

Rhinoceros was chosen as the visualization for this design tool. Each block type defined by the students are automatically defined as layers in the Rhino software. Rhino also properly trimmed all spaces to within their defined hull space and then recalculated their respective area and volume size. Visualizations of the submarine can be found in Figures 2 and 3. More details regarding location of systems within the layout could also be visualized after final concept definition by the students.

4.6 *Design output*

After visualization, the Excel file was then updated with the updated volumes and COGs computed by Rhino. The updated mass of each block was also computed. The lightship weight was then calculated by the tool. Regarding the variable weights, such as main ballast and trim tanks, students were expected to vary the fill factors of these tanks for different scenarios. A weight distribution table (Figure 5) was generated showing the relative weight of each block to the total weight. This table would help students check if relative weights are within proper ranges according to literature. The table shown in Figure 5 is for the parent hull form provided to the students, where the weight of many of the systems are not yet included. Additional systems could be added by the students and they would be included in this table.

5 LESSONS LEARNED

The authors have identified several lessons learned through this process. This section outlines feedback received from the students, and well as the instructors' lessons learned from both an educational and tool development perspective.

5.1 *Feedback from students*

Positive feedback:

- The tool setup was clear and helped with the precise placement of objects and tanks within the hull, greatly assisting with defining the layout.
- The 3D Rhino representation helped present the results clearly. Various 3D views of the design were possible once within Rhino.
- There was fast start-up via the use of Excel and Rhino as students have used both programs in the past. As one student wrote "A lot of the courses are very MATLAB focused and I think it is good showing excel can be used in such a way (since every company in the industry has access to Excel)."
- The open source code was appreciated by curious students who wanted to dig deeper into the inner working of the tool.
- Use of reference planes was a helpful addition to defining the locations. This offered flexibility in defining block locations, which some design teams chose to explore.

Areas for improvement:
- Weight and sizing estimates of secondary objects that were provided could be improved. This point will be continually updated in future years to ensure these estimates are as accurate as possible given openly available literature.
- Improve transparency regarding area and volume sizing calculations. This will give students more flexibility to study various load cases by better providing clearer insight into the impacts of object placement or tank capacity.
- The Excel/Rhino interface required too many steps for fast manual iteration, especially towards the end when every minor design change required running the script from the beginning.
- The Excel data sheets were data dense and not always clear. Future iterations will work to ensure spreadsheet layouts are clear and intuitive.
- Include various loading conditions in one sheet to facilitate the creation of the trim polygon. This

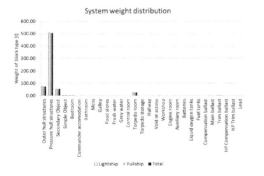

Figure 5. System weight distribution table for the parent hull as provided by the design tool. Note many systems are missing in the parent hull provided to students.

will not define the trim polygon automatically, but merely help facilitate defining the load cases which will then be calculated by the students themselves.

5.2 Educational lessons learned

- This tool was designed in part to support 3D layout generation and visualization, but not intended to replace general arrangements design. This subtle difference needs to be explicit when explaining the tool and objective of the design project to the students. Automated visualization of layouts does not (and should not) replace designer rationale of general arrangements.
- Calculating the trim polygon presented challenges. As this was explicitly left out of the design tool, some students still struggled with properly calculating it and integrating its implications into their design. There are no plans to implement this into the tool, as this is a new calculation to the students and is a fundamental aspect of submarine design.
- Students desire more automation in the tool. The instructors need to be very careful in which areas they automate to ensure it is not at the expense of the students learning and understanding of the iterative nature of the design process.
- Many final designs did not vary far from the parent hull generated by the default values of the tool. For example, most designs were semi-double hull because that was the default setting in the tool. This is not surprising, as most teams spent most of the effort getting to a converged design, as opposed to seeking creative solutions. To achieve a wider variation in designs concepts, or to force the students to work harder towards concept convergence, future parent hull may start with default values that are further away from a suitable converged design.

5.3 Tool development lessons learned

- The block based design tool method works well with students and is intuitive. Students found this breakdown easily approachable and suitable to generate a concept design.
- Good weight and cost estimates are necessary. Weight estimates are obviously necessary to ensure realistic and feasible designs, while good cost estimates are needed to ensure interesting trade-offs between on-board systems. Despite the sensitivity of this information from a submarine perspective, good first estimates were able to be gleaned from the literature.
- Students were able to adapt to tool updates that were given midway through the semester with minimal interruption. It is expected that in future years mid-semester updates will be minimized or will be entirely not necessary.

6 CONCLUSIONS

The development of a new submarine design tool for graduate education proved to be a valuable effort for the co-authors and provided added value to the course "Design of Complex Specials". If in future years the subject of the design project returns to a surface ship, the previous version of the design tool can be used again. This provides the instructors freedom in future years to change the vessel type for the design project. The underlying rationale of the tool proved to be sound, only requiring minor adjustments in future years. The authors believe that this tool did indeed help with teaching design of complex and specialized maritime vessels. Continual improvement of the tool will happen always with the objective of design education guiding the decisions on which way to move forward.

ACKNOWLEDGEMNTS

The authors would like to acknowledge Erik Takken and Etienne Duchateau, whose previous work built the foundation for this design education tool. The authors would also like to acknowledge all the students who provided constructive feedback on this tool outside of class time. Lastly, the authors would like to acknowledge the other instructors and guest lecturers that assisted in redeveloping this course.

REFERENCES

Andrews, D. 2017. Submarine Design is not Ship Design. *WARSHIP 2017: Naval Submarines and UUVs*, Bath, UK, 14–15 June. The Royal Institution of Naval Architects (RINA).

Burcher, R. & Rydill, L. 1994. *Concepts in Submarine Design*. Cambridge, UK: Cambridge University Press.

Friedman, N. 1984. *Submarine Design and Development*. Annapolis, MD: Naval Institute Press.

Jackson, H.A. 1992. Fundamentals of Submarine Concept Design. *SNAME Transactions* 100: 419–448.

Kormilitsin, Y.N. & Khalizev, O.A. 2001. *Theory of Submarine Design*. St. Petersburg, Russia: Riviera Maritime Media.

Rietveld, L.P.W. Optimization of a propulsion plant for a submarine based on first principles. MSc Thesis, TU Delft.

Smith, C.S. 1991. Design of Submersible Pressure Hulls in Composite Materials. *Marine Structures* 4: 141–182.

Stapersma, D. & de Vos, P. 2015. Dimension prediction models of ship systems components based on first principles. *12th International Marine Design Conference 2015* Vol 3: 291–405. Tokyo, Japan.

Van der Nat, C.G.J.M. 1999. A Knowledge-based Concept Exploration Model for Submarine Design. PhD Thesis, TU Delft.

Design methodology

Intelligent general arrangement

A. Yrjänäinen
Elomatic Marine Engineering Oy, Espoo, Finland

M. Florean
Cadmatic B.V., Groningen, The Netherlands

ABSTRACT: Naval architects create working and cost-efficient designs and also meet safety requirements in the early design phases of ship design projects. Specialised software is used for these tasks, after which the end result is usually published in the form of a 2D general arrangement. In this paper, we demonstrate how a general arrangement can be used to develop a ship concept in a 3D model, and how the concept is connected to supporting software and calculation aids. We have named this concept the Intelligent General Arrangement. The Intelligent General Arrangement allows a single naval architect to draw up a rough concept, which specialists in e.g. classification design and machinery arrangement can later add to simultaneously. It also allows the design to be optimised on a dedicated platform, as will be demonstrated in the HOLISHIP project (Holiship, 2018). The Intelligent General Arrangement contains internal modules, such as a weight calculation module and an analysis tool. It is also connected to external modules such as hydrostatic and hydrodynamic software and classification societies' rule-based calculators. The Intelligent General Arrangement can be easily modified. The aim is to utilise an optimisation platform to enable the rapid elaboration of ship concepts, thereby enhancing the entire design process and producing better end results.

1 INTRODUCTION

The General Arrangement of a ship is the principal document used to present the ship's basic dimensions and features. It is drawn up by a naval architect based on prior knowledge and experience. The design of a ship is a creative process, but it is often heavily based on an existing ship, a proven design, which is incrementally elaborated on in every generation of the vessel. It is also updated with every generation to respond to actual external parameters, such as fuel costs and the latest requirements of the authorities and classification societies. Naturally, the technical development of systems and equipment is incorporated in the vessel design.

The objective of the Holiship project and the Intelligent General Arrangement is to improve the traditional Evans' design spiral, where a ship concept's general arrangement (GA) is usually created in the form of a drawing. The goal is to rapidly create the first GA version, which is a functional and innovative concept. The Intelligent GA produces ship models that are optimised and that comply with safety regulations such as those of the IMO. It includes all aspects of ship design such as weight analysis, seakeeping, stability, strength, propulsion, etc. The amount of information in the ship model will be augmented. One of the goals is to improve the output to ensure that all the required 2D deck plans and other related documentation can be obtained from the model.

In order to improve the traditional ship design process and, in particular, the concept phase, we are developing a novel way of easily creating and handling ship models from the concept phase through to basic design and detail design. This will be achieved with the Intelligent General Arrangement.

The Intelligent GA is primarily a design and an analysis tool used to create, modify and analyse a ship's general arrangement. The tool supports qualified naval architects during the entire design process. Therefore, the Intelligent GA consists of several modules or supporting tools, such as a weight calculator tool and hydrostatic tools. The Intelligent GA not only consists of internal modules, but also interfaces with external modules such as strength analysis programmes and hydrodynamic tools for optimisation.

The tool allows a single naval architect to draw up a rough concept, to which specialists from different disciplines can simultaneously add more precise details. It allows the optimisation of the design on a dedicated platform; this will be demonstrated in the Holiship project.

In chapter 2, the term *concept* is defined, and the reasoning behind the development of the Intelligent GA is explained. A list of concept design stakeholders is also provided accompanied by an explanation of their interests and objectives in the ship design process. Chapter 3 outlines the general working principle and features of the tool. Chapter 4 introduces how these features are utilised in the core software tool, CADMATIC Hull. Chapter 5 presents the functionality of the Intelligent GA in the application case of a double-ended ferry.

2 CONCEPT DESIGN REQUIREMENTS

2.1 *Definition of concept design*

A concept may refer to fidelity stages of a design that are rather different. It may also refer to a simple sketch that demonstrates a novel ship solution, or a rather elaborate new design with a comprehensive design package.

The content and purpose of the different early design phases vary according to the vessel type and local shipbuilding traditions. Early design is associated with feasibility studies, concept design, initial design, early design, preliminary design, and contract design—all of which precede the basic design phase. There is no universal understanding of the scope of these activities. In this document, we consider the concept phase to be the first design phase. It is the design phase where feasibility is checked and where the main particulars are optimised. The concept phase includes a GA and other principal drawings, as well as the outline specification defining the main components. The concept phase accounts for only 1% of all design and engineering work. The impact on the design is, however, global and, therefore, crucial.

The concept design phase is followed by basic design where the concept design is fine-tuned. The following tasks are also performed while related documents are generated:

– classification drawings
– arrangements of all spaces
– system diagrams, required calculations and operational descriptions based on technology supplier documentation
– full specification
– all data required as the basis for detailed design

In some cases, ship contracts are signed based on concept-level documentation only, but may also include content from the basic design phase. This varies from one shipyard to another and according to the vessel type. As such, the content of the design contract may vary and does not necessarily exactly match the content of the design package.

2.2 *From 2D ship design process to 3D*

The concept phase design is usually created with a 2D software programme. Other software programmes are also utilised at the same time; these include NAPA, dedicated programmes for hydrostatic characteristics and hydrodynamics, and propulsion performance. Weight calculations are usually done with the use of spreadsheets, unless a dedicated software is preferred. The main concerns in this regard are data handling without automatic interfaces and the inflexible and time-consuming manner in which modifications are made. The malleability of traditional 2D design is generally poor. This problem area is also addressed by the Intelligent GA.

The ship design process is traditionally described using Evans' spiral, called the ship design spiral (Evans, 1959). It was developed in the 1950s to describe sequentially proceeding and gradually converting design processes. According to Lamb, the traditional design spiral is a rather ineffective method of designing ships. This is mainly due to the task structure that adheres to a design-evaluate-redesign logic. The problem with the traditional design spiral is that designers make an initial assumption based on reference ships, for example, about the general arrangement, after which they only seek to make improvements to the design. In other words, the design becomes 'stuck' in the initial configuration of the GA and, therefore, different design solutions are not explored (Lamb, 2004).

The Evans' spiral is a rather inflexible design process, which is not particularly suitable for the current demands on ship design processes. The current trend is to employ a simulation-based (SBD), modular 3D design building block (DBB) approach, which is a holistic and more flexible ship design process. This method takes advantage of 3D modelling already in the concept stage. The main benefits of this approach include the increased amount of data (3D model vs 2D) and the ability to conduct different simulations (virtual reality, evacuation simulations, seakeeping, etc.) (Tibbitts et al. 1993) (Andrews, 2006).

Figure 1. GA requirements and different stakeholders that have a major influence on the ship concept development and their dominant viewpoints. The naval architect confirms that all relevant requirements are considered in the ship concept.

3D modelling allows different tasks to be conducted simultaneously; a synthesis of a concept that is elaborated on to create the final vessel (see Figure 1). In its ultimate form, the ship concept is a *digital twin* of a ship that can be utilised in simulations to assist seafarers and improve safety during the lifespan of a ship.

In this project, the process is created and database structures developed based on the Intelligent GA. The process is developed in such a way that the tools work together to ease concept design and ensure design quality. It allows the effective and precise creation of malleable ship concepts. This supports the innovativeness and creativity of ship designers, which was identified as a major disadvantage of the traditional Evans' design spiral (Lamb, 2004).

2.3 Ship concept and stakeholder requirements

A ship concept is used for different purposes by various parties and stakeholders in the early design phase. It is necessary, therefore, that all the required information is based on accurate and up-to-date design materials, which form the basis of the general arrangement. It is also important that all the documents are based on the same information, calculations, and drawings and that no conflicts exist. The main stakeholders are the ship-owner (and operator if different), possibly the ship operator's customer, and authorities such as classification societies and flag authorities (see Figure 1).

The ship-owner (and the ship operator if not the same) have their own viewpoints of the vessel. However, they commonly represent the party who supplies the vessel and its capacity (or any other task) to the shipping market. Diverging viewpoints are also evident for persons in different organisational roles. Technical personnel have to be convinced of the feasibility of the mission and the functionality of the vessel, as well as safety factors. Seafarers' viewpoints also have to be taken into account. The main aspects illustrated by the GA are:

– The functionality of the vessel
– Manner in which cargo handling and other ship operations are handled
– Ship spaces; cargo areas, technical areas, service areas and accommodation areas; principal arrangements in these areas
– For cost estimation of a newbuilding, as a core part of the tender material defining the vessel in general

The shipyard needs the GA and other concept phase documentation as a basis for basic design. The concept will be verified and validated according to the shipyard's own experience, while possible elaborations will be made. This forms the basis for cost estimates of the ship contract process. It is important, therefore, that all parameters that influence costs are fixed for the cost estimate. Also, the performance of the vessel has to be indicated, but needs to be verified. Factors that affect the ship's building process are not included in the concept design.

Classification societies and flag authorities' need to ascertain whether the proposed ship design meets the rules and requirements. They do this is by inspecting principal drawings, including the GA. All other documents have to be compatible with the GA. Furthermore, many of the drawings are based on information from the GA (deck plans) and the traditional 2D drawing of the GA is, as such, still required.

2.4 Working with GA

The naval architect creates the initial illustration of the GA based on the vessel's task and ensures that the rules and regulations, as well as other aforementioned requirements are met. The entire package has to be competitive in the shipping market. The creative process is iterative, but also requires more flexibility to handle the design.

In addition to the functions required by the ship-owner and authorities, the naval architect needs the following requirements to be met in the general arrangement (model):

– Easy handling of the model
– Malleability and flexibility of the model for modifications and alternatives
– Working interfaces with other software programmes and calculation tools

In a 2016 survey, respondents were asked about their preferences regarding the most beneficial way the 2D drawing of a GA could be replaced by a 3D model, see Figure 2 (Jokinen, 2016).

Based on these results and further discussions, the functionality of the Intelligent GA was defined.

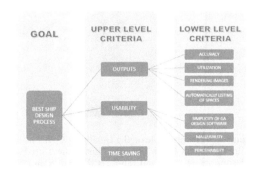

Figure 2. The criteria according to which the GA design process is evaluated (Jokinen, 2016).

3 INTELLIGENT GENERAL ARRANGEMENT

The Intelligent GA is a design tool that will assist naval architects to do their work. It is not an automated ship configurator. The Intelligent GA will be used to create innovative concepts, and to support naval architects in all design phases, from concept design to detail design.

The main features of the Intelligent GA are a new interface to easily sketch a ship's general arrangement plan and a model that communicates with external and supporting modules (see Figure 3). Some modules are internal e.g. the Weight Estimator Module, while others e.g. the Structural Strength Module, are external. This supports the naval architect in creating a feasible general arrangement with a single user interface.

3.1 Basic idea and functionality

The ship design process starts at the concept exploration stage where the mission/operational requirements (such as the required speed, seakeeping characteristics, cargo capacity, etc.) are defined by stakeholders (Tibbitts et al. 1993).

Once the mission requirement of a vessel has been defined (Mission Requirements module, external), the initial sizing of the ship model is defined (Initial Sizing, supporting module). The main dimensions and other ship-related parameters are stored in a data bank. This is combined with the model, drawings and output module to form the core of the Intelligent GA. In addition, the Intelligent GA consists of several independent internal and external (linked) modules that are organic parts of the Intelligent GA.

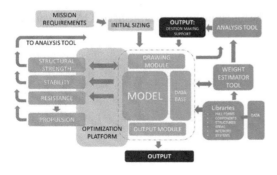

Figure 3. The principle of the Intelligent GA. The core of the Intelligent GA is presented inside the dotted line. The supporting tools and modules that are handled by the tool are on the ight side. The linked (external) modules are left side. The output is displayed below. All these elements together form the INTELLIGENT GA.

The drawing module is the naval architect's working tool in the Intelligent GA; it is the interface to the model. The intention is to keep the interface as simple as possible.

Different libraries are available to speed up modelling in the drawing module. These libraries include objects such as hull forms from existing reference vessels, 3D components/equipment and systems. The object information is stored in the data bank. This object-oriented hierarchy allows naval architects to select the hull forms and systems to be used in the drawing module. Objects can be modified in the drawing module and replaced, if necessary. This ensures the high malleability of the ship model, including topology and the GA.

The database not only supports naval architects in drawing, but also provides inputs for the analysis tool used in decision making (further improvements to the model). The data bank module is linked to external modules via the optimisation platform. These external modules utilise the data bank information (current status of the model) and provide the analysis/decision making tool with calculation results (strength, stability, hydrodynamics, etc.).

The output module within the Intelligent GA provides outputs for the stakeholders highlighted in chapter 2. The outputs include the 2D GA, tables and curves regarding ship characteristics (for example dimensions, strength, stability, performance and price) or other data.

A more detailed description of the modules in the Intelligent GA and the optimisation platform are provided in the following sub-sections.

3.2 Internal modules

Some of the features of the intelligent GA are integrated with the design tool as internal modules, while those that are considered to be linked tools are discussed later.

3.2.1 Weight estimator tool

The ship's lightweight (LW) estimate is one of the most important tasks in ship design, especially in the concept phase. In the early phase of ship design, the lightweight is estimated based on factors related to the main particulars, volumes, and areas of the ship. However, when the design proceeds and the main components are defined, more accurate calculations can be performed.

Different weight groups have diverging factors that affect the ship's weight. In this project, it is natural for the model and the decision making tool to be linked to the Weight Estimator Tool (WET) as presented in Figure 4.

The WET is based on the reference material of the selected vessels. These statistics are used in equations in order to calculate estimates for the

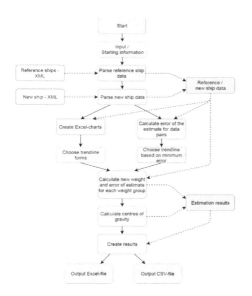

Figure 4. The structure of the Weight Estimator Tool. The blue colour indicates the function of the tool, orange is the input from the user, grey indicates input files, purple indicates data structures and green is the output. The solid arrows illustrate the execution order of the tool, while the dashed lines illustrate the information flow (Kahva, 2017).

different weight groups or sub-groups. The structure of the weight estimator tool is demonstrated in Figure 4. The result depends on the quality of the statistics and reference vessel selection.

The structural weight of the ship is by far the most important weight group. This is normally followed by machinery and auxiliaries. Other groups have a lesser impact on the ship's structural weight; however, they remain crucial in defining the total weight.

3.2.2 Libraries

The Intelligent GA consists of several libraries where, for example, different hull forms, 3D components (main engine, funnel, cabin modules, etc.) and steel structures are stored. These libraries supply a selection of predefined materials according to the naval architect's selection. The entities may be elements of given features, such as weight and dimensions, or may be subject to defined parameters.

These libraries serve the designer, but the data also includes weight information. The data and how it is transferred to the weight estimator tool and other modules need to be defined.

3.2.3 Powering the vessel

An internal tool is available to define the principal powering solution. It is used to define the main components of the propulsion train and its weight. This can be used to define the fuel consumption in conjunction with the resistance prediction.

3.2.4 Analysis tool

The analysis tool analyses and presents data for designers to support design work. The tool analyses the project ship's characteristics, but also compares these characteristics with data analysed in supporting databases. The analysis tool provides the naval architect with up-to-date information about selected ratios, figures and factors and compares this information with the reference vessel data. The top level ratios include L/B, B/T, and Fn, as well as space and volume analyses.

Furthermore, the analysis tool calculates the equipment number of the vessel. This value is used to define the related deck outfitting part and its weight. This is then used (with a certain factor for handling the rest of the weight of that weight group) as the initial value in the weight estimator module prior to the optimisation process.

The analysis tool supports the naval architect's decision making and keeps him/her on track to design a feasible vessel. All crucial deviations from the reference materials are brought to the designer's attention.

3.2.5 Output tool

As indicated earlier, different parties with diverging needs have to be satisfied with the model that replaces the general arrangement. However, the traditional general arrangement plan forms part of the output. Furthermore, other types of output can also be derived from the intelligent GA. The output module creates the required outputs from the model, which include:

– GA in 2D format, as a traditional general arrangement plan
– 3D model for visualisation purposes
– 3D GA for marketing purposes

All these output forms are necessary as outputs of the vessel concept. By using the intelligent GA, all these forms are based on the same model and therefore the risk of conflicts is avoided.

3.3 Linked modules

As previously indicated, the Intelligent GA tool is open to external software such as strength, hydrodynamics and hydrostatic analysis tools. These tools are commonly used in ship design, but in this case they are linked to the model and, therefore, all data remains up to date across the different software programmes.

3.3.1 Strength and structural weight

The strength of the vessel is calculated with a rule-based calculator, which is integrated into the Intelligent GA tool. It defines the scantlings for the first estimate of the vessel.

The model receives the unit weights as calculated from the main frame section and allocates these unit weights to other structural elements. This is rather simple for longitudinal structures, but may require more effort for other structures.

The main frame section defines the unit weights in the first phase. These unit weights are used for structures throughout the vessel. There are also some further requirements, for example, the collision bulkhead has to be reinforced according to the rules. The next step concerns the ends of the vessel. The goal is, however, to keep the process as simple as possible to calculate the steel weight rapidly and in the most efficient way.

3.3.2 *Hydrodynamics*

The resistance of the vessel is estimated with the *v-Shallo* panel code method developed by HSVA (HSVA). It is integrated with the Caeses optimisation platform (Harries, et al., 2017). Common regression analyses are also conducted based on resistance predictions as alternatives. Also, the results can be calculated with CFD analysis that is integrated with the optimisation platform by means of the surface response method.

3.3.3 *Hydrostatics*

The ship's stability is calculated with an external module linked via the optimisation platform. It receives the required information from the Intelligent GA and returns the calculated values. The required stability has to be met in order to achieve the minimum viable level for a concept. This module is a domino, which can be used if integrated with the optimisation platform.

3.4 *Optimisation platform integration*

The Intelligent GA is integrated with the optimisation platform that functions as the interface between the external/linked modules.

As presented in Figure 3, the optimisation platform requires a ship model that is imported from the Intelligent GA. This model includes the topology, weights, materials and costs of the structures and systems, and other ship related parameters. These parameters are used, for example, in hydrostatic calculations, as well as in calculations with hundreds of variations regarding performance and costs over the vessel's life cycle. These outputs are returned to the Intelligent GA to further improve the design.

The optimisation of the main particulars is based on the defined external operational scenarios of the vessel and its internal technical limitations and restrictions. The variations are automatically generated within given constrains. The vessel itself is designed by a qualified naval architect, but the optimisation is carried out by a computer in batch-mode without the manipulation of designers.

Examples of the optimisation platform are presented in chapters 0 and 5.

4 SOFTWARE DEVELOPMENT

One of the goals of the Holiship project has been the development of the Intelligent GA. For this purpose, a commercial software, Cadmatic, will be developed to fulfil the objective defined in chapter 3.

4.1 *Intelligent GA in CADMATIC Hull*

The Hull Structure module of the CADMATIC Hull system is at the core of the Intelligent General Arrangement tool. The Hull structure model is used for 3D modelling of the entire hull structure from basic design up to detailed design and production engineering of hull blocks, assemblies, panels and parts.

CADMATIC Hull offers topology that supports and eases the designer's work throughout the entire design process. A specific and dedicated 2D drafting functionality allows the creation and maintenance of general drawings and plans, while the Hull Shape Import Manager module allows the import of surface-based hull shape models directly into CADMATIC, based on different independent file formats. Shell development is done with an add-on module.

4.2 *Design procedure in CADMATIC Hull*

There are several factors or constraints that define the main dimensions of the ship model. These parameters (see Figure 5) are sets of values that can be defined as fixed values (e.g. length of ship is 130 m), or as a mathematical formulas. A parameter can also be defined with the use of another parameter (e.g. reference distance between decks) and can be modified at any given time. Changing these parameters will result, for instance, in a ship with different stability and damage calculations.

Figure 5. Example of parameters: based on fixed values or based on other parameters via mathematical formulas.

The ability to modify parameters throughout the entire project, ensures a high level of model malleability and thereby promotes the achievement of the Holiship project goals.

The theoretical model is created with the aforementioned parameters and is stored in the form of surfaces with properties, called reference surfaces. A reference surface is an "invisible" flat surface that is used as a topological reference structure to define plate construction. The actual steel structures refer to these reference surfaces, which means they are created by inheriting user-defined values such as thickness and material types. When the properties of the references surfaces are changed, the steel structures are updated accordingly. This is called property topology. Because the plane of a future plate and its common properties are predetermined by its reference surface, the process of creating a plate construction becomes a lot faster for the detail engineer.

The topology extends further than reference surfaces, up to creating a construction with a relation to another construction. When the related-to plate changes, the construction related to it follows the changes as well.

As a result, it is possible to define an entire system where modifying just one parameter changes every key construction in the ship, while automatically taking all the conditions into account.

The design procedure is as follows:

1. Define main characteristics in parameters such as ship length, main deck position, etc.
2. Create reference surfaces (such as decks and bulkheads) using parameters and present the locations of main structural elements
3. Create steel constructions inheriting the properties of the reference surfaces
4. Check the scantlings and calculate weight
5. Analyse the design in the optimisation platform
6. Update parameters with new values after the analysis
7. Replace hull with new hull form generated by the optimisation platform
8. Trigger recalculation of reference model
9. Trigger recalculation of steel model
10. Repeat until design is optimal

The above-mentioned design procedure is straightforward, and follows the working principle of the Intelligent GA presented in chapter 3.1.

4.3 Linked software programmes and API

The survey presented in subchapter 2.4 noted that the poor communication between different software programmes was seen as a major disadvantage in ship designers' daily work (Jokinen, 2016).

In the development of the Intelligent GA, the openness of 3D model data transfer is an important objective. The use of a "database" module in the Intelligent GA allows systematic data storage in the 3D model. The data includes the topology of the model, the weights and cogs of structures and equipment, etc. The data can be transferred between the Intelligent GA and other software programmes using APIs.

Two examples are presented below to illustrate how the communication between the Intelligent GA and other software programmes functions.

4.3.1 Scantlings analysis

A classification society's rule based calculator, Bureau Veritas' MARS2000 product in this demonstration is used to check the scantling of any transverse section or transverse bulkhead along the ship length.

The Intelligent General Arrangement tool is connected to scantlings analysis tool via a file-based system.

The information regarding cross-sections in the 3D model is exported to the Mars2000 module as an xml file. The xml file is read by Mars2000 to perform the scantling analysis.

Based on the results from the scantlings analysis, the engineer updates the 3D model with the necessary changes and with the correct unit weight. The

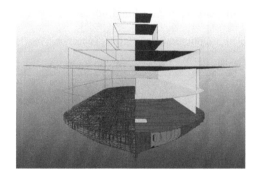

Figure 6. The model with visible hull structures. In the superstructure, the reference planes and structure locations defined are visible.

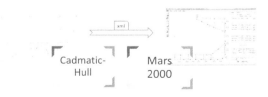

Figure 7. Schematic of the CADMATIC Hull and scantlings analysis tool interface.

Intelligent GA tool can then calculate the weight of all the items in the updated 3D model.

4.3.2 *Optimisation platform*

Caeses software, a platform for variable CAD modelling, CFD automation and shape optimisation, is used as the optimisation platform in the Holiship project. (Harries, et al., 2017) In general, Caeses allows any software which can be run in batch mode to connect to the platform and set up automated optimisation cycles.

In our case, the coupling between the systems gives Caeses access to the information from the Intelligent Arrangement Tool presented in Table 1.

Caeses performs the optimisation, generates a new hull form and updates the parameters. The Intelligent GA then replaces the existing hull form and parameters, and recalculates the 3D model to fit the new hull form and parameters.

4.4 *Future development*

A premise of the Intelligent GA design process is that a naval architect will be able to create 3D models in the same easy way as in 2D. Time saving was identified as a very important factor for ship designers in the afore-mentioned survey (Jokinen, 2016). Thus, 3D modelling cannot significantly increase the time used for design.

To speed up the design process, a new functionality needs to be added to allow the generation of sub-bulkheads and longitudinal bulkheads by drawing a simple line at the floor or deck level. The system will ensure that the 3D data is automatically generated.

To enhance simplicity and automation, an extra function needs to be added to the Intelligent General Arrangement tool to allow the creation of series of stiffeners with one command. The system will automatically calculate the length of the stiffeners based on the floors and decks and the stiffeners will be automatically connected to girders.

Since topology is the main factor that speeds up and automates steel creation, this property needs to be added to the equipment as well. This will ensure that the circle of automatic topological behaviour triggered by the parameters is applied to closing the loop. It will be done by saving the connection to the level view (i.e. level drawing) in which the equipment is added in the equipment data. Level views can already be related to reference planes and will thus be updated when the reference surface changes, as a consequence, updating the equipment as well.

5 APPLICATION CASE: DOUBLE-ENDED FERRY

Double-ended ferries are usually rather small, but complex vessels. In addition to ordinary ship systems and areas, they contain car and passenger areas like those found on ferries. Operation in both directions also results in special requirements for arrangements, hull shapes and propulsion. It forms, therefore, an interesting platform for showcasing how a parametrised ship could be designed and optimised on a dedicated platform.

Double-ended ferries are used widely in European waters, from the Greek archipelago to the Nordic countries, where they have a long tradition of connecting shorter routes over rivers and at sea. The operational areas and, therefore, the requirements vary heavily.

Double-ended ferries have been divided into three different size classes in the HOLISHIP project. These size classes are used as initial designs for further development. They are parametrised and can also be modified topologically, with different features such as motoring selections and public spaces as options.

A special characteristic of a double-ended ferry is the variety of powering options available. Currently, hybrid systems and a full range of electric solutions are widely considered as options for novel vessels. This is the case, especially, where short routes allow regular charging between voyages.

5.1 *The vessel*

The optimisation case was a double-ended ferry with three variations with dimensions as given in

Table 1. Optimisation platform, connection.

Information	Connection
Parameters	File-based, via an xml file
3D model, including weight	Direct communication, using APIs from Intelligent GA tool

Figure 8. Ice-going road ferry "Pluto", a type of double-ended ferry. © Uudenkaupungin Työvene Oy.

Table 2. The ferry in the application case, main particulars of 3 different ferries' initial values. Below the main particulars' indicative ratios (in cursive) are presented.

	Small	Medium	Large
cars	100	150	200
PAX	400	600	800
lanes	6	7	8
L	90	100	120
B	17,3	19,4	21,5
T	3,5	4	4,5
H	5	6	7
dwt	400	1000	1500
GT	1667	3333	6000
L/B	*5,2*	*5,2*	*5,6*
B/T	*4,9*	*4,9*	*4,8*
dwt/GT	*0,24*	*0,30*	*0,25*
LB	*1557*	*1940*	*2580*
H-T	*1,5*	*2*	*2,5*
LBH	*7785*	*11640*	*18060*

Table 3. Initial values for optimisation task.

	No	Unit
Round-trip	1	h
No of RTs	15	RT/day
Length of RT	10	nm
Loading/unloading	2 min + 3 s./car	

Table 4. The load cases used for the optimisation case and their respective shares of the round-trips.

	Trip 1	Trip 2	RTs
Load case 1	100%	20%	20%
Load case 2	50%	20%	40%
Load case 3	10%	0%	40%

Table 2. Each optimisation task is topologically identical. The variations can be optimised separately if needed. The topologically different solutions can be compared with each other in order to find the best solution.

The initial values for the different double-ended ferries are based on the regression of similar types of vessels.

5.2 The optimisation task

In order to allow the concept design to be optimised, it has to be integrated with the optimisation platform. This is the key point of the entire HOLISHIP project. Before starting the optimisation process, the ship concept has to be completed and it has to be topologically sound. A comparison of different solutions, such as vessels that are electrically driven or diesel engine driven, is done separately and the selection has to be done before entering the optimisation loop phase. However, both alternatives can be optimised and later compared with each other.

5.2.1 Optimisation environment

Optimisation is in principal done over the vessel's life cycle, meaning that it is done by taking the capital cost and the operational cost into account. In this demonstration, the values are taken from very conservative scenarios. The initial values for the operation are illustrated in Table 3.

The optimisation is done for estimated load cases as presented in Table 4. The figures refer to the number of cars and passengers at maximum capacity. The share of different load cases is also indicated. One round trip consists of trip 1 and trip 2.

5.2.2 Optimisation loop

The optimisation loop is based on the following procedure, and is repeated by the optimisation platform:

1. The *Intelligent GA* model is connected to Caeses (or better: the hull form comes from Caeses, structure from *Intelligent GA*). Every iteration of the optimisation produces a new hull form (L, B, T, C_b) in the *Intelligent GA*.
2. The *Intelligent GA* checks via scantlings analysis tool whether the scantlings are in order (and also not over-dimensioned). Thereafter, The Intelligent GA calculates the hull weight and returns the value to Caeses.
3. Caeses calculates the resistance for the hull form as given in item 1 for the lightweight as defined, taking into account item 2 (the flotation).
4. OPEX and CAPEX changes are estimated based on steel weight (CAPEX) and resistance (OPEX).

In this case, some simplifications have to be made for the optimisation, e.g. the longitudinal CoG of every load case is fixed and the ship navigates on an even keel (which can be realised with cargo handling).

The features or requirements that are not optimised, but may be affected by the optimisation process, have to be checked during the process, even if they were fulfilled in the initial design. These include at least the following:

- Intact stability of the vessel
- Load line requirements
- Cargo deck height in operational conditions (limits may be set)

Furthermore, the following requirements also have to be checked:

- Damage stability of the vessel.
- Equipment number, related to the deck fitting requirement

The consequences of the possible need for any changes to the initially-defined values have to be considered, at least with final variants of the optimisation reflected in the initial vessel.

The initial design, which is the basis of the main parameter optimisation, is based on a rule-based design process. No local optimisation is conducted on it. This results in a rather conservative concept, which may be heavier than a novel, locally-optimised counterpart.

The opportunity for locally-optimised structures could be taken into account with a negative weight reserve as the weight proposal of a sophisticated design. On the other hand, it should be compensated with a slightly more expensive structural unit price. The lower LW provides an opportunity for a positive effect on the fuel economy.

Operational tool integration with the optimisation platform has to be considered. It needs to be discussed, especially, whether it is inside the DE ferry loop, or whether it is done first for the initial design and then used as such. Also, timetable optimisation taking quieter times into account has to be considered.

6 CONCLUSIONS

In this paper, the functionality of the intelligent GA was presented. It illustrates how the integration of different tools can result in an enhanced and more compact design process. Furthermore, the model can be optimised with given parameters in order to create the most optimal vessel for the given initial values.

Model handling has been kept simple to allow for effective working; it is important that the model can be created and modified easily. The model also allows the optimisation of the GA within defined constraints and provides restrictions on a dedicated optimisation platform as demonstrated in the Holiship project.

The 3D model allows the GA to be used in diverse ways. The common model allows technical requirements to be demonstrated to technical personnel and other parties, and also keeps all related data up to date and uniform for all the parties involved in the project. The model is elaborated on during the later design phases and can be utilised after the construction phase of the vessel as a digital twin and as a database for data management.

REFERENCES

Andrews, D. (2006). Simulation and the design building block approach in the design of ships and other complex systems. In T.R. Society. doi:10.1098/rspa.2006.1728.

Evans, J. (1959). Basic Design Concepts. *Naval Engineers Journal*, 671–678.

Harries, S., Cau, C., Marzi, J., Kraus, A., Papanikolau, A., & Zaraphonitis, G. (Eds.). (2017). Software Platform for the Holistic Design and Optimisation of Ships. *STG 2017*.

Holiship. (2018). Retrieved from http://www.holiship.eu.

HSVA. (n.d.). Resistance Prediction by Panel Code Method. HSVA.

Jokinen, M. (2016). *Feasibility of 3 Dimensional Modelling for Design of Ship General Arrangement*. Espoo: Aalto University.

Kahva, E. (2017). *Weight Estimator Tool with Uncertainty Analysis for Ship Concept Design*. Espoo: Aalto University.

Lamb, T. (2004). *Ship Design and Construction*. SNAME.

Tibbitts, B., Comstock, E., Covich, P., & Keane, R. (1993). Naval Ship Design in the 21st Century. SNAME.

Vessel.js: An open and collaborative ship design object-oriented library

H.M. Gaspar
Department of Ocean Operations and Civil Engineering, NTNU, Ålesund, Norway

ABSTRACT: Open and collaborative JavaScript libraries are a key successful factor when developing web-based software, with many scientific applications already available, such as numerical calculations, data visualization and 3D graphics. *Vessel.js* intends to bring the object oriented, openness and web user interface characteristics from such examples to the ship design community. The library is organized as a compilation of core classes able to describe the ship entities and states during early stages of design. These classes allow ship knowledge to be gathered from key systems of the ship, such as a design building block, as well as access a library of pre-defined formulas and regressions, such as in parametric design. The combination of these bottom-up and top-down data can be manipulated, analysed and visualized via 2D/3D dashboards in HTML environment. The JSON object-oriented data format allows a flexible ship taxonomy while inputting data, with multiple tags for the same object as well as easy parametrization for traditional vessel data. *WebGL, D3* and other powerful support libraries are used for 2D/3D visualization and data analysis. Rather than promising a unique and unified ship design software, the *Vessel.js* library focus on a core tool that lays out a foundation to solve specific ship design problems. Examples in general arrangement, basic stability, design spaces and fuel consumption simulation are presented. A call for more development in JavaScript and other applications of the library concludes the paper.

1 VESSEL.JS: AN OPEN SOLUTION FOR DATA-DRIVEN SHIP DESIGN

Effective use of ship value chain data is a key factor for successful projects during the preliminary stages of design. Ship Data can be represented in two complimentary categories, top-down and bottom-up. The first comprises data from formulas, regressions, previous designs and parametric studies based on existing solutions, such as the one combining in Parsons (2011), most of Watson and Gilfillan (1976) and recently Roh and Lee (2018). Such top-down approach to collect data usually finds a practical and consensual solution at short time and cost but lacks in innovation and carries on the bias and out of date insights from previous designs. Bottom-up data is connected to specific key elements and subsystems that directly affects one or more phases of the value chain. Bottom-up data presents different taxonomies for vessel division and understanding, such as in methods like systems-based ship design (Levander, 2011) and design building block (Andrews, 2003). Bottom up data is the starting point for innovative and technological break-through designs, but suffers from the lack of knowledge connected to the uncertainty of one-of-a-kind projects.

Data-driven ship design methods (Gaspar, 2018) aims to connected both data categories, presenting a unique data structure able to consolidate top-down and bottom-up approaches during early stages of design, combining value chain data, from concept to operation. Commercial PDM/PLM software promises such integration, but it usually comes with a high integration cost and limited freedom to customize libraries and calculations.

Vessel.js is a JavaScript library for data-driven design combining conceptual ship design with a web-based object-oriented approach. *Vessel.js* represents the vessel as an JSON object, which is used to simulate different functionalities and behaviours. Currently, the library includes methods for hydrostatic and stability calculations, as well as parametric hull resistance and closed-form seakeeping equations, but it intends to grow to incorporate linear and non-linear hydrodynamics and structural models. *Vessel.js* has the ambition to be an open and collaborative web-based library toolset for ship design data, freely inspired by data-driven documents library (Bostock *et al.*, 2011), able to handle relevant information about the vessel systems and key components, as well as relevant parametric data when this is within the useful range, with a special flexibility for managing different taxonomies and attributes. The data-driven framework seen here is yet in its early stages and relies heavily on community to grow. The current paper presents the initial stages of *Vessel.js* development, introducing its design principles, basic functionalities, examples and a call for collaborators. The library is free, open and developed by the Ship Design and Operation Lab at Norwegian Univ of Science and Technology (NTNU) (http://vesseljs.org).

2 JAVASCRIPT FOR DESIGN AND ENGINEERING

2.1 *Why JavaScript*

The internet grew in the last years much faster than the engineering programming style (Gaspar, 2017). And so, it changed the routine on how to present and understand the results of analyses and simulations. Interactive dashboard and visual quantification of changes are not only a reality, but expected; pretty much every *App* in a smartphone has some sort of interactive page. Try, however, to create such engineering dashboards (Few, 2013), in a software like Matlab or Excel. Besides the cumbersome programing of outdated windows and user controls, it is practically impossible to cost-free share with others as well as control versioning without tremendous risk of loss of functionalities. JavaScript (JS), on the other hand, incorporates useful open source features, such as available code, traceability, reproducibility and versioning control. As engineering and design is an interactive process, to be able to track changes and fix bugs while testing and simulating are essential in modern programming. Moreover, JS is one of the core languages of the web, together with CSS and HTML, and most modern browsers support it without plugins (Flanagan, 2011). It was released in December of 1995, made originally to control dynamically webpages, but grew in the same fast pace as the internet, being used today in pretty much every online application available, from webapps for smartphones, server management, to video game development.

Gaspar (2017) claims five benefits for developing engineering applications in JS: speed, usage, compatibility, user interface and reliability, summarized as follow.

2.2 *Speed*

JS engines from web browsers, such as Google V8, are extremely optimized to performance in most of modern machines, and most of the cases provides a great cost-benefit in terms of simplicity versus computational power. A benchmark from 2015 (https://julialang.org/benchmarks/) shows that for some basic operations, such as *parse_int*, which parses a string from a user input or a table of values given in text file, and transform this string in an integer number, JS can be over 130 times faster than Matlab. Even more impressive, complex numbers calculations such as *Mandelbrot Set* and operations like the *pi sum* series are faster in JS than C, Fortran or Java. Given the pace of development and the fact that big companies like Goggle are behind powerful JS engines, it is not wrong to speculate that in few years the speed difference in other operations will be closer and closer to the C benchmark.

Speed to write and understand codes should also be considered. McLoose (2012) shows that JS requires in average 3.4 times less lines of code than C and, impressively, requires 6 percent less lines in average than an equivalent large task code in Matlab. Although both benchmarks show that JS is not the fastest among all languages to process algorithms or code typing, it is so efficient as other high-level languages (either open or proprietary), with the extra functionalities from the web when combined with HTML and CSS.

2.3 *Usage*

A study from 2017shows that JS is by far the most used language in Github (https://octoverse.github.com/), with over 2.3M active repositories, way ahead of the second place (Python, 1M repositories) and Java (986k repositories). Such large community means that many software, tutorials and examples made by thousands collaborators are available to re-use and contribute, with an extensive qualified community that shares its developments and results openly. Data Driven Documents library (D3 – https://d3 js.org/), for instance, keeps an impressively neat page of examples, tutorials and documentation available in Github, with ready-made codes for most of the visualizations presented.

2.4 *Compatibility*

Close to universal compatibility is the core of online applications. The idea that anyone with any modern browser can open, explore and run an engineering model in a few clicks, with no need to install or update anything is paramount (Gaspar, 2017). In other words, web applications are a key link to show and share academic results with the society and industrial partners without the complexity that usually follows simulation models.

The fact that a modern browser is the new standard for user interface means that developing in JS will most likely avoid future compatibility problems between versions and operating systems.

2.5 User interface

JS interactive graphical user interfaces (GUIs) can be visualized via mobile, tablets or PCs. Sliders are very intuitive to modify variables, and the real-time update recalculates automatically every plot. It requires no compilation, no run button, no external installation, runs direct from the browser, can be shared online (in a standard configured webserver) or private (with .HTML file and additional libraries). From the user's point of view, it requires almost no explanation when the GUI is made properly—sliders change variables, which changes the simulation and updates the plots.

A clean user interface is not a merit of JS, but of the powerful combination of HTML and CSS. Different from commercial software that have a GUI constrained by cells (e.g. Excel) or predefined windows and buttons (e.g. Matlab), the browser always start from a blank page, where text, image, video, buttons, charts and most of the interactive digital commands that we are used to can be placed with large freedom on style and interface. Pretty much every element of the page can be customized, from colours to fonts, graphs to buttons, text, and images.

The fact that everyone knows how to use a browser is another benefit. A clean dashboard requires little or no training. Sliders instead of buttons instigates interactivity and the real-time calculation eliminate the need of pressing the run button every time that a parameter is changed. The concept of buttons, sliders, tabs, hyperlinks are already part of the everyday life of not only engineers, but all stakeholders from a simulation model. In this sense, to create a GUI that is understood and can be interacted by a wider range of people is an advancement.

2.6 It just works

I am aware that this JS would not have been the language of choice to develop a ship design library 5–10 years ago. Most of the functionalities and libraries discussed were not available by then, and we did not have the powerful JS engines encoded in the browser as we have now. However, today, it just works. The fact that JS is a web language means that developers should take in consideration different devices, operating systems, browsers, versions and languages, and create standards and libraries that are functional for all of them. It is not wrong to speculate that the in the future we will not have to use a software that runs only in Windows XP with Service Pack 2. The reality is that we are already able to login via web in a powerful server in the cloud, allowing CFD calculations being made from a tablet and e-mailed back to you when finished, Gentzsch *et al.* (2016). As the scope of this paper is aimed to people like the author and his students, professional engineers while amateur developers, it is a relief to realize that a code just works, does not matter if open in the Windows 10 PC with Internet Explorer from the boss, or in the Macbook with Safari from the students (Gaspar, 2017).

3 VESSEL.JS LIBRARY

3.1 Overview—ship as an object

One of the fundamental concepts from object-oriented programming is the inheritance concept. Usually in a class-based programming language, objects are instances of classes, from which they can inherit properties and functions. In JavaScript objects inherit from special objects called prototypes. JS also allows encapsulation, enclosing all the functionalities of an object within that object so that the object's methods and properties are protected from the rest of the application, allowing particular properties and functions to be specific within the boundary of the object. These two concepts allow the build of applications with reusable code, scalable architecture, and abstracted functionalities expected in Vessel.js idea (Monteiro and Gaspar, 2016).

A vessel and its subsystems can be thus defined as objects (within objects), with general (predefined by the library) and customized (defined case to case) properties and methods. Such freedom in establishing a taxonomy is one of the key elements of *Vessel.js* to handle complex hierarchical structures as ships and its simulation. A well-defined object can be used in different calculations, for simulation and visualization of complex engineering models. Figure 1 presents the idea of a ship as an object, with properties like *Name* and *Length*, and methods such as *Sail( )* and *Anchoring( )*. This basic principle is incorporated in a generic virtual prototype model, detailed as follows.

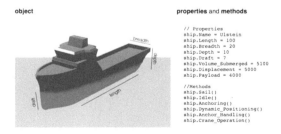

Figure 1. A ship as a JS object, with properties (e.g. name, length) and methods (e.g sail, idle) as basis for the Vessel.JS library structure (Gaspar, 2017).

3.2 Virtual prototype model applied to ship design

Inspiration for the fundamental elements required for an efficient web-based object-oriented virtual prototype library were taken from He *et al.* (2014). The authors states that a successful virtual prototype model is composed of other three subsequently models: entities (EM), states (SM) and processes (PM). *Entities* are related to the physical and logical components of the product, including 2D and 3D models, mostly reflecting the relation between basic structure and geometric information of each part; *States* are connected on how the entity behaves under changes in internal and external stimuli, which we interpreted as the main analyses; Lastly, *processes* incorporate entities and states in longer periods of time, that is, the accumulation of different state models for different conditions.

A preliminary application of this framework in ship design was sketched by Fonseca and Gaspar (2015) and recently updated (Fonseca *et al.*, 2018). The model considers the ship and its systems/parts as entities, while common analyses are tackled by a math library which updates the state model. Processes over time are considered missions, which can be as simple as a manoeuvring simulation or as complex as the fuel consumption simulation of the whole operational lifecycle. The interaction between these three models is observed in Figure 2.

The main structure of the *Vessel.js* library presented in the rest of this Section tackles mainly the EM and SM elements. Modelling processes (PMs) usually requires extensive external information, not always connected to the structure of the ship and its systems, and it must be organized case by case. A proposal for a stepwise PM, from a design space to a multibody simulation is exemplified in Section 5.

3.3 Vessel.JS structure

The basic structure of *Vessel.js* (Figure 3) incorporates so far three types of fundamental elements: *classes*, *fileIO* and *math*. These elements are able to model most of the ship related entities (EM) and analyse internal behaviour of a ship under different states (SM), such as fuel consumption and hydrostatics calculations.

classes is the heart of the library, with seven sub-libraries able to handle most of the ship design data including entities and internal states. *fileIO* contains functions for file exchange, such as load and download a pre-existing ship file. *math* aggregates a library of formulas and regressions that can be used for analysis and design, from simple assumptions like the sum of an area, to more complex non-linear formulas like hull resistance via *Holtrop* (Holtrop, 1984) and closed-form functions for seakeeping (Jensen *et al.*, 2004).

Moreover, the structure was designed with the open and collaborative mind-set, therefore it is flexible enough to allow the incorporation of other elements, such as new formulas, coefficients and regressions in *math*, new file formats in *fileIO* and ship elements in *classes*, as well as modification of the existing one, such as different assumptions for calculating GM in *Hull.js*.

3.4 classes

classes serves to properly handle different entities and states data from the ship. On the current version of the library (v0.13) there are seven key classes, as follow.

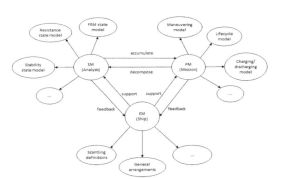

Figure 2. Virtual prototype model applied to ship design and simulation (Fonseca and Gaspar, 2015; adapted from He *et al.*, 2014).

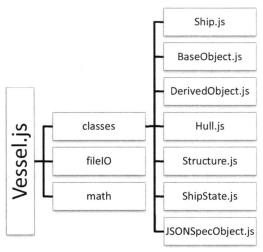

Figure 3. Vessel.js structure, with core *classes*, *fileIO* and *math*.

BaseObject.js: the image of every class object defined in the library, except the hull and structure. In other words, it constructs the library of physical entities (systems/subsystems) that construct the ship. It can be defined from a simple component (e.g propeller) to the whole machinery room. The main idea is that a base object can be served as basis for many derived objects. For instance, a base object *tank* defines a fixed height, volume, lightweight and 3D model for a given type of tank, as well as capabilities specific to a tank, such as pressure limit and manufacturer. Later, this tank can be copied several times in the ship, with different fluids and levels inside the ship. *BaseObject.js* can also contain a library of several engines, while only one will be instanced to the ship, adding standard attributes for this engine (area, volume, 3D model) as well as specific capabilities for engines (such as specific fuel consumption, fuel type and efficiency curves). A database of systems and components, with 3D models and internal methods, is expected to be available in future versions of the library.

DerivedObject.js: classes for individual ship component entities, derived from *BaseObject.js*. Every design is composed by a set of derived objects. In other words, every physical object considered in a *Ship{}* object must have a derived object. *DerivedObject.js* is the class that defines properties for individual component such as mass, center of gravity and tank filling. A vessel with six tanks based on one instance from *tank* from the base object, for instance, could have six derived objects (e.g. *tank01, tank02, tank03, tank04, tank05, tank06*), with each object having different specific properties, such as fluid density and deck position.

Hull.js: represents the geometric ship hull, currently based on a traditional table of offsets. Rather than dealing the hull as a derived object, this class handles hull as an object with very specific properties, containing internal methods to calculate most of its attributes, such as volume, displacement and common coefficients (e.g. C_B, C_P, C_M). This encapsulates most of hull data and methods inside the class, allowing easy access of its properties in other parts of the code. This class also includes methods for GM calculation and for truncating hull visualization according to a given draft. A database of hulls is expected to be available in future versions of the library.

JSONSpecObject.js: hidden base class for objects that are constructed from a literal object, (or optionally from a JSON string). Currently used to load and export objects in JSON format. The default implementation is now at preliminary stage.

Structure.js: defines structural specific objects, such as decks and bulkheads. It is in incipient stage of development but intends to include all relevant structural elements of a ship, including methods to handle longitudinal, transversal and local structural analyses. Future developments of this class intends to be connected to the *math* part to incorporate classification society rule-check and basic structural analysis. A database of structural elements and basic configurations is also aimed in the long term.

Ship.js: this is the ship itself, defined by a set of elements from other classes, that is, one hull (from *Hull.js*), one division of decks and bulkheads (from *Structure.js*) and all general arrangements (from *DerivedObject.js*). The Ship class will collect the analysis inherently to the vessel, performed by the library as they are implemented, such as GM considering hull (hydrostatic) and objects (weights). A database of ships with different hull and arrangements is expected to be available in future versions of the library.

ShipState.js: stores ship states for simulation calculations. This can be simple change in states, such as fuel level, heading and CG, to more complex state changes, such as powering under DP and demanding marine operations, checking simultaneous state of several derived objects, such as tanks, cranes and engine. *ShipState.js* currently accounts for load state, or the states of objects in the ship.

The object state assignments also have assignment rules based on groups and IDs, facilitating operations filtered by different taxonomies. A group of *baseObjects* could for instance be a category including all tanks that carry a given compound, regardless of their size and shape. The same goes for *derivedByGroup*. With this, there would be five types of assignments, where assignments of subsequent types override assignments of previous types:

- *common*: All objects.
- *baseByGroup*: Applies to every object that has its base object's property "group" set to the given name.
- *baseByID*: Applies to all objects that have base object consistent with the given ID.
- *derivedByGroup*: Applies to every object that has its property "group" set to the given name.
- *derivedByID*: Applies only to the object with given ID.

The caching and version control of ship states is incomplete at this stage but intends to control states at different frequencies for complex maritime simulation, for instance, *engine under DP* must be updated in 1000 Hz, while seakeeping converges in 100 Hz and fuel consumption can be checked for every 1 Hz. In this sense, complex time-domain simulations can be handled efficiently for every object of the ship.

3.5 *fileIO*

Class with pre-defined functions for handling file exchange, currently using JSON format. It is composed of three main methods:

Vessel.browseShip(): handy function for letting the user load a ship design from a local file)
Vessel.downloadShip(ship): very simple function to download of the specification of a given ship design in JSON format.
Vessel.loadShip(url, callback): function for loading a ship design from file

The *fileIO* class can be extended to allow exchange of files in any known open format, such as XML or comma-separated values (CSV).

3.6 *math*

math handles common ship-design functions used for calculations inside the core classes, but that are not necessary connected to one specific object. It consists of generic formulas such as moments of inertia calculation, as well as empirical formulas which requires inputs from different objects, such as parametric weight calculations. When one uses the work of Parsons (2011) or recently Roh and Lee (2018), most of the formulas presented there could be incorporated in this class. Currently, the class has the minimum required formulas to calculate basic hydrostatic and few parametric assumptions. Some of these formulas is exemplified as follows:

areaCalculations.js: calculation of sectional areas and moments of inertia.
combineWeights.js: returns mass and CG for a combination of weight arrays.
interpolation.js: linear and bilinear interpolation helpers. Also includes *bisectionSearch()*, a function that takes a sorted array as input, and finds the last index that holds a numerical value less than, or equal to, a given value. Returns an object with the index and an interpolation parameter *mu* that gives the position of value between *index* and *index+1*.
parametricWeightParsons.js: this function estimates the structural weight of the hull for simple draft calculations. This includes the weight of the basic hull to its depth amidships. It is based on Watson and Gilfillan (1976). This function exemplifies parametric equations.

vectorOperations.js: some small helpers for operations on 3D vectors. A vector is defined as an object with properties x, y and z.
volumeCalculations.js: volume calculation for vessel displacement based on numerical integration of sectional areas. Uses as input the table of offsets.

3.7 *Supporting libraries*

Diverse supporting JS libraries are used to create the examples presented in the next section. This exemplifies the power of developing in JS. 3D plots, for instance, are handled by *WebGL* via *Three.js* library, while change in parameters are easily handled via *dat.gui*.

A list of credited supporting libraries (used in examples only) follows:

Bootstrap: http://getbootstrap.com/ – used for creating responsive sites, that works well in PCs, tablets and mobile phones.
jQuery: https://jquery.com/ – used to simplify client-side scripting in HTML.
three.js: https://threejs.org/ – used for 3D plotting and 3D user interface,
dat.gui: https://github.com/dataarts/dat.gui/ – a lightweight controller library for changing design and object parameters.
renderjson: http://caldwell.github.io/renderjson – used to output design files in JSON.
D3: http://d3js.org – data-driven documents used to 2D plots and visualization of vessel data (previously applied in ship design by Gaspar et al., 2014).

4 VESSEL.JS EXAMPLES

4.1 *A very simple hull*

A very simple prismatic hull was created to verify the basic hydrostatic of the library. This exemplify the most basic top-down approach: a hull is given, and hydrostatic properties are calculated for each given draft and angle of list (Figure 4).

The only info required as input in this case is the table of offsets of the hull. In the case of the prismatic hull observed in Figure 4, the JSON object required for the calculation would contain simply:

```
"hull": {
          "attributes":  {"LOA":  20,    "BOA":  10,
"Depth": 4, "APP": 0,},
          "halfBreadths":   {"waterlines":   [0,1],
"stations": [0,1], "table": [ [0,0],[1,1]]},
          "buttockHeights": {}
      }
```

Figure 4. Prismatic hull and its hydrostatic properties calculated by Vessel.js.

The benefit of a HTML GUI can already be seen in this very simple example: the draft of the ship is changed by a slide in the top right corner of the screen, very intuitively. Automatically all hydrostatic properties are updated when the value is changed.

4.2 A very simple ship—from top to bottom

The next step in our simple case is to start to transform the hull into a ship, including physical entities to it. Each of these objects must have basic standard properties, such as positon, area, volume, weight and 3D shape. For the sake of exemplification, lets create a base object tank, and derive two ship objects from it, placing each of them in different coordinates and different decks.

```
baseObjects": [
{
"id": "TankL ",
"boxDimensions": {
        "length": 4,
        "breadth": 6,
        "height": 5
    },
"file3D": "tank1.stl",
"baseState": { "fullness": 0},
"weightInformation": {"contentDensity": 0,"volumeCapacity": 120,"lightweight": 15000}
}
"derivedObjects": [
{"id": "Tank1",
"baseObject": "TankL",
"affiliations": {"Deck": "WheelHouseTop", "SFI": "101"},
"referenceState": {"xCentre": 2, "yCentre": 0, "zBase": 4}
},
{"id": "Tank2",
"baseObject": "TankL",
"affiliations": {"Deck": "BridgeDeck", "SFI": "102"},
"referenceState": {"xCentre": 5, "yCentre": 0, "zBase": 2}
},
```

This is just an example, and more info is required as input for a proper object-oriented analysis, such as deck and bulkheads information (included in *Stucture.js*) as well as a fullness state for every tank in *ShipState.js*. A weight for the hull is also expected if the same hydrostatic properties observed in Figure 4 are going to be calculated. A

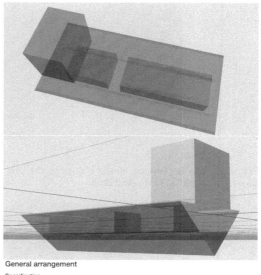

Figure 5. Adding blocks to the basic hull, exemplifying a possible transition to top-down to bottom-up.

simple GUI for adding block, as well as the effects in the 3D plot of a ship containing a prismatic hull and three objects is observed in Figure 5.

Note that, with hull and entities as elements, draft is no longer an input, but an output of the code, with objects (hull, blocks) as inputs. This may be the simplest example differencing between top-down and bottom up approaches. Adding objects starts to change the approach from a known behaviour *Draft is 2m*, to a question *What is the draft for this general arrangement?* given that, with weight, area and volumes, the draft is consequence of the equilibrium equations rather than initial input. This trigger other functions in the library to calculate the same hydrostatic properties and other simple analysis, given that VCG and LCG is known, such as GM calculation, as observed in Figure 6. The same objects could be encapsulated by a different hull, creating a new design, new GM.

4.3 PSV example

A more detailed platform supply vessel (PSV) is currently being used as example for more advanced calculations. The vessel consists of 106 derived objects. Different systems are incorporated in the

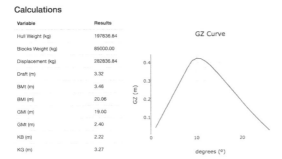

Figure 6. KG is provided by the objects in the ship, allowing a simplified GM calculation.

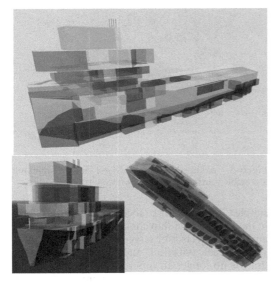

Figure 7. PSV composed of 106 derived objects.

design, such as propulsion, cargo, deck and accommodation. The vessel can also be organized in different taxonomies, with each of the entities being able to be included in one or more classification. In this way, the object *Tank1*, for instance, can be part of the *Cargo System*, as well as the *Midship Volume*, and the library would allow a colour representation of each of these different taxonomies.

Each of the objects is also able to receive one or more external files as attributes, for instance a 2D drawing, bill of materials or 3D file. In the current example, each derived object is connected to a library of 3D base objects, and they are rendered according to the information available in the library: if the object has an external 3D .STL file, this will be rendered in real time, otherwise a basic shape can be choose, such as cuboids and cylinders, to be plotted in the height and width defined. The example in Figure 7 shows the PSV, with the rendering of the 106 objects via their 3D .STL files or basic shapes.

5 VESSEL.JS FOR DESIGN AND SIMULATION

5.1 Creating a design space

Having one ship available is the starting point for more advanced analysis. A design space, for instance, can be created based in a unique ship, only by altering it objects' properties.

Fonseca *et al.* (2018) uses this concept for design space evaluation. The authors use *Vessel.js* to modify the existing hull from Figure 7 in 441 other vessels, with a step of 1% in the scaling ratio, but keeping all designs with the same displaced volume. The algorithm fixes one of the main dimensions (for instance length) and vary breadth and depth to achieve the same volume of the original vessel with a different configuration. Weights of hull, decks and bulkheads are scaled with their areas. Lightweights of tanks and compartments are assumed to remain constant. As capacities of tanks and compartments scale with volume, they also remain constant.

This design space consists thus of 441 different *Ship{}* objects, each of them with arrangements similar to the original ship, but with unique hull forms and main dimensions, as well as automatic modifications in the general arrangements, with rules for modify tanks dimensions and other derived objects that can be scalable.

Four examples of these 441 designs are observed in Figure 8, with the *Length, Beam* and *Depth* slides used for real-time 3D plot of each solution.

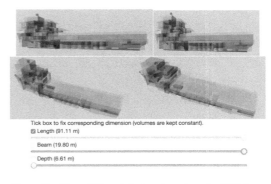

Figure 8. Four different designs based on the PSV from Figure 7: each design has the same displaced volume, but different main dimensions and arrangement; a total of 441 designs with the same displacement was produced based in a unique PSV.

5.2 Fuel-consumption simulation

Simulating complex ship design performance is a key objective for the properly development of *Vessel.js*. These cases are not part per se of the *Vessel.js* library, but uses it for solving specific problems, and the objective is to have a gallery of validated processes that feed the PM examples required for a full virtual prototype experience (Figure 2). These PMs can then be copied, modified and extended according the designer needs. Few cases are now being developed for simulate process, here exemplified by combining resistance, seakeeping and fuel consumption simulation.

Basic fuel consumption can be calculated including engine and propeller information in the *Ship{}* object. The *Holtrop* method is able to provide the resistance of the hull for each speed and draft inputed, and information from propulsion efficiency and specific fuel consumption curves from the engine can be used to estimate the fuel consumed for a given draft in a given speed for a given period of time (Figure 9).

Using as basis the simple process presented in Figure 9, Fonseca *et al.* (2018) starts from the design space from Section 5.1 to evaluate the fuel consumption of 441 different designs, for the same path and under same wave condition. Besides *Holtrop*, the method also includes a wave resistance factor, and uses closed-form functions for basic seakeeping (Jensen *et al.*, 2004). As input, the user enters a preferred speed, and a maximum vertical acceleration is selected as constrain, as well a base time step, which will be used to calculate behaviour at different frequencies. For instance, in the input observed in Figure 10 the wave encounter is calculated each ten seconds, while fuel level is updated every minute and draft changes, due to fuel usage, every hour. Each of the 441 vessels is then simulated following a pre-defined path, with different wave heading angles and amplitudes for each of the legs of the path. When the vertical acceleration reaches the maximum threshold, the preferred speed is diminished until the threshold is reached. As output, total average speed and fuel consumption of the route is analysed for all designs.

5.3 Real-time simulation

Real-time visualization of time domain simulations is also a focus of the project, heavily influenced by past successful JS development, exemplified by Chaves and Gaspar (2016). Figure 11 illustrates their work, which consists of a web-based simulator, using closed-form expressions to estimate wave-induced motion for mono-hull vessels. These expressions require only vessel main dimensions and basic hull form coefficients, being especially relevant for conceptual design, where little information about the hull form is available. The approach allows the designer to vary those parameters, and quickly assess their influence on the wave-induced motion. Seakeeping closed-form expressions were implemented in JavaScript code, and transposed to time domain. In this way, parametric real-time simulations are provided by the application, together with a 3D visualization of vessel's motion in regular waves (heave, pitch and roll – Figure 11).

The current case develops from the work from Chaves and Gaspar (2016) to standardize the input of the code as an object *Ship{}*, and adding the seakeeping methods in the *math* class. Real-time plot and 3D wave visualization is presented using the supporting libraries discussed in Section 3.4.

5.4 Real-time multibody simulation

The next challenge is to combine the simulation environment with the design space, in a multibody simu-

Figure 9. Basic fuel consumption calculation based on a *Ship{}*, varying speed and draft to calculate resistance and consequently fuel consumption.

Simulation Inputs			
Parameter	Value	Parameter	Value
Preferred ship sailing speed, knots:	11.00	Wave time step, seconds:	10
Vertical acceleration threshold, m/s²:	3.00	Fuel level time step, seconds:	60
Planar path coordinates, nm:	[0,0],[10,10],[0,0]	Draft time step, seconds:	3600
Base time step, seconds:	1		

Figure 10. Simulation inputs, including preferred speed, vertical acceleration limit, coordinates and time step data.

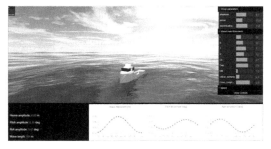

Figure 11. Chaves and Gaspar (2016) simulator, developed in JS.

Figure 12. Multibody real-time simulation: 441 design interacting in the same time-domain environment.

Figure 13. Testing the limits of the real-time plotting of multibodies—a little less than 10^4 are a reality.

lation. This research is yet under development, and the screenshot from Figure 12 presents the current stage, with a subset of the 441 cuboid-shaped bodies interacting in the same real-time environment. Each body is a different *Ship{}* object.

A magnitude of ±400 designs plotted in real-time in a modern computer did not presented lag for the real-time—each of the bodies movements in water were update within the 20–30 frame per seconds rate acceptable for human eyes, and waves seemed smooth. Therefore, to test the limits of current technology, a small research project was established to create the maximum of objects in the same environment using the current technology. The answer is a number a little less than 10^4 objects, each of them with unique geometry and material properties (Figure 13). This number can be raised if fewer unique instances are used, as well as computers with better graphical cards. On the other hand, the number is lower if we consider the sea and other details rather than cuboids for describing the ship.

6 A CALL FOR DEVELOPMENT— EXTENDING THE LIBRARY AND CREATING A COMMUNITY

I close this paper with a call for my colleagues and students to consider developing future engineering analysis and simulation in JS. The examples here presented are a working in process, and much of the library structure and methods will be improved in the years to come. The main point defended in this paper is that technology is not a bottleneck for collaborative ship design, exemplified by the current fast-paced stage of online web-development, neither the speed of the computer processors and memory size, but rather how efficient ship design data is able to be transferred from books and experience to useful reusable models. Every naval architecture is able to follow Holtrop (1984) instructions to create their own Excel spreadsheet with hull resistance, but although a good exercise for students, this is a process that should be done once, and from there the function would be freely available to all, much similar to the functions available in Excel or Matlab—we rarely start from zero for a function that calculates the standard deviation of a series or a multiplication of matrices; why then are we yet re-typing and re-assembly 30–40 year old ship design functions?

Efficient handling of ship design data also faces the multifaceted aspect of its objects, with different taxonomies and hierarchies for solving different problems in each of the ship value chain phases. Finding a common standard *one size fits all* is practically impossible, and compromises must be done even in an open and adaptable object-oriented data formats, such as JSON.

As for the development of real engineering application in JS, I recognize the power of scientific applications of other languages, like Python, or long-term usability as robust Fortran codes; but combined with web elements, I defend that no other framework will allow so much continuing collaboration and re-use of codes and libraries as JS. Even for more advanced applications JS is becoming a reliable option, with online compilers such as WebAssembly coming as standard feature soon in modern browsers, which will allow C and C++ compilation direct from the client, (http://webassembly.org/).

Regarding JS and the ship design community, other examples rather than the ones presented by this author are yet very seldom, and most of it is done by researchers connected somehow to the author. Commercial application is being tested by the ship design community in Norway (Gaspar, 2017), and collaborative research is done in cooperation the the Marine Group at UCL (Fonseca et al., 2018; Piperakis et al., 2018). The author is hopefully that in a matter of years much more will appear.

ACKNOWLEDGEMENTS

The author holds currently an associated professor position at the Department of Ocean Operations and Civil Engineering at NTNU (Ålesund, Norway), and has no commercial or professional connection with the software and companies cited. Statements on usability and performance reflects solely my opinion, based on personal experience, with no intention to harm or diminish the importance of current commercial state-of-the-art engineering tools. Moreover, *Vessel.js* is an open and collaborative library that could not be developed without the hard work from MSc students Elias Hasle, Ícaro Aragão, Alejandro Ivañez, Mateus Sant'ana and Olívia Chaves, and the given credits to each of their contribution is stated at the project page (http://vesseljs.org) – this paper would be possible without them.

REFERENCES

Andrews D.J., 2003. A Creative Approach to Ship Architecture. IJME No. 145 Vol. 3, September.

Bostock, M.; V. Ogievetsky; J. Heer. 2011. D3: Data-Driven Documents, IEEE Trans. Visualization and Computer Graphics.

Chaves, O.; Gaspar, H.M. 2016), A Web Based Real-Time 3D Simulator for Ship Design Virtual Prototype and Motion Prediction, 15th COMPIT., Lecce, Italy.

Few, S. 2013. Now you see it, Analytics Press.

Flanagan, D. 2011. JavaScript: The Definitive Guide, O'Reilly & Associates.

Fonseca, I.A.; Gaspar, H.M. 2015. An object-oriented approach for virtual prototyping in conceptual ship design, ECMS 2015, Varna.

Fonseca, I.A., Gaspar, H.M., Ryan, C., Thomas, G. 2018. An Open and Collaborative Object-Oriented Taxonomy for Simulation of Marine Operations. 17th COMPIT, Pavone, Italy (accepted).

Gaspar, H.M; Brett, P.O.; Ebrahim, A.; Keane, A. 2014. Data-Driven Documents (D3) Applied to Conceptual Ship Design Knowledge, 13th COMPIT Conf., Redworth.

Gaspar, H.M. 2018. Data-Driven Ship Design, 17th COMPIT, Pavone, Italy (accepted).

Gentzsch, W., Purwanto, A.; Reyer, M. 2016. Cloud Computing for CFD based on Novel Software Containers, 15th COMPIT Conf., Lecce.

He, B.; Y. Wang; W. Song and W. Tang. 2014. Design resource management for virtual prototyping in product collaborative design. Proc. IMechE, Part B: Journal of Engineering Manufacture, 1–17.

Holtrop, J. (1984), A Statistical Reanalysis of Resistance and Propulsion Data, Int. Shipbuilding Progress 31.

Jensen, J.J.; Mansour, A.E.; Olsen, A.S. 2004. Estimation of ship motions using closed-form expressions, Ocean Engineering 31/1, pp.61–85.

Levander, K. 2011. System Based Ship Design Kompendium, Lecture Notes, NTNU, Trondheim.

McLoose, J. 2012. Code length measured in 14 languages, Wolfram http://blog.wolfram.com/2012/11/14/code-length-measured-in-14-languages/

Monteiro, T.; Gaspar, H.M. 2016. An Open Source Approach for a Conceptual Ship Design Tools Library, 10th HIPER Conf., Cortona.

Parsons, M.G. 2011. "Parametric Design". In Ship Design and Construction vol. 1. SNAME.

Piperakis, A., Gaspar, H.M Pawling, R., Andrews, D. 2018. Designing Ships from the Inside Out - A Collaborative Object-Oriented Approach for the Design Building Block Method (unpublished).

Gaspar, H.M. 2017. JavaScript Applied to Maritime Design and Engineering 16th International Conference on Computer and IT Applications in the Maritime Industries, 2017.

Roh, M., Kyu-Yeul Lee, K. "Computational Ship Design", Springer Nature Singapore Pte Ltd. 2018, ISBN 978-981-10-4885-2.

Watson, D.G.M., & Gilfillan, A.W. (1976). Some ship design methods. Transactions of the Royal Institution of Naval Architects, 119, 279–324.

Exploring the blue skies potential of digital twin technology for a polar supply and research vessel

A. Bekker
Sound and Vibration Research Group, Department of Mechanical and Mechatronic Engineering, Stellenbosch University, South Africa

ABSTRACT: The SA Agulhas II is a South African polar supply and research vessel, which offers crucial research access to the Antarctic and Southern Ocean. In order to advance the scientific basis for ice-going vessels and ship-based ergonomics this vessel has been the subject of full-scale engineering measurements since 2012. The sensor infrastructure and advanced data analytics that have resulted position this ship as an ideal platform from which to explore a definitive trend in the future marine industry: digital twin technology. This is a digital, real-time, in-context, operational mime of an asset, which connects the digital and real word representations towards actionable insights. The technology readiness of the SA Agulhas II platform, is considered against the conceptual architecture required to implement digital twin technology. Furthermore, the advantages of digital twin technology are explored for stakeholders including the marine industry, the vessel owner and potential applications for advancing Antarctic science.

1 DIGITAL TWIN TECHNOLOGY

1.1 *Definition*

A digital twin is defined as a digital representation of the state and behavior of a unique, real asset or process in almost real time (Erikstad, 2017; Parrott & Lane, 2017) within its operational context. Datta (2016) refers to a digital twin as a software avatar that mimics the operation of a real asset or process.

Digital twin models are versatile and may be created in a wide variety of contexts to serve different objectives (Parrott & Lane, 2017). These models possess an integrated, holistic, and iterative quality of the physical and digital world pairing (Parrott & Lane, 2017) and can comprise various digital models and collections of information and processes (Erikstad, 2017).

Data can be in the form of 3D CAD models, dynamic and discrete simulation models, virtualized control systems and communication networks, analytical models, data models, sensor data, relationship data, as well as digital information including documentation and reports (Smogeli, 2017).

Erikstad (2017) defines five intrinsic characteristics of digital twins.

1. Identity: A digital representation of a single, real, unique asset.
2. Representation: The asset in its "as-is" state. This includes as-built models and all subsequent modifications.
3. State: A close to real-time representation.
4. Behavior: A digital representation of asset responses.
5. Context: Describing the context in which the asset operates.

The implementation of digital twin technology depends much on the asset or process type as well as the required accuracy, quality, availability and feasibility as governed by cost and technology readiness (Erikstad, 2017). At a minimum, a digital twin implementation must at least comprise:

1. Edge capabilities for the real-time observation of asset or process response. These may include sensors, data, analytics and integration, whereby information is communicated to a digital platform.
2. Digital twin core runtime, which is the twin model itself, using the input stream from the edge to render a (near) real-time representation of the asset.

According to Deloitte the architecture of digital twins should be designed for flexibility and scalability. Conceptually, the enabling technology is created through seven steps, which have been adapted from Parrot and Lane (2017).

1. Create: Outfitting the asset with sensors to obtain operational responses and environmental/context-specific data. Which are secured and potentially augmented with process-based information.

2. Communicate: Seamless, real-time, bi-directional integration/connectivity between a physical process and digital platform, potentially through network communication. This comprises three primary components including edge processing, communication interfaces and edge security.
3. Aggregate: Data is ingested into a data repository, processed and prepared from analytics. Aggregation may be performed on the premises or in the cloud.
4. Analyze: Data is analyzed and visualized.
5. Insight: Insights from data analytics are presented through dashboards with visualizations in real-time.
6. Digital twin: This is the digital mime, which can be generated from operational data or through models with outputs, which result from operational data inputs (for example Finite Element Model (FEM) representation from which stress localization can be interrogated).
7. Act: Actionable insights from previous steps are fed back to the physical asset or digital process to achieve the impact of the digital twin.

1.2 Enabling technology trends

Digital twins are relatively new to the maritime industry. Trends in the marine industry indicate that digital twin technology is likely a key competency, which will distinguish innovators in the future of this industry. This technology, which was named as one of the top ten technology trends in 2017 (Panetta, 2016), promises the contribution which is most prominently that the performance of the marine asset is monitored in operational conditions, and that this information is digitally represented in virtually real-time. Trends that will accelerate the adoption of this technology include:

1. The large scale of many marine structures challenges the ability to reproduce extreme thermal, mechanical, and acoustical loadings in a laboratory at anything more than the component scale. Therefore, computational simulation is required to identify and quantify of limit states (Glaessgen & Stargel, 2012). Digital twin technology increases the value of operational data by allowing the measurement and analysis of extreme operational loadings as if experiments were performed in an operational laboratory. As always, future generations of marine vehicles will require lighter mass while being subjected to higher loads and more extreme service conditions over longer time periods. This leads to increased demands on structural materials and requirements to decrease structural margins. Industries, with in-depth operational knowledge of their products will increasingly be able to shave off unnecessary structural margins whilst retaining safety-critical structural integrity.
2. The simultaneous development of several digital technologies together are enabling immense potential in digital manipulation and visualization of data. Specifically, 3D Laser Scanners have become more affordable and the resulting point cloud data can be processed with significantly less effort in software, which requires little training. Today, CAD software can handle enormously large point cloud datasets generated in shipbuilding at virtually little to no additional investment. The combination of these factors has recently led to the increased utilization of laser scanning in shipbuilding (Morais, Danese, & Waldie, 2016) and as such, powerful capabilities to progressively visualize ship structures in higher detail. The potential to maintain as-built models is increasingly feasible because of the shrinking additional costs and benefits to ship-owners who need to have a better understanding of their asset and insights, which will enable cost savings during operations.
3. Mutually reinforcing technologies such as Big Data analysis techniques are enabled by increased processing power through cloud storage and analysis. Furthermore, the increasing support of hardware and software interfaces for the IoT (Internet of Things) is now a reality, which increases the amount of data available and spurs the development of machine learning algorithms. Conditions are ripe for the utilization of data as a result of several technologies which have improved rapidly at the same time (Morais et al., 2016). The combination of these will place new tools and creative solutions at the disposal of the marine industry.
4. Advances in edge processing have eliminated bottlenecks, which increase the viability of digital twin technology. Edge interfaces connect sensors and process historians, processes signals and data from them near the source, and pass data along to the platform (Parrott & Lane, 2017). Edge processing translates proprietary protocols to more easily understood data formats whilst reducing network communication.

1.3 Advantages of digital twins

The ever-increasing ability to perform sophisticated data analytics and to visualize information presents industry with a data-rich layer, which is rife with potential. Deloitte (Parrott & Lane, 2017) emphasizes the real power of a digital twin is that it can provide a near-real-time comprehensive linkage between the physical and digital worlds. These are richer models that yield more realistic and holistic measurements of unpredictability

1. Visualization support for an "as-is" asset can allow for remote inspection or inaccessible locations, thereby reducing some of the effort, frequency and risk of physical inspections (Erikstad, 2017). In a 3D digital twin, users can navigate, perform measurements, calculate values, and display, select, filter, localize and annotate objects.
2. The accurate context of digital twins can be used for training—for example, training for inspection competence using a survey simulator or for real operations.
3. Digital twins can be used to simulate specific complex deployed assets and to monitor and evaluate wear and tear and specific kinds of stress as the asset is used in the field (Parrott & Lane, 2017). Such a model could provide accurate information, which informs service intervals and safety-critical fatigue problems.
4. A digital twin, fueled with sensor data, allows decision makers, to intervene or react, if not in real-time, then within a decision interval that enables actions that still have value (Smogeli, 2017).
5. As empirical information from sensor data accumulates, digital models will likely increase in predictive value, which will enable more pro-active vessel management, risk avoidance and increased profitability in operations (Smogeli, 2017).
6. Accurate hindsight and causality are enabled through the capture of operational data, coupled with simulation models. This delivers high quality design inputs for future builds or designs and can accelerate failure mitigation.
7. Digital twins are used for forecasting purposes. These models are named the so called 'probabilistic twins' whereby digital twins are coupled to risk models, thereby providing foresight (Smogeli, 2017).

1.4 Challenges for digital twin technology

1. Failures in the implementation of digital twin technology have proven that the successful implementation is not a one-company effort. A consortium comprising of DNV GL, the Norwegian University of Science and Technology, Rolls-Royce and SINTEF Ocean have initiated a collaborative effort to develop a new standard in marine digital twins. This standard will be an open-source platform, which is envisioned to include a digital library of generic product models, which can be accessed by any user in industry. Datta (2016) concurs that the rapid diffusion of digital twins calls for open source entity level models of sub-components. It is envisioned that the model owners of subcomponents, to create and contribute models to a common repository as the vast majority of users cannot deploy an army of engineers to create custom digital twins for their exclusive experiments.
2. The quality and reliability of data (veracity), especially when collected in large volumes, from a variety of platforms is challenging to maintain. The users of digital twin technology should further be cognizant of the applicability of digital models in specific situations. For instance, keeping in mind the fact that machine learning algorithms have decreased predictive capabilities when used outside the bound of the training data set (Erikstad, 2017). The emergence of complexly connected trends may be difficult to validate and interpret accurately.
3. Technical challenges exist especially related to sensor technology, decision-making intelligence, and system robustness (Heikkilä, Tuominen, Tiusanen, Montewka, & Kujala, 2017).
4. Increased sensor and communication capabilities have created new security issues. Common security approaches utilize firewalls, application keys, encryption, and device certificates. Contemporary security solutions are yet under development and solutions to safely enable digital twins will likely become increasingly pressing as assets are progressively IP enabled (Parrott & Lane, 2017).

This article explores the potential advantages of digital twin technology for a polar supply and research vessel, the SA Agulhas II (SAA II). A digital sister-ship of this vessel is envisioned, where model-based simulation, data analytics and visualization capabilities are connected in a cloud-based interface, with data from full-scale measurement sensors. The specific potential of digital twin technology for ship-based research and science is explored over and above the benefits for conventional stakeholders such as the vessel owner and marine industry.

2 THE SA AGULHAS II

2.1 Vessel background

The SA Agulhas II (SAA II) is a polar supply and research vessel owned by the South African Department of Environmental Affairs. She was manufactured by STX Finland in Rauma shipyard and measures 121.3 m between perpendiculars and is 21.7 m wide. She is propelled by four Wärtsilä 3 MW diesel generators that power two Conver Team electric motors, which are each, connected to a shaft with a variable pitch propeller. Accommodations are available for 44 crew and 100 passengers on annual research and re-supply voyages to South African research bases in Antarctica and the Southern Ocean.

In terms of research facilities, the SAA II offers eight permanent laboratories for marine, environmental, biological and climate research totaling 800 m². She is equipped with launching infrastructure for deep-water probes through an environmental hanger door on the starboard side. If the ship is operating in icy waters, an alternative launch porthole is provided through a 2.4 × 2.4 moon pool. A drop keel with transducers for the measurement of plankton density and ocean currents can be lowered 3 m below the hull surface. Furthermore, a hydraulic A-frame in the stern of the ship is provided to tow sampling nets and dredges.

The drive towards understanding and exploration of the oceans in globally perplexing matters such as climate change, places the SAA II in a strategic position for prominent research support (Moedas, Pandor, & Kassab, 2017). This is attributed to her annual research voyages to Marion Island, Gough Island, Antarctica and the Southern ocean, which are areas of immense interest for climatologists, oceanographers and marine biologists who are researching food security, global ice-cover and tipping points in the earth's ecosystem.

The global importance of these questions leads to the likelihood that the SAA II may well be funded for international research expeditions beyond her current utilization of 120 days per year for voyages to new geographical stations of interest, some of which will be in ice-covered waters.

An international research consortium comprising the Stellenbosch University, Aalto University, Aker Arctic, DNV GL, Rolls-Royce, STX Finland, University of Oulu, Wärtsilä and the Department of Environmental Affairs South Africa initiated a full-scale measurement program on the SAA II for her ice-trails in the Baltic Sea in March 2012 (Suominen et al., 2013).

These measurements included ice loads on the ship hull and propulsion system, ice-induced structural vibrations and noise, whole-body vibration comfort, ship dynamics in ice, global ice loads, underwater noise and mechanical and physical sea-ice properties.

The original aim of this work was to contribute to the scientific basis of ice-going vessels by providing

Table 1. A summary of full-scale measurements on the SA Agulhas II.

Measurement	Variables	Equipment	Number of channels	Sample rate
Ship context	Camera footage	Bosch underway camera system	~40	1 Hz
Wave	Height, direction	Visual observations	1	4 hours
Sea Ice	Thickness, cover, drift, temperature and salinity, structural properties, biogeochemical composition, light isotopes	GPS	1	1 Hz
		Underway ice camera	2	1 Hz
		Visual observations (Concentration, floe size, thickness)	3	10 Min
		Mobile and fixed polar laboratory (ship/land), Trace clean corer, clean lab (ship/land), CHN-IRMS light stable isotope (land)		
Ship vibration response (hull and super-structure)	Acceleration (Rigid body motion and flexure)	DC accelerometers,	10	2048 Hz
		ICP accelerometers	22	2048 Hz
Ship – shaft-line torsional and thrust vibration	Thrust, Torque,	Strain gauges, V-links and Quantum data acquisition units	2	600 Hz
	Bearing acceleration	Accelerometers to Quantum data acquisition units	6	2048 Hz
Ship – hull ice loading	Bow, bow shoulder and stern shoulder loads	Strain gauges, Central measurement unit	56 (+ 9)	200 Hz
Navigational parameters		Time, latitude, longitude, SOG, COG, HDT, relative wind direction, wind speed, depth		1 Hz
Ship machine control		Propeller motor current, speed and voltage for starboard propeller. Rudder order, position and pitch for port- and starboard shaft, rpm.	26	0.5 Hz
Ship – AIS data	Heading, coordinates, etc.	Various sensors and ship central measurement unit	10	1 Hz

Table 2. A summary of data products from full-scale measurements on the SAA II.

	Parameters	Methods	Sensors used
Human factors	R.m.s. acceleration, Vibration Dose Values	(ISO 2631-1, 1997)	Accelerometers in accommodation areas
	Motion Sickness Dose Value (MSDV)	(ISO 2631-1, 1997)	Accelerometers in accommodation areas
	Human response to wave slamming—Custom model from Vibration Dose Value and Survey responses	In-house model (Omer & Bekker, 2018)	Accelerometers throughout ship structure correlated to survey response data
Ship responses	Rigid body motion	A technique using rigid body dynamics and Kalmann filter (Sharkh, Hendijanizadeh, Moshrefi-Torbati, & Abusara, 2014).	DC-accelerometers
	Flexural modes—tracking	Operational modal analysis (based on Stochastic Sub-space Identification) using an in-house algorithm (Soal, Govers, Bienert, & Bekker, n.d.).	Accelerometers throughout ship structure and operational navigation and environmental data
	Shaft-line torque and thrust	In house algorithm using full-bridge strain relations and material properties (R. J. O. De Waal, Bekker, & Heyns, 2017).	Strain gauge bridges on the shaft-line
	Ice-loads on propellers	Inverse moment estimation (R. De Waal, 2017).	Strain gauge bridge for shear strain measurements on portside propulsion shaft.
	Ice-loads on ship hull	In-house algorithm using structural transfer function between strain measurements and theoretical hull load (Mikko Suominen, Kujala, Romanoff, & Remes, 2017).	Hull strain measurements of bow, bow shoulder and stern shoulder.
Environmental conditions	Relative wave height, wave direction, wave spectrum	Visual observations and sea state estimates using rigid body motion and strip theory.	DC accelerometer array, visual observations
	Ice thickness, ice concentration floe size	Visual observations and image processing from ice cameras	Video footage from ice cameras, visual observations

operational data and performance analyses during a three day ice trail in the Baltic Ocean (M. Suominen et al., 2013). Since then, Stellenbosch University and Aalto University have continued with a full-scale measurement campaign with the focus on human factors, structural dynamics and environmental conditions, which are experienced on research voyages with durations ranging between 14 and 78 days. A summary of full-scale measurement parameters is presented in Table 1.

Several research studies have focused on the analysis of the recorded full-scale data. Table 2 summarizes the data products and analysis algorithms, which have been developed through these efforts SAA II. These products relate to human factors, dynamic responses of the hull-structure and shaft-line and environmental conditions in which the vessel currently operates. Several measurement efforts since 2012 have resulted in an increased number of sensors and increased levels of expertise in determining the reliability of captured data.

An example of such growing expertise is the daily determination of data quality on research voyages. Accelerometer data was traditionally recorded by on-board accelerometers and stored in 5-minute data records. An LMS.Turbine Testing software and hardware combination is used to ensure truly continuous data recordings. Data was stored in LDSF format and extracted/converted to a processible format (.mat/.txt/.csv) about 15 days into a voyage. This required the start-stop of the measurement system. An onboard engineer manually checked the cable connections and measurement equipment on a daily basis to ensure system functionality. With experience, the measurement approach has changed to incorporate the start-stop of the system on a daily basis. The recorded data for this day can now be converted on a daily basis and run through a program to interrogate its veracity.

The software evaluates the statistical moments and distribution of the data channels as shown in Figure 1. This enables the identification of data which lies outside of what experience has taught to be normal levels. Researchers onboard thereby have an additional toolset with which to identify faulty or interesting measurements.

2.2 Readiness of the SAA II for digital twin technology

The full-scale measurement project on the SAA II is considered against the backdrop of Section 1.1 to assess its status against the ideals, software and hardware requirements to realize digital twin technology. The 'implemented' and 'required' elements are segregated and summarized in terms of 'haves' and 'have nots' in Table 3. From this analysis it is apparent that some promising architecture such as sensing and data analysis elements are already in place and that the promising applications, insights and associated requirements of digital twin technology remain to be defined and explored.

The realization of a digital twin concept for the SAA II will entail the on-board processing of sensor data from a variety of sources, in different formats and at different data rates and levels of reliability. Aside from overcoming these big data challenges, digital twin technology will require the transmission of reduced data and therefore on-board processing.

This potentially implies the requirement for data acquisition or sensing units that enable real-time data transmission and infrastructure for the aggregation and analysis of data prior to communica-

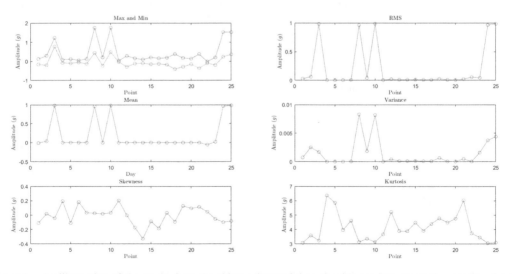

Figure 1. An illustration of the results from the object-orientated data algorithm to interrogate the veracity of data from 25 acceleration measurements. The minimum-maximum, root-mean-square (RMS), mean value, variance, skewness and kurtosis of the acceleration.

Table 3. An analysis of the readiness of conceptual architecture of the SAA II for the implementation of digital twin technology.

Define digital twin requirements	Implement conceptual architecture to enable digital twin model						
	Step 1	Step 2	Step 3	Step 4	Step 5	Step 6	Step 7
	Create operational sensing ability	Aggregate	Analyse	Communicate	Insight	Digital Twin model	Act
IMPLEMENTED	– Comprehensive sensor network has been installed (Summarized in Table 1).	– Data has been aggregated post several voyages.	– Value added data products have been developed from research outputs. Some data veracity checks are available.	– Sparse communication is possible in remote polar areas.	– Models and research results from operational data (as presented in Table 2).	– Some real-time as-built visualizations are possible.	
REQUIRED. The potential of digital twin technology for the SAA II is yet unexplored.	– No edge capability. Work is required to improve the robustness of sensor systems – Explore potential of radar imaging to improve environmental measurements (Kujala, 2017).	– Local ship network infrastructure installed to aggregate data. – Data storage infrastructure is required to retain data.	– Algorithms should be able to operate continuously and deliver results in real-time. Appropriate data reduction should be determined for communication.	– Strengthen signal coverage and internet connectivity between SAA II and the internet. Explore the installation of a high frequency radio channel with base stations in South Africa and Finland (Kujala, 2017). – Real-time data reduction cost must be determined. Perhaps explore real-time transmission of reduced data with full post-voyage updates.	– Full potential remains to be explored.	– Engineering product development models such as Computer Aided Design, Finite Element Models, Computational Fluid Dynamics, Machine control system models etc. reside with vessel manufacturers and are not at the disposal of the project. – Seek collaborators to contribute to digital twin platform.	– Desirable outcomes of digital twin technology are to be identified.

tion. Furthermore, support will be needed for for cheap and abundant data storage on the vessel and algorithms for real-time signal processing on the continuous data stream. It is clear that the demands of digital twin technology with regard to connectivity in polar environments will be an important factor, which governs the required data cost.

3 EXPLORING THE ADVANTAGES OF DIGITAL TWIN TECHNOLOGY FOR STAKEHOLDERS OF THE SAA II

The legacy of closely scrutinized, high quality data from the SAA II ideally positions this platform to prototype future trends in data technology. The question is: What advantages can be gained by embracing real-time digital twin technology? Despite obvious challenges, the proverbial blue skies benefits of digital twin technologies are conceptually explored for the SAA II. The perspectives of different stakeholders including the vessel owner, maritime industry and broader research community are considered.

3.1 *Benefits for the maritime industry*

1. An actual lifetime loading cycle can be accrued to determine the impact of Southern hemisphere storms and Antarctic ice on the structural health and propulsion systems, which could benefit future ship designs and cause-effect analyses.
2. Arctic Marine Transportation System remains demanding, dynamic and complex due to challenging hydro-meteorological conditions, poorly charted waters and remoteness of the area resulting in lack of appropriate response capacity in case of emergency (Heikkilä et al., 2017). This also applies to the Antarctic. The vast majority of marine casualties have their origins in human-related errors. The added advantage of digital twin technology includes safety benefits such as expert remote assistance, which could be offered in real time with the best available visualization and technical information.
3. Insight into the accrual of fatigue loading of the hull and propulsion structures will assist in the development of long-term strategies to operate and navigate ships for a prolonged service life.
4. In a more general sense, this pursuit can be motivated by the industrial drive towards the reduction of ship crew and new technologies for the remote sensing and control of ships. The ideal of automated ships, cannot be accomplished without sufficient high quality operational data to train and condition machine learning algorithms and prototype observations of ship responses to control algorithms. Digital twin technology creates the ideal environment for prototyping these ideas. The autonomous operation of ships with harsh voyage profiles will likely prove a significant engineering challenge. As such, the study of the voyages of the SAA II in harsh waves and ice could contribute to the training data required for autonomous operation algorithms.

3.2 *Benefits to the vessel owner*

1. Challenging ice conditions and rough open water are the factors that most degrade operational efficiency. The real-time processing of operational ship data will assist in aiding tactical judgments in ship handling to improve safe navigation and cost savings. This could result in direct economic benefits through fuel savings and increased scope for research with more efficient ship operations becoming possible for a given budget and available ship time.
2. Maintenance intervals could be optimized according to the factual requirements of the aging vessel structure. Accurate assessment of the fatigue life of the vessel as she nears the end of her service life would result on decommissioning based on facts which is sure to offer operational, safety and costing benefits over estimations based on the best engineering logic available.
3. The imperfections in engineering practice for ice-going vessels in the Southern hemisphere is emphasized by the fact that the SAA II is pre-disposed to prevalent stern slamming in even mild following sea states with 1 m swells, (Omer and Bekker, 2016). The pursuit of mitigating measures for slamming is governed by the underlying question: Will slamming vibration crucially reduce the operational lifecycle of the vessel? Public literature is largely devoid of recommendations to eliminate wave slamming through navigational strategies and full-scale measurements that elude to possible loading conditions and risks. The reason for this is the unpredictable and non-linear nature of slamming on full-scale ships, which implies that measurement campaigns can run for years without successfully capturing a slamming event.
4. Safer navigation in remote Antarctic ice is greatly benefited by a hull monitoring system because of the lack of ship response feedback to vessel operators. It is difficult for the crew to estimate the magnitude of the loads in harsh environmental conditions. Real-time, quantative measurements could provide information to avoid unintentional damage (Wang et al., 2001). The added advantage of digital twin technology includes safety benefits in that expert remote assistance could be offered in real time.

5. Digital twin technology could further be used as a ship "flight recorder" (Wang et al., 2001) to correlate accidents with the most recent measurement and simulated data available.

3.3 *Benefits to ship-based polar research*

1. Remote polar research could become a possibility through correct data reduction and transmission. This could reduce the pressure on berth space requirements. Participating researchers will not be required to be in situ, therefore also enabling the real-time involvement of researchers who are not healthy, or able to be away from shore for extended periods.
2. Digital twin technology would enable a land-based presence in Antarctic waters, which could reduce human impact in this sensitive natural environment, whilst enabling increased participation from academic supervisors and research collaborators.
3. The real-time availability of research data will result in the quicker delivery of results to land-based collaborators and shorter lead-times to research publications.
4. Adjustments or additional measurements could be requested by land-based scientists before the vessel progresses too far from her remote location which will increase scientific work quality and furthers the agenda of opportunistic science.
5. By embracing a digital, data-driven mindset the recording, modelling and reduction of various data will lead to cross-pollination of disciplines—observation of inter-relationships between more variables and new insights. South African research efforts will furthermore remain contemporary enough to contribute to international research efforts in developed nations who will embrace emerging technologies and rapid change through cloud-based digital technology platforms of their own.
6. The drive towards understanding and exploration of the oceans in globally perplexing matters such as climate change, places the SAA II and her operations in the spotlight (Moedas et al., 2017). This is furthered by the imperative of the South African government to build the South African Blue Economy. From this perspective, a digital twin of the SAA II has the potential to have immense showcasing potential whereby real-time vessel loads/oceanographic information could create a cloud-based museum. This could be show cased for the purposes of school tours or targeted events at the Department of Environmental Affairs offices at the iconic Victoria and Alfred Waterfront in Cape Town.
7. Research teams typically suffer from high turnover rates as post-graduate students graduate on a 2 to 5 year cycle. A digital twin of the SAA II offers immense training potential because of its remote accessibility. The consolidation of all available models and information could benefit knowledge retention and continuation of research efforts, which benefits long-term observations and trending which are essential to answer global research questions.
8. Real-time measurement and modelling places the vessel in an environmental laboratory, whereby the fidelity of digital response models can be interrogated and improved in their operational context in real-time. As such, operational data is truly utilized for the asset that it can be. High fidelity models could be of high value when considering the feasibility of chartered voyages to new, remote stations to support emerging research agendas.

3.4 *Challenges*

The successful implementation of digital twin technology will depend on the aspect of the vessel which is to be represented, and the associated requirements in terms of accuracy, quality, and time resolution. The ability to deliver such models will depend on technology readiness and cost of implementation. In the light of the foregoing discussion, some challenges/next steps can be highlighted towards the realization of a digital twin technology for the SAA II. It can be a daunting task to create a digital twin, which aims to deliver the blue skies advantages all at once. Deloitte (Parrott & Lane, 2017) recommends to start in one area, deliver value there, and continue to develop. Foreseen challenges include:

1. The versatile uses and many benefits of digital twin technology are numerous. It is important to focus and identify large benefits, which may be acquired through cost-effective means and little additional digital twin architecture.
2. Presently full-scale measurements on the SAA II are not obtained or aggregated through a centralized data acquisition network. A central information system is required and will enable concurrent data acquisition synchronized by a single time stamp. A robust sensing network with edge-processing (Mao, Hou, & Wu, 2008) capabilities is likely required to achieve the required data sensing capabilities (Erikstad, 2017).
3. Poor internet connections in remote Southern ocean environments will inhibit the streaming of live data and will require significant data reduction.
4. Utilizing the full potential of digital twin technology requires insight into the inter-disciplinary

research applications of the ship and her laboratories outside the ambit of engineering technology, science and research.
5. Some technology readiness challenges remain, especially related to sensor technology, decision-making intelligence, and system robustness in terms of ship-based measurements (Heikkilä et al., 2017).
6. To date the full-scale measurement project on the SAA II has not involved the development of simulation models. Insight with regard to stress concentrations and accrual of fatigue loadings will require access or creation of as-built digital models and vital partnerships with appropriate expertise in simulation and visualization.
7. In order to realize near-real-time results the processing of data or analysis models must be completed within the measurement time. Slow processing will introduce an ever-growing time lag between measurement and the digital twin model that at best is constant, and at worst continues to grow.
8. Not only does it require effort and skill to create a digital twin, but also to interpret and utilize the results it produces. Initial prototype systems may not be robust and will require expertise to debug. This implies that the successful implementation of this technology relies also on user readiness and training.
8. Data security concerns and secure data encryption are undeniable challenges of the future, which have not been addressed, nor explored in the contect of the SAA II.

4 CONCLUSIONS

Literature is rife with the potential applications of digital twin technology, which are clearly versatile. Implementations of digital twin platforms on ships have not yet been reported, however current trends indicate that such technology is best implemented in open source consortia and will be a distinguishing factor in the marine industry of the future. The SAA II polar supply and research vessel is a promising platform, which offers a mature sensing network and advanced data analytics, which can contribute immense value to the marine industry, the vessel owner and engineering and earth sciences if digital twin technology can be sufficiently harnessed.

ACKNOWLEDGEMENTS

The financial assistance of the National Research Foundation (NRF) towards this research is hereby acknowledged. The collaboration of Aalto University, Pretoria University and the Department of Environmental Affairs of South Africa is gratefully acknowledged.

REFERENCES

Datta, S. (2016). Emergence of Digital Twins Is this the march of reason? Retrieved January 9, 2018, from https://dspace.mit.edu/handle/1721.1/104429.

De Waal, R. (2017). *An investigation of shaft line torsional vibration during ice impacts on PSRV's*. Stellenbosch: Stellenbosch University. Retrieved from http://scholar.sun.ac.za/handle/10019.1/100973.

De Waal, R.J.O., Bekker, A., & Heyns, P.S. (2017). Bi-Polar Full-Scale Measurements of Operational Loading on Polar Vessel Shaft-Lines. In *Port and Ocean Engineering under Arctic Conditions*. Busan, Korea.

Erikstad, S.O. (2017). Merging Physics, Big Data Analytics and Simulation for the Next-Generation Digital Twins. In V. Bertram (Ed.), *High-Performance Marine Vehicles* (pp. 141–151). Durbanville, South Africa.

Glaessgen, E., & Stargel, D. (2012). The Digital Twin Paradigm for Future NASA and U.S. Air Force Vehicles. 53rd AIAA/ASME/ASCE/AHS/ASC Structures, Structural Dynamics and Materials Conference 20th AIAA/ASME/AHS Adaptive Structures Conference
14th AIAA, 1–14. https://doi.org/10.2514/6.2012-1818.

Heikkilä, E., Tuominen, R., Tiusanen, R., Montewka, J., & Kujala, P. (2017). Safety Qualification Process for an Autonomous Ship Prototype – a Goal-based Safety Case Approach. *Marine Navigation*, (August), 365–370. https://doi.org/10.1201/9781315099132-63.

ISO 2631-1. (1997). Mechanical Vibration and Shock—Evaluation of Human exposure to whole-body vibration. Part 1: General Requirements. Geneva, Switzerland.

Kujala, P. (2017). Ship monitoring and data transfer in extreme and varying environment near the pole (EMONITOR). Espoo.

Mao, S., Hou, Y.T., & Wu, M. (2008). Exploiting edge capability for wireless sensor networking. *Topics in Wireless Sensor Networking, IEEE Wireless Communications*, (August), 67–73.

Moedas, C., Pandor, N., & Kassab, G. Belem Statement on Atlantic Research Innovation Cooperation, 13 July § (2017). Lisbon: Atlantic Ocean Research Alliance. Retrieved from https://ec.europa.eu/research/iscp/pdf/belem_statement_2017_en.pdf.

Morais, D., Danese, N., & Waldie, M. (2016). Ship Design, Engineering and Construction in 2030 and Beyond. In V. Bertram (Ed.), *High performance marine vehicles* (pp. 297–310). Cortona, Italy.

Omer, H., & Bekker, A. (2018). Human responses to wave slamming vibration on a polar supply and research vessel. *Applied Ergonomics*, 67, 71–82. https://doi.org/10.1016/j.apergo.2017.09.008.

Panetta, K. (2016). Gartner's top 10 strategic technology trends for 2017. Retrieved January 9, 2018, from https://www.gartner.com/smarterwithgartner/gartners-top-10-technology-trends-2017/.

Parrott, A., & Lane, W. (2017). Industry 4.0 and the Power of the Digital Twin. *Deloitte University Press*, 1–17.

Sharkh, S.M., Hendijanizadeh, M., Moshrefi-Torbati, M., & Abusara, M.A. (2014). A Novel Kalman Filter Based Technique for Calculating the Time History of Vertical Displacement of a Boat from Measured Acceleration. *Marine Engineering Frontiers*, *2*(August). Retrieved from www.seipub.org/mef.

Smogeli, O. (2017). Digital twins at work in maritime and energy. *DNV-GL Feature*, (February), 1–7. Retrieved from https://www.google.co.za/url?sa=t&rct=j&q=&esrc=s&source=web&cd=6&cad=rja&uact=8&ved=0ahUKEwi22aKM473YAhWJuBQKHZalBpEQFgg2MAU&url=https%3A%2F%2Fwww.dnvgl.com%2FImages%2FDNV%2520GL%2520Feature%2520%252303%2520ORIG2b_tcm8–85106.pdf&usg=AOvVaw3fpRAEPnSt4BFOR.

Soal, K., Govers, Y., Bienert, J., & Bekker, A. (n.d.). System Identification and Tracking using a Statistical Model and a Kalman Filter. *Submitted to Mechanical Systems and Signal Porcessing*.

Suominen, M., Karhunen, J., Bekker, A., Kujala, P., Elo, M., Von Bock Und Polach, R., ... Saarinen, S. (2013). Full-scale measurements on board PSRV S.A. Agulhas II in the Baltic Sea. In *Proceedings of the International Conference on Port and Ocean Engineering under Arctic Conditions, POAC*.

Suominen, M., Kujala, P., Romanoff, J., & Remes, H. (2017). The effect of the extension of the instrumentation on the measured ice-induced load on a ship hull. *Ocean Engineering*, *144*(August), 327–339. https://doi.org/10.1016/j.oceaneng.2017.09.056.

Wang, G., Pran, K., Sagvolden, G., Havsgård, G.B., Jensen, A.E., Johnson, G.A., & Vohra, S.T. (2001). Ship hull structure monitoring using fibre optic sensors. *Smart Materials and Structures*, *10*(3), 472–478. https://doi.org/10.1088/0964-1726/10/3/308.

Combining design and strategy in offshore shipping

M.A. Strøm, C.F. Rehn, S.S. Pettersen, S.O. Erikstad & B.E. Asbjørnslett
Norwegian University of Science and Technology (NTNU), Trondheim, Norway

P.O. Brett
Ulstein International AS, Ulsteinvik, Norway

ABSTRACT: This paper presents the Design-Strategy Planning (DSP) procedure as a framework that integrates life cycle strategies of a ship into the early stages of the design process. We argue that understanding strategic, tactical, and operational aspects is essential when it comes to design of complex systems under uncertainty. Unfortunately, these are often neglected in ship design today. Using a Markov Decision Process Methodology, we demonstrate the insight gained from the concurrent exploration of system configurations and strategies, to better understand what to do when in the operational phase of the lifecycle. A case study is presented, where different tactical strategies of an offshore vessel are characterized. The results indicate that there are significant advantages in explicitly addressing ship owner strategy through DSP, when designing offshore support vessels that may be reconfigured in their lifetime.

1 INTRODUCTION

1.1 Motivation

Ship design is traditionally done with limited consideration of business strategy, even though acquisition of new ships is an important decision from that perspective. Ships are capital-intensive assets, meaning that design decisions will affect the overall financial viability and future business strategy of ship owners. Owners and designers will approach the same project with vastly different mindsets, meaning that there is a need to close the gap between overall strategic planning for the operational phase of the ship, and the decisions made by ship designers. In this paper, we attempt to close that gap.

1.2 Literature review

Early attempts at bridging the gap between shipping strategy and ship design include Benford (1967) who studied the connection between the initial sizing of cargo vessels, and maximizing the economic benefits given forecasts of cargo availability, while considering logistics. Erichsen (1989) and Wijnolst & Waals (1995) highlight the differences between the ship designer as an engineer concerned with the development of a ship description, and the ship designer as one someone who can translate business strategies into a ship description. Stopford (2009) also briefly consider ship design, suggesting that designers should understand the trades the vessel will serve, and the subsequent capacities, speeds and degree of flexibility. Lorange (2009) points out that shipping has moved significantly in the direction of specialization. Whereas before, shipping companies integrated across multiple activities such as owning, using, and operating ships, today's maritime business environment is more complex. He describes four strategic archetypes for actors in the maritime industry; owning steel, using steel, operating steel, and innovating around steel.

The connection between ship design and business strategy has also been elaborated on in previous International Marine Design Conference (IMDC) papers, including the work by Ulstein and Brett (Ulstein & Brett 2009; Ulstein & Brett 2012; Ulstein & Brett 2015). They specifically address the need for an interplay between technical, operational, and commercial considerations in ship design, signaling to shipowners that it is important to avoid too technical details at the early stages of the design process. They build on Brett et al. (2006), who introduce the Accelerated Business Development (ABD) process as a methodology for considering the link between shipping strategies, the shipping company value proposition, and the ship design process. The starting point for ABD involves four sub-processes in which the ship owner develops the business concept, with considerations of needs, expectations, risk and competitive positioning, building on the classical work of Porter (1979). Porter introduces five forces that drive industrial competition. The competitiveness of a firm is challenged by the strength of its suppliers, potential new entrants to the marketplace, the buyers of its products or services, substitutes, and current rivals. The outcome of the ABD process will be conceptual designs developed with basis in the ship owner strategy. Hence, the business proposition of the ship owner will strongly influence what functionality is sought, and consequently what concept the designers should iterate on.

With respect to the connection between the operational phase and ship design decisions, Erikstad et al. (2011) introduce the ship design and deployment problem (SDDP), as a mixed-integer programming model. Their model accounts concurrently for lifecycle deployment, and optimal design decisions. Further, Gaspar et al. (2012) provide added insights to the temporal aspects of the offshore ship design problem combining SDDP with Epoch-Era Analysis (EEA). For more information on EEA, see Ross & Rhodes (2008).

Uncertainty is an important consideration of the operational phase, which has received significant focus in the systems design literature. Ross et al. (2008) introduce the concept "value robust" to reflect the characteristics of a system that enables it to continue to deliver value throughout its lifecycle. What is valuable to a shipping company evolves as exogenous uncertainties resolve, and the strategies and tactics of the shipping company evolves as new aspects gain importance. Ships may be designed either to be able to statically deliver value as the context and owner strategies change, or ships can evolve through retrofits and reconfiguration to provide new functionality, as context and strategies change. The latter is often addressed by the term physical design changeability. Changeability is defined by Fricke & Schulz (2005) to be the superset of robustness, flexibility, adaptability and agility. The changeability concept has a strong link to the links between business strategy and ship design. Further, changeability has a strong link to real options, often characterized as the right but not the obligation to perform some action. This field of research has received attention also in the maritime industry. An overview of traditional real options research for managing risk in shipping is presented by Alizadeh & Nomikos (2009). Examples of options that have seen wide application in the shipping world include lay-up, the option to charter in additional capacity at peak demand, or the option to take on spot cargoes. Real options in the context of systems design has become a popular topic in recent years, as exemplified by de Neufville & Scholtes (2011).

For marine design applications, Niese & Singer (2014) introduce a methodology for assessing system changeability based on Markov decision processes (MDP). MDP is a structured method for modeling sequential decision-making under uncertainty, accounting for both the outcome of current decisions and future decisions opportunities (Puterman 2014). Previous research using MDP applied to ship design include analysis of ballast water treatment compliance (Niese & Singer 2013), energy efficiency (Niese et al. 2015) and emission control area regulation compliance (Kana & Harrison 2017).

1.3 Contribution

This paper argues for the importance of considering the strategic and tactical aspects of the operational phase of a ship lifecycle, already at the early stages of the ship design process. Further, this paper presents the design-strategy planning (DSP) framework that considers operational phase strategies and tactics of a ship at the initial stages of the design process. In addition, the DSP framework can support active management throughout the lifecycle. A case study is presented, where different strategies of an offshore vessel are characterized, and design characteristics valuable for each strategy are identified.

2 MANAGERIAL STRATEGY

2.1 Strategy as a plan

There is an abundance of definitions of *strategy*. Recognizing the multiplicity, Mintzberg (1987) presents five definitions (the five Ps) of strategy: as plan, ploy, pattern and position, for which this paper aligns with the first. As a plan, strategy is *some sort of conscious intended course of action, a guideline (or set of guidelines) to deal with a situation* (Mintzberg 1987). Thus, fundamental characteristics of strategy is that it is developed deliberately in advanced of being deployed. Other definitions following the idea of strategy as a plan are: *a careful plan or method: a clever stratagem (trick); the art of devising or employing plans or stratagems towards a goal* (Merriam-Webster). A specific plan of action to reach a particular objective (Mieghem & Allon 2015), and a *coordinated set of decisions* (Skinner 2009). In light of these definitions, we define *strategy as a plan to coordinate a set of decisions to reach a particular objective*. As stated by Andrews (1987): *Anything that is not planned is not a strategy, such that successful pattern of action that was not intended is not a be called strategy, rather brilliant improvisation or just plain luck*.

Strategy is often used to describe multiple managerial planning archetypes, while at the same time describing the highest planning level—the strategic level. The managerial strategy planning horizon are commonly divided into strategic, tactical, and operational levels, all terms referring to the use of the vessel in the operational phase of the lifecycle. For shipping applications, we define these terms as follows (Christiansen et al. 2007):

- **Strategic planning** refers to decisions with long-term implications, typically several years. For a ship owner, these decisions include acquisition, including ship design, sales and scrapping of vessels, as well as shipping network design.
- **Tactical planning** refers to decisions with medium-term implications, typically up to one year. For a ship owner, these decisions include chartering, deployment, lay-up, routing and scheduling.
- **Operational planning** refers to decisions with short-term implications, typically from days to months. Decisions at this level include speed optimization, and other detailed planning of marine operations.

Confusingly, the *operational phase* describes the entire time the ship is in operation. In a lifecycle perspective, the operational phase is everything that happens between production and disposal. The ship *design phase* is the process of finding a description of the ship to be built. Hence, ship design in itself is a strategic decision problem (Christiansen et al. 2007).

Figure 1 illustrates that there is a need for integrating the asset management philosophy used for the operational phase within the design process. To the left, the operational phase is decomposed from the strategic level, further to tactics and operations. To the right, the design process is described as an iterative mapping between function and form. The point of Figure 1 is that there needs to be an interplay between strategies for managing the ship in the operational phase of the lifecycle, and the ship design decisions in the conceptual design stage of the lifecycle.

For example, if the strategy of a ship owner is to operate a vessel in the platform supply North Sea spot (short-term) market, his ship design preferences will likely be different than if the newbuilding is intended for a long-term tender contract with a large oil company. A ship designed for the spot market would favorably be agile and be able to remobilize quickly, possibly with modular interfaces between the integral ship platform and topside equipment. In comparison, a ship designed for the tender contract may be less modular.

A parallel here can be drawn to "requirements elucidation", proposed by Andrews (2011). Where requirements elucidation favors that requirements are developed along with solutions, we here favor a strategy elucidation, where ship designers seek to critically understand the ship owner strategy when developing solutions.

There is an important difference between transportation shipping, such as bulk, tank or container shipping, and non-transport shipping, such as offshore service providing ships heavy lift and construction vessels. Christiansen et al. (2007) discuss aspects of planning for *transportation shipping*, which are not necessarily transferable to *non-transport shipping (offshore)*. Shipping strategy is also discussed by Lorange (2005), who also points out the important difference between commodity shipping and other types of shipping. He mentions several successful niche strategies in shipping, such as developing leverage niches, build niches and transform niches. What is of significant relevance for the strategies in these two segments is the competition. Lorange also emphasizes the *time scale* of competitive strategies, as where barrier to competition vanish in the long-run. However, human know-how and soft skills can be difficult to copy.

2.2 *Shipping strategies*

Christiansen et al. (2007) present examples of strategic planning problems. These including (not

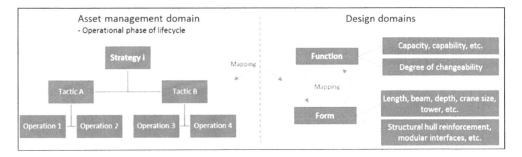

Figure 1. Connecting aspects of strategy, tactics, and operation to the traditional design problem.

limited to) market design and trade selection, *ship design,* network and transportation system design, and fleet size and mix decisions. Thus, ship design is fundamentally characterized as a strategic problem. The same holds for retrofits of the ship that may be done throughout the operational phase of the lifecycle.

In the context of fleet renewal and ship design, strategic planning can be connected to the business models of a shipowner. Business models in shipping are often classified in the following way:

- **Asset play:** Operational costs are not that important, as the main source of profitability is from the well-timed purchase and sale of ships (Lorange 2005). Hence, the owners will try to minimize capital expenditure.
- **Full ownership:** Long-term ownership is supported by operational cost minimization. This is similar to the "operations based strategy" by Lorange (2005). These actors care about technical and operational aspects, and will have strong ship design preferences.
- **Tonnage provider:** Focus is on buying and developing assets, for then to lease them on bareboat.
- **Other:** Depending on the specifics of the shipping case, one can also have combinations of one or more of asset play, full ownership and tonnage provider.

2.3 *Shipping tactics*

Christiansen et al. (2007) present examples of tactical planning problems. These include (not limited) fleet deployment, ship routing and scheduling, and ship management. Tactical planning decisions for offshore ships comprise contract and area selection. Contracts are of different length, and tactical decisions also involves the selection of operation in the spot or term market. Other aspects of ship management to consider at the tactical level is real options, such as lay-up and reactivation of ships, and expansion or equipment retrofits

2.4 *Shipping operations*

Christiansen et al. (2007) present examples of operational decision problems in shipping. These include, (not exhaustive) for example cruising speed selection, ship loading and environmental routing. Operational planning will not be covered in this paper.

2.5 *Strategy as a pattern*

As pointed out by Mintzberg (1987), in addition to defining strategy as a plan, one should consider the resulting stream of actions—the *pattern*. One definition of strategy as a pattern is *consistency in behavior, whether or not intended* (Mintzberg 1987).

Thus, while the plan creates an *intended strategy* (i.e. plan of action), only some of the intended strategy is *realized*. The d*eliberate* strategy represents the parts of the intended strategy that is retained, and *emergent* strategy represents strategy that becomes apparent along the way (Mintzberg & Waters, 1985). Therefore, while maritime decision makers attempt to develop strategic plans setting the direction for their operations, the plan is without value if the intended set of actions are not carried out.

2.6 *Design of strategic systems*

This paper proposes the term *strategic systems* to map the asset management domain and the design domain (Figure 1). The term strategic system refers to a specific design-strategy configuration, which will be used interchangeably. *Design* refers to the physical aspects of the vessel performing the operations resulting in stakeholder value, while *strategy* refers to the managers available options (both on an operational, technical and strategic level) to utilize the design. Using language of *real option*, this configuration encompasses a set of real *in* and *on* options (Wang & de Neufville 2005). While the real *in* options related to the physical design, the real *on* options relates to the management of the system.

This paper states that the objective of the conceptual design phase should be to create a strategic system. This extends the traditional view on design, from solely focusing on the physical configuration, to also considering how the physical configuration is an enabler for the strategic, tactical and operational decisions over the vessels lifetime. The strategic system should encompass a set of real options able to be aligned with the constant changes in context and needs, thereby creating a *sustained competitive advantage* and becoming *value robust*. In two extremes, the design configuration can either be perfectly aligned with the current context and needs, or not fit for all. The same goes for the strategies. Thus, the strategic systems are *unsuccessful* when neither the design or the strategy are fit to the current needs. On the other side, the strategic system is highly successful, and have a high competitive advantage, when both the design and strategy are aligned with the current context and needs.

In the process of adapting the strategic system to its environment, the key question to ask in the design domain is *how should the vessel be configured to have the functionality to meet the current market demands?* In the strategy/managerial domain, the key question to ask is *how should the vessel be utilized to gain competitive advantage?*

3 METHODOLOGY

3.1 *Design Strategy Planning (DSP)*

Design-strategy planning (DSP) is a systematic framework for supporting active management of exogenous uncertainty throughout the lifecycle of offshore vessels. As an iterative, four-step procedure, the framework consists of an (I) identification phase (II) development phase (III) implementation phase, and (IV) monitoring phase. The framework is presented in

Figure 2 note that, while the figure indicates a distinct sequential flow between the four-steps, this is not necessarily how it would play out. Especially the initialization and development phase consist of irregular activities, that are intertwined. Therefore, the procedure will often end up jumping back and forth between these phases. The feedback arrow illustrates this.

3.1.1 *Phase I: Identification phase*

The identification phase is a collaborative process between major stakeholders, for addressing strategic decisions and platform designs that together forms a "strategic system" able to deliver high stakeholder value and handling future uncertainty. The key objective is to get a shared understanding of the commercial, operational, and technical aspects of the design problem, to lay the foundation to find a design solution that fits with the business and operational domain.

First, major stakeholders are identified, and their objectives and resources clarified. Major stakeholders to include are, amongst others, designers, engineers, owners, operators, and analysts. The owners contribute with the commercial intent of the vessel, in addition to technical and operational expectations. The designers and engineers provide insight into feasible technical solutions, and the operators provide expertise in the vessel's performance and operational needs. It is crucial to ensure a joint understanding of the objectives, as this defines the criteria for the vessel's lifecycle success. Combining the different domain expertise from the very beginning is key for creating value robust solutions. Then, the internal assessment should focus directly on key stakeholders' resources related to operating the vessel. The focus should not only be on the tangible resource base, but also on intangible, such as knowledge, capabilities, attitude, and relationship to its network. Furthermore, only understanding the current stakeholders, objectives and resources is insufficient; it is essential to analyze how these potentially can develop over the vessels lifetime.

Secondly, major drivers of exogenous uncertainty must be identified, and, to the extend it is possible, quantified. The assessment should both consider the direct market environment and the wider contextual environment. Both the likelihood and the consequence should be assessed, to focus the process on the most high-risk aspects of the future. To be aligned with all aspects affecting the lifecycle of the design, all aspects from both the commercial, operational, and technical sides of the vessel should be analyzed.

Third, a set of platform designs should be identified. A good way to develop flexible engineering systems is to start from an existing set of platform designs, as it relaxes the computation burden of starting from scratch (Cardin 2014). The base designs will further be enriched by adhering to design principles for changeability. Modularity and redundancies are examples of design principles enabling changeability (Fricke & Schulz 2005). (Rehn et al. 2018) introduces the choice of changeability level, to illustrate that the ease of change by executing a change option (both in changing cost and time) can be controlled. The underlying hypothesis is that incorporating changeability becomes more relevant with increasing uncertainty, and for systems with longer planning horizon.

As the last point in the initialization phase, strategic decisions for mitigating vulnerabilities and exploiting opportunities inherent in the uncertain aspects should be identified and analyzed. As earlier pointed out, it is important to consider both strategic, technical and operational level in the strategy domain to grasp the full extent of how the vessel can adapt in the face of changes in context and needs to stay competitive.

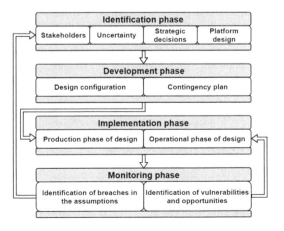

Figure 2. Illustration of the Design-Strategy Planning (DSP) framework.

3.1.2 *Phase II: Development phase*

In the second phase in DSP, the development phase, we want to iteratively develop and select a design configuration, and a contingency plan. The objective is thus to identify under which circumstances various strategic design and operational options should be executed. The underlying hypothesis implied by creating the contingency plan is that the future too uncertain for not having a pre-defined plan stating how to response to changes in the changes in context and needs.

Design configuration

The selected design configuration consists of a platform design, in addition to a set of selected principles of changeability and levels of changeability. Arguably, incorporating changeability is a means for the base design to better dealing with uncertainty. However, one key challenge is to strike the balance between the implementation and carrying cost of incorporating a changeability (referred to as the design for changeability level, or DFC level) and cost of executing the options, against the cost of executing the options without having it pre-installed in the design.

Contingency plan

The contingency plan states which real options that should be executed on a technical and/or operational level as a response to trigger information. Triggers are occurrences that require a response from the contingency plan to mitigate risks or take advantage of opportunities. Triggers can also result in a reassessment of the DSP as the underlying assumptions of the development phase are changed. Contingency planning recognizes that generating sustained value is not only about making solid design decisions in the early phase, but also a continuous managerial decision problem over the lifecycle of the vessel.

A well-developed contingency plan should be robust, meaning that a broad range of different futures should be considered, related to the technical, operational, and commercial domains of the vessel, both in the near future and in the end of the vessel's lifetime. Secondly, the contingency plan should be flexible, meaning that a broad range of tactical measures should be considered to find the best measures to handle the uncertain future. Third, the plan should be specific, stating which measure to implement under which situation. Also, it is of high importance to consider the ability of the manager of executing the planed procedures, and the resources available in the situation.

3.1.3 *Phase III and IV: Implementation and monitoring phase*

Following the development phase, some of the actions are immediately implemented in the production phase of the design. These actions are related to the building of the platform design selected. After the vessel is launched, in the monitoring phase other actions can be implemented in the operational phase of the lifecycle, but only as a direct response to trigger information.

In the monitoring phase, the environment of the vessel is monitored seeking for trigger information indicating vulnerabilities to mitigate and opportunities to exploit. If found, the contingency plane states which actions to implement.

The DSP process should be reassessed if the monitoring phase identifies major changes in the context and needs that breaches the underlying assumptions of the development phases. If so, the process would not start from the very scratch, this time the process starts off with a vessel design. Another reason for considering reassessing of the DSP process would if one deviates from the intended strategy outlined in the contingency plan. This could be a result of limited resources and/or capabilities of managing change. If that is the case, the contingency plan itself should be reassessed. However, another reason for not following the plan could be stakeholder's inherent resistance to change. If that is the case, one should seek to overcome this rigidity to change, rather than changing the plan.

3.2 *Markov Decision Processes (MDP)*

Markov decision processes (MDP) is a technique to quantitatively model and solve sequential decision problems. Since MDP can determine the "optimal" initial vessel design, and which real options to implement over the vessel's lifetime for maximizing long-term profit, it is able to support the development phase in the DSP framework.

Figure 3 illustrates a symbolic representation of a sequential decision problem. At a specific point in time, a system is in a state (or decision epoch), in which an action is to be made. The action is made from a set of available decisions, and is based on a decision rule, stating which action to make under

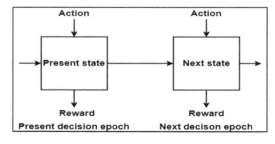

Figure 3. Symbolic representation of sequential decision-making problems (Puterman 2014).

Table 1. Illustration of a decision matrix.

		System space			
		s = 1	s = 2	...	s = |S|
Time space	t = 1	Act. II	Act. XI	...	Act. I
	t = 2	Act. I	Act. I	...	Act. IV
	...	...	...	...	...
	t = |T|	Act. XI	Act. X	...	Act. I

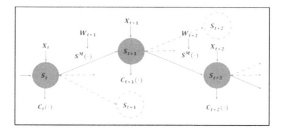

Figure 4. Markov decision process illustration (Strøm, 2017).

which circumstances. It is assumed that the action is made with a complete information of the system state. The consequence of the decision is two-folded: first, the decision maker receives an immediate contribution, secondly, the system transits to a new state. Which state the system transits into is both dependent on the decision made, and of some exogenous information revealed first after the action is made. After the new state is entered, the procedure is repeated. The procedure can last for a finite or infinite time. The objective of a sequential decision problem is to find the optimal policy which maximizes the contribution over the lifetime of the system. A policy is a sequence of decision rules, stating which action to make, for each future time step, under different circumstances. The optimal policy is often presented in a decision matrix (DM), as presented in Table 1.

More formally: At time t, the system is present in state $s \in S_t$, for which an action is made $x_t \in X_t$. Which action to make is stated in the policy, $\pi \in \Pi$. The policy is a function $\pi_t : S_t \rightarrow X_t$, that for each time step t, maps the current state to an action to make. After the action is made, the system receives an immediate contribution determined by the contribution function, $C_t(S_t, x_t)$, dependent on the current state and the action made. In the next time step, the system transits to a new state, S_{t+1}, determined by the transition function, S^M, as given in Equation 1.

$$S_{t+1} = S^M(S_t, X_t, W_{t+1}) \qquad (1)$$

The transition function can be dependent on the current state, the decision made, and the exogenous information revealed first after the decision is made, W_{t+1}, making the transition uncertain. The probability of transitioning from state S_t to S_{t+1} is given by the one-step transition matrix, $P(S_{t+1} | S_t, X_t)$, which depends on the current state and the current action made.

Extending Figure 3, Figure 4 illustrates a realization of a three-step system path. Depending on the nature of the problem, the sequence can continue for finite or infinite time, and the state, action and time space can be discrete or continuous. In addition, the contribution and the transition probability can be stochastic or deterministic. In this article, the focus is on stochastic, finite horizon processes, with discrete data, decisions, and time space.

The performance metric represents the performance of a specific policy. A traditional performance metric is the expected discounted contribution over the lifetime of the system, stating that receiving a contribution in the future is less of worth than receiving it immediately. By the discounted contribution performance metrics, the policy is evaluated as given in Equation 2.

$$\psi_i^\pi \equiv \mathbb{E}\left\{\sum_{t=0}^{T} \gamma^t C_t^\pi(S_t, X_t^\pi(S_t)) \mid S_0 = i\right\} \qquad (2)$$

Here, T is the length of the system's lifetime under consideration and γ^t is the discount rate. The expectation, E, is over the exogenous information affecting the contribution function. By using the discounted contribution metrics, the performance of a policy can be evaluated using the expression given in Equation 3.

$$\max_\pi \mathbb{E}\left\{\sum_{t=0}^{T} \gamma^t C_t^\pi(S_t, X_t^\pi(S_t))\right\} \qquad (3)$$

Despite its complexity, there are methods available for solving the Equation 3. To do so, let V_t be the value function, expressing the expected value of making the optimal decision, $x_t^*(S_t)$, in a given state, at a given time step. One way of expressing the value function is the standard form of Bellman's equation, as given in Equation 4.

$$V_t(S_t) = \max_{x_t \in X_t} \left(C_t(S_t, x_t) + \gamma \sum_{s' \in S} \mathbb{P}(S_{t+1} = s' \mid S_t, x_t) V_{t+1}(s') \right) \qquad (4)$$

By knowing the value function in each successive step, for all possible transition states, the optimal action is found by the argument of the maxima of the expression in Equation 5.

$$x_t^*(S_t) = \underset{x_t \in X_t}{argmax}\left\{C_t(S_t, x_t) + \gamma V_{t+1}(S_{t+1})\right\} \quad (5)$$

Following the equations presented above, there is a need for a method that calculates the value function in each state-time combination, such that the optimal set of actions can be found. There are several methods for doing so, one of which is approximate dynamic programming (ADP).

In Figure 5, a generic ADP algorithm is presented. First, in step 0, the value function is initialized to zero, and a starting state, S_o^1, is selected. Note that instead of using the true value function, a statistical estimate (i.e. approximation) after n iterations is used, $\bar{V}_t^n$. Here, n is the interation counter stating the number of times the algorithm is run. Secondly, in step 1, ω^n is the sample path the process follows at iteration n, representing how the stochastic information unfolds. Thus, $W_t(\omega^n)$ represents the realisation of the stochastic information at time t following specific sample path. Note that following a single set of sample realizations would not generate anything of value (as the same instances would occur each iterations), hence the procedure need a new sample path for each iteration. Third, in step 2, the algorithm loops over the time step of the system's lifecycle (t = 0,1,2,3...T). In each time step, a sample estimation, $\hat{v}_t^n$, of the value of being in state S_t^n is calculated using the approximation of the value function calculated in the previous iteration ($\bar{V}_{t+1}^{n-1}$). From this, action x_t^n is chosen to be the one that solves the maximization problem. The sample estimation is used to update the value function in the current iteration. Then, the system transits into a new state, before the process continues. The procedure is repeated for N number of iterations. After the final iteration, the approximated value function in each state-time combination is used to find the optimum decision with Equation 5. The set of optimal decisions comprises the optimal policy.

For instructional purposes, this paper outlined a generic form of ADP, while actually using the Q-learning algorithm (QLA) on the illustrative Case. QLA is one of the fundamental algorithms in ADP and reinforcement learning. We did do because the QLA is more comprehensive, and therefore encourage readers with interest in this field to Powell (2011) for a riche presentation of it.

3.3 MDP in support of DSP

The Markov decision process (MDP) methodology models the decision problem using the insights gained in the implementation phase of design-strategy planning (DSP). Following the notation presented in the former section: The time space represents the points in time over the vessel's lifetime in which decisions concerning the design-strategy configurations are made. The state space represents all states the strategic system can encounter. The decision space represents all actions stakeholders can execute to alter the system spaces. These actions are both related to the altering of the physical design configuration and the altering of the operational mode or strategy. The contribution function models the gains, or losses, from executing an action in a given state, depending on which state the system transits into. The transition function models how the strategic system evolves from one system state to another, which is dependent on the current state, the decision made and exogenous uncertain factors. The stochastic variable represents the exogenous uncertainty in the decision problem that makes the outcomes of every decision made (i.e. the contribution gained, and the state transitioned into) uncertain.

After having modeled the decision problem, the MDP methodology solves it. The output is a decision matrix (ref. Table 1) recommending decisions in every state, at every time step, to maximize the expected life cycle contribution of the strategic

---

Step 0. Initialization:

Step 0a. Initialize $\hat{v}_t^0$ for all states S_t

Step 0b. Choose an initial state S_0^1

Step 0c. Set n = 1.

Step 1. Choose a sample path ω^n

Step 2. For t = 0, 1, 2, ..., T do:

Step 2a. Solve

$$\hat{v}_t^n = \underset{x_t \in X_t^n}{max}(C_t(S_t^n, x_t) + \gamma \sum_{s' \in S} \mathbb{P}(s' \mid S_t^n, x_t))\hat{V}_{t+1}^{n-1}(s'))).$$

and let x_t^n be the value of x_t that solves the maximisation problem

Step 2b. Update $V_t^{n-1}(S_t)$ using

$$V_t^n(S_t) = \begin{cases} \hat{v}_t^n & S_t = S_t^n \\ V_t^{n-1}(S_t) & otherwise, \end{cases}$$

Step 2c. Compute $S_{t+1}^n = S^M(S_t^n, x_t^n, W_{t+1}(\omega^n))$.

Step 3. Let n = n + 1. If n < N, go to step 1.

Figure 5. Pseudocode for a generic approximate dynamic programming (ADP) problem using the one-step transition matrix (Powell (2011), p. 120).

system. The decision matrix can further be used as input in a life cycle simulation to for instance analyze the expected life cycle contribution, and gain insight from other metrics. This Markov decision process methodology is inspired by the approach proposed by Niese & Singer (2014) for assessing changeability. The decision matrix and the output from the life cycle simulation can then be analyzed to provide valuable insight to the development phase of the Design-Strategy Planning (DSP) problem.

4 CASE STUDY – OFFSHORE SHIP

4.1 Case description

Using an illustrative offshore case, this section presents how Design-Strategy Planning (DSP), supported by the Markov decision process (MDP) methodology, can be used to support the ship design process. The presented work is based on Strøm (2017).

4.2 Phase I: Initialization

In this illustrative case, the stakeholder is a shipowner seeking to build an offshore vessel targeted to operate in the offshore construction segment. The objective is to build a vessel with the highest expected discounted life cycle value. Thus, the case centers around a strategic level decision of choosing vessel design.

Initially, the vessel is to undertake a five-year offshore decommission contract in the North Sea. After the initial contract ends, the vessel is assumed to continue to operate in the North Sea. The time span of the analysis is 15 years from present, and since the first five years are determined, we analyze the subsequent 10 years thereafter.

A high degree of uncertainty affects the performance of the vessel over its lifecycle, where particularly the overall economic market state and operational requirements are of high importance. In addition, there is uncertainty related to whether the shipowner wins future contracts, and the dayrates for each mission.

One platform design is considered, with the following main dimensions: a length of 120 meter, beam of 25 meter and a depth of 10 meter. It has accommodation capacity for 250 persons, and main crane capacity of 400 tones. For more comprehensive design analyses, multiple platforms can be considered.

Tactical decisions to consider are selection of missions, associated contract duration, in addition to the options to lay-up or sell the vessel. In addition, the shipowner can alter the configuration of the vessel by altering the accommodation size, replace the main crane, add light well intervention equipment, remotely operated vehicles, cable laying equipment and a moonpool.

4.3 Phase II: Development

4.3.1 Modelling the system space

Combining the information found in the initialization phase with the MDP methodology, the state space comprises: the design state, strategy state, mission state, market state and technical state.

$$State\ space = (design,\ strategy,\ mission,\ market,\ technical) \quad (6)$$

Design state
The design state represents the set of possible vessel configurations under consideration, comprising both fixed and variable parameters. The fixed parameters represent the dimensions (length, beam, depth) of the platform design, and the variable parameters represent the design parameters that can be altered. Table 2 presents the set of design state variables.

Enumerating all the combinations of the design variables gives 216 unique design configurations, some of which are not feasible. Reducing the design space is crucial for lowering the complexity of the procedure. To reduce the design space, physical design feasibility, stability criterion and freeboard criterion were imposed.

The physical design feasibility constraint removes all designs with a deck area off less than zero square meters. The stability criterion removes all designs with an initial metacentric height (GM) less than 0.15 meter. The freeboard criterion removes all designs with a freeboard less than 1.5 meters. Imposing these constraints reduced the number of designs configurations to 12.

Strategy state
The strategy state represents the shipowner's available decisions, mainly concerning tactical options. From the implementation phase, there are four options availed. These options are whether to operate the vessel in the spot market, operating on one-year contracts or in the long-term

Table 2. Design state variables.

Design state variables	Units	Values
Accommodation	Persons	[50, 250 400]
Main crane capacity	Tonne	[0, 400, 800]
Light well intervention	Tonne	[0, 300, 600]
Remotely operated vehicle	[–]	[No, Yes]
Cable laying equipment	[–]	[No, Yes]
Moonpool	[–]	[No, Yes]

market, operating on three-year contracts. The vessel owner can, after the initial contract ends, also sell the vessel or lay it up.

Mission state

As presented in Table 3, eight missions are considered. It is assumed that all are available in the North Sea market at all times. However, which mission the shipowner takes is dependent on three factors: First, there are technical requirements associated with each mission that the vessel must comply with. These are dependent on the general requirement state in the market. Secondly, the vessel competes for the contracts with other vessels operating in the North Sea. The probability for winning a contract is a factor of the supply-demand ratio of vessels which depends on the market state. Finally, it is assumed that the vessel owner always takes the mission, of those available to him, with the highest day rate. The day rate is a stochastic variable, depending on the mission taken, the contract duration and the state of the market.

Market state & technical requirement state

The state of the market and the technical requirements represent the two major sources of exogenous uncertainty to the shipowner. Both are modelled as stochastic processes, with a discrete representation. The market state is modelled to follow a seven-year cycle, and the technical requirements are modelled as a stepwise, linear function representing the assumption that the difficulty of meeting the requirements will increase in the future. Table 4 presents the relative levels of the exogenous uncertainty. The overall activity in the market is represented by the "market state", which is assumed strongly correlated with the oil price. A high market state thus represents high activity levels, and a resulting strong demand for offshore vessels services. A strong demand side results in higher dayrates, everything else equal.

4.3.2 Starting state

Following Equation 6 modeling the state space, it is assumed that model initially starts off from the platform design (seq. 4.2), operating on short-term contract in a market with low technical requirements. The market state is uncertain, but with a higher probability of being in the lower end of the scale. Also, the initial mission is uncertain. The mission selected is the one with the highest dayrate of the missions the vessel can undertake under the current state of technical requirements.

4.3.3 Modelling decisions

The shipowner can alter the state of the strategic system by making one of the following decisions: the shipowner can change the design configuration and change which tactic to follow (i.e. taking short- or long-term contract, and which mission to take). A decision for each of these tree considerations, on whether to change or remain as before, must be made in each state. If the shipowner decides to retrofit the vessel, the switching time reduces the number of annual operational days in the subsequent period. If the decision only deals with which tactic to select, the vessel immediately starts the next operation. The decision is only made at the end of a contract. Hence, if there is a long-term contract, the strategic system remains unchanged constant over the length of that contract. For operations in the spot market, the frequency of decision-making is higher.

4.3.4 Modelling the transition function

The transition function is dependent on the current state, the decision made, and the exogenous information revealed to the decision maker after the decision is made. Thus, the transition function comprises one stochastic and one deterministic part. While the transition from one design state to another, between strategy and mission states, is fully dependent on the decision made and therefore deterministic, the transition between market states and technical requirement states are independent on decisions made and is therefore stochastic.

4.3.5 Modelling the objective function

The objective of the case is to evaluate vessels based on the net present value (NPV) of their lifecycle performance. Only monetary value is considered, assumed to only be dependent on building cost, operational revenues and switching costs.

Table 3. Mission states.

Mission	Abbr.
Subsea Installation and Construction	OSC
Inspection Maintenance and Repair	IMR
Light Well Intervention	LWI
Field Decommission Support	ODS
Offshore Accommodation	ACC
Offshore Cable Laying	OCL
Offshore Platform Supply	OPS
Offshore Aquaculture Support	OAS

Table 4. Exogenous information (market and technical) and discretized levels.

Exog. information	Level
Market State	[Low –, Low, Medium-low, Medium-high, High, High +]
Technical Req. State	[Low, Medium, High]

4.3.6 Results from the development phase

The MDP model was solved by approximated dynamic programming, using a Q-learning algorithm.

Table 5 presents an excerpt of the derived life cycle policy, stating which strategic action to take under each state-time combination. Exemplified, if the vessel, in year 11, has design configuration 2, operating in a short-term contract in a medium-low market, with a high technical requirement (i.e. currently in model state 63), the shipowner should exercise action 26, whose details are presented in Table 7.

Table 7 presents an excerpt of the action list, presenting tactical decisions made over the course of the vessel lifecycle. Continuing the example above, action 26 represents a change to design configuration 9, in addition to switching to a long-term contract continuing operating short-term contracts. Retrofitting to design configuration 9 increases the accommodation capacity to 400 persons.

Following the MDP methodology, the lifecycle policy is used in a lifecycle simulation for further analysis.

4.3.7 Results from the lifecycle simulation

The statistics of expected net present value of the analyzed vessel are presented in Table 6. Numbers are in million USD and are based on 1000 lifecycle simulations. The average number of design reconfigurations indicate that, in fact, the simulated ship usually undergoes some sort of retrofit during the lifecycle simulations, and switches design configuration.

Figure 6 presents the frequency in which (a) the market state, (b) strategy state, (c) mission state and (d) design state occurs. As seen in Figure 6 (a), the North Sea market is highly cyclical, indicating that the shipowner is to expect a low market state when the initial contract ends and a high market in the end of the period analyzed. Figure 6 (b) indicates that the shipowner will take short-term contracts in the first years, and then start taking long-term contracts. The vessel is never sold in this simulation. Figure 6 (c) indicates that the vessel normally continues to operate on the ODS contract after the initial five years. Then, the OPS, ACC and OAS contracts are taken most frequently. The description of these contracts is given in Table 3. Note that these are the "mission modes" which have the least technical requirements. Figure 6 (d) presents which design configuration that the vessel has, i.e. the equipment installed. After the initial contract, the vessel always keeps its initial design configuration (design 1). However, as time passes by, reconfiguration occur more frequently. After design 1, in declining order, ship design 11, 3 and 9 are most often changed into. The details of these vessel configurations are found in Table 5. Both design 11 and 3 have ROVs installed. Design 11 has an accommodation capacity of 400 persons, in contrast to the 250-person capacity of design 3. Retrofitting to design 9 only increases the accommodation capacity to 400 persons. This could indicate that it might be beneficial to have ROV capacity from the beginning, and that the shipowner also could consider increasing the initial accommodation capacity.

4.4 Phase III: Implementation and monitoring

4.4.1 Implementation in the design stage

Following the analysis above, the shipowner should build a vessel with an accommodation capacity of 250 persons, a main crane capacity of 400 tons, in addition to installing an ROV. Beside the ROV, the selected vessel configuration is similar to the base design.

Table 6. Expected net present value (NPV) of the life cycle simulations for the considered design, 1000simulations, numbers in million USD.

Characteristic	Value
Mean value	24.5
Standard deviation	16.8
Max. value	93.3
Min. value	−23
Average number of design reconfigurations	2.13

Table 5. Excerpt of life cycle policy for system state 61–64 (of 648).

System state						Year				
	Variable									
#	Des.	Strat.	Mkt.	T. req.		8	9	10	11	12
61	2	S	ML	L	Action #	4	10	5	10	35
62	2	S	ML	M		11	10	11	11	5
63	2	S	ML	H		1	34	11	26	11
64	2	S	MH	L		28	5	5	11	5

Table 7. Excerpt of the action list.

Act. #	Strat.	Des. #	Design configuration					
			ACC [Persons]	MC [Tonne]	LWI [Tonne]	ROV [–]	PC [–]	MP [–]
1	Short	1	250	400	0	No	No	No
4	Short	2	250	400	0	No	No	Yes
5	Long	2	250	400	0	No	No	Yes
10	Short	4	250	400	0	Yes	No	Yes
11	Long	4	250	400	0	Yes	No	Yes
26	Long	9	400	400	0	No	No	No
28	Short	10	400	400	0	No	No	Yes
34	Short	12	400	400	0	Yes	No	Yes
35	Long	12	400	400	0	Yes	No	Yes

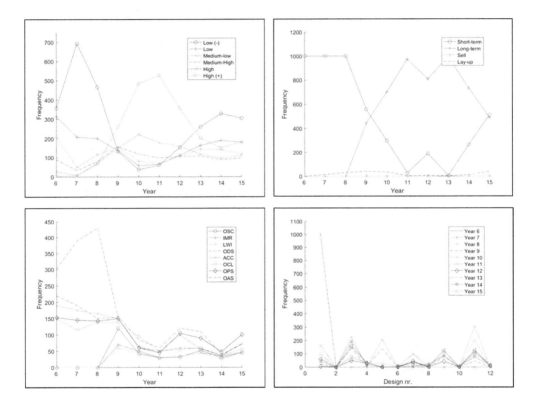

Figure 6. Top left (a) Market state, top right (b) Strategy (tactic) state., bottom left (c) mission state, bottom right (d) design state.

4.4.2 Monitoring phase/Implementation over the vessel lifetime

To illustrate the monitoring phase, one lifecycle simulation for the chosen vessel alternative was performed by following the contingency plan. The life cycle simulation is presented in Table 8. The table presents how the vessel would circulate between the implementation phase and the monitoring phase in the DSP framework. From the development of the uncertain information, the vessel experiences cyclical market, and a long period with low technical requirements, before the requirements are increased to medium in the end of the period analyzed.

To cope with the market dynamics, the policy states that the shipowner initially (i.e. after the initial five-year contract ends) should keep the initial design configuration (design 1) and operate on short-term OAS

Table 8. Example of one life cycle realization of the vessels lifecycle.

	Year	6	7	8	9	10	11	12	13	14	15
Uncertainty	Market	H	L	L(–)	L(–)	ML	H	H+	H+	H	ML
	Requirement	L	L	L	L	L	L	L	L	M	M
Decision	Design #	1	1	1	1	5	5	5	5	5	5
	Strategy	Short	Short	Short	Short	Long			Long		
	Mission	OAS	OAS	ACC	OAS	OPS	OPS	OPS	ODS	ODS	
Contribution [mill. USD]		1.6	1.9	2.2	2.9	15.3			2.1		

and ACC contracts. Then, in year 10, the shipowner would switch the design configuration to design 5, representing an increase in the accommodation capacity from 250 persons to 400 persons. In addition, the vessel will operate on long-term contracts for the remainder of the lifecycle, first on a three-year OPS contract, before ending with a three-year ODS contract. In this life cycle realization, the shipowner earned 44.8 million USD, in present values. Over 1000 lifecycle simulations, the vessel earned on average 35 million USD, with a standard deviation of 16.3 million USD. This is better than the initial base design analyzed, indicating that the analyses improved decisions made.

5 DISCUSSION

This article attempts to close the gap between overall strategic planning for the operational phase of the ship, and the decisions made by ship designers.

Opposed to most literature treating the strategic decision of ship design, this paper highlights the importance of considering the available managerial tactical and operational strategies for the vessel's life cycle utilization. We argue that a deep understanding of how managerial strategies and design functionalities interacts are important when creating value robust systems. Unfortunately, it seems like this managerial domain often are neglected in the ship design process, and consequently crucial factors of the ship owners' preferences are not reflected in the design models.

Supported by similar research, this paper argues that embedding flexibility increases the vessel performance by reducing the time and cost associated with adapting to changing circumstances, despite increase in initial building cost and carrying cost. However, despite often increasing the expected lifecycle value, increasing the building and carrying cost increases the financial risk. Therefore, if not utilized, the embedded flexibility ends up becoming an extra liability.

We have introduced the design-strategy planning framework with Markov decision processes to bridge the gap between shipping strategy in the operational lifecycle phase and ship design.

From the illustrative case, it is evident the DSP framework supported by the MDP methodology can support the conceptual ship design process. The framework supports the development of a flexible concept design and can be used to derive a contingency plan, stating under which circumstances (i.e. trigger information) the various tactical design and operational options should be executed. Using the contingency plan as input to a lifecycle simulation enables the use of metrics (such as average number of design reconfigurations) to gain valuable knowledge to support the development of strategic system. This is an area for further research. Note that although the case is based on relevant information from subject matter experts, the case presented is intended for illustrative purposes.

With the desire to form strategies for the future, we find ourselves in a tension between creating strategies that shapes the future, and simultaneously needing to realize that the future is uncertainty such that one after all end up with needing to change the strategy. In the perspective of Mintzberg & Waters (1985), the DSP framework creates an *intended* strategy (i.e. plan of action), however, as the future unfolds, only some aspects of the intended strategy is *realized*. The Design-Strategy Planning framework copes with this tension. The framework recognizes that the future is highly uncertain, and develops a contingency plan stating how the manager should alter the design and/or operational strategies to adapt to changes in context and needs. This planned adoption balances the opposite demands of the deliberate and emergent strategy, by having a formal process developing the plan, while still recognizing the range of multiple scenarios that can unfold, thereby providing the option to alter in response to changing scenarios. This stands in contrast to most traditional approaches for supporting uncertainty management that is based on a deterministic view of the future and does not pre-define how the manager should respond to changing circumstances. Generally, the core purpose of this framework is getting key stakeholders to exchange knowledge and ideas, establish lines of communication and coordinate all the activities taking place.

Incorporating flexibility in vessel design requires a more forward leaning approach in the management, by actively looking for opportunities to exploit and threats to avoid by utilizing the strategic options embedded in the contingency plan. This should be a constant process, where all levels in the organization—from top management to the vessel operator—interact to analyses future development and decide how to response to it. This increases the importance of what we call monitoring phase.

Despite having a contingency plan in place, there are many factors that hampers its use, some of which are the shipowners/managers inherent ability and willingness to utilize it. The ability is related to recognizing the emerging vulnerabilities and opportunities in context and needs, understanding of the strategic options available in the contingency plan, and the ability to select the best one, in addition to having the required resources (tangible and intangible) to do so. Strøm (2017) refers to these factors as manager *aptitude*. Note that this is not only related to the manager in charge, but also the organization as a whole. It is therefore related to organizations psychological and cultural factors that can either hamper or encourage change. In relations to its importance, this paper has not emphasized this issue. As manager aptitude directly affect to which degree the intended strategy is realized, it is important to recognize the managerial dimension in the development of the contingency plan.

With great flexibility, the MDP methodology captures the dynamic interaction between the system domain and managerial domain. Supported by MDP, the DSP framework can develop a more comprehensive contingency plan. However, recognizing the range of strategic options and the managerial dimension, increases the complexity of the already highly complex traditional ship design problem. It is questionable that increasing the dimensionality of the design problem increases the accuracy of the design solution, or making it more uncertain as there are more available decision paths for the strategist, therefore causing the need for more assumptions about the problem. As with all quantitate techniques for modelling the future, MDP relies on a trade-off between the realism of the model and its complexity.

There are several methods for solving MDPs, many of which falls under the umbrella term Approximate dynamic programming. In general, the Q-learning algorithm (QLA) applied is appropriate to use in problems with small state and action space. One of the reasons for using QLA is that it overcomes the need for the one-step transition matrix. This is important because it is impossible to probabilistically describe the outcome in environments characterized by a high degree of exogenous uncertainty.

Further challenges with the MDP methodology is that it to some degree is a black box. Thus, as the policy is based on millions of lifecycle iterations, it is hard, if not impossible, to fully understand the output of the model, besides some trivial relations. To trust the output results, it is important to trust the generic model and the input parameters. Unfortunately, it is difficult to create "realistic" models and difficult to find reliable data to base analyses on. For instance: What is the probability of winning a contract having different functionality installed? And, how much time does it take to increase the crane capacity by a certain amount? Not to forget, how do should you model future market and technological development? As pointed out by Stopford (2009) (pg. 608), the extremely small size of the market for non-cargo ships (special vessels) make it extremely difficult to analyze the market with any authority.

The ship design problem is characterized as a *wicked* (Andrews 2012) and *ill-structured* (Simon 1973; Pettersen et al. 2017). For several reasons, one could say that the attempt to develop a contingency is impropriate when dealing with problems of such characteristics, some of which are: The wickedness makes it difficult to understand the underlying drivers in the problem, therefore there is no definitive formulation of it. The ship design problem can be interpreting and defined in so many ways, and because one cannot get a complete understanding of it one might end up "paralyzed" in the analyses—unable to make pragmatic progress for real life decision making. Further, as stated in *Knagg's Law:* the more grandiose plan, the larger the chance of failure. Thus, attempting to create a comprehensive plan to manage the life cycle of offshore vessels may be like asking for trouble. In addition, as pointed out by Mason & Mitroff (1981): Generating a broad variety of alternatives in the design and operational strategies for coping with uncertainty will increases problem complexity, however it is essential for finding better quality decisions. Due to these aspects, we question whether the MDP methodology is the best tool for supporting the DSP framework, since we find it hard to apply such qualitative tool on such strategic and highly complex problems.

Scenario planning approaches of lower complexity could stand out as a better approach. In scenario planning, scenarios, rather than forecasts are developed to describe the future. These forms the basis for discussion of how to react do different plausible scenarios. This can be regarded as an approach of dividing the problem into sub-problems, before tackling them one-by-one. The solutions to each subproblem can then be combined into a cohesive whole, forming a contingency plan.

Despite the potential lack of authority in any strategic planning process focused on wicked problems, we still encourage design- and operational decision makers to perform analysis of this kind. We do this because the most value is not necessarily in the results itself, but in the insight gained by following a stepwise framework, and developing the models for supporting it. We especially highlight the important role of such frameworks as a mechanism of coordination, communication, and control in the conceptual design process. We believe improvement of these factors increase the likelihood of successful outcomes.

6 CONCLUSION

In conclusion, we state that embedding changeability in a design have the potential to increase its life cycle performance, by enhancing the ability to adapt to changing circumstances. However, embedding changeability increases the financial risk, such that if it is not realized it ends up becoming an extra liability.

We state that it is advantageous to explicitly addressing the options inherent in design and operational strategies to cope with the unforeseen changes in future operational context. Design-strategy Planning is of a framework supporting such a process.

Supported by the MDP Methodology, the DSP framework was found to be able to develop a comprehensive contingency plan. A key strength in the MDP methodology is that is can capture the dynamic interaction between the system domain and managerial domain. However, as with all models attempting to predict and future, we find it hard to rely on the analysis.

Still, we encourage decision makers to follow a step-wise procedure for analyzing the options inherent in the design and operational strategies for managing the future and support it with some sort of quantitative analyses (e.g. MDP or scenario planning). We do so because the real value is not necessarily in out output (which after all is unreliable), but in insight gained from performing the analyses.

Deeper insight into the strategic, operational, and technical aspects of the designs lifecycle is expected to enable decision makers to better handle uncertainty.

REFERENCES

Alizadeh, A.H. & Nomikos, N.K., 2009. Shipping Derivatives and Risk Management.

Andrews, D., 2011. Marine Requirements Elucidation and the nature of preliminary ship design. *Transactions of the Royal Institution of Naval Architects Part A: International Journal of Maritime Engineering*, 153(1), pp. 23–40.

Andrews, D.J., 2012. Art and science in the design of physically large and complex systems. *Proceedings of the Royal Society A: Mathematical, Physical and Engineering Sciences*, 468(November 2011), pp. 891–912.

Andrews, K.R., 1987. *The concept of corporate strategy*, Irwin. Available at: https://books.google.no/books?id=r5EoAQAAMAAJ.

Benford, H., 1967. On the rational selection of ship size. Pan American Congress of Naval Architecture and Maritime Transportation.

Brett, P.O. et al., 2006. A Methodology for Logistics-Based Ship Design. *IMDC 2006*, pp. 1–25.

Cardin, M.-A., 2014. Enabling Flexibility in Engineering Systems: A Taxonomy of Procedures and a Design Framework. *Journal of Mechanical Design*, 136.

Christiansen, M. et al., 2007. Chapter 4 Maritime Transportation. *Handbooks in Operations Research and Management Science*, 14, pp. 189–284.

Erichsen, S., 1989. *Management of Marine Design*, London, UK: Butterworths.

Erikstad, S.O., Fagerholt, K. & Solem, S., 2011. A Ship Design and Deployment Model for Non- Transport Vessels. *Ship Technology Research*, (October).

Fricke, E. & Schulz, A.P., 2005. Design for changeability (DfC): Principles to enable changes in systems throughout their entire lifecycle. *Systems Engineering*, 8(4), pp. 342–359.

Gaspar, H.M., Erikstad, S.O. & Ross, A.M., 2012. Handling temporal complexity in the design of non-transport ships using Epoch-Era Analysis. *International Journal of Maritime Engineering*, pp. 109–119.

Kana, A.A. & Harrison, B.M., 2017. A Monte Carlo approach to the ship-centric Markov decision process for analyzing decisions over converting a containership to LNG power. *Ocean Engineering*, 130(December 2016), pp.40–48. Available at: http://dx.doi.org/10.1016/j.oceaneng.2016.11.042.

Lorange, P., 2005. Shipping Company Strategies,

Lorange, P., 2009. *Shipping Strategy: Innovating for Success*, Cambridge University Press.

Mason, R.O. & Mitroff, I.I., 1981. *Challenging Strategic Planning Assumptions: Theory, Cases, and Techniques*, John Wiley & Sons Incorporated. Available at: https://books.google.no/books?id=EmaQAAAAIAAJ.

Mieghem, J.A. & Allon, G., 2015. *Operations Strategy: Principles and Practice*, Dynamic Ideas LLC.

Mintzberg, H., 1987. The Strategy Concept I: Five Ps for Strategy. *California Management Review*, 30(1), pp. 11–24.

Mintzberg, H. & Waters, J.A., 1985. Of strategies, deliberate and emergent. *Strategic Management Journal*, 6(3), pp. 257–272.

de Neufville, R. & Scholtes, S., 2011. *Flexibility in Engineering Design*, The MIT Press.

Niese, N.D., Kana, A.A. & Singer, D.J., 2015. Ship design evaluation subject to carbon emission policymaking using a Markov decision process framework. *Ocean Engineering*, 106, pp. 371–385.

Niese, N.D. & Singer, D.J., 2014. Assessing changeability under uncertain exogenous disturbance. *Research in Engineering Design*, 25(3), pp.241–258. Available at: http://link.springer.com/10.1007/s00163-014-0177-5.

Niese, N.D. & Singer, D.J., 2013. Strategic life cycle decision-making for the management of complex Systems subject to uncertain environmental policy. *Ocean Engineering*, 72(0), pp. 365–374. Available at: http://www.sciencedirect.com/science/article/pii/S0029801813003193.

Pettersen, S.S. et al., 2017. Ill-Structured Commercial Ship Design Problems: The Responsive System Comparison Method on an Offshore Vessel Case. *Journal of Ship Production and Design*, 0(0), pp. 1–10.

Porter, M.E., 1979. How Competitive Forces Shape Strategy. *Harvard business Review*, 57(2), pp.137–145. Available at: http://faculty.bcitbusiness.org/kevinw/4800/porter79.pdf.

Powell, W.B., 2011. *Approximate Dynamic Programming: Solving the Curses of Dimensionality*, Wiley. Available at: https://books.google.no/books?id=VBuZhne7pmwC.

Puterman, M.L., 2014. Markov Decision Processes: Discrete Stochastic Dynamic Programming, Wiley.

Rehn, C.F. et al., 2018. Quantification of changeability level for engineering systems. *[Working paper for journal]*.

Ross, A.M. & Rhodes, D.H., 2008. Using Natural Value-Centric Time Scales for Conceptualizing System Timelines through Epoch-Era Analysis. In *INCOSE International Symposium*. Utrecht, the Netherlands, pp. 1186–1201. Available at: http://s352047256.onlinehome.us/seari/documents/preprints/ROSS_INCOSE08.pdf.

Ross, A.M., Rhodes, D.H. & Hastings, D.E., 2008. Defining changeability: Reconciling flexibility, adaptability, scalability, modifiability, and robustness for maintaining system lifecycle value. *Systems Engineering*, 11(3), pp. 246–262. Available at: http://doi.wiley.com/10.1002/sys.20098.

Simon, H.A., 1973. The structure of ill structured problems. *Artificial Intelligence*, 4(3–4), pp.181–201.

Skinner, D.C., 2009. *Introduction to Decision Analysis: A Practitioner's Guide to Improving Decision Quality*, Probabilistic Pub. Available at: https://books.google.no/books?id=1Y9NAQAAIAAJ.

Stopford, M., 2009. *Maritime Economics*, Abingdon, UK, NY: Taylor & Francis.

Strøm, M., 2017. Design-Strategy Planning For Life Cycle Management of Engineering Systems. NTNU.

Ulstein, T. & Brett, P.O., 2012. Critical Systems Thinking in Ship Design Approaches. In *IMDC 2012*.

Ulstein, T. & Brett, P.O., 2009. Seeing what is next in design solutions: Developing the capability to build a disruptive commercial growth engine in marine design. In *IMDC 2009*. pp. 26–29.

Ulstein, T. & Brett, P.O., 2015. What is a better ship? – It all depends… In *12th International Marine Design Conference (IMDC)*. pp. 49–69.

Wang, T. & de Neufville, R., 2005. Real Options "in" Projects. In *Real Options Annual International Conference*.

Wijnolst, N. & Waals, F.A.J., 1995. *Design Innovation in Shipping: The only constant is change*, Delft University Press.

System engineering based design for safety and total cost of ownership

P. Corrignan, V. Le Diagon & N. Li
Bureau Veritas Marine and Offshore, France

S. Torben, M. de Jongh & K.E. Holmefjord
Rolls-Royce Marine, Norway

B. Rafine & R. Le Nena
Naval Group, France

A. Guegan
Sirehna, France

L. Sagaspe & X. de Bossoreille
APSYS, France

ABSTRACT: The increased complexity and automation of ships, very diverse operational profiles and critical operations they are involved in, call for an evolution of the processes and tools applied for their design. Safety, security, reliability, maintenance and life cycle cost of the systems should be addressed very early at concept design stage, and in collaboration between users, design teams, suppliers and classification society. Such a collaborative process, based on Systems Engineering, has been developed within the HOLISHIP project and is illustrated for the concept design of an Offshore Supply Vessel. It is supported by specific tools, developed or adapted to marine and offshore applications:
– a new system architecture tool is used to highlight functional chains of interest;
– model–based performance assessment tools, support CAPEX and fuel consumption assessment by exploring a large number of equipment alternatives and optimal modes of operation for power generation;
– a model based dynamic simulation tool for safety and reliability supports RAM analyses.

1 INTRODUCTION

1.1 Background

The level of complexity and automation of ships is increasing due to environmental regulations and economical concerns. This calls for an evolution of the design of complex ships equipped with many systems, operated in complex multiple operational profiles and involved in critical operations, where malfunctions would result in large impacts on human, asset or the environment. Key identified focus areas are:

- developing a closer collaboration between design teams, system suppliers and classification societies;
- ensuring and verifying safety and reliability;
- accounting for the systems maintenance and life cycle cost.

As a consequence, shipyards, naval architects and equipment manufacturers focus more and more on total cost of ownership (i.e. CAPEX plus OPEX) rather than on CAPEX only, as traditionally done, which gives freedom to propose to their customers various CAPEX & OPEX strategies. However, this should be addressed very early at concept design stage, where decisions on design solutions with large impacts on CAPEX are made.

These challenges are addressed in the HOLISHIP project, where the Model Based System Engineering approach used in aeronautical and aerospace industries are brought in, and adapted to ship design, in a collaborative environment.

1.2 Goal

Rolls-Royce Marine, Naval Group and Bureau Veritas have teamed up in the HOLISHIP project, with the additional support of BV's partner Apsys, to demonstrate how the collaborative methodology defined and the tools developed and/or adapted in this purpose can be applied to the concept design of an Offshore Support Vessel, in particular:

- a new system architecture tool meant to highlight functional chains throughout the system (System Architecture & Requirements management – SAR tool);
- model based performance assessment tools, dedicated to marine systems, allowing to assess CAPEX and fuel consumption by exploring a large number of equipment alternatives and modes of operation for power generation and distribution (Marine Power System Evaluation Tool – MPSET);
- a model based safety and reliability simulation tool, developed for performing dynamic RAM analysis in the aeronautic/aerospace industry and adapted to marine and offshore applications (SIMFIA).

The present paper gives a description of this collaborative methodology as well as an overview of the integrated tools.

2 PROCESS OVERVIEW

The global process is represented in Figure 1.

First, the overall, high-level architecture of the ship is modelled with the help of the System Architecture & Requirements management tool (the SAR tool). Machinery is modelled at a lower-level, in the SAR tool, based on sketches provided by the supplier of the machinery.

Once the architecture has been modelled, components are optimised. The Rolls-Royce MPSET tool is used to find the optimum sizing of the power system components based on the vessel's operational profile and chosen performance KPI's.

RAM analyses are conducted in parallel in order to validate the architecture with respect to RAM requirements. First of all, RAM experts outline the functional chains involved in propulsion & power. The SAR tool helps them ensuring comprehensiveness by highlighting both the functional chains and their interfaces with the rest of the ship. The components involved in these functional chains are then modelled in SIMFIA for further reliability analysis.

The MPSET tool provides system component information on equipment type, power rating and running hours, along with failure rates and time to repair to be used by the RAM analysis tool

Results from the MPSET tool and RAM analysis tool are collected and exploited in a Life Cycle Cost tool that is not presented in this paper.

3 POWER SYSTEM ARCHITECTURE

3.1 Operational profile

One of the most crucial data collections for generating the most efficient power and propulsion system is the operational profile. For this article, a very simplified operational profile is used, containing three main tasks typical for an Offshore Supply Vessel, more specific an Anchor Handling Tug Supply vessel (AHTS). These are Transit, Dynamic Positioning (DP) and Harbour operations. In Transit operation the vessel is sailing from harbour to the work-site, or opposite. In DP operation the vessel is performing its requested task. In Harbour operation the vessel is preparing for its next task.

An illustration of the operational profile is shown in Figure 2. This operational profile will vary from vessel to vessel. It will even vary from

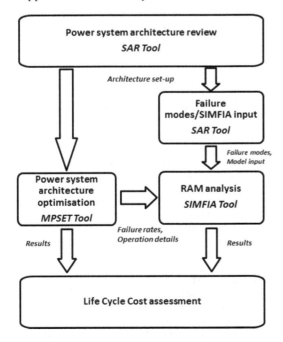

Figure 1. Power system architecture evaluation tool basic set-up and information flow.

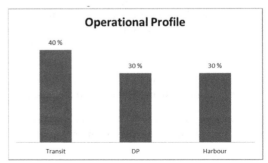

Figure 2. Simplified operational profile illustration.

owner to owner on similar vessels as different ship owners have different operating strategies. The flexibility to evaluate any type of operational profile is crucial, and the chosen profile seen in Figure 2 is purely to demonstrate the collaborative method between different tools as explained in the upcoming chapters. The operational profile will also contain details on the required power needed for hotel loading and required power and redundancy needed for propulsion thrust in the different operational tasks.

3.2 Machinery diagram

The collaborative optimisation process between the different tools allowing for both performance and RAM-analysis requires a certain amount of specifications to limit the possible total configurations. This was performed in the SAR-tool resulting in the first specification where the overall power and propulsion topology is defined for the OSV. This is defined to a Hybrid Shaft Generator-setup (HSG), which will allow for evaluations of diesel-mechanical, diesel-electrical and hybrid power modes depending on the operational strategy and dimensioning of the components.

The one line diagram seen in Figure 3 below illustrates the setup that investigated further for both efficiency dimensioning and RAM analysis. This is a more physical way of illustrating the setup than the SAR-tool and helps visualizing the topology of the Offshore Supply Vessel. None of the equipment have any specification or sizing on them, as that is part of the evaluation that will be performed in MPSET. Some of the components can also be changed in amounts, not just size, such as the auxiliary diesel generators.

The one line diagram is typically used to show ship owners the main setup for their vessel. It does not provide either information about dependencies between the ship and the system (unlike the SAR tool) or operational efficiency and modes (unlike the MPSET tool) or any reliability data (unlike the RAM tool).

3.3 Architecture diagrams

A novel Architecture Diagram Tool has been developed in the HOLISHIP project in order to model and handle the complexity that arises in ship design due to the numerous interactions between the systems on-board (Guegan et al. 2017). The tool is called "SAR" after "System Architecture & Requirements management"; the approach is inspired by established systems engineering methods and naval architecture standards. Work is ongoing to incorporate the tool in a wider design management framework (Guegan et al. 2018).

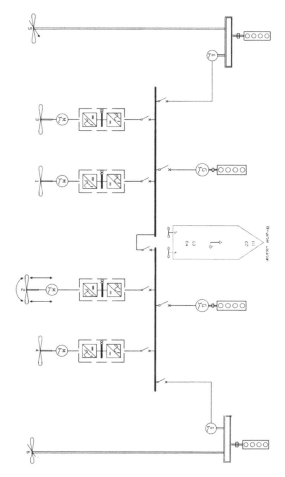

Figure 3. One line diagram.

The tool provides an exhaustive and hierarchical description of ship systems, sub-systems and components. Systems engineering models focus on interactions between these systems but they often fail to scale when the number of parts is high. The SAR tool has been designed to overcome this issue.

The OSV has been modelled with the Architecture tool. A model of the whole ship is displayed in Figure 4. Each square block represents an item at the highest level of the Product Breakdown Structure (PBS). A more detailed model of the machinery with all its components (at lower levels of the PBS) is displayed in Figure 5. Solid lines represent physical interfaces between components. Solid lines in Figure 4 have been inferred from physical interfaces in Figure 5, and generated automatically by the tool.

Architecture diagrams have the same hierarchical structure as the PBS: the model in Figure 5 is a child of the ship model in Figure 4, with the "machinery" block being its father. This induces

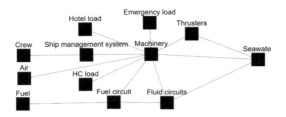

Figure 4. Architecture diagram at ship level.

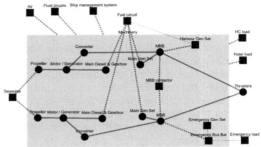

Figure 6. Components involved in "DP2, mechanical" mode.

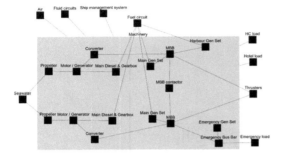

Figure 5. Architecture diagram at machinery level.

a relationship be-tween block interfaces at different levels: if components of a lower level are connected, then their "father" components are both connected as well.

Two complementary modelling processes may be used to draw architecture diagrams. In the present case, a top-down approach has been used to devise the product breakdown at top level (Figure 4). The detailed model of the machinery in Figure 5 was built up in a bottom-up approach in which the one line detailed diagram in Figure 3 served as a base to build up the architecture diagram.

3.4 Scope elicitation

The Architecture Diagram Tool provides a way to handle and visualize components and their interfaces at specific PBS levels. Although the total number of components in the ship may be high, only a limited number of components—and their interactions—are displayed at a time. This improves model browsing significantly with respect to the "flat" view in which all components are displayed at once.

An interesting feature in the Architecture Diagram Tool is that blocks can be assigned to a specific functional chain. This allows, e.g. safety analysts to high-light the components that support a specific safety function. The interfaces between this set of components and the rest of the system are highlighted as potential sources of perturbations to the safety function. Figure 6 shows the components and interfaces involved in the "DP2, mechanical" mode (round blocks, thick lines). In this mode, all diesel engines are running and the main bus breaker is open. The functional chain and its interfaces with the rest of the system (dashed lines) can be analysed further with advanced RAM techniques described in the following sections.

4 OPTIMIZATION OF POWER SYSTEMS

4.1 MPSET – Overview

The Marine Power System Evaluation Tool (MPSET) is being developed to provide an agile optimization and visualisation environment to determine the best customer solution, either directly with the customer or for a development team (Wilson et al. 2017). Today this is a manual, time-consuming and non-visualizing task needed for every new vessel and system design.

4.2 MPSET – Flexibility

The marine industry provides solutions for all types of vessels and load profiles, for instance Platform Supply Vessel (PSV), Anchor-Handling Tug Supply vessels (AHTS), Cruise vessel, Ferries etc. The electrical power system can be provided with an AC bus or DC distribution system, with energy storage in various locations and in electrical, mechanical or hybrid system configurations. The power source capacity (main engine, auxiliary diesel generator, battery, fuel cell etc.) and power flow control logics (PFC) must be selected and engineered to satisfy the various load profiles. To facilitate automatic power system operation, the algorithms are defined based on experience from marine engineers and classification requirements, which forms the fundamental to program the Power Flow Logic (PFC). It covers load-dependent auxiliary engines, PTI/PTO coordination with main engines, bus-tie breaker settings, energy storage functions and so on. Figure 7 illus-

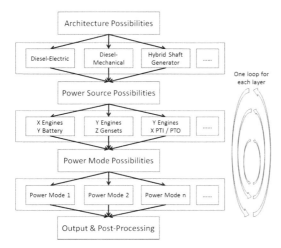

Figure 7. MPSET logic.

trates the main logic of MPSET where the Power Mode layer covers all these PFC logics, creating a matrix of operational power mode possibilities

4.3 MPSET – Model blocks

In summary, many power system configurations and control modes are possible. The first step is to define the inter-connection network with all system components separately defined as library elements that reflect the physical products. These components can be allocated to nodes in the architecture definition achieved with the SAR tool. The component function block library includes diesel engines, generators, frequency converters, batteries, inverters, motors, propellers and is built based on mathematical equations. The component library is built up in Matlab with well-defined inputs/output for each component, which can be connected to formulate different systems based on user's configuration.

4.4 MPSET – Screenshots

As illustrated in chapter 4.2, MPSET evaluates a large amount of operational and configurational scenarios. Table 1 below gives an example of one specific evaluation, where a specified architecture, power source combination and power mode have been simulated for a specific Transit scenario.

In this scenario, the vessel is transiting in a HSG-setup, where the main engines are supplying both the propulsion required for the thrust of the vessel, and electric power to vessels power system. This is typically a scenario designed for in an architecture sizing. One screenshot as shown in the simplified table below will in MPSET consist of a large list of

Table 1. Example of configuration for transit scenario.

Component	Operational parameter
Main Engines	1750 bkW & 205 g/bkWh
Aux. Engines	Off
PTI/PTO	PTO @ 275 ekW
Propellers	1475 bkW
All thrusters	Off

Figure 8. Fuel consumption overview.

operational parameters for all model blocks. This overview of thousands of operational scenarios will then be transmitted to Life Cycle Cost tool.

4.5 MPSET – Results

The main optimization logic performed in MPSET is related to fuel consumption. Figure 8 below shows an illustration of how 5 different configurations were simulated to investigate the efficiency of the setups. Configuration 2, 3 and 5 would then be further evaluated with RAM analysis.

5 RAM ANALYSIS

5.1 RAM analysis description

RAM modelling simulates the configuration, operation, failure, repair and maintenance of all equipment included in a system or a vessel. The inputs for a RAM modelling of a system include the physical components, equipment configuration, Mean Time Between Failures (MTBF) and Mean Time To Repair (MTTR), maintenance philosophy & logistics and operational profile. The outputs determine the operational performance of the system or vessel over the vessel's life cycle.

5.2 RAM analysis objectives

RAM modelling is part of the HOLISHIP project in order to evaluate the performance of alternative system architectures and compare them in order to get the best design in term of reliability and availability, i.e. exposure of risk.

In other words, RAM modelling is able to determine which system alternative is the most reliable, and consequently the one that will maximize the chance of success of vessels mission, increasing the vessels profitability and reducing penalties due to contract breach.

Then, the outputs of RAM simulation are also used as an input to evaluate the Operation Expenditures related to equipment failures and maintenance based on the association of the number of failures and repairs, quantity of maintenance activities (planned or unplanned) and maintenance utilities mobilizations performed during the life cycle to their respective costs.

With a detailed analysis of RAM results and more specifically about equipment contribution to operational losses, some design improvements can be pro-posed (e.g. equipment sparing). In this case, new alternative design configurations are created and can be introduced in the overall process starting with its evaluation by the MPSET tool (see section 4).

Furthermore, the RAM modelling is able to calculate the actual running time of each equipment that consumes combustible (Engines, Generators, etc.).

Finally, the decision of choosing the best design is done by calculating the total cost of ownership, i.e. its Life Cycle Cost (LCC). LCC of the system is determined by combining the overall OPEX and CAPEX. The design with the lowest LLC means that this design is the most cost-effective one. Figure 9 illustrates the interaction between the different tools used in HOLISHIP project in terms of inputs and outputs required for the RAM analysis and LCC evaluation.

5.3 RAM process in HOLISHIP

5.3.1 Scope definition

The scope of the systems to be modelled in the RAM tool is defined with the help of the SAR tool. The MPSET tool provides more detailed component information such as equipment specifications and operational modes based on the vessel operational profiles and power generation requirements.

The SAR tool helps to define the system configurations and functional links between equipment for each vessel operational mode, i.e. which equipment are required to be running and those that are in standby-mode (redundancy).

Figure 10 presents an example of the Power System configuration modelled in the RAM tool.

5.3.2 Criticality analysis

An essential step before starting modelling is the Criticality Analysis. This analysis consists in defining the impact of equipment failures on the system performance and further in the vessel's mission. This is done thanks to the SAR tool where it is possible to easily visualize the functional effects of equipment single point failures on the system. It is also possible to quickly highlight all the possible causes (equipment failures) that can lead to the

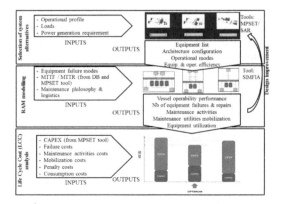

Figure 9. RAM analysis in the HOLISHIP process.

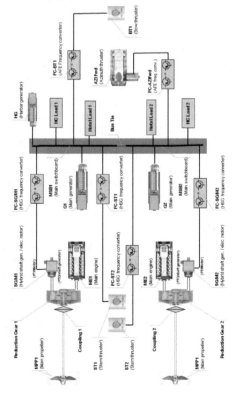

Figure 10. Scope of RAM model.

loss of vessel's mission for each operational mode and which alternative operational modes could be used in each case.

Thanks to the modelling flexibility and simplicity of the SAR tool, additional equipment that are not in the scope but that are somehow connected to the system (e.g. DP system, Auxiliary machineries, etc.) can be included in the SAR modelling. This "expanded" vision of the scope avoids missing some considerations that cannot be visualized in the basic scope—thus improving comprehensiveness, which is a pillar of RAM analyses. Figure 5 presents the scope of a power generation system and its auxiliary systems. As presented in the One Line Diagram (Figure 3), the Main Diesel Engine is not directly connected to the main switchboard (MBB). However, the SAR representation shows that the engines fuel circuit is connected to the MBB (Figure 5). We realize that a MBB failure can impact the Main Diesel Engine functionality because the engines fuel circuit cannot be power supplied. Without this "expanded" vision, this consideration could be missed and the model would probably not consider that the main switchboard could affect the main engine because they are not directly connected.

5.3.3 Reliability data collection

The main data to be collected for the RAM analysis are Mean Time To Failure (MTTF) and Mean Time To Repair (MTTR) for each equipment considered in the scope. Those data are collected from reliability data books like (OREDA 2009) or (IEEE 1984) and they are reviewed by partners based on their experience from similar projects. Those data are also documented in the MPSET Tool.

5.3.4 RAM assumptions

Assumptions should be established to define what considerations are taken in the case that will be modelled. Assumptions usually consist in operational and maintenance parameters and conditions used to simulate a realistic case. However, some aspects cannot be modelled, thus some assumptions have to be defined for simplifying the model (e.g. failure due to human errors are not considered in the model) or for adapting the model due to lack of information (e.g. in design phase, the spare parts strategy is not already defined, so it is considered in the model that spare parts are all available when needed). It is very important to use the same assumptions for RAM models of system alternatives; otherwise the results will not be comparable.

The nature of the assumptions depends on the system to be analysed and the degree of accuracy that is to be achieved. Here is a non-exhaustive list of assumptions that were made for the HOLISHIP RAM model:

- Failure considerations:
  - All simulated failures stop immediately and completely the failed equipment.
  - Failure on demand is applied to Main Engines and Diesel Generators.
- Maintenance philosophy:
  - Part of failures is repairable on-board and other part is repairable only at port.
  - Failures repairable on-board are immediately repaired by ship crew.
  - Failures repairable only at port are repaired as soon as the vessel arrives at port.
  - All repairs at port are performed in parallel.
- Operation philosophy:
  - Transit operation starts at Electrical mode (i.e. with only one Main Gener-ator operating; second Main Generator and both Main Diesel Engines in stand-by mode), DP2 mode starts at Mechanical mode (i.e. both Main Generators and both Main Diesel Engines operating at the same time).
  - When 1 propulsion is down, the vessel returns to port for repair using the other propulsion line with half speed capacity.
- Logistic delays/considerations:
  - Spare parts and maintenance crew at port and on-board are always available when needed.
  - If propulsion is completely down, a rescue ship is required to tow the vessel to port for repair.
- Preparation time for repair task:
  - No preparation has been considered.
- Sparing philosophy:
  - Spare parts at port and on-board are always available when needed.
- Preventive maintenance:
  - Preventive maintenance is not modelled (No impact of preventive maintenance because it is assumed that planned activities are performed when vessel is at port between missions).
- Operations sequence:
  - It was considered that the vessel will perform the following operation cycle:
    1. Harbour (3 days at port)
    2. Transit to off-shore (2 days trip)
    3. DP2 (3 days of operation)
    4. Transit return to port (2 days trip)

Vessel cannot leave the port (Harbour) if a piece of equipment is under repair.

- Lifecycle:
  - RAM model simulates the operation cycle during a period of 10 years.
- External factors:
  - External factors like weather and sea conditions, human factor, client delays etc. have not been considered.

5.3.5 *RAM modelling and simulation*

Once the scope and all assumptions is established, the system can be modelled in the RAM tool SIMFIA.

SIMFIA is a modelling tool based on AltaRica language. It enables to create a behavioural and/or functional mode of the system. Qualitative and quantitative dependability indicators can be computed from the model thanks to its module Simul (Dynamic simulation). This module allows modelling equipment of a system adding dynamic functional and dysfunctional behaviours: operational reconfigurations can be modelled depending on the vessel's operational modes, dysfunctions of equipment and logistic to repair failed equipment. By Monte-Carlo stochastic simulation, it generates sequences of events and calculates the resulting performance of the vessel, the system itself and each equipment.

A model is constituted of bricks, containing failure modes, and links between bricks to exchange flows of information, energy, etc. and propagate failure effects from one brick to another.

Models are well-structured thanks to the hierarchy notion. Bricks can model systems, sub-systems, items or components by nesting them into each other. The hierarchy structure and the number of levels are not limited.

Figure 11 presents the system modelled in SIMFIA, where bricks represent vessels operations and functional states, i.e., Harbour, Transits, DP2 operation, Failed at sea, Under repair at port (the 6 bricks at the top of the figure) and equipment, i.e., Main Engines, Generators, Switchboards, Propellers, Thrusters, Electrical consumers, etc. (interconnected bricks at the lower part of the picture).

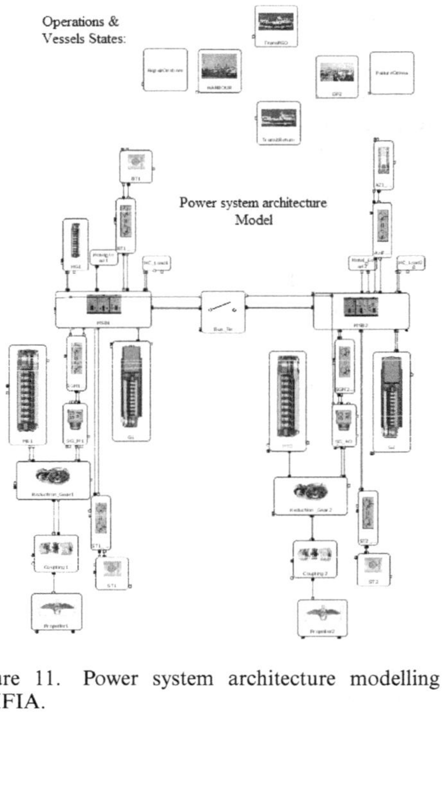

Building a model may be a long and collaborative process. At some point, it becomes necessary to check that the model represents effectively the system. To this end, SIMFIA offers functional analysis and interactive simulation. Step by step simulation enables to check that the model has the expected behaviour in typical scenarios, and is useful for presentation purposes.

Figure 12 represents the simulation of a scenario where one of the main generators fails during transit operation. It is possible to visualize that the second main generator –which was in standby mode—is now functioning.

Once the model is checked and validated, cycles of operation are simulated over the life cycle duration. The RAM tool will simulate the vessel performing the sequence of operations as per assumptions previously defined and also equipment failures and repairing based on the reliability data entered for each equipment.

As the tool performs the simulation, the impact of the sequence of failures and repairs on the vessels performance over its life cycle is progressively computed and measured.

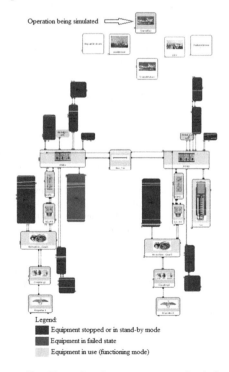

Figure 11. Power system architecture modelling in SIMFIA.

Figure 12. Example of power system simulation in SIMFIA.

By using the appropriate observers, RAM factors can be customized to produce the required indicators needed to compare the simulated system alternatives performance and to feed further the LCC analysis in the next step.

Thus, the main indicators extracted from RAM analysis for each system alternatives are:

- Number of ship missions completed with success;
- Number of ship missions completed with delay;
- Number of ship mission aborted due to system failure;
- Number of ship mission not started due to vessel unavailability;
- Number of repairs for each equipment;
- number of maintenance utility mobilizations (e.g. rescue ship, maintenance staff from shore if considered);
- Total time at port under repair;
- Occurrence of planned maintenance (if considered in the model);
- Total operational/functioning time of each equipment (overall value or per operation).

Secondary indicators can also be extracted in order to compare system alternatives performance:

- Vessel reliability, i.e. the chance of mission success when 1 cycle is started;
- Ship availability when required, i.e. chance of vessel to be ready to start a cycle;
- Ship operational availability, i.e. chance of vessel to be ready to start a cycle and succeed mission;
- Time the vessel was in operational or failed state;
- Time in stopped state and in failed state of each equipment.

It is to be noted that the indicators are calculated for the life cycle duration. It is very important to use the same life cycle duration for each simulation, otherwise the indicators cannot be used to compare the system alternatives performance.

Furthermore, SIMFIA is also able to compute sequence sets and FMECA to have a more precise analysis in order to define the equipment contribution to vessels mission failure.

All these outputs can be used to modify the system design and architecture and try to define new alternative cases with an enhanced performance.

5.4 *Life cycle cost calculation*

The final step consists in calculating the Life Cycle Cost (LCC) for each modelled alternative systems. The LCC is divided into 2 parts:

1. Capital Expenditures (CAPEX): costs related to investments on purchasing equipment and building the system. Investments dedicated to modifications and life extension of the system is not considered in this project.
2. Operating expenditure (OPEX):
   a. Costs related to activities or material used during system operation such as maintenance, logistics, fuel consumption, etc.
   b. OPEX also includes all costs related to penalties due to delays or unaccomplished missions due to vessels unavailability.

The following sections present how are respectively defined CAPEX and OPEX.

5.4.1 *CAPEX*

CAPEX is the sum all the following costs:

- Costs of purchasing of all equipment included in the scope including their specific auxiliary devices;
- Estimated cost of building the system (the more complex the system is, the more expensive the building is).

CAPEX information are extracted form MPSET Tool.

5.4.2 *OPEX*

OPEX are essentially calculated from the main indicators extracted from the RAM analysis described above.

These indicators associated with their respective costs form all elements necessary to calculate the OPEX

OPEX are the sum all the following costs:

- Cost of corrective maintenance: number of equipment repairs combined with equipment repair cost including materials, and man work.
- Cost of preventive maintenance: number of preventive maintenance combined with equipment preventive maintenance cost including materials and man work
- Cost of fuel consumption: total operational/functioning time of engines combined with fuel consumption rate and fuel cost (this calculation should be done in association with the MPSET tool).
- Cost of logistics (examples):
  o Number of rescue ship mobilisation combined with cost of hiring such ship;
  o Total time at port under repair combined with cost rate of being at port for repair;
  o Number of mobilization of maintenance staff from shore to vessel in case of repair at sea combined with costs of mobilization (e.g. helicopter hiring).
- Penalty costs (examples):
  o Number of ship missions completed with delay combined with penalties for delayed operation;
  o Number of ship missions aborted combined with penalties for not accomplishing the mission;
  o Number of ship missions not started due to vessel unavailability combined with loss of earning per unaccomplished mission;

o Penalties due to NOx emission could also be included by associating the NOx emission rate of the engines, total time of operation of each engine and costs of NOx emission.

It should be noted that all costs that are identical for all the alternative cases can be ignored in the LCC analysis. For example, the cost of engineering (CAPEX) is considered the same whatever the system architecture is. Similarly, it is assumed that the number of operators in the vessel does not change from an alternative to another. As the associated costs are the same for each alternative, they have no impact on the LCC findings.

5.5 *Final decision and conclusion*

At the end, the selected alternative is the one that presents the lowest LCC (Figure 9).

In some situations, the optimized design alternative in term of fuel consumption as determined by MPSET tool has poor performance in term of system dependability from SIMFIA simulation, and inversely.

By associating all costs of CAPEX and OPEX, the LCC study allows comparing all alternatives using unique criteria and determining the design with the best overall performance including fuel consumptions and dependability performances.

6 CONCLUSION

The present paper describes a System Engineering based collaborative ship design process aiming at addressing safety, security, reliability, maintenance and life cycle cost of the ship systems at concept design stage.

This process relies on specific tools that are being developed, or adapted, to marine and offshore applications, within the EU HOLISHIP project:

- a new system architecture tool used to highlight functional chains of interest;
- model–based performance assessment tools, supporting CAPEX and fuel consumption assessment by exploring a large number of equipment alternatives and optimal modes of operation for power generation;
- a model based dynamic simulation tool for safety and reliability supporting RAM analyses.

The application of this process to the concept design of an Anchor Handling Tug Supply vessel has started, and will be further developed within HOLISHIP.

Promising results obtained so far indicate that dependability analyses such as the ones performed in the aerospace industry are relevant to the assessment of ship design. They enable the shipyard and designer to manage risks better and account for total cost of ownership in an early stage of the design process. The system architecture tool eased cross-domain and cross-entities communication during collaborative design activities and it helped outlining the components of main interest with regard to performance and RAM analyses. The Marine Power System Evaluation Tool allows reducing the time for optimizing the power plant, based on customer requirements (operational profile), by exploring automatically a large number of design alternatives and sizing. The dynamic RAM simulation tool SIMFIA proved to be well adapted to model a marine power system and its various modes of operations. In particular functional analysis and interactive simulation capabilities make it possible to check the expected behaviour of the model for the defined scenarios.

The next steps will consist in completing this application exercise, enhancing the interconnection between the analyses and tools and consolidating the results to deliver expected KPIs (OPEX+CAPEX calculation). Making the whole process user-friendly for the naval architect will also be addressed.

ACKNOWLEDGEMENT

HOLISHIP has received funding from the European Union's Horizon 2020 research and innovation programme under grant agreement n° 689074.

REFERENCES

Guegan, A., Le Néna, R., Corrignan, P., Rafine R., 2018. Compliance matrix model based on ship owners' operational needs, *Proceedings of 7th Transport Research Arena TRA 2018*, Vienna, Austria.

Guegan, A., Rafine, B., Descombes, L., Fadiaw, H., Marty, P., Corrignan, P., 2017. A Systems Engineering approach to ship design, *Proceedings of the 8th Complex Systems Design and Management conference.* Paris, France.

IEEE 1984. Std 500-1984 Reliability data

OREDA 2009. Offshore Reliability Data Base 5th Edition, SINTEF

Wilson, G., McCarthy, J.F., Xiong J., Huan, Q., Liu, X., Tjandra, R. 2017. Use of Modelling and Simulation for Optimal Naval Ship Electrical System Design. *Proceedings of the International naval Engineering Conference.* Singapore

Optimization of ship design for life cycle operation with uncertainties

T. Plessas
Ship Design Laboratory, National Technical University of Athens (NTUA), Athens, Greece

A. Papanikolaou
Hamburgische Schiffbau Versuchsanstalt (HSVA), Hamburg, Germany
Ship Design Laboratory, National Technical University of Athens (NTUA), Athens, Greece

S. Liu
Nanyang Technical University, Singapore (former Ship Design Laboratory, NTUA)

N. Adamopoulos
Maran Tankers Management, Athens, Greece

ABSTRACT: In the present paper we demonstrate the optimization of alternative hull/engine/propeller setups for a defined operational scenario of a tanker that includes ship's operational profile in calm seas and representative weather conditions, as well as all relevant safety and efficiency regulatory and technical constraints. The developed procedure, which may be used as a Decision Support Tool (DST) for interested ship investors, includes a Life Cycle Assessment (LCA) model that take into consideration the uncertainties of the most dominant economic parameters in tanker ship operation, namely for the achieved freight rates and the fuel cost, whose accurate prediction for the whole investment period is practically impossible. The developed methodology and the associated software tool are herein based on empirical relationships allowing the fast exploration of the huge design space in the frame of a concept design optimization procedure.

1 INTRODUCTION

The most challenging part of the preliminary ship design optimization procedure is the handling of multidisciplinary complexities and uncertainties of related to ship's life cycle operation. The designer must find the optimum vessel in terms of some Key Performance Indicators (KPIs) that fulfills numerous stability, strength and safety criteria, and while taking into account various constraints in terms of main dimensions, cost, delivery time etc. Inevitably, various assumptions and simplifications are made at the early ship design stage in order to assess the different aspects of ship design which allow the designer to deal with the overall problem effectively. The traditional approach is to divide the whole ship design problem into sub-problems referring to the various disciplines of ship design and to solve it consecutively; this includes the definition of the hull, possibly as the result of a hydrodynamic optimization, the control of ship's stability, then the selection of the engine and propeller etc. In these design phases/steps, many assumptions are made regarding the economy of operation (freight rates, fuel cost etc.), technical ship characteristics (weights, stability, powering etc.), operational aspects (route, weather conditions etc.), while after completing a design step, it is often impossible to implement possible corrections backwards without interfering with the other design steps; thus, all steps need to be repeated until a convergence of results occurs (design spiral, J.H. Evans 1959 [see Taggart 1980]). This approach is not capturing the interactions of the various aspects of the design process in parallel and limits the potential of the optimization procedure and the design space. A holistic approach to ship design (Papanikolaou, 2010), which is also adopted in the EU funded project HOLISHIP, considers the multi-disciplinary optimization of ship as a mathematical multi-objective optimization problem with multiple regulatory and technical constraints, which is implemented in practice by the integration of multi-disciplinary software tools on a design software platform.

In the present paper we focus on the optimization of ship design for life cycle operation with uncertainties and demonstrate our approach with the investigation of alternative hull/engine/propeller setups for a defined operational scenario of a tanker that includes ship's operational profile in calm seas and representative weather conditions, as well as all relevant safety and efficiency regulatory and technical constraints. The procedure includes a life cycle assessment (LCA) model that take into consideration the uncertainties of the most dominant economic parameters, namely for the freight rates and the fuel prices, whose accurate prediction for the whole investment period is practically impossible. The developed methodology and the associated software tool are herein based on empirical relationships allowing the fast exploration of the huge design space in the frame of a concept design optimization procedure.

2 THE HULL PARAMETERS

In the examined scenario, a life-cycle assessment is performed for an investment that refers to the purchase and operation of a VLCC tanker of 300,000 tons DWT for twenty years life cycle. For the description of the hull, we employ 9 design variables: Length (L), Breadth (B), Draught (T), Depth (D), Block coefficient (CB), Midship Coefficient (CM), Length of Entrance (LE) and Length of Run (LR) of the Design Waterline (DWL), and Service Speed.

For the estimation of the fuel consumption of the vessel while operating in different wave and wind conditions, the total resistance of the vessel is divided into three basic components:

1. Calm Water Resistance,
2. Added wave resistance due to wind waves,
3. Wind resistance

2.1 *Calm water resistance*

The calm water resistance of the vessel is estimated by use of Holtrop's semi-empirical method (Holtrop, 1984) and is subdivided into the following components:

$$R_{total} = R_F \cdot (1+k_1) + R_{APP} + R_W + R_B + R_{TR} + R_A \quad (1)$$

where:

R_F is the frictional resistance according to the ITTC-1957 friction formula

$(1+k_1)$ is the form factor describing the viscous resistance of the hull form in relation to R_F

R_{APP} is the resistance of the appendages

R_W is the wave-making and wave-breaking resistance

R_B is the additional pressure resistance of the bulbous bow near the water surface

R_{TR} is the additional pressure resistance of the immersed transom stern

R_A is the model-ship correlation resistance

This type of calculations can be replaced by more accurate methods (e.g. by CFD and use of the Response Surfaces Technique, see Harriers et al, 2017); however, it is considered satisfactory for the demonstration of the proposed concept to use the simplified empirical method and a correction factor to adjust the total calm water resistance resulting from the semi-empirical method to available measurements of a baseline ship's sea trial report.

2.2 *Added wave resistance due to waves*

The added resistance due to irregular waves X_d can be determined according to spectral theory (Liu-Papanikolaou, 2016):

$$X_d = 2 \cdot \int_0^\infty \int_0^{2\pi} \frac{X_d(U,\mu',\omega')}{A^2} \cdot S_{\zeta\zeta}(\omega') \cdot D(\mu-\mu') d\omega' d\mu' \quad (2)$$

where:

$\frac{X_d}{A^2}(N/m^2)$ is the quadratic transfer function of the added resistance in regular waves and A is the wave amplitude,

$S_{\zeta\zeta}$ is the seaway energy density spectrum,

D is the spreading function of wave energy with respect to mean wave direction; for long-crested seaways it can be assumed that D = 1; in this case, the resulting added resistance X_d is multiplied with the reduction factor 0.9 to take into account the short-crestedness of the actual seaway,

ω' (rad/s) is the frequency of wave component,

μ' (rad) is the direction of the ship's forward speed.

$U(m/s)$ is the specified ship's forward speed.

The quadratic transfer functions of added resistance in regular head to beam waves $X'_d = \frac{X_d}{A^2}$ can be calculated as the sum of two parts, namely

$$X'_d = X'_{dM} + X'_{dR} \quad (3)$$

where,

X'_{dM} is the part of added resistance due to the motion (radiation) effect,

X'_{dR} is the part of added resistance due to the reflection (diffraction) effect of an individual wave component with frequency ω' and heading angle μ'.

The above components of the added resistnace are calculated using a semi-empirical formula proposed by Liu-Papanikolaou (2016a, 2016b) and

Figure 1. Definition of length of entrance and run of DWL (Liu-Papanikolaou, 2016, IMO-MEPC, 2016).

which was also submitted for consideration to IMO-MEPC (2016).

2.3 Wind resistance

The wind force and moment acting on the ship hull are estimated as follows:

$$R_{wind} = 0.5 \cdot \rho \cdot C_x \cdot A_T \cdot V_{res}^2 \quad (4)$$

where

$$V_{res} = \sqrt{(V_{wx} + U)^2 + V_{wy}^2} \quad (5)$$

is the resultant airflow velocity felt by the ship
V_{wx} is the x-component of the velocity of the wind compared to the ships direction and the
C_x coefficient is a function of ship's profile (here of a tanker) and of relative wind angle (Blendermann, 1993).

3 THE PROPELLER

The propeller is assumed to be of type Wageningen-B Series (Oosterfeld et al, 1975). There are four design variables that define the propeller and its characteristics: Diameter (DP), Pitch Ratio (P/D), Blade Area Ratio (AE/A0) and Number of Blades (Z). With the K_T-K_Q curves available for propellers of the Wageningen-B series type and the assumed diameter, the speed of advance and the required thrust known, the rotational speed of the propeller derives by finding the intersection of the $k_T - J$ curve with the parabola:

$$k_T = C \cdot J^2 \quad (6)$$

where

$$C = \frac{K_T}{J^2} = \frac{T/(\rho \cdot n^2 \cdot D^4)}{(V/(n \cdot D))^2} = \frac{T}{\rho \cdot V^2 \cdot D^2} \quad (7)$$

4 THE ENGINE

For the modeling of the engine, the model ensures that the operating point of the vessel lies within the engine limits regarding the power and the rotations per minute that are allowed by the manufacturer (MAN, 2017). The basic characteristics of various two stroke engines relevant to the study ship are imported into a database list, from which the optimization algorithm selects the appropriate engine that is closest to the required brake horsepower. In

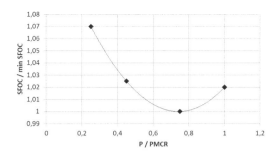

Figure 2. SFOC ratio vs. power ratio.

addition, the efficiency of the engine is determined by importing a simplified Specific Fuel Oil Consumption (SFOC) curve, from which the fuel consumption is estimated based on the ratio P/P_{MCR} (Figure 2); note that the minimum SFOC of the database MAN two stroke engines at 75% of MCR varies for the different engine types from 160.5 gr/kWh (MAN G90ME-C10.5 and G95ME-C9.6) to 165 gr/kWh (S70ME-C8.5).

5 THE ECONOMY MODEL

In order to evaluate a prospective new ship investment, an evaluation criterion is required that will allow the designer to compare the different designs and to select the optimum (Buxton, 1976). The selected economic criterion for this study is the Net Present Value (NPV). The cash flows of the overall investment are graphically depicted in Figure 3.

The NPV is calculated by the following formula:

$$\begin{aligned}NPV = &-30\% \: OF \: SHIP' SCOST \cdot \frac{1}{(1+i)^{Nc}} + \ldots \\ &(FREIGHT \: REVENUE - ANNUAL \: COSTS - \\ &LOAN \: PAYMENT) \cdot PWF(Nt,r)\frac{1}{(1+i)^{Nc}} + \ldots \\ &(FREIGHT \: REVENUE - ANNUAL \: COSTS) \cdot \\ &PWF(N',r) \cdot \frac{1}{(1+i)^{Nc+Nt}} + \ldots \\ &(SCRAP \: OR \: RESALE \: VALUE) \cdot \frac{1}{(1+i)^{N}}\end{aligned} \quad (8)$$

where:

$$r = \frac{1+i}{1+f} - 1 \quad (9)$$

$$PWF(N,r) = \frac{(1+r)^N - 1}{r \cdot (1+r)^N} \quad (10)$$

$$N' = N - Nt - Nc \quad (11)$$

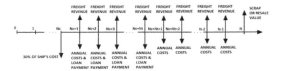

Figure 3. Investment cash-flows.

i: interest rate (set to 2% in the examined scenario),
f: inflation rate (set to 3% in the examined scenario),
Nc: construction time (1 year in the examined scenario),
Nt: total number of annual loan payments (loan repayment period is assumed to be 10 years),
N: investment period (the investment period is assumed to be 20 years).

The building cost of a VLCC, when this study was conducted (2017), was approximately 81 million dollars (Clarkson, 2017). The average daily operating cost of a VLCC vessel (excluding the cost of the fuel) was 9,950 $ (Moore-Stephens, 2017).

6 DEALING WITH UNCERTAINTIES OF FREIGHT RATE AND FUEL COST

In the present study the most crucial uncertainty parameters of the examined investment are the achieved freight rate on the income side and the fuel price on the expenses side. Instead of dealing with these parameters with a 'single value', namely with a deterministic approach, a uniform stochastic distribution will be introduced for a range of each of the above parameters.

6.1 Handling uncertainties with the Monte Carlo method

Monte Carlo methods are basically computational algorithms that use random sampling to obtain numerical results. With Monte Carlo methods, problems that might be deterministic in principle can be examined by using randomness and observing the statistical characteristics of the various different cases. In the present study a Monte Carlo method is used, where 1,500 random combinations of freight rates and fuel prices are simulated and the probability of having a negative NPV is estimated (Figure 4). This represents basically the 'failure probability' of an investment, namely the risk to invest in a non-profitable ship, which can be used in our optimization problem as an objective function to minimize the risk of the investment. In Figure 4, the investment 'failure probability' is visualized for a sample design. A clear boundary between the blue points (positive NPV) and the red points (negative NPV) representing the Required Freight Rate (RFR) and corresponding fuel price generating a zero profit/or no loss investment. By minimizing the failure probability, the number of designs with positive NPV increases and those with negative NPV decrease, thus minimizing the risk of having a non-profitable investment. Clearly, the lower the Required Freight Rate and Fuel Price, the more profitable will be an investment.

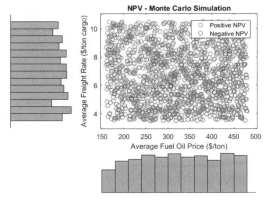

Figure 4. NPV calculation for random freight rates and fuel prices combinations.

7 MULTI-OBJECTIVE SHIP DESIGN OPTIMIZATION FOR LIFE CYCLE OPERATION UNDER UNCERTAINTY

7.1 Setting up the optimization problem

Assuming that we are splitting the investment period in annual intervals, the ideal vessel would be the one with:

1. The maximum cargo transferred annually (e.g. maximum freight revenue)
2. The minimum annual fuel consumption (e.g. this can be translated in minimum operating cost)
3. Minimum investment risk regardless the freight rate and fuel price fluctuations throughout the investment period (e.g. ensuring that in any possible scenario regarding the values of the freight rate and the fuel price, the ship will be profitable, thus minimizing the risk of the investment).

Therefore, a multi-objective life cycle optimization problem has been formed with three objective functions and fourteen (14) design variables that are summarized in Table 1 noting that the lower and upper bounds were defined based on existing vessels of similar size.

Table 1. Design variables.

Design variables	Units	Lower bound	Upper bound
Length (L)	m	314	324
Breadth (B)	m	58	60
Draught (T)	m	18	22.5
Depth (D)	m	28	30.5
Block Coefficient (C_B)	–	0.79	0.89
Midship Coefficient (C_M)	–	0.88	0.93
Length of Entrance (L_E)	% of L/2	20	50
Length of Run (L_R)	% of L/2	20	50
Service Speed	kn	12	14
Propeller Diameter (D_P)	% of T	40	50
Pitch Ratio (P/D)	–	0.6	1.4
Blade Area Ratio (A_E/A_0)	–	0.4	1.05
Number of Blades (Z)	–	3	7
Engine Selection*	–	1	27

*From a list of MAN two stroke engines, MCR 20,580 kW to 48,090 kW, MAN 2017.

In the assumed operating scenario, the vessel operates in a route with approximately one port call after every 5,000 nm. Regarding the weather conditions, it is assumed that the ship is traveling in 20% of its time in calm water, 60% in a representative weather scenario (IMO-MEPC, 2012) and 20% of its time in adverse sea/weather conditions (e.g. windforce 8 Beaufort and associated wave conditions). Note that the speed of the vessel in adverse conditions is assumed reduced to 9 knots. It is also taken into account that the ship is operating approximately half of its total voyage time in ballast condition.

7.2 Constraints

In the optimization study the following constraints were used:

– Geometry Related Constraints:
1. CB/CWL < 0.94: since we have design variables that determine both CB and CWL, we need to ensure that the produced geometries are meaningful.
2. CP < 0.9: again, since we define both CM and CB, we need to ensure that the resulting hull is meaningful and according to design practice.
3. The geometrical displacement ΔG = L*B*T*CB*ρ, must be equal to the sum of weights ΔW = LS + DWT, where LS is the light ship of the vessel calculated from empirical methods and DWT is calculated from regression analysis from existing ships based on ΔG (Papanikolaou, 2014). Thus, (ΔG-ΔW)/ΔG < 4%.

– Safety Related Constraints:
4. GM > 0.3 m. A minimum value of GM is required in order to ensure satisfactory stability and safety for the ship (Papanikolaou, 2014).
5. Freeboard Constraint (IMO-ICLL, 1988).
– Engine Limit Related Constraints:
6. 0.8 < PB/MCR < 1: we need to ensure that the operating point of the vessel's engine, when travelling in representative sea weather conditions in full load condition and at the service speed, is within engine limits and close to the engine's MCR.
– Propeller Related Constraints:
7. 0.8 < nprop/nMCR < 1: we need to ensure that we have the appropriate propeller for the selected engine. This constraint ensures that the propeller is designed to operate close to the MCR of the engine.
8. Cavitation check (Burrill et al, 1978).
– EEDI compliance (IMO-MEPC, 2011 and 2014)
9. $EEDI_{attained} < EEDI_{required}$

7.3 Optimization methodology

The optimization is performed in three stages. The first stage is the generation of 50,000 random designs that cover the whole design space uniformly. This large number of initial designs ensures that the optimization algorithm will not be trapped in a local minimum. The second stage encompasses the selection of a base of optimum designs which are set as the initial generation for the genetic algorithm that is used for the optimization procedure. The third and final stage is the generation of the Pareto front of favored designs (Deb, 2001). By starting with a strong initial population (the

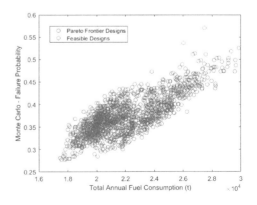

Figure 5. Annual fuel consumption vs. investment failure probability.
Blue circles: 1,375 feasible designs, but not EEDI compliant.
Red circles: 9 Pareto front designs, EEDI Phase I compliant.

best designs out of 50,000 randomly generated designs), the optimization algorithm converges very fast to the Pareto Frontier, namely by running the developed optimization code in MATLAB (Mathworks-MATLAB, 2017), the results for each design are obtained in less than 1 second on a conventional laptop.

7.4 Optimization results

The randomly generated 1,375 feasible designs and the 9 resulting Pareto Frontier designs of the conducted Multi-Objective Optimization Problem are presented in Figures 5, 6 and 7, while some finally selected, favorable designs are shown in Table 2.

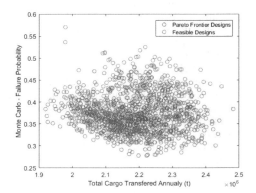

Figure 6. Annual cargo transferred vs investment failure probability.
Blue circles: 1,375 feasible designs, but not EEDI compliant.
Red circles: 9 Pareto front designs, EEDI Phase I compliant.

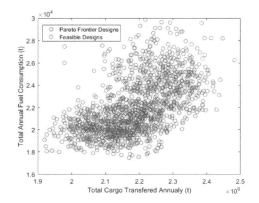

Figure 7. Annual cargo transferred vs annual fuel consumption.
Blue circles: 1,375 feasible designs, but not EEDI compliant.
Red circles: 9 Pareto front designs, EEDI Phase I compliant.

Table 2. Selected optimum designs vs. reference vessel.

Design variables	Reference vessel	D1 (Max. Cargo)	D2 (Min. Fuel)	D3 (Min. Risk)
L (m)	320.00	321.25	319.38	320.64
B (m)	60.00	59.63	58.45	58.91
T (m)	20.60	22.42	22.42	22.45
D (m)	30.50	29.62	29.72	29.69
C_B	0.816	0.808	0.794	0.803
C_M	0.890	0.906	0.906	0.906
L_E (m)	59.49	78.66	78.76	78.74
L_R (m)	68.82	58.81	68.03	65.21
$V_{SERVICE}$ (kn)	13	12.23	12.15	12.16
D_P (m)	10.3	10.84	10.78	10.83
P/D	0.851	0.766	0.752	0.755
A_E/A_0	0.657	0.821	0.742	0.740
Z	5	5	5	5
MCR (kW)	31,641	24,010	24,010	24,010
RPM	78	91	91	91
DWT (t)	290,927	313,043	299,850	307,100
V_{DESIGN} (kn)	15.6	14.5	14.4	14.4
Annual Cargo (mil. t)	2.25	2.27	2.17	2.22
Annual Fuel Cons. (t)	21,602	18,997	17,998	18,315
Failure Prob. (%)	32.5	28.4	28.1	27.7
NPV* (million $)	55.79	68.75	66.51	69.32
NPVI* = NPV / Purchase Cost (%)	68.72	81.3	81.0	83.1
RFR* ($/ton cargo)	6.50	6.21	6.19	6.16

*Fuel Oil Price = 320 $/ton, Freight Rate = 7 $/ton cargo.

8 THE DISCUSSION OF RESULTS AND CONCLUSION

By taking a closer look at the obtained optimization results (Table 2), it can be observed that the differences between the identified optimum designs are small. This is due to the large number of considered variables and implemented constraints, noting that especially the EEDI constraint proves to be a very stringent design constraint for the sample tanker ship. In Figures 5, 6 and 7 it is shown that only few designs (red points) could be identified fulfilling the current EEDI requirement for Phase 1; note that Phase 1 ends in year 2020, while an additional reduction of EEDI by 10% is required for Phase 2 ending in year 2025 and another 10% reduction for new ships built thereafter (Phase 3).

This outcome regarding the regulatory efficiency requirement of IMO-MEPC could have been expected, as similar results were obtained in the EU funded project SHOPERA (2013–2016) and are subject of long-lasting controversial deliberations at IMO-MEPC in recent years. It should also be mentioned that the reference vessel, fails to comply with the EEDI Phase 1 requirements, thus it could not have been built by today's regulations in force. The closeness of the data of the resulting optimum design ships with the reference ship also suggests that the implemented algorithms are working properly and this may serve as a simple, first round validation procedure for the developed software tool, though it remains to be also tested in other sample cases.

Despite the fact that the identified optimum designs are very similar in their general characteristics, it is interesting to observe the differences between the various designs (even if these differences are partly marginal), since they reveal the trend direction the algorithm has followed in order to achieve the set objectives:

1. The vessel with the maximum annual cargo transferred (D1) is the largest and fastest possible vessel that fulfills all the constraints.
2. It can be observed that the vessel with the minimum annual fuel consumption (D2) has the largest L/B ratio, smallest CB and slower speed compared to the others. Of course, a local hull form optimization (not conducted in this study) could further reduce the fuel consumption and lower EEDI (and CO_2 emissions).
3. The vessel associated with the minimum investment risk (D3) is basically an intermediate solution between the other two designs.

The presented methodology demonstrated some important aspects of the holistic approach in the life cycle ship design optimization process. While the vessel and its most crucial sub-systems need to be treated as a whole, it is at the same time important that at least the most crucial uncertainties for its life cycle assessment, namely those referring to the freight rates and fuel prices are treated as random variables with ample margins. This may be useful in the decision making of potential investors in a newbuilding, as it supports the identification of technical design solutions with lowest possible risk of investment.

ACKNOWLEDGEMENTS

The support of this research by the European Commission research project HOLISHIP under the European Union's Horizon 2020 research and innovation program under grant agreement n° 689074 is acknowledged. The European Commission and the authors shall not in any way be liable or responsible for the use of any knowledge, information or data presented, or of the consequences thereof.

REFERENCES

Blendermann, W., Parameter identification of wind loads on ships. Journal of Wind Engineering and Industrial Aerodynamics, 51, 339–351, 1993.

Burrill, L.C., Emerson, A. Propeller cavitation: further tests on 16 in. propeller models in the King's College Cavitation Tunnel. Trans. NECIES, 195, 1978.

Buxton, I.L., "Engineering economics and ship design", British Ship Research association, 1976.

Clarkson's Shipping Intelligence Network website, "sin.clarksons.net", Clarkson Research Services Limited, 2017.

Deb, K., "Multi-Objective Optimization Using Evolutionary Algorithms", Chichester, England, John Wiley & Sons, 2001.

Harries, S., Cau, C., Marzi, J., Kraus, A., Papanikolaou, A. Zaraphonitis, G: Software Platform for the Holistic Design and Optimisation of Ships, Proc. Annual Meeting STG, Potsdam, November 2017.

HOLISHIP (2016–2020), HORIZON 2020 – EU funded project, Grant Agreement n° 689074, www.holiship.eu.

Holtrop, J., "A statistical re-analysis of resistance and propulsion data", International Shipbuilding Progress 31 (363), 272–276, Delft University Press, 1984.

IMO-ICLL, "Resolution Msc. 143(77) Adoption Of The Amendments To The Protocol Of 1988 Relating To The International Convention On Load Lines", IMO, 1966.

IMO-MEPC.1/CIRC.796, "Interim guidelines for the calculation of the coefficient fw for decrease in ship speed in a representative sea condition for trial use", 12 October 2012.

IMO-MEPC 62/24/Add.1, "Amendments To The Annex Of The Protocol Of 1997 To Amend The International Convention For The Prevention Of Pollution From Ships, 1973, As Modified By The Protocol Of 1978 Relating Thereto", 15 July 2011.

IMO-MEPC 66/21/Add.1 "2014 Guidelines On The Method Of Calculation Of The Attained Energy Efficiency Design Index (EEDI) For New Ships", 4 April 2014.

IMO-MEPC 70/INF.30 "Air Pollution and Energy Efficiency: Supplementary information on the draft revised Guidelines for determining minimum propulsion power to maintain the manoeuvrability of ships in adverse conditions", 19 August 2016.

Liu S. & Papanikolaou A., "Fast approach to the estimation of the added resistance in head waves", Ocean Engineering, Vol. 112, pp. 211–225, 2016a.

Liu, S. & Papanikolaou, A., "Prediction of the Added Resistance of Ships in Oblique Seas", Proceedings International Offshore and Polar Engineering Conference, Rhodes-Greece, 2016b.

MAN Technical Papers/Brochures. "Marine Engine IMO Tier II and Tier III Programme 2nd edition 2017", MAN Diesel & Turbo, Denmark, 2017.

Mathworks. MATLAB, www.mathworks.com, 2017.

Moore Stephens, OpCost website, "www.opcostonline.com", 2017.

Oosterveld, M.W.C. & Oossanen, P. van, "Further Computer Analysed Data of the Wagenigen B-Screw Series", ISP, 22, July 1975.

Papanikolaou, A., "Holistic ship design optimization", Computer-Aided Design 42(11):1028–1044, 2010.

Papanikolaou, A., Ship Design- Methodologies of Preliminary Design, 628p, 575 illus., SPRINGER Publishers, e-book ISBN 978-94-017-8751-2, Hardcover ISBN 978-94-017-8750-5, September 2014.

SHOPERA (2013–2016) Energy Efficient Safe SHip OPERAtion, FP7-SST-2013-RTD-1, www.shopera.org.

Taggart, R., (ed) (1980) "Ship design and construction", SNAME Publications, New York.

Handling the path from concept to preliminary ship design

G. Trincas, F. Mauro, L. Braidotti & V. Bucci
Department of Engineering and Architecture, Section of Naval Architecture, University of Trieste, Italy

ABSTRACT: To manage the overall design process of complex systems like ships, particular attention has to be deserved to the connections and overlapping between the design stages. Therefore, the paper addresses the complex aspect of quickly moving from concept to preliminary design, so yielding an initial design in a decreasing time span. Concept design is here treated as a MADM synthesis process, where a huge number of feasible designs are randomly generated by an adaptive Monte Carlo sampling and ships' properties are assigned by means of metamodeling techniques. Non-dominated solutions are then filtered as a Pareto set identifying a short list of preferred solutions up to the "best possible" design characterized by top-level specifications (geometry, ship performance, capex, opex, etc.). This procedure has its core in the trade-off between technical and economic attributes also embracing uncertainty in a ship's lifetime perspective. A set of ship parameters are oriented to preliminary hull geometry and general arrangement definition. Hierarchical dependencies are used to match the concept hull form with modular internal spaces. The process is here applied to a case study, consisting in the initial design of a Compressed Natural Gas (CNG) ship as member of a fleet optimised for gas shipping from the Zohr field (Egypt) to the Adriatic Sea.

1 INTRODUCTION

Successful ship design is mainly a matter of fast and efficient decision making in a conflicting environment. A very competitive market in shipbuilding and shipping industry compels to improve design methods. This is especially true for concept ship design, which is the most important stage of the design process because it exerts the highest influence on the overall lifetime cost of the ship compared to the subsequent stages.

Worldwide experience indicates that successful innovative ships presuppose relevant innovation in design strategy. Designing ships is a complex system problem requiring rational integration of many disciplines of both technical and economic nature. Following the algorithm information theory, complexity grows along the design process since description of the ship contains more and more information moving from the concept to detailed ship design. Then, the design process can be structured through a hierarchical approach where the concept stage stays at the top of a lexicographic order. Hence, to define the ship as a system, manageable in spite of its complexity, the concept design stage has to require few information which nonetheless have to be highly relevant.

The main goal of this paper is to develop a design environment conceived to fully integrate concept ship design with preliminary design thus reducing the time needed to produce a pre-contract solution. To this end, the concept design stage is subdivided in two phases. The former assesses the main geometrical characteristics, principal performance and engineering economics of feasible designs by application of a multicriterial decision-making approach in order to select the "best possible" solution among a set of non-dominated candidates. The latter builds the hull form and preliminary general arrangement of the selected ship by defining the sectional area curve which complies with geometrical data and internal layout as yielded in the first phase by assembling modules of *primitive cargo units*. This way, accuracy of ship's cargo capacity is enhanced and a starting point for classical preliminary design is quickly provided.

The present work introduces an enhanced concept design process focusing the attention on optimal fleet composition, best ship selection, internal cargo space modelling, and concept hull form determination.

The overall process is illustrated with the aid of a specific example, considering the case of a CNG ship, starting from fleet optimisation up to preliminary hull form determination. The economic/financial and logistic scenario is referred to the compressed gas shipping from the Zohr field (Egypt) to the Adriatic Sea.

2 OVERVIEW OF MCDM METHODS

Ship is an expensive product and is very sensitive to uncertain operating and environmental conditions as well as to continuously changing market requirements, so compelling to assess the robustness of numerous viable alternative solutions since concept design stage.

Sorrowfully, ship design is not yet structured as an open process where decision makers can easily and quickly take under control the two primary factors affecting technical performance and economic viability of the final product, that is, cost and risk which may be high especially when departing from proven designs. Major weak points and ineffectiveness of the iterative single-point (Evans 1959, Nowacki et al. 1970) and innovating decision-based design approaches (Ray & Sha 1995, Sen 2001, Frye 2010) are mainly related to the poor integration of different subsystems and lack of the required interaction between technical decisions and economic evaluation.

Traditional descriptive design models perform initial stages by systematic variation of principal dimensions of the ship together with sectional area curve (SAC) of the underwater hull form or by conversion of an earlier ship design (Mandel & Leopold 1966, Gilfillan 1969) by introducing optimization procedures based on linear mathematical modelling without checking compliance of the design solution with constraints. A more systematic and rational approach was introduced when a set of multicriterial decision support methodologies, e.g. weighted criteria methods, random-search techniques, gradient approximations, etc., was introduced (Nowacki 2003).

As to concept ship design, many efforts have been devoted to apply optimisation techniques (Ray et al. 1995, Campana et al. 2007, Diez & Peri 2010), using multiobjective decision-making (MODM) schemes. But such an approach to the concept design is merely capable to quantify and aggregate a variety of factors influencing the design outcomes into a single objective function, often a single economic criterion of choice. It is a diffuse opinion that the MODM approach represents an ineffective simplification of the ship design process also because it still relies on a sequential descriptive paradigm even though the *sequential synthesis in ship design* (Andrews & Dicks 1997) helps in simplifying the process.

Poor success of this direction of research has been recognized by the authors themselves and attributed to non-acceptance of the MODM procedures by practical designers together with difficulty of generating objective functions fully describing technical properties and economic outcomes of ships simultaneously.

In other terms, the MODM approach cannot address both the inherent complexity of the ship and the complexity of the design process at the same time. Moreover, this methodology does not satisfy basic principles of design (Suh 1990) which require that for a good design the qualities are as much as possible uncoupled with respect to the parameters, and that the information content is simple and minimal.

Nevertheless, since three decades it is evident that MODM techniques, apart from the computational difficulties, are too inelastic to account for complex products like ships as a whole (Grubišić et al. 1988). Since then a bulk of scientific papers and applications have been yielded to describe the multiattribute decision-making (MADM) approach. An extensive review of the associated techniques is given by Trincas (2001).

The MADM procedure treats ship design as a whole requiring only simple evaluation and selection procedure, is advocated as the best method to tackle the concept ship design. It provides generation of a large number of concept solutions and implementation of an effective evaluation process to select the "best possible" concept design. In this respect, this term is preferred instead of *optimum* as no designer and/or decision maker knows how many other designs can be generated which could result better than the selected one in a different operating scenario.

Since now, the main drawbacks of the MADM approach have been the poor control of the accuracy of the hull shape and slow transfer of concept design outcomes to preliminary design. In the following a possible method to speed up the data transfer is described for the particular case of a CNG ship, following the scheme reported in Figure 1, where the optimal fleet composition, as a part of the *external task* selection (Pashin 1983), is at the heart of the concept design.

The MADM concept design was here used to build several databases of optimal CNG ships, each one depending on the type (full steel, steel liner wrapped with fibres, full composite) of Pressure Vessels (PV) and characterized by the required gas capacity. Then, the best fleet composition is determined, which concurrently provides the minimum shipping tariff, the number of sister ships, their capacity, and service speed, while respecting the logistic constraints.

After accomplishing the *external task*, the cargo space of the concept ship is modelled in detail. The number and length of PVs complying with the internal spaces is perfected by defining the Sectional Area Curve (SAC) at design draught as well as on tank top. These SACs allow the definition of a concept lines plan inferred from the geometry of ships previously stored in a database. The concept

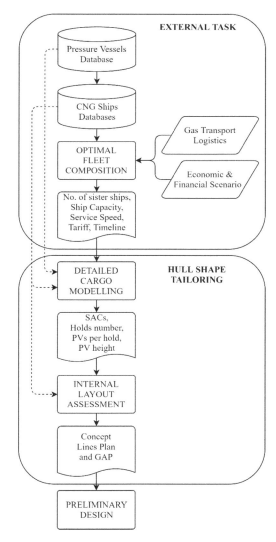

Figure 1. Conceptual design procedure.

General Arrangement Plan (GAP) is then established combining the concept lines plan with the internal cargo layout. The so determined lines plan and GAP lead to a *concept geometry* automatically respecting concept design attributes being constraint to the preliminary design stage.

3 MADM CONCEPT DESIGN SCHEME

At least 70% of the ship lifetime cost is decided in the concept design stage (Pugh 1991), thus making it the key element in the design process. In fact, possibilities for influencing economic success of a ship are very high during concept design and decrease in the following stages. This stage is particularly challenging since it encompasses a complex and uncertain decision-making environment. The selection of the "best possible" design results from a set of performance properties (attributes) and design constraints also reflecting norms and desiderata of the different design stakeholders. This set of controlling factors is not expected to vary substantially upon the subsequent design stages.

3.1 *Design methodology*

In the frame of the MADM approach, each design can be represented by a point in the design space spanned by the design variables. It can also be considered as a point in the attribute space spanned by the design attributes. The process of design generation is then separated from the process of design selection. Constraints in the design space bound the subspace of feasible designs. If sufficient density of non-dominated points is generated, a discrete inversion of design space on attribute space may be obtained. Otherwise, further solutions are generated in mini-cubes around the best non-dominated designs through fractional factorial design (FFD).

Information on the synthesis shell, on graphic representation of design and attribute space, as well as on organization of data structure was extensively detailed by Trincas et al. (1994). A mathematical design model (MDM) is built for efficient design description and assessment of design properties (attributes). The MDM is inserted into the generation phase of the shell after definition of the design space. It is composed of different modules which are associated to functional decomposition of the ship, where detailed requirements and functions are grouped into modules. It deals with problems of hull form, scantlings, light ship weight, stability, resistance and powering, deck area and subdivision arrangement, seakeeping, acquisition cost estimation, ship lifetime economy, etc. The MDM is driven by an adaptive Monte Carlo sampling, or alternatively by a genetic algorithm (GA) in the design space, which randomly generates a large set of designs around a baseline point. Feasible designs that overcome all constraints are filtered for dominance.

In the attribute space the following main tasks are performed:

- determination of attribute values to estimate the ship performance on the basis of empirical formulae and metamodels;
- definition of intra-attribute fuzzy functions based on fuzzy set theory for intra-attribute preference and inter-attribute preference matrix to account for subjective judgment across ship properties;

- feasibility evaluation of ship designs subject to crisp criteria;
- use of dominance algorithm to filter the Pareto set through metrics of attributes values' distance from the identified ideal point;
- extraction of the preferred solutions according to the established preference structure;
- refinement of design space around specific non-dominated designs (mini-cubes) via GA to identify robust solutions.

3.2 *Design model platform*

The MADM structure for concept design searches for main technical and economic characteristics of a set of alternative designs which have to satisfy aspiration values of attributes. The principle inspiring the formulation of the technical modules is to exploit the existing know-how, the experimental data and past experience gained by naval architects. It is designers' responsibility to establish the free variables, parameters, attributes and constraints of the MDM.

As there is no existing tool for creating a platform for concept ship design, particularly for a CNG ship prototype, a proprietary design model has been implemented which has been tailored to incorporate Type III pressure vessels (steel liner wrapped with composite fibres) of 2.35 m internal diameter at 250 bar, in a family of CNG ships stored in a database (Fig. 1).

3.2.1 *Free variables and parameters*

The design model is structured around a number of free variables (length, beam, draft, midship section coefficient, and vertical prismatic coefficient), which are sufficient to define the ship accurately and uniquely at the concept design stage. Some of the variables that include technical constants, the so-called parameters, are determined outside the model and consequently remain fixed for one design cycle. They can be classified as shape and size parameters, topological parameters and positioning parameters.

As the mathematical model has to represent the ship in a simple but meaningful manner, a Design of Experiments (DoE) was performed to reduce the number of significant design variables by running different sample runs (Trincas 2014).

In order to restrain the model from shifting to infeasible regions of the design space, design variables are not completely free since, in addition to lower and upper bounds, they may be constrained by dependencies with each other.

3.2.2 *Attributes and metamodels*

Design attributes are principal elements for decision making regarding performance and quality of the ship. Based on the attained level of the attributes, designs will be accepted or rejected from further analysis. Number of attributes may be as large as needed. Attribute values which usually have different units of measurement, are normalized via fuzzy sets before entering the dominance analysis. The upper and lower limits of satisfaction for each attribute are provided by the design team.

Most of the attributes are calculated via metamodels (Myers et al. 2008). Search and optimization algorithms were used to find the metamodels of ship response functions which are represented by multi-linear regression equations on the predictors.

The following attributes identified by metamodels refer to:

- resistance & propulsion
- intact and damage stability
- global strength & vibrations
- seakeeping & manoeuvring
- holds surface area
- gas consumption in a running cycle

3.2.3 *Hard and soft constraints*

To prevent the design from attaining some unwanted characteristics, all intermediate and final results are subject to constraints which are linear and nonlinear equations of the equality and inequality type. Some relationships between geometrical variables and parameters are used as min-max constraints. In principle, constraints are used for hard type of decisions to distinguish between feasible and unfeasible designs, verifying that the solution remains within allowable design space. Attributes, which are design performance measures, are in fact constraints since their values may be constrained; hence, they may be considered as soft (fuzzy) constraints.

4 SELECTION OF THE OPTIMAL CNG FLEET

LNG and pipelines are not the only options to transport natural gas. Marine transport of compressed natural gas (CNG) is poised to be the most viable solution to bring natural gas supplies to new markets, to satisfy small-demand markets and to monetize small reserves and stranded fields. CNG technology can be used for transporting gas from with small throughputs (from 1 to 4 billion of cubic meters per year) along short and medium seaborne distances (up to 2500 nm). Therefore, CNG marine transport has a market niche between subsea pipelines and production volumes and distances covered economically by LNG.

CNG does not require expensive liquefaction and regasification facilities and thus may offer energy consumers with a more cost effective alternative to LNG and pipelines in markets that are located within regional proximity to a gas supply. The relative low cost of CNG delivery terminals as well as short time-to-market will reduce the cost entry barrier to introduce new gas supplies. However, lighter weight of PV and less expensive gas containment systems are a precondition to establish the commercial viability of CNG marine transport. In this respect, the PVs of Type I (fully steel) cannot be competitive at all against LNG and pipeline transport modes. The CNG PVs of Type IV and even more the adsorbed natural gas (ANG) PVs are going to be the winner solutions.

Moreover, the scalability of CNG projects allows decision makers to begin the projects earlier on a smaller scale and then to expand the fleet capacity by deploying additional CNG ships into the fleet when required. On the other side, when supply volumes and production rates decline, CNG solution offers the ability to redeploy CNG ships in other areas, thus reducing the project risks of technical, commercial and political nature. In fact, approximately 85–90% of a CNG transport project's assets are deployable.

4.1 CNG Transport modelling

Although economics represents the main milestone in setting up any gas transportation project, a number of general factors always apply in the choice of any gas transport scheme, mainly related to primary ship performances and gas containment system.

Since CNG concept ships are prototypes, inadequate information about configuration and size of feasible designs is overcome by generating efficient solutions. After fuzzification of attributes' outcomes, the "best possible" design is extracted from the Pareto frontier (non-dominated designs) as that one having a minimal distance, measured by Čebyšev metrics, from the ideal design (Žanić et al. 1992).

4.2 Shipping

The CNG marine transport system is economically optimised based on a number of dedicated CNG ships which will carry a specified volume of compressed gas per year from loading terminal (both onshore and offshore) to destination terminals. The shipping scenario is described by distance to the market, stand-by time, connecting/disconnecting times, loading and offloading rates, and (possible) storage facilities.

The primary attributes of the optimal fleet, which has to provide the expected rate of return, are the number of ships in the fleet, the capacity and the service speed of each sister ship. Therefore, the optimal fleet composition is to be decided upon by appropriate economic criteria, e.g. shipping and infrastructure tariff, the latter being of minor importance. Special attention should be deserved to choose the loading and discharging rate by economic optimisation of the overall system.

4.3 Cost estimate methodology

A built-in economic model provides monetary value of each alternative CNG project.

Cost estimate is broken down by project elements and then by discipline. Each discipline shall contain details as to outline of the cost estimate breakdown structure, technical data sources, cost data sources, contingency.

4.3.1 Discounted cash flow analysis

The main tool for analysis of engineering economics is a discounted cash flow (DCF) model, which allows to determine the net present value (NPV) for an expected rate of return and project life. Discounting allows for the time value of money, which is an effective tactic for evaluating the lifetime value of a project in terms of today's money. Calculation of DCF involves identification of input parameters, such as capital amount, terms of project financing, operating expenses, tax and interest rates, depreciation rate, etc. Cash flows are calculated on an after tax basis and are assumed to occur on an end-of-year basis.

In the case study considered later, the following assumptions will be made: monetary values at a base value of 2017 US dollars and an inflation rate of 2% for revenues and expenses, including the cost of natural gas at wellheads; a 30% corporate income tax; linear depreciation of ships for a period of 15 years.

4.3.2 Financial parameters

Capital for ship building will be acquired through both equity and debt, at a rate of 30% equity to 70% debt. The loan term is assumed to be 8 years at an interest of 5.5%. Interest payments begin at the end of the project's initial year.

For the three years of the CNG ships' construction, the principal repayment is assumed to be 40%, 30%, and 30% of the total project cost. The expected rate of return for the CNG project is assumed to be 12.5%, a reasonably fairly standard value for the oil & gas companies

4.3.3 Capex and Opex

A cost estimate for building ships and facilities (capex) is performed with breakdown for main items. Transferring the gas from the production facility to the CNG containment system on board is technically straightforward and requires a minor capital investment for a short subsea pipeline, riser pipes and gas transfer buoys.

As regards the ship building cost, it is evaluated distinguishing between material costs and direct labour costs; it is assumed to be a function of ship size and capacity applying a CGT (compensated gross tonnes) factor. It is estimated from available productivity data from Italian shipyards.

Operating expenditure for ships and infrastructure facilities (opex) is assessed for the project life. Opex includes labour rates and material costs typical for the supply and market locations. It is escalated at 2% per annum starting from the first gas delivery.

4.4 Criterion for selecting the optimal fleet

As stated by Lamb (2003), ship success depends substantially on economic success. By hypersurface plotting of NPV as a function of ship variables, it is straightforward to determine the optimal fleet composition and main characteristics of the most profitable ship which exhibits the higher NPV, that is, the lower tariff for unit energy transport.

Resulting tariffs should be read in relative terms, since the building cost of each ship is evaluated on the basis of average hourly cost from Italian shipyards, whilst daily operating costs were derived from average data for LNG ships, fuel costs from present market information, and under assumed financial scenarios. The overall structure of the simulation model for identification of the optimal fleet is given in the upper part (the *external task*) of Figure 1.

5 SELECTION OF THE PREFERRED CNG SHIP

A database of eighteen CNG twin-skeg ships of different capacity ranging from 50 to 900 million of standard cubic feet (mmscf) has been made ready for fleet composition.

These ships have been developed as satisficing solutions in the first phase of the concept design via MADM approach to transport gas at various average supply rates per day. The ship has to be capable of sailing up to and connecting with a STL buoy and to a SAL system without assistance of tugs. The DP2 Class is preferred, which requires that the failure of any single active component (transversal thrusters, azimuth thruster, generators, switchboards, etc.) does not cause loss of position (Mauro & Nabergoj 2015). The CNG fleet shall guarantee a continuous delivery to the destination terminal, being flaring not allowed and re-injection not considered presently.

For the CNG marine transport scenario from Zohr giant field in Egyptian Sea to Adriatic Sea (Brindisi), a hub-and-spoke continuous-intermittent service scheme is considered here with 5 mmscm/d (1.75 billion cubic meters per annum) loading and offloading rate, where 4 mmscm/d is the utilized capacity.

For the CNG marine transport scenario from Zohr giant field in Egyptian Sea to Adriatic Sea (Brindisi), a hub-and-spoke continuous-intermittent service scheme is considered here with 5 mmscm/d (1.75 billion cubic meters per annum) loading and offloading rate, where 4 mmscm/d is the utilized capacity.

Figure 2 illustrates the feasible CNG fleets, showing interrelations of number of ships in the fleet, ship capacity and optimal service speed. The depicted lines show that the shipping tariff increases dramatically when the number of ships increases with corresponding decrease of ship size. Each point in a group (fleet) denotes a ship with a specific service speed (from 13 to 18 knots). Here, the best fleet is composed by four ships with about 350 mmscf capacity sailing at 17 kn. The main characteristics of the selected ship are provided in Table 1.

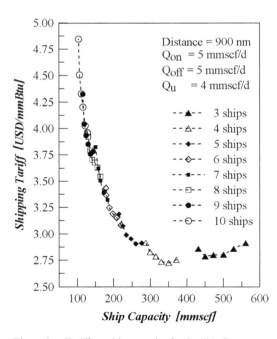

Figure 2. Tariff vs. ship capacity for feasible fleets.

Table 1. Main characteristics of the selected CNG ship.

Pressure vessel		Ship			
		Geometric characteristics		Performance attributes	
Number	316	Length overall	181.35 m	Service speed	17 kn
Diameter (internal)	2.35 m	Length btw PP	172.70 m	Main engines	2×8725 kW
Length	29.00 m	Beam	37.56 m	Propeller speed	124 rpm
Weight	42.20 t	Draft	7.02 m	Propeller diameter	4.8 m
Gas weight	20.50 t	Block Coefficient	0.703	Pitch ratio	0.960
Heel gas volume	3260 m^3	Vertical Prismatic Coeff.	0.857	Blades number	4
		Wetted surface area	7085 m^2	Area ratio	0.450
		Long. centre of buoyancy	81.25 m	Natural roll period	18.51 s
		Bulb area	17.40 m^2	Natural pitch period	7.84 s
		Transom area	3.85 m^2	Vertical acc. (at SS5)	0.58 m/s^2
				First vibration mode	82.90 cpm
				Capex	433.20 mm$
				Opex	7.10 mm$/y
				Tariff	2.69 $/mmBtu

6 DESIGNING THE CNG SHIP HULL FORM

In the generation step of the first phase of the concept ship design, a function-to-form mapping approach has been implemented in the mathematical design model. Since for whichever generated ship, to some extent length of STL and compressors room, azimuthal thruster space, engine room and cofferdams, are independent on ship size, the overall length of the hold spaces is derived subtracting these lengths from ship length. Then, a preliminary number of PV is easily determined. A database of PV provides the vertical length of the PV as a function of their number and required ship capacity. The PV maximum length is subject to a technological limit, currently assumed as 35 m. Finally, the hull particulars of the candidate design is drawn which is substantially dependent on PVs' number and distribution.

This design is simultaneously subject to two crisp constraints, e.g. compliance with IMO stability criteria and avoidance of roll resonance in a seaway. The best solution coming out from the first phase of concept design is intrinsically satisfying those two constraints. Then, the second phase (hull shape tailoring) of the concept design starts (Fig. 1) where the hull form and the cargo hold spaces at the "best possible" ship are improved.

6.1 Concept sectional area curve

Control and enhancement of hull form since concept design stage is essential for economic success of a ship. To manage the concept geometric parameters together with internal layout, control of sectional area curve (SAC) is mandatory. In fact, the SAC at design draught is including the information regarding the hull volume distribution, being the latter extremely important not only to assess hydrodynamic performance of the ship, but also to establish the internal spaces needed to install the required PVs.

Usually the data coming from concept design outcomes are not directly converted into a feasible SAC. On the contrary, since a database of hull forms is available, also a *concept SAC* (Fig. 3) can be determined by interpolating SAC curves of the ships stored in the database.

Not only is the SAC at design draught relevant for the project, but also the SAC at tank top height, being the ship volume distribution over the tank top essential for the cargo area determination. By doing that, it will be then easier to identify accurately cargo areas since concept design stage.

6.2 Primitive cargo unit

To evaluate the main dimensions of the ship by considering the cargo capacity as one of the most important characteristics for the project since the concept design stage, a modular approach has been used.

Whichever cargo ship is to be designed, cargo space and cargo type will influence the overall dimensions of the ship. Of course, since not all the cargo ships can be identified and modelled in the same way, it is impossible to find a unique universal parameter to directly identify different cargo ships.

By consequence, to have a more flexible approach suitable to be applied on a wide range of ship types,

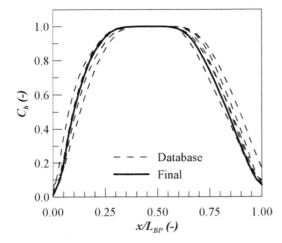

Figure 3. SAC as interpolated from database.

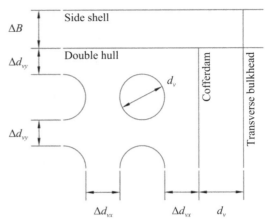

Figure 4. Boundaries around primitive cargo unit for CNG ships.

it can be handy to figure out whether cargo spaces can be identified by different sets of recursive volumes that can be used as basic unit for the cargo space modelling. This minimal unit can be defined as *primitive cargo unit*. As an example, for a container ship it is reasonable to define the TEU as the minimal cargo unit, for a car carrier a specific car size, or for a LNG carrier the liquefied gas tank volume.

In such a way the cargo space of the ship can be identified as a modular function of the minimal cargo unit, assuring that cargo space inside the ship will intrinsically have sufficient volume to comply with the required cargo capacity requested by logistic routines.

In the specific case of a CNG ship, the cargo area is occupied by cylindrical gas PV disposed vertically in multiple rows and columns inside dedicated hold spaces (Fig. 4). Typically, the PV diameters are constant and, according to the class regulations, also the minimal spaces between them shall be in certain predetermined range in both longitudinal and transversal direction. That is why it can be handy to consider the pressure vessel external diameter d_v as the minimal cargo unit for the CNG ship. In such a way, the entire hold dimensions can be defined in terms of d_v as follows:

$$l_h = m\left[d_v + \Delta d_{vx}(d_v)\right] + d_v \qquad (1)$$

$$b_h = n\left[d_v + \Delta d_{vy}(d_v)\right] + \Delta d_{vy}(d_v) \qquad (2)$$

where l_h is the hold length, b_h is the hold breadth, m is the number of PV columns in the hold and n is the number of PVs per column.

Longitudinal and transversal distances between vessels, Δd_{vx} and Δd_{vy} respectively, can be expressed as functions of d_v. As it can be seen in equations (1) and (2), the hold dimensions are including an elongation term, which is taken into account to comply with safety regulations. In fact, at least one d_v cofferdam space per hold should be considered in longitudinal direction as well as there must be another interstice with each hold side. In the present CNG model, Δd_{vx} and Δd_{vy} have been assumed as 600 mm and 200 mm, respectively.

This modelling is adequate for a hold located in the parallel middle body or in areas where the waterline at the hold bottom (tank top) is close to the full beam of the ship. To take into account reduction of available cargo space at fore and aft shoulders, a space-reduction coefficient C_h is defined as:

$$C_h(x) = \left.\frac{A(x)}{A_M}\right|_{T=h_{TT}} \qquad (3)$$

where $A(x)$ is the area of the transversal section at x position and A_M is the maximum sectional area, both at tank top height h_{TT}.

The space-reduction factor is acting on the b_h determination; so, to consider the effective number N_{pvh} of pressure vessels that can be installed in a hold, the following expression can be used:

$$N_{PVh} = \sum_{i=1}^{m} n_i = \sum_{i=1}^{m} \text{floor}\left[nC_h(x_i)\right] \qquad (4)$$

where x_i is the ordinate of each PV column centre. Then by adding N_{PVh} pertaining to each hold, the total number of pressure vessels N_{PV} is obtained.

6.3 Internal layout

In the generation process based on Monte Carlo approach, the main dimensions of the ship are taken as random variables inside the design space. Anyway, in order to generate a feasible solution since the concept design stage, it is necessary to model the internal spaces. In the specific case of a CNG ship, some specific spaces should be taken into account. In fact, besides cargo length and engine room length, other longitudinal spaces should be dedicated to equipment necessary to gas loading and offloading process. This is the case of the conical recess in the fore part of the flat of bottom for connection to the STL system and of the main compressors.

For such a reason dedicated space should be considered to install the above mentioned equipment, leading to a definition of cargo length L_C:

$$L_C = L - L_{ER} - L_{CR} - L_{STL} - \Delta L \quad (5)$$

where L_{ER} is the engine room length, L_{CR} is the compressor room length and L_{STL} is the STL compartment length. For a CNG ship, the aforementioned longitudinal dimensions can be assumed as constants: L_{STL} will be almost independent of ship size, L_{CR} will lightly be affected by the actual gas volume to load and offload as well as by the difference in pressure inside the PVs and onshore or storing equipment. In any case they are invariant for each specific set of generated ship, since the total capacity of the selected "best possible" ship varies with operative/economical profile only. ΔL is a safety length estimated as a percentage of L_C taken into account to have a suitable margin for the total length definition.

As mentioned above, for the hold breadth b_h, an easier formulation can be done:

$$b_h = B - \Delta B \quad (6)$$

where ΔB is the minimum double side width in accordance with class regulations.

The cargo length L_C shall be subdivided into a proper number of cargo holds n_h in order to comply with damage stability requirements. The number of holds shall be minimized to reduce the total cofferdam spaces and it can be assessed with a regression formula.

The maximum values of n and m are defined by modelling the holds as composed by primary cargo units; then, applying the space-reduction factor the total number of pressure vessels N_{PV} is defined.

A logistic scenario is univocally defined by a fleet composed by n_s ships, characterized by a cargo capacity C and a design speed V.

The capacity is function of type, number and length of pressure vessels, therefore, to assure that the generated ship satisfies the logistic scenario, the length of pressure vessels l_{PV} is defined as:

$$l_{PV} = \frac{C}{N_{PV} C_l} \quad (7)$$

where C_l is the capacity per length unit of the selected type of pressure vessel at maximum allowed pressure. The new resulting PVs' length is again subjected to the already mentioned technological limit; e.g. 35 m.

7 MOVING FROM CONCEPT TO PRELIMINARY DESIGN

With reference to ship hull shape, as already explained in the previous sections, at the end of the first phase of the concept design stage only the main hull parameters have been determined as a modular function of *primitive cargo unit*.

To establish the possibility to build a link between outcomes of the first phase of the concept design with a primary representation of the "best possible" ship, a dedicated procedure has been implemented. the target is to *quickly* visualize a concept hull form and GAP by considering the internal space modularity.

As already mentioned for SAC, interpolation methods can be used to determine a hull form compliant with the hydrodynamic performance and internal capacity assessed during the first phase together with economic attributes. However, the process is somewhat more complicated compared with the SAC case determination.

In fact, for an enhanced hull form definition, two constraints are taken into account, that is, the SAC at design draught and the SAC at the tank top draft which has been used to define the hold spaces.

Interpolating between selected numbers of similar ships makes it possible to determine each transversal section via a B-spline. Its area is then evaluated at design draught and tank top height in such a way to control whether the section is compliant with the predetermined SAC curves. The solution is reached by modifying spline curves nodes by means of target search algorithms (based on genetic algorithms) which are automatically finding a compliant solution within a certain confidence range.

By doing that, a suitable concept hull form can be found (Fig. 5). It is then straightforward to determine the desired waterlines at design and tank top drafts. So by considering the subdivision length given for cargo and main equipment areas, it is possible to figure out a concept GAP (Fig. 6)

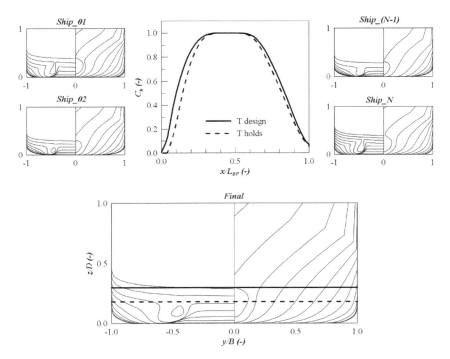

Figure 5. Generation of hull form from SACs.

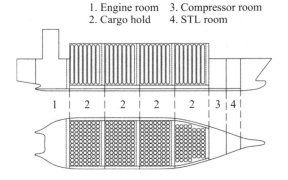

Figure 6. General arrangement.

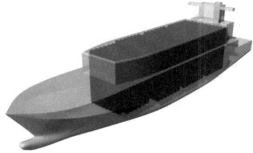

Figure 7. Rendering of the ship at preliminary design stage.

respecting the surface and volume constraints. For the specific case of a CNG ship, it is also possible to figure out the PVs' layout inside the hold spaces, respecting all the constraints related to the PVs' interstices as described in Section 6. The hull form and GAP built in the second phase of the concept design stage are then used to further proceed with the ship design process. In fact, these concept hull form and GAP are suitable for a direct use in the preliminary design stage, provided fairing and detailed optimisation processes are activated afterwards. The preliminary hull form is a good starting point for sub-optimisation of specific parts, mostly related to hydrodynamics issues.

Besides, availability of a preliminary GAP and a weight breakdown is also helpful for a preliminary analysis of ship damage stability and strength.

The path from concept to preliminary design is realized by a customized user interface which provides the lines plan and the GAP in a format readable by general purpose computer programs.

A perspective view of the CNG ship as it appears after derivation from the integrated concept design process is shown in Figure 7.

8 CONCLUSIONS

This paper has presented the synthesis procedure and methodology of decision making for the multiattribute ship design problem as the most appropriate at concept design stage, being superior to traditional design methodologies. The application has been described for a member of an optimal CNG fleet thus enabling a rational selection of the "best possible" design extracted from the Pareto frontier.

In order to speed up the design process, a new geometrical procedure has been developed to make the concept hull form description quasi-automatically transferred to preliminary design stage as an extension of the initial design.

By adopting a modular definition of the internal cargo area (*primitive cargo unit*) and resorting to a database of similar hull forms, it has been possible to derive a sufficiently accurate hull form and GAP to be earlier used as starting point in the preliminary design stage.

This procedure drastically reduces time required for hull form modelling and internal compartmentation resulting in a cost reduction of the design process while increasing the quality and reliability of its outcomes. Moreover, it makes easy the 3D visual representation of the ship.

Future research could include enhancement of the gas containment system design as well as implications of different loading/offloading systems and different propulsion systems (pods, fully electric).

REFERENCES

Andrews, D.J. & Dicks, C. 1997. The building block design methodology applied to advanced naval ship design. *Proceedings of the 6th International Marine Design Conference, IMDC'97*, Newcastle-upon-Tyne, pp. 3–19.

Campana, E.F., Peri, D. & Pinto, A. 2007. Multiobjective optimization of a containership using deterministic particle swarm optimization. *Journal of Ship Research*, Vol. 51, no. 3, pp. 217–228.

Diez, M. & Peri, D. 2010. Robust optimization for concept ship design. *Ocean Engineering*, Vol. 37, pp. 966–977.

Evans, H.J. 1959. Basic design concepts. *Naval Engineers Journal*, pp. 671–678.

Frye, M.C. 2010. *Applying set-based methodology in submarine concept design*. MIT, Cambridge.

Gillfillan, A.W. 1969. The economic design of bulk cargo carriers. *Trans. RINA*, Vol. 111.

Grubišić, I., Žanić, V. & Trincas, G. 1988. Concept design system for interactive optimisation of specialized vessels'. *Proceedings, CADMO'88*, Southampton, pp. 371–382.

Lamb T. 2003. Ship Design and Construction, Vol. I, SNAME.

Mandel, P. & Leopold, R. 1966. Optimization methods applied to ship design. *Trans. SNAME*, Vol. 74.

Mauro, F. & Nabergoj, R. 2015. Integrated station-keeping and seakeeping predictions. *Proceedings of IMAM 2015*, Pula, Croatia.

Myers R.H., Montgomery D.C & Anderson-Cook C.M. 2008. *Response surface methodology* – process and product optimization using designed experiments. Third edition, J. Wiley & sons, Hoboken, New Jersey.

Nowacki, H. 2003. Design synthesis and optimization—an historical perspective. In *Course Notes of 39th WEGEMT summer school*, Birk & Harries Eds, pp. 1–26.

Nowacki, H., Brusis, F. & Swift, P.M. 1970. Tanker preliminary design—an optimisation problem with constraints. *Trans. SNAME*, Vol. 78, pp. 357–390.

Pashin, V.M. 1983. Ship Optimization—system approach mathematical models. *Sudostroenie*, no. 3 (in Russian)

Pugh, S. 1991. *Total design: integrated methods for successful product engineering*, Workingham.

Raharjo H., Xie M. & Brombache, A.C. 2010. A systematic methodology to deal with the dynamics of customer needs, *Quality Function Deployment*, Vol. 10.

Ray, T. & Sha, O.P. 1994. Multicriteria optimisation model for a containership design. *Marine Technology*, Vol. 31, no. 4, pp. 258–268.

Sen, P. 2001. Communicating preferences in multiple-criteria decision-making: the role of the designer. *Journal of Engineering Design*, Vol. 12, no. 1, pp. 15–24.

Suh, P.N. 1990. *Principles of design*. Oxford University Press, New York.

Trincas, G. 2001. Survey of design methods and illustration of multiattribute decision making system for concept ship design (plenary paper). *Proceedings of MARIND 2001*, Varna, 2001, Vol. III, pp. 21–50.

Trincas, G. 2014. Optimal fleet composition for marine transport of compressed natural gas from stranded fields. *Proceedings of 2nd INT_NAM Conference*, Istanbul, pp. 25–38.

Trincas, G., Grubišić, I. & Žanić, V. 1994. Comprehensive concept design of fast ro-ro ships by multiattribute decision making. *Proceedings of 5th International Marine Design Conference, IMDC'94*, Delft, 1994, pp. 403–418.

Žanić, V., Grubišić, I. & Trincas, G. 1992. Multiattribute decision-making system based on random generation of non-dominated solutions: an application to fishing vessel design. *Proceedings of PRADS'92*, Elsevier Applied Science, Vol. 2, 1992, pp. 403–418.

A concept for collaborative and integrative process for cruise ship concept design—from vision to design by using double design spiral

M.L. Keiramo & E.K. Heikkilä
Royal Caribbean Cruises Ltd., Newbuilding and Innovation, Turku, Finland

M.L. Jokinen
Turku School of Economics, Finland Futures Research Centre, Turku, Finland

J.M. Romanoff
Aalto University, School of Engineering, Marine Technology, Espoo, Finland

ABSTRACT: In the cruise ship building industry, new product innovation has become increasingly challenging, driving companies jointly to create and implement smarter operation models to stay competitive. A collaborative innovation model amended by a collaborative leadership process is one way to meet this challenge. It is expected that this model would produce more opportunities to lead complex architectural design, technical development and new innovation projects by intertwining work streams and breaking silos. This paper, describes preliminary ideas for the development of a 'fit for purpose'-driven collaborative innovation model created for cruise ship conceptual process. This model synchronizes human and technical systems of innovation processes. The human system analyzes innovative thinking, partnership culture and leadership in the context of collaborative innovation process. The technical system focuses on architectural and technical aspects of new product design in a disciplined and structured manner. To research the development process, actions in innovation workshops will be documented and examined by ethnographic observations; the social network analysis will produce information on social and communication structures. In addition, in depth interviews will bring understanding of the stakeholders' prerequisites, expectations on gains, enablers and obstacles of sharing, open communication and building a joint partnership culture and a distributed way of leading. Outcomes and effects of the innovation process will be evaluated by interviews and questionnaires for the stakeholder groups. The research methods will mainly be qualitative descriptive methods and a case study of a specific development process.

1 INTRODUCTION

In today's global economy, organizations are collaborating more and more. Thus, organizations are engaging in new forms of highly collaborative mechanisms and networked structures capable of providing a competitive advantage by combining the best skills or core competencies and resources of two or more organizations, as well as customer knowledge of a product or a service to co-create a value proposition more compelling and relevant to the consumers' needs and expectations (Romero & Molina, 2009). New product innovation in the cruise ship building industry has become increasingly more challenging, driving companies to create and implement smarter operation models and processes to stay competitive. The collaborative design process amended by a distributed leadership model is one way to meet the challenge. Ships of tomorrow will evolve naturally in line with economic and technological trends. The long-term economic and ecological pressure for energy efficiency will inevitably lead to for example lower ship speeds and bigger fleets. At the same time, smarter design processes will look at power requirements in realistic operational scenarios to minimize yearly fuel consumption (e.g. Hochkirch & Bertram, 2012). Advances in simulation technology include using 3D ship product models for example fast finite-element modelling and computational fluid mechanics techniques, the as-is condition of a ship will be capable of being simulated at any time over the design and operational life spans as envisioned (Wilken et al., 2011). Thus, overall a new model or paradigm shift is necessary, from closed to open innovation paradigm (Chesbrough, 2003). Taking into account such complicated structure of innovation and high demand for breakthrough solutions, generating new meaningful ideas to lucrative concepts in collaboration with a supplier and client becomes more momentous than ever. For creating concepts for new products even in shorter time and with less resources, the smarter collaborative approach and sustainable processes are essential for the companies to stay in the business.

Over the years, the understanding of innovation has been connected with creating new products (Schumpeter, 1934) and technologies (Mishra & Srinivasan, 2005) that would satisfy the customer and open new markets. In a modern fast-developing world, creating the new product or service is not enough. The entire innovation system including business processes, core processes, product performance, services, channels, customer experience and brand could be designed in order to provide value for the customer (Doblin, 2012). Innovations do not come by accident any more than the future does. They result from various intentions. The fifth wave of industrial society development ended with the financial crisis and we are now in the beginning of the sixth (see Table 1). The central question of the sixth wave is whether we can generate enough very diverse innovations. That means new technologies and new business models, in other words, solutions that are based on more intelligent production, consumption, and distribution models (Wilenius, 2017).

In today's global economy, no organization is local any longer. The new ecosystem of global business, where individuals, organizations, governments, and economies are all networked and interdependent, we need a new innovation model. This model must be based on the platform where internal, external, collaborative, co-creative ideas can converge to create organizational and shared value. We label this new approach to innovation "Co-Innovation" (Lee et al., 2012). Some of the key enablers of effective innovation have been explored such as strategic vision, culture, direction and sense of urgency of the organization. Then, it follows that organizations should design proper DNA for sustainable innovation by balancing exploitative and explorative innovations and the innovation value infrastructure should be properly aligned with organizational strategy to create dynamic capability to operate in market environment (Teece, 2009). Today, there is no time to re-design the same product over several rounds, i.e. to follow the principal of classical ship design spiral. Furthermore, a shorter design and construction phase necessitates teams to design concepts that enable fit-for-purpose and first-time-right solutions and products. In this paper, we describe and sketch out foundation for a fit-for-purpose driven collaborative model created for cruise ship conceptual design process. The model synchronizes architectural and technical new product design including the cost aspect in a transparent, disciplined and structured manner.

2 RESEARCH OPPORTUNITY AND METHODOLOGY

Collaborative Networked Organizations (CNOs) show a high potential as drivers of value co-creation, allowing organizations access to new knowledge, sharing risk and resources, joining complementary skills and capacities which allow them to focus on their core competencies. In addition, collaborative networks induce innovation, and thus co-create new sources of value by confrontation of ideas and practices, combination of resources and technologies, and creation of synergies (Romero & Molina, 2011).

The latter also encompasses collaborative partnerships and their leadership capability to enhance a joint vision of how to help each other succeed in a more competitive business environment. It is

Table 1. The succession of development waves in industrial societies (Wilenius, 2017).

THE SUCCESSION OF DEVELOPMENT WAVES IN INDUSTRIAL SOCIETIES						
K-Waves	1st wave	2nd wave	3rd wave	4th wave	5th wave	6th wave
Period	1780–1830	1830–1880	1880–1930	1930–1970	1970–2010	2010–2050
Drivers	Steam Machine	Railroad Steel	Electricity Chemicals	Automobiles, Petro-chemicals	Digital communication technology	Inteligent resources efficient technologies
Prime field of application	Clothing industry and energy	Transport, infrastructure and cities	Utilities and mass production	Personal mobility and freight transport	Personal computers and mobile phones	Material and energy production and distribution
Human interest	New means for decent life	Reaching out and upwards	Building maintence	Allowing for freedom	Creating new space	integrating human, nature and technology

essential to have full commitment of the parties for hard work to build up joint processes wherein joint values and priorities, expectations and achievable goals including risks are defined. The concrete steps that organization can digest are crucial and mapping out the main challenges such as how to avoid resistance, how to get people to work together, how to build trust on shared values and how to maintain flexible and collaborative attitude. These elements enable organizations to achieve expected behavioural changes and reach a shared goal for the new product development.

A dynamic network of changing and learning actors is expected to discover new radical possibilities, ideas and be able to let go of old practices and behaviours. The concept design phase is the most discontinuous and challenging part of a ship building process. Therefore, different approaches and new ideas are highly welcomed for producing and managing incremental or radical innovations in collaborative environment between multiple partners. The model encompasses collaborative partnerships and leadership capability and an importance of understanding a joint vision of how flexibly support partners to succeed in highly competitive innovation environment.

The research of complex phenomena such as concept development in collaborative partnership contains rich elements of creative interpretations and participative elements such as observations. The approach is therefore multidisciplinary and combines different methods to capture the concept development process as a part of cruise ship planning phase.

The study aims to contribute to innovation and transition research by applying, in addition to case description, other methods such as network analysis and futures workshops. Empirical data is produced using three methods: social network analysis, content analysis and observations of workshops in addition to semi-structured interviews.

3 PRESENT PROCESS OF CRUISE SHIP CONCEPT DESIGN

Current practice in concept design usually start from existing concept and easily goes on adding new inventions to the old concept rather than having an early phase of growing ideas for a new concept. The difference between shipbuilding and many other industries that involve delivering finished goods to the marketplace for a customer to choose from is that the final design and construction of a ship does not start until the ship is sold. The production decision, which is therefore made by a customer rather than the manufacturer, is based on the outline of the ship. Consequently, in the shipbuilding industry the shipyards or design agencies do not create a product in advance to be offered to customers. Instead, shipping company looking to acquire a new ship can itself direct the design process. The design process often takes place simultaneously in several shipyards and private firms. In this way, several parties can share the workload or, alternatively, competing proposals can be drafted. However, in addition to the features specified by the customer, other features that the customer has not even considered are often offered by the supplier (Keinonen, 2010).

Traditionally, ship contracts have been compiled in such a way that when entering into one, shipyards are committed to deliver technically specified vessel within the required timeframe given only the operational and architectural demands, but with no detailed technical plans or drawings. This situation puts considerable pressure on the planning effort (Hellgren, 2016). The design should start from the mission specified for the ship describing the "musts and wants". The aim is reduce the number of loops needed to find a technically feasible and economically preferable solution as demonstrated by the classical design spiral of systems-based-design (Levander, 2000).

4 COLLABORATIVE AND INTEGRATED DESIGN PROCESS

Innovation driven concept design in a multi-partner environment is understood in this paper as the set of joint activities and operations that are performed in order to collaboratively and transparently bring fresh ideas, inventions and solutions to be part of cruise ship concept design. The first area is to leverage innovative ideas to introduce new products, services or even new ventures. This process requires collaborative efforts with internal and external partners. According to (Lee et al., 2012) value creation involves value chain innovation to make the architecture more efficient which in turn will cut the cost, improve quality, and/or increase the speed of the process. It also includes reinventing the concept of customer value. This area is especially fruitful for value co-creation with customers for a shared value. The traditional customer values of price, quality, speed, and customization are of course still essential. However, today's customer demands more than just these such as experience, emotional impact and the public good. Customers would like to be engaged in the process of experiencing the product or service, a sense of beauty or safety, and an opportunity to learn new things. However, producer activities play an important role in the cultural production process because producers' efforts to offer innovative products with diverse appeals to consumers also generate the impetus for meanings to be revisited and reassigned to physical artifacts. Product design and advertising are two producer activities that play a central role in the process of symbolic value creation (Ravasi et al., 2008).

Business models represent the approaches that the organization strategically selects to produce and deliver its goods or services to the customer. One of the areas of value creation is to expand the customer base. (Lee et al., 2012.) Creating customer value for all customers and also differentiated value for specific customer groups require innovation. Many conceptual design methods and processes are quite laborious and require extensive capabilities by the designers. Switching from one stage of the process to another often causes problems especially in the system-based design process because the goals and the perspective often change radically.

The main goal of designing a new process is to ensure that concept, design and construction of a cruise ship, especially in regards of prototype vessels, are guided by the first-time-right principle so that details are fit-for-purpose designed. The developed idea for a new design process is quality-driven collaborative cruise ship design process that synchronizes ship architecture and technologies in a disciplined and structured manner but that also leaves room for a creative approach. The aim of this collaborative and integrative development process is to reduce hours and redesign work and cost. Further to improve product quality, total design lead time, maximize creative time by reducing wasted time spent on searching, waiting and missing information. It synchronizes all necessary processes and enables transparent knowledge sharing among multiple partners. In addition, it standardizes roles and responsibilities, fosters two-way collaboration, transparent communication and learning between project teams and leadership. In addition, the structured and disciplined way of working with clear layers, process steps and defined quality gates reduces the stress level for all partners.

4.1 Towards collaborative and integrated cruise ship concept design

The logic of product development is an interminable and cyclic activity in a mature, continually operating, knowledge-accumulating business. Product development is usually organized as a series of linear projects, the reason is that it is easier to steer the activity in this way as stated by Routio (2007). The costs of product development tend to escalate sharply as the project unfolds and the same is true for the costs caused by any change in the design. The possibilities to affect the qualities of the product decrease at the same time as the process advances. This process is highly dependent on the experiences and insights of the skilled experts. Further, detailed design information is difficult to share, and design conflicts are resolved via a common effort by the design engineers during the downstream design stages. Meyer (2010) suggests that effective creating, sharing and use of knowledge is a principle factor of corporate competitiveness in today's global economy. The development of the concept design process is facing need to explore for user oriented design process (Purnomo, 2016).

4.2 Principles of participatory design

As for an introduction, the participatory design as explained by Purnomo (2016), is a relatively more democratic design process. Various stakeholders participate fully in the design process. The stakeholders are not only the users, but also the experts of various fields. In such case the designers themselves can be considered as one of those experts. Therefore, unlike the conventional design process, designer is considered not any more as the main subject of the process but only play as a part of the component of the design process. The participants engage on generating design alternatives but also participate in problem definition, sharing solution (design alternatives), development of ideas (generating design alternatives), evaluation of ideas (design alternatives) and deciding which is the best idea out of all the generated ideas in the process. Each cyclic consisted of the consecutive steps beginning from information gathering, generating ideas, followed by sharing ideas, evaluation and choosing the best alternative solution. This can be compared to Delphi method (Linstone, 2002).

The first problem of the method is related to the implementation where the mindset of the professional stakeholders that are reluctant in giving up the ability to control the project. Another difficult problem is in identifying the stakeholder or participant that accurately reflects the profile of the actual users (Webcredible, 2006). Even if we could identify such stakeholders, another problem is on how we could ascertain their participation and attendance in every step, non-professional stakeholders often do not appreciate their own knowledge. The solution that can be used to boost the spirit of participation among non-professional stakeholder is to bring them in the meeting room and let them have the feeling on how to seat on par with other stakeholder (Muller, 2007).

4.3 Collaborative and integrative design model

The user oriented and participatory design process, a joint envisioning and brainstorming for a new concept kicks off the entire process. The outcome of this phase is the full vision of principal elements of the concept that builds an aspirational foundation for a new cruise ship project. From the very beginning, the cost related questions as well other necessary studies (Routio, 2007) are pivotal to include in the routine of each working team.

One of the targets is to have a higher level of maturity in the concept design phase before the next design phase commences. This is the clear check-point process that aligns progress, architecture, structure, aspirations and cost. Instead of traditional linear "straight path", single spiral concept creation process, the new user oriented, collaborative and integrative process can include several quality gates (Cooper 1990; Aaron et al., 1993). The process is split into sub-processes with their subtasks. In order to take a work stream from a design cycle to another, each work stream has to pass a quality gate before entering to the next cycle. In this study, the proposed principal work streams for a cruise ship design are related to ship exterior, ship performance studies, principles for accommodation, principles for technologies, public venues and logistics flows, economics, safety, quality, risk and resources management related analysis.

The idea creation process of the defined design work streams are by disciplined manner pressed through the pre-defined quality gates in seamless, integrative and flexible multi-team collaboration. The preferred outcome of this very first phase in a new ship design is lucrative, aspirational and attractive vision for the entire ship concept ready to enter for the next design phase. Desirable is, that teams own collective vision of the concept options on each of the work streams in the beginning of the concept design phase. The content with increasing amount of layers, information and details for each work stream are refined during further design cycles and finalized before a start of ship production.

4.4 Introduction of double spiral model

The proposed double spiral model (Figure 1) is including the main elements of the traditional system-based design spiral. In the double spiral the concept phase design work is presented as cyclical process carried out by a joint effort of multiple teams consisting of technical specialists, architects and designers, futurists, suppliers and many other partners. The joint effort is "doing more by smarter and more agile approach" in several phase by small teams of specialists creating concept design evolutions, and working on close and transparent collaboration that enables swift information change between architectural design and naval architectural questions in regular basis.

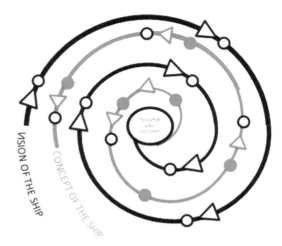

Figure 1. Principle of double-spiral guiding the integrative path from a vision to a concept (Keiramo et al., 2017).

The phase is well structured and categorized according to several layers of work streams such as architectural and technical design, safety, financial, risk, quality and resource related management streams. The results of each workstream are evaluated at regular intervals and they have to pass quality gate and criteria before the work can continue in the next cycle. The roles and areas of responsibilities for each team member are clearly defined with aligned understanding about the goals, main milestones followed by a joint acceptance of computer software systems for design and other work.

The design tools that are collectively used by each of internal or external design team member are playing key role in collaborative and integrative model. Selected tools enable efficient sharing of information in a way that the history and all changes of each design phase are accessible to every user. To support collaborative design, computer technology must not only augment the capabilities of the individual specialists, but must also enhance the ability of collaborators to interact with each other and with computational resources. The conceptual design needs to adopt a more pragmatic approach, through collaboration, support by artificial intelligence, and fuelled by information technologies (Wang et al., 2002).

Collaboration comes in a variety of forms. What used to be closed systems have given way to open systems emphasizing co-innovation focused on creating shared value. This has had impact in the form of leveraging innovative ideas to develop new products/services, relying on more efficient value chain architecture to apply process innovations. Focus on customer value provides new and better products/services for consumers, leading to broader customer bases, often supported through on-line purchasing. New business models enable organizations to produce and deliver these improved goods or services in more efficient ways (Lee et al., 2012), and designing a product right-the-first-time. Implementation of virtual 3D design station or studio could be an integrated collaborative design environment, allowin a distributed design team working together in harmony, as if they are in the same office. Required tools and technologies need to be selected after careful study and with caution (Wang et al., 2002).

4.5 Multi-level perspective on transitions in concept development process

The concept development process is complex and contains several layers of actions as described above. Geels et al. (2004, 2007, 2016) have used the multi-level perspective (Figure 2) to analyze transitions in especially in socio-technical systems. Transitions are seen as non-linear processes that consist from the interplay of multiple levels. Geels framework has three analytical levels: niches (the locus for radical innovations), socio-technical regimes (the locus of established practices and associated rules), and an exogenous socio-technical landscape. The MLP framework is meant to explain simultaneous and complex activities and changes in innovation systems and relative stability of existing regimes.

The MLP framework has been used in the context of sustainable technologies innovations and systems development as well as formal rules and institutions rather than normative and cultural-cognitive institutions. The criticisms and discussions on the MLP theory focus on (1) lack of agency, (2) operationalization of regimes, (3) bias towards bottom-up change models, (4) epistemology and explanatory style, (5) methodology, (6) socio-technical landscape as a residual category, and (7) flat ontologies versus hierarchical levels.

Geels response to criticism on lack of agency inspires this study to try to incorporate foresight systems in collaborative innovation networks in order to deepen understanding how especially collaboration mechanism interacts with other levels of niche innovations. Present model also aims to add dynamic futures perspective to MLP model, which concentrate on description of current patterns and mechanisms and their interplay by emphasizing anticipation, emergent development of interaction practices, and social contingency. This way we are able to incorporate interpretive and discursive dimensions to which have had less attention in MLP theory than formal rules and institutions.

4.6 Distributed leadership in the environment of collaborative design processes

The right to innovate inside the company has been possessed by a limited number of people such as managers, R&D departments, and top management (Chesbrough, 2003). However, the majority of great ideas are not discovered behind the closed door by an exclusive group of people. They are discovered by people who deal with everyday problems facing the customer and providing a service to different stakeholders. The notion is that employees from all levels of an organization can produce valuable solutions and by doing this bring competitive advantage to their companies (Kesting & UlhØi, 2010). One of the Heikkilä's interviewees stated (Heikkilä, 2018):

> "...and you have to create meetings which are everything except projects, which is linked to processes, continuous improvements, innovation, but mainly processes, which is very important. So you have to create forums of discussions, around these topics. If they don't exist, and you don't step out from the projects, you are not able to create the right environment. And, that's what we've tried to do on many occasions here, to focus very much on setting up the processes right."

The success of large-scale partnership projects is a prerequisite for smooth and seamless co-operation at the leadership level as well. Distributed leadership (DL) could be productive model for ensuring the success of design projects. Leadership and project management of each organization and team must be constantly on track in regards of the frequency and quality of communication between

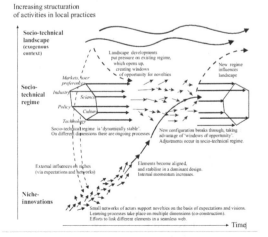

Figure 2. Multi-Level Perspective (MLP) on transitions (Geels 2011).

teams and their individual members. In the event of difficulties, the management must immediately react, providing the necessary support to the various parties so that problems can be resolved as quickly as possible to ensure non-stopping progress of the conceptual design. In addition, for reaching each quality gate on time, communication at all levels, but especially among the leaders, is a central element of timely decision making and key for successful passage through each gate.

In many collaborative projects with multiple partners, distributed leadership (DL) has become a popular "post-heroic" (Badaracco 2001) representation of leadership which has encouraged a shift in focus from the attributes and behaviours of individual "leaders" as promoted within traditional traits, situational, style and transformational theories of leadership (see Northouse 2007 for a review) to a more systemic perspective, whereby "leadership" is conceived of as a collective social process emerging through the interactions of multiple actors (Uhl-Bien 2006). From this perspective, DL works through and within relationships, rather than individual action' (Bennett et al. 2003).

In creative design teams, one of the management tasks is to provide space for experts to work and produce material, to ensure that teams have productive, meaningful and modern software tools, and common workplaces that meet the requirements of the project. Furthermore, the management responsibility is to monitor that teams remain on time and within budget, and produce plans that are in line with the common ambitions defined in the envisioning phase and strengthen during each leadership "talk and walkthroughs." In addition, the management's role is to act as a solicitor for possible conflict situations, and at the end, to provide time and an appropriate environment for the decisions that are leading the work for a joint vision of a concept at the end.

Complexity leadership theory divides required leadership actions for complex programs, such as a shipbuilding project, under three types of leadership:

- Enabling – committee's and PM (project oversight and management),
- Administrative – PMO (program management) and PM,
- Adaptive – processes such as concept development.

During the concept development phase of a shipbuilding project the importance of adaptive leadership, to challenge the thinking of the team members, is apparent. The role of adaptive leadership deserves a focus also due to its challenging nature. Complexity leadership theory suggests that instead of adaptive leadership coming from a single source it is better for it to emerge from systems, processes and social constructions. Also, the theory of distributed leadership discusses leadership as a systemic process rather than an individual act. By combining the research findings of these two research discourses we are able to critically learn and develop the leadership practices of the concept development phase (Heaslip, 2014).

5 FUTURES ORIENTATION AND FORESIGHT SYSTEM OPPORTUNITY IN COLLABORATIVE NETWORKS

As the futures are emerging and ever changing, the reflective practices and awareness of changing are critical components of every process that try to reach future needs or scan and analyze possible environmental developments. Futures orientation and images are vital in the process of identifying new possibilities or transformations to happen. The key question is how to form shared understanding, how valid futures information is produced and processed in order to analyze it and make useful conclusions from fuzzy sets of beliefs, assumptions, images and other inputs. The futures-oriented concept development model calls for a holistic approach to support radical innovation development, decision making and management of socio-technological systems.

The key question in the future-oriented concept development model is how to link together sociotechnical and partnership systems, which takes in consideration as well personal assets as managerial and leadership practices. Corporate foresight gives us a framework for functions of organized social process or interventions in and futures-oriented, context-driven and actionable knowledge creation (Piirainen & Gonzalez, 2015). The functions of foresight are linked to strategic decision making, improvement of long term planning, improvement of the innovation process and improvement of the speed to reach to environmental changes. Dominant logics of foresight activities are divided to expert-based foresight, model-based foresight, trend-based foresight and context-based open foresight. Context-based open foresight aims at shifting from 'trend-reactions-impact' logic to 'trends-context-strategy' logic by embracing softer values of foresight critical success factors. (Daheim & Uerz, 2008). According to Rohrbeck et al. (2015, 2):

> *"Corporate foresight permits an organization to lay the foundation for future competitive advantage. Corporate Foresight is identifying, observing and interpreting factors that induce change, determining possible organization-specific implications, and triggering appropriate organizational responses. Corporate foresight involves multiple stakeholders and creates value through providing access to critical*

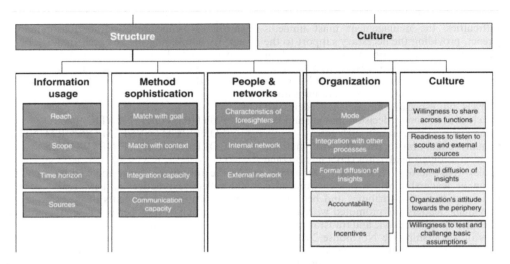

Figure 3. Elements of structural and cultural approaches (Rohrbeck 2011).

resources ahead of competition, preparing the organization for change, and permitting the organization to steer proactively towards a desired future."

There are several studies and theory on how foresight functions are organized within organizations, but less insights regarding how foresight processes run within networks of companies. (Rohrbeck, Daheim, 2015, 2008). Foresight has a strong connection to strategy which might explain why foresight is often a closed top management activity. We take as a starting point a corporate foresight model introduced by Rohrbeck (2011). His model covers structures and cultural aspects in corporate foresight process. This model will be modified and analyzed in partnership networks. Rohrbeck (2011) acknowledges the fact that national foresight still relies heavily on forecasting for identifying economic and social benefits in technologies, in contrast to the alternative futures oriented corporate foresight, which aims at identifying uncertainties and discontinuities. Foresight networks on local and national levels are implemented widely, while innovation networks on the other hand are extensively studied.

6 CONCLUSIONS

The design process of cruise ships differs from many other technology products that are delivered to customers as completed products to be sold. The cruise ship as a product is sold as a sketch and the actual design takes place during the design and manufacturing process among the network of several partners.

The specific ability of companies or alliances to work together is based on a shared understanding of the content and regulations of the activity and interaction between people. In such a complex and extensive network, a common open concept of future thinking, innovation and the creation of future concepts is particularly challenging because different actors can have both common and diverging goals. The interest in exploring and developing the conceptual design of cruise ships is particularly focused on improving the process so that the commonly set time, cost and quality objectives are achieved more flexibly both at the social side of the co-operation and the interaction between different actors at different times.

The most common way to describe the ship design process is by a spiral model, capturing its sequential and iterative nature. The task structure is "design-evaluate-redesign". The model easily locks the naval architect to the first assumption and will result in a patch and repair to this single design concept rather than generate alternatives. An approach that better supports innovation and creativity should be used (Levander 2000).

In this paper, the focus is on the conceptual design phase of a cruise ship and its further refinement. The aim of the study is to develop an approach that supports the creative process and development of innovations and to avoid the principal of system-based process: "design-evaluate—redesign", instead focusing on the process development that provides design right-at-the-first-time. In addition, the main goal is further to develop and pilot through case studies the collaborative and integrative model in which the architectural and technical concepts of a ship can be designed flexibly and uniformly by several teams and through multiple simultaneous work processes on the same schedule.

This operating model differs from the traditional, linear, mainly because it allows layering of work streams. Therefore, work is no longer done in "silos" but is seamless and can be continuously spread between different teams in a multi-professional environment. Further, the work can distributed through virtual design rooms that are enabling e.g. a-synchronized collaboration model especially between teams operating in different time zones. The purpose of this new collaborative and integrated approach is to ultimately ensure the production of more profitable and better quality concepts at the beginning of the project. In addition, it has a positive impact on the design quality, schedule and cost, furthermore having a positive impact on the construction phase, the overall quality of the end product and the end-user satisfaction. All in all, this research approach is part of the long-term continuous development of conceptual design of cruise ships and of this research.

The paper also described starting points and approaches for analyzing cruise ship concept development in real time and context. The aim was to build an analytical framework, which allows us to understand further and better the conceptual development process in stakeholder networks. The key research themes are how concept creation process is managed and linked to other parts of design and planning; how innovation networks are structured; and how creative partnerships are promoted. The project will examine leadership themes in collaborative partnership networks. The approach is multidisciplinary combining theoretical concepts of industrial product development, innovation and transition studies, collaborative networks, corporate foresight, and distributed leadership.

Finally, one of the most important research questions is how to combine and adapt the various fields in such a way that they ultimately enable creation of successful visions and concepts for a new ship design. For solving this, we still have to strive to build a link between open research questions and theoretical starting points. All in all, this research approach is part of the long-term continuous development of conceptual design of cruise ships and the purpose of this article is to present a research concept that will be further developed in the next phases of the dissertation project.

REFERENCES

Aaron, J. et al. 1993. Achieving Total Project Quality Control Using The Quality Gate method. *Whitepaper presented at the 1993 Project Management Institute Annual Symposium.*

Asaro, P.M. (2000). Transforming society by transforming technology: the science and politics of participatory design. *Accounting, Management and Information Technologies*, *10*(4), 257–290.

Badaracco Jr, J.L. 2001. We don't need another hero. *Harvard Business Review*, 79(8), 120–6.

Bolden, R. 2011. Distributed leadership in organizations: A review of theory and research. International Journal of Management Reviews, 13(3), 251–269.

Chesbrough, H. 2003. The Era of Open Innovation. *MIT Sloan Management Review* 44 (3): 20, 35–41.

Cooper, R. 1990. Stage-gate systems: A new tool for managing new products. *Business Horizons* May-June 1990. pp. 44–54.

Daheim, C., Uerz, G. 2008. Corporate foresight in Europe: from trend based logics to open foresight. Technology Analysis & Strategic Management, 20(3): 321–336.

Doblin, . 2012. Ten types of innovation framework. Available at: http://hbr.org/hb/article_assets/hbr/1105/F1105Z_A_lg.gif [Accessed: 24 January 2013].

Eyres, D.J. Bruce, G.J. 2012. *Ship construction*. Oxford: Butterworth-Heinemann.

Ferdows, K., Lewis, M., Machuca, J.A.D. 2014. Zara— The World's Largest Fashion Retailer. The Case Centre. 615-059-1. (Available at https://www.thecasecentre.org/corporate/products/view?id=130606).

Geels, F.W. 2004. From sectoral systems of innovation to socio-technical systems: Insights about dynamics and change from sociology and institutional theory. *Research Policy*, 33(6): 897–920.

Geels, F.W. 2010. Ontologies, socio-technical transitions (to sustainability), and the multi-level perspective. Research policy, 39(4), 495–510.

Geels, F.W. 2011. The multi-level perspective on sustainability transitions: Responses to seven criticisms. *Environmental innovation and societal transitions*, 1(1): 24–40.

Geels, F.W., & Schot, J. 2007. Typology of sociotechnical transition pathways. *Research Policy*, 36(3), 399–417.

Geels, F.W., Kern, F., Fuchs, G., Hinderer, N., Kungl, G., Mylan, J., ... & Wassermann, S. 2016. The enactment of socio-technical transition pathways: a reformulated typology and a comparative multi-level analysis of the German and UK low-carbon electricity transitions (1990–2014). *Research Policy*, 45(4): 896–913.

Harari, Y.N. (2015). What explains the rise of humans?. *TED, Internet, July*.

Heaslip, R.J. 2014. *Managing Complex Projects and Programs: How to Improve Leadership of Complex Initiatives Using a Third-Generation Approach.* Somerset, Massachusetts: Wiley.

Heikkilä, E. 2018. Exploring the relationship between distributed leadership and power. A case study from the shipbuilding industry. *University of Turku*.

Hiekata, K., & Grau, M. 2015. Shipbuilding. In *Concurrent Engineering in the 21st Century* (pp. 671–700). Springer International Publishing.

Himmelman, A.T. 2001. On Coalitions and the Transformation of Power Relations: Collaborative Betterment and Collaborative Empowerment. *American Journal of Community Psychology*, 29(2): 277–284.

Hochkirch, K., Bertram, V. 2012. Hull optimization for fuel efficiency—Past, present and future. 11th COMPIT conference proceedings, Liege: 39–49. (Available at http://data.hiper-conf.info/compit2012_liege.pdf).

Keinonen, T.K., Takala, R. 2010. *Product Concept Design: a review of the conceptual design of products in industry.* Springer Science & Business Media.

Kesting, P., Parm Ulhøi, J. 2010. Employee-driven innovation: extending the license to foster innovation. *Management Decision*, 48(1): 65–84.

Kim, W.C. and Mauborgne, R. 2005. *Blue Ocean Strategy*, Boston: Harvard Business School Press.

Lee, S., Olson, D.L. & Trimi, S. 2012. Co-innovation: convergenomics, collaboration, and co-creation for organizational values. *Management Decision* 50(5): 817–831.

Levander, K. (2000). *System based ship design.*

Linstone, H.A., & Turoff, M. (2002). *The Delphi method: Techniques and applications* (Vol. 18). Addison-Wesley Publishing Company, Advanced Book Program.

Meyer, J., (2010). Empirical perspectives on learning at work (Doctoral thesis, Jacobs University, Bremen, Germany).

Mishra, B.P., Srinivasan, R. 2005. A framework for technology innovation. *Journal of Advances in Management Research* 2(1): 61–69.

Muller, Michael J. (2007). Participatory Design: The Third Space in HCI, In J. Jacko and A. Sears (eds.), *Handbook of HCI 2nd Edition. Mahway NJ USA: Erlbaum*,10.

Piirainen, K.A. & Gonzalez, R.A. 2015. Theory of and within foresight—"What does a theory of foresight even mean?". *Technological Forecasting and Social Change* 96: 191–201.

Purnomo, A.B. (2009). Teknik Kuantitatif untuk Arsitektur dan Perancangan Kota.

Purnomo, A.B. (2016). User Oriented Design Process. *International Journal on Livable Space*.

Ravasi, D., and Violina R. 2008. Symbolic value creation. In: *Handbook of new approaches to organization*: 270–284.

Rindfleisch, A. & Moorman, C. The acquisition and utilization of information in new product alliances: A strength-of-ties perspective. Journal of marketing, 2001, 65.2: 1–18.

Rohrbeck, R. & Gemünden, H.G. (2011). Corporate foresight: Its three roles in enhancing the innovation capacity of a firm. Technological Forecasting and Social Change, 78(2), 231–243.

Rohrbeck, R., Battistella, C. & Huizingh, E. (2015). Corporate foresight: An emerging field with a rich tradition. *Technological Forecasting and Social Change* 101: 1–9.

Romero, D. & Molina, A. 2011. Collaborative networked organizations and customer communities: value co-creation and co-innovation in the networking era. *Production Planning & Control*, 22(5–6): 447–472.

Romero, D.; Molina, A. 2009. Value co-creation and co-innovation: Linking networked organizations and customer communities. In: *Working Conference on Virtual Enterprises*. Berlin, Heidelberg: Springer: 401–412.

Routio, Pentti. (2007). Arteology, the science of products and professions. http://www2.uiah.fi/projects/metodi/e00.htm, 2007.

Schumpeter, J.A. 1934. The Theory of Economic Development: An Inquiry into Profits, Capital, Credit, Interest and the Business Cycle. Cambridge: Harvard University Press.

Spinuzzi, Clay. (2005). The Methodology of Participatory Design, Technical Communication, Volume 52, Number 2.

Teece, D.J. 2009. Dynamic Capabilities and Strategic Management: Organizing for Innovation and Growth. Oxford: Oxford University Press.

Volker Bertram (DNV GL), Future of Shipbuilding and Shipping—A Technology Vision. Hamburg/Germany, volker.bertram@dnvgl.com.

Wang, L., Shen, W., Xie, H., Neelamkavil, J., & Pardasani, A. (2002). Collaborative conceptual design—state of the art and future trends. *Computer-Aided Design*, 34(13), 981–996.

Webcredible. (2006). User-centered design (UCD)-,http://www.webcredible.co.uk/user-friendlyresources/web-usability/user-centered-design.shtml.

Wikipedia. (2007). Participatory design, http://en.wikipedia.org/wiki/Participatory_design.

Wilenius, Markku 2017. *Patterns of the Future: Understanding the Next Wave of Global Change.* World Scientific.

Wilken, M., Eisen, H., Krömer, M., Cabos, C. 2011. Hull structure assessment for ships in operation, 10th COMPIT Conference, Berlin: 501–515. (Available at http://data.hiperconf.info/compit2011_berlin.pdf).

High-level demonstration of holistic design and optimisation process of offshore support vessel

M. de Jongh, K.E. Olsen, B. Berg, J.E. Jansen & S. Torben
Rolls-Royce Marine AS, Norway

C. Abt
Friendship Systems, Potsdam, Germany

G. Dimopoulos & A. Zymaris
DNVGL, Piraeus, Greece

V. Hassani
Sintef Ocean, Trondheim, Norway

ABSTRACT: Design of highly complex vessels involves a multitude of disciplines designing different modules and units of the vessel using different tools and methods requiring specialist competence in specific areas. The vessel may have a complex operational profile and multiple key performance indices. Sub-optimisation of each module for different KPIs without taking into account the interaction between the modules does not necessarily lead to an optimised overall performance of the vessel. A holistic approach is needed to achieve this. At the early design stage of a vessel, important parameters are defined having a huge impact on the performance of the vessel according to the KPIs. Changing these parameters at a later stage in the design process is difficult and requires a considerable effort from the multidisciplinary design team. This paper introduces a practical approach for holistic design approach at an early stage of the design process of an Offshore Service Vessel.

1 INTRODUCTION

Vessels with complex operational profiles have multiple key performance indices (KPIs), determining the operational and economic performance of the resulting vessel. Diverse modules and units in the ship contribute in each of these KPIs to some level. All the designers and manufacturers of different units in the maritime sector, particularly in offshore industry, are trying to optimise different KPIs, such as energy consumption, environmental impacts, operational windows, etc. However, the complexity and interconnection of the different modules in the system defy the concept of having efficient ship with maximal performance through optimisation of each module independently. As a matter of fact, improving the efficiency of a subsystem without considering its interaction with other models does not necessarily contribute to total efficiency of the system. Hence, a holistic optimisation framework is required to analyse the overall design in an early design stage. The current article not only seeks to propose sound concepts and techniques for designing vessels with optimised performance but also to define valid and practical approaches to assess the effect of each sub systems using integrated numerical simulations.

For Offshore Service Vessels (OSV), defining the main dimensions requires significant effort and review by a wide range of expertise. The choice of main design parameters results in a narrow bandwidth of performance expected for all of the KPIs. Significant adjustment of these parameters at later stage of the design process is often difficult, as this would result in going through the full design process again.

The ability to assess a wider range of main design parameters against the major vessel KPIs at an early concept definition stage, provides benefits in engineering time, operational efficiency and vessel cost in both CAPEX and OPEX.

As part of the HOLISHIP project (HOLISHIP, 2016–2020), which aims to bring Model Based System Engineering approach in ship design to a new level, Rolls-Royce Marine has teamed up with Friendship Systems, Sintef Ocean and DNVGL to establish an integration platform for the various design tools required for the high level concept development based on CAESES (https://www.caeses.com/products/caeses). These tools include:

- Hull lines import and transformation
- Stability
- Vessel motions
- Station Keeping and Dynamic Positioning
- Resistance and propulsion
- Lightship and steel weight
- CAPEX Cost
- OPEX Cost

These tools are often used at more detailed stages of the design process. However, the integration platform also allows the use of these tools at an early stage. The integration platform enables the various tools to be run, provide inputs and collect the outputs. It also provides the capability to perform a multi-parameter optimisation of the main design parameters based on the output of the various tools.

2 BACKGROUND

2.1 Software integration case study

To demonstrate the integration platform, a demonstration case has been defined based on an existing Subsea Construction Vessel (Fig. 1). The mission of the vessel is to perform subsea installation of heavy modules in ultra deep water using a subsea crane. So the main purpose of the vessel is to transport the heavy module from shore to the installation site, and to be a stable platform for the lifting operations over the side of the vessel using the subsea crane. The objective of the case is to find the combination of vessel size and crane type capable of performing the mission at lowest cost considering both CAPEX and OPEX.

The dimensions of a subsea construction vessel results from a large number of operational factors as well as chosen mission equipment. The development of fibre rope cranes using lifting ropes with neutral buoyancy, resulted in higher lifting capacities at larger depths. This may have an important influence on the dimensioning of the vessel and consequently on the CAPEX, OPEX and emission levels. Other dimensioning factors like operability, cargo deck dimensions and accommodation size must also to be taken into consideration in the design process.

2.2 Case implementation

Implementation of the holistic design tool for the case study is described in the following sections starting with the integration platform followed by a description of each step in the design process with the corresponding specialist tools and methods used. An overview of the holistic design process is shown in Figure 2.

The main steps are:

- Case setup/input
- Step 1: Definition of hull lines.
- Step 2: Steel weight estimates.
- Step 3: Hull verification/Stability.
- Step 4: Vessel motion calculations.
- Step 5: Station keeping calculations.
- Step 6: Resistance calculations.
- Step 7: Propulsion and machinery systems.
- Step 8: CAPEX and OPEX estimates.

When setting up the case, information about the mission and the crane type must be provided in addition to definition of the design space that shall be explored using a parametric hull model (Step 1). For each hull size, steel weight is estimated (Step 2) before stability calculations are performed taking into account the heeling moment from the crane operation. This is resulting in a go/no-go decision (Step 3). Vessel motion performance is calculated defining a limiting weather criteria taking into account the active heave compensation performance of the crane (Step 4). Station keeping calculations are performed to define the required thrust forces fore and aft of the vessel in Dynamic Positioning (DP) operation (Step 5). Resistance of the hull is calculated to dimension the main propulsion requirements for the vessel in transit operations (Step 6). Based on the thrust and propulsion needs, the propulsion and thruster units are selected (Step 7). Fuel consumption is estimated based on a simplified operational profile and a power system setup adapted to the selected propulsion and thruster units. High level estimates of CAPEX and OPEX can then be calculated (Step 8).

This process is iterated to explore the entire design space followed by an optimisation process

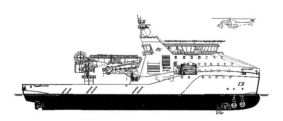

Figure 1. Subsea construction vessel.

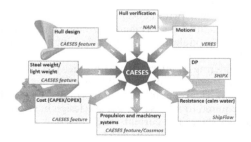

Figure 2. Overview of the holistic design process.

to identify the optimum hull. To compare the two vessel alternatives for the given mission based on fibre rope crane and steel wire crane respectively, the process is repeated twice with different set of input date. The final result is a relative comparison of the performance and the CAPEX and OPEX of the created designs.

3 INTEGRATION PLATFORM

3.1 *Holistic design tool architecture*

The architecture of the tool is quite basic in set-up with CAESES in a central position and the various specialist software tools arranged around it. CAESES uses a central database which is set up in a bottom up procedure, i.e. additional information can be added and detailing of methods can be performed at any time. The central database stores all direct input, like operational conditions and parameters to control the shape of the vessel or other relevant data that influence the performance of the design. In addition to direct inputs, relationships between inputs to assessment tools and methods are managed—as well as the methods for evaluating results of the integrated assessment tools. The involvement of time costly simulation methods, e.g. for predicting the power requirements by means of CFD, calls for a software architecture that only triggers an update of the individual results if any supplying data to the result has been changed (D. A. Watt, W. Findlay). This behaviour is realized by utilizing *lazy evaluation*. In the strictly hierarchical dependency, every object has a logical link to its suppliers and clients, except for those objects in the top level that only have clients and those in the lowest level that only have suppliers. The chosen approach allows for an economic management of computational recourses and ensures at the same time the integrity of the entire model.

3.2 *CAESES platform*

As the central hub to complex projects in a CAE context, CAESES comes with a variety of CAD interfaces, geometric modelling and pre-processing capabilities, probably the most prominent difference to other Process Integration and Design Optimisation (PIDO) tools (Ora Research LLC and intrinSIM LLC). The capability for defining new features and customised import and export routines inside CAESES allows the seamless integration of most tools that can be remotely controlled.

3.3 *Tools integration*

Running a tool and evaluating its results in an automated way requires three classes of information relating to input, execution and output:

– files that are made available to the tool,
– the location of the executable of the tool,
– files that contain the results of the tool execution.

This information is configured in the *Software Connector* in the platform which is also responsible for maintaining the consistency of data (Fig. 3).

Files that are made available to a tool usually contain data that depends on other objects, e.g. the shape of the vessel for resistance prediction, the vessel's speed and the type of crane or any other relevant information. These relationships are modelled in the software connector and managed by CAESES to ensure that the tool is executed anew, as soon as information from the output of the tool is requested, under the precondition that supplied data has also changed.

Figure 4 illustrates the dependency of two tools. Tool 2 requires an output from Tool 1, e.g. station

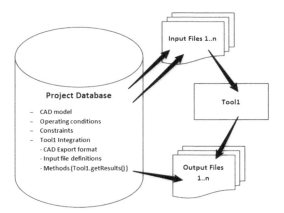

Figure 3. Tool integration.

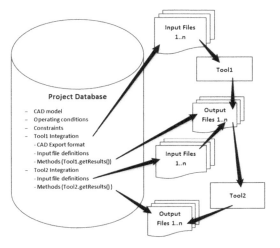

Figure 4. Dependency of tools.

keeping calculation using the environmental limits calculated by the seakeeping code. If an input to the seakeeping tool has changed, both tools will be executed in a sequence, if only an input to the station keeping code has changed, the latter will re-compute its results.

4 HULL LINES

This section refers to step 1 in Figure 2.

4.1 Tool application

In order to determine the optimum dimensions of the OSV considering the crane type and size, first a hull has to be created. From a baseline vessel hull, a variation of hull main dimensions is performed. Traditionally this was done by hand and this limited the amount of configurations that could be reviewed. This tool allows for automatic generation of hulls within a design space.

All tools integrated and described in the following paragraphs are preconfigured for an OSV vessel of a particular range, in the example illustrated this is in the range of 80–120 m in length.

Technically this is implemented by linking an empty container object of the geometry to the tools that require the hull-form for analysis.

To initialise a project, the baseline geometry is imported into the platform and an analysis feature is executed to adjust the settings of main particulars to the baseline (Fig. 5).

The setup of the project allows creating a parametric twin of the imported vessel (Fig. 6), i.e. if the preconfigured design variables are not changed, the imported, original geometry is analysed by all tools described. This analysis generates the datum design as a reference.

Instead of starting from the imported vessel, all design variables can be modified to define a new starting point and hence a new baseline design. For the envisioned design space exploration of its main particulars, the vessel's length, beam, height and draught are preconfigured as design variables in a given range, which can be adapted according to the specification of the study. Figure 7 shows a parametric variant of the vessel featuring a different length.

The objective functions CAPEX and OPEX call for minimum thruster sizes within the given constraints to achieve the required transit speed and to keep station during crane operation. During a design space exploration focusing on principal dimensions, additional shape characteristics can be utilized for minimizing required power in transit mode within a nested optimisation based on potential flow analysis. A set of design variables is provided in the set-up that allows changing properties of the vessel that are commonly used in standard hull form optimisation processes: Bulbous bow shape, shoulder location as well as longitudinal centre of buoyancy. The number of design variables is kept relatively low to reduce the overall computational effort, since for each design of new principal dimensions, a set of variants is assessed aiming for minimum resistance. Figures 8 and 9 show the bulb shape based on the baseline design and a modified variant respectively.

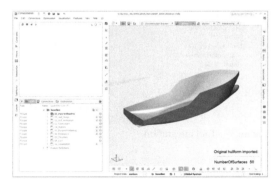

Figure 5. Imported baseline.

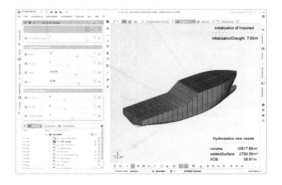

Figure 6. Parametric twin.

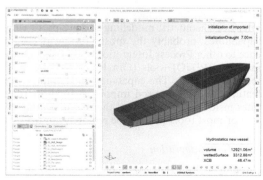

Figure 7. Parametric variant.

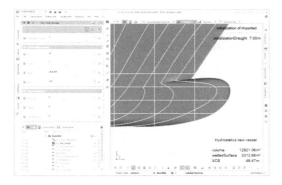

Figure 8. Bulb base line design.

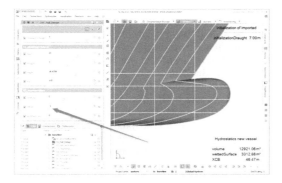

Figure 9. Bulb modified variant.

The new vessel geometry based on the shape modifications as described in this chapter are linked to export methods and specialized features that provide the required file format for the calculation and assessment methods integrated in the case study and don't need any a further manual processing.

5 WEIGHT ESTIMATIONS

This section refers to step 2 in Figure 2.

5.1 *Tool application*

For the stability and cost calculations, the weight of the vessel is of importance. Relative changes in weight compared to the baseline vessel are calculated. As indication of the change in lightship weight, estimations for steel weight and main equipment weight are calculated.

5.2 *Tool description*

The steel weight is calculated based on main dimensions and steel coefficients for the hull and deck-house. These coefficients are based on the reference vessel and can be adjusted in case other features like ice class or larger accommodation space are to be considered.

With regard to the lightship weight of the vessel, changes in main components are taken into consideration. The most important items are the crane and the lifting wire.

The centre of gravity of the lightship in vertical direction is also considered based on the changes in steel weight and equipment weight. The longitudinal position, as a simplification is considered to be at the longitudinal position of the centre of buoyancy. This because the actual loading conditions are not considered, due to the high level nature of the tool. It is assumed that adjustments can be made by ballasting the vessel. This is also the case for the vertical centre of gravity, where ballast is used to obtain suitable metacentric height values. However a check of the vertical centre of gravity position against the baseline vessel and the hydrostatics of the hull configuration, provides an additional quality check.

6 STABILITY CALCULATIONS

This section refers to step 3 in Figure 2.

6.1 *Tool application*

The stability tool is used to evaluate the feasibility regarding the vessel stability for each of the hull configurations generated with each crane type. Not all hull dimension combinations are viable regarding vessel stability.

6.2 *Tool description*

Stability calculations are performed in NAPA for Design (https://www.napa.fi/Design-Solutions). NAPA is a leading software for design stage development of hull shape and stability calculations. NAPA handles a variety of different hull definition methods, including import and export of graphic formats like IGES.

In this application, the hull shape is obtained from CAESES.

6.3 *Hydrostatics*

The hydrostatic model is set up by typical OSV parameters like height of working deck and length of buoyant accommodation above working deck. Refinement of the model, like introducing a moonpool or additional buoyancy can be easily controlled by CAESES parameters.

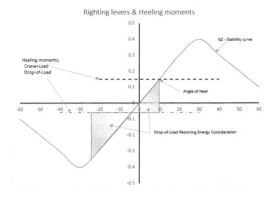

Figure 10. Crane operation stability—angle of heel & restoring energy.

6.4 Stability

Based on the hydrostatic model, together with the input from the weight estimation tool, the stability characteristics are verified against typical IMO requirements. These criteria investigate the extent of positive stability, the energy reserves against capsizing and the sensitivity to wind heeling moments.

For a vessel with a large crane for offshore lifting, additional requirements are introduced, like (Fig. 10):

– heeling angles, when subject to heeling moments from crane and crane load,
– heeling angles, as a result of a drop-of-load situation,
– consideration of restoring energy margins, following a drop-of-load situation,
– influence of counter ballasting.

7 VESSEL MOTIONS

This section refers to step 4 in Figure 2.

7.1 Tool application

The hull dimensions have an effect on the vessel motions in a given seaway. To determine the effect of the operability of the lifting operation, motion calculations are performed. From these calculations limiting wave heights and period are established based on the crane operational criteria on motions and accelerations. These limiting wave heights and periods are input into the station keeping evaluation. This to ensure that the station keeping performance is inline with the crane operability of each configurations.

7.2 Tool description

VERES is the vessel response plugin of SINTEF Ocean's ShipX workbench, which is applicable from the early design stage to the operational phase of a vessel (Fathi D.E. User Manual ShipX Vessel Responses (VERES) Plug-In). VERES is built upon strip theory and offers the ability to calculate ship motion and global loads in waves. The program calculates:

– motion (displacements, velocities and accelerations) transfer functions at arbitrary locations in six degrees of freedom for both, $Fn = 0$ and $Fn > 0$,
– motion transfer functions at specified points,
– unsteady global loads.

Mean drift forces and added resistance in waves can be calculated by either a pressure integration method or according to the radiated energy approach by Gerritsma and Beukelman (Gerritsma and Beukelman (1972)). Furthermore, global wave induced loads can be assessed. Beside the effect of the hull alone, VERES is capable of including the effects from the effect of bilge keels, moonpools, active and passive roll stabilizing tanks, rudder control, and other active and passive control surfaces (such as roll stabilizing fins, T-foils, and interceptors).

Moreover, the post-processor of the code can calculate short-term and long-term statistics, operability limiting boundaries and operability percentage for various relevant limiting motion criteria.

It is worth to mention that in VERES, the hull is defined by a set of body lines at freely selected longitudinal positions. Each of the cross-sections of the hull are specified by a number of offset points, which are further interpolated upon in VERES. In order to create the hull description in ASCII format for VERES from the parametric hull geometry in CAESES, a special module in CAESES has been developed. This module partitions the hull into strips by distributing the points on one half of the hull section which subsequently will be mirrored about the centre line plane by VERES to give a complete description of the hull section.

Figure 11 shows results of numerical simulations in which CAESES executes VERES for different wave headings, from head sea (0°) to beam sea (90°). For each heading VERES finds the maximum wave height (HsLim) for which a certain criteria on operability of crane operation is met. Furthermore, for each heading, CAESES generates an operability index for each heading that can be used as an operability index in the final optimization routine.

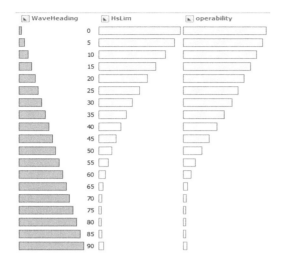

Figure 11. Maximum wave height (HsLim) and operability index for different wave headings.

8 STATION KEEPING

This section refers to step 5 in Figure 2.

8.1 Tool application

Another critical aspect of an OSV vessel performance is the station keeping ability and the required thrust in relation to the environmental conditions. The required thruster sizes are driving both CAPEX and OPEX, and is therefore part of the optimisation process.

The maximum wave height and wave length from the motions evaluation is used as input for the environmental conditions for the station keeping calculations. An additional scaling factor allows for operational margins and the wind speed is added based on the Beaufort scale. The current is based on IMCA North Sea industry standard values. (IMCA, Specification for DP capability plots, 2000)

8.2 Tool description

Station Keeping is another plug-in of SINTEF Ocean's ShipX workbench that enables analysis of the station keeping capabilities of a vessel in the early design process (R E. User Manual ShipX Station Keeping Plug-In). The results from station keeping analysis are of significant importance with respect to dimensioning, positioning and usage of the force generators on the vessel. It also allows study of the performance of the vessel in presence of faulty force generators. The station Keeping plug-in only addresses the static scenarios (steady state conditions); furthermore, static station keeping capability calculations are performed in the horizontal plane with a minimum amount of input data. While users can define their own force generator unit, the vehicle could be equipped seamlessly with any force generator model from the ShipX propulsion library developed by SINTEF Ocean. Similarly, users can develop their own thrust allocation algorithm or simply use any of allocation method developed by SINTEF Ocean and gathered in an allocation library. After configuring the vessel and defining different environmental conditions, the software can produce variety of performance indicators, including but not limited to the DP capability plot defined by IMCA (IMCA, Specification for DP capability plots, 2000), environmental regularity numbers (ERN) defined by DNVGL (DNV, Rules for Classification of Ships, PART 6 CHAPTER 7), and station keeping performance (SKP) defined by ABS (ABS, Guide for Dynamic Positioning Systems, 2014).

Figure 12 presents a typical IMCA DP capability plot in which station keeping capability of the vessel is illustrated in a polar plot. The objective of a DP capability analysis, using DP capability plots, is to determine the limiting environmental conditions (wind speed, wave height, and current) within which the vessel can maintain position and heading while using its DP capability. This is done by balancing the maximum obtainable thruster force against environmental forces due to wind, wave drift, current, and possible other loads.

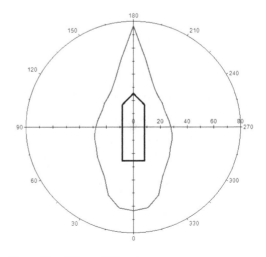

Figure 12. DP capability plot.

9 RESISTANCE

This section refers to step 6 in Figure 2.

9.1 Tool application

In order to determine the thrust and power requirement for transit operations, a resistance and propulsion evaluation is performed. The evaluation is performed for one speed, which is set as an input value.

Since the resistance is influenced by the shape of the bulb and hull details, an optimisation is performed, to find the lowest values for each hull configuration. Setup of calculations:

- baseline total resistance calculations by RANS,
- divide total resistance into viscous resistance and wave resistance,
- assume mechanical efficiency based on chosen propulsion solution, and total propulsion efficiency from available model test data.

9.2 Calculation approach

The potential flow solver SHIPFLOW XPAN by FLOWTECH International AB is utilized for iterative vessel design investigation (SHIPFLOW: www.flowtech.se/products/shipflow-basic). The best calculation results from XPAN are normally obtained for ship speeds in the range between $Fn = 0.2$ and $Fn = 0.35$.

In this high-level approach, the XPAN calculation results are used for relative ranking of the different design iterations, not for determining absolute resistance values.

9.3 Simplified approach

Initial hull is calculated in RANS code STAR-CCM+ (www.mdx.plm.automation.siemens.com/marine). The pressure resistance component of

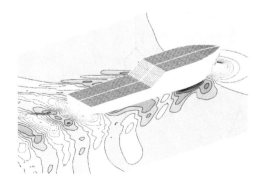

Figure 13. Calculated wave pattern for hull Loa = 92.05 m, B = 23.0 m at draught T = 6.0 m and ship speed Vs = 12kts.

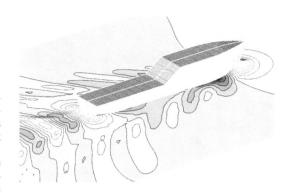

Figure 14. Calculated wave pattern for hull Loa = 92.05 m, B = 16.6 m at draught T = 6.0 m and ship speed Vs = 12kts.

the calculated total resistance for the initial hull is related to the CWTWC coefficient, which is a wave resistance value, obtained from a transverse wave cut method in the SHIPFLOW calculation. When ship speed or hull parameters (e.g. draught, length over all, and breadth) are modified, SHIPFLOW recalculates the wave pattern and a new CWTWC value is obtained. By comparing the CWTWC values of the baseline hull with the CWTWC value of the modified hull, it is possible to obtain a relative ranking of the resistance component contributed by the generated waves. This approach will show the effect of fore—and aft shoulder positions, bulb design and hull main dimensions on the wave-making resistance. Viscous resistance is calculated by a feature function following the ITTC 57 correlation line approach (ittc.info/media/2021/75–02–02–02.pdf Page 2 of 17).

9.4 Propulsion

Ranking of hull performance for transit operations is based on the resistance values. In a more detailed optimisation process, this would not necessarily be correct as propulsive efficiency can alter this. However, in this high-level approach this is reasonable. Propulsive efficiency assessment requires a detailed in-depth design analysis with experiments or advanced RANS runs.

The required propulsion power is directly related to calculated resistance through the total efficiency where the propulsive efficiency η_D is fixed and the mechanical efficiency η_M varies dependent upon main propulsion variant (ITTC symbol and Terminology List Version 2017).

10 MACHINERY SYSTEMS

This section refers to step 7 in Figure 2.

10.1 Tool application

With the performance of the hull configuration established, the cost impact has to be determined. With the machinery systems evaluation tool, required thrust for the main propulsion and manoeuvring thrusters is converted to installed power generation. Base on this a machinery set-up is generated and both fuel consumption and cost calculations are performed, feeding into the CAPEX and OPEX evaluation.

10.2 Tool description

The thrust requirements for both transit and DP operations are compared and the minimum required thrust levels for bow and aft thrusters are determined, based on a basic thruster configuration, which is an input for the tool.

A database with equipment information, consisting of equipment sizing, is used to determine which main equipment is to be installed to fulfil the operational requirements.

With the main thrusters selected, the total power requirement is calculated taking into consideration the electric losses, hotel load and heavy consumers. With this total power requirement the power generation machinery is selected. The machinery configuration is kept the same as the baseline vessel, but the elements are sized according the power supply requirement.

The machinery system of the vessel and the propulsion powertrain is modelled using the DNVGL COSSMOS framework (Dimopoulos et al. 2014). COSSMOS is DNVGL's in-house process modelling framework and consists of a library of reconfigurable generic models of ship machinery components. The models capture the steady-state and dynamic thermo—fluid / mechanical / transport phenomena / electrochemical behaviour of each component. The component model library is coupled with a graphical flow sheeting environment, in which the user can hierarchically synthesize system models of varying complexity. This process results in large systems of non-linear Partial Differential and Algebraic Equations (PDAEs), subject to initial and boundary conditions. The required numerical and optimisation solvers are incorporated to our framework in order to perform a wide range of model-based studies such as steady-state and dynamic simulations, parameter estimation, and non-linear, mixed-integer and dynamic optimisation. COSSMOS has been successfully applied to machinery studies in all ship segments, and with particular relation to this application in OSV vessels and LNG carriers with electric propulsion (Stefanatos et al. 2015, Dimopoulos et al. 2016).

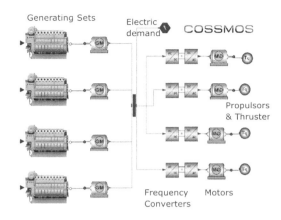

Figure 15. Propulsion powertrain model in COSSMOS.

In the current study, COSSMOS is used to model a diesel-electric propulsion powertrain at steady-state conditions. The developed system model is depicted in Figure 15. The propulsion powertrain consists of four diesel generating sets covering the total electricity demand from the propulsors and thrusters as well as and hotel and auxiliary electricity demand.

The base line vessel propulsion powertrain consists of:

- four RR B32:40 L6 generating sets each with power at 2880 kW, at 720rpm,
- two main stern azimuthing thrusters each with power at 3000 kW,
- two forward tunnel thrusters each with power at 1200 kW.

The system model is utilized as the fuel consumption, efficiency and related costs estimator, within the CAESES optimisation framework. The steady-state simulation model receives information from the other CAESES modules on propulsion and electricity demand and returns the fuel consumption and related costs. The COSSMOS propulsion powertrain system module is packaged as a stand-alone executable and it is interfaced with CAESES in a batch file mode.

The subsequent steps, with respect to machinery systems modelling, simulation and optimisation, consist of the development of a more generic system model suitable for design and operation optimisation.

11 CAPEX AND OPEX ESTIMATION

This section refers to step 8 in Figure 2.

11.1 Tool application

In addition to the fuel consumption, which is part of the operational expenditure (OPEX), the capital expenditure (CAPEX) is represented by the cost of the power and propulsion system elements, crane and steel costs. Values of CAPEX and OPEX relative to the baseline, shows the differences between the various possible configurations reviewed by the tool. The early design stage application of this tool allows this simplification.

11.2 Tool description

The costs for power and propulsion system are based on the thrust requirement for station keeping and sailing, as described in sections 8, 9 and 10. The costs of these elements are summed up to obtain a reference value representing the power and propulsion cost of the CAPEX.

The steel cost is based on the steel weight, as described in section 5 and is based on a price per ton of worked steel. Since the deckhouse size is kept as a constant in this case, no corrections for accommodation outfitting is taken into consideration.

Further parameters determining the CAPEX and OPEX levels can be added if deemed important for the comparison, by adding additional features in CAESES or linking additional life-cycle cost tools e.g. Reliability, Maintainability and Availability (RAM) tools for assessment of maintenance and repair cost.

12 DESIGN SPACE EXPLORATION

This section refers to CAESES as the hub in Figure 2.

The process of finding promising designs and hence optimising a product for one or several objective functions is typically initiated by running a design space exploration (DSE). The design space is defined by the given range of input parameters that influence the system's behaviour and its objective functions. The number of input parameters—the design variables—defines the complexity of the system and also gives an indication how many designs need to be investigated for developing an understanding of the design space. A good estimate for investigating an n-dimensional problem in the context of hydrodynamic performance is n^2 samples. Algorithms for populating the design space are e.g. the Latin Hypercube method or a Sobol Sequence (M. Cavazzuti, 2013). Both methods aim for distributing design candidates in the design space such that all areas of the design space are covered with as little evaluations as possible.

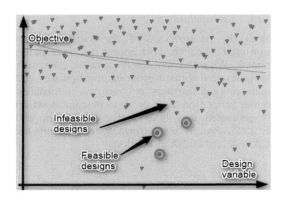

Figure 16. Constraint dominated design space.

In practical design tasks the design space is often strongly reduced by constraints. It can be easily understood that for an independent variation of length and beam as design variables, a constraint for the required displacement is violated frequently and feasible designs will not be distributed evenly in the design space. The aim of a DSE therefore is not only getting an understanding of the design space but also identifying its limitations by constraints. Figure 16 shows a design space that is dominated by constraints. Out of 100 variants only 3 feasible designs have been identified.

A design space exploration helps the design team to understand the system's behaviour, adapting the design space or to reconsider limits of constraints. As long as the constraints are not subject to costly simulations a-priori investigations of the variants is very helpful without launching any CFD simulation.

After having identified the most promising feasible designs, pre-defined optimisation algorithms are executed to squeeze out their potential, starting from one or several of these candidates.

13 RESULTS AND FURTHER WORK

The implementation of the early stage holistic design tool for the case study of a Subsea Construction Vessel is ongoing and the following has been achieved to date:

– definition of a case study applying holistic design synthesis at an early stage of a design process,
– establishment of a multi discipline design team involving domain specialists from four different organizations located in four different EU countries,
– establishment of a integration framework based on the CAESES platform from Friendship Systems,

- implementation of a parametric hull model enabling automatic design space exploration,
- integration of NAPA for stability calculations,
- integration of ShipX/VERES from Sintef Ocean for vessel motion calculations,
- integration of ShipX/Station Keeping from Sintef Ocean for DP capability assessment,
- integration of SHIPFLOW from Flowtech for resistance calculations,
- integration of COSSMOS from DNVGL for fuel consumption estimations,
- Implementation of dedicated features in CAESES for weight estimations,
- Implementation of feature in CAESES for thruster and propulsion selection from Rolls-Royce Marine product portfolio,
- Implementation of feature in CAESES for CAPEX and OPEX calculations.

Next steps in this development are:

- implementation of automatic exploration of the design space in CAESES,
- implementation of multi parameter optimisation in CAESES,
- implementation of user interface.

The end results of this holistic design tool is an early definition of the main dimensions of a hull considering the mission of the vessel and exploring different key options like selection of mission critical products, propulsion system configuration and power system type and architecture.

14 SUMMARY AND CONCLUSION

Decisions taken at the early design stage can have a significant impact on the final performance and life cycle cost of a vessel. For specialized vessels with a complex operational profile, defining the best trade-off between main design parameters satisfying the mission requirements of the customer, respecting constraints in rules and regulations and providing the optimum solution with regard to multiple KIPs as life cycle cost, emission is a challenging task. Multiple disciplines working at different modules or units of the vessel must interact to find the best solution. At an early design stage major decisions are taken, e.g.:

- main dimensions of the hull,
- selection of propulsion configuration,
- definition of power system type and architecture
- selection of mission critical components like cranes and other deck machinery systems

This paper describes a methodology based on a Model Based System Engineering approach. The design integration platform CAESES is used to connect several specialists tools in a design synthesis assuring a consistent state of the overall design during the design process. The tool can explore a user defined design space automatically and perform multi parameter optimisation of the results to find the optimum solution based on KPIs. This is a tool for a naval architect, enabling him to perform early design synthesis involving advanced specialist tools. A simplified user Interface for the naval architect is important to make this a practical usable tool. A web interface where the design case can be configured and tools can be executed is planned to make this a practical usable tool.

Domain specialists are still as important as before in this design methodology, but the interaction with the naval architect is structured in a new way with the aim of reducing design time and increasing the quality of the decisions taken at an early design stage.

This paper has described a case study on this methodology based on early conceptual design of a Subsea Construction Vessel. The implementation of this case is still ongoing in the HOLISIP project (HOLISHIP, 2016–2020).

ACKNOWLEDGEMENT

HOLISHIP has received funding from the European Union's Horizon 2020 research and innovation programme under grant agreement n° 689074.

REFERENCES

American Bureau of Shipping (ABS). *Guide for Dynamic Positioning Systems, 2013 (Updated July 2014)*.
CAESES. www.caeses.com/products/caeses/.
Cavazzuti M, Optimization Methods. *From Theory to Design*, Springer-Verlag Berlin Heidelberg 2013.
Det Norske Veritas (DNV). *Rules for Classification of Ships, PART 6 CHAPTER 7, 2013*.
Dimopoulos GG, Georgopoulou CA, Stefanatos IC, Zymaris AS, Kakalis NMP. *A general-purpose process modelling framework for marine energy systems*. Energy conversion and management. 2014;86(0):325–39.
Dimopoulos GG, Stefanatos IC, and Kakalis NMP. *Assessment of LNG Carriers electric propulsion configurations via process modelling*. 28th CIMAC World Congress, Helsinki, Finland, June 6–10 2016.
Fathi D.E. *User Manual ShipX Vessel Responses (VERES) Plug-In*. Internal Report SINTEF Ocean. 2017.
Gerritsma J. and Beukelman W. *Analysis of the Resistance Increase in Waves of a Fast Cargo Ship*, Netherlands Ship Research Centre TNO, Report 169S, April 1972.

Gianni D, D'Ambrogio A, Tolk A. *Modelling and Simulation-Based Systems Engineering Handbook*. CRC Press. 2015.

HOLISHIP (2016–2020). *Holistic optimisation of ship design and operation for life cycle, Project funded by the European Commission, H2020- DG Research, Grant Agreement 689074, www.holiship.eu.*

International Marine Contractors Association (IMCA).

ITTC 57. *ittc.info/media/2021/75-02-02-02.pdf Page 2 of 17.*

ITTC symbol and Terminology List Version 2017. ittc.info/media/7937/structured-list_2017_a.pdf.

NAPA. www.napa.fi/Design-Solutions.

Ora Research LLC and intrinSIM LLC. *DESIGN SPACE EXPLORATION MARKETS & OPPORTUNITIES. A collaborative market report, 2018.* Ringen E. *User Manual ShipX Station Keeping Plug-In. Internal Report SINTEF Ocean. 2017.*

SHIPFLOW. *www.flowtech.se/products/shipflow-basic).*

Specification for DP capability plots, IMCA M140 Rev 1, 2000.

STAR-CCM+. *www.mdx.plm.automation.siemens.com/marine.*

Stefanatos, IC., Dimopoulos GG, Kakalis, NMP, Vartdal BJ, Ovrum E, Sandaker K and Haugen BR. (2015). *Modelling and simulation of hybrid-electric propulsion systems: the Viking Lady case, 12th International Marine Design Conference—IMDC 2015, Tokyo, Japan.*

Watt DA, Findlay W. *Programming language design concepts*. John Wiley and Sons, 2004.

HOLISTIC ship design optimisation

J. Marzi, A. Papanikolaou & J. Brunswig
HSVA, Hamburg, Germany

P. Corrignan, L. Lecointre, A. Aubert
Bureau Veritas Marine and Offshore, Nantes, Cedex 3, France

G. Zaraphonitis
NTUA, Ship Design Laboratory, Athens, Greece

S. Harries
Friendship Systems, Potsdam, Germany

ABSTRACT: The present paper describes the HOLISHIP–Holistic Optimisation of Ship Design and Operation for Life Cycle project approach to ship design and operation and demonstrates a subset of its functionality on the basis of a case study. This refers to a RoPAX ferry optimisation for minimum powering requirements and maximum life-cycle economic performance in realistic operating conditions by use of concurrent engineering tools from different project partners operating in collaboration on a common design software platform. The impact of alternative operating/speed scenarios on case study ship's efficiency and safety is presented and discussed.

1 INTRODUCTION

Today's shipping industry operates in a complex environment with numerous economic, environmental and even social restrictions. Energy efficiency, safety and environmental protection are key requirements for a sustainable shipping industry and the means of transportation need to be adapted accordingly. This calls for significant changes in the traditional ship design process, which is a complex, multi-disciplinary and multi-objective task of both technical and non-technical nature. Likewise multifaceted is ship operation. A system approach to ship design and operation considers the ship as a complex system, integrating a variety of subsystems and their components, e.g. for energy/power generation and ship propulsion, for cargo storage and handling, accommodation of crew/passengers and ship navigation. Any state of the art design process inherently involves optimisation, namely the selection of the best solution (trade-off) out of many feasible ones for a given target function or transport task, depending on vessel type. Today, this trade-off or formalised optimisation increasingly involves life-cycle considerations and objective functions.

In practice often only parts of the ship design and even less of the ship's life-cycle are integrated in a common database and software platform. This typically results in less favourable selections of optimised sub-systems or components while the optimal ship would have been the result of a holistic optimisation of the entire ship system. It should be noted that the system ship is actually a component of the wider transport system, thus a holistic approach to ship design should actually also consider aspects of fleet composition and transport/mission scenario optimisation, which are not addressed in this paper. For a systems approach to ship design see, e.g. (Hagen et al., 2010) and (Guégan, A. et al., 2017) in the HOLISHIP project.

The approach chosen in the HOLISHIP project (www.holiship.eu) acknowledges the fact that, in practice, surrogate models need to be employed for several sub-systems and components to reduce computational/processing time and the complexity of the overall optimisation problem; also, the often conflicting constraints and requirements of the optimisation, which in turn result from contradicting interests of the various stake holders in the maritime transport chain, need to be optimally balanced. The volatility of market conditions and associated transport demand, the variability of the operational conditions over a ship's life-cycle, the cost of raw materials as well as energy cost during operation all need to be considered in compliance with continuously changing regulatory requirements regarding ship safety and the ecology of the marine environment.

The present paper addresses the topic of design and optimisation of ships and their operation by a holistic approach, as elaborated in (Papanikolaou 2010), constituting a multi-disciplinary and multi-objective problem. The implementation of this approach requires the coupling and integration of a series of software tools within a design software platform, sharing common data, as will be outlined in the next sections. For illustration purposes a representative application case covering important design aspects for a modern RoPAX ferry is shown. The HOLISHIP project will further address 8 other application cases for merchant and research vessels which will be shown in the future.

1.1 The EU-project HOLISHIP

To meet present and future challenges as outlined above, a large team of 40 partners, led by HSVA (Hamburgische Schiffbau-Versuchsanstalt) and NTUA (National Technical University Athens), set out to develop the concept of an integrated, holistic ship design platform and implement it in the context of the Horizon 2020 Research Project HOLISHIP – Holistic Optimisation of Ship Design and Operation for Life Cycle (2016–2010, www.holiship.eu). The project considers all relevant design aspects, namely energy efficiency, safety, environmental compatibility, production and life-cycle cost, which are to be optimised in an integrated manner with the aim to deliver the right vessel(s) for future transport tasks. To do so, HOLISHIP addresses different design steps, covering basic design and contract design of vessels as well as virtual prototyping for design and operational assessment. These are implemented in two platforms of which the first one, covering concept and contract design, is addressed in the present paper.

This HOLISHIP design platform is built on CAESES®, a state-of-the-art process integration and design optimisation environment developed by FRIENDSHIP SYSTEMS. It integrates first-principles analysis software from various disciplines relevant to ship design and combines them with advanced multi-disciplinary and multi-objective optimisation methods. Due to the complexity of several evaluations surrogate models are employed to limit computational effort.

Compared with traditional approaches the interplay of all design components in form of a design synthesis model—hosted via the HOLISHIP platform(s) – allows exploring a much wider design space and, finally, helps achieving superior designs in less time (see, e.g., an application to tanker design by Sames et al., 2011). Figure 1

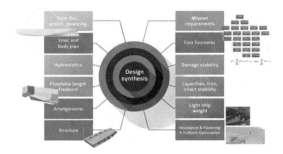

Figure 1. HOLISHIP design synthesis combining all relevant disciplines of ship design.

illustrates the holistic approach to tool integration which enables a concurrent analysis and optimisation of systems and components, contrasting the sequential approach associated with the idealised view of the classical design spiral.

During its first project phase HOLISHIP integrates a full range of disparate software tools into the design platform. This paper presents a snapshot after 15 months of development (late 2016 to 2017). It highlights a first application of the HOLISHIP platform and the tools coupled so far to the design of a modern RoPAX ferry. This utilises parametric models for the hull form, general arrangements, structural design, engine layout and energy simulation and life-cycle assessment. Simulation codes are used for hydrodynamic analyses, provided by HSVA and NTUA, as well as for intact and damage stability, realized by NTUA and engine room and energy simulation from BV. This application case is continuously enhanced as more tools will be added to the platform. By the end of the project nine different application cases, as diverse as a double-ended ferry and an offshore platform for an arctic environment, will have been worked on through the HOLISHIP platforms.

2 INTEGRATION APPROACH

2.1 The integration platform for concept and contract design

The HOLISHIP design platform is based on CAESES®, a general process integration and design optimisation (PIDO) environment developed and licensed by FRIENDSHIP SYSTEMS. It allows to couple any software which can be run in batch-mode and to set up process chains for automated design and optimisation studies.

The available coupling mechanisms are very flexible and based on template files for input and output of external s/w components. The templates are used to specify parametric data relevant for the optimisation. An elaborated example is given in MacPherson et al. (2016), for more background see Abt et al. (2009). Software tools to be coupled can be made available either locally (on the same computer) or remotely (e.g. within the same network), possibly combining different operating systems (Windows and Linux).

Within a CAESES project many different software connections can be used concurrently so that arbitrary process chains can be built. Moreover, CAESES supports the set-up of hierarchies so that every entity of a model knows on which data items it depends and which data items it serves.

2.2 Coupling tools for HOLISHIP

While the coupling mechanism within CAESES is very flexible the actual process needs both expert knowledge of the tool to be integrated and expertise in using CAESES itself. In order to enable a larger group of users to effectively exploit integrated software tools in standard applications a new functionality is introduced into CAESES that reduces complexity and makes it easier for the non-expert to run high-quality simulations. A novel wrap functionality for specific use cases offers technical APPs, short for applications, which provides customised expert knowledge for a given task.

2.3 Usage of surrogate models

Since several simulation tools require substantial computer resources (flow computations, structural analysis, parametric stability models) and special environments which may not all be available at the same time and to the same people, CAESES provides methods to pre-compute data for later usage and store them in response surfaces: the Surrogate Model. Here a design-of-experiment (DoE) is undertaken for a chosen set of free design variables, which form a task-specific sub-set of the total design space of interest to build a surrogate model in CAESES. To this end DAKOTA, an open-source optimisation kit by Sandia National Laboratories (dakota.sandia. gov), is embedded in CAESES. A simulation tool can then be run from CAESES separately and upfront to be subsequently replaced by a suitable response surface. A range of models are made available, for instance, Artificial Neural Networks (ANN), polynomial regression and kriging (Harries, 2010).

3 DESIGN DISCIPLINES IMPLEMENTED

3.1 Hydrodynamics

The hydrodynamic performance of a ship determines to a large extent the energy efficiency and—together with stability—a major part of its safety. The required propulsive power for a specified speed is a key contractual item for any new vessel as it determines fuel consumption and hence cost and emissions. Low resistance and high propulsive efficiency are fundamental prerequisites and optimising the hullform and the propeller/propulsor performance using different specific CFD tools is a must. A variety of further analysis tools for seakeeping performance, added resistance in seaways and due to wind, manoeuvring or the effects of hull appendages and energy saving devices up to the prediction of the effect of increased frictional resistance due to hull fouling form the basis for a complete hydrodynamic analysis.

The range of simulations applied to a specific design is adapted to its particular requirements. CFD predictions typically require substantial computational effort which is barely tolerable during an actual design optimisation process. Such analyses are successively implemented and generate response surfaces (surrogate models) which can be used during design and optimisation.

3.2 Ship stability

The safety of ships against sinking/capsize in case of loss of their watertight integrity is of prime interest to the maritime regulatory bodies, the maritime industry and to the entire society. The new probabilistic damaged stability regulation for dry cargo and passenger ships (SOLAS 2009) represents a major step towards the rationalisation of the procedure for the assessment of a ship's survivability in damaged condition. While the new regulation is more rational than the earlier deterministic approach (SOLAS 90), it requires the consideration of some hundreds of damage stability/flooding scenarios, which can only be studied by dedicated software tools (Papanikolaou, 2007). This effort is further increased when considering alternative arrangements and thus calls for specialised design software tools as an alternative to the traditional manual study of a few design/compartmentation alternatives. This is a crucial, yet very demanding task of contemporary passenger ship design. The EU funded project GOALDS (Papanikolaou et al., 2013) developed software tools for the parametric design and auto-mated multi-objective optimisation of RoPAX (and cruise ships), which are adapted to the new regulations and lead to vessels of enhanced survivability, while considering also building cost and efficiency in operation (Zaraphonitis et al., 2012). These software tools for the assessment of a ship's damage stability, along with corresponding ones for the intact stability, are now integrated in the HOLISHIP platform, allowing the concurrent optimisation of a ship's stability/safety with all other major design disciplines.

3.3 Energy systems simulation

Energy system simulation focuses on the way energy is produced and consumed on-board. In an approach to reduce overall energy consumption during ship operation, this complements traditional optimisation to improve propulsive power requirements which are largely based on hydrodynamics and (combustion) engine improvements. A ship is a highly complex system of sub-systems and components, e.g. propulsion, electricity, cooling, fresh water etc. with strong couplings between the different energy flows. This offers vast opportunities for simulation and optimisations which are typically performed using a Model Based System Engineering approach. The tool applied in the present context is SEECAT, an energy modelling and simulation software package developed by Bureau Veritas which takes into account the various operational and environmental conditions met during the life time of a vessel (Marty et al., 2012). This Model Based System Engineering approach has proven suitable to model complex energy flows while considering complex ship operational profiles. It has been successfully implemented for ship optimisation in previous research and industrial projects (e.g. Faou et al., 2015).

The innovative and challenging aspects developed in HOLISHIP concern the way such an approach can be integrated in the new design synthesis developed in the project. This covers the definition of the energy systems optimisation workflow as well as parametric models of the ship energy systems and the exchange mechanisms between tool and the design platform. In a first step SEECAT was connected to the HOLISHIP platform using a Python script whose parameters are modified by CAESES. The script drives the simulations and modifies parameters in the SEECAT environment using a COM interface (Component Object Module) thus allowing to perform energy system optimisation in-line with hull calculations. The approach offers the opportunity for a parallel execution of hull design and energy system optimisation which will change the traditional—sequential—workflow.

3.4 Cost assessment

Enhanced software tools for the evaluation of building and operational cost, annual revenues and eventually for the life-cycle assessment of alternative designs are currently developed in HOLISHIP. In the meanwhile, simpler tools, specifically developed for the application case presented here are used for the cost assessment aiming to close the design loop and to enable the demonstration of the potential of the adopted design procedure and of the developed optimisation platform. To account for some inherent uncertainty in the underlying cost data, differences in cost and Net Present Value (NPV) are used to compare with the baseline design rather than using absolute cost figures. The impact of design modifications on building and operating cost and annual revenues are calculated first and, based on them, the variation of the NPV for a specified life time is estimated.

3.5 Further HOLISHIP platform elements

The present study describes a snapshot of the HOLISHIP developments which will continue until 2020. During this period more tools will be added to the platform, as necessary for the conduct of the planned Application Cases. These mainly refer to structural design and life cycle assessment (LCA).

For ship's structural design and the generation of related data like structural weight/lightship/dis-placement, centroids and their effect on payload, stability etc. a variety of methods and tools are developed as appropriate for the various application studies. Structural design data for concept design are semi-empirical in principle or classification society rules-based, whereas contract design structural data are resulting from the application of advanced structural analysis methods like FEM. Such tools play an important role in the design of innovative vessels for which no empirical data are available, or when optimising vessels for minimum structural weight.

Life cycle aspects and their assessment (LCA) receive special attention in HOLISHIP. Future and better vessel designs need to adapt to changes of the operational profiles encountered during their life span. Assessment of the environmental, energy efficiency and economic performance of a vessel will be via suitable Key Performance Indicators (KPIs), e.g., Cumulative Energy Demand – CED, Global Warming Potential – GWP, Net Present Value – NPV. This assessment will include the evaluation of different operational profiles and maintenance strategies, allowing validation of the fit-for-purpose properties of the equipment and to extend warranty to the ship owner. A Decision Support System (DSS) will be developed allowing the identification of the most effective decisions/strategies to be assumed at any stage of the vessel's life cycle (as a function of the vessel's design features), while considering lifecycle uncertainties (e.g., fuel, chartering).

4 APPLICATION CASE

To illustrate HOLISHIP developments a realistic design example in form of a RoPAX ferry sailing

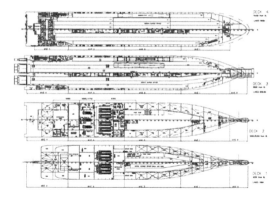

Figure 2. General arrangement—RoPAX ferry.

between Piraeus (mainland Greece) and Heraklion (Crete) was chosen. This representative example serves as both a testbed and a showcase to illustrate the potential of the design platform developed. An operational profile comprising a daytime trip of 6.5 hours at 27 kts and a night trip of 8.3 hours at 21 kts was specified.

4.1 Owner's requirements, operational profile and basic design

For the chosen application case a set of most important owner's requirements with regard to transport capacity have been selected on the basis of equivalent vessels. These are given in the following Table 1.

The above owner's requirements correspond to a baseline design initially developed by FINCAN-TIERI in the context of the EU-funded research project GOALDS and are further elaborated by the HOLISHIP partners. This is a twin screw RoPAX with mechanical propulsion, fitted with a main and an upper trailer deck and a lower hold. A hoistable deck is also fitted on the upper trailer deck. For loading and unloading of vehicles, the ferry is fitted with two stern ramps and side hinged bow doors with a bow ramp. The main characteristics of this vessel are given in Table 2.

The ship will be operated year-round, considering a high season of seven weeks with seven roundtrips per week, a medium season of twenty four weeks with five roundtrips per week, and a low season of twenty two weeks with three roundtrips per week resulting in total in 235 roundtrips per year. Appropriate occupancy rates for passengers, cars and trucks for each of these three periods have been assumed for the calculation of annual revenues. Since there are always limits in the demand for transport work, a gradual reduction of the occupancy rates for ships with larger transport capacity is assumed, when, for the purpose of optimisation,

Table 1. Owner's requirements.

Number of passengers	≥ 2,080
Number of passenger cabins	> 300
Lane length	≥ 1,950 m
Payload	≥ 3,500 t
Number of crew	120

Table 2. Main characteristics of baseline design.

Length between perpendiculars	162.85 m
Beam	27.6 m
Subdivision draught	7.10 m
Height of bulkhead deck	9.80 m
Gross tonnage (GT)	≈ 36,000
Deadweight (DWT)	5,000 t

parametrically varying ship's size. For example, for a 10% (resp. 20% or more) increase of transport capacity, compared to the baseline design, it was assumed that the annually transported passengers or vehicles increased by 7.5% (resp. 10%). This assumption ensures that larger vessels are only modestly exploiting the economy of scale.

4.2 Parametric design models

Various parametric models were built in order to undertake a first design and optimisation study:

- A flexible geometric model for the form of the bare hull within CAESES
- A comprehensive compartment model for spaces, including decks, bulkheads, tanks etc, within NAPA
- A weight model that estimates the weights and centres of gravity of key systems and components as functions of main dimensions
- A preliminary cost model (to be replaced in the future by more advanced life-cycle cost assessment).

The geometric model for the RoPAX hull uses main dimensions and relevant form parameters for hydrodynamics and stability, i.e., length, beam, draft, block coefficient, midship coefficient, centre of buoyancy, etc., along with local parameters for the bulbous bow. The model was set up in CAESES, allowing specific export of geometry to the coupled software tools in the formats required. For instance, NAPA receives the hull form as an IGES-file, the viscous flow solver *FreSCo+* would be fed by a watertight *stl-file* while the potential flow solvers, v-Shallo and NEWDRIFT+, obtain dedicated panel meshes of different topology (Harries et al., 2017).

A comprehensive parametric model for the watertight subdivision was developed within NAPA. It receives the hull form from CAESES as an IGES-file and subsequently creates all watertight compartments below the subdivision deck and on the main car deck. All openings and cross connections required for the damaged stability assessment are also created. Based on this model, the ship's light weight and the weight centre along with its transport capacity are calculated. A simplified procedure for the calculation of lightship is employed, which shall be later replaced by a more accurate external tool being currently developed within the HOLISHIP project. A series of loading conditions are defined for the evaluation of compliance with relevant intact and damage stability criteria. When changing the hull form all bulkheads and decks are "snapped" to the new shape, preserving the topology of the general arrangement. At this point all compartments are linearly scaled in longitudinal, transverse and vertical direction. This is a simplification and may become subject to a more elaborate treatment in the future.

4.2.1 Hydrodynamics

Hydrodynamic analysis requires precise knowledge of the actual hull geometry and a reasonably constructed parametric model of the hull. Using the parametric model described above, different hydrodynamic analysis tools have been employed to predict calm water resistance and power requirements as well as the effect of added resistance in a seaway.

For the *Calm Water* analysis a combination of HSVA's in-house tools v-Shallo (panel code, wave resistance) and FreSCo+ (RANS) was used. More information on the codes can be found in (Gatchell et al., 2000) and (Hafermann, 2007), the integration in the design platform is described in more detail in (Harries et al. 2017).

For the RoPAX design example two response surfaces for delivered power were established, one for the ferry's lower speed of 21 kts and the other for the top speed of 27 kts. Two design-of-experiments (Sobol) were run with v-Shallo, each comprising 360 design variants. Combining both v-Shallo and *FreSCo+* results to estimate power demand for all ferry variants during an optimisation, the response surface approach described in 2.3 was applied. Artificial Neural Networks were employed within CAESES and their accuracy was checked by comparing additional variants that were not contained in the training set with the corresponding results from direct simulations. A typical deviation of about 1% was found. Note that these hydrodynamic response surfaces can be viewed as a numerical hull series.

For the prediction of *Added Resistance* in waves NTUA's NEWDRIFT+ code (Liu et al., 2017) was employed. This 3-d panel code uses Green Functions to evaluate motions, wave loads and mean second order forces on ships in the frequency domain. The code is a further development of the original NEWDRIFT code by adding software tools for the calculation of added resistance of ships in waves based on the far field method, with empirical corrections for the short waves regime (Liu & Papanikolaou, 2016). It is fully integrated into the CAESES platform using a hull panelisation created in the CAD section of the platform.

Calculating the added resistance for a wave spectrum (here a JONSWAP spectrum with $h_s = 3$ m and $T_P = 7$ s) may require up to 20 minutes on a standard computer. Therefore, again a surrogate model in form of a response surface has been created for use in the optimisation. A comparison of added resistance results calculated by NEWDRIFT+ and estimates from the response surfaces are presented in Figure 3. Both models capture the relationship quite well, namely generally with an error of +/− 2.5%, which is considered much smaller than the accuracy of the ensuing seakeeping code (and of similar SoA codes in general).

4.2.2 Stability

According to SOLAS 2009, the ship's Attained Subdivision Index is calculated as the weighted average of partial indices at the deepest subdivision draught d_s, the partial subdivision draught d_p and the light service draught d_l (i.e. $A = 0.4A_s + 0.4A_p + 0.2A_l$). Each partial index is a summation of contributions from all damage cases: $A = \Sigma p_i s_i$, where i represents each group of compartments under consideration, p_i accounts for the probability that only this group of compartments may be flooded, and s_i accounts for the probability of survival after flooding. In addition, a Required Subdivision Index, is introduced, as a function of the number of persons on-board. The subdivision of a passenger ship is considered sufficient if the A-Index is not less than the R-Index

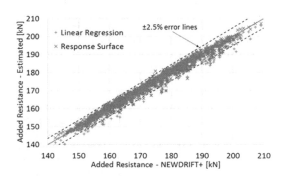

Figure 3. Comparison of added resistance in head seas at 27 kts calculated by NEWDRIFT+ and estimated by response surfaces and error bounds.

and if, furthermore, none of the partial indices (A_s, A_p and A_l) is less than 0.9 of the R-Index.

In order to speed up calculations during an optimisation campaign, the integrated models developed in CAESES and NAPA were used to carry out a series of preparatory calculations, to provide adequate data for the development of surrogate models for fast yet reasonably accurate estimation of the A-Index and the corresponding partial A-Indices. A comparison of the actual A-Index calculated according to SOLAS 2009 as amended and the estimated A-Index obtained using the response model is presented in Figure 4 and the error proves to be in general less than +/− 1%.

4.2.3 *Energy simulation*

A rather complete energy model of the RoPAX ship has been built by Bureau Veritas in its SEE-CAT simulation tool as indicated in Fig. 5.

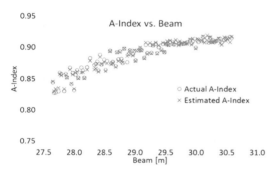

Figure 4. Actual and estimated A-Indices by response surfaces for parametrically varied ships vs. beam.

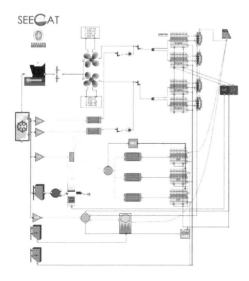

Figure 5. Energy system model.

The vessel's energy system comprises:

- a twin screw propulsion system, using one or two diesel engines per shaft line, depending on speed, and two controllable pitch propellers (CCP);
- An electric plant with three diesel generators and two PTO (Power Take-Off). The electrical power required at sea is 2.5 MW;
- Some auxiliary systems as shown in Fig. 5.

The necessary inputs for the energy simulation are the hydrodynamic resistance curve, as well as the thrust deduction factor t and the wake coefficient w. The platform integration assures the availability and consistency of the data and predictions parallel to hull computations without increase in effort.

The machinery optimisation focuses on the best configuration to operate the ship.

The machinery optimisation is done for a given operational profile as indicated before. A full round trip operational profile is defined as a time series of:

- Speed profile: 21 kts for 8.3 hrs and 27 kts for 6.5 hrs (daily rountrip requirement),
- Navigation mode (at berth, manoeuvring, at sea),
- Fresh water consumption: 5 m³/s and electrical power required: 2.5MW,
- Fuel type (MDO at berth or HFO when manoeuvring or at sea).

In this case, the fresh water consumption is considered constant. The optimisation is carried out considering 9 different configurations:

Config.	1	2	3	4	5	6	7	8	9
PTO	OFF	ON	OFF	ON	OFF	ON	OFF	ON	OFF
ME/shaft line	2	1	1	2	2	1	1	2	2
Rpm/ME	free	410	410	410	410	510	510	510	510

Rules and boundary conditions for the simulations include:

- The PTOs can't be used on the same electrical network as the electric plant, so at sea both cannot be used at the same time,
- Similar configurations for port and starboard side.

An example of the instantaneous fuel consumption at 21 kts and 27 kts for the configurations is shown in Figure 6. MCR has been set to 13 MW which means that only configurations with 2 engines per shaft line (#1,4,5,8,9) can reach 27 kts.

It appears that for 21 kts the best configuration is no 6 (PTO: on, 1 engine/shaft line, 510 rpm) and

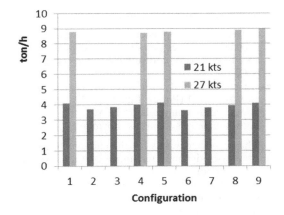

Figure 6. Instantaneous consumption at 21 and 27 kts, MCR = 13 MW/engine.

Table 3. Free variables and range.

Free Variable	Lower bound	Baseline	Upper bound
Length BP	155.0 m	162.0 m	170.0 m
Beam	27.6 m	27.6 m	30.6 m
Design draught	6.5 m	7.1 m	7.1 m

for 27 kts configuration 4 (PTO: on, 2 engines/per shaft line, 410 rpm). The interest of using PTOs is demonstrated as the simulated fuel consumption is reduced by 4% in average. The assessment of fuel consumption yields 86.8 tons of HFO per day/roundtrip for the optimal scenario. Different engines shall be compared in the next steps of the project.

4.3 Design optimisation set-up

With the vessel presented in Section 4.1 as a baseline, an optimisation study was undertaken to identify optimal RoPAX vessels fulfilling the owner's requirements. It should be noted that the FINCANTIERI baseline was originally designed according to the older SOLAS 2009 regulations. Any new design, however, should comply to the considerably more stringent damage stability requirements introduced by the IMO Resolution MSC.421(98), adopted in June 2017. It was therefore anticipated that, although sharing the same topology with the baseline, the outcome of the optimisation should be a significantly different design. In other words, the baseline, although being a valid RoPAX ferry when developed several years ago, would now have to be considered an *infeasible* design and, consequently, the design space was extended towards vessels of wider beam as can be seen in the following table:

In order to allow for many investigations and variants to be studied, the computational effort generally needs to be as low as possible. As discussed in previous sections, resource intensive simulations were first performed upfront (and at different sites) and afterwards replaced by dedicated response surfaces. Using these fast yet sufficiently accurate response surfaces, approximately 200 designs could be studied per hour on a standard desktop computer. For comparison, about one hour per design variant would have been needed if the simulations had to be undertaken directly with the various CFD tools for hydrodynamics and NAPA for damaged stability.

Suitable optimisation constrains were introduced so as to distinguish feasible and infeasible designs. The most important constraint required compliance with the intact stability requirements specified by the IMO Resolution A.749(18) as well as with SOLAS 2009 Part B, Reg. 6 and 7, as amended by the IMO Resolution MSC.421(98). As a temporary safeguard against possible inaccuracies in the GM estimation, suitable safety margins were introduced: The intact stability requirements should be met with a GM margin of 0.20 m, meaning that the actual GM in all loading conditions tested ought to be greater by at least 0.20 m than the one required by the intact stability criteria. For the A-Index and the three partial indices a safety margin of 0.02 was introduced, i.e., all feasible designs need to meet the inequality constraints $A - R \geq 0.02$ and $A_i - R \geq 0.02$, respectively. Additional constraints were employed to ensure adequate transport capacity in terms of lane length and DWT for each feasible design variant.

4.4 Selected results of first optimisations

Utilising the established synthesis model, the optimisation was conducted in two stages:

First, a design space exploration was undertaken in which 500 variants were generated within CAESES by means of a design-of-experiment (*SOBOL*). The hull forms were transferred to NAPA in order to create their watertight subdivisions and, following this, were evaluated using the tools and procedures described above. From these 500 designs only 3 proved feasible, emphasizing the challenge of finding acceptable let alone optimal designs.

Subsequently, a multi-disciplinary and multi-objective optimisation was carried out in which the Net Present Value of the designs was to be maximised while the fuel consumption per roundtrip was to be minimised. It is acknowledged that the minimisation of fuel consumption is inherently included in the first objective (i.e. the maximisation of NPV). However, it was decided to include this second objective in the optimisation to boost our search for designs of enhanced economic competitiveness and at the same time of minimal environmental footprint. The genetic algorithm NSGA II (Non-dominated Sorting GA II), available within CAESES, was used, resulting in 1130 feasible and 799 infeasible designs.

The aggregate of the results are presented in a series of scatter diagrams, see Figures 7 to 14 (for more clarity only feasible designs are shown). Figures 7 and 8 present scatter diagrams of the Net Present Value difference of each alternative design in comparison with the baseline (herein denoted as DeltaNPV) versus the ship's Length BP and Beam respectively. Note again that the baseline design is infeasible, since it fails to comply with the newly revised R-Index. These diagrams indicate that DeltaNPV generally increases with Length BP and decreases with Beam. This is due to the impact of length and beam variations on the propulsion power, and eventually on the fuel consumption.

A constraint was introduced in this study, according to which all feasible designs should have positive DeltaNPV. Because of this constraint, as shown in Figure 7 all feasible designs have a Length BP above

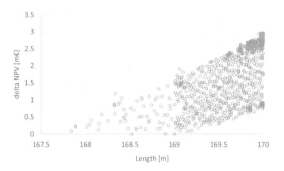

Figure 7. Increase of delta net present value vs. length BP.

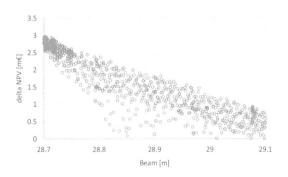

Figure 8. Decrease of delta net present value vs. beam.

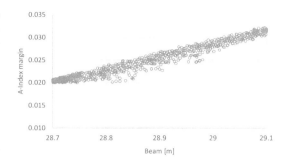

Figure 9. A-Index margin vs. beam.

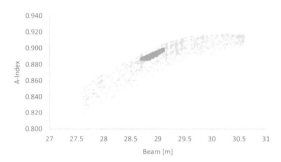

Figure 10. A-Index vs. beam.

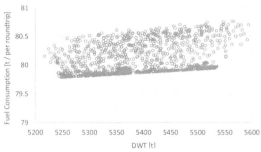

Figure 11. Fuel consumption (propulsion only) per roundtrip vs. DWT.

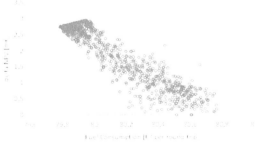

Figure 12. DeltaNPV vs. fuel consumption per roundtrip.

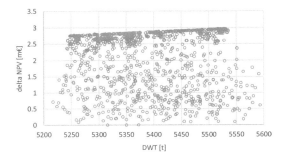

Figure 13. DeltaNPV vs. DWT.

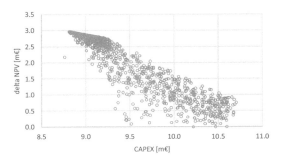

Figure 14. DeltaNPV vs. CAPEX.

167.8 m. The A-Index margin (i.e. the difference between the Attained and Required Subdivision Indices) is plotted in Figure 9 as a function of Beam. All feasible designs have a significantly increased Beam (at least 1.1 m larger than that of the baseline). Not surprisingly, this is due to the new damaged stability requirement (which the baseline had not had to comply to). A diagram of the A-Index vs. Beam is presented in Figure 10. In order to provide more insight on the impact of Beam on damaged survivability, both feasible and infeasible designs are included in this figure. The feasible designs are marked by full blue circles and can be clearly seen surrounded by a 'cloud' of infeasible designs.

The diagram in Figure 11 presents the relationship between the fuel consumption for the vessel's propulsion per roundtrip and DWT. Scatter diagrams illustrating the relationship between DeltaNPV and fuel consumption per roundtrip, DWT and CAPEX (i.e. the corresponding increase of building cost in comparison with the baseline) are presented in Figures 12, 13 and 14 respectively.

The most promising design, selected for further study was the one with the maximum DeltaNPV. This design has a length BP of 170 m, i.e., the maximum length considered, a beam of 28.7 m, i.e. the minimum beam for which the damaged stability requirement was fulfilled, and a design draught of 6.8 m. Its propulsion power at 21 kts and at 27 kts is equal to 14.7 MW and 40.3 MW respectively and its Net present Value and Building Cost are increased by 2.964 m€ and 8.814 m€ respectively, in comparison with the baseline.

5 CONCLUSIONS

HOLISHIP develops novel concepts for ship design and operation, which are implemented in versatile, integrated design platforms, offering a vast variety of options for the efficient development of alternative ship designs by use of tools for their analysis and optimisation with respect to all relevant (ship) design disciplines. An open architecture allows for continuous adaptation to current and emerging design and simulation needs, flexibly setting up dedicated synthesis models for different application cases.

The material presented describes a "snapshot" of on-going developments in HOLISHIP 15 months into the project. These cover elements of intact and damage ship stability, hydrodynamic performance in calm water and in a seaway, energy simulations and initial cost assessments. Together they largely determine two fundamental criteria of ship design, namely high safety and excellent efficiency. The material presented highlights the integration concept which will be further refined and extended to other design disciplines during the following project phase. The present status already allows demonstrating the effect of the holistic design and optimisation concept for the application case of a RoPAX ferry.

The chosen application case represents a realistic transportation scenario for a combined passenger and car ferry operating in European coastal waters. Starting from the definition of the transport demand for a specific route (and a baseline design that will be made available by the interested shipowner or be taken from a database) the most suitable main particulars of the ship are determined using advanced design-analysis methods, which already indicate that the traditional borders between concept and (preliminary) contract design will be blurred in the future.

The procedure applied in the present study led to feasible and good designs within very short lead time. The impact of a varying service speed on ship design, which is often an uncertain parameter, was clearly demonstrated and even if the results obtained may not be a surprise for an experienced designer, the speed, quality and extent of information generated by an automated, computer-aided procedure, examining hundreds of realistic variants before concluding on the best designs, is convincing. Extrapolating these first results onto further developments, namely also including other design disciplines like structural design, more striking design improvements may be expected, especially when considering higher complexity and flexibility of operation during the life-cycle of a vessel.

ACKNOWLEDGEMENT

HOLISHIP is being funded by the European Commission within the HORIZON 2020 Transport Programme.

Horizon 2020
European Union funding
for Research & Innovation

REFERENCES

Abt, C.; Harries, S.; Wunderlich, S.; Zeitz, B.: Flexible Tool Integration for Simulation-driven Design using XML, Generic and COM Interfaces, Int. Conference on Computer Applications and Information Technology in the Maritime Industries (COMPIT 2009), Budapest, 2009.

CAESES-Friendship; www.caeses.com

Faou, M., Roche, S., Jin, C.H., Corrignan, Ph., Marty, P., Wandji, C., Guerrero, C., 2015. Joint development of a new generation of LNG carrier, GASTECH 2015 conf., Singapore

Gatchell, S.; Hafermann, D.; Jensen, G.; Marzi, J.; Vogt, M.: Wave resistance computations—A comparison of different approaches, Proc. 23rd Symp. on Naval Hydrodyn. (ONR), Val de Reuil, 2000.

Guégan, A., Rafine, B., Descombes, L., Fadiaw, H., Marty, P., Corrignan, P., A systems engineering approach to ship design. *8th International Conference on Complex Systems Design & Management (CSD&M)*, 2017.

Hafermann, D.: The New RANSE Code FreSCo for Ship Applications, Proc. of Annual Meeting Schiffbautechnische-Gesellschaft, STG Jahrbuch, 2007.

Hagen, A., Grimstad, A., The extension of system boundaries in ship design. Int. Journal Marit. Eng. 152, 2010.

Harries, S.: Investigating Multi-dimensional Design Spaces Using First Principle Methods, Proc. 7th Int. Conference on High-Performance Marine Vehicles (HIPER 2010), Melbourne, Florida, USA, 2010.

Harries, S., Cau, C., Marzi, J., Kraus, A., Papanikolaou, A. Zaraphonitis, G: Software Platform for the Holistic Design and Optimisation of Ships, Proc. of Annual Meeting Schiffbautechnische-Gesellschaft, STG, November 2017, Potsdam.

Liu, S.; Papanikolaou, A.: Fast Approach to the Estimation of the Added Resistance of Ships in Head Waves, Journal of Ocean Engineering, Elsevier, Vol. 112, pp. 211–225, 2016.

Liu, S.; Papanikolaou, A.; Zaraphonitis, G.: NEWDRIFT+ Enhanced Version of the Seakeeping 3D Panel code NEWDRIFT for Calculation of the Added Resistance in Waves, Report Ship Design Laboratory, NTUA, Athens, Sep. 2017.

MacPherson, D.; Harries, S.; Broenstrup, S.; Dudka, J.: Real Cost Savings for a Waterjet-driven Patrol Craft Design Using a CAESES-NavCad Coupled Solution, Int. Conf. on Computer Applications and Information Technology in the Maritime Industries (COMPIT 2016), Lecce, 2016.

Marty, P., Corrignan, P., Gondet, A., Chenouard, R., Hétet, J-F, Modelling of energy flows and fuel consumption on board ships: application to a large modern cruise vessel and comparison with sea monitoring data, Proc. 11th International Marine Design Conference (IMDC2012), June 2012, Glasgow.

Papanikolaou A.: Holistic Ship Design Optimization, Journal Computer-Aided Design (2010), Elsevier, Vol. 42, Issue 11, pp. 1028–1044.

Papanikolaou, A. Review of Damage Stability of Ships—Recent Developments and Trends. Proc. PRADS 2007, Houston, October 2007.

Papanikolaou, A., Hamann, R., Lee, B. S., Mains, C., Olufsen, O., Tvedt, E., Vassalos, D., Zaraphonitis, G., GOALDS—Goal Based Damage Stability of Passenger Ships, Trans. SNAME, Vol. 121, 2013, pp 251–293 (SNAME Archival Paper).

Sames, P.C.; Papanikolaou, A.; Harries, S.; Coyne, K.P.; Zaraphonitis, G.; Tillig, F.: BEST Plus—Better Economics with Safer Tankers. Proceedings of SNAME Annual Meeting and Expo 2011, November 2011, Houston, Texas, USA.

Zaraphonitis, G.; Skoupas, S.; Papanikolaou, A.; Cardinale, M.: Multi-objective Optimization of Watertight Subdivision of RoPAX Ships considering the SOLAS 2009 and GOALDS s Factor Formulations, Proc. 11th International Conference on the Stability of Ships and Ocean Vehicles, 23–28 Sep. 2012, Athens.

A methodology for the holistic, simulation driven ship design optimization under uncertainty

L. Nikolopoulos & E. Boulougouris
Department of Naval Architecture, Ocean and Marine Engineering, University of Strathclyde, Glasgow, UK

ABSTRACT: The change of scenery in shipping has been evident over the past 20 years. The ever changing oil and fuel price, tough and cyclical market conditions, the constant societal pressure for a «green» environmental footprint combined with ever demanding international safety regulations create the new framework in which commercial ship designs are subject to. As a result of this current status of shipping commercial a change of attitude in the philosophy and process of ship design is required in order to shift towards new approaches where holistic approaches are deemed necessary. Apart from considering all the interrelationships between the subsystems that consist the vessel lifecycle and supply chain considerations are the key in successful and «operator oriented» designs.

The methodology herein presented is built and fully integrated within the Computer Aided Engineering (CAE) software CAESES that integrates in the design process CFD codes. It can be successfully used for the optimization of either of the basic design of a vessel or the operation of an existing vessel with regards to the maximization of the efficiency, safety and competitiveness of the final design. Stability, strength, powering and propulsion, safety, economics, operational and maintenance and in service management considerations are tightly integrated within a fully parametric model. This tight integration enables the user to simulate the response of the model in variations of the geometrical, design variables of the vessel (including its propeller) under conditions of simulation and uncertainty. For each of the potential design candidates, its operation is simulated based and assessed on a lifecycle basis and under conditions of uncertainty. The uncertainty modelling is extensive and in several levels including but not limited to Economic, Environmental, and Operational uncertainty as well an accuracy modelling of the methodology itself. The methodology is applied on the iron and coal seaborne trade and more specifically the case of large bulk carriers. The uncertainty models are based on Big Data statistical analysis, from the on-board real time monitoring systems of a fleet of 15 vessels for a period of more than 18 months on the examined trade.

1 INTRODUCTION

For centuries the backbone of global trade and prosperity has been international shipping, with the vast majority of transportation of raw material as well as manufactured goods being conducted through seaborne trade. While the 20th century saw the expansion of shipping coincident with the industrial revolution, the first decade of the 21st posed a series of challenges for commercial shipping. The economic recession combined with a fall in freight rates (due to tonnage overcapacity as well as a global economic slowdown in terms of growth per capita) has threatened the financial sustainability of numerous companies. At the meantime, following the Kyoto protocol and the societal pressure for greener shipping gave birth to a number of international environmental regulations legislated by the UN International Maritime Organization (IMO) and classification societies that set the scheme for future as well as existing ship designs.

Among others, future vessels' carbon emissions are controlled both by technical and operational measurements while the must also incorporate ballast treatment facilities to mitigate the risk reduced biodiversity (especially in sensitive ecosystems such as reefs) due to the involuntary carriage of evasive species inside water ballast tanks.

When focusing in the dry bulk cargo transportation, the carriage of major bulk commodities, i.e. iron ore, coal and grain the iron ore and coal dominate this market with 650 and 690 million tons respectively in 2005 as per Stopford [1]. This number grew significantly to 1,364 million tons of iron ore and 1,142 million tons transported by sea in 2017 in accordance with United Nations UNCTAD Report [2]. The total dry bulk seaborne trade in 2017 totaled at 4,827 million tons making iron ore and coal the dominant commodities with 28.3% and 23.7% of the total trade.

The rapid expansion of Chinese economy created a constant demand for both iron and coal. On the

Figure 1. Major iron ore trades.

other hand the major iron ore exporters are located in South America (primarily Brazil) and Australia. From the other hand, coal production in order of mil tons is concentrated in Indonesia, Australia and Russia with 383, 301, and 314 mil tons accordingly. Serving the supply chain and flow of iron ore and coal. The coal consumers are the Atlantic market consisted by Western European countries (Germany and the UK) and the Pacific market, which consists of developing and OECD Asian importers, notably Japan, Korea and Chinese Taipei. The Pacific market currently accounts for about 57% of world seaborne steam coal trade. For the past half century global bulk shipping has focused on providing tonnage to serve the above trade with vessels of considerable size due to limited size restrictions both due to ever expanding port terminals as well as to the absence of physical restrictions (e.g. Panama Canal). The present paper focuses on vessels intended for this trade which can be grouped in the Capesize/Very Large Ore Carrier (VLOC) segment of the shipping market.

The design of such and all bulk carriers in general for the past decade (2008–2018) focused on the increase of efficiency by two means: increase of cargo carrying capacity and decrease of energy demands. In most cases the optimization, if any, is based on a single design point in terms of both speed and loading condition (draft and thus displacement). This paper in turn proposes the herein developed and proposed holistic methodology intended for the optimization of the basic design of large bulk carriers based on their actual simulated operational profile, for their entire lifecycle and under conditions of uncertainty. The speed and trading profile is simulated for the entire economic life of the vessel and the optimization focuses on the minimization of all operating costs, maximization of income, minimization of internal rate of return (IRR) summarized by the Required Freight Rate (RFR) from one hand and from the other the minimization of the energy footprint of the vessel expressed by the Energy Efficiency Design Index (EEDI), simulated Energy Efficiency Operating Index (EEOI), lifecycle emissions as well as the minimization of the required water ballast amount for stability in order to minimize (or even eliminate) the energy and costs for the treatment of water ballast onboard. From the safety point of view the optimization targets on the minimization of the risk of structural failure without unnecessary increases of the lightship weight.

2 OVERVIEW OF THE HOLISTIC METHODOLOGY

Holism (from ὅλος holos, a Greek word meaning all, whole, entire, total), is the idea that natural systems (physical, biological, chemical, social, economic, mental, linguistic, etc.) and their properties, should be viewed as wholes, not as collections of parts. This often includes the view that systems somehow function as wholes and that their functioning cannot be fully understood solely in terms of their component parts. Within this context the authors have developed such methodologies in the Ship Design Laboratory of NTUA with use of the Computer Aided Engineering (CAE) software CAESES developed by Friendship Systems that can simulate ship design as a process in a holistic way. This approach has been applied in a variety of cases in the past such as tanker design optimization [Nikolopoulos, 9] as well as to containership design [Koutroukis, 13].

Holistic ship design

The methodology is holistic, meaning that all of the critical aspects of the design are addressed under a common framework that takes into account the lifecycle performance of the ship in terms of safety efficiency and economic performance, the internal system interactions as well as the trade-offs and sensitivities. The workflow of the methodology has the same tasks as the traditional design spiral with the difference that the approach is not sequential but concurrent.

Simulation driven design

The methodology is also simulation driven, meaning that the assessment of the key design attributes for each variant is derived after the simulation of the vessel's operation for its entire lifecycle instead of using a prescribed loading condition and operating speed (Nikolopoulos, Boulougouris [15]). The operation simulation takes into account the two predominant trade routes large bulk carriers are employed in and models the operation based on actual operating data from a fleet of large bulk carriers (Capesize and Newcastlemax). By employing such a technique, the actual operating conditions and environment with all uncertainties and volatilities connected to the latter is used to assess the merits of each variant of the optimization ensuring that the design will remain robust and attain its good performance over a range of different environments and for its entire lifecycle. The dimensioning of the principal components, e.g the main engine and propeller is based on the margin allowed from a limit state condition assumed in the analysis.

Design under uncertainty

A new novel approach with regards to uncertainty is introduced in the herein discussed version of this methodology. The entire methodology is evolved from deterministic to probabilistic by the introduction of various levels of uncertainties in the following levels:

a. Environmental Uncertainties
b. Market Uncertainties
c. Methodology Uncertainty.

Design and simulation environment

The environment in which the methodology is programmed and is responsible for the generation of the fully parametric hull surfaces is the CAESES CAE which is a CAD-CFD integration platform developed for the simulation driven design of functional surfaces like ship hulls, propeller and appendages, but also for other applications like turbine blades and pump casings. It supplies a wide range of functionalities or simulation driven design like parametric modeling, integration of simulation codes, algorithms for systematic variation and formal optimization. The offered technologies are:

- Complex fully parameterized models can be generated. Additionally, (non-parametric) imported shapes can be manipulated with parameterized transformations. Feature modeling, special parametric curve and surface types, as well as transformation techniques support those tasks.
- External simulation codes, be it in-house codes or commercial codes can be conveniently coupled in a multitude of ways: tool-specific coupling, coupling via a common data interface on XML basis, project based coupling with template files and communication via the Component Object Mode (COM) interface. Except for the first one, all interfaces can be set up by the user.

A range of different algorithms for systematic variation, single- or multi- objective optimization is offered from the so-called Design Engines.

The holistic methodology proposed has the following workflow:

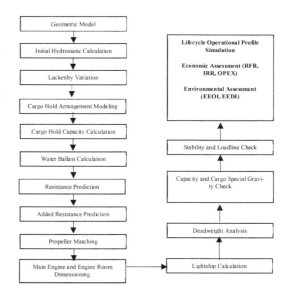

Figure 2. Workflow of the proposed methodology.

Picture 1. Finalized hullform after Lackenby variation.

2.1 *Geometric core*

The core of this methodology and any similar developed in a CAD/CAE system is the geometrical model (geometrical core). The original surface is produced as group of parametric sub-surfaces modeled in the CAESES.

2.2 *Initial hydrostatic properties*

The hydrostatic calculation aims on checking the displacement volume, block coefficient and center of buoyancy of the design. It is performed by an internal computation of FFW and for its execution a dense set of offsets (sections) is required as well as a plane and a mirror plane.

2.3 Lackenby variation

In order to be able to control the desired geometrical properties of the lines, and more specifically the block coefficient (Cb) and the longitudinal centre of buoyancy (LCB), the Lackenby variation is applied. This variation is a shift transformation that is able to shift sections aft and fore accordingly. Instead of applying quadratic polynomials as shift functions, fairness optimized B-Splines are used allowing the selection of the region of influence and the smooth transition as well. The required input for the transformation is the extent of the transformation which in this case is from the propeller position to the fore peak and the difference of the existing and desired Cb and LCB as well[9].

2.4 Cargo hold modeling

Using the output surface from the Lackenby variation, the cargo hold arrangement is generated with a feature of the Friendship Framework and its capacity is calculated.

The cargo hold surfaces and their respective parametric entity were realized within CAESES. Furthermore, the hydrostatic calculations of CAESES were used to calculate the capacity of the cargo holds, which is necessary for most of the computations. The parameters/variables controlling this area were the positions of the bulkheads, the position of the Engine Room bulkhead, the frame spacing as well as some local variables such as the hopper width and angle, the topside tank dimensions (width and height), the lower stool height and length and double bottom height.

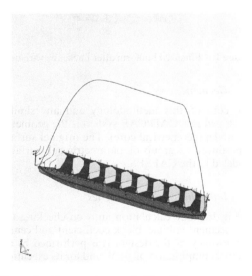

Picture 2. Parametric cargo hold surfaces.

The capacity of each tank is calculated by creating offsets for each one of the tank surfaces and joining them together. Afterwards, a hydrostatic calculation of the tanks takes place and the total capacity can be checked. Furthermore, a calibration factor derived from the parent hull is introduced in order to take into account the volume of the structural frames inside the cargo holds as well as a factor in order to derive with the Bale and Grain capacities.

The result of the parametric tank modeling can be also seen at the CAESES snapshot (picture [2]).

2.5 Resistance prediction

Calm water resistance

The resistance prediction of this model uses a hybrid method and two different approaches, depending on the optimization stage.

Initially, during the design of experiment and the global optimization phase, where a great number of variants is created there is a need for high processing speed and subsequently computational power. For this particular reason the Approximate Powering Method of Holtrop[4] is used that derives from editing statistical data and is a very fast method. Especially in bulk carriers it is very accurate too, since the wave making resistance as well as the viscous pressure resistance are very small fractions of the total resistance with the frictional resistance (direct function of the wetted surface) dominating all resistance components due to the dimensions and very small Froude number. The only inaccuracy of this method can be identified in the local viscous resistance effects and is common to all prediction methods.

However, in order to improve the prediction accuracy, especially for of design conditions such as the ballast condition, the coefficients for each component of the resistance used in Holtrop and Mennen methodology were recalibrated against the parent vessel model tests while the coefficients used for the powering prediction were calibrated both from model tests and analytical CFD calculations on the parent vessel (Nikolopoulos and Boulougouris [18]). In subject publication the constants and parameters from Holtrop and Mennen approximate power method were systematically varied with use of genetic algorithms with the goal of calibrating the method for minimum error against the statistical database used. The calibration database is consisted by the model tests (in both design, scantling and ballast loading conditions) of 7 different vessels with very similar geometric characteristics (full hull forms) and Froude number of the parent and target vessels. In total 111 points of power vs. speed for the Laden conditions and 61 points of power vs. speed for the Ballast conditions were assessed.

The calibration was performed by a systematic optimization approach. The optimization variables were the statistic coefficients as well as power values used in Holtrop methodology with a relatively big margin of variance as well as the introduction of some additional terms in existing equations. Then the methodology would be applied for each speed/power point of the model tests and the difference in powering would derive. The minimization of this difference is the optimization target of this particular sub problem. The applied algorithm for the optimization was the NSGA II with roughly 4000 variants being produced in two steps for each condition. The first step was the calibration of the equations for the calculation of the bare hull resistance and power (EHP-Effective Horse Power) while the second calibrated the equations for applying the self-propulsion problem and thus calculating the delivered horse power (DHP).The result was an average difference of −4.3% and −0.20% of the EHP and DHP respectively, for the Ballast Condition and −1.94% and −6.5% of the EHP and DHP respectively for the Laden Conditions with the Holtrop results being more conservative (over estimation) than the model tests. The standard deviation, variances as well as a full statistical analysis was produced and the prediction error of the methodology was modelled in the IBM SPSS with a non-linear regression method as a function of the vessels dimensions, block coefficient and wetted surface and subsequently programed in the methodology.

The entire Holtrop method is programmed within the Framework and is also generated as a feature for later use. Actual data from the geometric model is also used, such as the entrance angle, prismatic coefficients etc, making the process more precise and representing of the specific design.

Added resistance due to wind
The vessel's added resistance due to wind is calculated for two separate occasions in subject methodology. The first being for the assessment for sizing the main engine at a prescribed condition for the latter and second, within the simulation of the vessel's operation for each leg and stage of the simulated voyage route. The tool used for the resistance is the formula of Fujiwara et al [25] which is also used in the ISO15016-2015 [20] when doing corrections in the measurements obtained in sea trials. Subject method is considered as reliable, robust and accurate as the formula contains sensitivities and correlations with the hull and deckhouses geometry (via the use of projected surfaces).

Added resistance due to waves
The added resistance due to waves is similarly used in the two modules mentioned previously, namely main engine sizing and operational simulation. The tool used for the added resistance estimation is different depending on the stage of the optimization. For the initial stage, empirical formulae based on the Maruo far field are utilized while in a second stage, integrated panel codes using potential theory to solve the seakeeping motions problem and then through added mass calculate the added resistance.

For the first stage, after assessing the method of Kwon et al [14]. [15] as well as STAWAVE2 (as presented in ISO15016-2015 [20]) the new method of Liu and Papanikolaou [25] for the estimation of added resistance in head waves is chosen instead.

The method of Liu and Papanikolaou offer a fast and efficient calculation alternative to running a panel code, strip theory code or using RANS codes. The formula is based on best fitting of available experimental data for different types of hull forms. The formula, has been simplified to the extent of using only the main ship particulars and fundamental wave characteristics for the estimation of ship's added resistance.

The formula takes the below form:

$$R_{AW} = R_{AWR} + R_{AWM}$$
$$R_{AWR} = \frac{2.25}{2} \rho g B \zeta_a^2 \alpha_d \sin^2$$
$$E\left(1 + 5\sqrt{\frac{L_{pp}}{\lambda}} Fn\right) \left(\frac{0.87}{C_B}\right)^{1+4\sqrt{Fn}}$$
$$R_{AWM} = 4\rho g \zeta_a^2 B^2 / L_{pp} \overline{\omega}^{-b_1}$$
$$\exp\left[\frac{b_1}{d_1}\left(1 - \overline{\omega}^{d_1}\right)\right] a_1 a_2$$

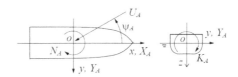

Fig. 1 Coordinate system.

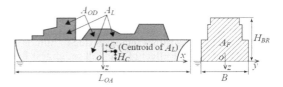

Figure 3. Coordinate system and input used in Fuiwara empirical formula for the estimation of added resistance due to wind [25].

where:

$$E = a\tan(B/2L_E)$$
$$\alpha_T = 1 - e^{-2kT}$$
$$a_1 = 60.3 C_B^{1.34}\left(\frac{0.87}{C_B}\right)^{1+Fn}$$
$$a_2 = \begin{cases} 0.0072 + 0.1676 Fn & \text{for } Fn < 0.12 \\ Fn^{1.5}\exp(-3.5 Fn) & \text{for } Fn \geq 0.12 \end{cases}$$
$$\overline{\omega} = \begin{cases} \dfrac{\sqrt{L_{pp}/g}\sqrt[3]{\dfrac{k_{yy}}{L_{pp}}}0.05^{0.143}}{1.17}\omega & \text{for } Fn < 0.05 \\ \dfrac{\sqrt{L_{pp}/g}\sqrt[3]{\dfrac{k_{yy}}{L_{pp}}}Fn^{0.143}}{1.17}\omega & \text{for } Fn \geq 0.05 \end{cases}$$

For $C_b < 0.75$:

$$b_1 = \begin{cases} 11.0 & \text{for } \overline{\omega} < 1 \\ -8.5 & \text{elsewhere} \end{cases}$$

$$d_1 = \begin{cases} 14.0 & \text{for } \overline{\omega} < 1 \\ -566\left(\dfrac{L_{pp}}{B}\right)^{-2.66} * 6 & \text{elsewhere} \end{cases}$$

For $C_b > 0.75$:

$$b_1 = \begin{cases} 11.0 & \text{for } \overline{\omega} < 1 \\ -8.5 & \text{elsewhere} \end{cases}$$

$$d_1 = \begin{cases} -566\left(\dfrac{L_{pp}}{B}\right)^{-2.66} & \text{for } \overline{\omega} < 1 \\ -566\left(\dfrac{L_{pp}}{B}\right)^{-2.66} * 6 & \text{elsewhere} \end{cases}$$

The L_E has been defined as the distance from the fore peak to the position where the maximum ship breadth is reached.

Fouling related resistance
The last environmental related added resistance factor taken herein into account both in the design modules (propulsion prediction and main engine selection) as well as input in the operational simulation module is that of marine biological fouling. More specifically, as the hull of the ship ages the average roughness values increases due to hull biological fouling. The effect of the hull roughness for the vessel's resistance can be calculated from the below formula (International [19]):

$$\frac{\Delta R}{R} = \frac{\Delta C_F}{C_T} = 0.044 * \left[(k_2/L)^{1/3} - (k_1/L)^{1/3}\right]$$

With k_2 and k_1. being the current and previous hull roughness respectively. The hull roughness increase on an annual basis is also estimated from [International [18]] which starts from an average of and continues on an exponential rate. Furthermore, in order to further enhance the lifecycle considerations, the dry docking recoating is taken into account in the 5, 10, 15, 20 and 25 year interval with a reduction of the roughness to a level 10% higher than the previous coating system (e.g roughness in 5 years is 10% higher than the newbuilding value, roughness in 10 years is 10% than the 5 year value etc). The starting roughness value at the delivery stage of the vessel is assumed to be aaverage value of 97.5 microns (derived from minimum 75 and maximum 120 microns).

The power increase corresponding to the above resistance increase is approximated by the following formula (International [19]):

$$1 + \frac{\Delta P}{P} = \frac{1 + \Delta R/R}{1 + \Delta \eta/\eta}.$$

With the increase on the propeller open water efficiency being:

$$\frac{1}{1 + \Delta \eta/\eta} = 0.30 * \left(1 + \frac{\Delta R}{R}\right) + 0.70$$

2.6 *Propeller model*

While the vessel's Propeller is not modelled geometrically at this current stage, it is assumed to be a part of the Wagenigen B-Series of propellers. All the Wagenigen polynomials are modeled within the methodology (Bernitsas [17]) so the open water diagrams of a propeller with a selected pitch, diameters, blade number and expanded area ratio can be derived. Following this, the self-propulsion equilibrium is conducted in the design speed in an iterative manner in order to derive with the final propulsion coefficients, shaft horse power, torque, thrust and propeller revolutions (RPM). This is in turn used for the propeller-engine matching and the propulsion plant dimensioning.

The optimal selection of the propeller parameters (diameter, pitch, blades) is also part of the global/preliminary design stage.

2.7 *Main engine and engine room dimensioning*

Main engine
After the propeller is dimensioned, the Main engine should be matched to that hull and propeller. In order to avoid the well-known (and rather recent) risk of underpowered vessels, instead of

employing a weather and fouling margin (typically 15%), a dimensioning condition was in turn used as determined by users. This condition is such that the vessel should maintain the full speed and corresponding engine load, power and RPM at head and beam waves corresponding to sea state 5, with adverse (head) current of 1.5 knots, roughness due to fouling corresponding to 4 years without cleaning and the corresponding head wind of sea state 5. In addition to the power requirements of the above an RPM of 10% (in accordance with MAN B&W requirements []) is imposed as well as an additional margin of 5% which is considered for derating the main engine and ensuring smaller Specific Fuel Oil Consumption (SFOC).

For the final requirements the main engine is matched with the existing "G-Type", "ultra-long stroke", engines available from MAN[6]. Firstly an "engine library" with alternative configurations is created, which is utilized in the selection module in combination with an internal iterative procedure ensures that the engine will have sufficient light running margin and that the layout point on the diagram is close to the L2 L4 line corresponding to bigger torque/MEP margins and smaller SFOC values.

From the above the final SFOC curve from 10% to 100% is produced and corrected for the actual engine layout.

All engines within the engine selection library are Tier III compliant in accordance with the MARPOL Annex VI, Regulation 13 as amended by the IMO MEPC 66 requirement [26] for ships built after the 1st of January of 2016. Additionally the engine library contains all three different available NOx abatement technologies, namely: Exhaust Gas Recirculation (EGR), High Pressure Selective Catalytic Reaction (HPSCR) and Low Pressure Selective Catalytic Reaction (LPSCR). The choice of which technology will be applied is one of the optimization variables. Furthermore, in future development, the engine library will be expanded also with Gas engines.

In addition to SFOC curves, curves of steam production from 20% to 100% are produced. These are used in turn as steam production curves in the operation simulation, in order to assess the potential load (if required) of the composite boiler to match the steam consumption requirements.

Diesel generators
The electrical balance analysis of the parent vessel is non-dimensionalized for each consumer and each condition respectively and the ratios are used within the methodology to determine the load of each consumer for the generated variants and thus the electrical load for a each condition.

The required alternator output is calculated based on this (after including a safety factor), while the prime movers (diesel generators) of the alternators are sized by assuming an 85% electrical efficiency.

Exhaust gas boilers
Similarly to the case of the electrical balance, the steam balance of the vessel is also non-dimensionalized. For applications of fuel tank heating (whether bunker or settling/service tanks) the steam consumption (in kg/h) is non-dimensionalized by the fuel tank capacity (calculated in intact stability module).

2.8 *Lightship weight prediction*

The lightship calculation follows the traditional categorization in three weight groups, the machinery weight, the outfitting weight and the steel weight.

Machinery weight
The machinery weight calculation is based on the average of two methods: the Watson-Gilfillan formula and the calculation based on the Main Engines weight respectively.

The machinery weight estimation is based on an empirical formula due to Watson-Gilfillan[5]:

$$Wm = Cmd * Pb^{0.89} \tag{1}$$

The average is used to balance out any extreme differences, and the coefficients of the Watson-Gilfillan formula are calibrated for low speed, two stroke engines based on statistic data available for a fleet of bulkers.

Outfitting weight
The outfitting weight is also based on the average of two independent calculations. The Schneekluth method is one and the use of empirical coefficients for sub-groups of that particular weight group is the other one.

Steel weight
During the initial design stages, and the selection of optimal main dimensions, it is necessary to identify the effect of the change of the principal dimensions of a reference ship on the structural steel weight. Thus, at first, an accurate calculation of the steel weight of the reference ship is conducted. Following this, the "Schneekluth Lightship Weight Method" was applied [Papanikolaou, 6]. Given that the steel weight for the parent vessel was available as derived from summing the individual steel block weights (from the shipbuilding process) a TSearch algorith was employed in order to vary the values of the statistical coefficients and constants of subject methodology with the objective of the

minimization of the difference between the actual and calculated values for the steel weight. The result was an accuracy of 0.3% which is more than acceptable within the scope of basic/preliminary design. The error was modeled also in the IBM SPSS as a function of the principal particulars and block coefficient.

2.9 Deadweight analysis

The deadweight of the vessel is comprised by subgroups such as the consumables, the crew weight and the deadweight constant. The Deadweight analysis is the prediction of the payload of the vessel based on the calculation of the consumables.

As mentioned before, the consumables for the machinery is calculated, namely the Heavy Fuel Oil for the main engines, and diesel generators, the Lubricating Oils of the engines and generators.

Furthermore, based on the number of the crew members (30), the fresh water onboard is calculated as well as the supplies and the stores of the vessel.

2.10 Stability and loadline check

The initial intact stability is assessed by means of the metacentric height of the vessel (GM). The centre of gravity of the cargo is determined from the capacity calculation within the framework while the centre of gravity for the lightship and consumables is determined from non-dimensioned coefficients (functions of the deck height) that derive from the information found in the trim and stability booklet of the parent vessel. All the above are calculated with the requirements of the IMO Intact Stability Code for 2008[3].

2.11 Operational profile simulation

This module is an integrated code within the methodology that simulates the actual operating conditions of the vessel for its entire lifecycle. Two trade routes are considered, the Brazil to China roundtrip and the Australia to China roundtrip. Each voyage is split into legs depending on distinctive sea areas.

For the Australia to China roundtrip the following legs are considered:

- Leg A: Sea Passage from W. Australia loading ports to Philippines being subdivided into 4 sub-legs.
- Leg B: Sea Passage from Phillipines to Discharging port being subdivided into 4 sub-legs.
- Leg C: Only for the ballast leg to Australia a stop in Singapore for bunkering is considered.

For the Brazil to China roundtrip the following legs are considered:

- Leg A: Sea Passage from the Brazilian Loading port to the Cape of Good Hope in South Africa. This leg is subdivided into 4 equal sub-legs.
- Leg B: From the Cape of Good Hope in S.Africa to Indonesia and is subdivided into 4 equal sub-legs
- Leg C: Sea Passage through the Malacca straight and Singapore including a port stay in Singapore for bunkering operations.
- Leg D: Sea Passage from Singapore through the Taiwanese straight into the discharging port of China. This leg is subdivided into to 2 sub-legs.

Input data
For each one of the legs (given distance in nautical miles) the average speed and added resistance curves are input as well as the loading of the generators, the maneuvering time. If the leg includes a discharging, loading or bunkering port the port stay in hours is also used. Based on this profile the voyage associated costs together with the fuel costs are calculated on a much more accurate and realistic basis.

The input variables of the operation simulation model for each model can be seen in the below Table 1.

Added resistance
For each leg, stage and corresponding time step the added resistance module is called from within the operational simulation module in order to calculate the added resistance. The final estimation is a probabilistic one, which means that the added resistance for different wave directios, wave heights and wave lengths is estimated and then a probabilistic figure is derived based on the probability distribution functions modeled from the onboard measurement data.

Environmental parameters modeling
The operating speed for which the added resistance (and thus added propulsion power) is calculated is also probabilistic.

Initially the uncertainty of the average operating speed per leg is applied. The probabilities of having a ±15% deviation from the estimated average of each leg are calculated from the probability density function derived from onboard data analysis. A probabilistic steaming speed is then produced from the weighted average of the higher and lower speeds.

Currents
The second source of uncertainty with regards to the operating speed is environmental and is related to the local currents. For each leg/sea area a statistical analysis from onboard collected data, reveals both the average as probability distribution of the current speed and current direction. In the simulation mod-

Table 1. Operational simulation input parameters.

Operational simulation input parameters	Unit
ISO corrected SFOC Curve	
Speed Power Curve – Calm Water	
Auxiliary Engines Power	kW
SFOC curve for auxiliary Engines	
Auxiliary engine load during cargo hold cleaning	%
Time for Cargo Hold Cleaning	hours
Main Engine SMCR	kW
Main Engine Load in Maneuvering	%
Cylinder Oil Feed Rate (normalized average)	gr/kWh
Electrical power required during normal sea going	kW
Blowers Electrical Power	kW
Required electrical power during maneuvering	kW
Main Engine SFOC during Maneuvering	g/kW
Sulphur Content in Fuel	%
Main Dimensions	
Length Overall	m
Length between perpendiculars	m
Breadth	m
Voyage Draft	m
Wind Profile	
Total Lateral Projected Area	m^2
Total Transverse Projected Area	m^2
Lateral projected area of superstructures above deck	m^2
Fujiwara Hc	m
Height of Superstructures	m
Added Resistance	
Wave length probability distribution function curve	
Entrance Angle Length	m
Fouling – Resistance Increase due to roughness	N
Propulsion	
Thrust Deduction Curve	
Wake Fraction Curve	
Propeller Diameter	
Number of Blades	
Expanded Area Ratio	m^2
Pitch over Diameter Ratio	
Propeller Shaft Mechanical Efficiency	
Relative Rotative Efficiency	
Speed – RPM Curve	
Loading/Discharging Port	
Auxiliary Engine Load during Loading	%
Time in Loading/Discharging Port	hours
Time for maneuvering	hours
Sea Passage Leg	
Distance	miles
Average Transit Speed	knots
Probability of Head Current	
Probability of Astern Current	
Low Current Velocity	knots
Mid Current Velocity	knots
High Current Velocity	knots
Sea Passage Leg – Singapore (additional)	
Maneuvering Time	hours
Port Stay for Bunkering	hours
Auxilliary Engine Load in Port	%

ule these calculated probability distribution functions are used in order to estimate the probability of encountering a high, medium and low current (their amplitude is determined from the minimum, maximum and average speed from the onboard data). The correction to the operating speed is positive for the cases of astern current and negative for ahead current. The ahead and astern currents are considered for an "operating envelope" of ±45 degrees both in the ahead and astern term, as the side currents will only yield deviation rather than speed loss.

From the above mentioned two corrections the probabilistic ship speed is derived based on which both the calm water required delivered power is calculated as well as the added resistance and power calculations takes place.

Fouling
The fouling margin, is also calculated depending on the age of the vessel in the respective simulation stage by calling the fouling resistance calculation module described previously.

2.12 *Economic model*

In total the code calculates the Operational Expenditure (OPEX), the Capital Expenditure (CAPEX), the Required Freight Rate (RFR), the Internal Rate of Return (IRR) as well as the IMO Energy Efficiency Operational Index (EEOI).

The Economic model also follows the principle of simulation driven design and design under uncertainty. The uncertainties in the economic model can be identified both in terms of the shipping market as well as the fuel prices which directly the fuel costs (burden to owners that operate in the tramp/spot markets).

The market uncertainty is predominately expressed by the uncertainty of the vessel's Earnings. Through the Clarkson's Shipping intelligence database (Clarkson's [21]), a probability distribution function for the Capesize earnings was produced based on the data from 1990 to 2015 which cover a typical vessel's economic (and engineering) lifetime. Based on the earnings the probability of high (150,000 USD/day TCE), mid (35,000 USD/day TCE) and low (5,000 USD/day TCE) were calculated and thus a probabilistic value for the vessel's annual as well as lifecycle (by applying the interest rates) profitability was derived. Apart from this earnings directly affect the other shipping markets, namely the acquisition market (both the S&P and Newbuilding market; for the case herein presented the second as well as the scrap market. For this particular reason and in order to further enhance the correlation to the vessel's design the newbuilding prices and scrap prices were expressed (after suitable adjustment) per ton of lightship and were correlated from the Clarkson's Shipping Intelligence database to the Earnings of the vessel with the following formulas:

$NBprice = 157.335 * Earnings^{0.269}$

and

$Scrap\_price = 25.648 * Earnings^{0.244}$

For both equations the value returned is USD/ton of lightship and serve as magnification factors for the acquisition and residual values of the vessel. Furthermore, the two last which are used for the CAPEX calculation, are also probabilistic by applying the same probabilities that are used for High, Mid and Low Earnings with the respective amounts introduced in the above presented formulas.

By this way, it is able to accurately depict the volatility of the market and the response of each design variant as well as the effect of its dimensions to its lifecycle economic performance.

This is further enhanced by the calculation of the Fuel Price cost which is outside the usual time charter provisions of bulker Charter Party agreements. The Fuel prices cost is also probabilistic with the probabilities for High (1500 USD/ton), Mid (450 USD/ton) and Low (150 USD/ton) prices being derived from the probability distribution function that was calculated from the Clarkson's Shipping Intelligence Database.

This is a key point of this methodology, namely to optimize the vessel's design under uncertainty as the produced designs correspond to a more realistic scenario and the dominant variants of the optimization have a more robust behavior over a variety of exogenous governing market factors.

The derived probabilistic values of RFR and the deterministic value of the EEOI are the functions/targets used in the optimization sequence later.

2.13 Energy efficiency design index calculation

The Energy Efficiency Design Index (EEDI) is calculated according to the formula proposed in the IMO resolution MEPC.212(63), using the values of 70% deadweight and 75% of the MCR of the engines and the corresponding reference speed:

The minimization of this index is one of the primary targets of the conducted optimization. The engine power is directly related to the resistance of the hullform, while the deadweight is also related to both the hullform in terms of displacement and to ship's lightship weight.

2.14 Modeling uncertainties from big data analysis

One of the novel aspects of this methodology has been the use of big data and the statistical analysis of the latter with the IBM SPSS toolkits for the creation of linear and non-linear regression formulas as well as probability distribution functions and descriptive statistical studies. The big data taken into account and analyzed (as already described in the various subcomponents of the methodology) are in two categories:

a. Onboard data (write about their origin) and production of PDF for environmental criteria.

The Onboard data were collected from two the installed Vessel Performance Monitoring (VPM) System of a fleet of Capesize and Newcastlemax bulkers that operate both in the Brazil and Australia trade routes. This VPM system collects real time data (30 sec logging and averaging into 5 minute intervals) of the vessel's Alarm and Monitoring System (AMS) and the vessel's navigational data from the Voyage Data Recorder (VDR) into an onboard server. This gathering, together with the use of signals from torque meters and flow meters provides an extensive database that is used for the statistical analysis with the IBM SPSS toolkit of the following parameters:

1. Operating Speed
Normal PDF with a Mean and Standard Deviation depending on the leg of the passage.
2. Wind Speed
Normal PDF with a Mean and Standard Deviation depending on the leg of the passage.
3. Wind Direction
Normal PDF with a Mean and Standard Deviation depending on the leg of the passage.
4. Current Velocity
Exponential with a scale of around 1 to 1.5 depending on the leg of the passage.

$$EEDI = \frac{\left(\prod_{j=1}^{M} f_j\right)\left(\sum_{i=1}^{nME} P_{ME(i)} * C_{FME(i)} * SFC_{ME(i)}\right) + \left(P_{AE} * CF_{AE} * SFC_{AE}\right)}{f_i * Capacity * V_{ref} * f_w} + \frac{\left\{\left(\prod_{j=1}^{M} f_j * \sum_{i=1}^{nPTI} P_{PTI(i)} - \sum_{i=1}^{neff} f_{eff(i)} * P_{AEeff(i)}\right) * CF_{AE} * SFC_{AE}\right\} - \left(\sum_{i=1}^{neff} f_{eff(i)} * P_{eff(i)} * CF_{ME} * SFC_{ME}\right)}{f_i * Capacity * V_{ref} * f_w} \quad (2)$$

5. Current Direction
Normal PDF with a Mean and Standard Deviation depending on the leg of the passage.

b. Clarkson's Ship Intelligence Database for the modelling of market conditions.

The Clarkson's Shipping Intelligence Database (Clarkson's [21]) has been used extensively for the market modeling and studying of the correlations for the following parameters:

1. Capesize Earnings (1990 to 2015).
   Lognormal PDF with Scale = 23194.925 and Shape = 0.830.
2. Fuel Price – IFO380 (1990 to 2015).
   Lognormal PDF with Scale = 246.930 and Shape = 0.711.
3. Fuel Price – MGO (1990 to 2015).
   Triangular PDF with min = 101.25, max = 1268.13 and mode = 120.65.

3 DESIGN CONCEPT

3.1 Large bulk carrier market

The focus of the present study lies within the large bulk carrier segment. The market for subject vessel size is positioned on the seaborne transportation of primary bulk commodities for industrial activities (iron ore, nickel ore and other major minerals) as well as for energy in the form of coal.

As already mentioned previously, the trade routes for the above mentioned markets are between Latin America and the Far East (China primarily and then Korea and Japan) as well as between Australia and again the Far East. The optimal vessel for the maintenance of an efficient supply chain in these two routes is the primary objective of this study.

Traditionally in such markets Capesize markets have been employed as well as Very Large Ore Carriers (VLOCs). During the last decade a new class of vessels has been emerged, known as Newcastlemax as they are the largest vessels that can enter and load in the Coal Terminal of Newcastle in Australia

3.2 Baseline vessel – 208k Newcastlemax

As in any ship design optimization case study it is imperative that a baseline is set in the form of the parent vessel used as a primary source of reference as well as calibration for the methodology and all the formulas/computations applied in the latter. For this particular reason it is necessary to have as complete data as possible for the parent vessel in order to achieve a better degree of accuracy as well as being able to make proper comparison during the analysis of the dominant variants of the optimization front.

The vessel chosen for this study belongs to the new category segment of Newcastlemax Bulkers and is a newly delivered vessel. The baseline parametric geometry has been adapted to fit the hull form lines available. As mentioned in the previous chapter the model test results of subject vessel were used to calibrate and better adapt Holtrop statistical methodology for the prediction of powering along the entire speed-power curve. The principal particulars of the vessel can be found in the below table:

3.3 Proposed design concept characteristics

A small Froude number (slow speed) and full hull form is herein proposed as the base hull for the global optimization. The absence of a bulbous bow is evident as it is a recent trend in bulk carrier design as such absence assists in the reduction of the vessel frictional resistance (primary resistance component) while the wave making resistance is not increased. The effect of the bulbous bow on the above as well as the added resistance are investigated in depth in separate study. In addition the use only of an electronically controlled Main Engine is considered and no Energy Saving Devices (wake equalizing duct, pre-swirl fin, bulbous rudder etc) are considered since there is no such device installed on the parent vessel and further to the above such devices and their effect is to be considered in a post analysis study.

<u>Simulation driven design, choice of hullform parameters</u>
The assessment of the design is derived from the simulation of the operational, economic and trading profile (as per methodology in chapter In other words instead of using only one design point (in terms of draft and speed) multiple points are used derived from actual operating data of a shipping company.

Table 2. Baseline vessel principal particulars.

Length over all	299.98
Lengthbetween perpendiculars	294
Beam	50
Scantling Draft	18.5
Deck Height	25
Cb	0.8521
Main Engine Specified MCR (kW)	17494 @ 78.7 RPM/ MAN B&W 6G70ME-C9.2
Deadweight (tons)	Abt 208,000
Lightship Weight (tons)	26,120
Cargo Hold Capacity (m^3)	224,712.1

Newcastlemax design concept
The maximum moulded dimensions (Length Over All and Breadth) for subject study in the optimization problem set also as optimization constraints are the maximum allowable dimensions in order to load in the port Newcastle in Australia.

3.4 Optimization target/goals

The target of any optimization procedure is always to achieve the most desiring values/properties for the set optimization objectives. The alteration of the designs and assessed entries is performed through the systematic variation of their distinctive parameters, while each one of the designs must comply with the set constraints, e.g. stability criteria/maximum dimensions or deadweight

The generic targets or objectives in almost any ship design optimization problem are:

Competitiveness
The market and economic competitiveness of a an individual vessel variant is the core of any optimization as a vessel will always be an asset (of high capital value) and can be expressed by the following indices:

1. *Required Freight Rate.*
The required freight rate is the hypothetical freight which will ensure a break even for the hypothetical shipowner between the operating costs, capital costs and its income based on the annual voyages as well as collective cargo capacity and is such expressed in USD per ton of cargo.

2. *Operating Expenditure (OPEX).*
The operating expenditure expressed on a daily cost includes the cost for crewing, insurance, spares, stores, lubricants, administration etc. It can indicate apart from the operator's ability to work in a cost effective structure, how the vessel's design characteristics can affect. The lubricant cost is based on actual feed rates used for subject engines as per the relevant service letter SL2014-537 of MAN [14].

3. *Capital Expenditure (CAPEX).*
The CAPEX is a clear indication of the cost of capital for investing and acquisition of each individual design variant. The acquisition cost is calculated from a function derived from actual market values and the lightship weight for vessels built in Asian shipyards, and more specifically in China.

Efficiency
The merit of efficiency is herein expressed by the IMO EEOI index. Although on the design basis in practice the IMO Energy Efficiency Design Index is used as a KPI and measure of the merit of efficiency in new design concepts as well as for any newbuild vessel, in this study the calculated Energy Efficiency Operating Index is used instead. The reason for this change is the use of the Operational Profile simulation module which contains from a wide statistical database of a bulker operator the daily average speed per each stage of each voyage leg (refer to par. 2.10) thus given the cargo capacity calculation (par. 2.4) the EEOI can be accurately derived, which can depict more accurately the efficiency of the design given the fact that it takes into account all operating speeds (instead of one design speeds) and all operating drafts (instead of the design draft) thus expressing the actual transport efficiency of each variant by a simple ration of tons of CO_2 emitted (direct function of the tons of fuel consumed) to the tons of cargo multiplied by the actual distance covered (in nautical miles). In addition to the above, each operational practice such as slow steaming is taken into a full account, also considering side implications (for example the use of two diesel generators in the normal sea going condition instead of one in order to cover the blower's electrical load). Furthermore, the minimization of the required ballast water amount for the ballast conditions is set as optimization target.

3.5 Design variables

From the below Table 3, one can identify the selected design variables of the subject optimization problem. The latter are in three categories;

Table 3. List and range of design variables of the optimization problem.

Design variable	Lower boundary	Upper boundary
Length between Perpendiculars	275	320
Length Overall	280	325
Beam	42	55
Draft	16.5	19.5
Deck height	24	27
Hopper Height	7	10
Hopper Breadth (m)	2.5	4
Topside Height (m)	5	9
Topside Breadth (m)	8	12
Inner Bottom Height (m)	2	3
Block Coefficient Cb	0.84	0.87
LCB (%Lbp)	0.49	0.55
Bilge Height (m)	2.4	8
Bilge Width (m)	2.4	8
Propeller Diameter (m)	8	10
Propeller Expanded Area Ratio	0.35	0.55
Propeller Pitch over Diameter	0.75	1.2

principal dimensions, hull form characteristics (Cb, LCB and Parallel Midbody) and cargo hold arrangement parameters. The more detailed design variables of the hull form arrangement for the detailed shape of the bulbous bow (if any), flair and stem shape as well as stern shape are going to be assessed in a separate optimization study with the use of integrated CFD codes.

3.6 Optimization procedure

The optimization procedure applied for this study follows the rational of any optimization loop in engineering as it is evident from Figure 4.

For each iteration of the same loop the design variables receive their input values from the «design engine» applied in the Friendship Framework. The design engine can either be a random number generator or an optimization algorithm depending on the optimization stage. The applied values then trigger the generation of a new variant from the holistic, parametric model that utilizes the developed methodology for that matter.

After the variant generation, the Design Objectives, which are selected as the measures of merit of each variant are logged and assessed accordingly while at the meantime the Design Constraints imposed are checked for compliance. The Design constraints chosen for this application were the calculated values for Deadweight, Cargo Specific Gravity and the Stability Criteria of the 2008 Intact Stability Code. The size restrictions (in terms of vessel's dimensions) were not used in constraints given the fact they were taken into account in the applied range of the Design Variables.

The optimization procedure described in this paper can be described as a stepped (multi stage) one. At first, it is necessary to explore and fully understand both the design space (potential for improvement with given constraints) as well as the sensitivity of the methodology by a Design of Experiments procedure, using a system available random number generator that follows the Sobol sequence procedure [30]. The sensitivity analysis is a very important, preparatory step in which it is ensured that no major, unreasonable manipulations occur. In addition to that it is important to see that the results are realistic both on a quantitative and qualitative basis, with the latter in need of particular attention since the design ranking and selection is the essence of optimization (the value of a favored design is not important than the relationship with all the other produced designs).

The following formal optimization runs utilize genetic algorithm techniques (NSGA II algorithm [28]). The formal optimization runs involve the determination of the number of generations and the definition of population of each generation to be explored. Then the generated designs are ranked according to a number of scenarios regarding the mentality of the decision maker. One favored design is picked to be the baseline design of the next optimization run, where the same procedure is followed. When it is evident that there little more potential for improvement the best designs are picked using the same ranking principles with utility functions, and are exported for analysis.

Both the SOBOL and NSGA II algorithms as well as a plethora of other variant generation and optimization algorithms are fully integrated and available within the Friendship Framework.

3.7 Design of Experiment (DoE)

The Design of Experiment has the primary purpose of the calibration, test and sensitivity check of the methodology from one hand as well as the investigation for the optimization margin. From the first indications, as anticipated, there is a strong scale effect which one can say that dominates this particular optimization problem. This effect is very common in ship design were the largest vessels usually dominate the smaller since the increase of cargo capacity does not trigger an equivalent increase in the powering requirements or the vessel's weight. In addition to the scaling effect it was observed as in the formal optimization algorithm that there was a strong linear correlation between the Required Freight Rate (RFR) and the EEOI, which was also anticipated since both functions use cargo capacity.

The feasibility index was in a very high level (above 90%). In total 250 designs were created.

3.8 Global optimization studies

In this stage of the formal, global design optimization the NSGA II algorithm is utilized. The latter is a genetic, evolutionary algorithm that is based on the principles of biological evolution (Darwin [10]). As in the biological evolution each design variant is an individual member of a population of a generation.

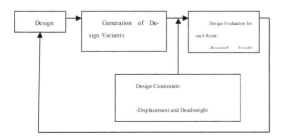

Figure 4. The optimization loop applied.

Each individual of the population is assessed in terms of the Optimization Objectives, as well as its relation to the desired merits. For the application in ship design optimization it is usual to apply a large population for each generation with an adequate number of generations. The large population combined with a high mutation probability ensures that the design space is properly covered, while the number of generations ensures that there is a push towards the Pareto frontier for each case of objective combination. For this particular application a combination of 17 generations with 100 variants population each was selected. The mutation probability was increased from the default value by CAESES of 0.01 to 0.05 in order to increase mutation events that trigger the variation of the design variables and thus have a wider design space.

In Figure 5, the scatter plot of the generated design population is depicted, with the RFR of each design on the x-axis and the respective EEOI on y-axis. A distinctive linear correlation between the EEOI and RFR is evident. This has been observed regardless of the use of uncertainty functions and is attributed to the direct linear correlation of the fuel consumed and CO_2 emissions (through the carbon conversion factors). We can see that the both the baseline as well as dominant variants are close to the middle of the straight cloud line comprised by the generated designs. It should be noted that the vessels with lower RFR has significantly increased OPEX and Required Ballast Water amount values making them thus less favored in the decision making process.

In Figure 6, the scatter plot of the RFR vs CAPEX is found. A clear Pareto frontier is formulated on which the decrease of CAPEX triggers in turn an increase in the RFR. This pattern can be attributed to the fact that these two objectives are contradicting. The RFR can be decreased by the increase of cargo carying capacity (and thus income) but this in turn will increase the vessel size and thus building cost. The CAPEX is comprsied by the acquisition (new building) cost and dry-docking costs both of which have been formulated as a non-lnear function of the vessel's lightship. Rather interestingly, the baseline design is far from the pareto frontier to an increased CAPEX compared to the dominant variants, which have the smallest CAPEX values.

The scatter plot of the RFR vs the OPEX (Figure 7), shows the same pattern as the previous plot of CAPEX. Again here, the relationship of

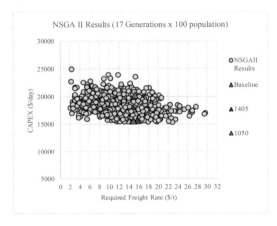

Figure 6. Scatter plot of the optimization results: RFR vs CAPEX.

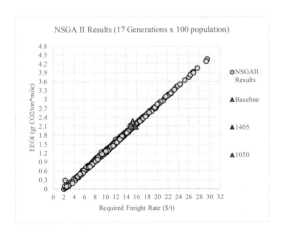

Figure 5. Scatter plot of the optimization results: RFR vs EEOI.

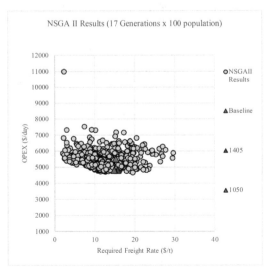

Figure 7. Scatter plot of the optimization results: RFR vs OPEX.

RFR to OPEX is antagonistic as the larger vessels with lower RFR values will have larger installed engines which will have significantly higher maintenance costs (non-linear function of vessel's SMCR) and require higher crewing and insurance costs (non-linear function of the vessel's GT). Like in the case of CAPEX the baseline design has a distance from the frontier, but in this case this is smaller due to the small OPEX of this vessel.

Lastly, an interesting and clear Pareto frontier is observed in the scatter plot between the Required Ballast Water Amount and the vessel's OPEX. Here, the increase of Required Ballast will also correspond to an increase of the OPEX, which is rather sharp. The front is therefore localized at the bottom left corner of the graph. The underlying mechanism between this relationship is that the Ballast Water amount required, determines the ballast pumps capacity and in turn the Ballast Water Treatment System (BWTS) capacity and both of them Auxiliary Engines rating. The running cost of the BWTS is a significant component of the OPEX, both due to the higher maintenance costs of the electric generating plant but due to the cost of chemicals both for treatment and neutralization. The same will also apply for the relationship of Required Ballast Amount with CAPEX since the cost of the installation of the BWTS system is significant and an exponential function of the Ballast Pumps Capacity which is calculated basis on the Required ballast amount and ballasting and de-ballasting time (constant).

3.9 Dominant variant ranking

One of the most critical steps during optimization of any system is the selection and the sorting of the dominant variants. For this particular reason it is necessary to follow a rational, rather than an intuitive, approach in order to consider in an unbiased way all trade-offs that exist. One such method is utility functions technique.

The optimum solution in our case would dispose the minimum EEOI, RFR, OPEX and CAPEX values. Instead of using fixed weights for the set criteria in the evaluation of the variants, we rather assume a utility function as following

$$U = w_{EEOI}*u(EEOI) + w_{RFR}*u(RFR) + w_{CAPEX}*u(CAPEX) + w_{OPEX}*u(OPEX) \quad (3)$$

The maximization of this utility function is the objective now, and the dominant variants of those 10 most favorable with respect to the 4 defined utility scenarios (Table 4) resulting in the identification and sorting of 40 designs with best performance according to each utility scenario.

From the above ranking (Figures 9 to 12) it is very interesting to observe that there is a certain repetition in the top three dominant variants from

Table 4. Weights used for the utility functions.

Maximum objective weight	U1	U2	U3	U4
RFR_	0.2	0.3	0.2	0.1
EEOI	0.2	0.1	0.1	0.3
OPEX	0.2	0.1	0.3	0.1
CAPEX	0.2	0.2	0.1	0.3
Required ballast water amount	0.2	0.3	0.3	0.2

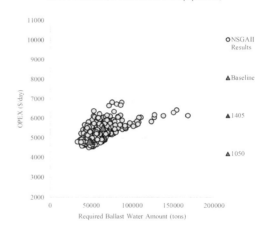

Figure 8. Scatter plot of the optimization results: Required Ballast Water Amount vs OPEX.

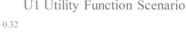
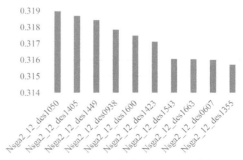

Figure 9. Ranking of dominant variants with U1 scenario.

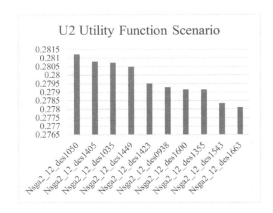

Figure 10. Ranking of dominant variants with U2 scenario.

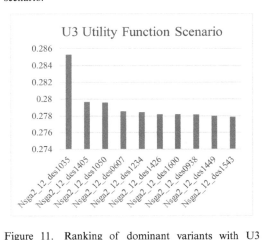

Figure 11. Ranking of dominant variants with U3 scenario.

the ranking procedure. Furthermore, for scenario U3 where there is an equal weight for all objectives, the three top dominant variants are the ones from scenario's U1 and U2. All the above illustrate that the peak on the observed pareto front is strong and apart from that, the dominant variants that can be selected (e.g 1405, 1050, 1035) perform better in a robust way under different assumptions and weights from the decision maker point of view. The characteristics of these three variants can be found in the Table 5.

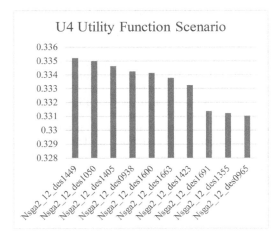

Figure 12. Ranking of dominant variants with U4 scenario.

Table 5. Principal particulars of baseline and dominant variants.

Particulars	Baseline	ID1405	ID1050	ID1035
Lbp (m)	294	275	276.1	277.8
Beam (m)	50	42.15	42.353	42.718
Deck Height (m)	25	25	25.26	26.53
Cb	0.8538	0.8599	0.8555	0.844
LCB	0.51986054	0.52	0.499	0.5480
LOA (m)	299.98	279	278	278.7
Draft (m)	18.5	16.59	17.02	16.93
Topside Breadth (m)	12	8.27	11.33	9.468
Topside Height (m)	9	5.15	7.71	5.024
Hopper Height (m)	10	9.98	9.046	8.529
Hopper Breadth (m)	4	3.25	3.42	3.412
Double Bottom Height (m)	2.5	2	2.85	2.14
Propeller Diameter (m)	9	9.27	8.87	8.05
Propeller P/D	0.9	0.942731	0.763	0.804
Propeller Expanded Area Ratio	0.55	0.516	0.4544	0.459
Bilge Height (m)	2.4	5.19	2.16	6.901
Bilge Width (m)	2.4	6.06	2.58	2.512

Table 6. Design objectives of the baseline vs the dominant variants.

Particulars	Baseline	1405	Difference %	1050	Difference %
RFR	15.22	15.69	3.09	14.88	−2.23
EEOI	2.25	2.11	−6.22	2.12	−5.78
OPEX	5520	4827	−12.55	4823	−12.63
CAPEX	20322	15771	−22.39	15648	−23.0
Required Ballast Water Amount	64244	49298	−23.26	47616	−25.88

4 DISCUSSION OF THE RESULTS – FUTURE PERSPECTIVES

From the table below (6), we can observe that for design 1405 an increase of the RFR of 3% was observed with a decrease however of the EEOI by 6%, of the OPEX by 12% and CAPEX and Required Ballast Water amount by 23%. Design I.D 1050 seems to be more promising as the improvements in EEOI, OPEX, CAPEX and Required Ballast Amount are marginally higher than these of the I.D 1405, however the RFR is 2.23% lower than that of the baseline. The marginal reduction of the RFR can be justified by the reduction of generally vessel size primarily in terms of beam and length (beam given the fact that these vessels are not stability limited) and thus the reduction of the initial capital cost, while in the meantime the cargo capacity has inevitably decreased, reducing thus the profitability of the vessel.

From the above discussion we can conclude that the novel methodology herein proposed for the simulation driven design with lifecycle, supply chain and the actual operating in service parameters can successfully trigger a reduction in the RFR and EEOI via systematic variation and advanced optimization techniques. However, this is a preliminary work restricted only into illustrating the applicability and potential of this method. The following work is planned for the next steps:

1. Integration of a Rankine panel code, for the vessel motions and added resistance calculation in irregular waves. This is developed at the moment and expected to finish within the next months.
2. Systematic variation of the modeled uncertainties and sensitivity analysis of the current model.
3. Move to a dynamic simulation, instead of quasi-steady state with a finer time grid of minutes. The data are already structured and processed and the aim is to depict transitional and dynamic phenomena.
4. Integration of calculation of both design and service structural loads (bending moments and shearing forces) as per IACS Common Structural Rules in the Steel Structural model.
5. Lifecycle assessment to include also a wastage model for the vessel's structure, such as the one proposed by Soares et al [31].
6. Use of dynamic energy functions similar to those developed by Chicowicz et al [27] for the modeling of the propulsion and auxiliary machinery plants.
7. Integration of equipment age degradation models for the main engine and auxiliary machinery (generators, boilers etc).
8. Expansion of optimization variables also to engine tuning, Tier III compliance (EGR, HPSCR, LPSR, Gas Engines) elements of which are already modeled.

ACKNOWLEDGEMENTS

The authors would like to express their deep gratitude to their mentor and teacher, Professor Apostolos Papanikolaou, from the Ship Design Laboratory at the National Technical University of Athens whose lectures on ship design and holistic optimization have been the inspiration for this paper. Furthermore, authors would like to acknowledge the help of Star Bulk Carriers Corporation in providing valuable data from their operating fleet and reference drawings.

REFERENCES

[1] Stopford M., "Maritime Economics 3rd Edition", Routledge, ISBN 978-0-415-27558-3.
[2] United Nations Conference on Trade and Development, "Review of Maritime Transport 2017", UNCTAD secretariat calculations, based on data from Clarksons Research, 2017d.
[3] FRIENDSHIP-SYSTEMS: FRIENDSHIP—Framework, www.friendship-systems.com.
[4] Papanikolaou A., Holistic ship design optimization. Computer-Aided Design, doi:10.1016/j.cad.2009.07.002, 2009.
[5] IMO, Intact Stability Code 2008
[6] Holtrop, J. Mennen, G.G.J., An approximate power prediction method, International Shipbuilding Progress 1982.

[7] Papanikolaou A., "Ship Design—Volume A: Preliminary Design Methodology", Publ. SYMEON, 1988/1989, 2nd edition: 1999 (Greece).

[8] MAN B&W, Marine Engine Program 2015.

[9] Nikolopoulos L. "A Holistic Methodology for the Optimization of Tanker Design and Operation and its applications", Diploma Thesis NTUA, July 2012.

[10] Darwin C *"On the origin of species by means of natural selection, or the preservation of favoured races in the struggle for life"*, London: John Murray. [1st ed.], 1859.

[11] Abt C., Harries S., "Hull Variation and Improvement using the Generalised Lackenby Method of the FRIENDSHIP-Framework", The Naval Architect, RINA © September 2007.

[12] [MEPC 62] – MEPC 62/24/Add.1, "Report of the Marine Environment Protection Committee on its sixty-second session", 26 July 2011.

[13] Koutroukis G., "Parametric Design and Multi-objective Optimization-Study of an Ellipsoidal Containership", Diploma Thesis NTUA, January 2012.

[14] Kwon, Y.J. (2008). 'Speed Loss Due To Added Resistance in Wind and Waves', the Naval Architect, Vol. 3, pp.14–16

[15] Lu, Turan, Boulougouris, «Voyage Optimisation: Prediction of Ship Specific Fuel Consumption for Energy Efficient Shipping», Low Carbon Shipping Conference, London 2013.

[16] MAN Diesel and Turbo, Service Letter SL2014–587 «Guidelines of Cylinder Lubrication», Copenhagen 2014.

[17] Nikolopoulos L., Boulougouris V. "Applications of Holistic Ship Theory in the Optimization of Bulk Carrier Design and Operation", EUROGEN 2015 Conference, September 2015 Glasgow.

[18] Nikolopoulos L., Boulougouris V. "A Study on the Statistical Calibration of the Holtrop and Mennen Approximate Power Prediction Method for Full Hull Form Vessels, Low Froude Number Vessels", SNAME Journal of Ship Production and Design, Vol., Month January 2018, pp 1–28.

[19] Bernitsas M.M., Ray D., Kiley P., "K_T, K_Q and Efficiency Curves for the Wageningen B-Series Propellers", University of Michigan, Report No.237, May 1981.

[20] [MEPC 68] – Publication of ISO15016:2015, Ships and marine technology—Guidelines for the assessment of speed and power performance by analysis of speed trial data, IMO, February 2015.

[21] International Marine Coatigs, Propeller No.15, January 2003.

[22] International Marine Coatings, Propeller No.16, August 2003.

[23] Clarksons Shipping Intelligence database.

[24] Fujiwara, T., Ueno, M. and Ikeda, Y. (2005), "A New Estimation Method of Wind Forces and Moments acting on Ships on the basis of Physical Component Models", J. of The Japan Society of Naval Architects and Ocean Engineers, 2, 243–255.

[25] Liu S., Shang B., Papanikolaou A., Bolbot V., "An Improved Formula for Estimating the Added Resistance of Ships in Engineering Applications", J. Marine. Sci. Appl. (2016) 15: 442.

[26] Marine Environment Protection Committee (MEPC), 66th session, 31 March to 4 April 2014

[27] Cichowicz J., Theotokatos G., Vassalos D., (2015), "Dynamic Energy Modelling for ship life-cycle performance assessment", Journal of Ocean Engineering 110 (2015).

[28] Deb K., Pratap A., Agarwaj S., Meyarivan T. (2002), "A fast and elitist multiobjective genetic algorithm: NSGA-II", IEEE Transaction on Evolutionary Computation, Volume 6, Issue 2, April 2002.

[29] Ulungu E.L, Teghem J., Fortemps P.H, Tuyttends D. (1999), "MOSA method: a tool for solving multiobjective combinatorial optimization problems" Journal of Multi-Criteria Decision Analysis, Volume 8, Issue 4, July 1999.

[30] Sobol, I.M. (1976) "Uniformly distributed sequences with an additional uniform property". Zh. Vych. Mat. Mat. Fiz. 16: 1332–1337 (in Russian); U.S.S.R. Comput. Maths. Math. Phys.16: 236–242 (in English).

[31] Y. Garbatov, C/Guedes Soares, "Corrosion wastage modeling of deteriorated bulk carrier decks", International Shipbuilding Progress 55 (2008) 109–125, DOI 10.3233/ISP-2008-0041, IOS Press.

Performance analysis through fuzzy logic in set-based design

H. Yuan & D.J. Singer
Department of Naval Architecture and Marine Engineering, University of Michigan, Ann Arbor, MI, USA

ABSTRACT: The United States Navy has currently mandated that Set-Based Design will be used as its preferred design method for all its future design activities. With this need in mind, the expansion of a previously developed Set-Based Design facilitation and negotiation method is presented. The original work demonstrated the ability to utilize expert opinion, in the preliminary design stage, as a means to successfully execute a Set-Based Design activity. However, the existing model focuses only on technical feasibility. This paper proposes an improved model through the addition of performance evaluation into the fuzzy logic method.

The extended method utilizes fuzzy logic state space mapping in conjunction with fuzzy logic controllers to enable the modification of design convergence so that factors such as production risk and development risk can be taken into account. The mathematical formulation of the methods as well as a design study is presented to demonstrate the research.

1 INTRODUCTION

Naval ships and marine structures are multi-purpose systems which require suitable and rigorous design. The area of research concerning different design concepts applied in the marine systems has been an advancing field. In order to give a clear classification to these concepts, the McKenney-Singer taxonomy (McKenney 2013) can be used to define them as one of the four categories, namely design approach, design process, design method, and design tool. This classification focuses on the scope that a proposed design concept covers.

Design approach covers the broadest scope and nowadays most of the ship designs follow the principle of current engineering design approach to some degree. In terms of the design process, System Engineering (Kossiakoff et al. 2011, Calvano et al. 2000) can be considered as a typical example in ship design. As for design method, iteration design method plays a role to find the final solution for a long time. However, the amount of variables in ship design problems is increasing rapidly and the combination of sub-systems is more and more complex, which is beyond the capacity of iterative design method. As a consequence, a new design method Set-Based Design (SBD) (Singer et al. 2009) has been introduced to ship design.

SBD was first developed in the automotive industry by Toyota (Ward et al. 1995) and then adopted by naval design (Singer 2003). This design method was used on the U.S. Navy Ship-to-Shore Connector in the program's preliminary design development (Mebane et al. 2011). Traditionally, a design solution is determined by optimizing results again and again in the feasible domain, which is regarded to be an iterative method. In contrast, SBD is a convergent method concentrating on eliminating infeasible and dominated points, which is more efficient to handle conflicting requirements and flexibility for trade-offs at early stages. As a result of the convergent method, SBD embraces a broad range for each design variable instead of single points. Those ranges indicate the possibility of both normal-performance and high-performance design schemes. Since it is a unique concept in marine systems, limited research has been conducted to the development of supporting design tools that can be used in SBD.

Design tool is a category that deals with the development of data that directly provides reliable information for decision making. Within the area of SBD limited research has been completed in the field of SBD tool development. Singer and Gray (Singer 2003, Gray 2011) developed a fuzzy logic SBD facilitation tool and Design Space Mapping (McKenney 2013) was utilized to evaluate the effect of cutting some regions of design space. All three authors utilized fuzzy logic, a popular design tool, in the development of their SBD research.

Fuzzy logic was initially introduced by Zadeh (Zadeh 1965). He defined membership values as a continuous membership function (MF) between 0 and 1. As stated, there were several attempts to utilize fuzzy logic theory. Singer (2003) and Gray (2011) proposed a hybrid agent Type I and Type II Fuzzy Logic SBD tool to facilitate the execution of SBD. In this research, several teams of designers who were responsible for distinctive aspects of ship design described their individual preferences for the negotiation of a particular parameter. They defined a fuzzy set in each team to demonstrate linguistic values of preferred, marginal, and unpreferred following four operations, which were fuzzification, linguistic fuzzy rules, fuzzy inference, and defuzzification, to obtain a common preferred set range. This range in Singer's model (Singer 2003) was modified further by Gray (2011). Gray tried three randomization methods to mimic uncertainty towards the range, which performed better under tight ship constraints.

The purpose of this paper is to build fuzzy logic models to elicit expert opinions and analyze the potential performance of a ship design beyond the basic feasibility requirements set by Singer (2003) and Gray (2011). In this paper, a final modified model is planned to build based on this foundation through three main steps. First, fuzzy logic systems are involved to model subjective knowledge on design variables like length and beam. Second, the application of design mapping method, similar to the method developed by McKenney (2013), is used to generate preference level to objective spaces that are functions of basic variables. Third, the same variable will be mapped back from different functions resulting in their corresponding curves and a fuzzy logic controller will determine the final preference level of a certain variable. The evaluation of performance can be considered during these steps in SBD. Section 2 will describe those methods in detail; section 3 will show a case study of ship design; and section 4 will report the conclusion and discussion.

2 METHOD

2.1 Type I fuzzy logic

The first operation of type I fuzzy logic is fuzzification. In a crisp set, MF of a certain element must equal to 1 if it belongs to the set or 0 if not. Whereas, fuzzy set owns a different rule which allows an element to stay in several sets at the same time and the value of MF represents the percentage of a certain set. This rule fits the linguistic evaluation of ship design variables. Three fuzzy sets: {Preferred}, {Marginal}, and {Unpreferred} are set up for experts to negotiate. Assume N agents determine the variables from their own perspectives independently so that there will be N membership functions for one variable. Figure 1 is one of the membership functions for X_1: $\mu_{preferred}(X_1 = 7) = 0.5$ and $\mu_{marginal}(X_1 = 7) = 0.5$.

The second operation is to establish linguistic fuzzy rules, which are in the form of "IF...AND...

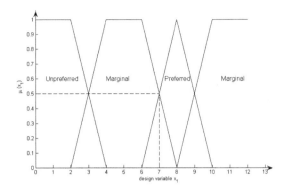

Figure 1. Fuzzy set example.

THEN" statements to determine the fuzzy set of the output. The number of agents also influences the linguistic fuzzy rule matrix and output membership. As this is an improved method of Singer's (2003) model, the same fuzzy rules as his previous work has been introduced. For example, Table 1 and Figure 2 explains the situation for N = 2. The shaded area in Table 1 indicates such a rule: if value x is considered as {Preferred} in Agent 1 and also {Preferred} in Agent 2, then its output belongs to {Emphasized}.

The third operation is to decide an appropriate fuzzy inference scheme. The fuzzy rules defined in the second operation include the AND logic. Such logistic rule is associated with correlation-minimum inference. All the membership function values are collected as inputs to find a minimum. Then the output membership function is clipped at this minimum value. For instance, continue the calculation at value x in the previous step and the inference result is from $\mu_{output\text{-}Emphasized}(x) = \min\{\mu_{Agent1\text{-}Preferred}(x), \mu_{Agent2\text{-}Preferred}(x)\}$. If the minimum value is 0.4, corresponding graph is shown in Figure 3.

The last operation is defuzzification. It calculates a crisp output based on multiple fired fuzzy rules and their clipped output membership functions. Centroid defuzification is suitable method in this paper. Its equation is described as follows:

$$\text{crisp output: } y = \frac{\sum_r a_r x_r}{\sum_r a_r} \quad (1)$$

where r is the number of fired rules; a_r is the area of the clipped output membership function; x_r is the centroid of the clipped output membership function.

The final result through these four operations is a joint preference curve (JOP) of each design variable. And the same procedure will repeat

Table 1. Fuzzy rules for N = 2.

Agent 1 \ Agent 2	Preferred	Marginal	Unpreferred
Preferred	Emphasized	Preferred	Trim
Marginal	Preferred	Marginal	Trim
Unpreferred	Trim	Trim	Trim

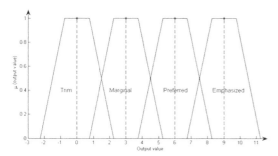

Figure 2. Output membership for N = 2.

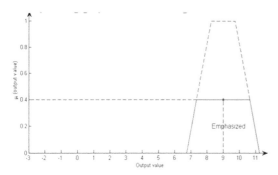

Figure 3. Clipped membership function.

when design agents evaluate vessel performance. At that time, the fuzzy sets are named as: {High}, {Medium}, and {Low} to represent different performance levels. Similarly, its corresponding result is a JOP of each objective function which is obtained from design mapping method.

2.2 Method of imprecision

The Method of Imprecision (MoI) is utilized here as the design space mapping techniques to support SBD. A tutorial about MoI was presented by Antonsson & Otto (1995) which presented imprecision calculations about deducing the preference level of design variables on performance variables. The work flow is expressed in the following steps.

Firstly, JOP curves of design variables $\{d_1, d_2 \ldots d_m\}$ are inputs as design variable preferences. For each design variable d_i, the range of x axis in JOP is its design variable space (DVS) X_i and the values of y axis is the preference level μ_{di} for X_i.

Secondly, define objective parameters as a function of design variables, such as length to beam ratio $R = f(L,B) = L/B$. These parameters serve as performance variables $\{p_1, p_2, \ldots p_k\}$ to evaluate performance. The valid ranges of p_j is the performance variable space (PVS), which should be computed later.

Thirdly, preference level of performance variables should also be mapped from DVS to PVS. This step can be accomplished via Zadeh's (1965) extension principle. Its definition for continuous-valued function is:

$$\mu(p_j) = \begin{cases} \sup \min\{\mu_{d_1}, \mu_{d_2} \ldots \mu_{d_m}\} \\ \quad if\ d_1, d_2 \ldots d_m : f_j(d_1, d_2 \ldots d_m) = p_j \\ 0\ if\ d_1, d_2 \ldots d_m : f_j(d_1, d_2 \ldots d_m) \neq p_j \end{cases} \quad (2)$$

An algorithm called Level Interval Algorithm (LIA) is used here to realize the computation of this extension principle. This algorithm was first introduced by Dong & Wong (1987) and was further modified by Wood & Otto (1992). The algorithm version here follows the steps in Wood & Otto's (1992) paper:

1. Each performance variable p is a function of vector $\vec{d}$ which includes N parameters $\{d_1, d_2 \ldots d_N\}$. Discretize their preference function into M α values ($\alpha_1, \alpha_2, \ldots \alpha_M$).
2. Determine the intervals for $d_1, d_2, \ldots, d_N$ at each α cut.
3. For each α cut, one design parameter owns two end points. Since there are N parameters in the performance objective function, 2^N permutations exist corresponding to 2^N p values.
4. Among 2^N p values of that particular α cut, determine the minimum and maximum to be the boundaries of that α preference level.

When coding these steps above, it is also necessary to pay attention to the prerequisites of preference functions:

1. Normality and convexity conditions
2. Continuous functions
3. No singularities

Performance evaluation will be conducted after obtaining the all the information of performance variables. Experts from various agents measure the performance through the whole PVS by dividing the ranges and adding linguistic labels. The purpose of this negotiation is to get JOP of each performance function.

2.3 Fuzzy logic controller

After performance evaluation, the next step is to map back to the DVS according to new JOP curves of performance variables. As a result, several new JOP curves of a certain design variable d_i may exist because d_i might be an independent variable in more than one performance functions. Figure 4 illustrates an example of the whole process and it also shows that two JOP curves are created for d_2 and only one JOP curve is generated for d_1 and d_3 after mapping back from PVS.

A fuzzy logic controller will be activated if MoI method generates more than one JOP curves for one design variable from multiple performance disciplines (e.g. d_2 in Fig. 4). In order to satisfy all the judgments from design agents, a minimum preference level method of fusion logic is chosen for this improved model. Its equation is expressed as:

$$F_s(\bar{x}) = \wedge_{j=1}^{J} F_j(\bar{x}) \qquad (3)$$

where $F_j(\bar{x})$ is the output from different disciplines; $F_s(\bar{x})$ is outcome of fuzzy logic controller.

Figure 5 is the framework of the Hierarchical Fuzzy Logic Controllers (HFLC).

A modified JOP curve for each design variable comes out at a result of fusion logic operation. And that is the reference which contains performance evaluation to determine a new range of the design variable for the next round of negotiation.

3 CASE STUDY

3.1 Model background

The example selected to demonstrate the method revolves around the design of a yacht carrier that will provide service within Mediterranean basin. The vessel will be required to transport a diverse set of yachts which is presented in Table 2 below.

3.2 Model variables

The primary step of design should focus on determining the dimensions and shape of the vessel. And this improved model for SBD has carried out at this point. Given the arrangement of yachts in the carrier, the initial ranges for design variables beam (B), length (L), draft (T), block coefficient (C_B) are defined in Table 3.

Each design variable is analyzed by several agents, which are listed in Table 4.

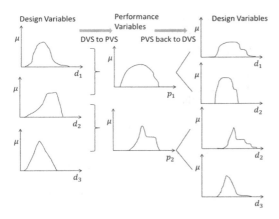

Figure 4. Map back from PVS to DVS.

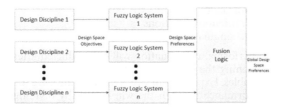

Figure 5. HFLC framework in engineering design space (Cuneo 2013).

Table 2. Yacht parameters.

Parameter	Large yacht	Small yacht
Total number	9	27
Displacement (MT)	225	40
Length (m)	40	22
Beam (m)	9	4.5

Table 3. Design variable ranges.

Design variable	Lower bound	Upper bound
Beam (m)	32	40
Length (m)	164	188
Draft (m)	4	10
Block coefficient	0.5	0.8

Table 4. Design variable and relevant agents.

Design variable	Design agent
Beam (m)	Stability, Resistance, Yacht arrangement
Length (m)	Strength, Controllability, Yacht arrangement
Draft (m)	Stability, Resistance
Block coefficient	Capacity, Resistance

3.3 Model results

The linguistic information is collected in the fuzzy sets and relevant JOP curves are created based on those fuzzy sets. Take beam as an example in Figures 6–9. Figures 6–8 demonstrates three agents providing their preference for a range of beam values relative to their role: Stability agent responsible for guarantying a stable vessel, Resistance agent responsible for providing resistance and powering guidance, and Yacht Arrangement agent responsible for the general arrangement design. The system then takes the agents' inputs and exports a JOP curve which is displayed in Figure 9.

Repeat the same procedure to define original JOP of length, draft and block coefficient, which are shown in Figures 10–12.

According to those JOP curves of each design variable, some ranges of the variable have been eliminated due to technical infeasibility. The new ranges are gathered in Table 5.

The next step is to define the three performance variables based on the individual value JOP curves.

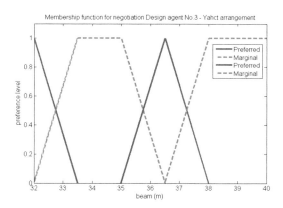

Figure 8. Beam fuzzy set—yacht arrangement.

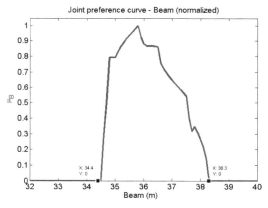

Figure 9. Beam JOP (original).

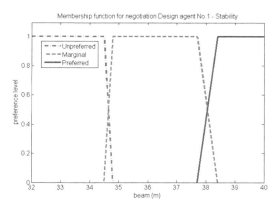

Figure 6. Beam fuzzy set—stability.

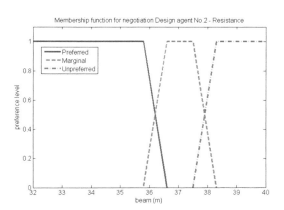

Figure 7. Beam fuzzy set—resistance.

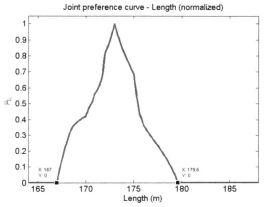

Figure 10. Length JOP (original).

249

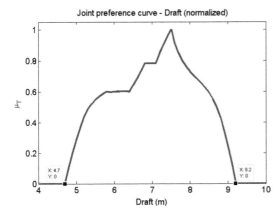

Figure 11. Draft JOP (original).

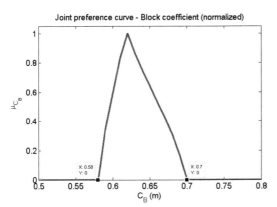

Figure 12. Block coefficient JOP (original).

Table 5. Design variable ranges (after design variable negotiation).

Design variable	Lower bound	Upper bound
Beam (m)	34.4	38.3
Length (m)	167	179.6
Draft (m)	4.7	9.2
Block coefficient	0.58	0.7

The performance variables in this case are the length to beam ratio p_1 = L/B, beam to draft ratio p_2 = B/T and displacement p_3 = C_BLBT. The initial preference level of these three performance variables can be obtained automatically through MoI shown in Figure 13, Figure 14 and Figure 15. And it is necessary to get the range of p_1, p_2, p_3 for further performance evaluation: $p_1 \in$ [4.36, 5.21]; $p_2 \in$ [3.75, 8.15]; $p_3 \in$ [15000, 44000].

Performance evaluation for p_1 (L/B ratio) is conducted between 4.36 and 5.21. Two agents are

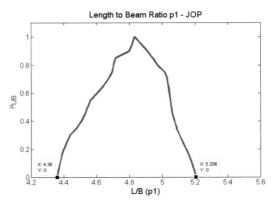

Figure 13. Length to beam ratio JOP (from mapping).

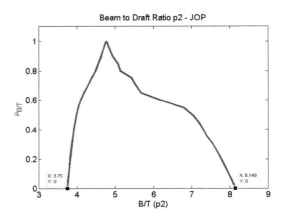

Figure 14. Beam to draft ratio JOP (from mapping).

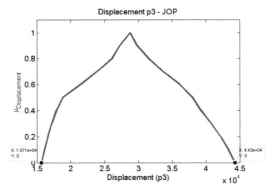

Figure 15. Displacement JOP (from mapping).

involved in its evaluation, which are speed and ship production. The goal of these agents is to provide input concerning the production cost within the range as well as economic feasibility of providing the desired speed need for the range of possible ship configuration. Corresponding membership function is shown in Figures 16 and 17 and the modified JOP for performance evaluation is provided in Figure 18.

As shown in Figure 18 the incorporation of additional performance information changes the L/B ratio JOP curve. According to the JOP of p_1, design parameter length and beam get a new JOP which is mapped back from p_1 respectively so that the impact of performance information can be related back to the basic variables used in design. Figures 19 and 20 demonstrate relevant results.

Performance evaluation for p_2 is concerned about speed and safety. Performance evaluation for p_3 in concerned about customer satisfaction and manufacture cost. As a result, all corresponding JOP curves are mapped back respectively and a modified JOP of each design variable is generated in the end. Figures 21 to 24 concludes the modified JOP of beam, length, draft and block coefficient.

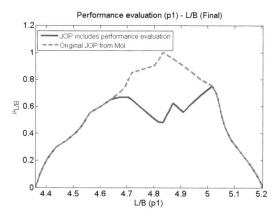

Figure 18. Length to beam ratio JOP (from performance evaluation).

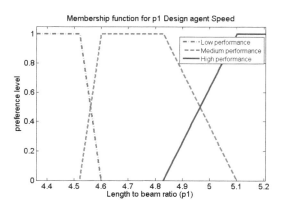

Figure 16. Length to beam ratio—fuzzy set—speed.

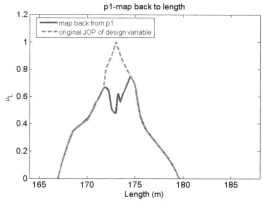

Figure 19. Length JOP (original & map back from p_1).

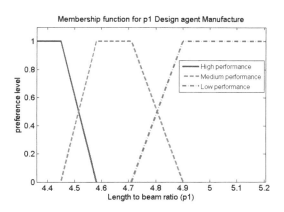

Figure 17. Length to beam ratio—fuzzy set—manufacture.

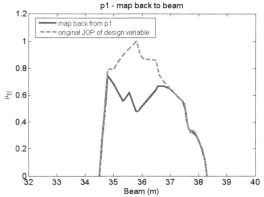

Figure 20. Beam JOP (original & map back from p_1).

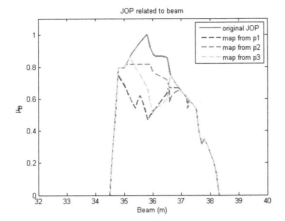

Figure 21a. Beam JOP (all related curves).

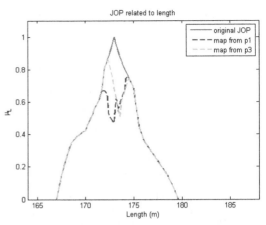

Figure 22a. Length JOP (all related curves).

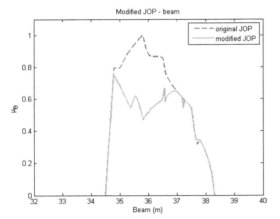

Figure 21b. Beam JOP (final modified curve).

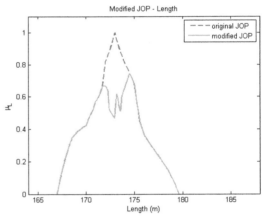

Figure 22b. Length JOP (final modified curve).

The green curves in the figures are the final modified JOP which is based on the lowest value of the JOP curves at every variable value.

As can be seen from Figures 21 to 24, there are various changes between modified curves and original ones. The modified JOP curve has the greatest difference against the original curve at the peak value. Beyond the maximum value difference between the original and modified JOP curves there is also a shift in the location of maximum preference as well as the shape of the curve. For beam, the best preference value shifts from 36 m to 35 m because the performance evaluation assigns different linguistic preferences to the range. The modified JOP of length owns two peaks compared to the original curve so that the emphasized areas are around 172 m and 175 m in the next negotiation round. The modified JOP of draft has the

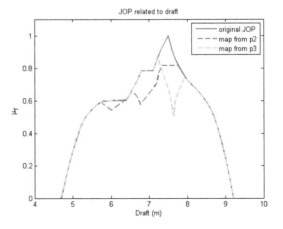

Figure 23a. Draft JOP (all related curves).

252

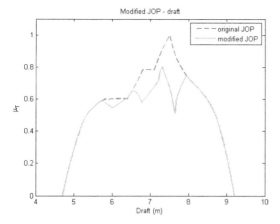

Figure 23b. Draft JOP (final modified curve).

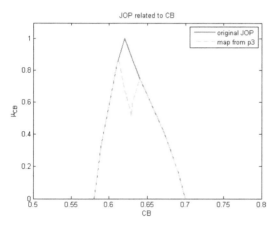

Figure 24a. Block coefficient JOP (all related curves).

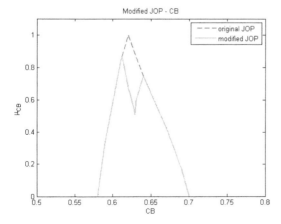

Figure 24b. Block coefficient JOP (final modified curve).

similar tendency as the previous one except for a drop around 7.7 m. And the modified JOP of block coefficient shows a significant drop at the peak.

4 CONCLUSION AND DISCUSSION

In this paper the authors have presented an extension and improvement from previous SBD fuzzy logic negotiation research through the inclusion of additional performance evaluations. The results show that most of the changes between modified JOP and original JOP occur at the peak value, which is an artifact of the example developed, but does demonstrate how the method successfully allows SBD teams to incorporate performance functions into the negotiation process. The change in JOP is a direct influence of adding performance analysis to design variables. Because sometimes high performance means pursing extreme design samples which may challenge the present technical evaluation. The shift of peaks provides more information to instruct different agents to conduct their simulations for the next negotiation round.

Another important observation is about the number of JOP curves of one certain design variable. Some design variables are the independent variable in several performance functions. As a result, there will be multiple JOP curves when mapping back from the PVS. Modified JOP curves of such design variables contain more information to predict potential performance.

However, there are still limitations and challenges in this modified method. First, it should be noted that it has restrict requirements to distinguish technical design agents and performance evaluation agents. If those two categories were set up repeatedly, the modified JOP would be less convincing. Second, it has been found that convexity is a crucial prerequisite to check before applying MoI to JOP curves which are from the technical negotiation round. For example, the case study in this paper includes four JOP curves (Figs 9–12) that need to be checked. Figures 10 to 12 satisfy the convexity requirement successfully while Figure 9 for beam shows a little inconformity between 37 m and 38 m. This small concavity has been ignored in the MoI procedure because it only has little influence on the mapping result. However, if there is a huge violation that could not be neglected, this modified method won't work well. In such circumstances, MoI might be used piecewise or other alternatives would be recommended to accomplish the mapping step.

Furthermore, more research should be done to make this performance analysis more robust. In the PVS space, if the area beyond certain preference value is quite large, it will indicate a greater possibility to maintain such performance in the next negotiation round. Otherwise, a performance variable peak value will disappear easily if its surrounding area is pretty small. A reward function of Markov Decision Process (MDP) is a feasible choice to demonstrate different area values, so a forward step may be integrating MDP into the presented model to realize robustness of performance analysis.

Overall, a primary problem in the preliminary design is the lack of comprehensive information. Both technical evaluation and performance evaluation from expert experiences are essential to improve the situation.

REFERENCES

Antonsson, E.K. & Otto, K.N. 1995. Imprecision in engineering design. *Journal of Vibration and Acoustics* 117(B): 25–32.

Calvano, C.N., Jons, O. & Keane, R.G. 2000. Systems Engineering in naval ship design. *Naval engineers journal* 112(4): 45–57.

Cuneo, B.J. 2013. *Development of a human centric multi-disciplinary design optimization method using fuzzy logic systems and controllers* (Doctoral dissertation). University of Michigan, Ann Arbor.

Dong, W.M. & Wong, F.S. 1987. Fuzzy weighted averages and implementation of the extension principle. *Fuzzy Sets and Systems* 21(2): 183–199.

Gray, A.W. 2011. *Enhancement of Set-Based Design practices via introduction of uncertainty through the use of interval type-2 modeling and general type-2 fuzzy logic agent based methods* (Doctoral dissertation). University of Michigan, Ann Arbor.

Kossiakoff, A., Sweet, W.N., Seymour, S.J. & Biemer, S.M. 2011. *Systems Engineering: principles and practice*. John Wiley & Sons.

McKenney T.A. 2013. *An early-stage Set-Based Design reduction decision support framework utilizing design space mapping and a graph theoretic markov decision process formulation* (Doctoral dissertation). University of Michigan, Ann Arbor.

Mebane, W.L., Carlson, C.M., Dowd, C., Singer, D.J. & Buckley, M.E. 2011. Set-Based Design and the ship to shore connector. *Naval Engineers Journal* 123 (3): 79–92.

Singer, D.J. 2003. *A hybrid agent approach for Set-Based conceptual ship design through the use of a fuzzy logic agent to facilitate communications and negotiation* (Doctoral dissertation). University of Michigan, Ann Arbor.

Singer, D.J., Doerry, N. & Buckley, M.E. 2009. What is Set-Based Design? *Naval Engineers Journal* 121(4): 31–43.

Ward, A.C., Liker, J.K., Christiano, J.J. & Sobek, D.K. 1995. The second Toyota paradox: how delaying decisions can make better cars faster. *Sloan Management Review* 36(2): 43–61.

Wood, K.L., Otto, K.N. & Antonsson, E.K. 1992. Engineering design calculations with fuzzy parameters. *Fuzzy Sets and Systems* 51(1): 1–20.

Zadeh, L.A. 1965. Fuzzy sets. *Information and Control* 8(3): 338–353.

Managing epistemic uncertainty in multi-disciplinary optimization of a planing craft

D. Brefort & D.J. Singer
University of Michigan, Ann Arbor, USA

ABSTRACT: This article introduces a multi-disciplinary optimization method for early-stage ship design. The limited knowledge available about a vessel in the early stages of its development limits the applicability of traditional optimization methods in such stages. This epistemic uncertainty stems from several sources, three of which are considered in this article. First, the uncertainty caused by limited information of a vessel model which is not fully defined. Second, the uncertainty which stems from the difficulty identifying precise design performance targets and comparing designs with similar performance. The third source of uncertainty relates to the difficulty identifying precise bounds on the validity of analysis tools. A multi-disciplinary optimization method is introduced which uses type-2 fuzzy logic to model the vague information inherent to early-stage design and incorporate human expertise directly into the optimization to handle this uncertainty. The method is illustrated on the optimization of a planing craft with respect to resistance and seakeeping.

1 INTRODUCTION

The design of ships and other large engineered systems involves generating knowledge for decision-making through time. A common method of generating such knowledge is mathematical modelling, which is defined as "an abstract description of the real world giving an approximate representation of more complex functions of physical systems" (Papalambros & Wilde, 2017). It attempts to capture and structure knowledge about a system into a form which is useable to make a decision. For instance, models are often used to understand the relationship between design variables and performance parameters (i.e. objective functions), and to understand trade-offs between multiple competing performance parameters. However, models are nothing more than the structured description of a system used to guide engineering decisions. The actual decisions result from the analysis and interpretation of the mathematical model's results by the human designer—what is commonly known as the engineer's judgement. It is the iterative interplay between information generation (i.e. using models) and human judgment, which leads to a better understanding of a design. The engineer's judgement comes from rich mental models developed over years of ones' professional life. It captures experience, personal values, and expert beliefs, which cannot be easily incorporated into product models. Instead, the engineer's judgement is typically used a posteriori to evaluate the output of a model, limiting designers' ability to use their expertise directly in the modelling process. In response, this paper proposes a method which uses fuzzy logic to incorporate human expertise directly into optimization methods. Fuzzy logic was inspired by a human's ability to converse, communicate, reason and make rational decisions in an environment of imprecision, uncertainty, incompleteness of information and partiality of truth. As such, it is an adequate tool to model human expertise and incorporate domain knowledge directly into an optimization model.

Recent advances in computational capability and market competitiveness have pushed organizations to develop computer-based tools capable of rapidly producing and evaluating a large number of design solutions (Gray et al. 2013). One of these tools is optimization, which uses a mathematical description of the product being designed to find the solution with the highest performance. The mathematical structure of the model allows a quick and effortless analysis of the product, helping the designer find design variable values that maximize a product's performance. Multidisciplinary optimization (MDO) has been particularly useful to the design of engineered systems, where interdependencies between diverse disciplines with conflicting requirements become a driver of the system's overall performance. MDO has found numerous applications in aircraft design (Henderson et al. 2012, Vlahopoulos et al. 2011, Allison et al. 2006) and in automotive engineering where it was used to design a vehicle chassis according to ride quality and handling (Kim et al., 2003). It has also found uses in propeller design where hydrodynamics and structural dynamics interact (Takekoshi et al. 2005, Young et al. 2010), and for the optimization of a ship's propeller-hull systems (Nelson et al., 2013).

Mathematical modelling and optimization; however, suffer from several limitations which keep them from being more widely used in professional engineering settings. The first major limitation stem from a model's inherent property of being a reduction of the actual problem being investigated. A model cannot capture every detail of a problem, so simplifications and abstract representations of the product are required. Properly modelling every aspect of a problem is impossible due to time and budget constraints, because some of the data is unavailable for legal or privacy reasons, and because some of the information cannot fit the rigid mathematical formulation required by optimization models. The second major limitation of optimization stems from the possible unacceptance or misunderstanding of the optimization results by the human user (Meignan et al. 2015). A complacent user might trust a bad solution, and a skeptical user might reject a good optimization solution because they don't trust the model.

Several approaches have already been proposed to address these limitations. The constant increase in available computing power has pushed back the limits on model detail and complexity. For naval design, Kassel (2010) discusses the need for a tool to "rapidly produce a full range of feasible ship arrangements from a basic shell of a ship" and analyze the "vulnerability implications of the sizing and arrangement of the ship." However, even the best mathematical models cannot accurately capture all aspects of a design problem. Interactive optimization recognizes this limitation and calls on the user to fill this modeling gap by providing feedback about the optimization results during the run, enriching the model and guiding the search process (Meignan et al., 2015). For example, Kim and Cho (2000) developed an interactive genetic algorithm to determine the attributes that make a piece of clothing fashionable. Aesthetics cannot be modelled mathematically, so at each iteration of the algorithm, a user ranking of several clothing designs replaced the traditional objective function used by optimization algorithms. However, this method is prone to tedium and error, and the population size of the algorithm is limited. Fuzzy logic has been used to incorporate human input a-priori into optimization algorithms, reducing the need for human evaluation of every solution. Fuzzy logic was originally developed as a method of programming computers using natural language (Zadeh, 1965), making it an adequate tool to formalize personal preferences and experiences into a computer algorithm. Steinberg (1993) developed an automatic aircraft carrier landing system for an F/A-18 using fuzzy logic to incorporate "elements of human pilot 'intelligence' with more conventional automatic control laws." In optimization, fuzzy logic has been used to rank preferences between performance criteria of a multi-objective optimization using linguistic statements, making it easier for the user to communicate the relative importance of problem objectives to the optimization (Yazdi, 2016). Additionally, fuzzy membership functions have been used in ship design to assign preferences to ranges of design variables, allowing users to formalize their preferences for the product. Singer (2003) used fuzzy membership functions to assign a designer's preferences for ranges of ship design variables with respect to multiple design disciplines, and negotiate the ranges of design variables to cut during a Set-Based design reduction process. Cuneo (2013) uses a similar method of modelling designer preferences through fuzzy membership functions, but uses them to classify the user's preferences on results of design analysis tools. Human designers interpret optimization results, allowing them to infer more knowledge than is captured in the model's objective function. Cuneo's method uses fuzzy logic to mimic human judgement and use it to increase the information content of an optimization run.

This article extends Cuneo's work through the inclusion of type-2 fuzzy logic into the multidisciplinary optimization to better handle epistemic uncertainty associated with linguistic reasoning used by people, and limited models. Although type-1 fuzzy logic systems have been successfully applied to many real world problem (Mendel, 2000), their use of crisp membership functions limits their ability to effectively handle uncertainty. This criticism of type-1 fuzzy sets was answered when Zadeh (1975) introduced type-2 fuzzy sets in which the membership functions used to describe linguistic variables are themselves fuzzy, allowing them to handle uncertainties where type-1 systems can't (Ozen et al., 2003). For instance, the meaning of words used in linguistic variables is uncertain, and can mean different things to different people (Mendel et al. 2002). This article proposes a type-2 fuzzy logic multidisciplinary optimization to handle three types of epistemic uncertainty associated with early stage design and human expertise. The first is the uncertainty associated with limited information when only preliminary models of a vessel's arrangements exist. The second is the uncertainty associated with the bounds on analysis tools validity. The third type of uncertainty originates from the uncertainty associated with the linguistic terms used to characterize a design. Human expertise is conveyed linguistically, so modeling this type of uncertainty is crucial to properly incorporating human expertise into MDO algorithms. This article is structured as follows. Section 2 reviews multidisciplinary optimization and introduces the method of incorporating human experience into the optimization model. Section 3 explains how this is done on a planing craft case study. Results are presented in Section 4, followed by concluding remarks.

2 METHODS

2.1 Optimization methods

Generally speaking, optimization is defined as the search for design variable values that minimize or maximize an objective function value, subject to constraints. First, the constituents' objectives are determined (i.e. the performance parameters used to evaluate a solution). Then, a mathematical model is built which defines the relationship between design variables and performance parameters. These models can range from low fidelity regressions of historical data to high fidelity simulations (Jouhaud et al., 2007; Yang et al., 1996). Finally, users run the models through an optimizer for a variety of test cases to find designs with good performance and gain knowledge about the design space.

2.2 Multidisciplinary optimization

Multidisciplinary design optimization (MDO) is particularly useful in complex engineering problems. In such problems, the management of interdependencies between diverse engineering disciplines with conflicting requirements becomes a driver of the design because "the performance of a multidisciplinary system is not only driven by the performance of the individual disciplines but also by their interactions" (Martins et al., 2012). MDOs provide a structured approach of coupling multiple disciplines through numerical optimization and coordinate the search of each discipline to find a consistent set of design variables (Hannapel, 2012). In hierarchical multidisciplinary optimization for example, a top level (i.e. system) optimizer sets targets for the design variables of lower level (i.e. discipline) optimizers. Each discipline independently searches for the set of design variable values that maximizes its own performance, subject to design variable targets set by the system level optimizer. Thus, the top level optimizer's main role is to minimize the difference between design variable values returned by the lower level optimizers.

2.3 Hierarchical fuzzy logic multidisciplinary optimization

In open ended problems like early-stage design, few precise mathematical model of the product exist. This limits the designer's ability to use classical optimization methods which have greatly benefited the aerospace and automotive industry. To bring these benefits to early-stage ship design, this article proposes a hierarchical fuzzy logic multidisciplinary optimization (FL-MDO) method which can better handle the uncertainty of early-stage design. The FL-MDO is used to formalize human expertise directly into an optimization framework and fill knowledge gaps without developing large complex models. The overall structure of the FL-MDO is shown in Figure 1.

The first step of the optimization is the discipline analysis, where each engineering discipline evaluates design variables with respect to several performance indicators of interest to them. These performance indicators can come from the objective functions of mathematical models, or can be from less precise models like rules of thumb. The second step consists of assigning preferences to performance indicator values through a fuzzy logic mapping of human expertise. In step three, the fuzzy output preferences $F_j(\mathbf{x})$ of each discipline j are aggregated into an overall system level solution through fuzzy intersection (Dubois & Prade, 1980) (Eq. 1) or through the mean of all discipline preferences (Eq. 2)

$$F_s(\mathbf{x}) = \bigwedge_{j=1}^{n} F_j(\mathbf{x}) \qquad (1)$$

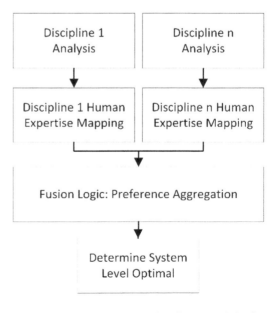

Figure 1. Fuzzy logic multidisciplinary optimization structure.

$$F_s(\mathbf{x}) = \frac{\sum_{j=1}^{n} F_j(\mathbf{x})}{n} \qquad (2)$$

In the final step, the design variable instance with the highest performance is found.

2.4 *Fuzzy logic systems*

The heart of the FL-MDO presented above lies in the human expertise mapping stage (step 2 of the overall FL-MDO), which allows expert opinion to be included directly into optimization models. This mapping is done using fuzzy set theory and fuzzy logic systems. Fuzzy logic was inspired by a human's ability to converse, communicate, reason and make rational decisions in an environment of imprecision, uncertainty, incompleteness of information and partiality of truth. Unlike classical set theory, where an element e is either a member of a set A (i.e. $e \in A = 1$), or an element e is not a member of the set A (i.e. $e \in A = 0$), fuzzy set theory allows the degree of membership of an element e to a set A to lie between 0 and 1. In classical set theory, a person must either be tall or not tall, but fuzzy set theory allows a person to be classified as partially tall and partially not tall, better modelling the vagueness and nuances associated with real life and human judgement.

Fuzzy logic systems mimic human reasoning by mapping a crisp input variable to a crisp output variable through a set of linguistic rules (Fig. 2).

The set of linguistic rules, or rule bank, is at the center of the fuzzy logic system. It consists of a set of IF-THEN statements that map a set of linguistic inputs (i.e. antecedents) to linguistic outputs (i.e. consequents). These rules represent learned patterns or mental models, and can be populated using expert opinion or extracted from data. A linguistic rule could be stated as follows:

If air moisture is high and temperature is hot, then chances of a thunderstorm are high.

To use the rule bank, crisp input variables must first be converted to linguistic variables. This is done through the fuzzifier, which uses a membership function to determine how similar a numerical variable value is to its linguistic description.

The fuzzy inference determines which rules are activated, and calculates their activation level. This is analogous to a human associating new environmental cues to similar past experiences to navigate an unknown situation.

The final step is the output processing block, which combines the outputs from activated rules into a single decision. A new situation will likely share traits with several past experiences, each of which can inform the decision maker.

In the context of the FL-MDO, a separate fuzzy logic system is used to assign preferences to design variables with respect to each design discipline. The discipline's fuzzy logic system takes crisp inputs from several objective functions related to the discipline's performance. The fuzzifier converts these values to a set of high, medium, or low linguistic preferences based on their objective function value. The linguistic variables are then passed through the fuzzy inference process to assign preference bounds to each design variable instance (Fig. 3). The process is illustrated in detail in Section 3.

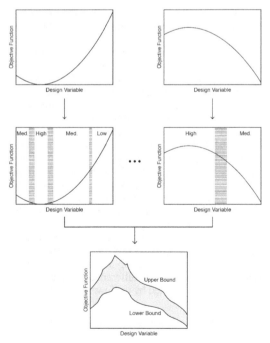

Figure 3. Human expertise mapping of a single discipline (adapted from Cuneo, 2013).

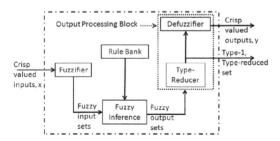

Figure 2. Type-2 fuzzy logic system (Mendel, 2000).

3 CASE STUDY

The early-stage multidisciplinary design optimization method presented in section 2 is applied to the design of a planing vessel. The vessel will be analyzed with respect to seakeeping and resistance performance. In early-stages, when details of the hull form are still uncertain, the hydrodynamics properties of a planing craft can be simply described by the vessel's length, beam, deadrise, and longitudinal center of gravity (Knight et al. 2014). These are the variables considered in this study. The optimization process begins with an arrangements generator, which determines the most likely location of the longitudinal center of gravity (LCG). This LCG is then used as a parameter in the seakeeping and resistance analyses. The optimization problem formulation is given as follows. (Note: the tildes in constraint g_2 represent fuzzy or approximate numbers.)

$$\text{Maximize} \begin{cases} \textit{Seakeeping Performance} \\ \textit{Resistance Performance} \end{cases}$$

With respect to L, B, β

Subject to: $g_1 : 2.5 \leq \frac{L}{B} \leq 7.0$
$\qquad\qquad g_2 : \tilde{3} \leq \tau \leq \tilde{7}$

Given: LCG
$\qquad\quad \Delta$
$\qquad\quad V$
$\qquad\quad h_{1/3}$

where: L = length
$\qquad\;\, B$ = beam
$\qquad\;\, \beta$ = deadrise angle
$\qquad\;\, \tau$ = trim
$\qquad\;\, LCG$ = longitudinal center of gravity
$\qquad\;\, \Delta$ = displacement
$\qquad\;\, V$ = speed
$\qquad\;\, h_{1/3}$ = significant wave height

3.1 Arrangements generator

The arrangements generator uses a fuzzy logic system to determine the most likely location of the longitudinal center of gravity based on the designer's vague mental model of the vessel's arrangements. The concept of the arrangements generator is illustrated below for a two compartment vessel. If the compartment is assumed to be a standard rectangle, and no additional information about it is available, then the location of its LCG can only be assumed to be in the middle of the box (i.e. 50% of the compartment length). A human designer with a mental model of the compartment's arrangements; however, can easily provide his or her belief on the location of the LCG.

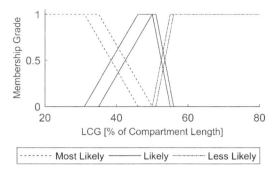

Figure 4. Compartment LCG belief membership function.

In this work, these beliefs are captured by type-2 fuzzy membership functions (Fig. 4). The membership function assigns likelihood to the LCG's location based on the designer's knowledge of the vessel's typical arrangements layout. In Figure 4, representing the engine room compartment, the LCG is likely to be at midpoint of the compartment (i.e. the center of the box). However, a human designer knows that the engine is typically located near the rear of the compartment, so assigns the LCG a higher likelihood of being near the rear of the compartment. Uncertainties associated with the selected propulsion system limit his or her ability to know the exact definition of the "likely' and "most likely" membership functions, and where the transition between the two occurs though. This uncertainty is captured by the parallel dashed and solid lines to the left of Figure 4, which provide a measure of dispersion associated with the designer's model and the linguistic terms he or she uses. The bottom lines represent the lower membership functions, or lower bound on the uncertainty, and the top lines represent the upper membership functions, or upper bound on the uncertainty. The large size of a water-jet propulsion system pushes the LCG forward compared to a propeller, for example, but the propulsion system might not be known until later in the design process, hence the large uncertainty on the engine room LCG model around of 40%.

Beliefs about each compartment's longitudinal center of gravity are aggregated into an overall belief about the planing craft's LCG through a fuzzy logic rule bank (Table 1). In a two compartment vessel for example, the vessel's LCG is the average of the two compartments' LCGs, and the designer's preference for this overall LCG depends on the designer's belief about each compartment's LCG location likelihood. If compartment 1 has its LCG at 30% (most likely membership grade of one) and compartment 2 has an LCG at 50% (likely membership grade of one), the vessel's overall LCG at 40% of the vessel length has a medium-high likelihood of being the true vessel LCG.

Table 1. Vessel LCG belief rule bank

LCG 2 belief \ LCG 1 belief	Most likely	Likely	Less likely
Most Likely	High Likelihood	Medium-High Likelihood	Medium Likelihood
Likely	Medium-High Likelihood	Medium Likelihood	Medium-Low Likelihood
Less Likely	Medium Likelihood	Medium-Low Likelihood	Low Likelihood

The fuzzy logic system used by the arrangements generator produces a model of the vessel's LCG location by aggregating the designer's beliefs about several compartments' arrangements layout. The output of this model is shown in Figure 5, where the uncertainty associated with the membership functions is propagated to the model. The vessel LCG selected for seakeeping and resistance analysis is chosen as the average of the LCGs with the highest likelihood.

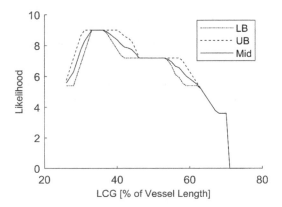

Figure 5. LCG location likelihood model.

3.2 Resistance discipline

The purpose of the resistance discipline is to minimize vessel drag subject to trim restrictions. The Savitsky method is used by the discipline analysis step of the FL-MDO to calculate the equilibrium trim and drag by summing the pressure drag, frictional drag, and dynamic lift acting on the vessel (Savitsky 1964). The total resistance is given by Equation 3.

$$R = \Delta \tan \tau + \frac{\rho V^2 C_f \lambda B^2}{2 \cos \beta \cos \tau} \tag{3}$$

The drag and trim values of a design are input into the human expertise mapping step. They are first converted to linguistic variables which convey the designer's preference for the design with respect to its drag and trim values (Figs. 6–7). In Figure 6, R^{min} is the smallest drag of the population of evaluated designs and R^{max} is the drag value of the design which is optimal with respect to accelerations. High resistance preference is one at R^{min}, and linearly goes to zero as drag increases to R^{mid} (i.e. the midpoint between R^{min} and R^{max}). The medium resistance preference increase from zero to one at the same rate. The exact transition between the high, medium, and low resistance preference is not known exactly, which models uncertainty in linguistic terms used to describe designs and the difficulty comparing two designs with similar performance. This uncertainty is captured by the parallel lines in the membership function, which provide the bounds on the membership grade.

The design's trim is used to penalize designs with trim outside of the Savitsky's method's bounds, which is only valid for trim values between 3° and 7°. However, it is unreasonable to believe that these

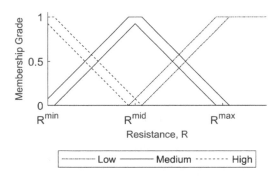

Figure 6. Resistance preference membership function.

bounds are crisp. That is that the Savitsky method is valid for designs with trim of 3.01° and invalid for designs with trim of 2.99°. The uncertainty associated with the fuzziness of model bounds is modeled by the membership function for trim shown in Figure 7. Designs with trim below 2.8 and above 7.2 are unpreferred with membership grade one. Between 2.8 and 3.2, the unpreferred membership grade linearly goes to zero.

Designer preferences about the design's drag and trim values are aggregated into an overall resistance discipline preference using a fuzzy logic system and the rule bank shown in Table 2.

3.3 Seakeeping discipline

Although intended for high speed operations, the performance of planing craft in semiplaning and displacement conditions must also be considered, especially in rough seas where planing conditions might not be achievable (Savitsky, 1985). The seakeeping discipline aims to minimize the craft's vertical accelerations in planing conditions, improve non-planing seakeeping performance, and limit slamming. Each of these performance characteristics are estimated in the seakeeping discipline analysis step. Vertical accelerations in planing conditions are given by Savitsky (1985) in Equation 4.

$$a = \frac{1}{C_\Delta} \left[\frac{0.0104\tau}{4} \left(\frac{h_{1/3}}{B} + 0.084 \right) \left(\frac{5}{3} - \frac{\beta}{30} \right) \left(\frac{V_k}{\sqrt{L}} \right)^2 \left(\frac{L}{B} \right) \right] \quad (4)$$

The length to beam ratio of the vessel is used as a surrogate for non-planing seakeeping performance. Vessels with length to beam ratio greater than five reduce the impact of acceleration in semi-planing conditions (Savitsky, 1985) so vessels with larger length to beam ratios are assumed to have better non-planing seakeeping performance. Finally, deadrise is used as a surrogate for slamming performance. A moderate aft deadrise around 15° is assumed to provide the best compromise between minimizing slamming and minimizing resistance. The membership values encoding these preferences

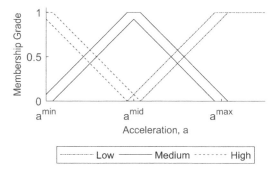

Figure 8. Accelerations preference membership function.

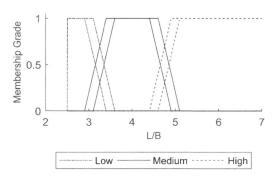

Figure 9. Length to beam preference membership function.

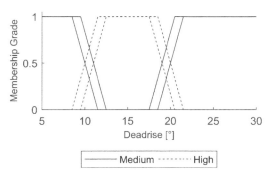

Figure 10. Deadrise preference membership function.

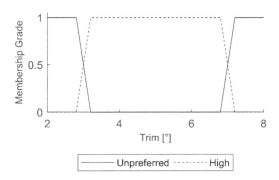

Figure 7. Trim preference membership function.

Table 2. Resistance discipline rule bank.

Trim Pref.	Resistance Pref. High Preference	Medium Preference	Low Preference
High Preference	High Preference	Medium Preference	Low Preference
Unpreferred	Unpreferred	Unpreferred	Unpreferred

Table 3. Seakeeping discipline rule bank.

L/B Pref. \ Acceleration Pref.	High Preference	Medium Preference	Low Preference
High Preference	High Preference	Medium Preference	Low Preference
Medium Preference	High Preference	Medium Preference	Low Preference
Low Preference	Medium Preference	Low Preference	Low Preference

High Deadrise Preference

L/B Pref. \ Acceleration Pref.	High Preference	Medium Preference	Low Preference
High Preference	High Preference	Medium Preference	Low Preference
Medium Preference	Medium Preference	Low Preference	Unpreferred
Low Preference	Low Preference	Unpreferred	Unpreferred

Medium Deadrise Preference

are shown in Figures 8, 9, 10, and aggregated using a fuzzy logic system and a rule bank (Table 3). The acceleration value a^{min} is the smallest acceleration of the population of evaluated designs and a^{max} is the acceleration of the design with the smallest drag value.

The rule bank allows designers to easily incorporate their expertise into the optimization model. Table 3 shows that the designer places more importance on vertical accelerations than length to beam ratio, since a high acceleration preference gives a high output preference for high and medium length to beam ratio preference. This rationale is explained because planing craft operate mostly in planing conditions, and planing accelerations are much more violent on the crew than non-planing accelerations.

4 RESULTS

The optimization uses the following parameters, with the LCG value being returned by the arrangements generator module.

Δ = 30 tons
V = 50 knots
$h_{1/3}$ = 1 meter
LCG = 0.35% of vessel length from aft

4.1 *Pareto analysis*

In multidisciplinary optimization, the performance objectives of the different disciplines are often conflicting. In such cases, the performance increase of one discipline comes at a performance cost to at least one of the other disciplines and the optimization results form a Pareto front. A design point $\mathbf{x}^* \in \Omega$ is said to be Pareto optimal (Coello et al. 2004) if for every $\mathbf{x} \in \Omega$ and $I = 1,2,...,k$ disciplines.

$$f_i(\mathbf{x}) = f_i(\mathbf{x}^*) \quad \forall i \in I \quad (5)$$

Or there is at least one $i \in I$ such that

$$f_i(\mathbf{x}) > f_i(\mathbf{x}^*) \quad (6)$$

According to Equation 3, the resistance increases with increasing beam and deadrise. According to Equation 4, vertical accelerations decrease with increasing beam and deadrise, making their objectives conflicting. Figure 11 shows the Pareto front of the planing craft seakeeping and resistance performance built with the fuzzy logic multidisciplinary optimizer detailed in Section 3. It clearly shows the conflicting objectives of the two disciplines, with resistance preference increasing as seakeeping preference decreases, and vice-versa. The uncertainty associated with membership function definition in section 3 is propagated to the Pareto front through the spread on preferences between the Pareto front lower and upper bound. The min ub best and mean of mean best system preferences are calculated using Equations 1 and 2.

Several points along the mean Pareto front where sampled and are listed in Table 4 (design one exhibits the highest resistance preference and lowest seakeeping preference, and design six exhibits the lowest resistance preference and highest seakeeping preference). These results show the tradeoff between seakeeping and resistance performance. The vessels with better resistance characteristics tend to be shorter with smaller deadrise angles. At high speed, friction drag becomes a large component of drag so vessels with large wetted lengths exhibit high resistance. Additionally, although deep V-hulls exhibit better seakeeping characteristics, they require more power to plane, and thus exhibit worst resistance characteristics. The designer's preference for high length to beam ratios for non-planing seakeeping performance is also captured in Table 4, with length to beam ratios increasing along the Pareto front from design one to six.

4.2 Fusion logic

As stated in Section 2, two types of fusion logic are used in this case study. The first is the fuzzy intersection, or min fusion logic (Eq. 1). The second is the average fusion logic (Eq. 2). Table 5 shows the optimal design found with each of these method. Both fusion methods, plotted in red in Figure 11, provide comparable results which have a good compromise between resistance and seakeeping.

Table 4. Pareto optimal design points.

Optimal	Pareto designs 1	2	3	4	5	6
Length [m]	9.5	11.6	15.1	19.1	21.7	26.5
Beam [m]	3.8	4	3.2	2.9	3.1	3.8
Deadrise [°]	5	5	5	5	14	25
Drag [kN]	36	37	39	40	45	59
Accelerations [g]	2.2	1.9	1.6	1.3	1.1	0.8

Table 5. System level optimals.

Optimal	Fusion type Min fusion	Mean fusion
Length [m]	21	20.9
Beam [m]	3	3
Deadrise [°]	9	5
Drag [kN]	43	42
Accelerations [g]	1.2	1.2

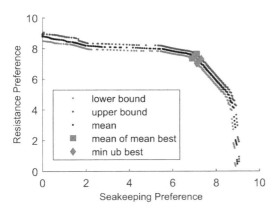

Figure 11. Planing craft Pareto front for V = 50 knots.

The system level preferences found using min fusion and mean fusion are plotted in Figures 12 and 13. These show a high preference for vessel lengths between 15 and 22 meters, which provides a compromise between resistance and seakeeping performance. Vessels shorter than 15 meters tend to favor the resistance discipline and vessels longer than 22 meters tend to favor the seakeeping resistance as illustrated in Table 4.

4.3 LCG location sensitivity

The arrangements generator provides a model of the vessel's LCG value according to the designer's belief about the arrangements layout. Figure 5 shows the most likely LCG location to be between 30% and 40% of the vessels length (measured from the transom). Optimal design values calculated using fuzzy intersection, or min UB, fusion logic (Eq. 1) are given in Table 6 for LCGs at 30%, 35%, and 40% of the vessel's length. Forward LCG

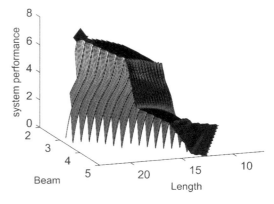

Figure 12. Min fusion system preferences for deadrise of 9°.

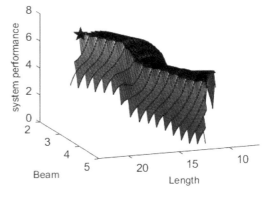

Figure 13. Mean fusion system preferences for deadrise of 5°.

Table 6. System optimal designs for varying LCG locations.

Optimal \ LCG	30%	35%	40%
Length [m]	22.8	21	19.5
Beam [m]	3.3	3	2.8
Deadrise [°]	9	9	9
Drag [kN]	45	43	42
Accelerations [g]	1.1	1.2	1.3

values reduce the vessel's trim, increasing vertical accelerations and decreasing resistance in planing conditions. Future work should include a robust optimization of the planing craft with respect to uncertainty of LCG values. The arrangements module provides bounds on likely LCG values, and the LCG with the worst system level performance between these bounds should be used as a parameter in the optimization.

5 CONCLUSIONS

This article presented a method of incorporating human expertise directly into a multidisciplinary optimization through a type-2 fuzzy logic MDO. Uncertainties associated with the use of reduced models in early-stage ship design make traditional optimization methods ill-suited for these applications. The incorporation of human expertise directly into the optimization allows designers to easily enrich models without needing high-fidelity simulation. However, human expertise models are uncertain themselves, due to the vagueness of linguistic terms humans use to reason and communicate. This uncertainty is handled by type-2 fuzzy logic.

The FL-MDO is demonstrated on a planing craft. Human mental models of the vessel's arrangements are first used to determine the most likely location of the vessel's LCG. This information is then input into the FL-MDO to optimize the craft with respect to seakeeping and resistance, where additional uncertainties are modeled. Uncertainty associated with the bounds on model validity, and uncertainty associated with drag and acceleration preferences are modeled. In addition, the seakeeping analysis shows the method's ability to aggregate planing, non-planing, and slamming performance into an overall seakeeping performance index, and to formalize vague information, like rules of thumb, into an optimization process.

The Pareto front shows conflicting objectives of the seakeeping and resistance disciplines, but the system level optimals returned by the optimization show a good compromise between the two disciplines. The sensitivity of the results with respect to LCG are given, but future work should include a robust optimization of the vessel with respect to all likely LCG values given by the arrangements module.

ACKNOWLEDGEMENTS

The authors would like to thank Ms. Kelly Cooper and the U.S. Office of Naval Research for their support. This work was funded by the Naval International Cooperative Opportunities in Science and Technology Program (NICOP) contract number N00014-15-1-2752.

REFERENCES

Allison, J., Walsh, D., Kokkolaras, M., Papalambros, P. & Cartmell, M. 2006. Analytical Target Cascading in Aircraft Design. *44th AIAA Aerospace Sciences Meeting and Exhibit, Aerospace Sciences Meetings*, Reno, USA.

Coello, C.C., Pulido, G.T. & Lechuga, M.S. 2004. Handling multiple objectives with particle swarm optimization. *IEEE Transactions on Evolutionary Computation*, 8.

Cuneo, B. 2013. Development of a Human Centric Multidisciplinary Design Optimization Method Using Fuzzy Logic Systems and Controllers. *Ph.D. Thesis*, University of Michigan, USA.

Dubois, D. & Prade, H.M. 1980. Fuzzy Sets and Systems: Theory and Applications. *New York: Academic Press*, New York City, USA.

Gray, A.W., Cuneo, B.J., Vlahopoulos, N., & Singer, D.J. 2013. The Rapid Ship Design Environment Multi - Disciplinary Optimization of a U.S. Navy Frigate. *In ASNE Day 2013 Proceedings*, Crystal City, USA, American Society of Naval Engineers.

Hannapel, S. 2012. Development of Multidisciplinary Design Optimization Algorithms for Ship Design under Uncertainty. *Ph.D. thesis*, University of Michigan, USA.

Henderson, R.P., Martins, J.R. & Perez, R.E. 2012. Aircraft Conceptual Design for Optimal Environmental Performance. *The Aeronautical Journal*, Vol. 116, 2012.

Jouhaud, J.C., Sagaut, P., Montagnac, M. & Laurenceau, J. 2007. A Surrogate-Model Based Multidisciplinary Shape Optimization Method with Application to a 2D Subsonic Airfoil. *Computers and Fluids*, 36, 520–529.

Kim, H.M., Rideout, D.G, Papalambros, P. & Stein, J.L. 2003. Analytical Target Cascading in Automotive Vehicle Design. *Journal of Mechanical Design*, 125(3), pp. 481–489.

Kim, H.S. & Cho, S.B. 2000. Application of interactive genetic algorithm to fashion design. *Engineering Applications of Artificial Intelligence*, 13(6):635–644.

Knight, J., Zahradka, F., Singer, D.J. & Collette, M. 2014. Multi-Objective Particle Swarm Optimization of a Planing Craft with Uncertainty. *Journal of Ship Production*, 30(3):1–7

Martins, J. & Lambe, A. 2012. Multidisciplinary Design Optimization: A Survey of Architectures. *AIAA Journal*, 51(9):2049–2075.

Meignan, D., Knust, S., Frayet, J.M., Pesant, G. & Gaud, N. 2015. A Review and Taxonomy of Interactive Optimization Methods in Operations Research. *ACM Transactions on Interactive Intelligent Systems*, 5(3).

Nelson, M., Temple, D.W., Hwang, J.T., Young, Y.L., Martins, J.R. & Collette, M. 2013. Simultaneous Optimization of Propeller-Hull Systems to Minimize Lifetime Fuel Consumption. *Applied Ocean Research*. 43:46–52.

Mendel, J.M. & John, B.J. 2002. Type-2 Fuzzy Sets Made Simple. *IEEE Transactions on Fuzzy Systems*, 117–127.

Mendel, J.M. 2000. *Uncertain Rule-Based Fuzzy Logic Systems: Introduction and New Directions*. Hoboken: Prentice Hall.

Ozen, T. & Garibaldi, M. 2003. Investigating Adaptation in Type-2 Fuzzy Logic Systems Applied to Umbilical Acid-Base Assessement. In *Proceedings of 2003 European Symposium on Intelligent Technologies (EUNITE)*, pp.289–294.

Papalambros, P.Y. & Wilde, D.J. 2017. Principles of Optimal Design. 3rd ed. Cambridge University Press, Cambridge, UK.

Savitsky, D. 1985. Planing Craft. *Naval Engineers Journal*, 97(2):113–141.

Savitsky, D. 1964. Hydrodynamic Design of Planing Hulls. *Marine Technology*, 1(1):71–95.

Singer, D.J. 2003. A Hybrid Agent Approach for Set-Based Conceptual Ship Design Through the Use of a Fuzzy Logic Agent to Facilitate Communications and Negotiation. *Ph.D. Thesis*, University of Michigan, USA.

Steinberg, M. 1993. Development and Simulation of an F/A-18 Fuzzy Logic Automatic Carrier Landing System. *IEEE International Conference on Fuzzy Systems*.

Takekoshi, Y., Kawamura, T., Yamaguchi, H., Maeda, M., Ishii, N., Kimura, K., Taketani, T. & Fujii, A. 2005. Study on the Design of Propeller Blade Sections using the Optimization Algorithm. *Journal of Marine Science and Technology*. 10(2):70–81.

Vlahopoulos, N., Zhang, G., & Sbragio, R., 2011. Design of Rotorcraft Gearbox Foundation for Reduced Vibration and Increased Crashworthiness Characteristics. SAE Technical Paper 2011-01-1704.

Yang, J.B. & Sen, P. 1996. Interactive Trade-off Analysis and Preference Modelling for Preliminary Multiobjective Ship Design. *Systems Analysis, Modelling and Simulation*, 26:25–55.

Yazdi, H.M. 2016. Implementing Designer's Preferences using Fuzzy Logic and Genetic Algorithm in Structural Optimization. *International Journal of Steel Structures*. 16(3):987–995.

Young, Y.L., Baker, J.W. & Motley, M.R. 2010. Reliability-Based Design and Optimization of Adaptive Marine Structures. *Composite Structures*, 92(2):244–253.

Zadeh, L.A. 1975. The Concept of a Linguistic Variable and its Application to Approximate Reasoning. *Information Sciences*, vol. 8, pp. 199–249.

Zadeh, L.A. 1965. Fuzzy Sets. *Journal of Information and Control*, 8:338–353, 1965.

Quantifying the effects of uncertainty in vessel design performance—a case study on factory stern trawlers

J.J. Garcia, P.O. Brett & Ali Ebrahimi
Ulstein International AS, Ulsteinvik, Norway
Department of Marine Technology, Norwegian University of Science and Technology, Trondheim, Norway

A. Keane
Ulstein International AS, Ulsteinvik, Norway

ABSTRACT: A quantitative evaluation of the effects of uncertainty surrounding the development of the next generation of factory stern trawlers is suggested as progression of state-of-the-art ship design practice. To better understand and take into consideration the uncertainties related to the technical, operational and commercial aspects influencing the solution space definition of a factory stern trawler is paramount. This paper discusses such a challenge and reviews ways in which current ship design practice and vessel design solutions developed thereof, can be improved and implemented in novel ship design approaches. The fifth generation of factory stern trawlers—focusing on improving energy efficiency and food product quality—is currently under development, drenched in a deluge of uncertainty. With an aged fleet with little renovation and renewal since the early 90 s, the need for greener and more commercially effective vessels has spurred a fleet renewal market trend. The evolution of technologies and their future benefits, new regulations regarding fishing quotas, fish quality or fish processing, or the future availability of fish among others, have demonstrated, historically, to play an important role on fishing vessel performance. We, therefore, propose new methods of quantification of their effects to improve the performance of the vessel design of the next generation stern trawler.

The research behind this paper is based on a methodology of structured Accelerated Business Development (ABD) workshops identifying uncertainty factors, which are contrasted with those found in state-of-the-art literature. A MATLAB-based simulation model to quantify their effects on the economic performance of the vessel is developed and reviewed in this paper. This model is presented and discussed. The paper argues that a better understanding of the effects of uncertainty factors in the design and operation of factory trawlers, and all other vessel types, for that matter, should support more effective decisions and a better vessel design work process. The paper presents, therefore, a tool to support decision-making under uncertainty during both, the conceptual design phase and in the operational phase of the vessel.

1 INTRODUCTION

It is expected that aquaculture will cover a major portion of the growth in sea food demand in the future, as it is said to have less effects on the reduction of fish stocks. But ocean fisheries are also expected to contribute. Firstly, by improvements in the exploitation of current fish biomass (by means of fish oil, fish meal and other products), and secondary by exploitation of biomass species (mesopelagic). Both factors could contribute to the growth of biomass food produced from the sea, while maintaining the level of captures and reproduction of resources.

The current fleet of factory trawlers, may be challenged when trying to fulfil such future needs and expectations. Low fish prices and quality reputation of trawled fish at periods, are the result of the stagnation of technology and fishing vessel arrangement solutions and equipment, and consequence of the additional supply spurred by aquaculture (OECD, 2016). This development has led to an ageing of the fleet, currently with an average of 34 years (IHS Fairplay, 2017).

The reduction of newbuilding prices, and the need for more effective vessels producing higher quality biomass, motivate a renovation and renewal of the fleet – a fifth generation of factory trawlers. This new generation should be characterized by fishing efficiency, with focus on fish product quality and a better exploitation of fishing captures, flexibility from number of products, species and waters or regions, and a best possible quota utilization.

Another factor spurring a fleet renovation is the poor energy efficiency and environmental footprint of the current trawler fleet, as compared to more modern pelagic trawlers or purse seiners (Ziegler et al., 2013). The replacement of environmentally harmless refrigerants, used by many of the vessels in the current fleet, or more fuel efficient designs, could reduce the carbon footprint of factory trawlers by up to 30% (Ziegler et al., 2013).

The new generation of trawlers is over-come by uncertainty, from the availability of fish stocks, to quota regimes and technological development. In this paper, we propose an approach to quantify the effects of uncertainty in the economic performance of a factory stern trawler, by means of simulation techniques. By doing so, designers and fishermen can be better prepared to take more informed design, technical, operational, and commercial decisions. Firstly, we identify uncertainty factors by performing a qualitative evaluation involving different stakeholders, such as designers, shipbuilders and fishermen. Secondly, this qualitative data is quantitatively converted and implement as input in a simulation practice – a one year's operation of a factory trawler. We use contribution margin (Gallo, 2017) as relevant performance benchmark. It is our proposition, that a shipowner will not invest in a new vessel if he or she cannot see an economic benefit from it. Finally, we present our results and discuss them, including some recommendations towards design betterment for the fifth generation of factory stern trawlers.

2 HISTORY AND INTRODUCTION TO FACTORY TRAWLERS

MV Fairtry and the rest of factory trawlers built during the 50 s and 60 s, characterized by the combination of a stern ramp and onboard processing, are known as the first generation of factory stern trawlers (Standal, 2008). The first generation would further evolve by the introduction of filleting machines, reducing the need for personnel onboard from 50 down to 30. During the 70 s the second generation of factory trawlers evolved, characterized by two-lane trawl deck and the independent operation of two trawls simultaneously. The flexibility of working with two trawls allowed continuous operation, without the inconvenience of stopping the operation when one trawl was damaged, as was the case in the first generation. During the late 80 s the third generation of factory stern trawlers were developed, which did not represent a major change. The focus was to build wider vessels, with larger deck areas, aimed at increasing the workability of the trawl deck and expanding the factory onboard. The contracting activity which spurred the development of the third generation was especially consequence of protraction in the Exclusive Economic Zone (EEZ) (Fernandez et al., 2014), and quota incentives from national governments such as the Norwegian (Standal, 2008). The fleet in Norway, among others, expanded from 14 to 25 vessels.

The turning point of the fleet of factory stern trawlers could be seen in 1989. During that year, the cod stock collapsed, and the quotas were set to historical minimum, catching many owners with recently delivered expensive vessels. As a result, many companies went bankrupt, while others could sell their vessels to operators in regions such as Russia or New Zealand. Additionally, as a result of social pressure, the vessels were obligated to land part of the quota for processing further inland, as a measure to create land-based post-processing jobs. After this period, as depicted in Figure 1, the contracting activity was halved, to an average of approximately 40 vessels per year during the 90's and less than 10 during the 2000's. Those vessels are here considered as the "third+" generation, as they represent a continuation of the technology developed in the late 80 s and not a new generation. Low contracting activity during this period has also been a consequence of the growing interest towards Oil&Gas investments. Owners, shipyards and designers, triggered by the attractive offshore market moved to the background and often postponed any previous engagements in regards to activities pertaining to fishing vessels.

After the financial crisis in 2008, driven by high fuel oil prices, fishing units were at risk, and a need for a fleet renovation started growing, as a mean to improve competitiveness, energy efficiency, onboard activity optimization, safety and ergonomics (Fernandez et al., 2014). The vessels built after 2010 are considered the fourth generation of factory stern trawlers. With a strong focus aimed at fuel efficiency, the vessels inherent to fourth generation are characterized by engines boasting up to 20% lower specific fuel oil consumption than those of the third generation (Fernandez et al., 2014), representing a major improvement of a vessel's economy and emission footprint. According

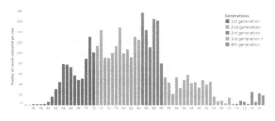

Figure 1. Historic contracting activity of stern trawlers 1954–2017 (IHS Fairplay, 2017).

to (CRISP, 2015), the newer tonnage is a 50% more efficient in terms of unit fuel consumption used per kilo captured fish/biomass and resulting fishing business, which is 5–15% more profitable. The same article highlights some of the improvements from this generation, such as hull shape design and factory processes, including robotized freezers and storage. However, methods for fish handling are still inadequate and arguably little evolution has taken place over the past 60 years (CRISP, 2015), resulting in more than 15% mortality rate before the fish is further processed.

We advocate, therefore, that the design of the next generation of factory stern trawlers, the fifth generation, should focus on *fishing efficiency*, aiming for maximizing revenue and profit though improved quality of the end biomass product. Moreover, flexibility should be demonstrated through an ability to produce an increased number of end products, utilize species and waters (regions) a better extend than today, and exhibit the best possible quota utilization. It therefore follows, that the integration of technical, operational, and commercial perspectives are paramount, and that a better collaboration from participating stakeholders during early stages of the ship's design process is vital.

3 FACTORY TRAWLER DESIGN, OPERATION AND COMMERCIALIZATION

The design, operation and commercialization of factory stern trawlers varies, among others, with regards to operational region and targeted species to be caught. Modern factory stern trawlers can catch cod in the Barents Sea, pollock off the coast of Alaska, shrimp in Islandic waters, hake outside Argentina or krill and other meso-pelagic biomass in the Antarctic. While a vessel operating in the Southern Atlantic (FAO 41) may sail 7,000 NM (nautical miles) from Spain for a six-month campaign (Fernandez et al., 2014), a vessel catching cod in the Barents Sea will typically, not sail more than 500 nm. Hence, ship owners and ship designers may have to prioritize different design parameters or consider future changes in operational conditions in early design stages.

Factory stern trawlers are fishing vessels that catch the fish with one or multiple trawls and then, with help from onboard facilities, process and store the catch in different products for consumption. From a holistic perspective, one can designate eight main system functions onboard a factory stern trawler. Each of the systems has a specific purpose, and they are interconnected as shown in Figure 2. The operation of the different systems could be a value chain, where a fault of one system will be propagated to the overall vessel's performance. A breakdown of one machine in the factory could halt the entire processing line, creating a bottle neck in the holding tanks which may require the termination of further trawling, which subsequently hinders any further activity.

The trawling operation can take from around one hour, if there is a large school of fish, up to several hours, when the availability of fish is lower. In case of the latter, the trawl will be recovered after six or seven hours to reduce the effects of fish squeeze in the trawl. When the trawl is full, it is hauled onboard the vessel, traditionally over a stern ramp. Once onboard, the net is emptied into the holding tanks (also known as RSW tanks—Refrigerated Sea Water). From those tanks, the fish is transferred into the factory onboard, where it will be stunned, headed, gutted and processed into the final fish products before going into the cold storage room. The RSW tanks are used as an intermediate storage of live fish between catch and the processing of the fish, allowing for an increasingly uniform workload in the factory, contrary to peak loads generated by each individual haul.

A simplified version of the activities onboard the factory is presented in Figure 3. At the entrance of the factory, the fish is electrically stunned, reducing the stress on the fish as it is further processed, improving the final quality of the product (Digre, 2013). After a cleansing stage with cooled seawater, the fish is sorted and graded, by species and

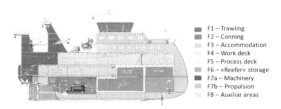

Figure 2. Factory stern trawler subdivided in eight system functions and their interconnections.

Figure 3. Diagram of fish processing onboard a factory trawler. Based on (Digre, 2013).

size. Depending on seasonality, fish prices, fish size or the capabilities of the vessel, the fish is further processed as fillets or is frozen as H/G (headed and gutted) fish. If further processed, the fish will be filleted, boned and frozen. Raw materials such as heads, guts, skin and bones will be further processed as fish meal or fish oil, or in some cases packaged as standalone products, such as cod liver. A better utilization of fish resources, by means of fish oil and fish meal, could potentially generate an increase of 13% for biomass and 16% in revenue, based on 2017 prices and typical level os Norwegian Sea catch.

From an operational perspective, a key factor on factory trawlers is to decide which species to fish, where to do it and what type of final product produce and ship to the markets. Those are decisions to be taken before the vessels leave port, as the type of trawls and the crew required onboard will depend on such decisions. They are also decisions, which must be taken at a conceptual design level, as a vessel designed without filleting machines cannot fillet fish without a costly retrofit, unless the vessel has been prepared for it (Rehn et al., 2017). Seasonality, plays an important role in this decision, as the availability of fish and its commercialization continuously influence a continually the decision of which species is to be targeted.

Figure 4 includes a monthly distribution of catches per species. Species such as cod and pollock are largely caught during the first months of the year, while less valuable species like hake and hoki are left to the end of the year. This shows how fishing companies look to maximize use of their quotas and how the new design solutions must be able to respond and support such dynamics. Their strategy, in most cases, looks for the full quota utilization of the most valued species, leaving the quota of other species that are less profitable to the end of the year, see Figure 4 and Figure 5.

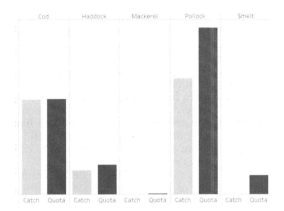

Figure 5. Quota and total catch by species of a Norwegian company during the period 2014–2016 (Based on data from Fiskeridirektoratet, 2017).

4 UNCERTAINTIES SURROUNDING THE FIFTH GENERATION OF FACTORY TRAWLER DESIGN

Uncertainty mitigation, is an important part of today's shipping industry, from the design phase through the construction and the operation of the vessel, the latter being where most literature has focused (Erikstad and Rehn, 2015). With regards to design of special vessels, uncertainty has been present in typically technical, commercial and operational areas of the decision making process (Garcia et al., 2016).

In much of the fishing ship design literature, little has been said and discussed as to how these technical, operational and commercial aspects together influence the overall design strategies and tactics of factory stern trawler solutions (Gates, 1984). Uncertainty has been considered at a macro-perspective in some cases, as a key factor in the management of world fisheries, but not taken into the detailing of a vessel design solution process. One example is the paper from (Davies, Mees and Milner-Gulland, 2015), who study the use of scenario planning techniques for the management of tuna fisheries in the Indian Ocean. A similar example, is the overview of the current status of global fisheries by (Sumaila, Bellmann and Tipping, 2016), who evaluate challenges and opportunities and the uncertainties regarding the fishing industry in the future. From another perspective, (Mangel and Clark, 1983) evaluate the effects of uncertainty considering the location of schools of fish in relation to the financial performance of fishing vessels, and develop a model to measure the effect of search in such uncertainty. Similarly, (Millar and Gunn, 1990) assess the impact of myopia and catch rate on the cost-performance of a fleet of trawlers. The broadest study regarding uncertainty in the fishing indus-

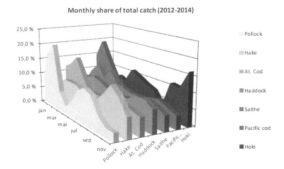

Figure 4. Seasonal distribution of catches by specie (Havfisk, 2017).

try may be the one carried out by (Gates, 1984), who proposes and studies six types of uncertainties: catch rates, equipment failure, prices, weather, quality of inputs, and institutions. (Gates, 1984) suggests that, together with fish prices, fish stock represents the most important and least controllable factor in business performance. One may argue that, up to certain level, the negative relationship of those two factors may reduce the overall effect on uncertainty in the vessel's economic performance.

Today, this proposition is not applicable. The industry has proactively evolved to become less dependent fish stock and fish prices. Firstly, by becoming more independent of fish availability, future fishing vessels are expected to become more flexible with regards to targeted species and waters in which to operate within. As such, a factory trawler can easily change from whitefish operation to shrimp, or even target medium depth or pelagic species. Newer sonar technologies and the better understanding of migratory patterns have facilitated the identification of fish schools. Secondly, to become less dependent on fish prices through a broader production portfolio of product, so that the total revenue is less dependent on variations in raw fish price. The vessel can decide whether to produce H/G fish, fillets, fish meal and/or fish oil. Hence, when fish prices are very low, the operator may decide to just produce H/G fish, reducing manning costs almost by half, as less personnel are needed onboard, in compensation for a lower revenue. Vessel operators can also focus on fish quality and increase their revenue by providing a highe quality product.

For this study, we have carried out a workshop to identify the principal uncertainties surrounding the design and operation of the fifth generation of factory stern trawlers. The workshop was part of a broader ABD initiative (Ulstein and Brett, 2009), with the purpose of discussing the needs of a fifth generation of factory stern trawlers, and further generating a competitive solution to enter the market. With this initiative, a large sample of stakeholders throughout the value chain were involved, both, directly and indirectly. Ship owner, fishermen, ship designers, ship builders and suppliers, among others. The different uncertainties captured during the workshop have been further classified in three groups, technical, commercial and operational as proposed by (Garcia et al., 2016), see Table 1.

Two technical factors are considered in the model, i) the performance of the hauling system and ii) the performance of the factory. The performance of both systems here represents two aspects. Firstly, relating to the quality of the outcome, the fish or fish products; secondly, the reliability of the system. The former aspect of uncertainty is the result of betting for innovation, looking for improving current practices. The lat-

Table 1. List of critical uncertainty factors.

Uncertainty groups	Current uncertainty factors	Literature support
Technical	Performance of hauling system	(Gates, 1984)
	Performance of factory	(Gates, 1984)
Operational	Fish stock	(Mangel and Clark, 1983; Gates, 1984)
	NOx taxation	(Thanopoulou and Strandenes, 2017)
Commercial	Fish and products prices	(Gates, 1984)
	Quota system	(Standal, 2008)
	Fuel price	(Jafarzadeh et al., 2017)

ter, although related to innovation aspect as well, it focuses on the quality of materials, machines and its maintenance. In this paper, we will focus of the former aspect, and relate it to the quality of the final fish products.

The operational aspects relate to Gates' work. How much fish can the vessel bring home, if it is fishing for a period "t"? Excluding the time factor, one could say that the amount of fish one vessel can carry back, it is a combination of its hold capacity, fish storage arrangement and fish processing. Hold capacity is in many areas an exogenous factor, as regulations typically limit the maximum size. Contrary to this, the storage arrangement (e.g. cartoon boxes or tubs) and the type of product processed (e.g. H/G or fillet), are endogenous factors, giving fishermen partial control of how much fish they can carry. By-catch can be a limiting factor here, as by-catch may represent less valuable fish species that will fill in the hold while providing less value. Being able to reduce by-catch or to refine it as alternative products, such as fishmeal or fish oil, will reduce the potential effect on the vessel's hauling value.

The latter aspects relate to the commercial operation of the vessel. Three aspects are here considered: a) fish and products prices, b) quota level and c) fuel price. The three factors are intrinsically exogenous, although fishermen have certain control on how those factors affect the performance of their operations. Fishermen cannot control fish prices, but they have control over the quality of the fish produced, therefore influencing their income from the sale of fish.

5 MODELLING AND DATA ANALYSIS

In order to measure the effect of uncertainties in the economic performance of a factory stern trawler, we have developed a MATLAB model to simulate a one-year commercial operation of a vessel. The goal of the model is to understand the influence of technical, commercial and operational uncertain parameters in the financial performance of the vessel business case.

Our models builds on the research of (Millar and Gunn, 1990; Inoue and Matsuoka, 2003), and expand on their models.

The vessel was modelled based on nine parameters: hold capacity, no. and size of the nets, hauling technique, crew, fuel type, newbuilding price, autonomy and installed power. Similarly, the operation of the vessel is described based on the time required to perform each activity; load and discharging of fish in port, net hauling, net launching, and the distance to shore. It is also modelled here, the process inside the fish factory. Building on the design parameters and the operation of the vessel, catch volumes are calculated per trip, considering seasonality effects and operational factors such as number of nets or fish processing technique. The last aspect is the economic performance of the vessel. The revenues from the different trips are calculated based on monthly prices for the different fish products. Vessel expenses are calculated following the model proposed by (Stopford, 2009). Finally, the contribution margin is presented and used as reference to evaluate the impact of uncertainty factors in the economic performance of factory stern trawlers.

The model is developed in a way that looks for the optimum exploitation of the given quota structure, from an economic perspective. Building on the model proposed by (Inoue and Matsuoka, 2003), Equation (1), we expand it by considering more than one fish species, multiple trawls and seasonal effects. Our model is presented in Equation (2).

$$C = F_c N = qEN \qquad (1)$$

where C represents the catch, F_c the fishing capacity, as a function of q, gear efficiency and E, effort, and N representing the stock of fish.

$$C_{(l,i)} = F_c N_{(l,j)} = qnEN_{(l,j)} \qquad (2)$$

where n represents the number of trawls, and l, i and j account for fish species, trip number and months respectively. Hence, the catch ratio of the vessel will depend on the gear selected, the operational region and the time of the year. We model fish stock (N) as the average monthly catch rate per vessel, based on the information available of fish landing per specific by stern trawler in Norway. Three years of data (2014–2017) are used as a basis. Fishing effort (E) is used to correct the fish stock (N) based on vessels size, GT as reference.

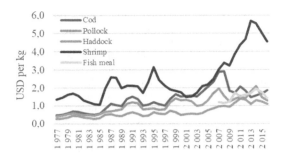

Figure 6. Annual fish and fish meal prices in Norway (1977–2016). (Based on data from (SSB, 2017)).

Finally, q and n are modeled by the amount and size of trawls used in each trip. We assume a fixed number of trawls per trip, although in reality this may vary from haul to haul.

Vessel revenue is calculated as described in Equation (3), considering the catch per fish type and trip $C_{(l,i)}$, the monthly price of each fish product $P_{(l,j)}$, and the type of product p, fillet, H/G, meal or oil. A reduction for by-catch (α) is also considered. Further, by-catch is used for meal and oil is added to the total revenue.

$$Revenue = \sum_{i}^{\infty}\sum_{1}^{\infty}\left(C_{(l,i)}(1-\alpha)pP_{(l,j)}\right) \qquad (3)$$

As fish prices, we use the monthly average price based on the prices for the individual fish species for the past 8 years. Data for historical annual prices over a 40-year period (Figure 4) has been used to reflect the variance of prices and the relevance of the 8-year period selected. Monthly prices for the last 8 years are quite representative looking at the 40-year period. Prices for shrimp may be perceived as high compared to the long-term run. Still, as it represents a sub-market for the vessel, it will not have a large impact in the vessel's economic performance overall. Similarly, prices for fish meal and fish oil are based on monthly average prices for the period 2007–2017.

A field study as described by (Lurås and Nordby, 2014) was carried out to get a better understanding of the operations, so the model could better simulate the real operation of a factory trawler. Further, the model was tested on a real design case and calibrated based on the feedback from the fishing company.

6 RESULTS

Our reference benchmark of financial performance for the vessel design solution is the average contribution margin generated per day, based on a year of operation. The base case represents a factory stern trawler operating in the Norwegian coast producing fillets. It is considered that the vessel owns three structural quotas to catch cod, haddock and pollock respectively. A structural quota is the right to catch a fixed amount of fish defined annually by the relevant political authorities of each operating region. The vessel has an innovative hauling system, which is assumed can improve the quality of the fish product and provide up to a 10% higher fish price. Current prices for fish, fish products and fuel are assumed. Considering our default input parameters, the contribution margin of the base case vessel is of 8,988 USD per day, see Table 2.

A lesser performance than what is to be expected for the hauling system, would require the use of a conventional ramp, hence, losing the potential additional revenue and reducing vessel's contribution margin by 20%. Potentially, and if this aspect wouldn't have been considered during the conceptual design, the vessel could have had a higher impact on its financial performance, as it wouldn't be able to bring onboard the catch.

Fish stock and fish prices are the factors with highest influence in the economic performance of the vessel. Those factors have a negative effect on the contribution margin of 23% and 36% respectively. A reduction on quota has a lower impact, as the vessel has the flexibility of operating as shrimp trawler when its quota level for white fish is fulfilled.

Political factors such as the taxation level on NOx emissions influence considerably the economic performance of the vessel. Hence, doubling the current taxation level would have a negative contribution of 16% into the vessel's contribution margin. Such an effect would be lower for a vessel operating with liquified natural gas (LNG) as fuel, as the level of NOx emissions is lower. The effect in this case would be only 2%.

7 DISCUSSION

The development of the fifth generation of factory stern trawlers focuses on fishing efficiency and retaining high quality catch. Fishing efficiency relates both directly and indirectly to the efficiency and performance of the individual vessel. High fish quality focuses on the reduction of fish stress during reception and processing. The control of fish temperatures throughout the entire process is also delicate in the final product quality. A vessel's performance is influenced by uncertainty factors such as fish stock, fuel price, quota levels and regulations, as demonstrated in our analysis. Hence, such uncertainty factors should be catered for in the design process of this fifth generation of factory stern trawlers to ensure the efficiency of the vessel.

Designers and operators can influence and reduce the negative influences of uncertainty factors. As example, by increasing the number of fish products produced, the vessel can indirectly increase its flexibility and improve its robustness towards negative changes in fish prices. In our analysis, we show that fish and product prices have the highest impact on economic performance. We have considered that the price of all fish species and fish products decreased by 10% compared to the base case. This is a realistic assumption for a vessel operating only with one fish type, but more improbable for those vessels catching multiple species. As an example, a reduction of 10% in cod prices would only contribute to a reduction of 18% in vessel's contribution margin. The effect is halved when considering that the prices of other species and fish products remains stable. Price reduction for non-core fish species or products will have a lower influence, as they have a lower contribution to the overall vessel revenue. Similarly, by choosing LNG as fuel, the vessel becomes less dependent of, for example, the effect of changes in taxation of NOx emissions.

Table 2. Quantified effects of uncertainty factors in the contribution margin of the vessel.

Current uncertainty factors	Case	Contribution margin (UDS/day)	Difference (%)
Initial considerations	Base case	8 988	0%
Performance of hauling system	Conventional ramp	7 218	−20%
Performance of fish factory	H/G	8 143	−9%
Fish stock	(−)10% catches	6 876	−23%
NOx taxation (Regulations)	x2 NOx tax	7 574	−16%
Fish and products price	(−)10% all fish and product prices	5 743	−36%
Quota system	(−) 10% quota	7 303	−19%
Fuel price	(+)10% fuel prices	8 410	−6%

Our overall conclusion is that the performance of factory stern trawlers will be challenged by uncertainty factors that are not yet under the control of the ship owner or designer. Stakeholders can influence on the effect of such uncertainty factors in the overall vessel's economic performance by spending more time and resources in analyzing, estimating and understanding the causes and effects of different influential performance factors. We here propose here measures to mitigate such uncertainties and quantitatively measure the effect in terms of vessel's economic performance. Building and applying on more appropriate performance yield models, like suggested in this paper, better vessel design solutions will appear and will be brought into use in our industry.

Vessel economic performance simulation studies as carried out in this study, prove to be a robust method in handling uncertainty related to the design of more effective factory stern trawlers.

REFERENCES

CRISP (2015) 'Quality improvement', *Crisp Annual Report*.

Davies, T.K., Mees, C.C. and Milner-Gulland, E.J. (2015) 'Second-guessing uncertainty: Scenario planning for management of the Indian Ocean tuna purse seine fishery', *Marine Policy*. Elsevier, 62, pp. 169–177. doi: 10.1016/j.marpol.2015.09.019.

Digre, H. (2013) 'Automatic fish handling system onboard'. Bergen, Norway: SINTEF Fisheries and Aquaculture.

Erikstad, S.O. and Rehn, C.F. (2015) 'Handling Uncertainty in Marine Systems Design - State-of-the-Art and Need for Research', in *International Marine Design Conference (IMDC)*. Tokyo, Japan, pp. 324–342.

Fernandez, A.S. et al. (2014) 'ARALFUTUR PROJECT: Energy Efficiency of Deep Sea Trawlers for South Atlantic Fisheries', in *Third International Symposium on Fishing Vessel Energy Efficiency E-Fishing*. Vigo, Spain.

Gallo, A. (2017) 'Contribution Margin - What It Is, How to Calculate It, and Why You Need It', *Harvard Business Review*, pp. 1–3.

Garcia, J.J. (2016) 'Handling Commercial, Operational and Technical Uncertainty in Early Stage Offshore Ship Design', in *Conference on System of Systems Engineering*. Kongsberg, Norway.

Gates, J.M. (1984) 'Principal types of uncertainty in fishing operations', *Marine Resource Economics*, 1(1), pp. 31–49.

Havfisk (2016) *Industry Handbook - The White Fish Industry 2015*. Ålesund, Norway.

IHS Fairplay (2017) 'IHS Maritime World Register of Ships Q3 2017'. IHS Global Limited.

Inoue, Y. and Matsuoka, T. (2003) 'Distribution of catch per haul in trawl and purse seine fisheries: Implication for reduction of fishing capacity', in Pascoe, S. and Gréboval, D. (eds) *Measuring Capacity in Fisheries*. Rome, Italy: FAO, p. 314.

Jafarzadeh, S. et al. (2017) 'LNG-fuelled fishing vessels: A systems engineering approach', *Transportation Research Part D: Transport and Environment*. Elsevier Ltd, 50, pp. 202–222. doi: 10.1016/j.trd.2016.10.032.

Lurås, S. and Nordby, K. (2014) 'Field Studies Informing Ship's Bridge Design at the Ocean Industries Concept Lab', in *International Conference on Human Factors in Ship Design & Operation*. London, United Kingdom.

Mangel, M. and Clark, C.W. (1983) 'Uncertainty, search, and information in fisheries', *Journal du Conseil Permanent International pour l'Exploration de la Mer*, 41, pp. 93–103.

Millar, H.H. and Gunn, E.A. (1990) 'A simulation model for assessing fishing fleet performance under uncertainty', in *Proceedings of the 1990 Winter Simulation Conference*, pp. 743–748.

OECD (2016) *The Ocean Economy in 2030*. Paris, France. doi: 10.1787/9789264251724-en.

Rehn, C.F. et al. (2017) 'Quantification of Changeability Level for Engineering Systems', *Pending final approval*.

SSB (2017) *Fisheries, Quantity and value of catch from Norwegian vessels*.

Standal, D. (2008) 'The rise and fall of factory trawlers: An eclectic approach', *Marine Policy*, 32(3), pp. 326–332. doi: 10.1016/j.marpol.2007.07.003.

Stopford, M. (2009) *Maritime Economics*, Routledge. Abingdon, UK: Taylor & Francis.

Sumaila, U.R., Bellmann, C. and Tipping, A. (2016) 'Fishing for the future: An overview of challenges and opportunities', *Marine Policy*, 69, pp. 173–180. doi: 10.1016/j.marpol.2016.01.003.

Thanopoulou, H. and Strandenes, S.P. (2017) 'A theoretical framework for analysing long-term uncertainty in shipping', *Case Studies on Transport Policy*, 5(2), pp. 325–331.

Ulstein, T. and Brett, P.O. (2009) 'Seen whats is next in design solutions: Developing the capability to develop a commercial growth engine in marine design', in *International Marine Design Conference (IMDC)*. Trondheim, Norway.

Ziegler, F. et al. (2013) 'The Carbon Footprint of Norwegian Seafood Products on the Global Seafood Market', *Journal of Industrial Ecology*, 17(1), pp. 103–116. doi: 10.1111/j.1530-9290.2012.00485.x.

The role of aesthetics in engineering design—insights gained from cross-cultural research into traditional fishing vessels in Indonesia

Richard W. Birmingham
Marine, Offshore and Subsea Technology Group, School of Engineering, Newcastle University, Newcastle upon Tyne, UK

I. Putu Arta Wibawa
Politeknik Perkapalan Negeri Surabaya, Jl. Teknik Kimia, Kampus ITS Sukolilo, Surabaya, Indonesia

ABSTRACT: The design of a small fishing vessel for the coastal waters of Indonesia is a straight forward technical challenge, however some initiatives to introduce newly designed vessels with more advanced technology have been unsuccessful. The observation that the appearance of rejected craft contrasts markedly with the traditional vessels operating in the area has led to the consideration of the importance of aesthetics in engineering design that is reported in this paper. Definitions of beauty, their relevance to engineering design, and alternative approaches to decisions regarding form and style are discussed, with examples taken from the marine domain. A proposal to make explicit the place of aesthetics in one theoretical model of the design process is followed by a description of the design of a fishing vessel for eastern Java, this being an example of cross-cultural design in practice, where aesthetics was given a prominent role.

1 INTRODUCTION

The products of engineering design, due to their size or location, can have a significant impact on the visual experience of those who are operating them, and on those living or working in relatively close proximity. Despite this the role of aesthetics in the design process is not always considered in any detail, and can even be ignored entirely. In the maritime sector a range of cases can be found, from the design of a luxury yacht at one extreme, where style can be elevated to the most important design driver, to the design of an offshore oil platform at the opposite extreme, where appearance might be considered an irrelevance. However even in an entirely functional artefact, such as a cargo ship, anecdotal evidence suggests that there is value in a product that is pleasing to the senses, one example (described informally by an ex-mariner, Dr Kayvan Pazouki, 2016) being the preference of a company's seagoing personnel to work aboard the oldest ship in the fleet, despite its poorer reliability, simply because its more traditional lines and style gave them greater pride in their work. Despite this evident pleasure afforded to the operators working aboard a vessel that had acknowledged aesthetic merit, to include such a subtle and esoteric benefit into calculations, and to establish the added value in cash terms, would be an almost impossible task.

Fishing vessels would conventionally be put into the same category as cargo ships, in terms of the unimportance of aesthetics in a design task that is driven by functionality. However in the course of research (funded by the Indonesian Ministry of Research Technology and Higher Education in the form of an overseas post graduate studies scholarship) into the design of a sustainable fishing vessel, it became apparent that the visual characteristic of the design had greater significance than anticipated. The vessel was to be used for operation in the inshore fisheries of eastern Java, where the traditional boats are striking in their dramatic shape and ornamentation (Figure 1). As an exercise in engineering design a low technology fishing boat of under 15 meters length for operation in the

Figure 1. An example a traditional Indonesian fishing vessel of eastern Java.

developing world (implying a low labour cost economy), is a straight forward challenge for a naval architect, and has little technical complexity.

However despite the apparent simplicity of the task, evidence from earlier initiatives indicated that new vessel designs were not readily accepted by the operators. The research revealed that although the technical challenge involved in designing a new fishing vessel was straightforward, there was a parallel challenge associated with the cross-cultural nature of the task that could only be resolved by giving prominence to the appearance of the vessel. Exploring the issues relating to aesthetics for this specific case led to insights into their role in the general context of engineering design.

2 PRIORITISING AESTHETICS IN ENGINEERING DESIGN

At the outset of a design exercise the list of requirements that could be considered and prioritised might start with cost, and then take in such things as safety, weight, size, efficiency, reliability, ease of production, and ease of maintenance. The requirements might also include elements relating to human factors such as limits on temperature, noise and vibration, or others relating to habitability and comfort. But for a design exercise that is driven by functionality, aesthetics need not be mentioned—after all, what unambiguous and verifiable metric can be used to specify a requirement for beauty?

This neglect of the physical appearance of the designed object is not the case for all engineered products. The field of industrial design is dedicated to the interaction of the product with people, with both ergonomics and aesthetics being central to the success of the design. For products that are to be mass produced and sold into a competitive market, an elegant form can be perceived as an indicator of quality, and the desirability of the product enhanced by the tactile and visual pleasure experienced by the user. The Apple Corporation, with its range of products derived from the i-phone, are a remarkable example of the power of aesthetics for a mass produced product, and in the automotive industry style is a significant driver of a design.

This paper is reporting on research undertaken in the marine context, so the relevant designed products are ships and boats of all types and sizes. Although the vast volumes associated with the production of smart phones and cars are not found in this sector, recreational craft can be produced in the hundreds and so, just as with cars, successful sales may depend on attractive styling. Even with much larger vessels that are made in small batches of two or three, if they are being sold into a highly competitive market the advantage of product differentiation derived from a visual appearance that is associated with a distinct brand can be significant. This can be seen in recent years with the success of the X bow concept, a patented hull form that has been adapted for a variety of vessel types (as found on the Innovation pages of the Ulstein Group website in July 2017), one example of which is shown in Figure 2. Claims are made for the performance advantages of this bow form, but even disregarding this technical aspect of the design it is evident that a marketing success has been achieved due to the strikingly different aesthetics which distinguish these vessels from other functionally similar craft with conventional bow shapes.

The examples given above all have appearance as an important element of the design, but in every case this is an element of the marketing strategy, and the purpose of enhancing the visual appeal is to gain a competitive advantage. There are however one-off products where appearance can dominate all other considerations for different reasons, and examples of these can again be found in the maritime sector. Luxury mega yachts, which are high powered floating recreational palaces for the extraordinarily rich, are ordered and owned essentially as a demonstration of wealth, and so the visual impression and onboard experience drives a designer to ensure that this statement is made boldly and clearly. As a result appearance becomes more important than many other considerations. Similarly passengers on cruise liners, even if not at the extreme end of the wealth scale, are in part looking for confirmation of their success, and so both the external impression and the visual impact of the internal accommodation and recreational facilities have to be considered in detail. In the case of cruise liners stylists and interior designers are contracted to work alongside naval architects and marine engineers to ensure that aesthetic and engineering decisions are linked (Montgomery 2015). With mega yachts the roles may be even be reversed, so that it is the engineers who are contracted in to provide support to the project. Design credit is given to those who are primarily responsible for

Figure 2. An offshore supply vessel with patented X Bow, a style that has created a strong brand identity (Designed by Ulstein Group, www.ulstein.com).

the external and internal appearance of the vessel (as evidenced in "The Fifty Most Beautiful" [2015]), these being designers who often do not have a formal education in naval architecture, but who have a background in industrial design or other creative disciplines, including fine art.

The discussion above demonstrates that examples can be found where the ultimate users of the product dictate that appearance must take a high priority. But what of the many engineering products where the design is driven by cost, and this is itself derived from efficiency and effectiveness? In some cases the client providing the design requirement may not be the operator directly engaged with the product, but desk based and remote from the built artefact itself, and so have no personal interest in the issue of appearance. Examples of such cases, again from the maritime sector, are cargo ships such as bulk carriers or oil tankers, dredgers, offshore supply vessels, and fishing boats. In such examples of engineering design, the visual impact of the design is only considered informally while optimising the explicitly stated requirements. In the formal procedures aesthetics are neglected, as evidenced by the established models of the design process found in the theoretical texts on engineering design. There is no spoke on the common presentations of the design spiral, nor a box in the established higher level models of design philosophy (see summaries in Birmingham et al, 1995), that is labelled 'aesthetics'.

3 BEAUTY IN ENGINEERING DESIGN

Beauty is a word that can make engineers uncomfortable. It is not just that there is no metric by which to measure it, after all 'engineering judgement' is used to make decisions based on experience rather than hard data. The difficulty is that most engineers would consider themselves untrained and unqualified to make a judgment as to whether the design 'pleases the aesthetic senses, especially the sight' (*Oxford Dictionary* 2017), or 'exalts the mind or spirit' (*Merriam-Webster* 2017), these both being elements of formal definitions of 'beauty'. Where such judgement is an important element of the design process the decisions are contracted out, as indicated in the examples discussed above where interior designers or stylists collaborate with the engineers. In other cases the engineer can fall back on the long established principle that 'form follows function'. This phrase was first coined by Sullivan (1896) when referring to the natural world in the context of the architectural form of sky scrapers. The concept permits the designer to abdicate responsibility for the appearance of the product, the justification being that if the product is functionally successful then its form is inherently correct too.

The idea that the appearance of an object will be pleasing if its shape (form) is dictated by what it has to do (function) has been extended into the concept of 'functional beauty' (Sheridan 2014). The analysis of functional beauty suggests that it has two parameters which in essence are the degree of refinement of appearance, and the degree of refinement of function, although Sheridan uses longer terms, the former parameter being 'Knowledge of function', and the latter being 'Purity in aim and elegance.' Any design can be mapped into the design space shown in Figure 3 (which extends the thinking of Sheridan [2014, 74]) with designs that do have 'functional beauty' being both familiar and efficient. Inspection of the diagram shows that designs could be unsuccessful with respect to functional beauty due to being insufficiently or excessively refined in either of the parameters of function or appearance:

- Insufficiently developed functionality is self-explanatory, but at the extreme the object just doesn't work.
- Excessive refinement of function is less intuitive, but it is possible that an object is extremely effective, but at such cost that it is no longer appropriate. It is not fit for purpose as it is not optimally efficient. It is interesting that Sheridan terms such a failing as 'elegant', although 'over engineered' is more fitting.
- Insufficient attention to appearance could result in a design that is so far from the expected form that the object is unrecognisable.
- Excessive refinement of appearance can result in the standard form being abstracted (or 'codified' in Sheridan's terms) to such an extent that it is unusable. An amusing example of this type of failing is the hotel shower fitting that looks striking, but proves impossible to discover how to make it work.

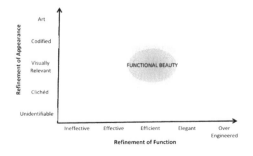

Figure 3. The parameters of functional beauty (after Sheridan 2014).

While Sheridan's analysis of functional beauty draws on philosophy and psychology using language that is unfamiliar to the engineer, the concepts are readily grasped if reinterpreted with the more accessible terminology introduced in Figure 3, and the conclusion that a good design is both recognisable in terms of function and is fit for purpose, is obvious to the engineer. However this analysis does also provide additional insights into how a design can fail aesthetically, and so takes the engineer beyond the platitude that 'form follows function.'

4 CONFLICTING CULTURAL NORMS

The concept described above need not cause any difficulty for the engineer, as it provides straight forward guidance that points the way to a satisfactory solution that has 'functional beauty'. Like the philosophy that form follows function, it makes no reference to style, nor does it require judgements as to whether as solution will 'please the aesthetic senses'. In fact, like many theories of design, this concept is as much describing practice as providing guidance. Many successful engineering designers who have never heard the term 'functional beauty', would say that the diagram in Figure 3 simply indicates what they strive to do. However the research underlying this paper, in part based on fieldwork in Indonesia, has led to a questioning of the adequacy of this approach in some situations, especially when there is a cross-cultural element to the problem.

The objective of the research was to develop a sustainable fishing vessel for the small scale fishers of Indonesia, this country being second to China in the scale of its capture fisheries (FAO 2016). It is estimated that there are over 640 thousand fishing vessels with 2.7 million fishers (MMAF 2014) directly employed aboard these vessels, and many more people working in related onshore activities. Amongst the fleets of locally built wooden vessels there are a small minority of craft that are built of fibre reinforced plastic (FRP), many of which have been provided to the communities in a series of initiatives by the Indonesian government or international agencies. However the success of these projects has been variable and further plans to continue such projects fiercely debated (Wibawa 2016, 5–6). Surveys undertaken by the authors demonstrated why this type of support for the fishing communities has been so controversial as it was observed that while some of these donated vessels were widely used, others were quickly abandoned. These rejected vessels could be found unused in remote corners of fishing harbours, even when they were only a few months old. The fact that both successful designs, and those that were not adopted, had significant technical variations from the locally built vessels indicated that it was not the imposition of unfamiliar technology that was the barrier to acceptance. The most obvious example of the introduction of an innovative technology was the use of fibre reinforced plastic (FRP) as the construction material. FRP inevitably introduces problems for the operators as it is difficult to modify and repair when compared to the traditional wooden vessels. Yet despite this difficulty examples could be found of FRP craft which were widely adopted (Figure 4) as well as examples where they had been rejected (Figure 5). While it is possible to identify elements of the failing designs where the operators' requirements have only been partially met, of greater significance is the fact that the successful designs emulate the shape and form of the traditional vessels, while the unsuccessful designs contrast strongly and present an aesthetic style that can be characterised as 'modern' or 'western'. The authors recognise that the failure to adopt innovative technology in the development context could be due to many issues including those associated with training, maintenance, ownership, and infrastructure,

Figure 4. Fibre Reinforced Plastic (FRP) boats of a type initially provided as aid after the 2004 tsunami have been widely adopted, and continue to be built on a commercial basis.

Figure 5. An example of a relatively new FRP fishing vessel that has been abandoned.

and also recognise that ways to address these issues could include ensuring that appropriate technology is employed and that stakeholder engagement is strong. However in this paper it is the significance of the appearance of newly introduced technology that is being considered.

The observed lack of acceptance of 'western' looking vessels contrasting with the enthusiastic adoption of vessels that followed the Indonesian style led the authors to consider the role of aesthetics in the context of technology transfer, and to propose the hypothesis that a cultural mismatch can be a significant barrier to the adoption of technology. Set in the maritime context, with the design of fishing vessels as a specific example, the argument can be expanded as follows.

If a consultant from the developed world were engaged to design an improved vessel for the developing world, the focus of the naval architect would tend to be on the economic, operational and technical requirements. The issue of appearance might be ignored on the premise that form follows function, however despite this there will still be an unconscious bias toward a solution that looks 'right' to the consultant. The result will be one that presents the consultant's own preferences of what is aesthetically pleasing but, more significantly and almost unavoidably, it will reflect the norms dominant in the consultant's own culture. This will not be a considered decision, it will simply embody the consultant's belief as to what a fishing vessel should look like. Imagine however if the geographical direction of the flow of expertise were reversed. Imagine trying to persuade fishermen from the north of Scotland (or Norway, or Canada) that a boat with the most up to date technology was available to them, but it looked like the vessel in Figure 1. No matter how big the subsidy offered, even 100% of the cost, it is highly improbable that the fishermen would be willing to take ownership of such a vessel. The cultural gulf demonstrated by the appearance of the vessel is just too great to bridge. This however is what is being done when an expert from the developed world proposes to introduce a 'better' solution into a developing world context—the cultural gulf, the mismatch, can be such a significant barrier as to make the new technology unacceptable.

5 OVERLAPPING CULTURES

The discussion above simplifies cultural differences, presenting an artificial case where two cultures are entirely alien to each other. In practice, in the digitally connected and in many aspects globally unified world of the 21st century there is an interchange of cultural values, and an overlap of norms. Indonesian boats can be seen decorated with the insignia of the football clubs of the European leagues, Figure 6, just as in many parts of the developed world *satay*, the Indonesian dish, is enjoyed when dining out. In addition in many situations designers do not have to conform to cultural norms, but make it their purpose to change perceptions, the world of fashion being the prime example of this. Marketing in any field is partly about discovering what the customer desires, and partly about convincing them that an alternative is even more desirable. However, in the case of promoting economic development through the transfer of technology, trying to drive a change in aesthetic values (either deliberately or unconsciously) is unnecessary and may obstruct achieving the primary objective. So rather than creating an additional potential obstacle to the successful introduction of new ideas, the designer should try to align the appearance of the proposed design with the prevailing culture. The difficulty of achieving this when the engineer is an outsider, operating in the context of an unfamiliar culture, should not be underestimated. The important cultural elements are not necessarily the obvious flam-

Figure 6. Evidence of cross cultural influences, here European football club insignia being used as ornamentation.

Figure 7. Examples of ornamentation that have regional or religious significance.

boyant ones, but subtle and obscure details that are difficult to identify. In addition, while a consultant may be unaware of how their own culture could be influencing and impacting on their decision making, so the customer may be unable or unwilling to express the cultural imperatives of their world. This could be because the specific details are in their eyes so obvious as not to be worth remarking on, or it could be because they are religious or spiritual in origin (Parastu, Sudamarwan, and Budiarta 2013), as shown in Figure 7, and any explanations might be considered difficult or inappropriate.

6 BALANCING THE COSTS AND BENEFITS OF BEAUTY

The insights, discussed above, into the role of aesthetics in engineering design emerged from research into the design of fishing vessels in the developing world, where it was realised that unless proper attention was given to the appearance of the design the proposal was at risk of being rejected by the operators however good the technology might be. Other benefits of raising the priority of visual appearance were then recognised that were universally applicable, including the potential for gaining a marketing advantage by generating a brand identity, and the greater satisfaction and loyalty generated in the operators if they could have pride in their vessel or equipment. It could even be argued that such pride might be translated into an enhanced attitude to health and safety.

Although it is clear that there are benefits from considering aesthetics in a design, in a commercial context this has to be balanced against cost. However in many cases there is no need for a good looking design, even a beautiful one, to be more expensive than a utilitarian or ugly one. To understand this the concept of 'satisficing', first proposed in 1969 by Simon (1996), needs to be revisited. Simon pointed out that although optimisation techniques are used throughout the design process, the final result is not an optimal one, but one that satisfies the design requirements. The optimising process stops once the requirements are met, as to continue would be an unnecessary expenditure of resources. However if Simon's idea of satisficing is considered further it can be seen that this process leads to an unexpected conundrum. Unlike the result of optimisation, satisficing does not lead to the inevitable single solution, but to one of a multitude of solutions all of which should be equally acceptable, as all would satisfy the design requirements. Although in theory all such solutions are equally good in practice the customer, if given a choice, would be able to indicate a preferred design.

Such a preference is not captured in the design requirements, but it would reflect the customer's priorities. When making design decisions compromises are traded. All of the many possible solutions balance these compromises differently, but given a choice the customer would be able to recognise which most closely reflects their values. For the designer to develop a design that will respond to the customer's values, it is necessary for them to be aware of the customer's priorities. If the design process starts with the elements that are most important to the customer, and the customer's priorities are considered every time a decision that requires compromise is made, then the result of the satisficing process will be one that reflects the customer's values. Optimisation processes are often explained by the analogy with climbing a hill, where the objective is to find the summit. Satisficing only requires that a predetermined altitude be reached, but the point where the optimising search process reaches that contour is defined by the point at the bottom of the hill where the climb starts. If the designer can start in the right place, then the result is more likely to respond to the customer's priorities and values.

While this principle applies to all the technical elements of a design, the customer will also have aesthetic values derived from personal taste and from the norms of their culture. As with technical aspects, if the style of the product is established at the outset, and appearance set alongside other considerations when each design decision is made, then the satisficing process will produce a result that reflects an aesthetic preference. While aesthetic considerations should not disrupt function or safety (Brewer, 1994) in many cases the technical design decisions do not relate to geometry and appearance, or at least only in a general way, so this aspect of the decision can be guided by aesthetic preferences with no impact on the technical outcome. If all else is equal, and if the design requirements have been satisfied then all else really is equal, the customer would prefer a design that is in their eyes beautiful.

7 EMBEDDING AESTHETICS IN ENGINEERING DESIGN THEORY

In developing a sustainable fishing vessel for operation in the waters of Indonesia, the authors' research led them to recognise that the vessel's appearance could be crucial to acceptance of proposed technical innovations. The technology had to be packaged in a form that was familiar, even appealing, to the operators. Responding to this concern became a significant part of all stages of the design process. Aesthetic considerations were integrated into all of the following: the requirement elicitation process; the interpolation of data from existing 'basis' vessels; and the evaluation of proposed designs (by referring to focus groups of fishing vessel skippers, as described in detail below). This extended process resulted in a design that contained all the technology identified as appropriate for a sustainable fishing vessel in the Indonesian context, but also one that would look at home in the fishing ports of the region, and so would be admired and desired by the fishers who would operate it.

Reflecting on this practical implementation of a design process, where aesthetic considerations have been given a high priority, can provide suggestions as to how this often ignored aspect of engineering design could be formally embedded into the design process in other situations. In exploring the role of aesthetics we can follow the terminology of formal optimisation as defined by Sen and Yang (2012, 18). In this interpretation of design the criteria are stated as objectives, each of which links an attribute to a required direction. For example the attribute 'cost' must be low, so the direction of the design process is to reduce this, while the attribute 'stability' (for a fishing vessel) must be high, so the direction is to increase this. If there is a specified threshold value to be achieved for the result to be accepted then the objective is considered a 'constraint', however if it is simply an aspiration to achieve the best possible result the objective is termed a 'goal'. In the examples just mentioned, stability is a constraint if it is specified that it must meet the requirements of regulations, while cost is a goal if it is simply required to be as low as possible. In these terms if appearance is included in the criteria for a design then this objective can be categorised as a goal, specifically to make the design as aesthetically pleasing to the customer as possible.

Design theorists usually resort to diagrammatic representations of the design process, and these models are as numerous as there are theorists. While not intending to introduce another model, it was noted earlier in this paper that aesthetics are neglected in many such models for engineering design so it is interesting to consider how this element could be incorporated. A widely accepted graphical interpretation of design at the strategic level is that of a spiral, indicating that each sub-problem in the design process has to be returned to iteratively until the requirements are satisfied. Versions of this model exist for different engineering sectors, and in the marine field such a visualisation of the process was first proposed by Evans (1959, 671–678). Other authors subsequently devised modified proposals to emphasise specific elements of the process such as the economic evaluation (Buxton 1987, 78), or to accommodate particular vessel types such as yachts (Larsson and Eliasson 2014, 5). While acknowledging that the spiral model is a huge simplification of the complexity of the activity of design, it has proven its value in communicating the nature of design to students and aspiring designers. As the spiral effectively facilitates a greater understanding it may be helpful to identify where aesthetics could be explicitly indicated in this model of the process.

The design spiral has two components, the circular loops indicating one complete cycle of the design process, and the radial spokes indicating sub problems that have to be addressed. Visualisation of the proposed design is an integral part of many of the sub problems, as producing a graphical representation of most elements of the design is an essential part of formulating a solution. However creating a drawing does not automatically imply that aesthetics have been taken into consideration.

Figure 8 presents a simple design spiral with aesthetic decision making explicitly identified in the design process. As can be seen aesthetics are considered in two ways. Firstly there is a dedicated 'spoke' added to the spiral at an early stage. This indicates that the 'style' and overall impression of the product should be explored (in sketches) at the very start of the synthesis process, and that this initial conjectured solution should be revisited in subsequent iterations as more detail is generated. It is interesting to consider the positioning of aesthetics in this respect, and to contrast it with other design goals, such as safety and cost. While all three, (aesthetics, safety and cost) provide a 'direction' for design decisions throughout the process, the latter two are essentially evaluated at the end of each iteration of the process, while aesthetics is considered at the beginning, so providing a visual template into which other decisions try to

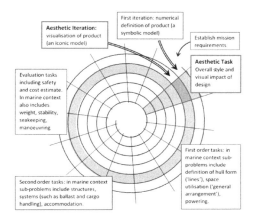

Figure 8. A generalised design spiral, with aesthetics explicitly indicated.

fit. Secondly Figure 8 shows that one of the loops of the process could also be considered as an aesthetic iteration. Although the concentric loops of the spiral are not usually explicitly identified, the outermost one is dedicated to establishing the value of the principal parameters of the design, so producing a symbolic model that defines the product only in terms of numbers (things like size, weight, capacity, power etc.). The next iteration is a visualisation one, where the symbolic model is turned into an iconic model in the form of sketches of what the product could look like. This is where aesthetics can again be seen to be considered explicitly, ensuring that of the many possible geometric forms that could satisfy the numerical requirement, a geometry is selected that also conforms to the aesthetic preferences of the customer.

8 AESTHETICS IN ENGINEERING DESIGN PRACTICE

Designers sketch possible solutions at the very earliest stages of the design process, conjectured from imagination and prior experience, or by adapting established solutions. While sketching every designer will be making choices based on what is considered to be the desired appearance, though it is possible that the designer will perceive the result an inevitable outcome of form following function. However in the light of the discussion above it is possible to consider the way that these decisions regarding appearance can be taken, as this is affected by the cultural context of the design activity.

In most cases the designer will be operating within a culture that is entirely familiar, which implies that the aesthetic values of the designer (their taste) aligns with that of the customer. Even without any explicit discussion as to appearance, the designer's instinct will be met with a favourable response. However even in this situation the designer does have a choice, which is either to stay safely within the conventional norms, or alternatively to step outside accepted solutions and introduce an imaginative or innovative proposal. In the former case the designer could be said to be following fashion. In the marine context this is achieved by consideration of the geometry of a number of basis vessels, then by scaling the dimensions using methods such as those suggested by Larsson and Eliasson (2014), to arrive at a proposal that is entirely in keeping with other vessels of a similar type. However in the latter case, where the designer proposes to introduce a design that is in appearance at variance from the established ones, the designer must have confidence that they are so familiar with the product, and so knowledgeable of the customer's aspirations and ambitions, that they can propose a solution that the customer will recognise as being just right despite its unusualness. If this is successful the client will be delighted that the proposed design exceeds expectations, and is evidence that the designer is so in tune with trends that they can lead fashion rather than follow it. This is clearly a risky strategy, but where marketing suggests that product or brand differentiation is beneficial, then this is necessary.

The contrasting situation, and the one which instigated this exploration of the role of aesthetics in design, is where the designer is working in an unfamiliar cultural setting. The design will provide an innovative technical solution in an area where the designer has recognised expertise, but the product will be operated in a cultural setting greatly contrasting from the designer's own. In this situation the risk of providing a visually innovative solution is great, as the designer may unwittingly present something that is at the very least unappealing to local taste and alien to cultural norms, and at worst offensive to religious or cultural sensibilities. Innovation theory suggests that new ideas should be introduced gradually in order to minimise the risk of failure (Abernathy 1988), so while technological innovation may be the purpose of the design exercise, introducing an innovative aesthetic adds unnecessary risk, so the designer should as far as possible maintain the appearance of the existing solutions. This situation is an extreme case of the method described above as 'following fashion', but here the designer must set aside their own preferences and tastes, and follow un-critically the norms identified from the existing solutions.

In the case of designing a fishing vessel for Indonesia, it was necessary to first select from the many contrasting vessels and regional variations (Samodra 2009) an appropriate vessel type that was extensively operated in the relevant area. In this case the *paying,* a boat type commonly operated out of Muncar and other ports of eastern Java was selected. The fleet was analysed in detail by considering the geometry of a selection of basis vessels and from this the expected dimensions and proportions of the craft were identified. This numerical analysis of the geometry included all significant visual features such as shape and angle of the bow and stern, the curvature of the deck line, the position and proportions of the cabin, and position and height of the mast. In addition relevant stakeholders, such as skippers, crews and owners, were engaged with at the outset of the process by using questionnaires and interviews (Figure 9).

It is quite possible to process the data from surveys of existing designs and still arrive at something that many would agree is ugly. However there are fundamental aesthetic considerations which are explored in the fine arts and in graphic and industrial design and which reference ancient cultures as their sources. These aesthetic principles are rarely explored in the

Figure 9. Questionnaires and in-depth interviews were conducted with stakeholders in their own environment, here with a fishing vessel skipper and crew member aboard their boat.

Figure 10. A focus group of stakeholders using scale models to evaluate alternative desigmn proposals.

engineering design context, however in the marine field the work by Guiton (1971) provides some analysis of what is effective in the context of ships and yachts. While it is not proposed to enlarge on this here, it can be noted that in considering the overall impression of the vessel Guiton emphasises the importance of conformity of lines (converging on focal points for example), and of the shape and proportions of the visual envelope, or overlapping envelopes, in which the design is contained. When developing the fishing vessel for Indonesia the authors used these principles in conjunction with the numerical data derived from basis vessels and the qualitative information elicited from the stakeholder interviews. (In doing this the authors assumed that these aesthetic principles are not a uniquely western convention, but are universal and can be applied in any culture, however it is recognised that further interdisciplinary research is needed to explore this.)

The stakeholders were referred to again toward the end of the design process when focus groups were organised to provide a critical commentary on the proposed designs (Figure 10). All the stakeholders participating in these groups brought valuable experience and knowledge to the sessions, however recognising that many had limited formal education, and as a result limited literacy and ability to interpret technical drawings, scale models were used to communicate what was proposed and to facilitate discussion, as can be seen in Figure 10. This close involvement with the operators, combined with the analysis of the geometry of basis vessels, was considered essential in order to avoid introducing or omitting visual elements to the proposed solution that, although unremarked by the designers, were of significance to the operators. The entire design process is detailed by Wibawa (2016) with the resulting design shown in Figure 11, and a computer generated visualisation shown in Figure 12.

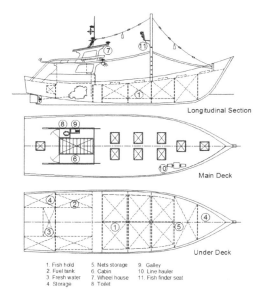

Figure 11. Drawing of the final design for a sustainable fishing vessel for the small scale fisheries of east Java, Indonesia.

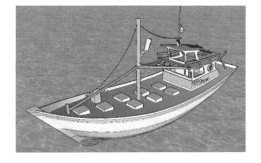

Figure 12. Visualisation of the fishing vessel design.

9 CONCLUSION

In engineering design the effort to develop a product that is efficient and profitable, with multiple design drivers that can include ease of manufacture, ease of maintenance, a minimum carbon footprint, maximum recyclability, and conformance with relevant regulations, it is unsurprising that appearance is often considered irrelevant, implying that aesthetics has no significance in the design process. However insights into the role of aesthetics was an unexpected outcome of combining research into the traditional fishing vessels of Indonesia with the challenge of designing a small sustainable fishing vessel for operation in the coastal waters of eastern Java. During the research it was observed that in a cross cultural context simply producing a good technical solution did not guarantee acceptance by the intended operators. If the appearance was not sympathetic to cultural norms and to the fishers' expectations of what looked appropriate then this would create a barrier that could result in appropriate technological innovations being rejected. Reflecting on this it was realised that benefits can be gained in other situations if a design is visually pleasing, such as market or product differentiation and more highly motivated operators. However of greater significance was the realisations that aesthetic decisions are continually being made during the design process, even if these are not conscious decisions but simply a reflection of the designer's personal preferences and cultural conditioning.

If it is accepted that aesthetic decisions are unavoidably being made, even if unconsciously or by default, and also that good aesthetic design can have direct benefits, then it is important to recognise that enhancing the appearance of a proposed design need not have resource implications. In many cases there are numerous geometries that will provide equally good technical solutions, so identifying a geometry that evokes pleasure, or even one that 'exalts the mind or spirit' (to quote again the dictionary definition of beauty [Merriam-Webster 2017]) need not cost more. Establishing what is pleasing to the eye in a cross-cultural context can be a significant challenge, overcome by close engagement with all the relevant stakeholders. In a more conventional setting identifying the stylistic and visual preferences of the client should be part of the requirement elicitation process, with the result that from all the possible solutions that would satisfy the technical requirements, the final proposal is one that also visually delights.

REFERENCES

Abernathy, William J. 1988. "Mapping the Winds of Creative Destruction" in Readings in the Management of Innovation, edited by M.C. Tushman. USA: Ballinger Publishing.

Birmingham, Richard, Graham Cleland, Robert Driver, and David Maffin. 1995. Understanding Engineering Design – Context, Theory and Practice. London: Prentice Hall.

Brewer, T. 1994. Understanding Boat Design. Maine: International Marine.

Buxton, I.L. 1987. Engineering Economics and Ship Design. 3rd ed. British Maritime Technology.

Evans, J. Harvey. 1959. "Basic Design Concepts." Naval Engineers Journal 7 (4).

FAO (Food and Agriculture Organisation). 2016. The State of World Fisheries and Aquaculture 2016.

Guiton, J. 1971. Aesthetic Aspect of Ship and Yacht Design. London: Adlard Coles.

Larsson, Lars, Rolf E. Eliasson. 2014. Principles of Yacht Design. 4th ed. London: Bloomsbury Adlard Coles

Merriam-Webster Dictionary. 2017. Merriam-Webster. https://www.merriam-webster.com/dictionary/beauty

MMAF (Ministry of Marine Affairs and Fisheries). 2014. Marine and Fisheries in Figures 2014. Indonesia.

Montgomery, Angus. 2015. "Designing the Interiors for a 141000-ton Cruise Ship." Design Week, January 8. https://www.designweek.co.uk/issues/5-11-january-2015/designing-the-interiors-for-a-141000-ton-cruise-ship/.

Oxford Living Dictionaries. 2017. Oxford University Press. https://en.oxforddictionaries.com/definition/beauty.

Parastu, P, A. Sudamarwan, and Budiarta. 2013. "The Ornament of Fishing Vessels in Perancak Village, Jembrana District, Indonesia." E-Journal Universitas Pendidikan Ganesha.

Samodra. 2009. "Traditional Boatbuilding in Indonesia: a Social and Technological Study of Current Practice and a Proposal for Appropriate Future Development." PhD Thesis, Newcastle University, UK.

Sen, Pratyush, Jian-Bo Yang. 2012. Multiple Criteria support in Engineering Design. London: Springer-Verlag.

Sheridan, Jonathan Andrew. 2014. "Synthesis of Aesthetics for Ship Design." MPhil Thesis, University of Southampton.

Simon, Herbert A. 1996. The Sciences of the Artificial. 3rd ed. Cambridge, Mass: MIT Press.

Sullivan, Luis H. 1896. "The Tall Office Building Artistically Considered." Lippincott's Magazine, March. https://ocw.mit.edu/courses/architecture/4-205-analysis-of-contemporary-architecture-fall-2009/readings/MIT4_205F09_Sullivan.pdf.

"The Fifty Most Beautiful Superyachts Ever Built." 2015. SuperYacht World, January 3. http://www.superyachtworld.com/yachts/the-most-beautiful-superyachts-ever-built-5841.

Wibawa, I.P.A. 2016. "Sustainable Fishing Vessel Development by Prioritising Stakeholder Engagement in Indonesian Small-scale Fisheries." PhD Thesis, Newcastle University, UK.

When people are the mission of a ship—design and user research in the marine industry

M. Ahola & P. Murto
Aalto University School of Arts, Design and Architecture, Helsinki, Finland

S. Mallam
University College of Southeast Norway, Vestfold, Norway

ABSTRACT: The role of non-engineering design disciplines (such as industrial design and interior design) and their applications in the marine industry is vague. People, as "users" of a ship, are commonly utilized as a source of design input for non-engineering design disciplines. Differing types of people inhabit a ship simultaneously and have differing roles, expectations, experiences and purposes for the onboard environment and its design features. This paper reviews the role of design methods and user-centred design research in the marine industry. Focusing on passenger ship design concepting, it explores how the two main users of passenger ships, (i) the paying customers/passengers and (ii) the onboard crew can be brought into the ship design and development process. This paper discusses the significance of a user-centred design research approach in the marine industry and the challenges of integrating non-engineering design disciplines into marine engineering and shipbuilding processes. Several perspectives will be discussed through case examples, including topics on design management, transition from 2D to 3D modelling in concepting phases, design for passenger safety and crew work efficiency. As such, this paper examines a new direction for research and practical applications in marine engineering and shipbuilding.

1 INTRODUCTION

Ships are designed based on their mission. According to Levander (2004) ship design process starts with a definition of the customer requirements, also called Mission statement. In this first phase ships': (i) task, capacity, performance demands, range and endurance, (ii) rules, regulations, and preferences, and (iii) operating conditions, like wind, waves, currents, ice are defined. Passenger ships are designed according to primarily function as a passenger carrier, whereas the main function of cargo ships is to carry cargo. In this paper, we concentrate on how people are considered as users of the ship in passenger ship design process. Passenger ships can be divided into two categories: ferries, which carry transportation cargo as their main income. Ferries typically have accommodation and amenities for passengers, similar to cruise ships that are designed only according to their mission of passengers' leisure-time voyages where the ship itself and its amenities are part of the experience. The cruise experience is an offering supported with products and services included in the cruise setting and that the cruise experience is impossible without the context (Ahola, 2017). Fundamentally, cruise vacations are a prototypical experiential product: a combination of a floating resort hotel, sightseeing vessel, gourmet restaurant, food court, nightclub, shopping centre, entertainment complex, and recreation facility (Kwortnik, 2005).

Although a broad human-centred research foundation exists for the above-mentioned individual aspects in other contexts, human-centred research for passenger ships' unique environment has remained scant. In general, research focusing on the human aspects of design has shown that engineering and non-engineering designers typically take fundamentally different perspectives in product development. It is common for engineering design to take an "inside-out" perspective and to work on the techno-economical aspects of products while non-engineering designers (such as industrial or interior designers) often employ an "outside-in" perspective where the human interfaces of products are taken as a starting point for development (see e.g. Walsh et al., 1992, Cross, 2008). In a ship design context, such difference in perspective could, for instance, manifest itself as a naval architect's focus on the economy of ship operation on one hand, and the interior designer's focus on creating a pleasurable ambiance within the ship. While such differences in perspective are generally handled well in shipbuilding practice, recent studies positioned in the intersection of engineering and non-engineering design in shipbuilding suggest that better integration of the two could provide competitive benefits in passenger ship design (see Ahola, 2017, Mallam, 2016, Murto, 2017).

Ship design follows International Maritime Organization's (IMO) general standards for safety, security and environmental performance in international shipping rules. In order to ensure that this vital sector remains safe, environmentally sound, energy efficient and secure, the IMO provides a framework to standardize diverse aspects of international shipping—including ship design, construction, equipment, manning, training, operation and decommisioning (IMO, 2017). Although, passengers appear in the frameworks for safety, energy efficiency, and security and are the main mission of the passenger ships, their needs, desires, and hopes are not sufficiently discussed in naval architecture literature. In other words, there's a limited understanding how to design the special environment of passenger ship from the passengers' perspective. This paper sheds light to this important issue of how to reveal user's insight and how to utilize the information in passenger ship design process through three case studies.

2 HUMAN-CENTRED RESEARCH METHODS

Non-engineering design knowledge could contribute to better passenger understanding in shipbuilding. Human-centred research approaches highlight, for example, different types of inhabitants of the passenger ships, their culture, experiences, and professional background, which all have a significant impact on how they experience and interpret the ship. Whereas, interpretations have significant impact on how people will behave in different situations (Hirschman & Holbrook, 1982). For example, modern cruise ships carry thousands of passengers with large age ranges: 25% are in the 30–39 age group, 25% in the 50–59 age range, and 25% in the 60–74 age group (CLIA, 2015). This means that there are thousands of individuals perceiving safety in different ways, consuming in different manners, and experiencing the ship differently. Furthermore, the onboard service staff and operational crew who work to ensure safe and pleasurable passenger experiences must cohabit the same physical structure. These onboard workers require that the built environment of ship support their work demands and optimize safe and efficient work practices. Thus, a cruise ship has differing types of "users" who have considerably differing expectations, requirements and experiences of a ship and it's design.

The methods used by non-engineering designers (e.g. industrial and interior designers) and in human-centred design are not always straightforward to integrate to engineering practice. For instance, non-engineering designers often make use of rich qualitative data, such as observation and interviews to understand the experience that people have with products or systems (see e.g. Ulrich & Eppinger, 2000, Polaine et al., 2013). The knowledge generated by these results is rarely quantifiable or generalisable in comparison to, for instance, the calculative methods commonly used in naval architecture and engineering. Moreover, the use of visual ideas, such as mood and image boards (see e.g. Murto et al., 2014), visual reference material (see e.g. Pasman, 2003), visual research stimuli (see e.g. Ahola and Mugge, 2017) or shared boundary objectives (see Mallam, 2016) as a central tool of design and collaborative processes may seem elusive from an engineering perspective (as discussed by e.g. Cross, 2008). While these challenges are partially a matter of communication and familiarity between non-engineering and engineering designers, our view is that there also exists untapped potential in better integration of nonengineering design methods in shipbuilding. In the following section, we present three cases that illustrate the challenges and benefits of better integration between non-engineering and engineering design in shipbuilding.

3 CASE EXAMPLES

3.1 *Design for innovative passenger experience*

It is generally accepted that the physical setting of a service influences how customers experience the services themselves (Bitner, 1992). Hence, the interior design, architecture and general arrangement of a ship has important implications for passenger onboard satisfaction. This case example builds on Murto's (2017) recent case study on the use of nonengineering design in the development of the Viking Grace cruise ferry, with a specific focus on outlining the challenges related to designing for innovative passenger experiences in shipbuilding.

There are two examples in Murto's (2017) study regarding the interior design of Viking Grace that are particularly relevant for understanding the integration of non-engineering design methods in shipbuilding practice. First, the separation of ship cost definition and interior design in the Viking Grace development process made it very difficult for the interior designers to deliver an innovative experience that the ship owner was after. For instance, the budget for lighting in the ship contract (and reference level) was initially inadequate for using LED-lighting in the ship—one of the key elements of experience innovation in the Viking Grace project. Hence, when aiming to design innovative passenger experiences, exploring interior design options in more detail prior to defining costs in the ship contract could help later in the process during the actual design. In practice, this could mean that the visual references used by non-engineering designers in outlining future passenger experiences should be brought to ship contract item development in more detail. Second, material use is an important means of defining how passengers experience the interior of a ship. However, ships have high demands for interior materials in terms of, for example, insulation and fire safety properties (Byun, 2006), reducing the amount of possibilities in interior design significantly. In the Viking Grace project, the ship owner used a new-to-the-industry interior design office in order to facilitate the development of a completely new interior experience, but faced difficulties in educating the designers on what materials can be used in a marine context. Thus, fast and effective dissemination of material knowledge is required in shipbuilding projects when using the ability of interior design to differentiate and innovate in shipbuilding projects. Based on the Viking Grace project, generating some kinds of means of disseminating material knowledge could also facilitate better transfer of information across projects and contractors working in ship engineering. For instance, some materials pushed by the interior designers in the Viking Grace project were resisted at first by interior contractors because they were perceived as being previously never used on ships—contrary to the fact that they had been used in previous ships but by different contractors. From the perspective of human-centred design methods, dissemination of material knowledge should also focus on the experiential qualities of the materials (such as visual and tactile properties) rather than e.g. cost, fire safety and weight only.

The two recommendations above provide some initial clues of how better integration of non-engineering design in passenger ship development could take place. Better integration across different phases of a project could mean the integration of interior designers earlier to development projects or adding of detail to interior design considerations prior to the ship contract. Better dissemination of knowledge between engineering and non-engineering design, and dissemination of knowledge across projects could be facilitated by the development of educational materials, material banks and databases that contain information of material requirements in shipbuilding. These suggestions can also have drawbacks: earlier integration of interior designers would likely increase project cost, while more detailed interior design prior to contract might slow down the development process. In terms of dissemination of knowledge, educational materials and material databases would require updating and investment as the ship interior material industry develops. Hence, in taking these actions, ensuring that the passenger experience is truly innovative is crucial for success. Based on the Viking Grace project, it can truly be so: Viking Line profits jumped from 0,9 million euro to 27,5 million following the release of Viking Grace while its market share on the Turku-Stockholm route rose c. 10% during the first year when Viking Grace started operating.

3.2 *Design for passenger safety experience*

Passenger ship safety design basis on safety regulations, which currently does not highlight the users (passengers) needs and desires (Akyuz & Celik, 2014, Le Coze, 2013). Design for passengers' safety in passenger ships lacks human-centred design approach. This shortage is critical, because it is apparent that passengers interpret ship environment and its safety state subjectively (Ahola, 2017). Perceptions are based on the surrounding design and perceptions influence on passengers' decision making and further behaviour (Sagun et al., 2014). However, safety regulations basis mainly on quantitative analysis, which in their current stage cannot consider in sufficient level the unpredictable nature of human behaviour. The IMO's Assembly, (Resolution A.947(23), 2003) has highlighted this lack of understanding and requested more attention to this research perspective. Although, shortage in safety regulations has been identified already in 2003, only few studies have addressed this issue to date.

With a human-centric approach to ship safety design the different user groups and their interactions with the ship environment can be better understood, and for example, possible misunderstandings of the design intentions, i.e. navigation and functionality of the safety equipment could be avoided. Passengers are not always able to recognize if their safety is guaranteed to an acceptable level. Instead they trust on their subjective interpretation of the safety state (Ahola, 2017). Therefore, it is important to go beyond objective safety and investigate what effects on people's subjective safety (Van Rijswijk et al., 2016). User-centric research approaches could help to identify the passengers' subjective perspective on safety. In Ahola et al. (2016) it has been identified that passengers' interpretations of the safety equipment and design intentions following the safety regulations may differ. This yields about the conflicting design emphasis: design emphasis has focused on issues that passengers don't necessary find important and instead they may concentrate on issues not considered in safety design. This possibility of misunderstanding is especially critical for evacuation situations. Furthermore, visibility of some safety equipment in the ship environment may trigger the unsafe feeling among passengers', whereas visibility of some safety equipment enhances the positive feeling of safety. Furthermore, passenger ships are designed for pleasurable living and to provide positive cruise experiences for the passengers. Safety is a critical element of building the positive experiences (Ahola, 2017). In other words, safety needs to ensure before people can have other 'more' positive experiences. Characteristics affecting on the interpretations can be identified and design of these characteristics improved. For example, in Ahola & Mugge (2017) it has been identified that ceiling and wall design having view to outside and curved shapes has a positive influence on passengers' safety perception, which in turn may affect their comfortability. More research is needed to reveal these conflicting issues.

Human-centred approaches in safety design comes with strengths and weaknesses. These approaches require significant effort in user-studies, in which thorough observation, interviews, and other methods for revealing users (passengers) insights. It is recommended that user-studies are conducted in real situations and environments. Revealing how real users (passengers and crew) use ship environment for different purposes design can be improved. However, must be noted that the thousands of people in modern passenger ships comes with thousands of individuals with diverse perceptions. Therefore, if a passenger ship community is considered as whole, only generalisations can be made and the most critical design issues identified. When the crew, as a group of professionals, is considered more detailed knowledge can be achieved. This is because crew is usually familiar with the ship,its equipment and repeating similar tasks. Whereas passengers' might be first time travellers and sometimes totally unfamiliar with the ship environment, marine culutre and interpret unfamiliar environmental characteristics, all of whichform their safety experience. Through userstudies these aspects can be revealed under normal operations. Furthermore, it is recommended that user studies are done in situ and therefore hazard situations are impossible to observe and interviewing accident participants may come with ethical issues. One solution to study realistic accident situations could be to demonstrate accidents and emergency situations in virtual environments to test design solutions with the users (See e.g. Ahola et al., 2014).

3.3 *Design for onboard service staff and operational crew*

Although the primary mission of a passenger cruise ship is to provide paying customers with a positive and enjoyable experience, a diverse onboard work staff are responsible for delivering varying customer-oriented services and amenities, as well as the overall operation and safety of the ship structure itself. The onboard work staff of a cruise ship have a wide range of job tasks, responsibilities and specializations. They consist of two distinct groups: (i) the onboard service staff, who interface with the customers and provide customer-oriented services, including hotel staff, waiters, cleaners, food services and travel agents; and the (ii) operational crew, who are trained and licensed seafarers that operate the ship structure and its various systems, including the navigation officers, marine engineers and various seafarer ratings. As both the onboard service staff and operational crew inhabit the same physical structure as the paying customers, a unique combination of conflicting expectations and demands from a ships'design emerges between the varying cruise ship "users".

Optimizing crew work tasks and onboard operational demands are typically not the focus of ship design. The general lack of attention for human element issues in marine structure design is reflected in naval architecture education, design methodologies and design regulations (Mallam, 2016, Mallam & Lundh, 2013, Orosa & Oliviera, 2010). However, human error is the main contributor of marine accidents (Hetherington et al., 2006, Rothblum, 2000). Furthermore, design deficiencies of ship work environments, layouts and equipment have shown to contribute to human error in marine accidents (Kataria et al., 2015). Inadequate design solutions of onboard workspaces and equipment can also influence crew to adopt unsafe work practices and "work-arounds" in order increase work efficiency, even at the expense of safety (Mallam & Lundh, 2016, Mallam et al., 2015). The crew themselves may even retrofit a work space or equipment after a ship is constructed and put into operation in order to better adapt the design to their work demands (Mallam, et al., 2015). Due to the influential role of human operators in marine accidents, and the impact that design plays in the safety and efficiency of crew operations, onboard personnel and their work tasks need to be addressed in the ship design process from both technical engineering and non-engineering perspectives (Graveson, 2002, Mallam, 2016).

Traditional ship design processes and naval architecture methodologies are typically engineering-focused. Unfortunately, contemporary ship designers generally do not have the experience or knowledge of even basic onboard ship operations or work tasks necessary to support crew needs (Chauvin, et al., 2008, Mallam et al., 2017). As marine related jobs, systems and operations are highly specialized, it is difficult for non-seafarers to conceptualize the demands required for a marine work system and its design. Crew are rarely involved in ship design or construction, and as such, their specialized knowledge of onboard operations and design requirements are often absent in ship design development. Thus, introducing a participatory design methodology into ship design aims to elicit the knowledge and experience of the end-users (i.e. ship crew) and utilize it throughout the development process. Participatory design reallocates power in collaborations, where designers do not "present" finalized or near-finalized solutions to other partners, but rather work together to create and develop solutions as a team (Carroll & Rosson, 2007).

Participatory design methodologies generally engage end-users and designers through group discussions, collaborative design exercises, scenario building and problem-solving tasks. Common, shared objects facilitate communication across disciplinary and social boundaries in order to create a common language, share knowledge and enable storytelling between stakeholders (Broberg et al., 2011). These objects may vary widely, however in design contexts often take the form of paper-based design sketching, scaled or full-size physical mock-ups, 2D and 3D Computer-Aided Design modelling, mapping scenarios, and analyzing documents or processes. Ultimately, the purpose of introducing participatory design processes is to create a collaborative framework to elicit and utilize highly specialized "expert" enduser knowledge. This aims to facilitate communication between designers and end-users in an effort to develop more user-centered, user-friendly design solutions onboard ships.

4 CONCLUSIONS

In this paper, we have presented three case studies that underline how human-centred design and its methods can add value to traditional ship design and building processes. Overall, the work carried out by engineering and non-engineering designers in ship-building often have different perspectives (inside-out vs. outside-in), which makes their integration challenging. Through a better understanding of how human-centred design methods could be used in ship-building, this paper aims to support the bridge between the different research traditions of human-centred design and naval architecture.

While being beyond the scope of this paper, some insights on how to integrate non-engineering design in shipbuilding practice are worthy of mentioning. According to Murto (2017), it is beneficial to utilize human-centred design expertise throughout a passenger ship design and building project: using human-centred design to define and clarify passenger and crew needs and wants is equally relevant in the very early phases of shipbuilding projects when the general arrangement is being defined and in defining the final interior and exterior design in greater detail. In practice, this is perhaps best achieved by integrating non-engineering designers with naval architects in charge of shipbuilding projects in all phases of the project and in different organisations taking part in the project. This does not mean using the same non-engineering designers in all phases of the project, but rather the use of such *type* of designes throughout the project. For instance, in the Viking Grace project, the passenger experience design of the ship was developed cumulatively and gradually over several years, with inputs coming from various different organisations and increasing in the amount of detail as the project moved forward—as opposed to being centrally planned and executed by a single organisations responsible for human-centred design in the project (for more detail, see Murto, 2017).

The main benefit of strengthening the bridge between human-centred design methods and naval architecture is in bringing the subjective experiences of the people who operate and "use" ships in differing ways to the development process. As displayed in Cases 3.1 and 3.2, the use of visual design research methods closer to naval architecture can aid in ensuring better experiences and hence, provide benefits for the ship owner's bottom line. Furthermore, as outlined in Case 3.3, the use of participatory design methodscan elicit users' perspectives, expectations and experiences to optimize the system, which would otherwise go unutilized. Thus, although the knowledge generated by human-centred design research methods is often subjective in nature, it may have very concrete and "objective" consequences in the real world.

REFERENCES

Ahola, M. 2017. *Tracing Passenger Safety Perception for Cruise Ship Design.* Aalto University publication series doctoral dissertations 3/2017. Helsinki: Aalto ARTS Books.

Ahola, M. & Mugge, R. 2017. Safety in Passenger Ship Environments: The Influence of Environmental Design Characteristics on People's Perception. *Applied Ergonomics* 59: Part A, 143–152.

Ahola, M., Salovuori, H. & Lehtonen, M. 2016. Safety Perception as a Sociotechnical Network. Proceedings of the *13th International Symposium on Practical Design of Ships and Other Floating Structures (PRADS' 2016).* 4–8 September 2016. Copenhagen.

Ahola, M., Magica, R., Reunanen, M. & Kauppi, A., "Gameplay Approach to Virtual Design of General Arrangement and User Testing", *International Conference on Education and Professional Development of Engineers in Maritime Industry, April 15th–16th 2014*, Busan, Korea, pp. 99–107.

Akyuz, E., & Celik, M. 2014. A hybrid decisionmaking approach to measure effectiveness of safety management system implementations on-board ships. *Safety Science* 68: 169–179.

Bitner, M.J., 1992. Servicescapes: the impact of physical surroundings on customers and employees. *The Journal of Marketing*: 57–71.

Broberg, O., Andersen, V., & Seim, R. (2011). Participatory ergonomics in design processes: The role of boundary objects. *Applied Ergonomics* 42(3): 464–472. DOI: 10.1016/j.apergo.2010.09.006.

Byun, L.-S., 2006. Peculiarity of interior design materials for accommodation areas of cruise ships: A state-of-the-art review. *Ships and Offshore Structures* 1(3): 171–183. https://doi.org/10.1533/saos.2006.0112.

Carroll, J.M., & Rosson, M.B. (2007). Participatory design in community informatics. *Design Studies* 28, 243261.

Chauvin, C., Le Bouar, G., & Renault, C. (2008). Integration of the human factor into the design and construction of fishing vessels. *Cognition, Technology & Work* 10(1): 69–77. DOI: 10.1007/s10111-0070079-7.

CLIA. 2015. *2015 Cruise Industry Overlook*. Retrieved April 12, 2016 from About the industry: http://www.cruising.org/docs/defaultsource/research/2015-cruiseindustry-outlook.pdf?sfvrsn=2.

Cross, N., 2008. *Engineering design methods: strategies for product design* (4th ed). Chichester, England: Hoboken, New Jersey: John Wiley.

Graveson, A. (2002). Human Factors in Ship Design and Operation. In *Proceedings of the International Conference on Human Factors in Ship Design and Operation, October 2nd–3rd, 2002*, London, United Kingdom.

Hetherington, C., Flin, R., & Mearns, K. (2006). Safety in shipping: The human element. *Journal of Safety Research*, 37(4), 401–411. DOI: 10.1016/j.jsr.2006.04.007.

Hirschman, E., & Holbrook, M. 1982. The experiential aspects of consumption: Consumer fantasies, feelings, and fun. *Journal of Consumer Research* 9(2): 132–140.

IMO-Assembly, *Resolution A.947(23) Human element vision, principles and goals for the Organization*. International Maritime Organization, London, 2003.

IMO. (2017). *About IMO (International Maritime Organization)*. Retrieved December 12, 2017 from http://www.imo.org/en/About/Pages/Default.as px

Kataria, A., Praetorius, G., Schröder-Hinrichs, J.U., & Baldauf, M. (2015). Making the case for Crew-Centered Design (CCD) in merchant shipping. In *Proceedings of the 19th Triennial Congress of the International Ergonomics Association, August 9th–14th, 2015*, Melbourne, Australia.

Kwortnik, R.J. 2005. Shipscape influence on the leisure cruise experience. International Journal of Culture. *Tourism and Hospitality Research* 2: 289–311.

Le Coze, J.C. 2013. Outlines of a sensitising model for industrial safety assessment. *Safety Science* 51(1): 187–201.

Levander, K. 2004. Passenger Ships. In T. Lamb *(Ed.), Ship Design and Construction* (pp. 1–39). New York: Society of Naval Architects and Marine Engineers.

Mallam, S.C. (2016). *Distributed Participatory Design in Multidisciplinary Engineering Projects: Investigating a Sustainable Approach for Ship Design & Construction* (Doctor of Philosophy Dissertation). Chalmers University of Technology: Gothenburg, Sweden. ISBN: 978-91-7597-462-0.

Mallam, S.C. & Lundh, M. (2013). Ship Engine Control Room Design: Analysis of Current Human Factors & Ergonomics Regulations & Future Directions. In *Proceedings of the Human Factors & Ergonomics Society 57th Annual Meeting* 57(1): 521–525. DOI: 10.1177/1541931213571112.

Mallam, S.C., & Lundh, M. (2016). The Physical Work Environment & End-User Requirements: Investigating Marine Engineering Officers Operational Demands & Ship Design. *Work* 54(4): 989–1000. DOI: 10.3233/WOR-162365.

Mallam, S.C., Lundh, M. & MacKinnon, S.N. (2017). Evaluating a digital ship design tool prototype: Designers' perceptions of novel ergonomics software. *Applied Ergonomics* 59(Part A): 19–26. DOI: 10.1016/j.apergo.2016.08.026.

Mallam, S.C., Lundh, M., & MacKinnon, S.N. (2015). Integrating Human Factors & Ergonomics in Large-Scale Engineering Projects: Investigating a Practical Approach for Ship Design. *International Journal of Industrial Ergonomics* 50: 62–72. DOI: 10.1016/j.ergon.2015.09.007.

Murto, P., 2017. *Design integration in complex and networked product development: A case study of architectural design in the development process of a greener passenger ship*. Aalto University publication series doctoral dissertations 185/2017. Helsinki: Aalto ARTS Books.

Murto, P., Person, O., Ahola, M., 2014. Shaping the face of Environmentally sustainable products: image boards and early consumer involvement in ship interior design. *Journal of Cleaner Production* 75: 86–95. doi:10.1016/j.jclepro.2014.03.078.

Orosa, J.A., & Oliviera, A.C. (2010). Assessment of work-related risk criteria onboard a ship as an aid to designing its onboard environment. *Journal of Marine Science & Technology* 15(1): 16–22. DOI: 10.1007/s00773-009-0067-0.

Pasman, G., 2003. *Designing with precedents*. Delft: Delft University of Technology.

Polaine, A., Løvlie, L., Reason, B., 2013. *Service Design: From Insight to Implementation*, 1st ed. Brooklyn: Rosenfeld Media.

Rothblum, A. (2000). Human Error and Marine Safety. Presented at *The Maritime Human Factors Conference 2000*. Retrieved December 8th, 2017 from: http://bowles-langley.com/wp-cotent/files_mf/humanerrorandmarinesafety26.pdf.

Sagun, A., Anumba, C.J., & Bouchlaghem, D., 2014. Safety issues in building design to cope with extreme events: case study of an evacuation process. *Journal of Architectural Engineering* 20, 05014004.

Ulrich, K.T., Eppinger, S.D., 2000. *Product design and development*, 2nd ed. Boston: Irwin/McGraw-Hill.

Van Rijswijk, L., Rooks, G., & Haans, A., 2016. Safety in the eye of the beholder: individual susceptibility to safety-related characteristics of nocturnal urban scenes. *Journal of Environmental Psychology* 45: 103e115.

Walsh, V., Roy, R., Bruce, M., & Potter, S., 1992. *Winning by design: technology, product design, and international competitiveness*. Oxford: Blackwell Business.

Human-centered, collaborative, field-driven design—a case study

E. Gernez & K. Nordby
The Oslo School of Architecture and Design, Oslo, Norway

Ø. Seim
PON Power AS, Skedsmokorset, Norway

P.O. Brett
Ulstein International AS, Ulsteinvik, Norway

R. Hauge
Ulstein Design and Solutions AS, Ulsteinvik, Norway

ABSTRACT: How can we design engine rooms that cater to the needs of their human operators? How can we do this in a design process that involves multiple companies and competences? We report on a design case where we facilitated a human-centered, collaborative design process crossing two companies. We present the methods used and the challenges experienced at each step of the process. We discuss what this process might enable for the designers, the engine room, and the ship crew. Based on our analysis, we argue that there is a need to (1) facilitate the collaboration between the companies involved, (2) collect qualitative data about the needs of the ship crews on board ships during operation, and (3) define the engine room as a human-centered working environment where the needs of human operators can be catered to. We argue that this process opens innovation venues by assisting collaborating companies in focusing on human-centered design solutions crossing the boundaries of their businesses, traditional roles and responsibilities.

1 INTRODUCTION

We present arguments for and solutions to adopting a human-centered design perspective on (1) the integration of the needs and challenges of the human operators of the ship in the ship design process, and (2) the facilitation of the human collaboration between the different companies involved in ship design processes in the maritime industry.

Across stakeholder groups involved in the ship design process such as designers (ship designer, sub-contractors, ship yard) and the end-users of the design object (ship owner, ship manager and operator, ship crew), the involved stakeholders have different levels and directions of expertise. Because of this, frameworks for understanding the separate parts of ship design can be hard to share across disciplinary gaps. This is especially important for the gap between the technical expertise of the ship designers (design) and the operational experience of the end-users (operation).

The gap between design and operation is a serious challenge since miscommunications and non-inclusive design processes can lead to suboptimal or even unsafe ship design solutions. Reviewing 29 published ship design processes, Ulstein & Brett argue that there is a need for new competences and new approaches to address this challenge (Ulstein and Brett, 2012). To address this need, we propose human-centered design methods, commonly used in industrial design, that enable the capture and exchange of the different needs of the stakeholders involved. This paper presents recent findings and results developed in the ONSITE project led by the Oslo School of Architecture and Design (AHO) together with the Ålesund branch of the Norwegian University of Science and Technology. The objectives of the project are to (1) introduce human-centered design methods to fill the gap between ship design and operation and to (2) study how this might contribute to the innovation processes of the different stakeholders involved.

The ONSITE project has generated design cases that involve a ship designer and ship builder (Ulstein group), an engine room integrator (Pon Power AS), and a Classification company (DNVGL). In this article, we present a case of engine room design that involves the ship design part of the Ulstein Group (Ulstein Design Solutions – UDS) and Pon Power. We address the following research questions:

1. How can we design engine rooms that cater to the needs of their human operators?
2. How can we do this in a design process that involves multiple companies and competences?

2 BACKGROUND

Ship design is commonly described as a decision-making process (Nowacki, 2009). Kuo observed that communication is one of the main design activities (Kuo, 2003). Ulstein & Brett argued for the need to secure "undistorted communication and equal roles in the dialog among stakeholders of the ship design process" (Ulstein and Brett, 2012). Erikstad stated that "gaining insights into the structure of the decision problem is at least as important as finding solution data" (Erikstad, 1996). It is, however, challenging to find research that proposes and reviews methods to facilitate these decision-making, communication, and insight-creating processes from a human collaboration perspective. The vast majority of the ship design research methods reviewed in the IMDC State of the Art Reports (Andrews et al., 2009, 2012; Andrews and Erikstad, 2015) represent the ship design process as a succession (or combination) of design steps, described in terms of the task that needs to be carried out before going to the next one, with the exception of Andrews's work (Andrews, 1986). In most cases there are no mentions of which stakeholder should be involved and executing each task and what other stakeholders might be consulted. The technical parameters information to be passed on from one step to another is sometimes represented (for example, main ship dimensions) but never in terms of how the information should be exchanged. In summary, the need to approach ship design as a human-centered design activity is often mentioned, yet there is a lack of proposed methods to facilitate human collaboration.

Van Bruinessen et al. observed how ship designers deal with innovation in the ship design process. They reflected that "further research is required, but exploring this social dimension is complex: it requires research-skills related to the social sciences, but sufficient knowledge is required to understand the subject matter" (Van Bruinessen, Hopman and Smulders, 2015). DeNucci attempted to develop a tool that could help "capture a design rationale" because of its potential to assist with the documentation, validation, evaluation, and communication of design decisions between design stakeholders (DeNucci, 2012). DeNucci pointed out that this is a "human-centered" challenge and that human-centered methods were required for this task, hence limiting his ability to research this topic.

Andrews introduced a method designed to facilitate the collaboration between the naval architect and the ship owner in the preliminary design phase (Andrews, 2003, 2011). The "Requirement elucidation" method helps to synthesize user needs into an initial design brief. Andrews pointed out that at this early stage, the ship designer needs to deal with requirements set by the "requirements owner" that are often contradictory, incomplete, and change with time. In the "Accelerated Business Development" process (ABD process) developed by Ulstein & Brett, the ship designer holds a workshop with the ship owner to help list out all the requirements for a new ship and rank them in order of importance (Ulstein and Brett, 2012). Recent approaches to ship design based on multi-objective optimization all refer to the need to capture how different stakeholders perceive and value "what is a good design" (Gaspar, Hagen and Erikstad, 2016) to be able to model it into the optimization problem. These examples show the need to use methods that can assist ship designers in translating the needs of their customers into their design processes.

Using qualitative research methods such as interviews, Solesvik observed and documented how different ship design companies deal with information sharing inside the company and externally with their customers. She gives a description of the stakeholders involved, their needs for information exchange, and how the tools they use enable them to exchange information (Solesvik, 2007, 2011). Although Solesvik provides a detailed observation of human collaboration in the ship design process, she does not propose methods to facilitate this collaboration.

In terms of innovation, the facilitation of information sharing between design stakeholders is also important. Levander criticized two prominent ship design methods (the Ship design spiral and the System engineering approach) for not enabling the exploration of innovation potentials lying at the meeting points between different design steps executed by different design stakeholders (Levander, 2003). The Nautical Institute publication "Improving ship operational design through teamwork" proposed the concepts of "operational design" and "operation driven innovation," arguing for the need to include the operational experience of seafarers in the design process to drive innovation in the ship design process (The Nautical Institute, 1998). The authors pointed out that there is an inherent barrier to such innovation due to the compartmentalization of the technical and commercial departments in most ship owning companies.

The need to include operational considerations in the design process is based on the argument that the safety and efficiency of a ship depends largely on the human operators' ability to take full advantage of its capabilities. Ship accidents database analyses can back up this argument (Grech, Horberry and Smith, 2002; Kataria et al., 2015; Praetorius et al., 2015). This being the case for operational safety, it is fair to assume that operational efficiency also needs to be addressed with a human-centered perspective. This is documented, for example, in an energy management study that found that "soft measures are the lever for realizing energy savings" (Kühnbaum, 2014).

According to the ISO 9241standard definition, human-centered design "aims to make systems usable and useful by focusing on the users, their needs and requirements, and by applying human factors/ergonomics, usability knowledge, and techniques. This approach enhances effectiveness and efficiency, improves human well-being, user satisfaction, accessibility and sustainability; and counteracts possible adverse effects of use on human health, safety and performance" (International Standards Association, 2014). There have been works related to the introduction of human-centered engine room design. For instance, Mallam & Lundh (Mallam and Lundh, 2013) reviewed the current regulations for the use of human-centered design in ship design guidelines by the International Maritime Organization (IMO) related to engine room and engine control room design. They concluded that although the IMO supports this approach, it currently lacks a regulatory framework to implement it. Mallam explored methods to collect insights from engine room operators to transfer them to the ship design process (Mallam and Lundh, 2014; Mallam, Lundh and MacKinnon, 2015, 2017).

Despite this, we see only sporadic application of human-centered design methods in ship design processes. The EU project Cyclades looked into "promoting the increased impact of the human element in shipping across the design and operational lifecycle." The project developed the concept of "crew-centered design" (The Nautical Institute, 2015; van de Merwe, Kähler and Securius, 2016) for ship design processes, highlighting the need to design for, and with, the end-users of the ship. The project documented some operational requirements for different design stakeholders and end-users, but it did not investigate what specific design activities could be used to facilitate the transfer of those operational requirements into the ship design process. The Cyclades project also documented the lack of practical seafaring experience of the design stakeholders, but it did not prioritize putting the designers in direct contact with operations onboard a ship.

There has also been increased attention toward including human-centered design competence in naval architecture and maritime engineering education (Abeysiriwardhane et al., 2015, 2017). Yet there is little evidence that these competences have transcended into professional practice to any serious degree. This is not surprising considering the current and increasing complexity in ship design processes (Gaspar et al., 2012). This makes the introduction of human-centered perspectives not a matter of individual competence but, rather, a matter of building a shared understanding among all involved stakeholders in a ship design process.

In the industrial design and human–computer interaction practices, human-centered design expands the notion of usability with participatory methods that may help to design objects that cater to users' needs in an inclusive and collaborative, co-constructed manner (Bødker and Buur, 2002; Buur and Bødker, 2000). In this tradition, design processes are thought of as innovation processes that typically follow three steps: insight collection, analysis, and prototyping. These three phases are carried out with a high degree of user involvement, using human-centered design methods. We used three such methods in the present case, which are briefly presented below.

The *field study* method originates from ethnography and anthropology (Blomberg, Burrell and Guest, 2009). The goal of this method in a design process is to enable the designer to personally and physically experience the context for which he or she is designing (referred to as "context of use," (Beyer and Holtzblatt, 1997), as well as to interact with the users he or she is designing for in their living and working context. This experience is expected to enrich the designer's judgment capacity (Lurås and Nordby, 2014), which is an important foundation for the designer's ability to deliver creative and innovative solutions (Nelson and Stolterman, 2003). A *workshop* is one method that enables a group of people to work out a problem together (Sanders and Stappers, 2012). We used workshops as a part of the field study process to work out, validate, and expand the field findings in a collaborative way (Millen, 2000). Finally, *prototyping* is a central activity in human-centered design processes that enables visualizing a concept and testing it with potential users in order to criticize it and improve it in a subsequent iteration (Buxton, 2010; Rogers, Sharp and Preece, 2011; Wensveen and Matthews, 2015). Prototyping occurred throughout our whole design process in the forms of sketching, use-scenario enactment, and 3D modeling.

3 RESEARCH APPROACH

In order to study the introduction and facilitation of human-centered, collaborative, field-driven design processes, we created real cases together with the project partners where we introduced human-centered methods that were collaborative and field-driven, and we then reviewed how the cases unfolded, what they created, and what the implications for the partners' design processes might be. In doing so, the researcher assumed two roles: a participant in the case and an observant of how the case unfolded. This type of approach is referred to as Participatory Action Research (Whyte, 1991).

The present case was initiated by Pon Power to better understand the experience of their end-users working in engine rooms: *How can we design for better experiences of ship engine rooms?* The case then followed a standard, open-ended, exploratory innovation process in which the exact content and outcome of each step was not known in advance: insights collection, insights analysis, and prototyping. The design methods used to implement this process were a field study, a workshop, and prototyping through the modeling of operational use-scenarios in a 3D CAD model. The case is summarized in Table 1 and presented step by step in Section 4. Throughout the case, a variety of data material was gathered, as presented in Table 2.

We reviewed the case outcomes in light of the research questions put forward in this article. We analyzed (1) how the design methods we used in the case captured the information describing the needs of the engine room human operators and (2) how this information was shared and dealt with in the design process across the different stakeholders involved. We based this analysis on the "actor centric mapping technique" developed through the ONSITE project (documented in an upcoming guide, (Gernez, 2018). It visualizes a process along a timeline, showing the stakeholders involved in the design process and information related to their contributions to the design process, for example, their roles or what activities they carry out, throughout the different steps of the design process.

Finally, we discuss the potential impact of this work on the engine room design process and its outcomes for the stakeholders involved. The discussion is based on the informal interviews of project participants throughout the project, as well as two half-day seminars with all the project partners. During the seminars, the status of the case was presented and used to collect the partners' feedback on what is important for them in the produced research and what should be prioritized further.

Table 1. Case summary.

	Field study	Workshop	Modeling of use-scenarios
Insight collection with	Ship crew, during ship operations, onboard ship	Engine room integrator, ship designer	Engine room integrator, engine mechanic, yard construction supervisor
Insight analysis with	Engine room integrator	Engine room integrator, ship designer	Engine room integrator
Prototyping steps	Early concept	Co-designed, refined concept	Prototype version 1.0

Table 2. Case data.

Data material category	Description
Visual maps	Visual notes from meetings and interviews (2D, A4 pages)
Hand-drawn concept sketches	Documentation of concept process in 2D, on paper
Rendered concept sketches	Produced by industrial design student
3D model prototypes	Produced by Pon Power
Meeting notes	From internal meetings at AHO and with Pon Power and Ulstein
Project logs	Time-based documentation of project progress
Field notes	Observation and reflection notes from the field study, workshop, and prototyping session
Field media	Photo, video, and audio material collected during the field study, workshop, and prototyping session
Presentation material	Presentations in Power Point format used to facilitate discussions with project participants during the field study, workshop, and seminar
Seminar interviews	Notes from seminar
Informal interviews	Notes from discussions with project partners

4 DESIGN CASE

4.1 Field study

The field study took place in the North Sea in December 2016 onboard a Platform Supply Vessel designed by Ulstein and built in China, with engine room systems provided by Caterpillar and Pon Power. It was carried out by one field researcher from AHO. The researcher was on board for 5 days. The objective of the field study was to collect first-hand data about the experiences of users of Pon Power engine rooms. This fed directly into the innovation process that drove the case: "*How can we design for better experiences of ship engine rooms?*"

From the start, the field study was designed to be human-centered by focusing on crew activities and by structuring field information about ship systems from the perspective of the use of the systems by the crew. The field study focused on: (1) the different tasks the crew members performed during the different phases of the ship operation, (2) the systems the crew members used to perform these tasks, and (3) the experience of the crew when using these systems while performing these tasks.

The field study was composed of different activities that needed to be performed in sequence to be effective (the produced results are useful for the design process informed by the field study) and efficient (a minimum of resources is used to produce these results). The activities are presented in Table 3.

The background and planning began with building an initial list of tasks and systems that we expected to be able to observe on board the ship and how we might go about the observations. To do this, we interviewed a researcher expert in field observations in engine rooms. Using the interview, we built an observation guide that indicated who to talk to and what location on the ship offered the best context for the conversation. The guide also indicated the best moments to carry out specific observations in regard to the ship's different operation phases. This was important because the field study needed to take place on top of current ship operations without disturbing the operations and respecting the recovery and recreational time of the crew when they were not on a shift. Then, we interviewed Pon Power employees that work with engine room modeling, system integration, and service. We visited their production site. This enabled us to understand their design process and adapt the field study to connect to it. Finally, we produced a detailed field study plan that was communicated to and approved by Pon Power, the company owning and managing the ship we studied, and the captain and crew. This helped create a shared understanding among all the stakeholders involved and secure their full participation.

Table 3. Field study activities.

Activity	Role
Scoping	Specifies what type of information needs to be collected
Background and planning	Specifies how to collect this information
Execution	Collects field information
Analysis	Structures field information, derives conclusions
Presentation	Communicates main findings and conclusions

Table 4. Field study process.

Activity	Outcome	Method
Field study scoping	Common understanding of field study goals	Face-to-face meeting and phone and email conversations
Background research	Observation guide: what to observe and how to observe it Map of Pon Power's design process	Interview with engine room observation expert; Interview and production site visit
Field study planning	Communication of field study goal, scope, and methods to all stakeholders	Plan drafted, then circulated to and approved by stakeholders
Field study execution	Pictures, videos, audio capture, hand-written notes, sketches, typed-up observations, and reflections	1 field study researcher followed the field study plan and adapted it to the ship operations taking place during the study
Field study results analysis	Selection of annotated photos and videos. List of observations and reflections	2 researchers reviewed, sorted, structured, and annotated the field data
Presentation to and analysis of results with Pon Power	Summary of observations in a Power Point document; draft plan for next phase	Summary document was presented and discussed during a face-to-face meeting

The observations collected during the field study consisted of: photos, videos, audio capture of interviews, and hand-written notes. The most significant observations took place at the very end of the field study when a maintenance intervention (changing the oil filters on one engine) was carried out on the way back from an oil platform to the logistics base on land. When back to shore, the data was reviewed by two researchers from AHO. The data was sorted and structured into observations and reflections related to the field study goal and objectives. We prepared a Power Point presentation summarizing the field study experience, illustrated with relevant media (photos and videos). We presented this summary to Pon Power and analyzed the findings together. The goal was to agree on what identified problem areas should be targeted and how to reframe them as innovation opportunities. The process is summarized in Table 4.

4.1.1 *Findings*

Field observations (Figures 1–4) showed that the crew was exposed to safety-critical risks (head injuries, slips and falls, burns) and that there were

Figure 1. Absence of working surface on the engine. The mechanic needs to place the tools and spares directly on the engine.

Figure 2. Service tools are stored in small boxes inside a larger box, making it inconvenient and time consuming for the operator to find the required tool. The lid of the tool box fell over the mechanic´s head during the operation.

Figure 3. Risk of head injury, trip, and fall, as well as oil spill. While carrying used oil filters in a receptacle full of oil, the mechanic needs to climb down stair steps while bending his back and knees to avoid hitting a beam with his head.

Figure 4. Constrained space around the engine, lack of body support and non-ergonomic body position. The mechanic is resting his elbow on pressure and temperature gauges that are not designed for carrying weight. Both his feet are only halfway resting on the flooring.

ergonomic issues (lack of body support and non-ergonomic body positions) as well as efficiency issues (tools and spares spread in different places, no protection nor recovery measures for important tools). We also observed that the crew made their own ad-hoc tools and their own solutions for routine cleaning and routine checks, which is evidence of a system design that does not entirely satisfy the needs of its users. The problem areas were summarized as:

"The engine as a working place": the engine needs to be seen as the central element of a working place where human operators need to carry out work tasks every day.

"Engine integration in the engine room": the engine integration in the whole engine room needs to enable the human operators to carry out their work tasks in the most safe and efficient way.

The problems were reframed in terms of innovation opportunity: by delivering safe and efficient working spaces, Pon Power can differentiate itself from its competitors. Because the ship designer designs the whole engine room, we agreed that the next step needed to be an innovation workshop with the ship designers from Ulstein Design Solutions.

In summary, the field study enabled Pon Power to articulate their challenges and innovation opportunities in terms of human-centered engine rooms and to advocate for a human-centered approach to the collaborative design process of engine rooms.

4.2 *Innovation workshop*

The objectives of the workshop were to (1) identify use-cases or design problems that repeatedly take place in engine room design activities, (2) identify entry points or ways into these problems, and (3) sketch opportunities for innovations.

Before the workshop, we recruited the personnel that had the mandate and competence to explore problems and implement solutions in their design process. From Pon Power, we recruited the technical director and an engineer in charge of modeling individual systems and their integration in engine rooms. Both worked all along the product chain, with sales and service teams on each end. From Ulstein Design Solutions, we recruited two engineers involved at the concept design stage, both specializing in machinery integration, as well as an engineer working with detailed design downstream in the design process. Still before the workshop, we briefed the workshop participants using a 5-minute video documenting a service intervention from the field study that showed several safety and efficiency issues. During phone interviews with each participant, we asked what they thought about the service intervention, what problems they saw, what might be the root of the problem, and what possible solutions they could think of. This enabled starting the workshop with an already established, common understanding of the problem at hand and a list of questions that could be addressed collectively:

1. How much do we know about engine room use, for example, like service scenarios? How can we find information about it?
2. How can we visualize such scenarios in 2D drawings and 3D models so that they can be used as input to the design process?
3. How can we manage the collaboration among the engine room integrator, the ship designer, and the yard to make sure the engine room is built according to the final design drawings and models?

Before the workshop, we also prepared visual concepts to synthesize the ideas we had discussed and developed so far with all the stakeholders. The visuals were also produced to support the workshop conversations by referring to specific ideas and to trigger further ideation processes by criticizing and improving the ideas. We sketched a concept of a human-centered engine room (Fig. 5) that included specific space requirements, such as space for circulating around the engine, flat working spaces, tool and spares storage, and space to manipulate the tools and spares used to service the engine. The space requirements were visualized as volumes on a

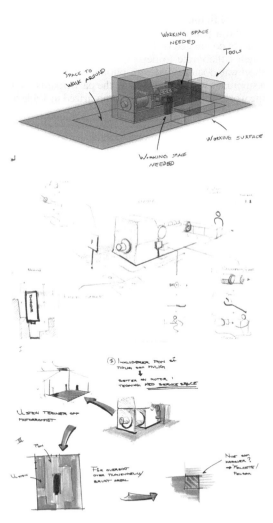

Figure 5. Concept sketches for a human-centered engine room. Top: concept for a 3D model of an engine that visualizes the space required for access around the engine, working surfaces, and tool storage as volumes of different colors. Middle: concepts for information that should be captured in the space requirements: body of the operator, tools, engine parts. Bottom: Color codes signaling the design responsibility of different design stakeholders in the engine room.

3D model with a color code indicating which stakeholder had the responsibility of the design in each area and which areas should be kept free of anything that might come in the way of the human operator.

The workshop itself took one day, including breaks, lunch, and transportation time. Two facilitators ran it, one facilitating the discussions and the other taking notes and visualizing ideas and concepts. The participants worked with the list of questions presented above, sharing their ideas on Post-its on the wall. Toward the second half of the workshop, the findings were summarized by one facilitator, and a plan was laid down that detailed a new concept co-created by the participants and how they might collaborate to develop it further. The workshop was documented with photos and videos, and a summary was shared with all the participants after the workshop. The process is summarized in Table 5.

4.2.1 *Findings*

The participants commonly agreed on the need to design engine rooms that enable service operations in good conditions. They collaboratively defined who the design stakeholders were that produced the information that informed this design process, as well as the design stakeholders that used this information in their design process. The groups of information producers and users were found to be overlapping: engine room integrator, concept designer, arrangement engineer, yard engineer, engine room crew, and machinery specialist working on land for a ship owner, as well as Class or verification authority and third-party service provider. Following, the workshop participants collaboratively agreed on the end-user needs: ergonomic body position for working, good access to engine parts, room around the engine for operation-maintenance and repairs, proximity to storage of spares, and service-friendly design while the ship is in operation. They agreed on examples of scenarios that might be used to qualify and quantify the required space: engine maintenance (pull out and exchange cylinders, change oil filters, tap and drain oil), operation (cleaning engine, performing temperature and other readings), and repair (dismantling, spare transport and lift, spare assembly). Finally, the workshop participants agreed that the main motivation was to improve the safety of the users, engine systems, and tools used to perform maintenance and repairs.

The participants continued exploring their own design processes and the challenges created when their respective design processes intersect. They mentioned that misunderstandings and wrong interpretations of design documents are common. This is a problem not only for the end-user but also for any stakeholder that produces the design of a ship part that ends up being built differently, as it impacts the quality of the final product and the image and reputation of the design stakeholder. They explained that the main reason for misunderstandings is that the information about space requirements in engine rooms is dispersed in different documents, handled by different design stakeholders, and presented in formats that are not able to carry the correct information. For example, the engine integrator produces user manuals for the engine systems in text formats, from which the ship designers select information they need to produce their own user manuals that are delivered to the yard. The engine integrator also produces visual descriptions in 2D drawings, but they typically only indicate the space taken by engine parts on the drawings and not the space required for the body of the human operator. When the ship designer or the yard turn the 2D drawing into a 3D model, there can again be information loss or misinterpretation.

The initial concept of the human-centered engine room modeled in 3D that was prepared before the workshop was expanded by the workshop participants. They added several layers of information to the model, each one displaying the space requirements for one use-scenario. They

Table 5. Innovation workshop process.

Activity	Outcome	Method
Mobilize workshop participants	List of confirmed workshop participants	Phone calls
Edit field data video	5 min video showing one service intervention on board	Software to edit several video clips and blur the face of the mechanic performing the service intervention
Discuss video with participants	Common understanding of the problem	Phone interview with each participant
Prepare inspiration visuals	Handmade and digital sketches	Industrial design student sketching
Run workshop	Co-created concept and implementation plan	2 facilitators. 3 main questions. Post-its on wall.
Share workshop documentation	Summary of workshop insights	Photo and video documentation

proposed ways to quantify space requirements for each scenario using videos from field studies, videos made by mechanics onboard the ship or on land in the workshop where the engines were being assembled, or on interviews with mechanics. They proposed using the model to check if other systems around the engine collided with the service space, enabling a test of the engine room design before it is built. The concept and its collaborative design process are visualized in Figures 6 and 7.

In summary, there were two main outcomes of the workshop. First, the participants managed to create a common understanding of the engine room design requirements when built from the perspective of the end-users and the design stakeholders in their respective contexts of use. Second, the participants managed to sketch a collaborative design process and a collaborative format supporting this process that were adapted to their own respective design processes.

4.3 Prototyping: Modeling of use-scenarios

The objective of this phase was to start prototyping the concept developed through the field study and the innovation workshop. A 3D model with a human avatar next to an engine was made by Pon Power using Teamcenter NX software. We organized a session at Pon Power's office where different engine interventions were filmed. A service mechanic showed how the intervention is done, what tools are used, and what steps are usually challenging to perform. Two sets of color tapes were laid on the floor around the engine at 0.5 m and 1 m to the engine center line to give distance indications in the video recording. We found and built props on the spot to approximately reproduce the sizes and shapes of different engine parts and servicing tools. Three examples are shown in Figures 8–10.

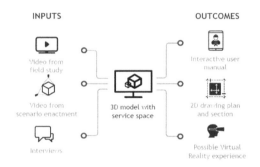

Figure 6. Concept of a 3D model of service space. The model can be built using different types of inputs, and can be used to generate different outcomes.

Figure 8. Space requirement capture for use-scenario: oil filter change. Note the room taken by the body of the mechanic when enacting the movement of pulling a filter out of the engine.

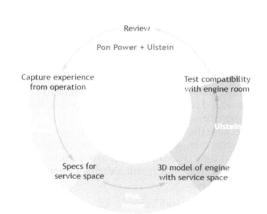

Figure 7. Collaborative process to produce a 3D model of design space, between the engine room integrator (Pon Power) and the ship designer (Ulstein). The process is designed to be iterative, hence its circular shape.

Figure 9. Space requirement capture for use-scenario: crankshaft removal. The crankshaft need to be pulled out entirely, requiring a lot of space.

The filming session took approximately two hours, with a total of five scenarios filmed. The session involved a project engineer, mechanic, and yard supervisor. It gave them the opportunity to share their experience with engine modeling, service intervention, and construction challenges.

The space requirements were then integrated into the 3D model. Each component of the engine that can be removed during a service intervention was movable, and the space needed by the human operator to perform the intervention was displayed in the model, both in 2D and 3D, using volumes. An early prototype is shown in Figure 11.

Figure 10. Space requirement capture for use-scenario: genset adjustment. The mechanic demonstrates the use of the tool required for this operation: a key with a long lever arm. To manipulate the key, the mechanic is standing one meter from engine.

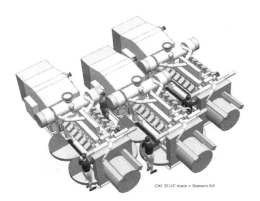

Figure 11. Early prototype of the 3D service space digital model with a human avatar and a color code: grey and light grey areas need to be kept clear of any other system to guarantee for service space. In this scenario, the avatar is in the position for removing an oil filter. Having 3 engines enables displaying this information on each side of the engine, including between two engines located side by side, as is often the case in compact engine rooms.

5 ANALYSIS

5.1 Design process

To analyze the design process followed in the case, we asked:

– What participants were involved?
– How did they work together?
– How are the needs of human operators captured and transferred into the design process across the different design stakeholders involved?
– How different is that compared to common practice?

We started the analysis by looking at how a ship design process would take place without any specific human-centered design intervention. We mapped such a process using the "actor centric mapping technique" based on data collected previously (Gernez, Nordby and Sevaldson, 2014). The result is shown in Figure 12.

We made two main observations from this mapping. First, each stakeholder interacts with at least two other stakeholders at every step of the process. In addition, the role of each stakeholder changes at almost each design step, which means that different teams with different competences and responsibilities need to be involved at different steps.

This means that there are also complex interactions inside each stakeholder company. Second, the end-users, such as the engine room users, are not involved in the design process. This means that the stakeholders designing and constructing systems for them do not know how usable the systems are or whether, how, or to what extent they address the users' needs. They do not have the opportunity to benefit from the ideas the end-users might have to make the systems better.

These observations showed that there are two important needs to address:

1. The interactions between the engine integrator, ship designer, and yard on the one side, and the end-users on the other side are not currently part of the design process. They need to be added to allow the design process to benefit from operational feedback.
2. Because the design process is built upon numerous and complex interactions between stakeholders, there is a need to facilitate these interactions from both the information exchange and human perspectives.

In Figure 13 and Table 6, we map and analyze how the design process we contributed to might have addressed these two needs.

Table 6 and Figure 13 show that both the inclusion of end-users and the facilitation of the collaboration of design stakeholders were addressed in

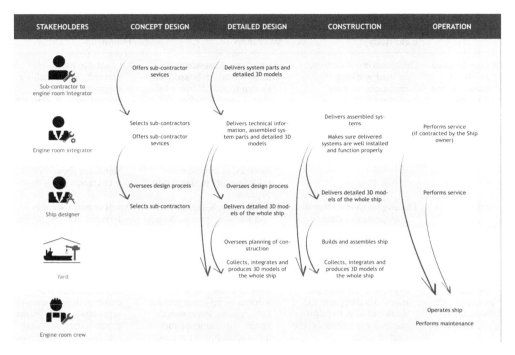

Figure 12. Actor centric mapping of a generic ship design process at a high level.

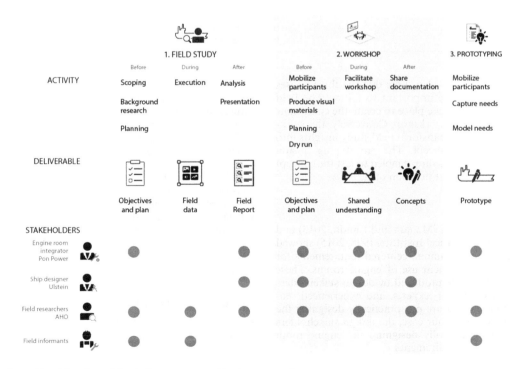

Figure 13. Mapping of the design process used in the presented case.

Table 6. Analysis of addressed needs.

	Field study	Workshop	Follow-up
Addressing the need for "involving end-users into the design process"	Observations of end-users in their working environment were captured, analyzed, synthesized, reframed, and shared back with the engine room integrator.	(1) a video of use-scenario as observed on board was shown to and discussed with ship designers before the workshop. (2) the workshop lead to a concept for how human operations can be captured during field studies and then modeled in engine room models, early in the design process.	(1) a service mechanic showed to a modelling engineer what is important to design for when designing for service scenarios. (2) the modeling engineer created a 3D model with a human avatar that simulates these service scenarios.
Importance	The engine room integrator now possesses data documenting engine room use experiences and has alternatives for ways to intervene.	The field data is shared with and worked upon with ship designers: operation data is one step closer to the early design phase.	Another end-user was brought into the design process: a service mechanic working with service intervention.
Addressing the need for "facilitation of design stakeholders collaboration"	Engine room integrators expressed the need to meet ship designers to work out how to improve the user experience of the engine room.	Engine room integrators met with ship designers and worked on co-constructing descriptions, definitions, needs, requirements and innovation opportunities.	A service mechanic, modelling engineer and yard construction supervisor, representing 3 different departments, met and exchanged perspectives.
Importance	The field data was used as evidence for the importance to have this meeting.	Together, these two design stakeholders from two different companies found a way forward, that also has potential to improve their own processes internally: time saving, increased knowledge sharing	The three design stakeholders from Pon Power learned how to arrange such a knowledge sharing session. They now have a digital format to capture and share the results of such a session.

our process. They also show that the three phases of the project are important to follow. The field study needed to take place to create the conditions for a productive workshop. Conversely, the workshop enabled to transform field study insights into an innovation concept. The prototyping session following the workshop enabled to test the concept and produce a first iteration of it.

5.2 Impact

Mallam & Lundh (Mallam and Lundh, 2013) and others (The Nautical Institute, 1998, 2015) showed the existence of human-centered requirements for the safe and efficient use of engine rooms. These requirements are produced by design stakeholders such as Class, HF experts, and experienced sea-going crews that are not practically designing the engine room. In our case, the design stakeholders that were practically designing the engine room defined these requirements.

The concept of human-centered engine rooms developed in this case has the potential to enable more efficient maintenance and service interventions for the end-users of the engine rooms, which reduces the risks for injury, system failure, and operational downtime.

For the design stakeholders involved, the use of human-centered design methods has the potential to improve the detection of design flaws and, consequently, to reduce the risk of additional design iterations to correct these flaws. Without these methods, it is challenging to consider parts of the ship from a dynamic perspective. We are, however, aware that the commercial perspective rules the main dimensions of a ship, and spaces and areas that are not directly involved with the main commercial functions of the ship are kept at a minimum.

5.3 Complementarity with ship design methods

The approach we propose is designed to be complementary to the way ship design processes are currently carried out. In the case described in this article, the approach is used on the design of engine rooms, but the approach is applicable the design of other ship systems as well. We visualize in Figure 14 how the human-centered design methods we have used can be placed in a simplified design process that connects the human-centered design components of the design process (what we refer to "Ship operations") with the technology-centered components (what we refer to "Ship architecture"). Similarly to starting a design process by looking at existing ships, we recommend to start the design process by a field study on a similar ship to map the working and living conditions of the end-users of the ship, as well as how they are performing ship operations currently. Using these field insights, we then recommend to analyze how the existing systems on board the ship enable its human operators to use the ship, and identify design problems that might impact the safety and efficiency of ship operations. From this analysis, we recommend to sketch what architectural solutions might enable the human operators perform their work in better conditions.

5.4 *Limitations*

We presented only one case in this paper, which limits the generalization of the impact of human-centered design methods. We have applied the same approach to two other cases in the ONSITE project. Referring to one of these cases, Ulstein Vice President Per Olaf Brett commented in a project seminar: "From now on, we will never do a new ship design project without using the field study methodology."

As illustrated in Figure 13, there is a high dependency on the design researcher to facilitate the use of human-centered design methods. The design researcher intervenes at each step of the process, as a field researcher in the field study, a facilitator in the workshop, and a contributor to the concept ideation and development throughout the whole process. In other ONSITE cases, the project partners have carried out field studies and run workshops on their own.

Looking at the cost of our approach, the field study took approximately 100 hours for the field researchers from initiation to conclusion. The innovation workshop took approximately 50 hours for the field researchers from initiation to conclusion. These costs need to be transferred to the design process costs. Considering their potential impact on the design process and the ship operations, these costs are negligible in comparison to, for example, the cost of one design iteration or one day of ship operational downtime. As expressed by Per Olaf Brett during a project seminar in November 2017: "Cost [of field studies] is not an issue. (…) Field studies can be very valuable for the downstream design process."

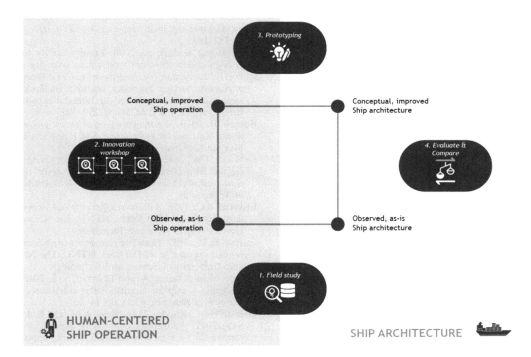

Figure 14. A design framework for human-centered, collaborative, field-driven design processes.

In terms of proximity to the end-users and their working environments, there are no alternatives to performing a field study. However, once a field study is performed, its results can be used over several design projects if there is a similarity in scope. The experience of the designer during the field study is also relatively transparent to the type of ship.

6 CONCLUSION

Using human-centered design methods in the design of engine rooms brings stakeholders together on common, shared issues and responsibilities that they can solve together to improve their own and collaborative design processes and, as a result, improve the quality of the outcomes for end-users. The requirements for implementing this approach are to (1) facilitate the collaboration between the companies involved, (2) collect qualitative data about the needs of the ship crew on board ships during operation, and (3) define the engine room as a working environment where human needs can be catered to.

These requirements are similar to human-centered design projects in other industries, for instance the Oil & Gas industry, where the involvement of end-users is a more common practice. Further research is needed to analyze how the uptake of human-centered design methods can be better facilitated, with regards to training, multidisciplinary collaboration, and the combination of user-centric, qualitative data with existing Computer Aided Design systems and ship design processes.

ACKNOWLEDGEMENTS

The authors would like to thank the ONSITE project partners, the Norwegian research council for funding, and Jon Fauske for the help in conceptual work, facilitation, and illustration.

REFERENCES

Abeysiriwardhane, A., Lützhöft, M., Petersen, E.S. and Enshaei, H., 2015. Future ship designers and context of use: setting the stage for human centred design. In: *Proceedings of The International Conference on Marine Design*. International Conference on Marine Design. The Royal Institution of Naval Architects, pp.1–16.

Abeysiriwardhane, A., Lützhöft, M., Petersen, E.S. and Enshaei, H., 2017. Human-centred design knowledge into maritime engineering education; theoretical framework. *Australasian Journal of Engineering Education*, 21(2), pp.49–60.

Andrews, D.J., 1986. An integrated approach to ship synthesis. *Transactions of The Royal Institution of Naval Architects*, 128, pp. 73–102.

Andrews, D.J., 2003. Marine design - requirement elucidation rather than requirement engineering. In: *Proceedings of IMDC 2003*. 8th International Marine Design Conference. Athens, Greece.

Andrews, D.J., 2011. Marine requirements elucidation and the nature of preliminary ship design. *Transactions of the Royal Institution of Naval Architects Part A: International Journal of Maritime Engineering*, 153 (Part A1).

Andrews, D.J., Duchateau, E.A.E., Gillespe, J.W., Hopman, J.J., Pawling, R.G. and Singer, D.J., 2012. IMDC State of the Art Report: Design for Layout. In: *Proceedings of IMDC 2012*. 11th International Marine Design Conference. Glasgow, UK.

Andrews, D.J. and Erikstad, S.O., 2015. IMDC State of the Art Report: Design Methodology. In: *Proceedings of IMDC 2015*. 12th International Marine Design Conference. Tokyo, Japan, pp. 89–105.

Andrews, D.J., Papanikolaou, A., Erichsen, S. and Vasudevan, S., 2009. IMDC State of the Art Report: Design Methodology. In: *Proceedings of IMDC 2009*. 10th International Marine Design Conference. Trondheim, Norway.

Beyer, H. and Holtzblatt, K., 1997. *Contextual design: defining customer-centered systems*. London, UK: Academic Press.

Blomberg, J., Burrell, M. and Guest, G., 2009. An ethnographic approach to design. In: *Human-Computer Interaction*. Boca Raton, USA: CRC Press, Taylor and Francis Group, pp. 71–94.

Bødker, S. and Buur, J., 2002. The design collaboratorium: a place for usability design. *ACM Transactions on Computer-Human Interaction (TOCHI)*, 9(2), pp.152–169.

Buur, J. and Bødker, S., 2000. From usability lab to "design collaboratorium": reframing usability practice. In: *Proceedings of the 3rd conference on Designing interactive systems: processes, practices, methods, and techniques*. DIS '00 Designing Interactive Systems. Brooklyn, NY, USA: ACM, pp. 297–307.

Buxton, B., 2010. *Sketching user experiences: getting the design right and the right design*. San Francisco, USA: Morgan Kaufmann.

DeNucci, T.W., 2012. *Capturing Design: Improving Conceptual Ship Design Through the Capture of Design Rationale*. PhD Thesis. TU/e Delft, Delft University of Technology.

Erikstad, S.O., 1996. *A decision support model for preliminary ship design*. PhD Thesis. NTNU, The Norwegian University of Science and Technology.

Gaspar, H.M., 2012. *Handling complexity aspects in conceptual ship design*. PhD Thesis. NTNU, The Norwegian University of Science and Technology.

Gaspar, H.M., Hagen, A. and Erikstad, S.O., 2016. On designing a ship for complex value robustness. *Ship Technology Research*, 63(1), pp. 14–25.

Gaspar, H.M., Rhodes, D.H., Ross, A.M. and Erikstad, S.O., 2012. Addressing complexity aspects in conceptual ship design: a systems engineering approach. *Journal of Ship Production and Design*, 28(4), pp. 145–159.

Gernez, E., 2018. *Guide: Actor centric mapping.* Unpublished.

Gernez, E., Nordby, K. and Sevaldson, B., 2014. Enabling a Service design perspective on Ship design. *Transactions of The Royal Institution of Naval Architects Vol 156 Part C1 2014.*

Grech, M.R., Horberry, T. and Smith, A., 2002. Human error in maritime operations: Analyses of accident reports using the Leximancer tool. In: *Proceedings of the human factors and ergonomics society annual meeting.* The human factors and ergonomics society annual meeting. Los Angeles, CA: Sage Publications, pp. 1718–1721.

International Standards Association, 2014. *ISO 9241-210:2010 – Ergonomics of human-system interaction —Part 210: Human-centred design for interactive systems.*

Kataria, A., Praetorius, G., Schröder-Hinrichs, J.-U. and Baldauf, M., 2015. Making the case for Crew-Centered Design (CCD) in merchant shipping. In: *19th Triennial Congress of the IEA.* 19th Triennial Congress of the International Ergonomics Association. Melbourne, Australia.

Kühnbaum, J., 2014. *DNV GL Energy Management Study 2014.* DNVGL.

Kuo, C., 2003. A management system based approach to ship design. In: *Proceedings of IMDC 2003.* 8th International Marine Design Conference. Athens, Greece.

Levander, K., 2003. Innovative Ship Design. In: *Proceedings of IMDC 2003.* 8th International Marine Design Conference. Athens, Greece.

Lurås, S. and Nordby, K., 2014. Field studies informing ship's bridge design. Human Factors in Ship Design and Operation. London, UK: The Royal Institution of Naval Architects.

Mallam, S.C. and Lundh, M., 2013. Ship engine control room design analysis of current human factors & ergonomics regulations & future directions. In: *Proceedings of the Human Factors and Ergonomics Society Annual Meeting.* SAGE Publications, pp. 521–525.

Mallam, S.C. and Lundh, M., 2014. Conceptual ship design, general arrangement & integration of the human element: a proposed framework for the engine department work environment. In: *Proceedings of the 11th International Symposium on Human Factors in Organisational Design and Management & 46th Nordic Ergonomics Society Conference.* Copenhagen, Denmark, pp. 829–834.

Mallam, S.C., Lundh, M. and MacKinnon, S.N., 2015. Integrating Human Factors and Ergonomics in large-scale engineering projects: Investigating a practical approach for ship design. *International Journal of Industrial Ergonomics,* 50, pp. 62–72.

Mallam, S.C., Lundh, M. and MacKinnon, S.N., 2017. Integrating Participatory Practices in Ship Design and Construction. *Ergonomics in Design,* 25(2), pp. 4–11.

van de Merwe, F., Kähler, N. and Securius, P., 2016. Crew-centred Design of Ships–The CyClaDes Project. *Transportation Research Procedia,* 14, pp. 1611–1620.

Millen, D.R., 2000. Rapid ethnography: time deepening strategies for HCI field research. In: *Proceedings of the 3rd conference on Designing interactive systems: processes, practices, methods, and techniques.* ACM, pp. 280–286.

Nelson, H.G. and Stolterman, E., 2003. Design Judgement: Decision-Making in the 'Real'World. *The Design Journal,* 6(1), pp. 23–31.

Nowacki, H., 2009. Developments in marine design methodology: roots, results and future trends. In: *Proceedings of IMDC 2009.* 10th International Marine Design Conference. Trondheim, Norway: Tapir Academic Press, pp. 47–80.

Praetorius, G., Kataria, A., Petersen, E.S., Schröder-Hinrichs, J.-U., Baldauf, M. and Kähler, N., 2015. Increased Awareness for Maritime Human Factors through e-learning in Crew-centered Design. *Procedia Manufacturing,* 3, pp. 2824–2831.

Rogers, Y., Sharp, H. and Preece, J., 2011. *Interaction design: beyond human-computer interaction.* John Wiley & Sons.

Sanders, E.B.-N. and Stappers, P.J., 2012. *Convivial toolbox: Generative research for the front end of design.* BIS.

Solesvik, M.Z., 2007. A Collaborative Design in Shipbuilding: Two Case Studies. In: *Industrial Informatics, 2007 5th IEEE International Conference on.* IEEE, pp. 299–304.

Solesvik, M.Z., 2011. Collaborative knowledge management: case studies from ship design. *International Journal of Business Information Systems,* 8(2), pp. 131–145.

The Nautical Institute, 1998. *Improving Ship Operational Design: Through Teamwork.* London: The Nautical Institute.

The Nautical Institute, 2015. *Improving Ship Operational Design.* The Nautical Institute.

Ulstein, T. and Brett, P.O., 2012. Critical Systems Thinking in Ship Design Approaches. In: *11th International Maritime Design Conference (IMDC).* Glasgow, UK: University of Strathclyde.

Van Bruinessen, T., Hopman, J.J. and Smulders, F.E., 2015. Controlled innovation of complex objects. In: *COMPIT 2015.* Conference on Computer Applications and Information Technology in the Maritime Industries. Ulrichshisen Germany.

Wensveen, S. and Matthews, B., 2015. Prototypes and prototyping in design research. *Routledge companion to design research,* pp. 262–276.

Whyte, W.F.E., 1991. *Participatory action research.* Sage Publications, Inc.

Seeing arrangements as connections: The use of networks in analysing existing and historical ship designs

R.J. Pawling & D.J. Andrews
University College London, London, UK

ABSTRACT: A growing trend in computer aided ship design, particularly in the early stages, is the utilisation of approaches and numerical methods developed in other disciplines. Examples include genetic algorithms, financial methods of risk assessment and the use of network science. Networks can provide an abstract mathematical representation of many types of connected features, properties and information, such that the associated network analysis metrics and approaches can offer new ways of investigating and evaluating ship designs. This paper reports on ongoing UCL investigations into the application of network science in assisting human analysis of the general arrangements of existing ship designs. This work includes designs of complex service vessels (research vessels) as a comparison with naval ships and makes use of freely available network analysis software. This project makes use of the experience in naval vessel concept design at UCL by enabling a comparison of expert judgement and interpretation of designs with the quantitative network metrics. This paper describes the network analysis approach adopted, the findings for the arrangements analysed, and also discusses the future work required to further the approach.

1 INTRODUCTION

1.1 Background to the study

The Marine Research Group, part of the Department of Mechanical Engineering at UCL (2018) conducts research into various aspects of maritime design and technology, in both Naval Architecture and Marine Engineering. One long-running theme, of particular interest to the authors, is ship general arrangements, including arrangements evaluation methods, architecturally-centred design methods that integrate configuration into the earliest stages of design (Andrews, 2003), and the problem of how to effectively teach arrangements design to undergraduate and postgraduate Naval Architecture students (Pawling et al, 2015).

This paper describes progress to date on an ongoing US Navy Office of Naval Research (ONR) sponsored collaborative international project to investigate various aspects of arrangements design. Previous collaborative outputs of this project have included an IMDC State-of-the-Art report (Andrews et al, 2012) and joint papers on the subject of style in design (Pawling et al, 2013, 2014) and a new taxonomy for describing distributed systems in ship design (Brefort et al, 2017).

The various partners in the project are undertaking independent research projects, with significant cross-pollination of ideas, and one area that UCL is investigating is the application of networks to arrangements design, with a particular emphasis on designer-centred processes, allowing the designer to "see" the general arrangement in a new, non-geometric, way. The ongoing UCL investigation of the application of networks to investigate developed general arrangements draws inspiration from four sources; a notable series of "comparative naval architecture" papers, previous UCL considerations of the meaning of "style" in ship design, considerations of topology and connectivity in the field of architecture and urban design, and the significant past work carried out by the NICOP project partners in this area.

1.2 Comparative naval architecture

Comparative naval architecture has its origins in the 1970s Cold War, with the need for NATO to understand the capabilities of Soviet warships without having access to reliable technical information. A type of reverse engineering, it assumes that the designers of the ships under investigation made rational decisions using the information available to them; but that those decisions may not be consistent across a range of international designs The practice was carried out using primarily numerical analysis (Kehoe, 1976) which included an attempt to re-design a US Navy vessel using Soviet practices and style, and also more holistic analysis including some internal arrangements (Kehoe et al 1980a, 1980b). As more information became available on Soviet vessels at the end of the Cold War such comparisons have been undertaken at a more detailed level, such as that by Brower & Kehoe (1993). Most recently this approach has also been applied to passenger vessels (Sims, 2003), with a particular emphasis on how safety was considered.

Generally these studies only included limited information on the general arrangement, such as the high-level breakdown in volume allocation shown in Figure 1 (Kehoe et al, 1980b), or simple profile views.

The UCL arrangement study described in this paper draws on existing ship designs, with the aim of examining the potential for networks to be used in comparing arrangements of vessels differentiated by nationality, era, and role.

1.3 Style in ship design

"Style" was introduced as a conceptual component of design methodology by Simon (1970), and first applied to ship design by Brown and Andrews (1981) in their "S5" summary of the naval architects considerations in ship design; Speed, Seakeeping, Stability, and Strength and Style. Andrews (2012) provided a listing of topics classed as "Style", which is repeated as Table 1.

These stylistic issues were proposed to be subtly different to other aspects of ship performance in that they bring a collection of disparate wholeship design issues, incorporating engineering sciences, managerial and user-focused issues. A novel definition of style was proposed, that it is a type of design information possessing several key characteristics; that it is cross-cutting, groups information, and is able to accommodate uncertainty (Pawling et al 2013). These characteristics are illustrated in Figure 2.

Style was proposed to be "cross-cutting", in that a given style decision will impact across multiple performance areas. The implication of this is that style in design is also a way of grouping information about design decision that have such predominately cross-cutting impacts. Finally, it was proposed that design solutions occur at the conceptual intersection of technical performance requirements and stylistic decisions (Figure 2d). Regarding the use of networks to analyse arrangements, the objective of the UCL study is to investigate whether networks can detect "style", in the form of network characteristics and metrics.

1.4 Ways of thinking about space

Large inhabited structures such as ships, building and even cities present problems of understand what "space" and "arrangement" mean. Although a ship is of course a 3D construct in Cartesian space, the internal arrangement can be viewed in different ways. A simple 3D model, such as that in Figure 3 is technically correct, but is not always straightforward to understand—the implications of this in education having been considered by Pawling et al (2015) and Collette (2015).

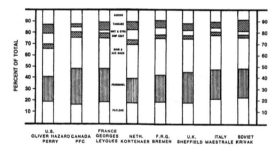

Figure 1. Volume allocation in NATO and Soviet frigates (Kehoe et al, 1980b).

Table 1. Aspects of style in naval ship design (after Andrews, 2012).

Stealth	Protection	Human factors	Sustainability	Margins	Design issues
Acoustic signature	Collision resistance	Accommodation standards	Mission duration	Space	Robustness
Radar cross-section	Fire-fighting	Access policy	Crew watch policy	Weight	Commercial standards
Infrared signature	Above water weapon effect	Maintenance levels	Stores level	Vertical center of gravity	Modularity
Magnetic signature	Underwater weapon effect/shock	Operation automation	Maintenance cycles	Hotel power	Operational serviceability
Visual signature	Contaminants protection	Ergonomics	Refit philosophy	Ship services	Producibility
	Damage control		Upkeep by exchange	Design point (growth)	Adaptability
	Corrosion control		Replenishment at sea	Board margin (upgrades)	Aesthetics

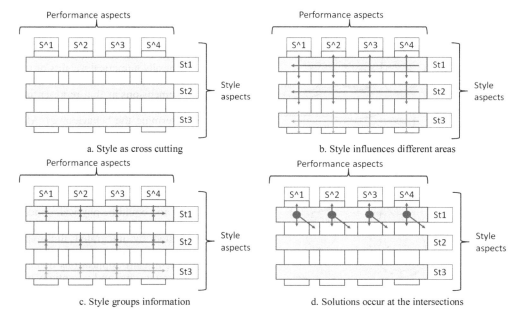

Figure 2. The main characteristics of style as a type of information in ship design as proposed by Pawling et al (2013).

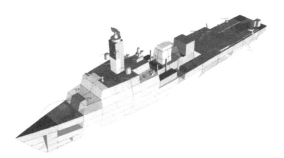

Figure 3. A 3D spatial model of a ship design.

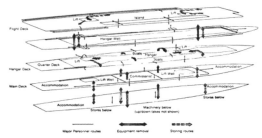

Figure 4. The internal flows in an aircraft carrier—an example of a 2.5D layout representation (Honnor and Andrews, 1982).

"2.5D" representations of the ship design (originally defined as "2D+" by the first author, (Pawling, 2007)) are frequently used, which show the design as a series of stacked decks with interconnections. These are also frequently used to effectively convey the logic behind a design, such as the internal flow in an aircraft carrier, shown in Figure 4.

A more abstract definition of ship arrangements is the contact diagram, which has been used to develop layouts for accommodation spaces (Cain & Hatfield, 1979). Figure 5 shows a typical contact diagram for the superstructure of a cargo vessel. Contact diagrams allow the layout to be built up from functional requirements, through the topology, to the contact diagram, and then to a geometry (Klem, 1983). More recently they were used by Dicks (1999) in the prototype demonstration of the Design Building Block approach, and Andrews (2003) incorporated some numerical comparison of relationships.

This type of abstract thinking about the nature of complex spaces has seen greater use in other fields. The seminal paper by Alexander (1965) used set and graph theory to examine the structures of notional cities, making a (social) argument for a lattice, rather than tree-type underlying structure. With the increasing application of CAD to (land based) architecture, the capability to consider topology and geometry in different ways has become more practical, e.g. Medjdoub & Yannou, (2000). Although it still frequently has a highly conceptual, theoretical basis, e.g. Savaskan (2012).

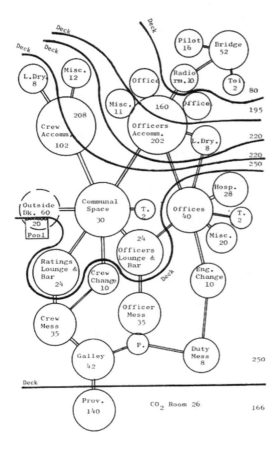

Figure 5. A contact diagram showing proximity relationships. The numbers are the required areas (Cain & Hatfield, 1979).

uni-directional. Networks can also be represented in matrix form, and Figure 6 illustrates a simple directed network and its associated matrix.

The nodes and edges can support additional information, typically numerical weights or textual data, which can be stored in additional matricies with the same dimensions as that representing the connections in the network. Early applications of networks to determine the behaviour of AI were known as Semantic Networks and described a decision making tree, as shown in Figure 7 (Sowa, 1992).

Networks are suited to application to problems that can be described in terms of connected entities. The entities can be represented by the nodes, or by the edges, and could be physical, conceptual or operational. Network methods can be made more sophisticated by integrated multiple networks representing the same system. One of the more recently significant applications of network theory is in understanding (and potentially manipulating) social structures as the "connectedness" of modern society increases (Easley & Kleinberg, 2010).

2.2 Network analysis

Newman (2010) and Mrvar (2018) provide a detailed description of the various quantitative methods that have been developed to analyse net-

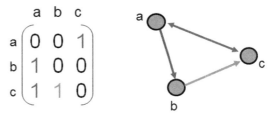

Figure 6. Matrix representation of a directed network of three nodes (after Collins et al. (2015)).

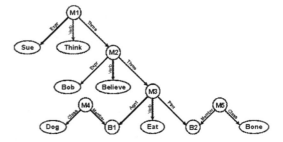

Figure 7. A semantic network representing a simple sentence, "Sue thinks that Bob believes that a dog is eating a bone" (Sowa, 1992).

Work at the Bartlett School of Architecture at UCL has led to the development of the concept of "Space Syntax", which applies aspects of network theory to existing and proposed urban landscapes, both to determine underlying emergent structures (Al-Sayed et al, 2012) and to guide proposals for development (Hillier, 2009). Notably this abstraction is suitable for application at a range of scales, from buildings to cities, (Hillier, 2014), and an early version was applied to ship design by Andrews (1984).

2 NETWORKS

2.1 Networks and matrices

A network is a collection of points, called vertices or nodes, joined by lines, called edges or arcs, Newman (2010). The edges can have no direction or be

works, both at the whole-network level, and with regards to individual nodal properties. Many of these metrics are best used in a relative manner, comparing nodes to one another, rather than as absolute values. The metrics used in the UCL study are described in Section 4, but they can be broadly summarized as relating to degree, centrality, or communities (modules).

Degree refers to the number of connections of the node, and can incorporate directionality, as shown in Figure 8. Any weighting on the edges can be incorporated in degree metrics.

Centrality measures relate to the connections between nodes and come in three broad groups; degree, closeness and betweeness. Degree centrality is a measure of the number of direct connections a node has. Closeness centrality is a measure of how close a node is to all other nodes in a network. Betweeness centrality evaluates the extent to which a node lies in the shortest path between pairs of nodes in a network, as shown in Figure 9. Eigenvector centrality is a variant of closeness centrality, in which the centrality of surrounding nodes influences the centrality value assigned.

Communities or modules are clusters of nodes with more arcs within the cluster than between clusters, as illustrated in Figure 10 in which there are more arcs inside the cluster than among clusters.

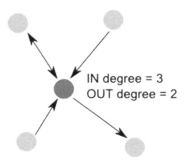

Figure 8. The degree of a node.

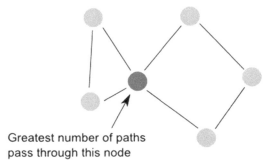

Figure 9. Betweenness centrality.

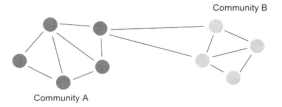

Figure 10. Communities in a network.

2.3 *Applications of networks to ship design*

The earliest application of network science to ship design was by MacCallum (1982), who used them to represent and explore relationships between ship design characteristics in computerized models, with a particular emphasis on understanding the interactions and influences between the parameters. Similarly Parker and Singer (2013, 2015) and Shields et al (2015) describe the application of modern network models to investigate ship design models and the flow of information in the design process.

More recent applications have focused on the potential applications of network science to arrangements design, as has been the case in land-based architecture. Gillespie (2012) used networks to examine emergent design drivers that could be detected from databases of arrangement preferences (i.e. without generating layouts first), this work making particular use of numerical methods to detect communities as shown in Figure 10 (Gillespie et al, 2013). Kilaars et al (2015) combined networks with automated approaches to layout generation, which are capable of producing a large number of possible arrangements, needing subsequent downselection, with Roth (2017) examining networks metrics as a possible method to differentiate between design options.

Another recent application of network science is in the modelling and analysis of distributed systems, with various levels of abstraction. Rigterink et al. (2014) applied community detection methods to ship hotel services, such as electrical systems. Of particular interest in the design of naval vessels is the use of networks to evaluate survivability of distributed systems (Shields et al, 2016, 2017).

3 PREVIOUS UCL APPLICATIONS OF NETWORKS TO SHIP DESIGN

UCL has engaged in a variety of investigations of the application of network science to ship design, including submarine concept design, layout preference analysis, surface ship concept model analysis and ship survivability.

3.1 Submarine concept design

Collins et al (2015) described ongoing PhD research into the use of networks to address issues of knowledge and uncertainty in the integration of new technologies, applied to submarine design. Submarine design is traditionally very conservative and the objective of this work is to improve understanding of the relationships and interactions in submarine concept design by representing the design model as a network of connections between variables. Numerical network metrics can then be used to determine the significance of various parameters. This has the aim of providing earlier identification of design features and parameters that will be disrupted by the addition of a new technology to the submarine.

3.2 Layout preference analysis

Pawling et al (2015) described the use of network analysis to investigate layout preferences in warship design. A database was populated with pairwise arrangement relationships (i.e. space A related to space B) and the NodeXL Excel plug-in (NodeXL, 2014) used to construct a network and conduct an analysis. This then illustrated those spaces (represented as nodes) were most significant in the network, and thus were afforded the greatest importance in that particular designer's view of arrangements design. Figure 11 illustrates the betweeness centrality ranking of the nodes in the database of one designer's preferences. The high closeness centrality of the Damage Control (DC) deck and the need for spaces to be split, indicates that this designer has a heavy preference for survivability in layout.

3.3 Surface ship concept model analysis

Pawling et al (2016) described a similar analysis to that of Colins et al (2015), with the network analysis software Pajek (Mrvar, 2018) used to investigate the significance of various parameters in the UCL MSc in Naval Architecture concept ship design

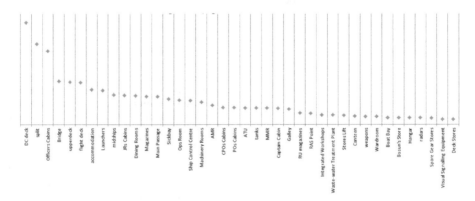

Figure 11. Ranking of spaces and arrangement features by closeness centrality (Pawling et al, 2015).

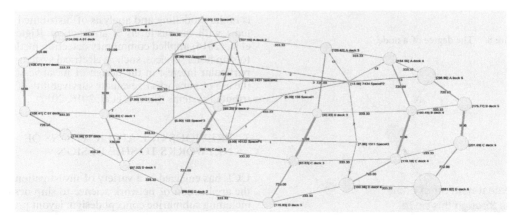

Figure 12. Section of a blast propagation network, including interconnecting ship systems.

model. This directed network, was used to examine influence within the concept design model via the proximity prestige metric, indicating that the most influencing parameter (node) changed as the concept design progresses, something not explicitly stated in the design documentation.

3.4 *Survivability*

Pawling et al (2016) also described the use of networks, again in Pajek, as a possible proxy for the modelling of blast effects after an explosion within the ship, due to a weapon impact. This was a comparative exercise, using a UCL model for internal blast developed as an MSc project (Edwards, 2015) as a baseline. The rationale for applying network methods to this problem was that blast is a phenomenon that propagates through connections (bulkheads) between entities (spaces) and thus can be represented as a network. In addition to the comparison of analytical capabilities, networks were considered as a possible means to visualise blast effects, as shown in Figure 12, where the relative size of the nodes represents the blast overpressure in a compartment and the thickness of the connecting edges represents the value of the failure criteria of the structure between them.

4 THE CURRENT UCL STUDY

4.1 *Introduction*

The latest UCL investigation of the application of networks to ship design is in the analysis of historical vessel designs, using a database of general arrangement drawings obtained from various sources. At the time of writing, this is an ongoing project and so this paper describes progress to date.

4.2 *Method*

4.2.1 *Encoding*

The first step in the analysis was converting the general arrangements into a network model. Watertight and non-watertight doors, hatches, ladders and stairs were included in the model. Evacuation scuttles (which are only used in extremis) and other openings, such as serving hatches, were not included. Where it could be clearly identified from the general arrangement drawing, spaces with the same function connected by an arched opening (i.e. a doorway with no door in it) were treated as a single space.

In addition to the connections themselves, their direction was recorded (vertical or horizontal). For the spaces, some additional parameters were entered into the Excel databases:

I. Functional Group: Based on the breakdown described by Andrews and Pawling (2008) – Float, Move, Fight (i.e. main role), Infrastructure and Access.
II. UCL function: A more detailed functional breakdown based on the UCL MSc Ship Design Exercise weight break down system (WBS).
III. UCL weight group: A slightly more detailed version of the UCL function. For example, the "sanitary" function does not differentiate between showers and heads, but the weight group does.

For general arrangements not in English, the first authors' extremely limited knowledge of French, German and Finnish was supplemented by Google Translate (Google, 2018). This is noted here as machine translation has little sense of context and so significant interpretation of the results was sometimes required, so this task requires a naval architect and alternative interpretations may exist.

An important note is the approach to external spaces, such as the upperdeck and superstructure decks. They were included in the baseline database for each ship, and a subsequent down-selection process removed all but the minimum required to connect the operational spaces in the vessel. This construction of an "external" and "internal" version of the network was both to investigate the impact of the upperdeck and to prevent "short cuts" appearing along the length of the ship that would be unrealistic in operation (i.e. one would not climb up the superstructure when one could use a passageway). The impact of this is discussed under the results (Section 5).

4.2.2 *Analysis*

This analysis, so far, has used the Gephi freeware software (Gephi, 2018). This decision was primarily determined by the relative ease with which networks can be transferred from Excel to this tool via Comma Separated Variable (CSV) file, and metrics generated. The Pajek software is capable of more sophisticated analysis (hence its use in previous work) and it is likely to be used in further work based on progress so far. Additionally, Gephi has an easy-to use Graphical User Interface (GUI) and this is of great utility to occasional users. The NodeXL Excel plug-in also used in previous work was not adopted here due to compatibility issues with the latest versions of Microsoft Windows and Office software.

All networks were un-directional and no weighting was applied to the edges or nodes, although Gephi represents multiple connections between two nodes as a weighting on a single edge.

Noteworthy is the fact that even on an obsolete computer (3.4 GHz dual core Pentium D and 32 bit operating system) the numerical analysis was effectively instantaneous; the greatest processing time was demanded by the layout algorithms used to generate visualisations of the network Thus a frigate-sized network took approximately 30–60 seconds to remove the majority of overlap between edges in the visualisation.

4.2.3 *Metrics*

Numerical metrics were generated for the overall networks and the individual nodes.

4.2.3.1 Overall network metrics

Table 2 summarises the numerical metrics measured for the overall network of each general arrangement (with and without external access).

4.2.3.2 Individual node metrics

Table 3 summarises the individual node metrics examined in this study. Some relate to the overall network metrics.

Table 2. Overall network metrics.

Number of nodes and edges

The number of unique spaces (nodes) and connections (edges) in the general arrangement network.
Average degree:
The average number of edges per node.
Average weighted degree:
The average sum of the weights of the edges of nodes; this will account for multiple connections between two nodes.
Network diameter:
The maximum distance, in terms of intermediate edges, between any pair of nodes in the network.
Network radius:
The minimum eccentricity of any node in the network.
Graph density:
The ratio of actual connections between nodes to the potential connections between nodes.
Modularity:
A measure of the strength of subdivision of the network into modules (communities). Higher values indicate a higher ratio of connections within modules to those between them.
Number of communities:
The number of communities (modules) in the network. The calculation of modularity and communities features a random component and so the exact values for both will vary between calculation runs.
Average clustering coefficient:
A measure of the tendency of the nodes to cluster together (separate from modularity).
Average path length:
The average number of steps along the shortest paths for all possible pairs of nodes in the network.

4.2.4 *Visualisations*

In addition to the numerical metrics, one objective of the study is to explore possible network visualisations that could be of use in investigating the general arrangement. It is possible to visualize the complete network, and several numerical approaches, known as force-directed layout algorithms, are available to arrange the many nodes in some layout, generally one that minimises the number of edges that cross. Figures 13a and 13b compare the same arrangement network visualized using the Gephi imple-

Table 3. Individual node metrics.

Degree:

The total number of unique connections to the node.
Weighted degree:
The total number of connections to the node (space) including multiple connections between spaces. This will be equal to the degree for most spaces.
Eccentricity:
The maximum distance from a node to the most distant node.
Closeness centrality:
An aggregate measure of the means distance from a node to all other nodes.
Betweenness centrality:
A measure of the extent to which a node (space) lies on the paths between other nodes. Most relevant for access routes.
Modularity class:
The module (community) in which the node lies.
Eigenvector centrality:
A measure of centrality where the centrality of each node is proportional to the sum of the centralities of its' neighbours. This indicates if a node is in a well-connected region of the network.

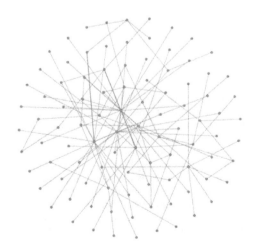

Figure 13a. Fruchterman Reingold visualisation of a layout network.

mentations of Fruchterman Reingold (1991) which represents the nodes as masses and the edges as springs, so tending to place the nodes an equal distance apart, and Force Atlas 2 (Jacomy et al, 2012), which distorts the locations of nodes in an attempt to spatialise the connections between them.

Further to these direct visualizations of the network, ways of exploring the distribution of network properties by node (and the associated characteristics of the space, such as functional group) were investigated. Figure 14a shows an example visualisation, where each diamond represents a space, arranged from left to right in order of decreasing normalized centrality (or some other network metric as indicated where appropriate), and banded by Functional Group (FLoat, ACCess, MOve, FIght, INfrastructure, ACcOmmodation and SToRes), with certain spaces highlighted.

Figure 14b shows the same information in a cumulative line graph, where the coloured lines represent the cumulative entries of the Functional Groups, and the dotted line shows the decreasing value of the centrality measure. Comparative visualisations such as this are considered to be of more use, due to the relatively abstract nature of some of the network metrics, in that their absolute values do not have a direct physical meaning.

Figure 13b. Force Atlas 2 visualisation of the same layout network.

4.3 The example ships

The selection of example ships to be analysed was largely driven by the availability of complete, labelled

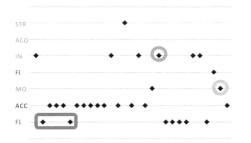

Figure 14a. Banded and ranked visualization of node centrality.

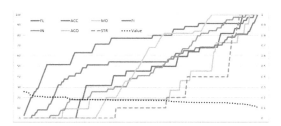

Figure 14b. Cumulative visualization of node centrality.

Table 4. Vessels examined in the study to date.

Name	Type	Year	Built	Nation	Displ. (tonnes)	Overall length (m)	Overall beam (m)	Accom.
DDL	Guided missile destroyer	1970	No	UK for Australia	4200	129.6	14.6	263
GPF	Guided missile frigate	1962	No	Canada	3300	121.3	14	240
DDH	Anti-submarine frigate	1965	Yes (after changes)	Canada	3800	129.1	14.7	244
Aconit	Anti-submarine frigate	1970	Yes	France	3870	127	13.4	232
DD-692 Long Hull	Anti-submarine destroyer	1947	Yes (after changes)	USA	2220	114.6	12.5	309
EPV Louhi	Patrol vessel	2007	Yes	Finland	3450	71.4	14.5	40
Falkor	Research vessel	1981	Yes	Germany	2260	82.9	13	42
Meteor	Research vessel	1986	Yes	Germany	5125	97.9	16.6	63
Armstrong	Research vessel	2015	Yes	USA	3204	72.5	15.2	44

general arrangement drawings, with the intention of examining a selection of drawings from a range of eras, nations and roles. Of particular interest is the comparison between warships and research vessels, as both are complex service vessels. Table 4 summarises the vessels that have been encoded into network form so far, along with their principal particulars. Some of the naval vessels are designs that were proposed but lost out to competitors, or were developed further prior to construction.

5 RESULTS

5.1 Overall network metrics

Table 5 summarises the overall network metrics for the designs analysed for this paper. The suffix "EXT" means that all external decks are included, "INT" means that only the minimum are included. For some ships it was not possible to remove the decks and leave a viable network, so for those vessels only a single network was created.

Examining these results, we see that the average degree is larger than we might expect, indicating that large numbers of spaces in real ship designs have more than one connection with another space. Examination of the distribution of degree for the designs shows that between 55 and 70% of spaces have only a single connection. It is the case that the statistical average is pulled up by a small number of access routes with a large number of connections. However, in each design there are some highly-connected non-access routes, as illustrated by Figure 15, which shows, for each vessel the 40 spaces with the highest degree.

Table 5. Summary of the overall network metrics for the designs analysed.

Design	Network size		Average degree		Network dimensions			Graph density	Modularity	Communities	Average clustering coefficient
	Nodes	Edges	Unweighted	Weighted	Diameter	Radius	Average path length				
DDL INT	269	282	2.097	2.134	16	8	6.831	0.008	0.846	16	0.017
DDL EXT	270	289	2.141	2.193	15	8	6.286	0.008	0.833	17	0.04
GPF INT	199	215	2.171	2.191	12	6	5.279	0.011	0.816	13	0.101
GPF EXT	204	223	2.186	2.225	12	6	5.264	0.011	0.808	12	0.098
DDH INT	205	217	2.117	2.127	12	6	4.877	0.01	0.825	13	0.123
DDH EXT	207	221	2.135	2.155	13	7	4.959	0.01	0.816	11	0.117
Aconit INT	221	231	2.09	2.109	18	9	7.388	0.01	0.848	13	0.078
Aconit EXT	225	240	2.133	2.16	15	8	6.73	0.01	0.829	14	0.075
DD-692 INT	145	152	2.097	2.124	21	11	8.841	0.015	0.815	11	0.044
DD-692 EXT	150	170	2.267	2.373	11	6	5.237	0.015	0.78	10	0.062
EPV	97	108	2.227	2.268	11	6	5.027	0.023	0.736	10	0.134
Falkor INT	126	138	2.19	2.222	16	9	6.716	0.018	0.785	10	0.137
Falkor EXT	133	148	2.226	2.271	15	9	6.8	0.017	0.789	12	0.138
Meteor INT	200	224	2.24	2.32	15	8	5.942	0.011	0.805	10	0.093
Meteor EXT	204	230	2.255	2.333	15	8	5.907	0.011	0.802	14	0.095
Armstrong	106	116	2.189	2.283	12	6	4.804	0.021	0.751	9	0.094

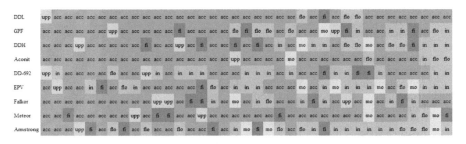

Figure 15. The Functional Groups of the spaces with the 40 highest degree in each design; external networks used in each case (colours indicate the Functional Group for the highlighted spaces – See Section 4.2.4).

Notable in Figure 15 are the DDL; which has two parallel passageways along the length of the vessel, with consequentially very high levels of connection; the DD-692, which has a large number of highly connected infrastructure spaces; and the research vessels, which have highly connected labs and working decks (under the "Fight" functional group in this study).

Turning to the network dimensions section of Table 5, it may be possible to see the significance of external access routes in the designs. The dimensions of the network for the DDL, GPF and DDH are far less affected by the removal of the upperdeck than Aconit or the DD-692 designs. In the latter case, the removal of the upperdeck edges and nodes significantly changes the network properties, and this is to be expected; the DD-692 design was developed late in the Second World War, during the US Pacific Campaign, whereas the other warships were designed for (or by designers who normally worked with) North Atlantic navies preparing for a conflict involving nuclear weapons, where ensuring good access inside the ship was paramount.

What can also be seen is that the overall parameters of the networks for the research vessels are less effected by the removal of the upperdeck. This may be a reflection of the provision of vertical stair towers in civilian designs, providing good internal access.

5.2 Metrics by node (space)

The three nodal metrics of initial interest are betweenness, closeness and eigenvector centralities. As betweenness centrality tells us how many paths pass through a node, it can potentially be used to evaluate the significance of a particular passageway to overall circulation. Figure 16 illustrates the effect of switching from Closeness Centrality to Eigenvector Centrality. The relative ranking of some functions, such as access, is strongly affected.

Closeness and eigenvector centrality both tell us how well connected a node is to the rest of the network, which may be used to evaluate the accessibility of a space from the rest of the ship. Eigenvector centrality is of interest as it effectively "weights" the spaces by their proximity to other well-connected spaces.

5.3 The DD-692 design

Figures 17a, b, c and d show the functional banding of the distribution of betweeness, closeness and eigenvector centrality for the DD-692 "Long

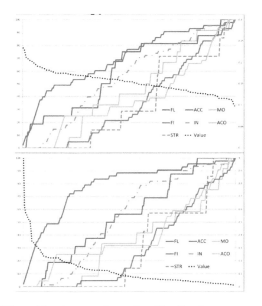

Figure 16. A comparison of ranking for closeness centrality (top) and eigenvector centrality (bottom) for the Aconit design.

Figure 17a. Betweenness centrality, including upperdeck, with main crew accommodation highlighted.

Figure 17b. Betweenness centrality, excluding upperdeck, with main crew mess decks highlighted.

Figure 17c. Eigenvector centrality, including upperdeck, with crew accommodation, galley (green) and CIC (red) highlighted.

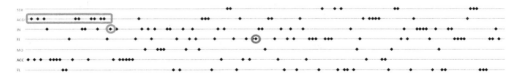

Figure 17d. Closeness centrality, including upperdeck, with crew accommodation, galley (green) and CIC (red) highlighted.

Hull" destroyer design. In these figures, the highest values are to the left of the diagram.

The first thing that can be seen is that several crew accommodation and mess decks (green box) have high values of betweenness centrality. Removing the upperdeck access, between Figures 17a and 17b, increases the relative significance of the mess decks to the overall accessibility of the ship. The highest betweenness centrality in a non-passageway space is assigned to a crew berthing space, highlighted in Figure 18. This is signifies the importance of non-dedicated passageways to overall access in ships of this period; examination of Figure 18 reveals a large number of watertight and hatches doors providing main longitudinal and vertical access through accommodation spaces.

Figures 17c and 17d also indicate that the galley is in a location affording easy access to the rest of the ship. Given the significance of this space to daily operations, this is not unexpected. Another contrast between Figures 17c and 17d is the change in relative position of the CIC/Operations Room, from which the ship is fought. As it is in the superstructure, its closeness centrality is reduced, however the eigenvector centrality may better capture the impact of its proximity to several other important spaces, such as the captain's cabin and radio rooms.

Although these banded diagrams may indicate some trends in arrangement, the cumulative representation is required for further insight. Figure 19 shows the cumulative eigenvector centrality for the design (including upperdecks). The convex nature of the lines for ACCess and ACcOmmodation indicate that there are a smaller number of spaces with high centrality (connectivity). The FIght functional group, however, has a concave line, indicating that many spaces have low centrality. Inspecting the general arrangement, it is notable that the DD-692 has several weapon control

Figure 18. Crew berthing space in the DD-692 design with very high betweenness centrality.

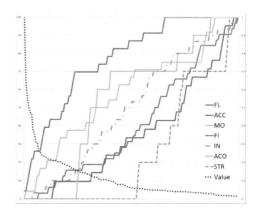

Figure 19. Cumulative eigenvector centrality.

spaces and magazines deep in the ship, with consequently limited access.

5.4 Comparing destroyers

Figures 20a, b and c show the eigenvector closeness distributions for the DDL, DDH and GPF designs (without upperdeck, given their post-war design). These can be compared with Figure 17c. Notable is

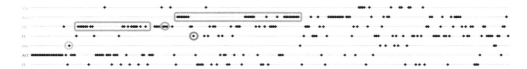

Figure 20a. DDL internal spaces in order of eigenvector centrality, with officers cabins (green rectangle), offices and admin spaces (green rectangle), galley (green circle), operations room/CIC (red circle) and machinery control room/SCC (yellow circle) highlighted.

Figure 20b. DDH internal spaces in order of eigenvector centrality, with spaces highlighted as above.

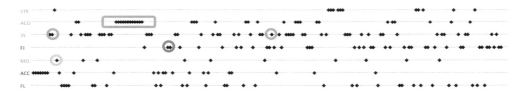

Figure 20c. GPF internal spaces in order of eigenvector centrality, with spaces highlighted as above, but with the dining halls highlighted instead of offices.

the apparent high priority given to officers accommodation, which has a higher centrality than in the DD-692. These ships also have more administrative spaces which are located in easily accessible locations, as shown by the high centrality of the offices.

The increase in both size and importance of accommodation and infrastructure spaces in post-war naval vessel design has been discussed in Brown (1988) and Brown & Moore (2003) and manifests itself in this network analysis. Comparing these with the older design (DD-962) also shows the a change in style to concentrated groups of accommodation spaces, in contrast to a smaller number of large messes in the later designs.

A key difference between the DDL and the other designs is that it has two longitudinal passageways on No.2 deck, and this is reflected in the large number of access spaces with high centrality values. The galley in all three ships has "middling" connectivity in the layout, but the GPF places the dining spaces in a well-connected location, effectively displacing, as the most connected spaces, the offices and admin compartments in the other two designs.

The contrast in relative ranking of the SCC can potentially be explained by examining its immediate connectivity. In the DDL, the SCC is on No.2 Deck and so benefits from the double passageways. In the DDH, it is again on No.2 Deck, but effectively in a cul-de-sac. The GPF locates the SCC lower in the ship, but retains watertight doors below the damage control deck providing access from the SCC to the machinery spaces. This drives the SCCs centrality higher. This practice was rapidly discontinued in post-war warships, with the SCC moved above the damage control deck (No.2 Deck) and no watertight doors fitted below the damage control deck. Although increasing survivability this has some impact on accessibility, and this appears to be reflected in the network analysis.

The apparently lower accessibility of the CIC in the DDH design corresponds to its location high in the superstructure—CIC location being one of the "stylistic" issues discussed by Kehoe et al (1980a, 1980b). Although CIC location in Figures 20a–20c does correlate with increasing height (by being in the superstructure), it is important to note that these are only relative within a single design, so smaller differences in centrality score, such as between the DDL and GPF, are not reliable indicators.

319

Figure 21a. Ranking of eigenvector centrality for Aconit including upperdeck, officers cabins, galley, CIC/operations room and SCC/machinery control room highlighted.

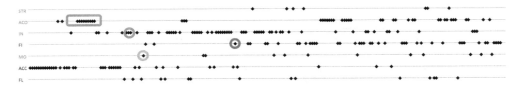

Figure 21b. Ranking of eigenvector centrality for Aconit exluding upperdeck, officers cabins, galley, CIC/operations room and SCC/machinery control room highlighted.

5.5 A non-NATO warship: FS aconit

France withdrew from NATOs' command structure in 1966, but rejoined in 2009. Although the 1980's saw several joint warship projects between France and other NATO nations, eventually leading to the Horizon frigates (Brown & Moore, 2003). This separation may have had some consequences for warship design and development, and differences in national approaches have been documented by Ferreiro and Stonehouse (1994) and Ferreiro and Autret (1995).

Aconit differs from the other post-war warships in that it does not carry a helicopter, and relies on guns for air defence. At the detail level, several layout features were noted that contrasted with UK, US and Canadian practice, including; the location of some officers accommodation aft; the provision of a large number of small messes acting as entrance lobbies for accommodation areas; and large duplicated conversion machinery for the sensors and weapons on No.2 Deck (the latter is possibly due to a large number of 1960s' era weapons carried on a small frigate). Figure 21a and 21b illustrate the ranking for eigenvector centrality for external and internal spaces in the Aconit design.

Comparing these two figures indicates that the availability of the upperdeck has minimal impact on the accessibility of Aconit, and indeed the arrangement shows airlocks and decontamination areas, indicating that she was designed to be fought under conditions of NBCD contamination. Comparing the results with the destroyers, we see similar results; the galley and SCC, located on No. 2 Deck, are well connected, while the CIC/Ops Room, located in the superstructure behind the bridge, is less so. A difference between the designs is the large number of non-accommodation infrastructure spaces that have high centrality in the Aconit design. This is driven by the large number of small messes, which generally lead onto the main passageway or vertical access, and Figure 22 illustrates this with the betweenness centrality, highlighting these small messes.

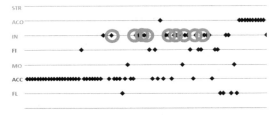

Figure 22. Betweenness centrality for Aconit, including upperdeck, with messes highlighted.

5.6 Research vessels

Research vessels were chosen as another type to be investigated, as they are service vessels, and their general arrangements are relatively readily available. Figures 23a, b and c illustrate the ranking of eigenvector centrality for the vessels, Falkor Meteor and Armstrong, respectively, with certain key spaces highlighted.

In all three designs, the scientist cabins have a higher centrality than the crew cabins, in common with officers on warships, and the tendency to concentrate cabins into groups can be seen. The working decks aft have high centralities, with multiple connections via labs (even if single passageways

Figure 23a. Eigenvector centrality ranking for Falkor, with main mess (M), galley (G), working spaces and labs (red) and machinery control room (yellow) highlighted.

Figure 23b. Eigenvector centrality ranking for Meteor, with main mess (M), galley (G), working spaces and labs (red) and machinery control room (yellow) and workshops (blue) highlighted.

Figure 23c. Eigenvector centrality ranking for Armstrong, with main mess (M), galley (G), working spaces and labs (red) and machinery control room (yellow) highlighted.

are used elsewhere). This concentration and accessibility of the fight functional group (representing "research" in this case) can be contrasted with the less accessible Fight spaces in the warships, which are be dispersed to meet survivability requirements. As has been previously noted, the mess spaces are generally better connected than the galleys.

These vessels follow civilian practice in having the machinery control room/SCC low in the hull, close to the engine room. Falkor features watertight doors connecting the SCC to the surrounding spaces, this significantly increases its accessibility, at the cost of potential vulnerability after a collision. Notable on Meteor, the largest and most capable of the three vessels, is the provision of several workshops, which have a high relative centrality.

5.7 EPV: The smallest network

The EPV Louhi was included as the smallest network so far investigated. Given the comparative nature of the visualizations and metrics, problems may occur with their applicability to networks with small numbers of nodes. Figure 24 illustrates one issue; the small number of spaces leads to a highly discretized numerical range, with many plateau.

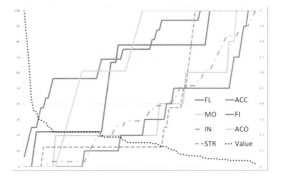

Figure 24. Cumulative eigenvector closeness for the EPV.

Figure 25 illustrates the eigenvector centrality ranking for the EPV, with certain spaces highlighted. The main capabilities of patrol and environmental protection vessels are usually in the working deck and boats, and in the EPV the latter have a high centrality, having access directly off of the space with the highest betweenness centrality (a stairwell in the superstructure), shown in Figure 26. The EPV has an arrangement similar to an Offshore Support Vessel, with most functional spaces concentrated in the superstructure forward, and this leads to high centrality values for the bridge, mess and galley. The SCC is less accessible, however, being low in the ship.

It is notable that the working deck has a lower centrality than the working decks in the research vessels. However it has a high degree with seven connections to other spaces. The low centrality may be a result of the concept of operations (in research vessels the working decks are a regularly used transit route for equipment and samples to and from workshops and labs) but in an environmental protection vessel this flow is not present, so operationally could be seen as a "cul-de-sac", where crew only occasionally go to carry out certain tasks, with far less requirement for connectivity.

6 THE SHAPE OF ARRANGEMENTS

Returning to the introductory sections which discussed conceptual models of ship arrangements, it is possible to visualize the complete network in a single image, as shown in Figures 13a and 13b so this can be used to examine the "shape of ships arrangements". At the time of writing, only simple illustrations produced using the "Force Atlas 2" (Jacomy et al, 2012) approach had been generated. Unfortunately the implementations of the graph layout codes in Gephi do not produce identical graphs each time they are run, and this makes comparing images a little difficult. It is possible to apply graphical effects to the nodes and edges based on numerical properties, such as colour-coding modules or a colour scale to indicate centralities.

Figure 27 illustrates the complete network for the EPV. It is perhaps notable that, despite the relatively small number of nodes in the EPV network, some distinct shapes are visible; the overall arrangement

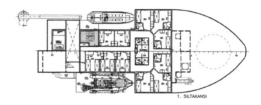

Figure 26. Partial GA of the EPV, showing the space with the highest betweenness centrality (Modified from Segercrantz, 2008).

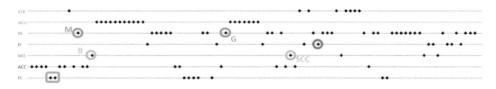

Figure 25. Ranking of eigenvector closeness for the EPV, with mess (M), galley (G), working deck (red), bridge (B), SCC and boat bays (blue) highlighted.

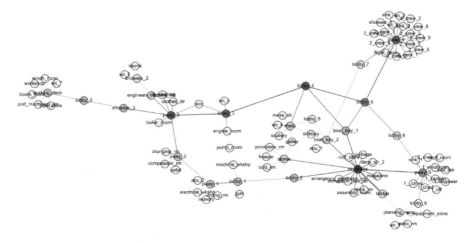

Figure 27. The complete network for the EPV, shaded by betweenness centrality.

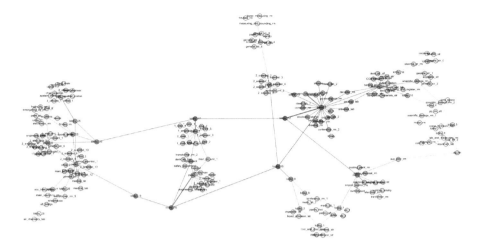

Figure 28. The complete layout network for Meteor, shaded by closeness centrality.

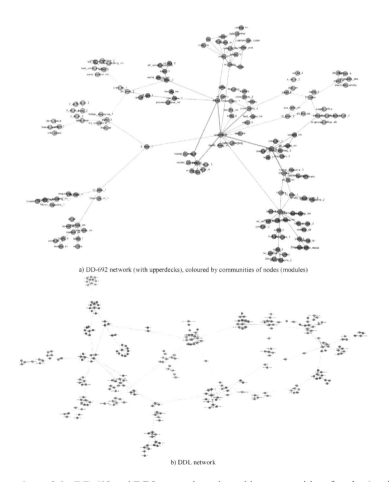

a) DD-692 network (with upperdecks), coloured by communities of nodes (modules)

b) DDL network

Figure 29. Comparison of the DD-692 and DDL networks, coloured by communities of nodes (modules).

is characterized by multiple loops, which branch off into star-shaped geometries. The colour scheme, in which nodes with a higher betweenness are shaded more darkly, highlights the main access routes.

These geometric features are more pronounced in larger vessels, for example the research vessel Meteor, shown in Figure 28. This graph features multiple loops and several stars, consisting of both accommodation and other functional spaces.

Figures 29a and b compare the complete networks for the oldest and the most recent designs in the analysis conducted so far. These have been coded with contrasting colours randomly assigned to the different modules (communities) in each network. A stylistic change is apparent in the reduced complexity of the star-shaped geometries, and the loops in the older design are smaller, composed of less spaces. This visual change reflects the move to more dedicated access routes in post war vessels. The interleaving of the communities in the DD-692 network is a result of this increased interconnection in this older design.

7 CONCLUSIONS

This paper describes progress to date on an ongoing UCL project to investigate the use of network science as a new way for designers to "see" general arrangements. As such, the work has examined ways of visualizing a relatively limited set of numerical network metrics, with analysis conducted using freeware software, and visualizations produced in Excel.

Several methods of visualizing numerical metrics in a comparative manner have been investigated: colour- and function-coded charts comparing different designs for a single metric by space; a functionally banded ranking visualization for various metrics in the designs; a cumulative plot of the same metrics; and a simple visualization of the overall network using established layout techniques.

The visualisations have allowed the authors to detect certain aspects of the "style" of the vessels, which correlate with the history of changes in warship design throughout the latter part of the twentieth century. Importantly, the visualisations retain a connection to the individual spaces and geometric layout of the design, as it is possible to identify which data point represents which space. It is considered that it is primarily this connection that allows the network metrics, when appropriately visualized, to be used as a new way to explore general arrangements, as the abstraction of network metrics alone would reduce their utility to the designer.

There are several possible criticisms of the analysis presented in this paper. Firstly, that the small number of examples makes drawing definite conclusions difficult. Secondly, the correlations described are dependent upon the authors' interpretation, and thirdly; correlation does not imply causation.

The small number of designs that have been addressed to date are being added to, to make the database more comprehensive—only a limited set could be fully explored at the time of writing. These include further patrol vessels and survey vessels.

It is not clear whether the interpretive nature of the analysis is a flaw particular to the investigations carried out so far; as one intention of this work is to develop new ways for the designer to examine the design, so some aspect of interpretation will always be present. The difficulty is in ensuring that such interpretation proceeds on a rational basis, and this is where the third criticism has significance. Considering the small number of arrangements examined so far, and the highly complex and emergent nature of ship arrangements in general, the possibility of illusory correlations being detected by the authors' interpretation cannot be ruled out.

Future work is partly directed at attempting to address these criticisms: the number and variety of ships will be increased; the range of numerical network metrics will be expanded, as only a limited set were used here; additional post-processing (such as statistical data) of the network metrics may also be of use in increasing confidence; and finally the functional meta-data assigned to spaces (functions and weight groups) may be incorporated into the analysis in a more numeric manner, as these could potentially form their own networks. These could then exist in parallel with the spatial connectivity networks which have been presented here.

Another key area for future research is in the potential application of this analysis in new ship designs. The analysis presented is considered to be best used as a tool to structure designer comparison of options, in a similar way to the land-based applications reported in the references. To develop a more extensive analytical approach network model metrics could be compared with other forms of simulation and analysis, such as the personnel movement simulations described by Andrews et al (2008), as this will assist in understanding correlations between network metrics and how the arrangement affects the ship's performance in specific scenarios.

ACKNOWLEDGEMENTS

This study was funded as part of the Preliminary Ship Design General Arrangements NICOP project and its successor NICOP, both funded by Ms. Kelly Cooper from the US Navy Office of Naval Research, and this support is gratefully acknowledged.

REFERENCES

Al-Sayed, K. Turner, A. & Hanna, S. 2012. Generative Structures in Cities. *Proc. 8th International Space Syntax Symposium*. Santiago de Chile.

Alexander, C. 1965. A City is Not a Tree. *Architectural Forum*. Vol 122. No 1. April 1965. pp 58–62.

Andrews, D.J. & Pawling, R.J. 2003. SURFCON – A 21st Century Ship Design Tool. *International Marine Design Conference (IMDC)*. May 2003. Athens, Greece.

Andrews, D.J. & Pawling, R.J. 2008. A Case Study in Preliminary Ship Design. *IJME*. Vol.150 Part A3, 2008.

Andrews, D.J. 1984. Synthesis in Ship Design. Ph.D. Thesis. London: University of London.

Andrews, D.J. 2003. A Creative Approach to Ship Architecture. *International Journal of Maritime Engineering (IJME)*. Vol.145, No.3, London: RINA

Andrews, D.J. 2012. Art and science in the design of physically large and complex systems, *Proc. R. Soc. A.* Vol. 469, pp.891–912.

Andrews, D.J. Casarosa, L. Pawling, R.J. Galea, E. Deere, S. & Lawrence, S. 2008. Integrating personnel movement simulation into preliminary ship design. *IJME*. Vol.150 Part A1, 2008.

Andrews, D.J. Duchateau, E.A.E. Gillespe, J. Hopman, J.J. Pawling, R.J. & Singer, D.J. 2012. State of the art report: Design for layout. *IMDC*. October 2012. Glasgow, UK.

Brefort, D. Shields, C. Habben Jansen, A. Duchateau, E. Pawling, R. Droste, K. Jasper, T. Sypniewski, M. Goodrum, C. Parsons, M.A. Yasin Kara, M. Roth, M. Singer, D.J. Andrews, D. Hopman, H. Brown, A. & Kana, A.A. An architectural framework for distributed naval ship systems. *Ocean Engineering*. 147 (2018), pp. 375–385, https://doi.org/10.1016/j.oceaneng.2017.10.028

Brower, K.S. & Kehoe, J.W. Captn(USN) 1993. Fast Attack Craft – A Comparative Analysis. *Naval Engineers Journal*. May 1993. ASNE.

Brown, D.K. & Andrews, D.J. 1981. The design of cheap warships. *J. Naval Science* 7/2. pp.81–95.

Brown, D.K. & Moore, G. 2003. Rebuilding the Royal Navy. London: Chatham Publishing.

Brown, D.K. 1988. The Naval Architecture of Surface Warships. *Trans. RINA*. Vol. 130.

Cain, J.G.D. & Hatfield, M.R. 1979. New Concepts in the Design of Shipboard Accommodation and Working Spaces. *Trans. RINA*. Vol. 121.

Collette, M. 2015. Studying Student's Experience of the Marine Design Synthesis Problem. *IMDC*. May 2015. Tokyo, Japan.

Collins, L.E. Andrews, D.J. & Pawling, R.J. 2015. A New Design Approach for the Incorporation of Radical Technologies: Rim Drive for Large Submarines. *IMDC*. May 2015. Tokyo, Japan.

Dicks, C.A. 1999. Preliminary Design of Conventional and Unconventional Surface Ships Using a Building Block Approach. Ph.D. Thesis. University of London

Easley, D. & Kleinberg, J. 2010. Networks, Crowds and Markets: Reasoning about a highly connected world. Cambridge University Press.

Edwards, G.W.B. 2015. Rapid Vulnerability Assessment Model. MSc Thesis. University College London.

Ferreiro, L. & Autret, G. 1995. A Comparison of French and US Amphibious Ships. *Naval Engineers Journal*. May 1995. ASNE

Ferreiro, L.D. & Stonehouse, M.H. 1994. A comparative study of US and UK frigate design. *Trans. RINA* Vol. 136.

Fruchterman, T.M.J. & Reingold, E.M. 1991. Graph Drawing by Force-Directed Placement. *Software – Practice & Experience*. Wiley, 21 (11): 1129–1164. doi:10.1002/spe.4380211102.

Gephi. 2018. https://gephi.org/

Gillespie, J. 2012. A Network Science Approach to Understanding and Generating Ship Arrangements in Early-Stage Design. PhD thesis. University of Michigan.

Gillespie, J.W. Daniels, A.S. & Singer, D.J. 2013. Generating functional complex-based ship arrangements using network partitioning and community preferences. *Ocean Eng.* 72, pp.107–115.

Google. 2018. https://translate.google.co.uk/

Hillier, B. 2009. Spatial Sustainability in Cities. Proc. *7th International Space Syntax Symposium*. Stockholm, Sweden.

Hillier, B. 2014. Space Syntax as a Theory as Well as a Method. *21st International Seminar on Urban Form*. Porto, Portugal. 3–6 July 2014.

Honnor, A.F. & Andrews, D.J. 1982. HMS INVINCIBLE - The First of a New Genus of Aircraft Carrying Ships". *Trans RINA*. Vol. 124.

Jacomy, M. Heymann, S. Venturini, T. & Bastian, M. 2012. ForceAtlas2, a Continuous Graph Layout Algorithm for Handy Network Vizualisation. 1st August 2012. http://medialab.sciences-po.fr/publications/Jacomy_Heymann_Venturini-Force_Atlas2.pdf

Kehoe, J.W. 1976. Warship Design – Ours and Theirs. *Naval Engineers Journal*. February 1976. ASNE

Kehoe, J.W. Graham, C. Brower, K.S. & Meier, H.A. 1980a. Comparative Naval Architecture Analysis of NATO and Soviet Frigates – Part I. *Naval Engineers Journal*. October 1980. ASNE.

Kehoe, J.W. Graham, C. Brower, K.S. & Meier, H.A. 1980b. Comparative Naval Architecture Analysis of NATO and Soviet Frigates – Part II. *Naval Engineers Journal*. December 1980. ASNE.

Killaars, T. Van Bruinessen, T.M. & Hopman, J.J. 2015. Network science in ship design. *IMDC*. May 2015. Tokyo, Japan.

Klem, P. 1983. Accommodation Design. Translated by Ellefsen, A. Edited by Benford, H. Department of Naval Architecture and Marine Engineering, University of Michigan, Ann Arbor. MI. USA.

MacCallum, K.J. 1982. Understanding relationships in marine systems design. *IMSDC*. London.

Medjdoub, B. & Yannou, B. 2000. Separating Topology and Geometry in Space Planning. *Computer-Aided Design*. 32(2000) 39–61.

Mrvar, A. 2018. Pajek: analysis and visualization of large networks. http://mrvar.fdv.uni-lj.si/pajek/

Newman, M.E.J. 2010. Networks: An Introduction, Oxford University Press.

NodeXL archive page: https://nodexl.codeplex.com/

Parker, M.C. & Singer, D.J. 2013. The impact of design tools: Looking for insights with a network theoretic approach. *International Conference on Computer Appli-*

cations and Information Technology in the Maritime Industries (COMPIT). 15–17 April. Cortona, Italy.

Parker, M.C. & Singer, D.J. 2015. Comprehension of design synthesis utilizing network theory. *IMDC*. May 2015. Tokyo, Japan

Pawling, R. 2007. The Application of the Design Building Block Approach to Innovative Ship Design. PhD Thesis. UCL

Pawling, R.J. Andrews, D.J. Piks, R. Singer, D. Duchateau, E. & Hopman, H. 2013. An Integrated Approach to Style Definition in Early Stage Design. *COMPIT*. 15–17 April. Cortona, Italy.

Pawling, R.J. Morandi, R. Andrews, D.J. Shields, C. Singer, D. Duchateau, E. & Hopmann, H. 2014. Manifestation of Style and its Use in the Design Process. *COMPIT*. 12–14 May. Redworth, UK.

Pawling, R.J. Piperakis, A.S. & Andrews, D.J. 2015, Developing Architecturally Oriented Concept Ship Design Tools for Research and Education. *IMDC*. May 2015. Tokyo, Japan.

Pawling, R.J. Piperakis, A.S. & Andrews, D.J. 2016. Applications of Network Science in Ship Design. *COMPIT*. 9–11 May. Lecce, Italy

Rigterink, D. Piks, R. & Singer, D.J. 2014. The use of network theory to model disparate ship design information. *Int. J. Naval Architecture and Ocean Engineering*. 6/2. pp.484–495.

Roth, M.J. 2017. Analysis of General Arrangements Created by the TU Delft Packing Approach making use of Network Theory. MSc thesis. Delft University of Technology.

Savaskan, D. 2012. Perception of Space in Topological Forms. Architecture Thesis. Syracuse University School of Architecture. Syracuse, NY, USA.

Segercrantz, H. 2008. New Design Keeps Baltic Clean. *Ship and Boat International*. May/June. London: RINA

Shields, C. Rigterink, D. & Singer, D. 2017. Investigating physical solutions in the architectural design of distributed ship service systems. *Ocean Eng*. 135, 236–245.

Shields, C. Sypniewski, M. & Singer, D. 2016. Understanding the relationship between naval product complexity and on-board system survivability using network routing and design ensemble analysis. *Practical Design of Ships and Other Floating Structures (PRADS)*. pp. 219–225. September 2016. Copenhagen, Denmark.

Shields, C.P.F. Brefort, D.C. Parker, M.C. & Singer D.J. 2015. Adaptation of path influence methodology for network study of iteration in marine design. *IMDC*. May 2015. Tokyo, Japan

Simon, H.A. 1970. Style in design, 2nd Ann. Environmental Design Research Association Conf. (EDRA), Pittsburgh, pp.1–10.

Sims, P. 2003. Comparative Naval Architecture of Passenger Ships. *Trans. SNAME*. Vol.111. 233–256

Sowa, J.F. 1992. Semantic Networks. http://www.jfsowa.com/pubs/semnet.htm

UCL Mechanical Engineering homepage. 2018. http://mecheng.ucl.ac.uk/

Process-based analysis of arrangement aspects for configuration-driven ships

K. Droste, A.A. Kana & J.J. Hopman
Department of Maritime and Transport Technology, Delft University of Technology, Delft, The Netherlands

ABSTRACT: Configuration-driven ships distinct themselves from other ships as their ability to perform their function is strongly influenced by the layout of the vessel. Exploring the balance between the performance of the vessel and its functions is typically done during early stage ship design. However this is more complicated for configuration-driven ships as evaluating the performance also requires a layout. A previously developed concept exploration method using the TU Delft Packing approach is used, but requires extensions to enable concept exploration of configuration-driven ships. Because of the limited and uncertain design information during these early design stages and the complicated interrelations within the problem, a solution based on stochastic modelling is proposed. The paper concludes with a Poisson-based metric to evaluate the layout of configuration-driven ships.

1 INTRODUCTION

This paper presents work done on improving concept exploration studies to balance the performance and the required functions for configuration-driven ships, such as amphibious assault vessels, aircraft carriers, or passenger ships. The large influence of the configuration-driven ships' layout on their performance make it difficult to predict this relation up front. With the methods currently available during early stage design the influence of the layout is hard to evaluate adequately. On the one hand, methods which manually generate designs require a lot of time and labour to fully explore the concept space. On the other hand, more automated methods struggle with creating 'good' layouts while maintaining design diversity. These methods require creative input from their users and with their results often aim to support the creative process of the designer (Schneider et al. 2010). Especially while also accounting for the specific aspects affecting configuration-driven ships. This can also be seen in the concept exploration method used here, which is based on the TU Delft Packing approach.

The introduction will first elaborate about what a configuration-driven ship is and present the problem in a system engineering view. Section 2 will elaborate about the challenges faced with this type of problem. Then Section 3 presents a method used for concept exploration and elaborates on the required changes that are applicable to configuration-driven ships. The section will conclude with initial results of a evaluation metric based on a Poisson process. Lastly Section 4 will present the conclusions of this paper and Section 5 will briefly present future work.

1.1 *Configuration-driven ships*

Configuration-driven ships refer to a certain category of ships where the performance of the vessel is dominated by the arrangement of spaces. For these ships the design process is often initiated by positioning the major spaces directly supporting the primary functions of the design (Andrews et al. 2012). Although it can be argued that most ships are driven by the configuration of the spaces to some extent, certain type of ships can be distinguished. For these ships, the configuration of the internal spaces and the connections between those spaces is driven by the operational processes. Yet the effectiveness of the design is strongly influenced by the resulting configuration. This Typical examples of these ships are amphibious assault ships, aircraft carriers, or passenger ships. For these ships the location of the primary spaces such as, the landing dock and garage facilities, the hangar and flight deck, and the accommodation and public spaces, have a strong influence on the initial sizing of the vessel and the required budget. For example, in the case of an aircraft carrier, one of the main goals of the vessel can be defined as to be able to deploy aircraft. The ability of the vessel to do this is strongly influenced by the configuration of the flight deck, the hangar, and the ammunition stores. Besides their configuration, the manner in which doors, stairs, and elevators connect these spaces is even of greater influence. These connections by themselves consume a considerable amount of space and therefore influence the layout extensively. The need for these connections depends on the arrangement of spaces, yet the arrangement of spaces is strongly influenced by the connections. This complex interdependency makes this a special category of ship design problems.

1.2 *System engineering view to the problem*

In order to understand how the operational and functional requirements influence the performance and effectiveness of configuration-driven ships, it is necessary to study the configuration of spaces and systems aboard. If the relation between the operational and functional requirements, and the ship's performance are understood this can be used to improve the designs by balancing the requirements set up front with the required performances. Figure 1 shows a system engineering V-diagram where the upper left corner shows how the primary function and operational processes provide input for the requirements and these requirements have influence on the configuration and performance on the right side. The focus of this V-diagram is on how the primary function relates to the effectiveness of the design by looking into the operational processes and subsequent supportive functions.

For configuration-driven ships it is these functions, and the processes by which they are implemented that make a configuration-driven ship unique. The need for elevators, staircases, and corridors to facilitate those processes depends on both the relative location of the spaces, the absolute location aboard the vessel, and on the speed and method used to execute these processes, for instance an elevator or crane. The space required for these transportation systems is of significant influence for the layout of the ship. However, the location and quantity of these systems also dictates the level of performance of the functions for which they are required. Therefore for a configuration-driven ship the influence processes have on the requirements and thereby on the performances can only be evaluated using the configuration. These layouts are hard to generate because they not only depend on the processes and systems required but they also dictate partly which systems are required. Gaining a more thorough understanding of the layout options during the early stage ship design supports better design decisions throughout the whole design effort and contributes to a better design. Before focussing on the challenges with the concept exploration of configuration-driven ships, a overview of the early stage design phase is provided.

1.3 *Early stage ship design*

The ship design process can be divided in three phases: early stage ship design, contract design, and detail design. The focus of the latter two is on increasing the level of detail of a single design, initially with the goal of drafting a contract and eventually building the actual vessel. The early stage ship design phase deals with the initial steps of the ship design process. It is during these initial steps that a full understanding of the design problem needs to be acquired and a rough idea of a solution. It is also in this phase that the initial configuration of spaces is formed. In the subsequent phases there might be changes, but the major decisions in the layout are taken and changing those will have both significant financial and scheduling consequences. To this extent the early stage ship design process can be divided in three different sub-processes: concept exploration, concept studies, and concept design.

Concept exploration tries to explore the possible solutions for the design problem and is often referred to as requirements elucidation. Andrews (2011) states that this is the only time designers can be truly radical and divergent in their thinking as this is the phase in a design process new ideas are ought to be formed. Besides the creative aspect, one could argue that the main goal of concept exploration is to improve the understanding of the design problem and identify its challenges and possible solutions.

Concept studies use the outcome of the concept exploration and further develop some of the

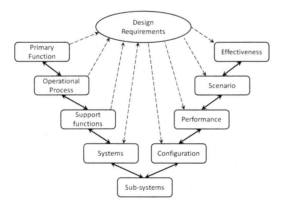

Figure 1. The systems engineering V-diagram shows how the primary function relate to the operational processes, support functions, and systems. These are integrated in the configuration and result in the performance and effectiveness of the design. Modified from van Oers (2011).

possible solutions and ideas to evaluate feasibility and to learn the characteristics of these solutions. Eventually concept design will make use of the new acquired knowledge of the design problem and possible solutions to create a concept design. van Oers et al. (2017) describes this process as executed by The Netherlands Defence Material Organisation (DMO) where they specifically distinguishes two phases, the concept exploration phase and a concept studies phase. The final concept design is selected from the designs studied in the concept studies phase. No matter how it is formulated, the early stage design process has three key elements:

1. Searching for a divers set of solutions in a concept exploration study.
2. Evaluating a set of more desirable solutions in higher detail.
3. Selecting/creating the concept design.

These three elements all serve the goal of gaining a thorough understanding of the design problem and formulating an initial solution in the form of a concept design. For this study the focus is on improving the concept exploration step in the early stage design process. The first step is to explore a broad range of possibilities before studying a selected number in more depth. Therefore the initial focus is on improving the concept exploration step of the early stage ship design process.

2 CHALLENGES

2.1 Configuration-driven ships

As mentioned before, the specific challenges in configuration-driven ship design originate from the fact that the relation between the functions and their performances is both influenced by the processes used to incorporate these functions and how these processes require additional systems and spaces in the layout. Especially the required connections such as staircases, elevators, or ramps enabling transportation between different decks is challenging. The requirements for these connections follow from the location of the systems and spaces in the ship while they themselves consume considerable space in the ship and therefore influence the layout.

2.2 Concept exploration

There are two main challenges faced during the concept exploration phase. The first one is with respect to the design requirements. Both the improved understanding of the solution and the external factors such as tighter budget constraints, new regulation entering into force, or newly available technology can cause changes in the original design requirements. These changes in design requirements often have difficult to predict consequences for the solutions space.

The second challenge can be found in the many operational processes possible to comply with the ship's primary functions, which in turn can be assigned to numerous combinations of systems and components. The systems engineering V-diagram in Figure 1 shows this relation between the primary function, operational processes, and eventually systems and components. Furthermore the figure shows how studying the potential solutions on different system levels contributes to the design requirements. There are numerous ways one can translate a primary function in multiple possible operational processes and there are multiple systems which can deliver those functionalities and support those processes (Duchateau 2016), (van Oers 2011), (Andrews 2003). This results in many different options for solutions at multiple levels of the system-of-interest and therefore tends to explode into a combinatorial problem.

2.3 Layout design for configuration-driven ships

For a configuration-driven design it has been stated that the layout is very important, yet designing the layout proves to be challenging. Three main challenges have been identified in the design of a layout.

1. The first challenge is related to the amount of rationale used to create a layout. This means that there are a lot of rationale that have influence on the layout varying from rationale about comfortability standards or safety regulations or even rationale related to logistical processes. Not all of these rationale are explicitly expressed and some maybe based on experience. DeNucci (2012) developed a method to capture and store those rationale to enable future usage. His study showed the diversity of designer rationale and the conflicts between them. This difference is further demonstrated on an organisational level by comparing rationale from US Coast Guard and DMO designers.
2. The second challenge is related to finding the proper balance between the individual, sometimes conflicting, requirements. This is especially challenging as this balance is not fixed for all problems or the entire design process but instead tends to change frequently (Shields et al. 2017). As a part of finding a balanced set, the relation between the requirements and the performance of the vessel becomes important. Specifically this relation is what is often studied during concept exploration as this relation can be complex.

3. The third challenge of particular relevance to configuration-driven ships is in defining the connections between individual spaces. As the connections depend on the locations of spaces and vice versa the location of spaces is influenced by the need for a connection to another space. In configuration-driven ships typically a lot of those connections are present as can be seen in Figure 2 with the example of a amphibious assault vessel. The figure shows both internal connections between spaces which are required to facilitate the movement of people and equipment inside, as well as the external connections to enable connecting to elements in the environment. There are two ways in which this is commonly approached. First the connecting elements are fixed in shape and location, and the other spaces are positioned around them after which the performance is evaluated. The other way is that the physical connections are generated based on the connections required using a topology defined up front. The evaluation of the layouts then focusses on evaluating the overall flow of goods and people through the physical connections or the rate of connectivity. These evaluations provide feedback on the defined topology. The second way of decomposes the problem uses a topology to capture which connections are desired and then solves the geometric part to validate the topology and evaluate the performance.

An overview of research into ship general arrangement methodologies is provided in Han et al. (2014). The overview shows that the layout of ships is a research problem approached from many different angles already. The more recent literature mentioned in the paper illustrate a trend to use the available computational power to increase information generation during the early design stages. One specific method doing that is the inteligent ship arrangement methodology (Nick 2008). Here the previously identified challenges regarding the large set of rationale and their conflicts are approached by the application of fuzzy logic evaluation of the constraints and objective function. Although this method is able to deal with large and conflicting setsof rationale, it is hard to relate the performance of these layouts to the primary functions of a configuration-driven ship. The large set of rationale required tends to converge to a single solution rather than more general insight into the relation.

The design building block approach (Andrews 2003) supports the generation of configuration-driven ships by initially focusing on the primary spaces involved in the ship's main functions and then iteratively adds more spaces to the model to progress the design in more detail. This way the rationale and their trade-offs are done manually and the tool focusses on analysing the design and informing the designer. This makes it hard to fully explore the design space as every solutions needs to be manually and sequentially generated.

Gillespie (2012) uses a network representation of the layout to study the relations between different spaces. Amongst the different networks analysed one looks into the passage ways where the relations are physical connections between spaces. Additional networks represent a fire fighting system where the relations are the plumbing of the system. This analysis proves very powerful as combining these individual network representations enables studies into the interactions between those individual systems, yet this method focusses on analysing the layouts instead of generating them.

Although the layout problem has been approached from multiple ways, creating a sensible configuration of spaces remains hard. Especially when combined with a method focused on automatically generating layouts to support concept exploration efforts. To approach this problem the usage of spaces is considered by taking into account the primary processes aboard the ship. The processes and functions provide the requirements for the interior layout of the ship, but they also can be used to evaluate the performance. Therefore the next section will look into processes and how they can be used for the generation and evaluation of layouts.

2.4 *Process analysis*

Given the system engineering view presented up front, the processes are the set of steps required to implement the primary functions on the left side in the system V-diagram of Figure 1. In order to be able to study the relation between the functions and the performance of the vessel a way to overcome the challenges presented by analysing the processes needs to be found. Within processes three main challenges have been identified.

1. Starting with the processes themselves, where there are many processes aboard a ship which are all required to make sure the ship is able to function. Both the number of processes, as well as the difference importance and in impact between the processes present a challenge.
2. The second challenge is the interaction between processes themselves and the layout. One process might depend on the result of another process to be able to start or multiple process might require the same space or system at the same time. Because of these interactions simply studying one process misses information and might lead to conflicts later on such as under-/overcapacity or congestions.

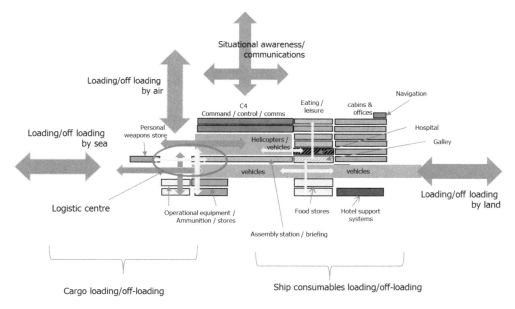

Figure 2. Primary spaces of an amphibious assault vessel with arrows decribing the main processes dicating the layout (Hopman 2013).

3. The third challenge is more of a practical nature. A full simulation of processes often requires extensive information regarding the layout, crew and processes, information which might not be available yet or uncertain at the start of the concept exploration as part of that information depends on the outcome of the concept exploration. Therefore applying advanced methodology such as multi-agent based methods during concept exploration is a mismatch in detail and will require more detail regarding the design.

In order to deal with the numerous amounts of processes and sub-processes, task-network theories as those presented in Nickerson (1995) and Baron et al. (1990) have been explored. Here the individual steps of the processes, called tasks, are used to model a structure of tasks. This structure can be used to organise the information regarding the individual tasks such as the required resources, conditions, and performance parameters as time to complete. Using task-networks the numerous processes can be ordered and structured.

In an effort to integrate human factors and personnel movement aspects into the preliminary design stage, Casarosa (2011) integrated a multi-agent personnel movement software package into the UCL's Design Building Block approach. In order to perform these analysis a lot of additional information is added to the concept design with respect to the connectivity and the layout. And also regarding the crew where an actual Watch and Station Bill is used. The need to add all this information makes this method more applicable to concept design and concept studies rather then concept exploration.

The application of eigenvalue and eigenvector analysis to the ship centeric Markov decision process (SC-MDP) enables an analysis of people movement during an emergency situation (Kana and Droste 2017). The application is especially suitable for preliminary design as it only requires a low detail general arrangement because of its network-based probabilistic approach, and very limited information concerning the crew composition. The used Markovian method enables a policy optimisation based on a reward function. The decisions are related to the rewards and the uncertainty. The probabilistic approach of the method predicts the likeliness of an individual actor moving through the network of spaces. Therefore it provides more insights into the evacuation potential of a layout than it does about the evacuation time of an individual. The interpretation of the results can be challenging as compared to a physics-based multi-agent method, as the outcome here is provided in possible routes and chances of occurring morethen in absolute numbers.

The above mentioned method uses Markov theory. From that same field originates queuing theory. Queuing theory knows many applications from computer networks (Gebali 2015) to waiting queues and scheduling in hospitals (Marynissen and Demeulemeester 2016, Helm and Van Oyen 2014). One particular interesting application of queuing theory is to factory production lines. The factory production lines can be modelled as a network of queues, with single machines representing a server in the network with a service time and a buffer. When the process time, down time due to failures or maintenance, or the chance rework are provided as stochastic distributions a more realistic model is obtained. Using these variabilities for the analysis of production lines, less obvious interrelations between different machines in the production line can be found (Hopp and Spearman 2011). This then allows the operational analysis of a physical system in a network representation. Similar analysis can be applied to the logistical network of a configuration-driven ship.

Both the Markov and the queuing example present a way of modelling the performance of a physical system using stochastic process algebra. A stochastic approach enables the analysis of a layout using processes as it addresses the challenges identified above. The use of stochastic distributions allow to cover wider variety of processes. Meanwhile the simultaneous analysis of multiple processes, interactions and interdependencies can be found. Thirdly using a stochastic method requires less data and detail compared to deterministic models. This enables application in the concept exploration phase. Brinksma et al. (2001) states that:

> The complexity of the situation, influences and conditions makes it impossible to describe such phenomena by deterministic models. Complicated interrelations, the lack of detailed information and some basic indeterminacy in the physical world make such processes to appear random. Nevertheless, measurements show that, although individual behaviours and different events are unpredictable, many statistical regularities can be observed and modelled by means of stochastic processes.

Therefore a solution using stochastic process algebra is proposed in this research as the stochastic process algebra allows the analysis of processes in a layout during the concept exploration phase. The analysis of these processes can be used to analyse the design and further understand which design operates more effective. Once it is possible to analyse the layouts the information gained can be used to improve the generation of the layouts. The improved analysis and generation of layouts enable designers to study layout features of configuration-driven ships using a concept exploration method.

3 CONCEPT EXPLORATION METHOD FOR CONFIGURATION-DRIVEN SHIPS

The previous section concluded that stochastic process algebra would enable concept exploration of layout features for configuration-driven ships. Therefore this section will explain the concept exploration method developed at the University of Technology Delft and highlight the additions required to allow the concept exploration of configuration-driven ships. In the second part of this section an idea of how a Poisson counting process can be used to analyse the logistical network of a layout is presented.

The interactive evolutionary concept exploration method is based on the TU Delft's Packing approach. The initial development of Packing focused on creating a tool which enabled the fast creation of a large set of varied designs (van Oers 2011). Being able to create such a set of designs, enabled concept explorations as one could quickly generate a diverse set of solutions for a specific design problem. This research was continued in the development of the interactive evolutionary concept exploration method by Duchateau (2016), which incorporated an interactive feedback element to the generation of designs by the packing tool. This allowed the designer to process results and adjust the exploration process based on the gained insight. Thus far the concept explorations done required certain spaces to have their global or relative position constrained in order to get a technical feasible design. For the ships studied until now this worked fine as they where not driven by the internal arrangement. However the quality of the internal arrangements was poor and for configuration-driven ships they are insufficient. Therefore current research focusses on improving the layout and thereby enabling new types of problems to be studied.

3.1 *Method structure*

The method addresses the concept exploration in seven steps and is visualised in Figure 3. The first three steps are preparing the input and creating the ship synthesis model. The fourth and fifth step focus on the generation of the ships and the data. Step six and seven use the obtained data either to create more insight and do additional calculations or to continue the design process. The seven steps will be further elaborated next:

1. The first step focusses on studying the current conditions and the operational needs to define the scope of the design problem and the type of solution preferred.

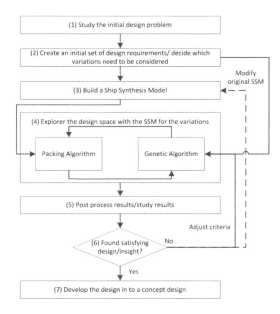

Figure 4. The positioning space (left) is used to position the blocks (middle) representing the spaces and systems in the ship synthesis model to create a design (right). Adopted from (Dyckhoff 1990).

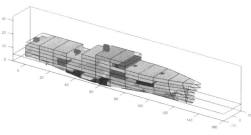

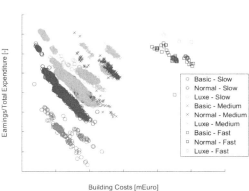

Figure 3. Concept exploration model, adopted from Duchateau (2016), initially presented in Droste (2016).

2. Step two defines an initial set of requirements and variations which will be studied in the concept exploration. For example for a cruise ship these variations can be the different luxury levels and operational scenarios.
3. In step three the variations and requirements are used to create the ship synthesis model. This model has a description of the set of spaces and systems required to study the variations. A more thorough description of this step can be found in (Droste 2016), (Duchateau 2016) & (Zandstra et al. 2015).
4. Step four starts the creation of designs using the Packing tool and the previously defined ship synthesis model. This tool uses a bin-packing algorithm to position the spaces and systems defined in the ship synthesis model into a positioning space, the ship's hull. A simple bin-packing example is provided in Figure 4. After positioning, various particulars and performances are determined such as lightweight, center of gravity, stability, resistance and powering, and costs. These performances can then be used to calculate objective values to pass on to the genetic algorithm, which uses the objective values to guide the search over the design space. The loop between the genetic algorithm and the packing algorithm continues for a pre-defined number of iterations. This results in a large set of designs.

Figure 5. Results from the cruise ship concept exploration study from (Droste 2016). On top a 3D visualisation of a single design, below a plot of some financial performances for the whole population of designs.

5. Step five post processes the set of designs generated in step four. As part of the post processing, for example, additional performances are calculated, visualisations are created, or the data is analysed using clustering algorithms (Jaspers 2017).
6. After the analysis step six evaluates whether sufficient insight is gained to decide on the best combination of variations defined up front or whether more searches with additional constraints are required or the original model needs to be modified.
7. If sufficient insight is gained step seven continues the development into a concept design.

The results generated in step four consist of a large set of roughly 20 000 ship designs with defined performance metrics. An example of these results is shown in Figure 5 where an individual 3D design is plotted and a plot of numerical data is provided. Each entry in this graph corresponds to an individual 3D design. Using this information the cost of different combinations of variations defined in step two can be analysed and specific solutions can be studied.

Although the data comes in great quantities, the quality of the design with respect to the layout is poor. When taking a closer look into the interior layout in Figure 6, it can be seen that there are various empty spaces, while other spaces cross watertight bulkheads. Furthermore the location of some of the spaces doesn't make too much sense, for instance the emergency control room in the middle of the accommodation on deck 6. This is amongst others caused by the fact that the only 'rationale' in the interior layout comes from constraints on the global location of systems or the relative locations towards other systems. These constraints are kept to a minimum to support variation in the resulting designs. Yet this also shows that for the concept exploration of configuration-driven ships these layouts are insufficient and improvements need to be made. To incorporate this the aim is on two specific areas of improvements, the generation of designs and the analysis of the designs. This will be achieved by using the stochastic process algebra in the concept exploration method to develop better analysis tools and enable process related input to improve layout generation. The first focus is on improving the analysis of the current layouts and establish a base line of the layout quality. Initial results of an analysis method are presented in the next section.

3.2 Stochastic algebra

To improve the analysis of the layouts for the concept exploration of a configuration-driven ship, a new metric is proposed based on a Poisson counting process (van Mieghem 2014). The chosen Poisson counting process can evaluate the design based on both the connectivity and the relative positions of spaces in a layout. The Poisson metric evaluates the layout in a probabilistic way enabling an analysis of the ability of the layout to support logistical processes. The mathematics of the Poisson counting process are presented in Equation 1. The equation can be used to calculate the probability of x occurrences during an interval $b - a$ when the average rate of occurrence is λ. In this research the Poisson process is used to evaluate the likely amount of activity on a certain connection in a ship's internal layout. The amount of activities on a connection is represented by x, the amount of movements on that connection. The interval $b - a$ is defined as the city-block distance between two spaces in the layout. What remains is the rate of occurrence λ, for now this has been defined based on the betweenness centrality of the connection between two spaces.

$$Pr(x) = \frac{(\lambda(b-a))^x * e^{-\lambda(b-a)}}{x!} \quad (1)$$

The betweenness centrality as defined in Equation 2 is the probability that a space is on the shortest path between any two random spaces in a layout (van Mieghem 2014). It therefore provides an average amount of activity of a space. In the equation v represents the node for which the betweenness is calculated, σ_{st} is the number of shortest paths between source s and target, t and $\sigma_{st}(v)$ is the number of shortest paths between s and

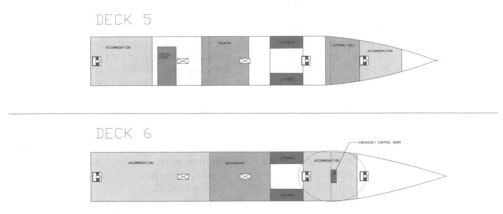

Figure 6. A general arrangement drawing based on a design generated with packing, but adopted for better visualisation (Droste 2016). The sub-optimal location of the emergency control room is highlighted in the drawing.

t visiting v. The N in the second term is the number of nodes in the network, this term normalizes the betweenness values to return values between zero and one.

$$g(v) = \sum_{s \neq v \neq t} \frac{\sigma_{st}(v)}{\sigma_{st}} * \frac{1}{(N-1)(N-2)} \qquad (2)$$

To compute this a network representation of the design is used as presented in Figure 7. The network is defined with nodes for every space and edges for the connections. The connections are defined in a way that every space is at least connected to one of the staircases present in the layout or an adjacent space, with exception of tanks and exhaust stacks. These staircases can be distinguished by the star-like clusters in the network. This originates from the static definition of the staircases in the original Packing model. The staircases are defined as a single space of five decks in height and because of that they only represent one node in the network. Using the betweenness centrality as the rate of occurrence in the metric couples the average activity to the probability of being visited. But the units of this formulation don't match yet, normally λ is defined per unit of time or length as the interval $b - a$ is a time or length. Yet the betweenness centrality is a probability with values on the domain [0–1] and unit-less. The definition of λ therefore is still active work which will be further discussed in Section 5.

3.3 Preliminary results

The betweenness centralities from Equation 2 and the network with distances are used in Equation 1 to calculate the results as presented in Figure 8. Each line represents a connection between two spaces in the layout. The X-axis shows the number of activities x occurring on that edge, while the Y-axis gives the corresponding probabilities. Two lines are coloured and marked, one starting high on the left in the graph (red, squares) and one with its maximum value in the middle of the graph (blue, circles). The red line shows a high probability that only one activity takes place on that edge, while the second, blue, line shows an edge where it is more likely that multiple activities take place on the edge. Another way to visualize the results is to take the maximum probability for each edge and the number of activities corresponding to it, and plot that on the network representation of the layout. Figure 9 shows these results for the probabilities and the activities individual and the product of the two.

The product between the maximum probability and the activity give a score which increases in value if an edge is more likely to experience more activity. This evaluation enables the analysis of the likeliness of a layout to be busy. The higher the score is, the more likely it is that

Figure 7. Network representation of the sideview of a ship, Roth et al. (2017).

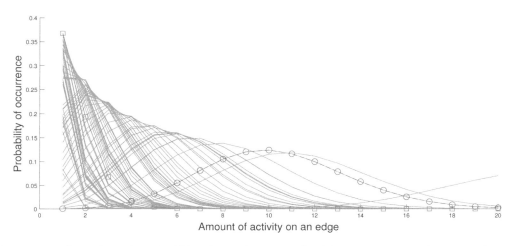

Figure 8. The figure shows the probabilities of a certain number of activities occurring on a given connection. The red, square marker line highlights a connection between a crew accommodation and the mudroom. The blue, circle marker line highlights a connection between one of the galleys and the second staircase. The blue line indicates a connection which is more likely to be busy, compared to the red line.

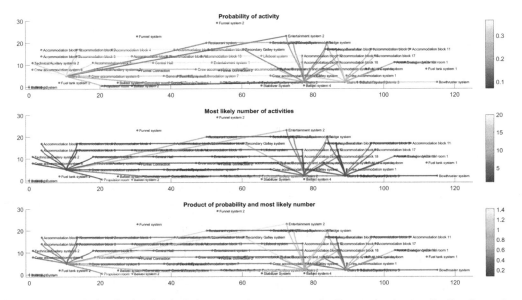

Figure 9. Network visualisation for the Poisson counting process results. From top to bottom the first figure shows the probability of the most-likely activity, the second figure shows the most-likely amount of activity, and the third figure provides the product of the former two.

multiple activities are occurring simultaneously on a connection in the layout and therefore a possible area to be concerned about. When the values of the individual edges in the bottom network of Figure 9 are summed together a score can be obtained to rate the complete layout. The probabilistic nature of the presented method limits its applicability to a relative comparison of the set of designs rather then absolute answers. Furthermore the level of detail of the layouts, for example how the staircases were modelled as described previously, limits the relevance of the results. The current results generally show a lot of likeliness of activity towards the staircases as no hallways are considered and the staircase is only a single object stretching multiple decks. Lastly the method only considers whether there is an activity or not and does not yet associate any flow or capacity to the activity.

4 CONCLUSIONS

This paper identified the challenges related to the evaluation of configuration-driven ships. The interdependencies resulting from the relation between spaces and the required connections make the design of configuration-driven ships difficult. The current tools available are not able to sufficiently deal with those interdependencies. Therefore this research proposes a method based on the operational processes of the ship. The initial results of a method were presented, demonstrating the ability to differentiate and evaluate the layouts based on a probabilistic approach. Although the method needs further development, the initial results presented look promising and it can be concluded that the use of stochastic modelling is a viable path forward.

5 FUTURE WORK

This paper presented the status of the research at the time of writing. Two extensions to the original concept exploration method were identified, one with respect to analysing the layouts and one regarding the generation of layouts. The paper provides preliminary results for a solution of one of them. Therefore the first thing is to continue the analysis presented using the Poisson metric. This involves both finishing the definition of the lambda variable and a further analysis of the results. After that, the focus shifts toward improving the generation of the layouts.

ACKNOWLEDGEMENTS

This research is funded by the Dutch Defence Materiel Organisation. The authors would also like to thank Ms. Kelly Cooper from the Office of Naval Research for her support.

REFERENCES

Andrews, D.J. (2003). Marine Design - Requirement Elucidation Rather Than Requirement Engineering. In *International Marine Design Conference 2003*.

Andrews, D.J. (2011). Marine Requirements Elucidation and the nature of preliminary ship design. *Transactions of the Royal Institution of Naval Architects Part A: International Journal of Maritime Engineering 153*(1).

Andrews, D.J., E.A.E. Duchateau, J. W. Gillespie, J. J. Hopman, R. Pawling, & D. J. Singer (2012). State of the Art Report Design for layout. In *International Marine Design Conference 2012*.

Baron, S., D.S. Kruser, & B.M. Huey (1990). *Quantitative Modeling of Human Performance in Complex, Dynamic Systems*. National Academy Press.

Brinksma, E., H. Hermanns, & J.-p. Katoen (2001). *Lectures on Formal Methods and Performance Analysis*. Berg and Dal, The Netherlands: Springer.

Casarosa, L. (2011). *The integration of human factors, operability and personnel movement simulation into the preliminary design of ships utilising the Design Building Block approach*. Ph. D. thesis, University College London.

DeNucci, T.W. (2012). *Capturing Design Improving conceptual ship design through the capture of design rationale*. Ph. D. thesis, Delft University of Technology.

Droste, K. (2016). *A new concept exploration method to support innovative cruise ship design*. Ph. D. thesis, Delft University of Technology.

Duchateau, E.A.E. (2016). *Interactive Evolutionary Concept Exploration in Preliminary Ship Design*. Phd thesis, Delft University of Technology.

Dyckhoff, H. (1990). A typology of cutting and packing problems. *European Journal of Operational Research 44*(2), 145–159.

Gebali, F. (2015). *Analysis of Computer Networks* (Second ed.). Springer.

Gillespie, J.W. (2012). *A Network Science Approach to Understanding and Generating Ship Arrangements in Early-Stage Design*. Ph. D. thesis, University of Michigan.

Han, D., H. Abdelkhalek, & M. Chen (2014). Design Methodologies of Ship General Arrangements-A Review. *Advanced Materials Research 936*, 2344–2351.

Helm, J.E. & M.P. Van Oyen (2014). Design and Optimization Methods for Elective Hospital Admissions. *Operations Research 62*(6), 1265–1282.

Hopman, J.J. (2013). Course on Naval Ship Design.

Hopp, W.J. & M.L. Spearman (2011). *Factory Physics* (3thd ed.). Waveland Press.

Jaspers, T. (2017). *Elucidating families of ship designs using clustering algorithms*. Master thesis, Delft University of Technology.

Kana, A.A. & K. Droste (2017). An early-stage design model for estimating ship evacuation patterns using the ship-centric Markov decision process. *Journal of Engineering for the Maritime Environment 0*(0), 1–12.

Marynissen, J. & E. Demeulemeester (2016). Literature review on integrated hospital scheduling problems. Technical report, KU Leuven, Leuven.

Nick, E.K. (2008). *Fuzzy Optimal Allocation and Arrangement of Spaces in Naval Surface Ship Design*. Ph. D. thesis, University of Michigan.

Nickerson, R.S. (Ed.) (1995). *Emerging Needs and Opportunities for Human Factors Research*.Washington, D.C: National Academy Press.

Roth, M.J., K. Droste, & A.A. Kana (2017). Analysis of General Arrangements Created by the TU Delft Packing Approach. In *Compit 2017*, Cardiff, pp. 201–211.

Schneider, S., J.-r. Fischer, & R. König (2010). Rethinking Automated Layout Design: Developing a Creative Evolutionary Design Method for the Layout Problems in Architecture and Urban Design. In J. S. Gero (Ed.), *Design Computing and Cognition*, Stuttgart, pp. 367–386. Springer.

Shields, C.P.F., J. Knight, & D.J. Singer (2017). THE DESIGN PROCESS AS A COMPLEX SYSTEM. In *ASNE: Technology, Systems & Ships*, Arlington, USA.

van Mieghem, P. (2014). *Performance Analysis of Complex Networks and Systems*. Cambridge: Cambridge University Press.

van Oers, B.J. (2011). *A Packing Approach for the Early Stage Design of Service Vessels*. Ph. D. thesis, Delft University of Technology.

van Oers, B.J., E.H. Takken, E.A.E. Duchateau, R.J. Zandstra, S. Cieraad, W.V.D.B.D. Bruijn, & M. Janssen (2017). Warship concept exploration and definition at The Netherlands Defence Materiel Organisation Introduction: The Netherlands. In *US Society of Naval Engineers: Set-based design workshop*.

Zandstra, R.J., E.A.E. Duchateau, & B. J. van Oers (2015). Concept Exploration for a Mine-countermeasures Vessel Using a Packing-based Ship Synthesis Model. In *World Maritime Technology Conference (WMTC)*, pp. 1–12.

A design space generation approach for advance design science techniques

Jason D. Strickland, Thomas E. Devine & Jonathan P. Holbert
Naval Surface Warfare Center Carderock Division, Maryland, USA

ABSTRACT: The focus of this paper is to introduce a novel approach to support ship conceptual design efforts. This approach leverages a baseline set of hull lines and systematically varies this hull to create a design space. This subsequent design space is evaluated for resistance and powering, stability, and seakeeping. This approach develops a robust design solution space. Evaluation of the space allows for the development of objective, quantifiable information in advance of a final hull form selection. This data directly supports multiple design science approaches such as Set-Based Design (SBD), Design Building Blocks (DBB), and Bin Packing Algorithms. These approaches to ship design provide solutions to design problems which are often subject to a high degree of uncertainty or late stage design changes. An attractive aspect of broad design space evaluation is the ability to concurrently develop design information at varying levels of fidelity using communication variables and set negotiations. This then allows the designer to assess and reduce the design space based on feasibility and dominance supported by objective evidence.

1 INTRODUCTION

The development of the design space to be analyzed was a multistep process that heavily leveraged the scripting of existing software packages, within an integrated team environment, augmented with various concurrent engineering heuristics. The scripted portion of this process consisted of four basic analytical steps that developed a portion of the database that was ultimately evaluated. This automatically generated data set was then appended with additional fidelity from traditional "hand-touch" operations and other assorted engineering applications. Section 2 DESIGN SPACE DEVELOPMENT of this document outlines the development, under lying assumptions, and presents some typical results. Section 3 APPLICATION of this document outlines the how the outline process can be applied to a larger design effort and coupled with other leading design approaches.

2 DESIGN SPACE DEVELOPMENT

The automated design space was developed in a three-step process. This process follows the subsection outline of this section. The initiation of this process was the development of a nominal "seed" point. A seed point or parent hull form is effectively the centroid of the excursion space. It is a converged design point that represents one possible solution to the design problem under evaluation. For this effort four seeds were identified. It is important to understand that these seeds need not be fully viable under the evaluation of perfect or more complete information. This is merely a process initiation point. This aspect will be highlighted in following sections. These seeds were selected in order to understand the implications of the inherited machinery architectures, and hull lines. All seeds were developed in accordance with the same baseline requirements.

2.1 Identification of baseline "seed" parameters

In order to initiate this analysis, the first order of business was to define a nominal stack up length for the seed. This stack up is the gross order of magnitude for the vessels overall length with consideration to the functional arrangement of the vessel's topside (Garner et al., 2015) or major machinery systems (McCoy, Smoot, Kuseian, Marshall, & Collins, 2013). Therefore, for a surface combatant a nominal stack up would proceed as follows from the stern towards the bow: Flight Deck → Hangar → Boat Bays → Deck House → Missile Compartment → Deck Gun → Anchor/Line Handling Area. This concept can be applied to any class of vessel, for example an OSV would likely resemble the following: Cargo Deck → Crane/Cargo Handling Space → Boat Bays → Deck House → Flight Deck. At this juncture, once a nominal stack up length has been determined one should revert to traditional heuristics of Length to Beam (L/B) and Beam to Draft (B/T) in order to develop a set of principal dimensions that are consistent with other ships of the appropriate class/type under evaluation. The dimensional ratios typically employed for these heuristics fall between 4–10 and 1.8–4 (SNAME, 1967). These are only an initial data point since the dimensions will be modified independently thus modifying these ratios. The approach below does not exclusively produce geometrically similar hull forms.

The principal dimensions and dimensional ratios of the four seeds evaluated for this effort are present in Table 1.

Since the stack up of these seeds were developed with a common set of requirements it is not surprising that the longitudinal length of the four seeds, are nearly identical. However, the variation of the hull shaping is evidenced by the fluctuation in the beam, depth, and displacement. Once nominal hull particulars were derived, a 3D surface model of each seed was developed in Rhino™ (Robert McNeel & Associates, 2018). Figure 1 – Figure 8 depict an isometric perspective and a profile elevation of each of seeds developed, respectively. The

Table 1. Design space initiation seed's principal dimensions and parametric data.

	Seed 1	Seed 2	Seed 3	Seed 4
L (m)	147	150	150	144
B (m)	16	19	18	20
D (m)	10	12	11	11
T (m)	6.0	7.2	6.6	6.8
L/B	9.2	7.8	8.3	7.3
B/T	2.7	2.7	2.7	2.9
Δ (MT)	5,400	6,700	6,500	6,500

Figure 1. Seed 1 isometric rendering.

Figure 2. Seed 1 profile rendering.

Figure 3. Seed 2 isometric rendering.

Figure 4. Seed 2 profile rendering.

Figure 5. Seed 3 isometric rendering.

Figure 6. Seed 3 profile rendering.

Figure 7. Seed 4 isometric rendering.

Figure 8. Seed 4 profile rendering.

color scheme represented in the following renderings is consistent for all following figures within this document. Therefore, in all following graphics Seed 1 will be represented with green data points.

These seeds were specifically chosen to evaluate the influence of the inherited machinery architecture and seakeeping of the relevant seeds. The machinery architectures are as follows: Seed 1 has two main machinery spaces with longitudinal separation, Seed 2 investigates a hybrid electric plant with adjacent main machinery spaces, Seed 3 has a single main machinery space, and Seed 4 has two adjacent main machinery spaces,. The baseline machinery architectures for each of the representative seeds is delineated in Table 2. The combination of the nominal hull forms and the baseline machinery architectures allows for the development of a converged design seed.

At this point, a nominal seed has been developed complete with main subdivisions allocations compliant with 3 compartment floodable length

Table 2. Initial machinery line up for the design space initiation seeds.

	Seed 1	Seed 2	Seed 3	Seed 4
Propulsion Train	Twin Shaft CODAD (2) 5 m CPP	Twin Shaft CODLAG (2) 5 m CPP	Single Shaft COGAG (1) 5 m CPP	Twin Shaft CODOG (2) 4 m CPP
Prime Movers	(4) MTU 20V 8000 Diesels	(2) 3.7 MW ELE Motors (1) GE LM 2500+ GT	(2) GE LM2500+G4 GT	(2) FM PA6B Diesels (2) GE LM2500+GT
Genset	(4) MTU 12V 4000 SSDGs	(4) MTU 20V 4000 SSDGs	(2) RR MT5 SSGTGs	(3) CAT 3608 SSDGs

Table 3. Seed parametric space range for principal dimensions of Length, Beam, and Depth. Draft is equal to 60% of total hull Depth. Displacement is calculated in salt water at the corresponding draft.

	Seed 1	Seed 2	Seed 3	Seed 4
L range (m)	118.0–177.0	120.0–180.0	120.0–180.0	115.0–170.0
B range (m)	13.0–19.0	15.0–23.0	14.0–22.0	15.0–24.0
D range (m)	8.0–12.0	9.6–14.4	8.8–13.2	9.0–13.6
T range (m)	4.8–7.2	5.8–8.6	5.3–7.9	5.4–8.2
Δ range (MT)	3,600–12,200	4,600–11,500	3,300–11,200	3,300–11,200

scenarios with an assumption of shell to shell flooding. Further, a nominal General Arrangement was developed in order to develop an understanding of the baseline seed with regards to available area/volume and weight centroids. This body of work was the entering argument to the Design Space Exploration Script (DSES). This script marks the debarkation point from traditional point design to Design Space Exploration.

2.2 Systematic design space exploration

The Design Space Exploration Script (DSES) was created using the Octave™ toolset (Arag, 2010). Its intended goal is to develop sets of data by systematically exploring the design space around initial seed points. The script has multiple functions which include the ability to perform hull form offset manipulation, assess intact and damage stability, generate a resistance estimate, match machinery plants, and calculate fuel requirements. Each of these functions are discussed in more detail below.

Various hull form characteristics are required as input, including offsets. The offsets are taken from the Rhino™ (Robert McNeel & Associates, 2018) surface models depicted in Figure 1 – Figure 8 and are input into the code as a set of points in a traditional coordinate system. Following import of the seed offsets, the script then performs linear scaling modifications on the parent hull form in one, two, and all three dimensions. A data set of morphed hull forms is then created. The baseline modification is (+/–) 20% in all three principal dimensions in 5% increments. This sampling scheme yields 729 unique hull forms for each seed. The range of principal dimensions developed with this perturbation scheme is outlined in Table 3 below.

2.3 Concurrent analysis

The aforementioned hulls were then directly passed to a Stability, Resistance and Powering, Seaway Motion, and Endurance Fuel analysis model for further evaluation. Again, the goal of this scripting was to develop a large data set that can be systematically perturbed and evaluated from a variety of disciplines in order to determine feasibility and dominance within the design space. The following subsections outline each of the topical areas. The approach applied and the results applicable to each of the discipline domains will be discussed.

2.3.1 Stability

An interface to the Ship Hull Characteristics Program (SHCP) was developed. SHCP is the US Navy's standard evaluation tool original developed in the 1960's for the evaluation of intact, damage and wind heel (M. Rosenblatt & Son, 1967). This program also produces stand curves of form and general hydrostatic data. Following the generation of the data set of modified hull forms, the offsets are then exported into a SHCP input file. The script then calls SHCP to perform intact and damage stability analysis. The analysis is run for each hull form at drafts ranging from 40% to 70% of depth. The results are then appended to the hull form database. In addition to the stability data, general hydrostatic values are recorded, which will be used as inputs for other steps in the analysis. Table 4 delineates the end of service life (EOSL) conditions with corresponding displacements, drafts, and vertical center of gravity to allowable vertical center of gravity ratio (KG/KGa). While the KGa is a direct function of the hull geometry, the individual KG for each hull was scaled from the seed designs using methodologies outlined by the Society of Allied Weight Engineers (Society of Allied Weight Engineers, 2001, 2007). It should be noted that not all of these conditions are ballast free. Seeds 1 and 3 both required ballast in the minimum operational (MINOP) condition in order to maintain intact stability. Seed 4 would require further analysis since it is at or near the stability limit.

Transverse metacentric height divided by the vessels beam (GMT/Beam) is a traditional heuristic for stability that has been employed for almost a century, if not longer. The first reference that could be located was from the 1920's (Hiatt, 1920). A nominal value of 5% has been used as a minimum for a multitude of craft types since that time (Faltinsen, 2006; Webster, Bates, Phillips, & Haag, 1920; Yun & Bliault, 2000). Figure 9 displays the design space developed with a hull prismatic parallelepiped versus the GMT/Beam ratio. It should be noted that not all of the hulls produced have a 5% GMT/Beam value. The hulls with a negative GMT to beam ratio would be unstable. The hulls between 0–5% have been identified as marginal and potentially too tender. Hulls above 20% have been identified as marginal and potentially too stiff. Subsequent analysis was isolated to a GMT/Beam range of 5–20%.

2.3.2 Resistance and powering

Hull resistance was approximated through the use of the Transom Stern regression (S.C. Fung & Leibman, 1995; Siu C. Fung, 1991). This calculation method was integrated in a similar method as in the Advanced Ship and Submarine Evaluation Tool (Kassel, Cooper, & Mackenna, 2010; Naval Sea Systems Command, 2012). This method provides a residuary resistance prediction for each hull form at a series of speeds. The frictional resistance

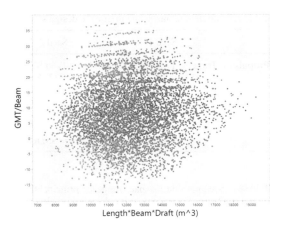

Figure 9. Composite surrogate stability analysis for the developed design space.

Table 5. Powering configurations considered during this design space evaluation.

Plant configuration	Auxiliary engine	Main engine
COGAG-2	LM 2500 +	LM 2500 +
COGAG-1	RR MT 30	RR MT 30
CODAG-2	FM PA6B STC	LM 2500 +
CODAG-1	MTU 20V 8000	RR MT 30
CODAD	MTU 20V 8000	MTU 20V 8000
CODOG-2	FM PA6B STC	LM 2500 +
CODOG-1	MTU 20V 8000	RR MT 30
COGOG	RR MT 7	RR MT 30
CODLAG	3.7 MW Elec Mtr	LM 2500 +

coefficient is predicted using the ITTC-1957 frictional correlation line (Morrall, 1970). Appendage drag was estimated as a ratio of total resistance based on barehull and appended model test data of ships with similar appendage arrangements. Air drag was calculated per DDS 051 (Naval Sea Systems Command, 1984), with the windage area scaled off of the seed design. Using these data sets, the total resistance and required effective power are calculated for the entirety of the design space. The required power is the amount of power necessary to drive the hull at a given speed. Once the 2,916 hulls were paired with plausible powering solutions, outlined in Table 5, the design space blossomed to 37,250 design points. This larger space is displayed in Figure 10.

Figure 10 depicts the prismatic volume which can be correlated to a nominal full load displacement of the converged design point on the axis of abscissas, with the corresponding EHP at the design speed as the ordinate. The resultant trend

Table 4. End of Service Life (EOSL) particulars and stability heuristic for design space seeds in the Minimum Operating (MINOP) loading condition.

	Seed 1	Seed 2	Seed 3	Seed 4
ΔEOSL (MT)	6,300	6,500	7,000	7,000
Draft (m)	5.6	5.1	5.9	5.8
KG/KGa	1.0	0.94	1.0	1.0

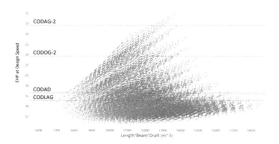

Figure 10. Expanded design space of hull and power option permutations.

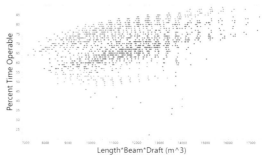

Figure 11. Design Space percentage time operable for transit and patrol mission in the Gulf of Alaska as a function of prismatic volumen.

is largely intuitive. The larger the vessel is with respect to displacement the more power is required to propel the vessel at a given speed. Additionally, for analysis purposes, iso lines of developed power have been added to the figure. These lines represent the powering configurations detailed in Table 5 above.

2.3.3 Seaway motion analysis

Seaway motion analysis was conducted with the Ship's Motion Program (SMP) (Meyers, Applebee, & Baitis, 1981). This program was utilized to generate roll amplitudes, pitch amplitudes, vertical accelerations, lateral accelerations, deck wetness instances, and hull slamming instances in a seaway conforming to Bretschneider spectrum (Bretschneider, 1959) for sea state 5 at the most probable modal period and the maximum significant wave height. Each hull machinery combination was evaluated with in the designated seaway. While these results were evaluated to the NATO STANAG 4154 (North Atlantic Treaty Organization, 2000) requirements for several missions, a desire to have a singular measure of performance for seakeeping necessitated the use of percentage time operable (PTO). The PTOs are calculated by applying mission criteria to a vessel response with an equal probability of occurrence for each speed, heading, and sea state combination. Figure 11 graphically displays the PTO for the Transit and Patrol Mission in the Gulf of Alaska. While this is a notably harsh environment, it can be seen that the majority of the design points fall between 60 and 80 percent operable.

These are the same hull-machinery combinations that are plotted in the preceding figures. One can begin to see the potential of multi domain analysis of a common data set. This type of analysis is essential in informing the variable negotiations of Set Based Design (Singer, Doerry, & Buckley, 2009; Singer, Strickland, Doerry, McKenney, & Whitcomb, 2017; Sobek II, Ward, & Liker, 1999), and is directly attributable to system level defini-

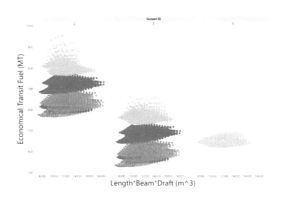

Figure 12. Design space of endurance fuel loading as influenced by machinery selection.

tion that is required to realize a Design Building Block (Andrews, 2006, 2012; Andrews, McDonald, & Pawling, 2010) or Bin Packing (Depetro & Hoey, 2013; Duchateau, 2016; van Oers, 2011) approaches.

2.3.4 Fuel loading calculation

An estimate for the economical transit fuel loading was calculated per DDS 200-1 (Naval Sea Systems Command, 2011). Power demand and electrical loading was estimated from previous efforts, however the effect of variations in the electrical plant was investigated. Figure 12 illustrates the effect of various electrical generator variants. This three-part graphic depicts the influence of generator selection. Once again, the x-axis is the prismatic hull volume, however, in this instance the vertical access corresponds to the economical transit fuel load required for an endurance range. The first column uses two Rolls Royce MT5 (Rolls-Royce, n.d.) aero-derivative engines as prime movers in a generator configuration combined with the main propulsion engines outline in Table 5. The cloud centroid of the first column is approximately 650 MT. The second column is the same data set with four MTU 12v4000 (Rolls-Royce, 2018) diesel engines, with a corresponding centroid of approximately 400 MT. The final column utilizes four MTU 20v4000 diesel engines (Rolls-Royce, 2018). This column is centered at approximately 450 MT due the common electrical loading referred to earlier. However, this plant line-up was developed for a combined diesel electric and gas turbine (CODLAG) arrangement. This machinery architecture had less commonality with the other seeds resulting in a sparse design space depicted.

3 APPLICATION

The developed approach creates a design space that is analyzed from a traditional hull centric perspective. Intact and Damage Stability, Resistance and Powering, Seakeeping and Seaway Motions, and Endurance Fuel Requirements are all direct functions of the hull geometry. This systematic approach is a departure from the traditional design spiral of ship design because no selection has been made at this point. The design space was developed to understand trends and interdependencies inherit to this convoluted problem of ship design.

This approach leverages the philosophy of Set Based Design of decisional deferral. While it is true that an initiation point was selected, one can effectively argue that all design endeavors: "Begin with the end in mind" (Covey, 1990). While the exact realization of what that end may be is undefined the general product type is known at the onset of the activity.

Further while it is not the focus of this paper, structural section requirements, electrical power loading, heating ventilation and cooling, general arrangements, and detailed weight estimation occurred concurrently with the approach documented herein. These efforts were examined in detail for the seed location and scaled accordingly or discretely calculated for the points within the cloud.

This is where the Design Building Block or Bin Packing approaches have an application point. Since these both focus on the development of functional blocks and understanding the internal compartmentation of the vessel via an inside-out design practice. This would consume the bulk of the effort outlined in the preceding paragraph and directly support the development of weight centroid data. One can see how this outside-in approach could benefit from trade space development or automation of the internal components and compartments of a vessel.

Further since the DSES utilized spiral development, the volume and weight of the auxiliary systems continued to improve as additional information became available. It is rather easy to envision an instance that as the weight centroids continued to evolve that a feasible hull becomes unfeasible from a stability perspective. Much of this could be mitigated with the employment of inside-out approaches such as DBB or Bin-Packing approaches. Another such learning point could be volumetric constraints. As system detail is developed and refined a hull becomes volumetrically deficient to support the desired subsystem arrangement. Again, this is where an approach that focuses on the detailed development of sub compartment would lend tremendous dividends. Under the traditional design spiral approach, it would be necessary at this point to start over with a new set of principal dimensions.

However, the value of this approach is proven in this exact circumstance. Since a multitude of hulls have been evaluated for critical parameters, if a stability or volumetric constraint is active and drives the preferred hull into an unfeasible space. There are 728 other hulls of that exact format and 2,915 hulls of a compatible format that are already at the same level of analysis available to default towards.

Joining this scripting approach with another process that focuses on the development of higher quality auxiliary system design and automation of weight accounting would be a powerful step for naval architectural conceptual design endeavors.

4 CONCLUSION

Under this effort, a surface vessel design space was developed and analyzed from multiple domains in a robust fashion. This process consisted of four basic analytical steps that developed a portion of the database that was ultimately evaluated. 2,916 individual hull forms were evaluated from a seakeeping, stability, seaway motions, and endurance fuel perspective. This automatically generated data set was then appended with additional fidelity from traditional "hand-touch" operations and other assorted engineering applications.

Insights developed during the execution of this initial tasking were extremely helpful for scoping the global size and performance of a notional vessel. This effort was helpful for narrowing the trade space and facilitating preliminary design efforts. Deficiencies that were uncovered during the execution of this tasking included several things that would support interdisciplinary, collaborative activities. Some of these deficiencies would include the ability to perform set negotiation, and the preemptive development data structures.

There is also a need for the development of sufficient understanding of sub-blocks and system level details. This understanding can be supported by DBB and Bin-packing algorithms, since they focus on this level of detail.

Further work could be performed on the development of structural systems, overall scoring metrics, and visualizations. While these activities have been handled outside of the initial synthesis loop, one can easily see the benefit of their inclusion. Finally, since this is a conceptual design tool, some assessment of risk would be greatly beneficial. This is currently accomplished via expert opinion. A means to objectively score the design risk, associated with the entirety of the space would potentially yield intriguing insights and should be evaluated further.

REFERENCES

Andrews, D.J. (2006). Simulation and the design building block approach in the design of ships and other complex systems. *Proceedings of the Royal Society A: Mathematical, Physical and Engineering Science*, *462*(2075), 3407 LP–3433. Retrieved from http://rspa.royalsocietypublishing.org/content/462/2075/3407.abstract.

Andrews, D.J. (2012). Art and science in the design of physically large and complex systems. *Proceedings of the Royal Society A: Mathematical, Physical and Engineering Sciences*, *468*(2139), 891–912. https://doi.org/10.1098/rspa.2011.0590.

Andrews, D.J., McDonald, T.P., & Pawling, R.G. (2010). Combining the Design Building Block and Library Based Approaches to improve Exploration during Initial Design. In V. Bertram (Ed.), *9th International Conference on Computer and IT Application in the Maritime Industries COMPIT* (pp. 290–303). Gubbio, Italy. Retrieved from http://www.ssi.tu-harburg.de/cgi-bin/webseiten/compit/privat_0/index_0.pl.

Arag, F.J. (2010). Gnu Octave. Retrieved from https://www.gnu.org/software/octave/.

Bretschneider, C.L. (1959). Wave Variability and Wave Spectra for Wind-generated Gravity Waves. *United States Beach Erosion Board, Technical Memorandum*, (118), 192.

Covey, S. (1990). *The Seven Habits of Highly Effective People, Powerful Lessons in Personal Change. Free Press (Simon and Schuster)*. …. Turtleback Books. Retrieved from http://scholar.google.com/scholar?hl=en&btnG=Search&q=intitle:Seven+Habits+of+Highly+Effective+People:+Powerful+Lessons+in+Personal+Change#0

Depetro, A., & Hoey, R. (2013). Rapid Generation and Optimisation of Ship Compartment Configuration based on Life Cycle Cost and Operational Effectiveness. *Iceaaonline.Org*. Retrieved from https://www.iceaaonline.org/awards/papers/2013_Methods_Models-2_paper.pdf

Duchateau, E. (2016). *Interactive evolutionary concept exploration in preliminary ship design.* TU Delft. https://doi.org/10.4233/uuid:27ff1635-2626-4958-bcdb-8aee282865c8

Faltinsen, O.M. (2006). *Hydrodynamics of High-Speed Marine Vehicles.* New York: Cambridge Unversity Press.

Fung, S.C. (1991). Resistance and Powering Prediction for Transom Stern Hull Forms During Early Stage Ship Design. *Transactions of the Society of Naval Architecture and Marine Engineers (SNAME)*, *99*, 29–84.

Fung, S.C., & Leibman, L. (1995). Revised Speed-Dependent Powering Predictions for High-Speed Transom Stern Hull Forms. In *Fast Sea Transportation* (pp. 151–164). Retrieved from http://scholar.google.com/scholar?hl=en&btnG=Search&q=intitle:Revised+Speed-Dependent+Powering+Predictions+for+High+Speed+Transom+Stern+hull+Forms#0.

Garner, M., Doerry, N., Mackenna, A., Pearce, F., Bassler, C., Hannapel, S., & McCauley, P. (2015). Concept Exploration Methods for the Small Surface Combatant. In *ASNE Day 2015 Proceedings*. American Society of Naval Engineers. Retrieved from http://www.navalengineers.org/Resources/Product-Info/productcd/AD2015.

Hiatt, F.M. (1920). Relation of Beam to Height of Metacenter. *International Marine Engineering*, *25*(1), 569–572.

Kassel, B., Cooper, S., & Mackenna, A. (2010). Rebuilding the NAVSEA Early Stage Ship Design Environment. In *Naval Surface Warfare Center Carderock Division.*

M. Rosenblatt & Son. (1967). Damge Stability System for the Ship Hull Characteristic Program. New Yor.

McCoy, T.J., Smoot, M., Kuseian, J., Marshall, H., & Collins, M. (2013). *Naval Power Systems Technology Development Roadmap.* Retrieved from http://www.navsea.navy.mil/Media/Naval Power Systems Technology Development Roadmap – Distribution A – 14 May 2013 – Final.pdf%5Cnpapers3://publication/uuid/F7DC9133-8312-4CC7-8DD1-C4EC17652F3B.

Meyers, W.G., Applebee, T.R., & Baitis, A.E. (1981). *User's manual for the standard ship motion program, smp* (Vol. 392234). Bethesda.

Morrall, A. (1970). 1957 ITTC Model-ship Correlation Line Values of Frictional Resistance Coefficient. In *Ship Report* (p. 16). National Physical Laboratory, Ship Division.

Naval Sea Systems Command. (1984). Prediction of Smooth-Water Powering Performance for Surface-Displacement Ships. DDS 051-1.

Naval Sea Systems Command. (2011). Calculation of Surface Ship Endurance Fuel Requirements. DDS 200–1 Rev. 1, (October), 21.

Naval Sea Systems Command. (2012). Ship design manager (sdm) and systems integration manager (sim) manual. Washington.

North Atlantic Treaty Organization. (2000). *Common procedures for seakeeping in the ship design process: stanag 4154.* North Atlantic Treaty Organization.

Robert McNeel & Associates. (2018). Rhincerous. Retrieved from https://www.rhino3d.com/.

Rolls-Royce. (n.d.). Gas turbines. Retrieved from https://www.rolls-royce.com/~/media/Files/R/Rolls-Royce/documents/customers/marine/gas-turbines.pdf.

Rolls-Royce. (2018). General Purpose Diesel Engines / 4000. Retrieved from http://www.mtu-online.com/mtu-northamerica/old-mountpoints/diesel-engines-overview/general-purpose-diesel-engines/4000/.

Singer, D.J., Doerry, N., & Buckley, M.E. (2009). What Is Set-Based Design? *Naval Engineers Journal, 121*, 31–43. https://doi.org/10.1111/j.1559-3584.2009.00226.x.

Singer, D.J., Strickland, J.D., Doerry, N., McKenney, T.A., & Whitcomb, C. (2017). *Sname T&R 7-12; Set-Based Design*.

Sname. (1967). *Principles of naval architecture (revised)*. (J.P. Comstock, Ed.). New York: Socity of Naval Architects.

Sobek II, D.K., Ward, A.C., & Liker, J.K. (1999). Toyota's Principles of Set-Based Concurrent Engineering. *Sloan Management Review*, 67–83.

Society of Allied Weight Engineers. (2001). *Weight estimating and margin manual for marine vehicles; rp14*. Los Angeles. Retrieved from https://www.sawe.org/technical/rp/sawe_rp-14_2001.

Society of Allied Weight Engineers. (2007). *Marine Vehicle Weight Engineering*. (D. Cimino, Ed.). Los Angeles: Society of Allied Weight Engineers.

van Oers, B.J. (2011). *A Packing Approach for the Early Stage Design of Service Vessels*. TU Delft. Retrieved from http://www.vssd.nl/hlf.

Webster, F.B., Bates, J.L., Phillips, S.M., & Haag, A.H. (1920). *Shipbuilding Cyclopedia*. (F.B. Webster, J.L. Bates, S.M. Phillips, & A.H. Haag, Eds.). New York: Simmons-Boardman. https://doi.org/10.1016/0003–6870(73)90259–7.

Yun, L., & Bliault, A. (2000). *Theory and Design of Air Cushion Craft*. Oxford: Elsevier.

An optimization framework for design space reduction in early-stage design under uncertainty

L.R. Claus & M.D. Collette
University of Michigan, Ann Arbor, Michigan, USA

ABSTRACT: Early stage design is marked by a low level of design definition, leading to high levels of uncertainty around design decisions made at this stage. Many techniques have been proposed to make better decisions given this uncertainty, including robust optimisation approaches and reliability-based design optimisation. Drawing inspiration from set-based design, this work presents a different approach. Instead of making a final decision with a margin for uncertainty, the procedure allows for a gradual reduction of the design space in a manner that maximises the designer's remaining flexibility. Two measures are first defined—the complexity of the remaining design space, and the regret, or potential loss of performance resulting from deciding at that time. The procedure solves for the Pareto front in the trade space between complexity and regret. To generate the Pareto front, the method uses two optimisers with one nested inside of the other; both the inner and outer optimisation problems are solved using a genetic algorithm. The outer optimiser is multi-objective with the complexity, or size, of the reduced design space and the resulting regret as the two objective functions. This method was used to solve a structural design problem, and the results are presented here.

1 INTRODUCTION

Decisions made in early-stage design have a significant impact on the quality of the final design. Unfortunately, these decisions must be made when the design is still loosely defined. This results in uninformed decisions in the early design stages of large design projects. These are often due to uncertain parameters where the final value is not known yet. In the exploratory initial design stage, there could still be requirements with unknown values, and parts of the design that will most likely change but are not under the designer's control. There are existing methods that can find an optimal solution while accounting for this uncertainty, but typically they are computationally expensive or provide a suboptimal solution for the majority of possible realisations of the uncertainty. Some of these existing approaches, such as robust optimisation and reliability based design optimisation, tackle this problem by adding a design margin which sacrifices the performance of the final design. Set-based design is a recently proposed method but computationally efficient tools to implement set-based design are currently lacking.

Robust optimisation is characterised by the goal of finding an optimal solution such that the feasibility of the design is minimally affected by parameter uncertainty (Bertsimas, et al., 2011). Many different objective functions have been suggested in the literature such as minimising the worst-case regret and minimising the worst case objective function, known as min-max robustness (Ehrgott, et al., 2014). Another approach is to formulate the problem as a multi-objective optimisation problem with the mean and standard deviation of the objective function as the two objectives (Wang & Shan, 2004). One of the major limitations of robust optimisation is that the solutions are often suboptimal for many of the possible realised values of the uncertain parameters (Ehrgott, et al., 2014). Another limitation of robust optimisation is the lack of established methods for multi-objective problems (Ehrgott, et al., 2014).

Reliability-based design optimisation (RBDO) is another approach that can optimise problems that have uncertain parameters. RBDO methods approach uncertainty by optimising an objective function and constraining the reliability of the problem. This is done by using a reliability assessment to calculate the probability that constraints will be satisfied under uncertainty (Wang Shan, 2007, Yao, et al., 2011). Traditionally, RBDO methods require a known distribution for uncertain parameters, but there has been work into alternative methods that also include interval uncertainty (Du, 2012, Huang, et al., 2017). The solution from an RBDO analysis has been shown to be more reliable than the solution from deterministic optimisation (Maglaras, et al., 1996). As with robust optimisation, a significant drawback of RBDO is that the method is designed to give a single solution, or solution set, that is feasible, but suboptimal, for the interval.

An existing method that informs the designer with the results of a trade study between performance and uncertainty is presented by Liu (2016). The method uses a multi-objective optimisation framework with the objectives being interval uncertainty and worst-case design performance. To efficiently solve the framework a new variable fidelity optimisation framework was developed; to improve the accuracy of the surrogate model multiple local Kriging surrogate models were built and updated online in the optimiser (Liu & Collette, 2014).

Another existing approach for early-stage design under uncertainty is set-based design (SBD). Unlike the previously mentioned methods, which are point based design methods, SBD is a convergent design method (Mckenney, 2013). In SBD solutions are found by eliminating infeasible and highly dominated solutions from an initial broad set of design values (Mckenney, 2013). This process allows design decisions to be delayed until later in the design process when there is less uncertainty. Fundamentally, SBD and traditional optimisation methods approach the problem in two different ways; SBD finds the best solutions by eliminating the inferior ones, and traditional optimisation methods iteratively check points for optimality.

A new method for early-stage design under uncertainty is presented here. While the proposed method is not a tool for SBD, it has been inspired by SBD and the idea of waiting for adequate information to make design decisions. Both RBDO and robust optimization find a single solution with the current uncertainty values. The proposed algorithm instead assumes that uncertainty will reduce as the design progresses. The algorithm seeks to give the designer a trade space between fixing design variables now and the freedom to adapt to new information in the future that reduces the problem uncertainty. The goal of the presented method is to explore the trade-space between the size of a reduced design space, and the loss in optimality that will result due to the design space reduction.

2 METHODOLOGY

The aim of this method is to find a reduced design space for the controllable parameters that will minimise the regret of the solution, a measure of sub-optimality, for any value the uncontrollable parameter may take. To accomplish this, problems are formulated to have two sets of design parameters, the controllable and uncontrollable parameters. This decomposition allows for the parameters to be handled differently and more appropriately. The output of the algorithm is the trade-space between the size of a reduced design space, here called complexity, and a measure of fitness compared to the global optimum, here called regret.

2.1 *Design under uncertainty*

As previously stated, the design variables are separated into two groups, the controllable and uncontrollable parameters, and each type is treated differently. The controllable design parameters, $x_i \in X$, are the design variables which the designer is selecting values for and are handled much like the design variables in a traditional optimisation method. The uncontrollable design parameters, $a_i \in A$, are the uncertain parameters; these parameters are modelled with interval uncertainty, as interval uncertainty is a valid method to handle both variability and ignorance (Ferson & Ginzburg, 1996). In an interval uncertainty model only an upper and lower bound on a value is known, and there is no information regarding distributions within the range.

2.2 *Algorithm structure*

The algorithm structure is comprised of two optimisers with one nested in the other. The inner optimiser is contained within the objective function evaluation for the outer optimiser. Both optimisation problems can be solved using a genetic algorithm (GA); the outer optimiser is multi-objective, and the inner optimiser is single-objective. The objective function of the inner optimiser is the fitness of an individual design much like a traditional optimisation problem; the objective functions for the outer optimiser are regret and complexity.

The chromosomes of each individual of the outer optimiser specify a sub-space, X_{range}, that is used as the variable limits for the inner optimiser; the inner optimiser is part of the objective function evaluation of the outer optimiser. The inner optimiser is used to find the optimal solution for each possible realisation of the uncertain parameter within the specified subspace. This list of solutions is then used to calculate the regret and complexity objective function values of the outer optimiser. Figure 1 depicts the interaction between the inner and outer optimiser.

As stated earlier, the objective function of the inner optimiser is the fitness of the design which is problem dependent. The outer optimiser's two objective functions are regret and complexity. The regret objective function, F_R, is a measure of the distance between the individual's fitness value and the optimal fitness value, here called the regret.

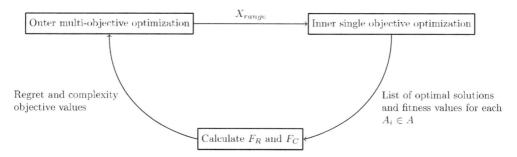

Figure 1. Interaction between outer multi-objective optimiser and inner single objective optimiser.

The average difference between the global optimal fitness value $f(\tilde{x}_t)$ and the optimal fitness value in the reduced space $f(x_i)$ for each $i: a_i \in A$ is computed using Equation 1. Since the objective is assumed to be minimised, the numerator will always be non-negative. The denominator, n, is the number of points in the A-space that have been evaluated.

$$F_R(\bar{x}) = \sum_{i=1}^{n} \frac{f(\bar{x}^i) - f(\tilde{x}^i)}{n} \quad (1)$$

The complexity objective value is a measure of the size of the reduced design space and an indicator to the resulting computational cost associated with size. In this application, the size of the reduced design space is the product of the number of possible values for each parameter that are used in the solution set $\bar{x}$. The complexity can be calculated using Equation 2. Figure 2 shows a visual representation of the complexity calculation.

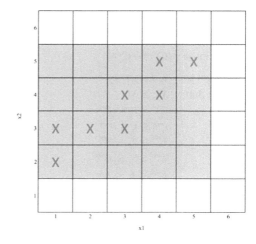

Figure 2. A simple visual representation of the complexity objective function calculation. The red X's are the optimal designs that are used in a solution to cover the uncertainty and the rectangle of the shaded area is the complexity value..

$$F_C(\bar{x}) = \prod_{j=1}^{n} s_j$$
$$s_j = \sum_{i=1}^{k_j} \begin{cases} 1 & \text{if } X_j^i \text{ is in } \bar{x}_j \\ 0 & \text{otherwise} \end{cases} \quad (2)$$

It should be noted that this is calculated for the solutions given back by the inner optimiser, not the subspace given to the inner optimiser as it is possible for the former to be smaller, or less complex than the later.

The outer optimiser has as many variables as there are controllable design parameters, x_i, for the problem. Each of the discrete variables of the outer optimiser corresponds to the range of a design parameter with a predefined list of parameter ranges.

The inner optimiser also has as many variables as there are controllable design parameters, x_i, for the problem. The optimiser solves for the optimal combination for these variables and calculates the fitness of the optimal solution in the given subspace. The optimal solution and its fitness value are then used to calculate the regret and complexity objective functions of the outer optimiser.

3 EXAMPLES

3.1 Box girder

The design of a simple t-stiffened box girder resembling a ship hull girder was used to validate the approach described above. The design problem was to select the scantling—plate thickness, stiffener spacing, and stiffener size—while not yet knowing the bending moment the box girder must resist.

The 3 m × 3 m box girder made of 6061 aluminium is symmetrical about both the x- and y-axes. The length, beam, and depth of the box girder were fixed to 5 m, 3 m, and 3 m respectively. A sample cross section is shown in Figure 3. The design of the box girder is described by six controllable design variables, the plate thickness, number of stiffeners, and stiffener size for the vertical and horizontal plates. This simple problem had a single uncontrollable design parameter along with the six controllable design parameters. The uncontrollable design parameter is the required bending moment of the box girder, and there are eight possible realised values for the requirement. The discrete possible values for all problem parameters are shown in Table 1.

The problem is as follows:

$$\text{minimise: } C_p^{norm} + C_w^{norm} \quad (3)$$

with respect to: $t_h, t_v, s_h, s_v, b_h, b_v$

such that: $M_z \geq M_{req}$

$t_h^{min} \leq t_h \leq t_h^{max}$

$t_v^{min} \leq t_v \leq t_v^{max}$

$s_h^{min} \leq s_h \leq s_h^{max}$

$s_v^{min} \leq s_v \leq s_v^{max}$

$b_h^{min} \leq b_h \leq b_h^{max}$

$b_v^{min} \leq b_v \leq b_v^{max}$

3.1.1 Box girder calculations

The design of the box girder is constrained by the required bending moment. The maximum bending moment of the design is calculated by Equation 4 where σ_{min} is the ultimate compression strength of

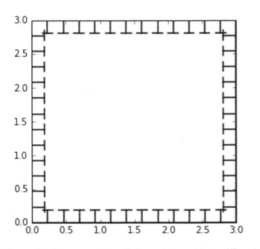

Figure 3. Cross section of box girder with t-stiffened panels.

Table 1. Variables for box girder example.

Variable		Possible Values
Required bending moment (TNm)	A	1, 5, 10, 15, 20, 30, 40, 50
Horizontal/vertical plate thickness (mm)	t_h, t_v	3, 4, 5, 6, 7, 8, 9, 10
Number of stiffeners on horizontal/vertical plates	s_h, s_v	8, 9, 10, 11, 12, 13, 14, 15
Stiffener size on horizontal/vertical plates	b_h, b_v	3 × 8.625, 4 × 11.5, 5 × 17.5, 6 × 25, 7.5 × 25, 9 × 35, 10 × 37.5, 12 × 45

the weakest panel. If a design is infeasible, in that the required bending moment is not met, a constraint violation is added to the objective function of the weight and cost. This constraint violation is given by the absolute difference between the box girder maximum bending moment and the required bending moment for ultimate compression strength calculation (Paik & Duran, 2004).

$$M_z = \frac{\sigma_{min} I_{NA}}{y} \quad (4)$$

The production cost, C_p, of the box girder is calculated with the method originally presented in (Rahman & Caldwell, 2012) and adapted in (Liu, 2016) and (Temple, 2015) using Equation 5. This cost calculation method includes six cost components to include the material and production costs. The six cost components, $C_{i,j}$ are:

$C_{i,1}$ = Materials cost for plate i
$C_{i,2}$ = Material cost for longitudinal stiffeners for plate i
$C_{i,3}$ = Welding cost for longitudinal stiffeners for plate i
$C_{i,4}$ = Preparation cost of brackets and joints for plate i
$C_{i,5}$ = Electricity cost and electrodes for plate i
$C_{i,6}$ = Fabrication cost of longitudinal stiffeners for plate i

$$C_p = \sum_{i=0}^{n} \sum_{j=0}^{6} C_{i,j} \quad (5)$$

The cross-sectional weight of the box girder, C_w, is simply calculated as the sum of all cross-sectional areas for each plate and stiffener. Each box girder design is evaluated for production cost, cross-sectional area, and ultimate strength fitness. The fitness of each design is the sum of the normalised production cost and cross-sectional

Table 2. Possible ranges given to inner optimiser for box girder example. The chromosome for each individual contained three label indicators, one for each range.

Range label	Range for t_h and t_v (mm)	Range for s_h and s_v	Range for b_h and b_v (T depth)
0	3–6	8–11	3–6
1	7–10	12–15	7.5–12
2	5–8	10–13	5–9
3	3–4	8–9	3–4
4	5–6	10–11	5–6
5	7–8	12–13	7.5–9
6	9–10	14–15	10–12
7	3–10	8–15	3–12
8	3	8	3
9	4	9	4
10	5	10	5
11	6	11	6
12	7	12	7.5
13	8	13	9
14	9	14	10
15	10	15	12

Table 3. Grouping of uncertain parameters to exploit the ordered nature of the uncertain parameter.

Grouping label	Grouping of points in A	Solution used for all points
0	$[A_1 - A_8]$	A_8
1	$[A_1 - A_4]$	A_4
	$[A_5 - A_8]$	A_8
2	$[A_1, A_2]$	A_2
	$[A_3, A_4]$	A_4
	$[A_5, A_6]$	A_6
	$[A_7, A_8]$	A_8
3	$[A_1]$	A_1
	$[A_2]$	A_2
	$[A_3]$	A_3
	$[A_4]$	A_4
	$[A_5]$	A_5
	$[A_6]$	A_6
	$[A_7]$	A_7
	$[A_8]$	A_8

Table 4. Run parameters for regret-complexity optimiser.

$|P|$	N_G	p_c	p_m
60	75	0.8	0.001

area objectives and the heavily weighted constraint violation determined by the bending moment.

The uncertain design variable in this design example has an ordered nature, and this was exploited as another way to reduce the number of designs in the design space. The possible required bending moments were numbered such that A_1 was the smallest and A_8 was the largest; this makes it that if a design is feasible for A_i, that same design will also be feasible for A_j if $j \leq i$. Given this, an additional grouping of solutions was employed as a variable of the outer optimiser. For example, with grouping label 1 the optimal solution would be found for A_4 and A_8 and the optimal solution for A_4 would be used as the solution for A_1, A_2, and A_3 as well. There were four possible groupings shown in Table 3.

3.1.2 Results

The box girder design problem was solved using a MOGA based on the NSGA-II sorting algorithm for the outer optimiser and a brute force table search for the inner optimiser (Deb, 2002). A brute force table search was utilised for the inner optimiser as the limited range of discrete variables kept the table to a tractable size and computational effort to a minimum. The run parameters for the outer MOGA are shown below in Table 4.

The Pareto front from the final generation of the outer optimiser is shown in Figure 4. The regret axis has been normalised and depicts the percentage of regret compared to the optimal objective values. The complexity objective value will always be an integer, which results in a seemingly sparse Pareto front. The rightmost point of the Pareto front, with a complexity of one, is the robust solution where one design is selected that is feasible for all possible realisations of the uncertain parameter. The leftmost point of the Pareto front, with a complexity of 100 and regret of zero, is the solution where eight distinct solutions are kept, one for each possible realisation of the uncertain parameter. The other points along the Pareto front correspond to solutions that have reduced the design space but have excluded some of the optimal solutions for each realisation of the uncertain parameter thus having some regret. The shape of the Pareto front informs designers where reductions of the design space will create a significant or negligible increase in regret. The Pareto front is steepest from the points with a complexity of 100–60 and 40–24; at these points, the design space is reduced by 40% and the average regret increases by only 0.06538%, a negligible amount for the significant reduction to the design space. The Pareto front is shallower as the complexity is smaller and closer to 1; this informs the designers that reducing the design space past a complexity of 4 will result in significant regret from even a small complexity reduction.

3.2 Cantilever tube

The design of a cantilever tube from Du (2007) is examined next. The objective of the problem is to design the tube such that the weight is minimised while keeping a sufficient reliability index when the angle of two applied forces is unknown. The geometry with the applied forces is shown in Figure 5. The controllable design variables are the average thickness and diameter, μ_t and μ_d, of the tube and the uncertain parameters are the angles the forces act on, θ_1 and θ_2.

The design problem is formally defined by:

$$\text{minimise: Area} = \frac{\pi}{4}\left(d^2 - (d-2t)^2\right) \quad (6)$$

with respect to: μ_t, μ_d
such that: $\beta \geq 3$
$t_{min} \leq \mu_t \leq t_{max}$
$d_{min} \leq \mu_d \leq d_{max}$

The possible discrete values for each parameter in X and A are presented in Table 5. The random variables and their parameters are shown in Table 6.

3.2.1 Cantilever tube calculations

Established methods in classical structural mechanics, shown in equation 7, were used to calculate the stress in the tube (Du, 2007).

$$\sigma_{max} = \sqrt{\{\sigma_x^2 + 3\tau_{xy}^2\}} \quad (7)$$

$$\sigma_x = \frac{P + F_{1sin}\theta_1 + F_{2sin}\theta_2}{A} + \frac{Mc}{I}$$

$$M = F_1 L_1 \cos\theta_1 + F_2 L_2 \cos\theta_2$$

$$A = \frac{\pi}{4}\left[d^2 - (d-2t)^2\right]$$

$$c = \frac{d}{2}$$

$$I = \frac{\pi}{64}\left[d^4 - (d-2t)^4\right]$$

$$\tau_{xy} = \frac{Td}{4I}$$

All reliability simulations to calculate β were calculated using the PyRe (Python Reliability) module (Hackl, 2013) and the maximum stress σ_{max}. The possible ranges given to the inner optimizer are shown in Table 7.

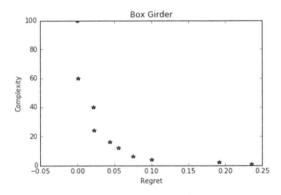

Figure 4. Pareto front for the box girder example.

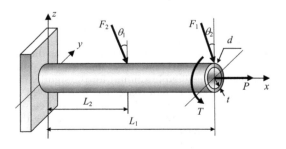

Figure 5. Geometry of cantilever tube from (Du, 2007).

Table 5. Possible variable values for cantilever tube example.

Variable	Possible values
θ_1	0, 1, 2, 3, 4, 5, 6, 7, 8, 9, 10
θ_2	5, 6, 7, 8, 9, 10, 11, 12, 13, 14, 15
μ_t	3, 4, 5, 6
μ_d	38, 39, 40, 41, 42, 43, 44

Table 6. Random variables for the cantilever tube example's reliability calculation.

Variables	Parameter 1*	Parameter 2*	Distribution
t (mm)	μ_t	0.1	Normal
d (mm)	μ_d	0.5	Normal
L_1 (mm)	119.75	120.25	Uniform
L_2 (mm)	59.75	60.25	Uniform
F_1 (kN)	3.0	0.3	Normal
F_2 (kN)	3.0	0.3	Normal
P (kN)	12.0	1.2	Gumbel
T (Nm)	90.0	9.0	Normal
S_y (Nm)	220.0	22.0	Normal

*For uniform distributions parameter 1 is the lower bound and parameter 2 is the upper bound. For all other distributions parameter 1 is the mean and parameter is the standard deviation.

3.2.2 Cantilever tube results

The small size of this problem, of only 64 individuals, suits it to be solved by enumerating all individuals instead of using an evolutionary algorithm to find the Pareto front. Many of these individuals did not produce feasible designs and therefore are severely penalised. Similarly to the previous example, the resulting Pareto front has limited points due to the integer complexity objective value and the small design space. The Pareto front is shown in blue in Figure 6 with the points in and near the front.

The steep Pareto front means that the designer can significantly reduce the design space without a significant increase in regret. Specifically, the designer can reduce the design space by 80% with very little regret. The endpoint of the Pareto front with a complexity of 20 and regret of 0 is a solution that includes the optimal design for all 121 possible combinations of θ_1 and θ_2; this solution has a design space that has been reduced by 28% but does not eliminate any optimal solutions for the possible realisations of the uncertain terms. The other endpoint of the Pareto front with a complexity of 1 is the robust solution with a single solution that is feasible for all possible realisations of the uncertain parameter.

4 CONCLUSION

A novel method has been developed for design optimisation in early stage design with uncertainty. The method uses nested optimisers to analyse design space reductions for the resulting regret and complexity, which are metrics for sub-optimality and the size of the design space. Since the regret and complexity of a reduced design space are competing objectives the method gives the Pareto front in the trade space of the two objectives.

Two examples have been presented for proof of concept of the method. These two examples were the design of a box girder with a larger design space and a single uncertain parameter, and the design of a cantilever tube with a smaller design space and multiple uncertain parameters. Both of these examples showed that this method is a valid approach to evaluate design space reduction options while minimising complexity and regret for small problems. Given this information, designers will be able to make more informed design space reduction decisions in early stage design.

With further exploration of the presented method, it was found that the problem becomes intractable for larger problems than those presented. The nested structure of the algorithm results in the computational effort exponentially increasing with an increasing design space. Future work is needed to restructure the algorithm for computational efficiency allowing larger problems to be solved.

Current methods for uncertainty in early stage design reduce either the complexity or the regret and in turn have a large, computationally expensive, design space or a sub-optimal final design. By minimising complexity and regret simultaneously, the presented method is able to explore the trade space between these two existing design approaches.

Table 7. Possible ranges given to inner optimiser for cantilever tube example. The chromosome for each individual contained two label indicators, one for each range.

Range label	Range for μ_t (mm)	Range for μ_d (mm)
0	3	38
1	4	39
2	5	40
3	6	41
4	3–4	42
5	5–6	43
6	4–5	44
7	3–6	38–44

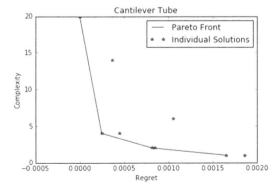

Figure 6. Pareto front for cantilever tube example.

REFERENCES

Bertsimas, D., Brown, D., & Caramanis, C. (2011). Theory and Applications of Robust Optimization. *SIAM Review*, 464–501.

Deb, K. (2002). A Fast and Elitist Multiobjective Genetic Algorithm: NSGA-II. *IEEE Transactions on Evolutionary Computation*, 6(2), 182–197.

Du, X. (2007). Interval Reliability Analysis. *33rd Design Automation Conference, Parts A and B*, 6(3), 1103–1109.

Du, X. (2012). Reliability-based design optimization with dependent interval variables. *International Journal for Numerical Methods in Engineering*, 1885–1891.

Ehrgott, M., Ide, J., & Schöbel, A. (2014). Minmax robustness for multi-objective optimization problems. *European Journal of Operational Research*, 17–31.

Ferson, S., & Ginzburg, L.R. (1996). Different methods are needed to propagate ignorance and variability. *Reliability Engineering and System Safety, 54*, 133–144.

Hackl, J. (2013). *PyRe 5.0.2: Structural Reliability Analysis with Python*. Retrieved from https://github.com/hackl/pyre

Huang, Z.L., Jiang, C., Zhou, Y.S., Zheng, J., & Long, X.Y. (2017). Reliability-based design optimization for problems with interval distribution parameters. *Structural and Multidisciplinary Optimization, 55*(2), 513–528.

Liu, Y. (2016). Surrogate-assisted unified optimization framework for investigating marine structural design under information uncertainty. *PhD Thesis*.

Liu, Y., & Collette, M. (2014). Improving surrogate-assisted variable fidelity multi-objective optimization using a clustering algorithm. *Applied Soft Computing, 24*, 482–493.

Maglaras, G., Ponslet, E., Haftka, R., Nikolaidis, E., Sensharma, P., & Cudney, H. (1996). Analytical and experimental comparison of probabilistic and deterministic optimization. *AIAA Journal, 34*(7).

Mckenney, T.A. (2013). An Early-Stage Set-Based Design Reduction Decision Support Framework Utilizing Design Space Mapping and a Graph Theoretic Markov Decision Process Formulation. *PhD Thesis*.

Paik, J.K., & Duran, A. (2004). Ultimate Strength of Aluminum Plates and Stiffened Panels for Marine Applications. *Marine Technology, 41*(3), 108–121.

Rahman, M.K., & Caldwell, J.B. (2012). Ship structures: improvement by rational design optimisation. *international shipbuilding progress, 42*(10), 61–102.

Temple, D. (2015). A Multi-Objective Collaborative Optimization Framework to Understand Trade-offs Between Naval Lifetime Costs Considering Production, Operation, and Maintenance. *PhD Thesis*.

Wang, G.G., & Shan, S. (2004). Design Space Reduction for Multi-Objective Optimization and Robust Design Optimization Problems. *SAE Transactions, Journal of Materials and Manufacturing*, 101–110.

Wang, G.G., & Shan, S. (2007). Wang, G. Gary Shan, S. *Journal of Mechanical Design, 129*(4), 370.

Design for resilience: Using latent capabilities to handle disruptions facing marine systems

S.S. Pettersen, B.E. Asbjørnslett & S.O. Erikstad
Department of Marine Technology, Norwegian University of Science and Technology, Trondheim, Norway

P.O. Brett
Ulstein International, Ulsteinvik, Norway
Department of Marine Technology, Norwegian University of Science and Technology, Trondheim, Norway

ABSTRACT: This paper explores how the resilience of marine systems against disturbances can be improved by considering latent capabilities. With resilience is meant the ability of a system to recover and return to an acceptable, stable state of operations after a disruption. Latent capabilities are distinct from capabilities intentionally designed for, and often remain unrecognized even during the operational phases. Our proposition is that these capabilities can be uncovered after designing, and be used in the operational phase to restore system operation after disruption, or to answer to emergencies in the marine environment. Drawing on fundamental theories in design, we illustrate how the function-form mapping can be adapted in response to these needs. Examples from marine transportation and marine service providers will be given in support of our arguments.

1 INTRODUCTION

Ships and other engineered ocean structures are cost-intensive systems often operating in a highly volatile environment. Marine transportation is a critical part of global supply chains, whose disruptions constitute substantial expenses to a multitude of stakeholders (Berle, 2012). Marine operations are similarly critical. Contractors in the offshore industry plan complex operations months in advance, relying on specialized, typically unique vessels determined, for example by tenders developed by the oil companies. If a vessel experiences a failure rendering it temporarily inoperable, the operation may therefore be delayed, resulting in large increases in overall project cost. Maritime safety has gathered loads of attention, due to the disastrous consequences of accidents to people and the environment. For example, the accidents of the Titanic, the Herald of Free Enterprise, the Estonia, and the Exxon Valdez significantly shaped regulation and consequently changed ship design.

For these reasons, the maritime industry have been one of the prime drivers of reliability and safety engineering. In fact, the word "risk" itself is derived from a Greek navigation term referring to the need to avoid difficulty at sea, later regarded as the risk of losing ships in renaissance Italy (Heckmann, Comes, & Nickel, 2015). The modern offshore industry has taken the drive for safe and reliable operations to a new extreme, with the unintentional consequence that the industry became less robust to changes in market conditions (Garcia, Brandt, & Brett, 2016). As good times turn to bad times, many stakeholders realize that the drive for safety and reliability may have come at the expense of effectiveness. This is not to say that we should accept unacceptable risks, rather we should realize that no amount of redundancy would remove every threat. We will only achieve robustness against disturbances we have been able to model (Woods, 2015). Instead of spending resources attempting to remove such risks, our approach favors resilience. Hence, we need to answer how we should respond once the hazard materializes: How can we recover to a stable level of operations after a disruption?

This paper proposes that we consider latent capabilities to address this issue. Latent capabilities imply that the ship or its subsystems can exhibit behaviors that can be used for functions that the ship or the subsystem was not intended for at the time it was designed. Latent capabilities can therefore prove to be useful for addressing failure modes that evolve in the ship or in the super-system (or system-of-systems, i.e. fleet) that the ship is a part of. In a recent paper, Pettersen, Erikstad, & Asbjørnslett (2017) further elaborate upon the concept of latent capabilities for engineering systems.

The remainder of the paper will be structured as follows: We review literature on resilience, to get a grasp on what that term really means. Second, we review the engineering design literature and its links to naval architecture. We then introduce latent capabilities with a review of similar concepts in the systems literature and functional modelling. Four illustrative cases from actual ship operations are then introduced to show that latent capabilities have applicability in the real world, and that accounting for such capabilities have value. Finally, we discuss what is needed for such approaches to work, with specific emphasis on the constraints to the usefulness of latent functions.

2 DEFINING RESILIENCE

Resilience is an increasingly popular concept within decision-making disciplines ranging from supply chain management, to systems engineering, safety science and emergency response. Due to its use in a variety of research fields, it has taken on a number of different meanings, leading to considerable semantic confusion regarding its exact meaning (Woods, 2015). Regardless of the specific definition chosen, resilience is an opposite of vulnerability, and different from robustness. Vulnerability is defined as the characteristics of a system that limits its abilities to endure threats and survive accidental events (Asbjørnslett & Rausand, 1999). Robustness is the ability of a system to resist and retain its stable situation through an event, while resilience implies that the system should have the ability to restore its functioning to a sufficient level of performance after the event.

The performance level profile through a disruption is shown for resilient versus robust systems in Figure 1. Whereas the performance of the robust system stays the same through the disruption, the resilient system is allowed to be disrupted, returning to a new stable state after some time. This figure does not consider the additional investment required to make the system robust against this specific event, nor does it specify whether this additional investment protects against other events. A similar figure is also presented by Boulougouris & Papanikolaou (2013), who consider the vulnerability and recoverability of naval combatants from enemy attacks. They distinguish primary performance degradation from secondary performance degradations beyond the primary loss of functionality. Their recoverability concept is limited to the implementation of damage controls, and hence does not cover efforts to repair the ship. Their concepts of vulnerability and recoverability provide, in sum, very similar results as what we refer to as resilience. Additional central definitions of resilience are given in Table 1.

Resilience is one among many desirable system lifecycle properties that we should strive for in design and operation (de Weck, Roos, & Magee, 2011). Resilience evolved as an "-ility" within supply chain management in the aftermath of a number of high-profiled detrimental events in the late 1990's and early 2000's. Common examples include the 1995 earthquake that destroyed the Port of Kobe, the terrorist attacks on 9/11 which lead to a shutdown of air traffic and spurred new security regimes at ports, and the 2000 fire at a New Mexico plant producing parts for both Nokia and Ericsson (Sheffi, 2005). These events made it apparent that consequences propagate in unforeseeable manners throughout integrated structures and supply chains, leading to the perspective that striving for resilience may lead to a considerable competitive advantage (Sheffi, 2005). Supply chain risk management has been a particularly fruitful field for the topic of resilience, first lead by initiatives at Cranfield University (Christopher & Peck, 2004) and Massachusetts Institute of Technology (Rice Jr. & Caniato, 2003).

Resilience engineering has emerged as a new paradigm for safety management (Hollnagel, 2010). Resilience engineering builds on previous experiences in safety management, by arguing that safety may be achieved by making more things go right, rather than making sure that fewer things go wrong (Hollnagel, 2010). It constitutes a shift of focus from considering failure in systems as a breakdown or malfunction, to seeing it as the inability of the organization or system to adapt to, and cope with current conditions (Madni & Jackson, 2009). The definition given by Dekker, Hollnagel, Woods, & Cook (2008), provided in Table 1 above, relates resilience to adjustment of functioning as a response to change or disturbance. This is a perspective that can easily be understood in terms of function-form mapping in engineering design: As a failure mode is observed, another component in the system can take over the function previously performed by the failed component, given that the component has the ability to perform that function, possibly without the intent of the designers.

Within systems engineering, resilience is seen as closely related to survivability (Castet & Saleh

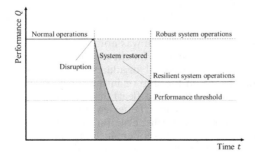

Figure 1. Comparing resilient and robust system performance through a disruption.

Table 1. Some definitions of resilience.

Reference	Resilience definitions
(Holling, 1973)	"… a measure of the persistence of systems and of their *ability to absorb change and disturbance* and *still maintain the same relationships* between populations and state variables."
(Wildavsky, 1988)	"… the capacity to cope with unanticipated dangers after they have become manifest, learning to bounce back."
(Foster, 1993)	"… the ability to accommodate change without catastrophic failure, or the capacity to absorb shocks gracefully."
(Asbjørnslett & Rausand, 1999)	"… a system's ability to return to 'a new stable situation after an accidental event'."
(Rice Jr. & Caniato, 2003)	"… an organization's ability to react to an unexpected disruption, such as one caused by a terrorist attack or a natural disaster, and restore normal operations."
(Christopher & Peck, 2004)	"… the ability of a system to return to its original state or move to a new, more desirable state after being disrupted."
(UN/ISDR, 2005)	"The capacity of a system, community or society potentially exposed to hazards to adapt, by resisting or changing in order to reach and maintain an acceptable level of functioning and structure."
(Dekker, Hollnagel, Woods, & Cook, 2008)	"A resilient system is able effectively to adjust its functioning prior to, during, or following changes and disturbances, so that it can continue to perform as required after a disruption or a major mishap, and in the presence of continuous stresses."
(Richards, 2009)	"… the ability of a system to recover from disturbance-induced value losses within a permitted recovery time."
(Chalupnik, Wynn, & Clarkson, 2013)	"… the ability of a system, as built/designed, to do its basic job or jobs not originally included in the definition of the system's requirements in uncertain or changing environments."

2012), especially in a context where the aim is to design military systems (Richards, 2009). Castet & Saleh (2012) see survivability as a measure of the loss of performance due to a disruption, seeing resilience as a superset of survivability and recoverability. (Richards, 2009) on the other hand relates survivability and resilience by seeing resilience enhancement as one of several design strategies for creating survivable systems. Resilience enhancement as a design strategy aims to maximize "the recovery of value-delivery within a permitted recovery time" (Richards, 2009).

A variety of design principles for resilience are proposed by Jackson & Ferris (2013) who propose to relate resilience to four attributes; capacity—the ability to survive threats, flexibility—the ability to adapt to threats, tolerance—the ability to degrade gracefully, and cohesion—the ability to act as a unified whole in the face of threats. Based on these four dimensions they propose fourteen, partially opposing design principles, like designing for functional and physical redundancies, and complexity reduction. Erden et al. (2008) argue that functional redundancies are different from the redundancies we normally find in reliability engineering, which are based on addition of redundant components to increase reliability. Functional redundancies instead work by covering the function with a component with a different primary function. Due to the increase in cost, and complexity implied by physical redundancy (adding redundant components),

Figure 2. The resilience trilemma. It is impossible to address all three dimensions without compromising one of them.

heuristics for systems design commonly favor use of functional redundancies (Jackson & Ferris, 2013). Functional redundancies are either included by intent, or result from system evolution over time, as in the case of fleet renewal (Uday & Marais, 2015).

For the purpose of this study, we take resilience to mean the ability of systems to recover from disruptions, regaining a stable, acceptable state of operational performance within a permissible timeframe, at an acceptable cost. A reasonable limitation of resilience, is then to not include human activities like repair of the system (Boulougouris & Papanikolaou, 2013), as repair activities will require that operations are shut down. This corresponds well with most attempts at quantifying the resilience of systems (Ayyub, 2015; Hosseini, Barker, & Ramirez-Marquez, 2016; Zobel, 2011). The result may be resilience as a trilemma, in which focusing extensively on cost will have a negative impact on the recovery time, or focusing extensively on minimizing disruption time will lead to a lower than desired post-disruption performance. The resilience trilemma is shown in Figure 2.

3 TOWARDS RESILIENCE IN DESIGN

3.1 Engineering design theory and ship design

Engineering design sees the task of designing as the process of creating artifacts meeting a stated need. The process becomes one of mapping statements about what the artifact should achieve, to some physical representation of the artifact, hence mapping between function and form. The product of the design process is hence a representation of the system which, in its manufactured form, will be able to perform a set of functions through conversion of material, energy or signals (Pahl & Beitz, 1996).

Pahl & Beitz (1996) describe four steps in design processes; clarification of task, conceptual design, embodiment design, and detail design. Task clarification includes information gathering about requirements and constraints to be met. Conceptual design includes derivation of functional structures, and search for solution principles for these, defining overall physical forms. Embodiment design and detail design further specifies the form of the system, to a level of detail needed for production of the system. It is evident from practical experience, and theoretic insights that the conceptual design phase is an iterative process. Evans (1959) took the functional requirements as given for the design spiral, and derived detailed descriptions of ship form on this basis. Asimow (1962) see the design process as spiraling between "analysis", "synthesis", and "evaluation", to increasing levels of detail. Analysis refers to the processes of seeking knowledge, the domain of natural sciences and traditional engineering analysis. Synthesis refers to the process of producing new representations of systems, based on the knowledge about the function to form relationships. Comparing system performance with required or intended performance, is in itself a process of analysis in which we derive new knowledge by testing the system representation. Later uses of the design spiral in marine design have more strongly emphasized synthesis, as exemplified by engineering economics (Buxton, 1972) and creative ship design (Andrews, 1981).

Suh (1990) introduces two axioms for design: i) keep functional requirements independent by ensuring a one-to-one mapping between function and form, and ii) keep information content (hence complexity) minimal. Further, he describes the iterative analysis-synthesis-evaluation process as zig-zagging between the functional and physical spaces: For example, if the overall mission statement is to "provide transport", the functional requirement on the next level in the functional hierarchy will be very different depending on the overall concept chosen. A transport ship will have very different functional requirements than a cargo airplane. While some recent ship design practice has gone in the direction of "solution-independent" functional specifications believed to open the solution space, these practices have been inefficient in dealing with the structural and behavioral complexity of ships, and iteration remains important (Andrews, 2011).

Equation (1) describes the common mapping between function and form, which underlies common theories of engineering design. FR refers to functional requirements, while DP refers to design parameters under the control of the designer. In accordance with common practice in decision sciences, DP could more accurately be described as design variables, as they describe the physical form resulting from the design process. a_{11} and a_{22} refers to elements of a design matrix A, whose form in the case below indicates an uncoupled design corresponding with the independence axiom of Suh (1990), as each FR corresponds exactly to one DP.

$$\begin{Bmatrix} FR_1 \\ FR_2 \end{Bmatrix} = \begin{bmatrix} a_{11} & 0 \\ 0 & a_{22} \end{bmatrix} \begin{Bmatrix} DP_1 \\ DP_2 \end{Bmatrix} \qquad (1)$$

According to axiomatic design, designs corresponding to this function-form mapping are preferable. Suh (1990) claims that coupled design is the reason for many failures, as this makes it difficult to adjust one design parameter, without affecting multiple functional requirements. Using this insight, one should attempt to make coupled designs decoupled, so that adjustment of design parameters can take place, as long as some ordered sequence is followed. While the design axioms seem more difficult to achieve in complex systems like ships than in product design settings, they provide some guidance as to what should be seen as a better design.

Modern approaches to ship design partially build on the generic design methodologies outlined above. System-based ship design (Levander, 2012) straightens the design spiral to illustrate that systematic design (Pahl & Beitz, 1996) starting with task clarification is valid also in the maritime domain.

In system-based ship design, a mission determines functions that are mapped onto a catalogue of known volumes representing systems that perform these functions. Iteration starts once these functional volumes are to be integrated by synthesis into constituting an overall ship form. Andrews (1998) suggests an architectural based on design building blocks. Design building blocks constitute the basic structures that represent functions that are combined into overall ship form in early-stage design.

3.2 *"Design for X"-perspectives*

Recent advances increasingly see ship design from a holistic systems perspective (Andrews & Erikstad, 2015; Ulstein & Brett, 2015), drawing on system theoretic insights and consideration of contextual factors that affect the value delivery from the marine system. The difficulties of capturing what a ship owner really will care about was recently discussed by Pettersen et al. (2018) who used epoch-era analysis, first introduced by Ross & Rhodes (2008a) to understand value generation through the lifecycle of an offshore ship. "Design for X" addresses the need for goal-based design methodology in ship design, where the "X" refers to some specific objective. Andrews, Papanikolaou, & Singer (2012) pointed out that this approach really is a multi-objective problem, where a set of objectives enter into the decision-making process. Ulstein & Brett (2015) point to three industrial "Design for X" perspectives; "Design for Efficiency" focuses on technical, operational and commercial aspects, while "Design for Effectiveness" integrated the "Design for Efficiency"-aspects with safety, and environmental friendliness. "Design for Efficacy" considered flexibility, agility and robustness as examples of design properties to ensure future abilities to deliver value.

Design for safety is another example of the "Design for X" perspective. Risk-based ship design (Papanikolaou, 2009; Boulougouris & Papanikolaou, 2013) proposes that safety be entered as an objective to the design process, rather than a constraint. This represents a deviation from previous approaches, which refer to safety as given by a set of regulatory requirements to the design problem. Under this new regime, specific regulations need not be followed as long as the design is proven to be equally safe or safer than the safety level provided by the rule-based standards. Goal-based standards and risk-based design processes indicate that safety considerations are handled in the functional space, rather than the form space (Bergström, Erikstad, & Ehlers, 2016). This opens the solution space for new innovative solutions that may be more cost-efficient than those adhering to prescriptive standards, keeping the door open for adaptation (and resilience) rather than static compliance.

Similar to the perspectives given above, "Design for Resilience" can be the objective of a process for identifying novel ways to use a designed system. We will here consider further how the resilience of systems, providing the ability to recover from disruption and provide emergency response can be supported by integrating this perspective in operational planning. Designing for resilience requires that designers revisit the function-form mapping, abstracting away from current ways to deliver functionality and uncover new capabilities, adapting to new ways of functioning. Our perspective on design will hence also include the design of concepts of operations for the ship. We will here advance the perspective that the ship has the potential to deliver value by performing functions that it was not intended for, and performing its functions in novel ways not intended. In this respect, planning for such operational procedures are perhaps beyond the common task description of the traditional naval architect, but remains within the task description for system managers tasked with developing the system and its concept of operations throughout its lifecycle.

4 LATENT CAPABILITIES

4.1 *Latent functions*

Latent capabilities are capabilities of a system that were neither intentionally designed for, nor recognized by the designers during the design process (Pettersen et al., 2017). Latent capabilities are useful because a system designed for a specific set of functional requirements, may need to meet other requirements in the future, or it may need to meet the same set of functional requirement in a different manner due to system-internal failures. We now introduce latent functions by reviewing previous work, study how latent functions provide resilience, and propose how latent functions can be identified, evaluated and utilized. Our arguments are also explored in the context of engineering design in Pettersen et al. (2017). We also refer to that paper for further expositions of connections between latent capabilities, and common theories in engineering design.

The process of conceptual design completes when a description of a conceptual design solution has been developed that meets the specified functional requirements. At this point, the system designers have a sufficiently good overview of which functional requirements the design should meet, and the "means" to achieve these requirements. This perspective does not account for the fact that users may perceive that these possible uses provide a value, without this being properly articulated (Ross & Rhodes, 2008b). Mekdeci, Ross, Rhodes, & Hastings (2011) also briefly address latent capabilities.

They use the concept to separate between what a system is designed to do, and what it can do.

The term "latent functions" originates in the social sciences. Merton (1968) introduces a functional taxonomy to study the effects of policy, in which the distinction between policy effects were outlined. First, the manifest functions refer to the desired transformations produced by the policy. Second, the latent functions refer to those effects of a policy that were "neither intended nor recognized", here limited to those effects with a positive impact. We here do not consider dysfunctions, which are the latent functions that have a negative impact.

Crilly (2010) reviews several additional function theories that attempt to distinguish between the functions artifacts should serve according to their designers, and the functions the artifact will be capable of fulfilling. The difference between intended and possible functionality can for example be attributed to differing stakeholder mentalities. System users often use systems differently from the uses the designer intended, and hence value is derived not only from the intended functionality. A similar argument is suggested in a review paper on functional modelling by Erden et al. (2008), who finds that the physical behaviors of an object do not only support the intended functionality. The object can also provide other, unintended behaviors. In their treatment of maintenance in relation to function modelling, they propose that "in a failure situation, another component, rather than the faulty one, can perform the function, perhaps in a less efficient way." That statement captures much of the essence of what we introduce in this paper for improvement of the resilience of marine systems in operations. An updated view of function can be represented by a move from statements of the type: "the function of system S is role R", to statements of the type: "the function of system S is role R in context C" (Crilly, 2015). Hence, a completed conceptual design phase does not imply that the designers are in possession of a complete inventory of capabilities. As the full set of possible system contexts is unknowable, designers and users may still discover additional system uses, even though these are neither intended nor recognized at the closure of the design process.

Finding latent capabilities is therefore not part of the design process per se, but rather an additional capabilities assessment to address vulnerabilities and future system uses. The difference between the design processes, and the proposed capabilities assessment is visualized in Figure 3. Design processes are characterized by the focus on synthesis. For this purpose, there is first a need for an analysis structuring the functional requirements, then to synthesize on that basis, and subsequently to evaluate the results of synthesis, and repeating until a satisfactory design has been found (Asimow, 1962). There is a jump to the

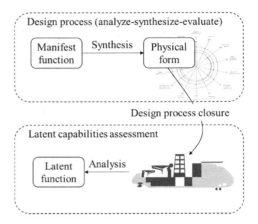

Figure 3. Assessment to identify latent capabilities in a designed system, as seen in relation to the process of designing the system.

additional capabilities assessment after the design process, where a mainly analytical process takes place, in which new useful ways of functioning should be identified.

4.2 *Identifying and implementing latent capabilities*

As latent capabilities are per definition not active at the closure of the design process, steps must be taken to identify and plan for latent capabilities. These steps constitute what we refer to as the "Latent capabilities assessment" in Figure 3. The latent capabilities assessment can consist of the following steps:

1. **Identification of disruptive context changes and failure modes:** How can the system context change and create a need for serving new functions? How can system components fail in such a way that continued operations become impossible, ie. what failure modes exist? Figure 4 shows the span of settings where latent capabilities can improve system resilience. In that figure, FR refers to functional requirements, and DP refers to design parameters.
2. **Identification of latent functions:** How can the system continue to deliver its intended functionality using different existing system components? How can the system deliver other functionality, given emerging operating context, using existing system components? Methods for identification of latent capabilities include design catalogues to identify new function-form relations (Pahl & Beitz, 1996), or simulation of the emergent behaviors of the system to determine whether functional requirements are met by latent functions. This step by itself may

follow the structure of design (analysis, synthesis, evaluation).
3. **Develop contingency plans using latent functions:** What operating procedures allow for the use of the system in a manner not intended by the designers? Development of operating procedures must take place in close collaboration between management, operators, and crew. Successful use will depend on organizational aptitude towards altering function-form mapping. For example, is the organization able to learn from previous successful adaptation of operational mode? Here, there is a strong link to be explored towards organizational theories that have investigated resilience, such as high reliability organizations and resilience engineering (Dekker et al., 2008).

5 CASE EXAMPLES

We here introduce some marine examples of latent capabilities for addressing disruptions. Latent capabilities for restoration of system operation after a breakdown is an addition to redundancies and margins included in the system design due to IMO regulations and class rules. Second, we study some cases where latent capabilities make it possible for a ship to perform vital operations in an emergency context. Accordingly, Figure 4 shows that latent capabilities are applicable as a response to both system-internal and system-external disturbances. An overview of the cases is found in Table 2.

5.1 Icebreaker operation in thick ice

Icebreakers are meant to operate in ice-infested waters. Examples of icebreakers include escort vessels for transport vessels, and research vessels. Facing thick ice, icebreakers commonly get stuck, indicating a temporary disruption from performing their transportation function. Auxiliary functions for breaking free from the ice have therefore been developed.

Sodhi (1995) states that merchant ships that got stuck in ice in earlier times, would hoist cargo by their crane, and swing the cargo sideways to help free the ship. Swinging heavy cargoes produces a roll movement that would wiggle the ship free from the ice, and reduce the need for powering to break through the ice ahead. The observation that this roll movement represented useful functionality, lead designers to include heeling tanks to provide this rolling movement in icebreakers (Sodhi, 1995). This hence serves as an example where an important function was performed by attributing a new function to the crane onboard, for which the crane was intended. Latent functionality was taken into the manifest design of icebreakers, and hence constitutes an example of learning from previous operational experience.

Even though heeling tanks are now intentionally used for operational recovery from getting stuck,

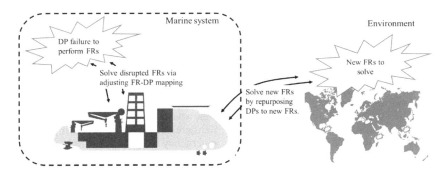

Figure 4. Range of situations where latent capabilities are useful. These include both the need to address failure modes internally in the system, and response to emerging needs in the system environment. As earlier, FR refers to functional requirements, and DP refers to design parameters.

Table 2. Case examples summarized.

Case	System	Manifest function	Latent function
5.1.	Crane on icebreaker	Lift cargo	Wiggle free from ice
5.2.	Anchor handler vessel winch	Hoist anchors	Tow icebergs, or tow vessels in distress.
5.3	Advanced offshore vessel	Various offshore construction tasks	Assist well-cap operation
5.4.	Platform supply vessels	Transport containers, wet and dry bulk cargoes	Search and rescue, accommodation

in some cases there is still a need for a more heavy-handed response to delays caused by very thick ice. There are reports of icebreakers that have needed to consecutively use heeling tanks and crane maneuvers to provide this required roll movement to break free from ice (Mielonen & Woolley, 2014). In such examples, the latent capabilities offered by the crane offers an opportunity to recover to normal operations.

5.2 Offshore ice management

As offshore oil and gas exploration and production commenced in the High Arctic, ship owners faced a novel operational problem. Whereas most other regions for offshore oil and gas are mostly without ice, this constitutes a huge challenge for development of fields in the High Arctic. A particularly critical area are the Grand Banks, offshore Newfoundland. That area is in the so-called "Iceberg Alley", where icebergs from Greenland drift south along the East Coast of North America. A report on iceberg management outlines several operational procedures for handling icebergs that drift in the direction of offshore installations (McClintock, McKenna, & Woodworth-Lynas, 2007). These are single vessel towing, dual vessel towing, water spraying, and propeller washing. Single and dual vessel towing are aimed at most large icebergs (some are too big), while spraying water and propeller washing are commonly used for smaller icebergs, including those termed "bergy bits" and "growlers" (smaller size icebergs up to 1000 tons). We see that all four procedures are examples in which equipment is utilized for functions outside their ordinary scope of operations, ie. they represent latent capabilities.

Towing operations is the most common deflection method for icebergs. This is commonly done using anchor handling vessels that normally are equipped with sizeable winches, and sufficient bollard pull. McClintock et al. (2007) suggest that the minimum bollard pull of offshore vessels involved with iceberg towing are in the range 70–140 tons. The efficiency of the towing operation will naturally be dependent on the size of the iceberg. Borch & Batalden (2015) investigate the management of offshore support vessels in high-turbulence, volatile environments. They point to this iceberg towing as a challenge to crews that operate offshore vessels in ice-infested areas. Doing fieldwork and interviewing crew on offshore vessels, they found that iceberg towing operations put significant strain on quality management systems. New operational procedures had to be invented to cope with the change in operations due to utilization of latent capabilities.

We also provide a short account of water spraying and propeller washing as ice management methods, due to their reliance on latent capabilities:

Fire monitors and nozzles are installed on many offshore vessels to fulfill tender requirements or regulations that require offshore support vessels to be outfitted with a firefighting system (FiFi). Besides the primary firefighting function of this equipment, water spraying have been found through testing and experience to be useful for ice management in several ways (McClintock et al., 2007). First, the water sprayed on the ice directly moves the iceberg. Second, the water sprayed onto the sea surface can induce a current that moves the iceberg. Third, the water sprayed on the ice leads it to break up, or melt. Naturally, using water to deflect ice masses will be most effective in warm weather.

Propeller washing works using the propellers to produce a backwards thrust that deflects smaller ice objects (McClintock et al., 2007). The vessel first backs up slowly to a smaller ice mass like a "growler" or a "bergy bit", to avoid the object to move in the direction of the vessel. Thereafter, the vessel accelerates to provide a wake that pushes the object in the opposite direction. This mode of operation is far from the intended functionality of the propulsion system, and results in a high fuel consumption and extensive wear on machinery (McClintock et al., 2007). There is also a high risk of contact between the vessel and the ice mass. Hence, there are high costs (fuel and wear) and high risks (damage due to impact) associated with utilizing this latent propeller function.

5.3 Deepwater Horizon

The Deepwater Horizon semi-submersible drilling rig exploded and sank after a blowout when operating the Macondo Prospect in the American section of the Gulf of Mexico, in April 2010 (Deepwater Horizon Study Group, 2011). The oil spill in the aftermath lasted for 87 days before the well was capped, and the leak stopped. The accident created an enormous, temporary demand for emergency response services that far surpassed the dedicated emergency response infrastructure, typically emergency response and rescue vessels (ERRVs). The resulting response effort consisted of a "fleet" of vessels, ranging from small privately owned boats, to fishing vessels, to US Coast Guard vessels, and to advanced offshore service vessels, working towards the common goal of reducing the impact of the accident (Graham et al., 2011, Mileski & Honeycutt, 2013).

Vessels of opportunity (VOO) were engaged to provide clean up of the oil spill, and complex special vessels equipped with ROVs and advanced offshore capabilities were used for tasks like damage assessment and capping the well. The contributions of these two classes of vessels are interesting as they delimit the type of operations:

Vessels of opportunity: Fishing vessels and small, privately owned boats that provide very simple services. These are primarily employed in the efforts due to the need for a large number of vessels, rather than their ability to perform advanced operations. An insufficient number of ERRVs would not have been able to cover all the affected areas.

Offshore support vessels: Vessels capable of performing advanced offshore operations. The functionality these vessels can offer go beyond the role they were designed for, and they can perform other advanced tasks. These operations are normally too complicated for the dedicated ERRVs.

Figure 5 differentiates emergency response efforts that took place after the Deepwater Horizon accident, in terms of both scope and scale of the operations. Scope refers to the complexity of the operation. Scale refers to the number of vessels needed to cover the entire oil spill-affected area. The dedicated emergency response infrastructure is sufficient to address events of a foreseen scope or scale. It is often not cost-beneficial to address accidental events that are perceived to be of an extremely low probability (like the Deepwater Horizon accident). However, when these events do occur, they still need to be addressed. This shortcoming of emergency response planning has been addressed in the emergency management literature (Altay & Green, 2006; Simpson & Hancock, 2009).

Common to both categories outlined above, is that the Deepwater Horizon accident constituted a business opportunity to ship owners who manage advanced service vessels: Once the accident materialized, the demand for services increased tremendously, from the side of the oil companies and responsible governmental agencies. British Petroleum and Transocean were liable to pay for the damages. As of 2015, the fines paid by BP exceeded 50 billion USD (British Petroleum, 2016). From the perspective of these companies, reduction of the consequences of the accident hence became a question of limiting their economic losses. A cost-benefit analysis from this side, could ascertain the exact rates the liable companies would have to pay to charter in advanced offshore support vessels from the spot market. We conjecture that the demand for response-capable vessels in this situation, and the subsequently high time charter rates offered to ship owners willing to assist, led to favorable business outcomes for ship owners willing and able to provide advanced offshore vessels for the emergency response. Hence, identifying how the ship owning organization can prepare for using their vessels for operations outside the intended operating profile constitutes a business opportunity.

5.4 *Platform supply vessels in Operation Triton*

Operation Triton is the European Union operation to resolve the Mediterranean migrant crisis. This operation started due to the need for increased border control to cope with the influx of migrants, and is supported by the efforts of numerous non-governmental organizations that cater to the need for humanitarian response (Cusumano, 2017).

To support these efforts, several offshore support vessels have been sent to the region. As their contribution to Operation Triton, the Norwegian government chartered the platform supply vessel Siem Pilot on a six months contract starting in June 2015 (Flaaten, 2015). The chartering rate agreed upon was 99.000 NOK, or around 12.000 USD at the time. At this rate, the daily loss to the ship owners amounted to 50.000 NOK, at least enough to cover the operational expenditures. However, the alternative modes of operation at the time, due to the downturn in the offshore markets, was either to put the vessel into lay-up, or to attempt to compete in the North Sea spot market. At the time, the spot charter rate for platform supply vessels could go as low as 20.000 NOK. Lay-up would also have a cost, without providing any opportunity for increased operational experience, or the boost in public relations that could be expected from participating in humanitarian services.

An interview with the captain on the platform supply vessel Bourbon Argos chartered by Doctors Without Borders reveals further details on how the vessel characteristics support search and rescue operations (Bourbon Offshore, 2015), and hence provide latent capabilities. The captain interviewed states that the "vessel perfectly matches the needs of this type of mission, it has a very low deck and is highly maneuverable, for fast rescues." He goes on to cite some minor adaptations that were needed, the largest of which was the outfitting of a hospital unit. The low deck characteristic eases the process of getting the people being rescued onboard the vessel. Maneuverability is important due to the need for a swift response needed to rescue people from an overloaded boat at risk of sinking.

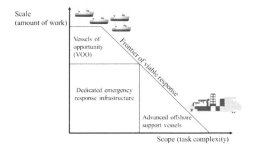

Figure 5. Vessels for the resource pool during Deepwater Horizon response, outside the scope and scale of tasks of the dedicated emergency response infrastructure.

6 DISCUSSION

The practical examples have several implications for latent capabilities. First, the cases show several advantages of exploiting latent capabilities with respect to enhancing the resilience of marine engineering systems. Second, the cases hint at the emergence of additional challenges, including constraints and risks that need to be addressed.

6.1 Advantages of latent capabilities

Latent capabilities are shown to; i) provide a way to address unforeseeable events that perturb system functioning, ii) provide operational recovery from a failed state, and iii) provide a business opportunity.

With respect to i), all the four cases addresses events whose first occurrence lead operators to derive functioning outside what was intended. Later operations, or later generations of the designed system may have taken the latent capabilities into the manifest design, as the latent capabilities were shown to be effective when dealing with these events. Hence, the system managers learn from previous experience, and accounts for the possibility that the unforeseen events can happen again.

With respect to ii), the three first cases show that desirable functioning can be recovered by the use of latent functions. Latent capabilities address both disruptions that limit system functioning (5.1), and disruptions that perturb the functioning of higher-level systems (super-systems), like the oil and gas production systems outside Newfoundland (5.2) and the Gulf of Mexico (5.3). Latent capabilities also address disturbances that are completely system-external, but occur in the maritime environment, as shown by the Operation Triton case (5.4).

With respect to iii), latent capabilities provide business opportunities. We see this especially in the cases of the Deepwater Horizon (5.3) and Operation Triton (5.4). We consider it likely that the ship owners that provided advanced response in the aftermath of the Deepwater Horizon accident received a considerable revenue. For Operation Triton we have economic data that illustrates that even though offshore vessels involved in the search and rescue operation were unable to cover capital expenditure, participation reduced their losses compared to the alternative modes of operation in the spot market or lay-up. For the case of ice management (5.2), this business opportunity argument may be framed as follows: The ship owners that are able to adapt to new operating profiles required in response to ice management needs, will be favored in bidding for tenders, compared to ship owners that are unable to adapt to the ice management needs.

As latent capabilities may provide novel business opportunities, there should be further exploration of other settings in which latent capabilities have value, beyond resilience enhancement. One example could be latent capabilities as a marketing argument for selling existing vessel designs, or for selling vessels in the secondhand market. These aspects of the concept needs further exploration.

6.2 Challenges of latent capabilities

There are several challenges with considering latent capabilities in practice. There will be significant differences in how well ship owners organize for resilience. They will vary with respect to the extent they implement training schemes, and facilitate knowledge sharing across crew members, so that latent capabilities can be taken advantage of. For example, Borch & Batalden (2015) found evidence that existing quality management procedures were challenged by engagement in iceberg towing operations that were outside previous operational profiles. The organizational aptitude towards resilience enhancement, and taking advantage of latent capabilities should be investigated further.

In a previous paper (Pettersen et al., 2017), the relationship between latent capabilities and complexity was studied. They found that using latent functions were partially in opposition to Suh's (1990) design axioms. Latent capabilities may alter the function-form mapping to a less desirable, more complex state, compared to the intended state. Hence, an issue is the possibility that reliance on latent capabilities for recovery from disruption, comes at an unacceptable cost, or introduces new risks. First, it is natural that re-arranging the function-form mapping to take advantage of latent capabilities will reduce the capacity of a system component to perform its primary function. For example, propeller washing as an iceberg deflection technique will cause significant wear on vessel machinery (McClintock et al., 2007). The increase in maintenance costs from this should be accounted for, when comparing to other methods for iceberg deflection. Second, some uses of latent capabilities may cause accidents, due to equipment being used outside their intended scope of operation. For example, iceberg towing induces several new risks, especially relating to interaction with icebergs, and the possibility of entangling tow lines into the propeller (McClintock et al., 2007).

An analogy can be made with risk-based ship design (Papanikolaou, 2009). The move from rule-based standards to goal-based standards for ship design is characterized by allowing that designers deviate from prescriptive rules that work in the physical domain, if they can show that their deviating concept has at least an equivalent safety level. Similarly, a requirement for developing operational procedures for latent capabilities should be improvement of overall resilience.

7 CONCLUSION

This paper has discussed the concept of latent capabilities for the marine design audience. We have shown how latent capabilities can contribute to increase the resilience of ships and other marine systems. There is a clear path between the theoretical constructs presented in the engineering design theory, such as functional modelling, to the application of latent capabilities to recover from disturbances in marine systems. We stress the importance of planning for use of such capabilities, as even though ships without intent possess latent capabilities; there does not in general exist operational procedures or concepts of operations that normally will allow their use. Hence, development of procedures for latent capabilities in ship operations should be a priority for ship owners and ship designers.

REFERENCES

Altay, N., & Green, W.G. 2006. OR/MS research in disaster operations management. *European Journal of Operational Research*, 175(1), 475–493.
Andrews, D.J. 1981. Creative Ship Design. *Trans RINA*, 123, 447–471.
Andrews, D.J. 1998. A comprehensive methodology for the design of ships (and other complex systems). *Proceedings of the Royal Society A: Mathematical, Physical and Engineering Sciences*, 454, 187–211.
Andrews, D.J. 2011. Marine Requirements Elucidation and the Nature of Preliminary Ship Design. *Trans RINA*, 153(A1), 23–40.
Andrews, D.J., & Erikstad, S.O. 2015. The Design Methodology State of Art Report. In *IMDC 2015*. Tokyo, Japan.
Andrews, D.J., Papanikolaou, A.D., & Singer, D.J. 2012. State of the Art Report: Design for X. In *IMDC 2012*. Glasgow, UK.
Asbjørnslett, B.E., & Rausand, M. 1999. Assess the vulnerability of your production system. *Production Planning & Control*, 10(3), 219–229.
Asimow, M. 1962. *Introduction to Design*. Prentice-Hall.
Ayyub, B.M. 2015. Practical Resilience Metrics for Planning, Design, and Decision Making. *ASCE-ASME Journal of Risk and Uncertainty in Engineering Systems, Part A: Civil Engineering*, 1(3), 1–11.
Bergström, M., Erikstad, S.O., & Ehlers, S. 2016. Assessment of the applicability of goal- and risk-based design on Arctic sea transport systems. *Ocean Engineering*, 128, 183–198.
Borch, O.J., & Batalden, B.-M. 2015. Business-process management in high-turbulence environments: the case of the offshore service vessel industry. *Maritime Policy & Management*, 42(5), 481–498.
Boulougouris, E., & Papanikolaou, A. 2013. Risk-based design of naval combatants. *Ocean Engineering*, 65, 49–61.
Bourbon Offshore. 2015. Bourbon Argos: Mediterranean rescue operations. Retrieved November 10, 2017, from http://www.bourbonoffshore.com/en/offshore/bourbon-argos-mediterranean-rescue-operations

British Petroleum. 2016. *Annual Report and Form 20-F 2015*. Retrieved from https://www.bp.com/content/dam/bp/pdf/investors/bp-annual-report-and-form-20f-2015.pdf.
Buxton, I.L. 1972. Engineering Economics Applied to Ship Design. *Trans RINA*, 124, 409–428.
Castet, J.F., & Saleh, J.H. 2012. On the concept of survivability, with application to spacecraft and space-based networks. *Reliability Engineering and System Safety*, 99, 123–138.
Chalupnik, M.J., Wynn, D.C., & Clarkson, P.J. (2013). Comparison of Ilities for Protection Against Uncertainty in System Design. *Journal of Engineering Design*, 24(12), 814–829.
Christopher, M., & Peck, H. 2004. Building the Resilient Supply Chain. *The International Journal of Logistics Management*, 15(2), 1–14.
Crilly, N. 2010. The roles that artefacts play: Technical, social and aesthetic functions. *Design Studies*, 31(4), 311–344.
Crilly, N. 2015. The proliferation of functions: Multiple systems playing multiple roles in multiple supersystems. *Artificial Intelligence for Engineering Design, Analysis and Manufacturing*, 29, 83–92.
Cusumano, E. 2017. Emptying the sea with a spoon? Non-governmental providers of migrants search and rescue in the Mediterranean. *Marine Policy*, 75, 91–98.
Deepwater Horizon Study Group. 2011. *Final Report on the Investigation of the Macondo Well Blowout Deepwater Horizon Study Group*. Retrieved from http://ccrm.berkeley.edu/pdfs_papers/bea_pdfs/DHSGFinalReport-March2011-tag.pdf
Dekker, S., Hollnagel, E., Woods, D., & Cook, R. 2008. *Resilience Engineering: New directions for measuring and maintaining safety in complex systems*. Lund, Sweden. Retrieved from http://citeseerx.ist.psu.edu/viewdoc/download?doi = 10.1.1.105.742&rep = rep1&type = pdf
de Weck, O.L., Roos, D., & Magee, C.L. 2011. *Engineering Systems: Meeting Human Needs in a Complex Technological World*. Cambridge, MA: The MIT Press.
Erden, M.S., Komoto, H., van Beek, T.J., D'Amelio, V., Echavarria, E., & Tomiyama, T. 2008. A review of function modeling: Approaches and applications. *Artificial Intelligence for Engineering Design, Analysis and Manufacturing*, 22(2), 147–169.
Evans, J.H. 1959. Basic Design Concepts. *A.S.N.E. Journal*, 71(4), 671–678.
Flaaten, G. 2015. Siem Pilot taper 50.000 hver dag i Middelhavet. Retrieved November 10, 2017, from https://sysla.no/maritim/siem-pilot-taper-50–000-hver-dag-i-middelhavet/
Foster, H.D. 1993. Resilience theory and system evaluation. In *Verification and Validation of Complex Systems: Humans Factors Issues*. Berlin, Germany: Springer.
Garcia, J.J., Brandt, U.B., & Brett, P.O. 2016. Unintentional consequences of the golden era of the Offshore Oil & Gas industry. In *ICSOS 2016*. Hamburg, Germany.
Graham, B., Reilly, W.K., Beinecke, F., Boesch, D.F., Garcia, T.D., Murray, C.A., & Ulmer, F. 2011. *Deep Water: The Gulf Oil Disaster and the Future of Offshore Drilling*. Washington, D.C.
Holling, C.S. 1973. Resilience and Stability of Ecological Systems. *Annual Review of Ecology and Systematics*, 4, 1–23.

Hollnagel, E. 2010. Prologue: The Scope of Resilience Engineering. In P. Jean (ed.), *Resilience Engineering in Practice: A Guidebook*. Ashgate Publishing Ltd.

Hosseini, S., Barker, K., & Ramirez-Marquez, J.E. 2016. A review of definitions and measures of system resilience. *Reliability Engineering and System Safety*, 145, 47–61.

Jackson, S., & Ferris, T.L.J. 2013. Resilience Principles for Engineered Systems. *Systems Engineering*, 16(2), 152–164.

Levander, K. (2012). *System Based Ship Design*. Trondheim, Norway: Norwegian University of Science and Technology.

McClintock, J., McKenna, R., & Woodworth-Lynas, C. 2007. *Grand Banks iceberg management*. St. John's, Canada. Retrieved from http://nparc.cisti-icist.nrc-cnrc.gc.ca/npsi/ctrl?action = rtdoc&an = 12328812

Mekdeci, B., Ross, A.M., Rhodes, D.H., & Hastings, D. 2011. System architecture pliability and trading operations in tradespace exploration. In *SysCon 2011*. Montreal, Canada.

Merton, R.K. 1968. *Social Theory and Social Structure*. New York, NY: MacMillan Publishing Co.

Mielonen, M., & Woolley, N. 2014. Finnish researchers testing themselves in severe ice conditions in Antarctica. Retrieved October 30, 2017, from http://www.helsinkitimes.fi/finland/finland-news/domestic/8880-finnish-researchers-testing-themselves-in-severe-ice-conditions-in-antarctica.html

Mileski, J.P., & Honeycutt, J. 2013. Flexibility in maritime assets and pooling strategies: A viable response to disaster. *Marine Policy*, 40, 111–116.

Pahl, G., & Beitz, W. 1996. *Engineering Design* (2nd ed.). London, UK: Springer.

Papanikolaou, A.D. 2009. *Risk-Based Ship Design*. (A.D. Papanikolaou, ed.). Berlin, Germany: Springer.

Pettersen, S.S., Erikstad, S.O., & Asbjørnslett, B.E. 2017. Exploiting latent functional capabilities for resilience in design of engineering systems. *Research in Engineering Design*, 1–15.

Pettersen, S.S., Rehn, C.F., Garcia, J.J., Erikstad, S.O., Brett, P.O., Asbjørnslett, B.E., Ross, A.M., Rhodes, D.H. 2018. Ill-Structured Commercial Ship Design Problems: The Responsive System Comparison Method on an Offshore Vessel Case. *Journal of Ship Production and Design*, 34(1), 72–83.

Rice Jr., J.B., & Caniato, F. 2003. Building a secure and resilient supply network. *Supply Chain Management Review*, 7(5), 22–30.

Richards, M.G. 2009. *Multi-Attribute Tradespace Exploration for Survivability*. Massachusetts Institute of Technology.

Ross, A.M., & Rhodes, D.H. 2008a. Using Natural Value-Centric Time Scales for Conceptualizing System Timelines through Epoch-Era Analysis. In *INCOSE International Symposium*. Utrecht, the Netherlands.

Ross, A.M., & Rhodes, D.H. 2008b. Using Attribute Classes to Uncover Latent Value during Conceptual Systems Design. In *SysCon 2008*. Montreal, Canada.

Sheffi, Y. 2005. The Resilient Enterprise: Overcoming Vulnerability for Competitive Advantage. Cambridge, MA: The MIT Press.

Simpson, N.C. & Hancock, P.G. 2009. Fifty years of operational research and emergency response. *Journal of the Operational Research Society*, 60, 126–139.

Sodhi, D.S. 1995. Northern Sea Route Reconnaisance Study: A Summary of Icebreaking Technology.

Suh, N.P. 1990. *The Principles of Design*. New York, NY: Oxford University Press.

Uday, P. & Marais, K. 2015. Designing Resilient Systems-of-Systems: A Survey of Metrics, Methods, and Challenges. *Systems Engineering*, 18(5), 491–510.

Ulstein, T., & Brett, P.O. (2015). What is a better ship? – It all depends... In *IMDC 2015*. Tokyo, Japan.

UN/ISDR. 2005. International Strategy for Disaster Reduction Hyogo Framework for Action 2005–2015: Building the Resilience of Nations.

Wildavsky, A. 1988. *Searching for Safety*. New Brunswick, NJ: Transaction Press.

Woods, D.D. 2015. Four concepts for resilience and the implications for the future of resilience engineering. *Reliability Engineering and System Safety*, 141, 5–9.

Zobel, C.W. 2011. Representing perceived tradeoffs in defining disaster resilience. *Decision Support Systems*, 50(2), 394–403.

Design for agility: Enabling time-efficient changes for marine systems to enhance operational performance

C. Christensen, C.F. Rehn, S.O. Erikstad & B.E. Asbjørnslett
Norwegian University of Science and Technology (NTNU), Trondheim, Norway

ABSTRACT: In this paper, we propose a model for quantifying the value of operational agility in shipping, i.e., the value of being able to exploit possible profitable market opportunities quickly. For a system operating in a dynamic context, the ability to be able to adapt and change is essential. However, for real-world applications, exploiting this flexibility comes with a time delay. If we are not taking the time delay into account, we may be biased towards estimating a higher value of flexibility than what is realizable, as well as failing to properly design the system to be able to change within an adequate time span. A real option valuation model based on Monte Carlo simulation is proposed, where we consider the time delay as a model parameter. The proposed methodology is applied to a bulk shipping case. Bulk fleet capacity expansion is currently achieved mainly through newbuilding or the 2nd hand market, but designing versatile and reconfigurable ships and fleets are also an alternative. The results indicate that significant value can be enabled by being agile, and potential design solutions enabling agility are proposed, both for fleets and single ship cases.

1 INTRODUCTION

1.1 *Motivation*

Shipping markets are often characterized as uncertain and volatile, introducing financial risks for capital-intensive investments (Alizadeh & Nomikos 2009). This risk is notably two-folded, as it can result in both payoffs and losses. In this context, the ability to adapt to changes in a quick an efficient manner could be the difference between failure and success. Over the recent decades, there has been an increasing focus on ilities in systems engineering (De Weck et al. 2012), and in marine systems design (Erikstad & Rehn 2015). However, although timing is often considered as the most critical factor in shipping (Cullinane 2005; Lorange 2005), little research is directed to reflect temporal aspects of the execution of the flexibility of a system.

1.2 *Literature review*

A variety of definitions of agility exist in the systems engineering literature, most reverting around the aspect of the time or frequency of change. As far as change is concerned, changeability represents a collective term for system properties that relate to the ability to change, such as flexibility, adaptability, or agility. Agility is recognized as an essential aspect of changeability, representing a system's ability to be changed rapidly (Fricke & Schulz 2005), or the ability of a system to make changes quickly (Ross et al. 2008). Ross et al. (2008) introduce the agent-mechanism-effect framework to describe a change, including path-enablers which are features that enhance changeability. Path-enablers enable easier change, making real options quicker and less costly to exercise. Haberfellner & De Weck (2005) clarify the difference between agile "systems engineering" and "agile systems" engineering, where the first phrase concerns the design and product development phase, and the latter concerning systems that can change during the operational phase of the lifecycle. In this paper, we focus on the latter. Agility has been identified as a vital system property in multiple domains. For example, agile supply chains through fast fashion (Cachon & Swinney 2011), where changes in consumer demands are identified and met quickly. In finance, market inefficiencies and short arbitrage opportunities can be exploited through high-frequency trading (Budish et al. 2015).

Flexibility in engineering design is discussed in detail by de Neufville & Scholtes (2011). They provide a practical framework for identifying, analyzing and implementing flexibility in a broad range of engineering systems. For engineering systems, real options can be divided between "in" and "on" options (Wang & de Neufville 2004). In contrast to "in" options, "on" considers technology as a black box. "On" options can be thought of as traditional managerial decisions, in line with traditional financial options. "In" options involve either: 1) flexibility for redesign or physical change of systems, such as the case of airport expansion, or 2) explicit consideration of "built-in" flexibility from the design stage, analogous to multi-functionality, which is often characterized as *versatility* (Chalupnik et al. 2013). Examples of the latter can be multi-purpose offshore vessels. Proper design and analysis of "in" options require technical insights beyond what is needed for pure "on" options analyses.

Managerial flexibility is often addressed as part of the traditional real options literature (i.e. "on" options). Real options are typically referred to as the right, but not the obligation to undertake some action, with some sort of irreversibility (Trigeorgis 1996). For assets such as ships, managerial options can be entry, exit, layup, or reactivation (Dixit & Pindyck 1994; Alizadeh & Nomikos 2009), or other embedded options such as chartering out a ship (Bendall & Stent 2007). However, time lags are not frequently addressed in the real options literature, and most of this literature consider markets where the underlying variable is modeled with a positive drift and do not consider cyclical markets typical for shipping (Sødal 2006). Among these is Majd & Pindyck (1987), who conclude that increased uncertainty increases the importance of time lag. Bar-Ilan & Strange (1999) extend the entry-exit model of Dixit (1989) with the influence of time lag, resulting in a reduced spread between the investment triggers. In cases with an underlying positive trend, the time lag is not necessarily a negative element. Instead, the increased uncertainty strengthens the incentives for capacity expansion with a time lag, as increased volatility gives a higher potential upside (Bar-Ilan et al. 2002; Aguerrevere 2003).

Adland & Jia (2015) discuss "sticky newbuilding prices," connecting the term to the difference in the time lag between the delivery of a ship from the second-hand market and the newbuilding market. They find that newbuilding prices are less volatile than second-hand prices. This can be connected to the significant time lag that comes with acquiring a newbuilding, which can be between two and four years. In other words, shipowners can be willing to pay a premium to buy a ship from the second-hand market due to the more instant delivery. This phenomenon can thus be used to quantify the value of agility in shipping implicitly. For example, during the boom in 2007–2008, second-hand prices of vessels were significantly higher than newbuilding prices (illustrated for the dry bulk Capesize market in Figure 3). Another insight to be drawn from the shipping markets is the option value of being in the spot market. By operating in the spot market, a shipowner is positioned to potentially quickly take advantage of upturns in the market, rather than being inflexible from having vessels locked into long-term contracts (Axarloglou et al. 2013). However, there is also a downside risk from being in the spot market in contrast to operating on a secure long-term contract. Thus, one can observe that the degree of agility preferable for a shipowner depends on their strategy of operation. For example, what is their risk attitude, and how is the fleet positioned compared to the competitors' fleets?

Sull (2010) discusses three distinct types of agility that help companies compete: strategic, portfolio and operational. Sull further provides an excellent example of strategic agility in the maritime industry: "*Carnival, for example, entered the cruise business in 1972 but didn't build any new ships until the late 1970s, when CEO Ted Arison recognized that airline deregulation would reduce the price of flying to Miami just as the television series The Love Boat was serendipitously educating consumers on the merits of cruises. As Carnival commissioned the industry's first new ship in nearly a decade, the industry leader, Royal Caribbean, enlarged two existing ships by carving them in half with welding torches and inserting a new midsection*".

As we can see from the reviewed literature, time lag and agility are addressed in multiple domains related to the shipping industry. However, there is little focus on designing agile maritime systems (ships), or system-of-systems (fleets). Although the timing is essential, we do not state that we will better predict the future to better time investments. By focusing on systems design, we address how systems can be designed to be more agile—enabling managers to make timely decisions more efficiently due to reduced time lags.

This paper is in part based on the MSc thesis by Christensen (2017), and the reader is advised to see this source for more detailed material.

2 DESIGN FOR AGILITY

2.1 *What is "design for agility"?*

The term *design for agility* represents design principles that enable rapid changes for a system and can hence be seen as a subset of the more general term *design for changeability*, as introduced by Fricke & Schulz (2005). This is because agility is considered one aspect of changeability, explicitly addressing the temporal aspects of change. For a design to be financially viable, we also note that for the agility to be relevant, it should be cost-effective. That is, even though agile solutions always can be designed for, they are not relevant unless they have reasonable concerns regarding costs. For practical applications, we are mainly interested in the ability of a system to change operational modes.

2.2 Two approaches enabling agility

There are mainly two approaches for achieving operational agility by design. That is either versatility or physical changeability.

First, one can achieve operational agility by *versatility*. Examples of versatile ships can be a multi-functional offshore ship or an oil-bulk-ore (OBO) ship. These can rapidly change between markets without having to perform any retrofits.

Second, one can achieve operational agility by the change of form, i.e., physical design change. This can be achieved by *reconfiguring* or *retrofitting* an asset. For example, ships can be elongated to increase their capacity, or equipment can be retrofitted on offshore vessels enabling them to take on missions they otherwise would not be capable of.

In general, there are many words describing a system's ability to change. A more neutral way of describing it is by "operational changeability without physical change," or "operational changeability by physical change." For more information on this topic, the reader is advised to see Rehn et al. (2017).

2.3 Fleet vs. single ship agility

The agility of a system is highly dependent on what is within the system boundaries. This brings up the critical separation between fleets of ships and single ships. Fleets of multiple units are often described as system-of-systems (SoS). Following the discussion from Section 2.1, a fleet can to a much higher degree be operationally changeable, without any need for reconfiguring the individual systems.

2.4 Characteristics of volatile markets

In terms of characterizing the operational context in which agility may be a valuable system property, we separate between two central market states, with different characteristics: growth and non-growth. This is illustrated in Figure 1, characterizing the establishment of an industry with the S-curve.

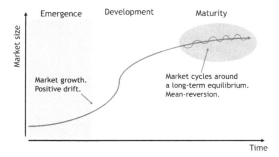

Figure 1. Two essential characteristics of context scenarios and agile properties: growth (drift) and non-growth but with mean-reversion.

Growth markets are characterized by drift on average and are typically represented in the first part of the S curve—the transient phase. Relevant questions to address in these markets include how quickly and how much the market will grow before stabilizing. A good example from 2017 is the situation with electric cars taking a significant position in the car market. As of late 2017, Tesla sees an extreme demand for their new mass-market car, Model 3, but ramping up production takes time. The time lag between early reservations as of 2016 and deliveries are several years. Quicker production response capabilities are valuable but difficult to achieve. At the same time, other established car companies are responding to the electric shift as quickly as they can.

Non-growth markets are characterized by no drift on average, but there may be significant volatility and cycles. This behavior is typical for mature industries, that are in the latter part of the S curve—the steady-state phase. Traditional shipping, such as bulk or tank, falls in this group. This can, for example, be seen with descriptions of cyclical markets (Stopford 2009). Thus, these markets have mean-reverting properties, and the relevant question, therefore, becomes: how quickly can we enter or exit? This question becomes highly relevant for the decision on whether to invest or not. Associated decisions that can be made are how agile the system should be, for example separating between strategic, portfolio and operational (Sull 2010). The case in this paper belongs to the non-growth segment.

However, we obviously recognize that mean-reverting properties also can occur in markets with an underlying growth. For our classification of growth and non-growth to make sense, we consider the development on a relative timescale. That is, relevant to the case and agility-level at hand. Nevertheless, we argue that in terms of modeling, the differentiation makes sense.

2.5 Enhancing agility through path-enablers

We are interested in understanding how we can design more agile systems, as that may be of significant value. For engineering systems to adjust to market changes, they need flexibility. For engineering systems to adjust to market changes in a rapid way, they need agility. As flexibility generally can be incorporated into systems as real options, agility can be achieved explicitly by introducing real options that could be exercised with a short time lag. How can one design a system that rapidly can exercise real options? For this, path-enablers are introduced. Path-enablers can be seen as variables that enhance changeability (Ross et al. 2008). Path-enablers are equivalent to what Fricke & Schulz

(2005) call *principles* enhancing changeability. Path-enablers can enhance changeability either by enabling versatility or by making it easier to change the form. An operational path-enabler for versatility can be a multifunctional offshore ship, while a physical design path-enablers can be structural reinforcement (margins) and modularity for equipment retrofit, or elongation. The main point of the path-enablers is that they enable easier change, i.e., cheaper and quicker. For considerations of agility, we are mainly interested in the temporal aspects of path-enablers.

3 METHODOLOGY

The methodology is inspired by a value-focused approach to design decision making (Keeney 1992). This means that we first explore the value of different levels of system agility, and then proceed by exploring solutions with different levels of agility.

We follow a four-step approach as illustrated in Figure 2. First, we define the problem, i.e., the market, the intended architecture of the design solution and the strategy of operating the asset in the volatile market. Step two involves analyzing the temporal aspects of the market. Step three involves building a model to value the system agility, for example, the value of entering the market as a function of the time lag. The result from step three is ideally a plot of the value (of flexibility) as a function of the time lag (agility). Fourth, and last, depending on the characteristics of the value function, we explore and identify potential physical solutions that can enable different levels of agility. The conclusion will generally be that if the cost of enabling systems for a specific level of agility is lower than the expected additional value it results in, it should be included in the design. This is the same as to say that the enabled agility has a positive expected net present value.

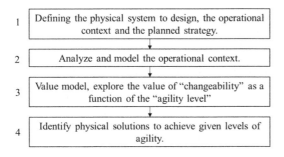

Figure 2. Methodology for valuing agility in the operating context and investigating solutions to enhance system agility.

4 CASE STUDY

This case study concerns the bulk shipping market, as illustrated in Figure 3. Following the methodology outlined in Chapter 3, the case study is structured the following way: First, we investigate the value of agile fleet capacity change in the Capesize segment, before we discuss design solutions that enable different levels of agility. The bulk market is chosen as an illustrative case due to two main reasons: i) the cyclicity in this market is significant, and the market is well known to be close to a "perfect market," and ii) there exists structured data that which gives the possibility to calibrate realistic valuation models. However, the procedure followed is generic, and can equally well be applied to other maritime and non-maritime systems design cases.

4.1 Problem definition

For a hypothetical shipowner in the Capesize dry bulk market, we are interested in investigating the value of the option of being able to add capacity to the market rapidly. The value of agility is to be determined on a per tonnage basis, per time lag unit representing the level of agility. This enables us to gain insight into designing more valuable fleets, being able to adapt to changing market conditions rapidly.

4.2 Context modeling

The dry bulk market is a well-established market, with transparency in both vessel earnings and vessel values. It is also profoundly affected by cycles; thus, agility and utilization of peak periods is a success factor in this market. For the analysis, we consider data of daily earnings in the Capesize market from 1990 to 2017. From the graphical presentation of both second-hand values and earnings in Figure 3, we can observe the link to the mature market presented in Figure 1, with cyclical behavior around a long-term mean. This long-term mean will, according to traditional economic theory for perfect markets, be the marginal cost of a transported dwt. Market imbalances will then make

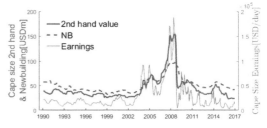

Figure 3. Cape size second-hand value (5yr) and earnings from 1990 to 2017 (data from Clarksons).

spikes, as we especially notice around the period 2007-2008.

In Figure 3, we can observe cyclic market behavior, with different periods. These periods could be investigated further by Fourier-analyses. However, this is not performed in this analysis, partly due to limitations in the length of the time series. An interesting observation from the data is the significant premium that was paid for 5-year-old second-hand vessels, compared to newbuildings, during the peak in 2007-2008. In this peak period, the delivery time of newbuildings was in the range of three to four years. Thus, the observed premium can be interpreted as a value of agility, as it reflects the willingness to pay for immediately entering the market.

To model the market, an underlying mean reversion stochastic process is assumed. The nature of mean reverting stochastic processes is that they allow for short-term fluctuation around a more long-term equilibrium. For applications to market modeling, the equilibrium is often considered as a marginal cost of production (or freight), as market players can adjust supply to meet demand (Dixit & Pindyck 1994). This is backed up by shipping research (Taib 2016; Koekebakker et al. 2006; Benth et al. 2015; Sødal 2006).

To assure that the modeled market value is non-negative and that the volatility is proportional to the given state, we further proceed with a geometric mean reversion (GMR) model, originating from the Ornstein Uhlenbeck process (Uhlenbeck and Ornstein, 1930). This is in line with Tvedt (1997). The geometric mean reversion stochastic differential equation is expressed in Equation 1, representative for the incremental change dX_t.

$$dX_t = \rho(\hat{x} - \ln(X_t))X_t d_t + \sigma X_t dZ_t \quad (1)$$

Here, ρ represents the speed of mean reversion, $\hat{x}$ is the mean reverting level, σ is the standard deviation of the change and dZ is the Wiener increment. We follow the process described by Rollins & Insley (2005) and get the fit of the model parameters for our dataset, as given in Table 1. To check the dataset for stationarity, the Augmented Dickey-Fuller test was performed. This hypothesis test implies that our data set follows an autoregressive alternative, hence the data is stationary and with mean reverting properties.

Table 1. Parameters for geometric mean reversion for the analyzed cape earnings data.

	$\hat{x}$	ρ	σ
Earnings	23 284,4	0.0825	0.2972

4.3 Valuation of theoretical market agility

4.3.1 Description of value

We take a real options approach to evaluate the possibility of entering the market with new capacity, for a given a time lag. This estimated value will be *design-solution-neutral*, as it is purely based on changes in operation and underlying market rates. The process to find the option value, for a given time lag, is a two-step process: i) Find the option value by the established framework, and ii) adjust the option value according to a given cost of time lag.

For step ii), the cost of time lag (CTL) is modeled as the opportunity cost of not staying in the market, much in line with newbuilding delivery lag cost phenomenon (Adland & Jia 2015), and the cost of holding excess capacity (Bar-Ilan et al. 2002). This opportunity cost is represented by the free cash flow from operation, i.e., the daily earnings (E) minus fixed operation costs (FOPEX). Earnings will be voyage freight rate less voyage cost (bunker, port cost) while the fixed operational is assumed to not vary with the operations of the ship, including crew cost and insurance. We write the cost of time lag (CTL) as given in Equation 2.

$$CTL(t) = e^{-rt}\int_t^T E(t) - FOPEX(t) \quad (2)$$

Here, r is the discount rate, and earnings are represented by a geometric mean reversion process.

4.3.2 Option value with timing flexibility

For the option valuation, we extend the Least Squares Monte Carlo framework (Longstaff & Schwartz 2001) and include a new parameter, time lag τ. This framework allows for the early exercise of an American option, and by this cover the element of timing, which is crucial in cyclical markets.

We further use that the vessel value is equivalent to the value of all discounted future cash flows from the vessel, and we use a vessel value corrected for time lag as underlying process for the option valuation. The evolution of the vessel value is based on the evolution of the earnings, replicated with the GMR-process, and transferred to vessel value through a regression formula of second-hand values on vessel earnings. Thus, the option value is the expected value from the distribution generated by the Monte Carlo simulation, as presented in Equation 3. This is an extended version of the valuation formula presented by McDonald (2003).

$$V(S_0, 0, \tau) = \frac{1}{n}e^{-rt}\sum_{i=1}^{n} V(S_t^i, t, \tau) \quad (3)$$

Here, the option value V (or payoff) is a function of the underlying variable S (net cash flow from the project), time *t*, and investment lag τ. The start state for each simulation is drawn from the distribution of Capesize earnings. Each simulation is then corrected for a time lag in the range from 0 to 3 years (0 to 1095 days in Figure 4).

The results from the agility evaluation model developed is presented in Figure 4. Here, we can see the option value (of adding tonnage to the market) as a function of the time lag it takes between the decision is made to the tonnage enters the market. As expected, the value is decreasing with increasing time lag. The option value will foremost describe how such a value depreciate with time, as the model always exercises the option, whether it is "in-the-money." The initial discount rate is found by the weighted average cost of capital, with a debt ratio of 60%, debt cost of capital of 2.3% and equity cost of capital of 17.4%, which reflect alternative investments in the Norwegian OBX-index.

From Figure 4, we can infer that for a typical Capesize bulk ship of 180 000 dwt, the value of being able to enter the market immediately is $10.8 million. With a time lag of one year, this option value is significantly reduced to $4.3 million. At the limit of three-year lag, the option value is reduced to $0.2 million. Since the typical time lag of a newbuilding is approximately three years, this result is consistent with efficient market theory, as there should be no apparent "free lunch" in the market (using established market mechanisms).

The range of results from Figure 4 (0 to $10,8 million), seems to be consistent with the observable market premiums between newbuilding and second-hand values in Figure 3, which is up to $50 million at maximum, but on average much less. The option value cannot be negative, as the option will not be exercised if it is not profitable.

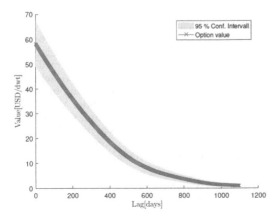

Figure 4. Option value as a function of time lag and a discount rate of 8%.

Table 2. The option value as a percentage of initial option value with no time lag, and for a Capesize 180 k dwt vessel with different lags (discount rate 8%).

No lag	6 m lag	12 m lag	24 m lag
60 USD/dwt	65%	40%	12%
180k dwt	$6.6 m	$4.1 m	$1.2 m

4.4 Identification of agile design solutions

Identification of solutions for improved agility is discussed both at the fleet level (system-of-systems) and at the individual ship level (system). This particular case study only investigates agility for capacity expansion.

The general notion for these analyses is that if the cost of adding the agile path-enabler is lower than the expected added value from agility, then the agile system design should be considered. We will not go into detail in terms of looking at the cost-benefit tradeoff, only discuss relevant principles enabling capacity expansion agility.

4.4.1 Fleet level

At the fleet level, we are interested in identifying solutions that can improve the speed of increasing the capacity of a fleet, other than going to the second-hand market. This section focuses on versatility.

By having a large fleet of multifunctional ships, one can quickly change its composition. For example, by having a combined tank and bulk fleet, together with oil-bulk-ore (OBOs), the OBOs can quickly switch between the two relevant markets (in addition to achieving higher utilization due to triangulation voyages). This is often used in the industry today. This approach would be characterized as "in" options approach to enable changeability, as it must be explicitly designed for. Thus, adding OBOs to a fleet thus represents a path-enabler to more quickly change the fleet from one composition to another.

An alternative "on" options approach to enable agility for a fleet can be to utilize contractual rights to have the option to charter in ships from other players in the market at a given time lag.

4.4.2 Ship level

At the ship level, we are interested in identifying solutions that can improve the speed of increasing the capacity of a ship, other than going to the second-hand market. A single ship can be versatile, such as the case for the OBO vessel discussed at the fleet level. This enables a ship to switch between wet and dry markets. However, there are potentially other markets a ship can switch between, including bulk. This section will instead focus on physically changeable ships.

All ships can, in theory, be converted into any other ship; it all depends on the conversion cost and time (Rehn et al. 2017). Without becoming too philosophical, there are specific characteristics of the alternative markets to switch between, in addition to bulk, which enable a starting point with more reasonable switching cost and time. If the market characteristics are significantly different, then the conversion cost and time will be high.

A general procedure for identifying change mechanisms, and path-enablers, is to compare the two designs we want to change between (de Neufville & Scholtes 2011). What are the main differences, in addition to the added capacity? Are there any costly retrofits that will follow from the capacity change? That is when we want to perform a change to the design, how does the change propagate through the design? Change propagation research is a field in the literature (Clarkson et al. 2004), and a general strategy to reduce the change propagation is by adding margins (Eckert et al. 2004).

4.4.3 *Bulk vessel expansion*

Within the bulk market, there are two types of expansions possible. One is to expand to switch between bulk size markets, such as from a Panamax sized (60 k – 80 k dwt) to a Capesize (180 k + dwt). The other is to scale the ship size within one of these segments, such as increasing from a 180 k dwt ship to a 250 k dwt ship within the Capesize segment.

A bulk vessel can theoretically be expanded in three dimensions: length, breadth, and height. The conventional and the most realistic way to expand a vessel is by its length, by inserting a new midship section. This has been performed for several vessels, such as cruise vessels, RORO-vessels, and container vessels. Elongation of ships is also discussed by Knight & Singer 2012 and Ericson (2014). The expansion of a ship in its breadth dimension has been performed, among others by Rederei NSB for MSC, in line with the widening of the Panama Canal.

The process to elongate a vessel is quite straightforward, as a vessel is docked, split amidships, and an extra midship section is inserted. Though, requirements for the new dimensions must be followed. To build the initial design with an expansion length in mind could be to implement path-enablers, features that enable a more rapid change. In the following we present two possible path-enablers that can be built into a bulk ship in the initial design phase, to enhance an agile transition. Both are purely for illustrative purposes, and we do not investigate the effects of these regarding operational cost and hence earnings. We do neither consider, e.g., how additional strength will affect the light weight of the vessel.

Table 3. Minimum plate thickness requirements.

	Length 150 m	Length 200 m
Bottom & sides	11 mm	12 mm
Keel	14.5 mm	17 mm
Strength Deck	8.5 mm	9.5 mm

Figure 5. Visibility requirements.

Structural reinforcement
Structural reinforcement can be implemented in the initial design phase, to ensure that the scantlings will satisfy the current rules for newbuilding of the possible length after expansion. Example of this is presented in the following table, with a basis in rules from DNV GL (2013).

Thus, by building the initial design with thickness for plates for the length after the expansion, the expansion will be less costly and more time-efficient.

Visibility
For vessels that enter the Panamax class, the rules for visibility is stricter than others, due to the sailing through the Panama Canal. From the conning positions, the view of the water surface is required not to be disturbed by more than 1x LOA forward the ship for laden vessels, according to DNV GL and *New Requirements for Panama Canal transit from 2016*[1]. Thus, path-enablers to ensure that a retrofit of the bridge is not necessary is to build the bridge in compliance with new requirements.

What has not been considered in terms of deciding whether to elongate or not, is the opportunity cost from the market that the ship operates in before switching. For example, elongation of a smaller bulk vessel will involve taking the ship out of the market during elongation, which involves adding an even higher opportunity cost, especially if the bulk market, in general, is good.

5 DISCUSSION

Fleet expansion in the dry bulk market usually relies on either newbuilding or acquisition in the second-hand market. With a delivery lag of 2–3 years for a newbuilding, an excellent initial myopic

---
[1]https://www.dnvgl.com/news/new-requirements-for-panama-canal-transit-58190 [07.12.2017].

decision could be reduced to a bad one due to market uncertainty resolving during the delivery process. To reduce this exposure, one could invest in second-hand vessels with shorter delivery time. However, second-hand vessels can be expensive compared to newbuildings—especially in good markets. In this case, we have discussed alternatives ways to expand the capacity of a fleet beyond what is commonly used today. Both conversions or combination (OBO) carriers could be used to move tonnage from one market to another, but we emphasize that also physical design changes could be an exciting alternative, such as elongation.

5.1 General shipping applications

For general shipping applications, what is relevant is the potential to switch assets between markets and to scale within a market to increase exposure. Both cases are to be able to exploit temporary arbitrage opportunities. Some ships can do this by inherent multifunctionality (versatility), while other need conversion or retrofit. The degree of value from switching between two markets would, in theory, be highly dependent on their correlation, or if the difference is modeled by a mean-reverting model, it would be highly dependent on the speed of mean reversion between the two markets.

A standard, versatile architecture in shipping, enabling rapid switching between transporting goods in various markets, is containerization. Container vessels can in theory switch between most markets relatively quickly. However, the degree of modularization enabled by containerization also has a high cost for transporting commoditized goods such as traditional dry bulk. Furthermore, container ships generally have the property of being volume constrained and not weight constrained, in contrast to bulk ships, and physics thus generates barriers for efficient switching. Further, the maximum weight capacity of a container vessel is significantly lower than bulk ships of the same length.

The case presented in this paper is a bulk shipping case. However, the same analysis can be applied to other cases of high relevance. Two segments that would be interesting to investigate further are:

- **Cruise shipping.** Agility in the cruise industry is identified as of particular relevance. This can also be seen with several elongation cases in the industry[2], in addition to the case discussed by Sull (2010).

- **Offshore shipping.** Here agility is especially crucial in the spot market, where mobilization and contract duration are short-term. However, the market in offshore is considerably different compared to traditional shipping, as it is heterogeneous and of more binary structure—as in multiple companies may compete for the same contract.

5.2 Agility valuation difficulties

Agility evaluation is not straight-forward. The approach presented has focused only on commercial value alone. We do not consider other aspects of valuation than monetary and evaluate the change in operation by accumulated earnings, drawn from the market rates. An interesting facet of this approach is that we obtain a *design-neutral* value. This enables system designers to have economic boundary conditions for their work. However, agility evaluation could also include other aspects, from customer satisfaction to safety performance. This will make the analysis more complicated. Such factors have also different importance between shipowners, but by a monetary scale where all players have the same information about historical values, fair values could be obtained without taking a stand in a supply and demand discussion, and risk appetite could be adjusted with different discount rates.

5.3 Value of agility and the forward curve

The value of agility can be related to forward freight agreements (FFAs) in the bulk market, although they are significantly different. FFAs mainly represent instruments used for hedging and locking in future trades, and historical spot rates have traded with a premium over FFAs. The agility valuation curve (Figure 4) is in a way the opposite: It represents the value of a rapid expansion or entry, which can be interpreted as a *risk premium* of the spot rates. This will then represent a value as the shipowner can await more information before decisions are made, while FFA is to make future decisions here and now—which the buyers have to pay a premium to do.

6 CONCLUSION

In this we paper, we discuss several aspects that can enable agility both for single vessels and fleets. Common for these aspects is that they are *design solutions*, either enabling versatility (no change of design), or more agile change of physical design. Thus, solutions that are *designed for agility* are explicitly made to reduce the time lag between the decision is made, and the change is completed. Many decision-makers admit that they cannot

---

[2]http://www.20thcenturyliners.com/ol_stretched.htm [15.11.2017].

predict the future in detail. With solutions designed for agility, they can await more information than conventional design solutions before making decisions, or respond to unforeseen events in a quicker way. This can be highly valuable for systems operating in markets with high volatility.

ACKNOWLEDGMENTS

We are grateful for the valuable advice and for exciting discussions with Professor Stein Erik Fleten at NTNU and Professor Roar Ådland at NHH.

REFERENCES

Adland, R. & Jia, H., 2015. Shipping market integration: The case of sticky newbuilding prices. *Maritime Economics and Logistics*, 17(4), pp. 389–398.
Aguerrevere, F.L., 2003. Equilibrium Investment Strategies and Output Price Behavior: A Real-Options Approach. *Review of Financial Studies*, 16(4), pp. 1239–1272.
Alizadeh, A.H. & Nomikos, N.K., 2009. Shipping Derivatives and Risk Management.
Axarloglou, K., Visvikis, I. & Zarkos, S., 2013. The time dimension and value of flexibility in resource allocation: The case of the maritime industry. *Transportation Research Part E: Logistics and Transportation Review*, 52, pp. 35–48.
Bar-Ilan, A. & Strange, W.C., 1999. The Tming and Intensity of Investment. *Journal of Macroeconomics*, 21(1), pp. 57–77.
Bar-Ilan, A., Sulem, A. & Zanello, A., 2002. Time-to-build and capacity choice. *Journal of Economic Dynamics and Control*, 26(1), pp. 69–98.
Bendall, H.B. & Stent, A.F., 2007. Maritime investment strategies with a portfolio of real options. *Maritime Policy & Management*, 34(5), pp. 441–452.
Benth, F.E., Koekebakker, S. & Taib, C.M.I.C., 2015. Stochastic dynamical modelling of spot freight rates. *IMA Journal of Management Mathematics*, 26(3), pp. 273–297.
Budish, E., Cramton, P. & Shim, J., 2015. The high-frequency trading arms race: Frequent batch auctions as a market design response. *Quarterly Journal of Economics*, 130(4), pp. 1547–1621.
Cachon, G.P. & Swinney, R., 2011. The Value of Fast Fashion: Quick Response, Enhanced Design, and Strategic Consumer Behavior. *Management Science*, 57(4), pp. 778–795.
Chalupnik, M.J., Wynn, D.C. & Clarkson, P.J., 2013. Comparison of Ilities for Protection Against Uncertainty in System Design. *Journal of Engineering Design*, 24(12), pp. 814–829.
Christensen, C., 2017. *Agility by Design: A Real Option Approach to Identify and Value Time-efficient Changes in Marine Systems*. NTNU.
Clarkson, P.J., Simons, C. & Eckert, C., 2004. Predicting Change Propagation in Complex Design. *Journal of Mechanical Design*, 126(5), pp. 788–797.

Cullinane, K., 2005. *Shipping economics*.
de Neufville, R. & Scholtes, S., 2011. *Flexibility in Engineering Design*, The MIT Press.
De Weck, O.L., Ross, A.M. & Rhodes, D.H., 2012. Investigating Relationships and Semantic Sets amongst System Lifecycle Properties (Ilities). *Third International Engineering Systems Symposium CESUN 2012, Delft University of Technology, 18–20 June 2012*, (June), pp. 18–20.
Dixit, A. & Pindyck, R., 1994. *Investment under uncertainty*, Princeton University Press.
Dixit, A.K., 1989. Entry and Exit Decisions under Uncertainty. *Journal of Political Economy*, 97(3), pp. 620–638.
DNV GL, 2013. Classfication Notes—Conversion of ships., (8).
Eckert, C., Clarkson, P.J. & Zanker, W., 2004. Change and customisation in complex engineering domains. *Research in Engineering Design*, 15(1), pp. 1–21.
Ericson, M., 2014. LENGTHENING OF A SPECIALIZED REEFER SHIP Is it economically viable?
Erikstad, S.O. & Rehn, C.F., 2015. Handling Uncertainty in Marine Systems Design—State-of-the-Art and Need for Research. In *12th International Marine Design Conference (IMDC)*. pp. 324–342.
Fricke, E. & Schulz, A.P., 2005. Design for changeability (DfC): Principles to enable changes in systems throughout their entire lifecycle. *Systems Engineering*, 8(4), pp. 342–359.
Haberfellner, R. & de Weck, O., 2005. Agile SYSTEMS ENGINEERING versus AGILE SYSTEMS engineering. *INCOSE International Symposium*, 17(2), pp. 1–17.
Keeney, R.L., 1992. *Value-Focused Thinking: A Path to Creative Decisionmaking*, Cambridge, MA: Harvard University Press.
Knight, J.T. & Singer, D.J., 2012. A Real Options Approach to Evaluating Flexible Architectures in Ship Design. *Imdc 2012*, (November 2015).
Koekebakker, S., Adland, R. & Sødal, S., 2006. Are spot freight rates stationary? *Journal of Transport Economics and Policy*, 40(3), pp. 449–472.
Longstaff, F.A. & Schwartz, E.S., 2001. Valuing American Options by Simulation: A Simple Least-Squares Approach. *Review of Financial Studies*, 14(1), pp. 113–147.
Lorange, P., 2005. *Shipping Company Strategies*.
Majd, S. & Pindyck, R.S., 1987. Time to build, option value, and investment decisions. *Journal of Financial Economics*, 18(1), pp. 7–27.
McDonald, R.L., 2003. *Derivatives Markets*, Addison Wesley.
Rehn, C.F. et al., 2017. Quantification of changeability level for engineering systems. *[Working paper for journal]*.
Rollins, K. & Insley, M., 2005. On solving the multirotational timber harvesting problem with stochastic prices: A linear complementarity formulation. *American Journal of Agricultural Economics*, 87(August), pp. 735–755.
Ross, A.M., Rhodes, D.H. & Hastings, D.E., 2008. Defining Changeability: Reconciling Flexibility, Adaptability, Scalability, Modifiability, and Robustness for Maintaining System Lifecycle Value., 11(3), pp. 246–262.

Stopford, M., 2009. *Maritime Economics*, Abingdon, UK, NY: Taylor & Francis.

Sull, D., 2010. Competing through organizational agility. *McKinsey Quarterly*, (1), pp. 48–56.

Sødal, S., 2006. Entry and exit decisions based on a discount factor approach. *Journal of Economic Dynamics and Control*, 30(11), pp. 1963–1986.

Taib, C.M.I., 2016. Forward pricing in the shipping freight market. *Japan Journal of Industrial and Applied Mathematics*, 33(1), pp. 3–23.

Trigeorgis, L., 1996. *Real Options: Managerial Flexibility and Strategy in Resource Allocation*, MIT Press.

Tvedt, J., 1997. Valuation of VLCCs under income uncertainty. *Maritime Policy & Management*, 24(2), pp. 159–174.

Uhlenbeck, G. E. and Ornstein, L. S. (1930). On the theory of BrownianMotion.

Wang, T. & de Neufville, R., 2004. Building Real Options into Physical Systems with Stochastic Mixed-Integer Programming. In *Options*. pp. 1–35.

Design for Decommissioning (DfD) of offshore installations

Chengi Kuo & Calum Campbell
University of Strathclyde, Glasgow, Scotland, UK

ABSTRACT: Many offshore oil/gas installations are at the end of their useful lives and require decommissioning. The process involves shutting down wells, cutting, lifting, moving components and these activities can cause adverse effects on the environment. Today the operators of an installation must perform an Environmental Impact Assessment (EIA) and have plans in place for decommissioning. A most useful approach would be to incorporate decommissioning requirements at the design stage or adopt the concept of Design for Decommissioning (DfD) methodology. This paper gives the main features in applying this methodology. These features include a full examination of the installation in question, before a removal method is selected. An EIA is then conducted to determine the risk of each hazard present during the decommissioning operations, the critical ones would be selected for risk reduction by through each phase of the installation's lifecycle. It is at the design phase that fresh methods and barriers would be introduced to enhance the care of the environment during decommissioning operations.

Keywords: Design, Decommissioning, Risk, Hazard, Environment

1 INTRODUCTION

Seeking oil offshore in the North Sea began in the 1960's and production was realised in the early years of 1970. The stage has now been reached that many offshore installations have come to the end of their useful lives and require decommissioning. The process involves shutting down wells, cutting, lifting, moving components, and these activities can cause adverse effects on the environment. Today the operator of an installation must prepare an Environmental Impact Assessment (EIA) and have plans in place for decommissioning. This information must be submitted to the relevant authority for approval. There are several approaches available for decommissioning and these depend on the types of instillations involved.

A most useful approach would be to incorporate decommissioning requirements at the design stage or adopt the concept of design for decommissioning (DfD) methodology. This paper is aimed toward adopting a DfD approach. It begins by highlighting offshore instillations before reviewing present decommissioning methods. The concept of design for something is introduced and the proposed DfD methodology outlined.

A risk assessment is performed by identifying the environmental hazards and classified into intolerable, tolerable and negligible risk levels. The intolerable risks would then be reduced by design methods and implemented through the various phases of the life-cycle. An illustrative example of the use of the DfD would be given.

2 HIGHLIGHT OF OFFSHORE INSTALLATIONS

There are four main types of offshore installations which have been used for oil and gas production and require decommissioning. It is useful to provide brief highlights of each type.

2.1 *Normally unmanned structures*

These are structures which are relatively small and fixed to seabed, see Fig. (1) for a typical example. They are used mainly for producing gas in shallow waters of southern regions of North Sea. There are no facilities for supporting people living on board and personnel go onto these installations

Figure 1. Normally unmanned gas installation with offshore support vessel (Source: Rigzone).

for maintenance purposes. There are several types of Normally Unmanned Structure, ranging from Type 0 structures, which are larger and are fitted with helipads and cranes, to Type 4 structures, which typically only consist of one small deck producing from one well (Nielsen 2016).

2.2 Jacket structures (Steel)

In offshore fields with large reserve of oil and gas, large steel structures are used and given the name "Jacket". Fig. (2) gives a typical example of such a structure. It can be seen the jacket is used to support topside facilities for oil/gas production and these include drilling equipment, accommodation for supporting people working on board, storage, etc. These installations are generally used in "harsher" environments, where bending moments and hydrodynamic forces are higher. They are also limited to relatively shallow water.

2.3 Floating production units

The concept of floating production unit was first promoted when oil was discovered in fields with limited reserves. It was not cost effective to install structures fixed to seabed in the long term. The first solution was to use a jack-up structure, see Fig. (3), where the legs are lowered onto seabed

Figure 3. "Jack-up" in operation (Source: Greatship Global).

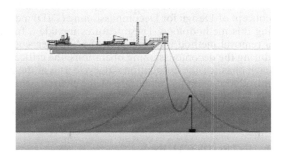

Figure 4. Diagram of an FPSO with steep-S riser configuration (Source: Offshore Energy Today).

when drilling takes place, and jacked up and unit towed away once the assignment was completed. Later, more efficient methods were developed and today for offshore oil production in even larger fields and operations in deeper water, the use is made of FPSO (Floating Production Storage Offloading), See Fig. (4). Semi-submersible units, as well as Tension Leg Platforms (TLP), can also be used for deep water, although TLP's can be limited in terms of maximum operational depth due to their use of tensioned mooring lines and top-tensioned risers. SPAR platforms are another option for floating platform design, with several variants possible (e.g. truss, classic, hybrid).

2.4 Subsea installations

As the search for oil and gas continued, more fields with limited reserve were discovered, with the basic characteristic that they were close to each other. The concept of having a core or main unit on the seabed and have satellite units feeding to the core unit was adopted. In this way, a more efficient system was used, see Fig. (5). There are many flow lines on the seabed around the core unit. Subsea installations can be fed directly to an existing

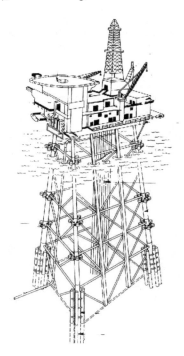

Figure 2. Illustration of typical Jacket structure (Source: ESRU Strathclyde).

Figure 5. Illustration of a subsea installation arrangement (Source: Offshore Energy Today).

Figure 6. Manifold arrangement (Source: KL4220 Bawah Laut).

pipeline, or can be sent back to a topside via risers. This depends very much upon the location of the well(s), as some locations have better pipeline infrastructure than others (i.e. Gulf of Mexico). A closer diagram of a typical manifold arrangement is shown in Fig. (6), with subsea trees and flowlines clearly illustrated.

It will be noted that each of the four methods would require their own approach to decommissioning.

3 BASIS OF DECOMMISSIONING PROCESS

The process of decommissioning an offshore oil/gas instillation involves many tasks and activities. It is useful to outline the basis of this process and highlight some of key issues involved.

It should be realised that decommissioning is a "big" business for the UKCS (Continental Shelf) as there are over 470 installations, 15 onshore terminals, 25,000 km of pipeline, 5000 platform wells and 6000 subsea wells. These all need to be decommissioned. The forecast for the period 2014–2023 is £14.6 billion.

Decommissioning is guided by regulatory agreements at international, regional, national and sub-national (local) levels. A typical agreement is OSPAR, which was derived after international meetings held in Oslo and Paris. The actual decommissioning programme (DP) starts after it has been established that the instillation in question has achieved all the economic benefits and is ready to cease production. The key steps are: cessation of production approved by government (in UKCS this involves the Dept. of Business Energy and Industrial Strategy) and there is a guideline report for operators, operator proposes an option and submits a DP for approval. The operator will also consult stakeholders. In developing a DP, the factors to be taken into account include the following:

a. *Asset data gathering*: assist work scope
b. *Technical studies*: feasibility, methodology
c. *Safety investigation*: risk levels of hazards in an activity
d. *Environmental assessment*: estimate impact level
e. *Option cost estimated*: for use in comparisons
f. *Project management*: implement DP in practice
g. *Stakeholder engagement*: seeking their views
h. *Societal understanding*: note other users of the seas

In doing the work of decommissioning of the four types of instillations, all required some of the following basic functions;

- *Surveying*: The underwater environment of the installation's location using remotely operate vehicles
- *Cutting*: Above and below water lines using tools that include: burning, saw, diamond wire, abrasive waterjets, hydraulic shears
- *Lifting*: Removing parts of installation after cutting using crane barges with various lifting capabilities
- *Handling*: By external facilities such as tugs

An excellent overview of decommissioning and examples can be found in the presentation of (Hoare 2017).

4 THE CONCEPT OF DESIGNING FOR SOMETHING

The term design is popularly used to explain some form of generating fresh ideas or for putting focus on a specific goal. It should be recognised that design belongs to a group of entities called "non-absolute'. Everyone has his or her own idea how to design and there is no standard way of performing the design tasks, so long as the objective(s) is met. There is no single definition of the term, while there are those which cover broad aspects and other narrower areas.

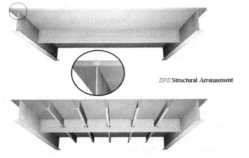

Figure 7. Sandwich plate system.

There is also a tendency to treat design as a method of generating data for hardware, but in practice this is a very restrictive view point as design outputs can take many forms ranging from information for hardware and software to organisational functions and human factors.

Some examples of design definitions can be found in references such as (Watson 2002), (Papanikolaou 2014), (Andrews 2015), (Milland 2011). For this reason, design is often associated with its aim. The earliest task-orientated design is "design for production" or more specifically design for the ease of production. This is best illustrated using Sandwich Plate System (SPS) in place of traditional plate and grillage approach, see Fig. (7) and reference (Kennedy 2010).

Other task–orientated design would cover the other phases of a product's life cycle and include design for operation, design for commissioning, design for maintenance and design for decommissioning (DfD).

DfD presents a greater challenge for two reasons. First, it is the last phase of a product's life cycle and there is less flexibility. Second, it is not so straight forward due the presence of many uncertainties associated with caring for the marine environment.

In addition to the task-oriented design concept outlined above, there are more global design concepts such as design for safety and design for cost-effectiveness. These concepts are associated with specific phases of a product's life-cycle and instead they can be involved in all phases. It will be noted that the overall objective would be same but the hazards identified could be different as well as their risk levels and methods of risk reduction.

5 THE PROPOSED DFD METHODOLOGY

The methodology adopted in the research includes the following steps:

5.1 Step 1: Examine installation

The installation selected for decommissioning would be examined in order to acquire a full understanding how it will be decommissioned and dismantled.

5.2 Step 2: Select removal option

In general, the installation will be broken into smaller units to ensure that they can be best lifted. The above water platform or topside would be separated from the main structure and lifted onto a barge. The main structure would be cut at appropriate depth underwater before removing by lifting or reverse-installation (Atitkar 2017). For subsea instillations, the wells would be plugged before separating the flowlines main and satellite units. The various items would then be removed from the seabed.

5.3 Step 3: Identify hazards in operation

In the process of implementing Step 2, there will be hazards that affect the caring of the environment. A full list of hazards would be identified using popular methods such as HAZOP and FMEA, (Kuo 2007), as well as the experience of the operators.

5.4 Step 4: Assess risk levels of hazards

A risk assessment would then be performed in order to classify the hazards into intolerable, tolerable and negligible risk levels. The methods suitable for use include the application of risk matrix approach, (Kuo 2007). The focus of attention would be on the hazards with intolerable risks.

5.5 Step 5: Trace hazard back to design phase

The determined hazard with intolerable risk would be traced back to the design phase via the other life cycle phases while identifying any contributions which can affect the risk levels from these phases.

5.6 Step 6: Risk reduction via design

It is at this step that the intolerable risk level would be reduced by design methods. Scope available includes risk reduction by probability of occurrence (P), by consequences (C) and by a combination of both P and C. In addition, attention should be given to introducing barriers as part of design solutions (see Section 8 for further discussion).

6 RISK ASSESSMENT OF ENVIRONMENTAL HAZARDS

The main tasks in doing a risk assessment are as follows. The first task is to identify hazards associated with the performing of a specific decommissioning operation. A hazard is defined as something that can lead to undesired effects or harm in the process of meeting the objective. In general, there are several types of hazards caused by one or more sources. For example, accidental damage of a flow line could lead to leaks and disturbance to seabed. The second task is to assess the risk levels of the identified hazards. Risk is a term that is often confused with the term hazard and used interchangeably as if they have same meaning. Risk has two parameters; the likelihood of hazard occurring (P) and how severe (C) it will be when compared with some reference standards. It is meaningful to classify risk into three regions—intolerable, tolerable and negligible levels. The third task is to reduce intolerable risk to tolerable levels by addressing both P and C. For further details, see (Kuo 2007).

Some decommissioning activities that will generate hazards with various risk levels include (see Table (1)).

7 AN ILLUSTRATIVE APPLICATION EXAMPLE

The example selected is the decommissioning of a subsea satellite unit. This is illustrated in Figures 8(a) and 8(b).

Table 1. Decommissioning activities and associated hazards.

No.	Decommissioning activities	Typical hazards
H1	Shutting down wells	High pressure, leaks
H2	Work near satellite unit	Noise, disturb seabed
H3	Disconnect flow lines, pipes	Leaks, pollute seabed
H4	Removing flow lines, pipes	Collisions, leaks
H5	Cutting structural components	Noise, pollute seabed
H6	Waste removing	Disturb seabed, noise
H7	Towing structural components	Collisions, disturb seabed
H8	Loading components to barge	Damage to barge

As an example, disconnecting flow lines and pipes from a satellite unit is selected. The hazards are outlined in Table (2).

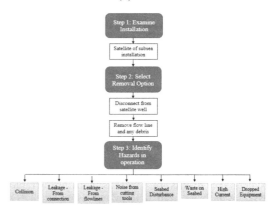

Figure 8a. Steps 1–3 of DfD methodology.

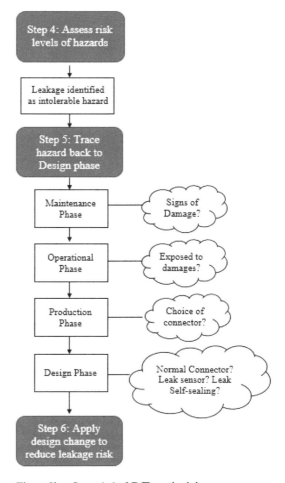

Figure 8b. Steps 4–6 of DfD methodology.

Table 2. Examples of hazards associated with disconnecting flowlines/pipes (H3).

No.	Hazards	Risk level of hazard
H3.1	Collision—ROV and unit	Tolerable
H3.2	Leaks—From connection	Tolerable
H3.3	Leaks—From cut flow lines	Intolerable
H3.4	Noise—From cutting tools	Tolerable
H3.5	Seabed disturbance—Work	Tolerable
H3.6	Waste on seabed—Cuttings	Tolerable
H3.7	High current—Unexpected	Negligible
H3.8	Damage from dropped items	Negligible

8 MAKING USE OF DFD RESULTS

When the results of hazards with intolerable risk levels have been determined using the DfD method, the information would be taken back to the design phase. The efforts would then be used to reduce these intolerable risks to tolerable by design methods.

Ideally, fresh and innovative methods should be sought and applied. In practice this requirement can be very difficult to achieve because the innovative ideas must be first generated, and then go through a number of "tests" to establish their applicability and robustness. Even when these steps are completed, the proposed solutions have to be "sold" to colleagues and the organisation's management. Generally, this means that the process can be very time consuming and the likelihood of quick implication is small.

The alternative to the ideal solution is to introduce barriers to reduce risk. There are a number of barriers which can be used and the main ones include the following:

a. *Physical barriers*: This barrier is usually associated with hardware and used in engineering solutions. For example, erecting a panel to stop or reduce noise from transmitting from one location to another. In practice, several layers of barriers can be installed in series. However, there is no guarantee they will be fully successful. J. Reason, in his book on human errors (Reason 1990), showed that such an arrangement can be like a block of Swiss cheese with holes in it and would not prevent errors from penetrating the barriers.
b. *Organisational barriers*: This barrier is based on software and is popular with organisations. Examples include introducing working procedures and giving special training to those involved in the decommissioning activities. While this is a very valuable barrier, it has to be "independent" and robust during the usage. When checking the application of this barrier it should be done by persons who are independent from the designing of the procedure and not by a member of the same team.
c. *Human factors barrier*: It should be remembered that humans play a crucial role in the design and application of activity and system. Humans make errors and it is also humans who can prevent errors. For this reason, in the designing of solutions the influence of humans should be given due attention.

It will be noted that these types of barriers can be used either individually or together as appropriate for the situation.

9 DISCUSSION

There are a number of aspects which deserve brief discussion.

9.1 Aspects needing further research

So far, the present study has focused on the principles and illustrated application, but there is scope for more detailed investigations. Typical examples will include hazards related to environmental effects, but also the effort needed to recover from any damage that has occurred.

Noise reduction during cutting and minimizing seabed debris are other areas for research, as well as improving lifting capabilities in more adverse weather conditions. Collection of data on damages/corrosion could also be considered.

9.2 Speedier feedback system

Since we are dealing with non-absolute entities, a management system is needed. This in turn requires systematic feedback at all times, so that modification and amendment can be readily made. It would be helpful in view pf the advances in computer technology. Some form of Artificial Intelligence methodology could be introduced.

9.3 Lifecycle changes

Design phase is the starting point, and in the process of going through various phases of the lifecycle there will be changes. These are noted but not necessarily integrated into modified design features. As a result, the final situation can be different to what was originally planned. This is particularly true in the case of decommissioning. For example, the initial weight and the final weight can be quite different. This will affect the lifting. Since there is no simple way of knowing what the final weight is, one tends towards over capability. It could also affect the variability of lifting vessels, and their lifting capabilities.

10 CONCLUDING REMARKS

Firstly, there are four main types of offshore instillations needing to be decommissioned and the proposed DfD methodology offers a systematic way of incorporating design features into the decommissioning process.

Secondly, for the decommissioning activities it is essential to use a management system based approach to identify hazards and assess their risk levels before returning to the design phase for determining how risk can be reduced cost-effectively.

REFERENCES

Andrews, D. 2015. Ship design and Naval Architecture. International Congress, Cartagena.
Atitkar, Hrishikesh. 2017. Decommissioning of Offshore Jackets by Reverse Installation. MSc thesis, University of Strathclyde
Gorman, D.G. et al. (eds.). 1998. Decommissioning Offshore Structures. Springer-Verlag London Limited.
Hoare, D. (2017). Offshore oil and gas decommissioning seminar given at workshop in the course: Decommissioning of offshore oil/gas instillations. University of Strathclyde.
Kennedy, S. J., Martino, A.E., Kong, J. 2010. Development of Rules for Designing Steel-Elastomer Sandwich Plates for Ship Structures. 9th International Conference on Sandwich Structures ICSS 9.
Kuo, C. 2007. Safety Management and its Maritime Application. The Nautical Institute.
Milland, A.F. (ed.). 2011. Marine Engineering Reference Book: A guide to ship design.
Nielsen, A. 2016. Unmanned Wellhead Platforms—UWHP Summary Report. Ramboll.
Papanikolaou, A. 2014. Ship design: methodologies of preliminary design. Springer.
Reason, J. 1990. Human Error. Cambridge University Press.
Watson, D.G.M. 2002. Practical ship design. Elsevier Ocean Engineering series.

Understanding initial design spaces in set-based design using networks and information theory

C. Goodrum, S. Taylordean & D.J. Singer
Department of Naval Architecture and Marine Engineering, University of Michigan, Ann Arbor, Michigan, USA

ABSTRACT: The US Navy has adopted Set-Based Design (SBD) as the preferred design method for addressing increasingly complex naval products. SBD requires functional groups to create and negotiate design spaces, as well as to manage convergence of the intersection of these regions over time. While there has been significant research conducted on negotiated spaces, little has been done to understand the creation of design spaces from the outset of the SBD process. Understanding initial design space landscapes is crucial to effectively implement SBD. The creation of design spaces is predicated by the design tools used to generate design knowledge. This concept paper investigates the creation of design space information from the outset of the SBD process to understand whether employed tools predicate design spaces, if proper negotiation channels are open between design groups, and if all appropriate parties are involved in set negotiations. Previous research is leveraged to create an integrated design space network, based on representative design tools. This network is analyzed using information theory to determine inherent hierarchical network structures, which provides insight to the landscape of design spaces using the limited information present in early design stages.

1 INTRODUCTION

1.1 *Motivation*

Set-Based Design (SBD) is a convergent design method that is better suited to tackle complex engineering design problems as compared to traditional methods (Singer 2003, Bernstein 1998, Sobek 1997, Liker et al. 1996, Ward et al. 1995). While traditional design methods have been used to successfully design ships in the past, they are inadequate to handle the increasing complexity of modern naval design problems. This is because traditional methods, based on the design spiral approach (Evans 1959), attempt to develop a 'best' design satisfying all design requirements at each design iteration. While this method often produces converged designs, it falls apart in the face of complex design landscapes, where external, internal, and temporal factors mean the design problem can never be fully understood (Shields 2017a). This is due to both the path-dependent nature of generating design knowledge, and time and budget limitations (Singer et al. 2009, Shields 2017a).

On the other hand, SBD requires the creation of broad sets of design parameters which are used to define tradeoff information as it relates to the design. These sets are developed concurrently, and are gradually narrowed with increasing information fidelity until a more globally optimal design point is revealed. This removes the requirement to search for 'the best' feasible designs at each point in the process and instead aims at discarding designs which are infeasible or dominated. SBD delays decision making until later design stages when there is more information to make an informed decision. This process has enabled more flexibility in adapting to design changes, and removed constraints early on in the design process which often predicate solutions (Singer et al. 2009).

The benefits of applying SBD to complex design activities have led to the US Navy utilizing the SBD method for a number of activities, including the ship-to-shore-connector and requirements evaluation (Mebane et al. 2011, Doerry & Fireman 2009). SBD has been well advocated for in regard to future naval ship designs (Keane et al. 2006, Kassel et al. 2010, Eccles 2010, Doerry 2009, Sullivan 2008).

There have been a number of challenges to widespread adoption of SBD in the US Navy. One major problem has been transitioning design tools that were successful in the traditional framework to enable success in SBD. Historically, there was a large push in traditional design approaches to remove the human from the design process through the use of automated codes. Improved computational power spurred the creation of large integrated toolsets and synthesis models. These have increasingly been used to select designs, instead of being used as a design tools (McKenney et al. 2012), and are ripe sources of design bias (Sypniewski & Singer 2017). These tools have had extensive investment, research, and time dedicated to their creation, and are not likely to be replaced. As such, it is difficult to fit many tools which were effective in traditional point-based design approaches into the new SBD framework. The structure of the analysis tools and widespread removal of the designer from the design process has drastically affected the way design space negotiations are made and has had huge implications of the success of SBD.

To date, an extensive amount of research has been conducted on the way designs are negotiated between design groups (Parsons et al. 1999, Singer 2003, Gray 2011, Cuneo 2013), but little has been done to understand the role these tools play on structuring the SBD process. Significant information can be gained by mapping the relations between variables used in design tools to a network framework (Parker 2014). This framework provides insight to the design process, and uncovers variables important for internal and negotiated design spaces. It also provides a novel way of understanding the impact the approach to an engineering analysis has on the ability to negotiate between design groups.

This paper outlines a novel framework of investigating design tools and how they relate to the design method. The paper introduces the framework, how it can be used to analyze a design process, and how it can be used to enable SBD. Section 2 familiarizes the reader with the network terminology used throughout the paper. Section 3 describes how the variable network was created, and Section 4 outlines how hierarchical network structures can be elicited through the created network. Section 5 presents a representative case study to demonstrate the method, and the results are presented in Section 6. Section 7 expands the discussion to the impacts this method has on structuring existing tools to be applicable for SBD.

1.2 Related research

Significant work has been done on understanding the impact design tools have on shaping the naval design process (McKenney 2013, Parker 2014, Parker & Singer 2015). McKenney (2013) identifies that the overreliance on design tools leads designers to treat tools as 'black boxes' and can cause a misunderstanding of the entire design problem. Additionally, he highlights the need for tools that facilitate the exchange of information. Parker (2014) utilizes a multi-partite network framework to elucidate the structure of design tools and inspired much of the work conducted in this paper. Parker defines three node types in his formulation: variables, functions, and disciplines. Variables are related to one another through the function layers, and are prescribed a design discipline dependent on the nature of the function. Parker's approach enables a mapping of variable interactions through functions, to design disciplines which led to a larger understanding the structure of naval design.

One shortcoming of the existing research is the requirement to prescribe what functions lend themselves to which disciplines ahead of time. While in some cases these allocations are obvious, many times functional memberships are not totally discrete. The framework presented in this paper enables the mapping of variables to disciplines organically through the structure of the network itself. The proposed method enables the disciplines to be uncovered without the need to define them a-priori. This research also provides insight as to how groups should be organized to maximize the efficiency of information flow, and provides insight as to which variables should be used to negotiate between groups.

2 NETWORK TERMINOLOGY

Networks are utilized throughout this work to uncover the structure of design tools and a design process. *Networks* (or *graphs*) are powerful and versatile tools which abstract representations of a system using points (*nodes*) and lines (*edges*). Nodes represent entities, and edges represent relationships between them. Edges can be directed or undirected, depending on the nature of the relationship. Networks' ability to represent and utilize information of varying fidelity makes them powerful tools to study design. Abstracting design tools to network space allows the structure of the tool to be studied independent of how it is implemented. The study of how these nodes and edges relate to one another enables the study of the network's *structure*, which can be quantified in a number of ways. As such, this section briefly outlines key network terminology used throughout the remainder of the paper. For an in-depth study of networks, see Newman (2010).

A key metric used to understand a network's structure is the degree distribution. A node's *degree* is the number of edges connected to it. For directed graphs, additional information can be gained by considering *in-degree* and *out-degree*, which measure the number of inbound and outbound connections a node has, respectively. These are node-centric measures of structure, and are the basis for more global structural perspectives, such as *centrality* and *communities*.

Additional information about the network structure can be gained by considering paths through the network. A *path* is a sequence of nodes such that every consecutive node in the sequence is connected by an edge. A *random walk* is a path across the network created by taking repeated random steps between nodes by moving along connected edges. Beginning at an initial vertex, an edge is selected at random to create the next step in the path, and the process is repeated. Random walks allow for nodes and edges to appear multiple times the sequence defined by the path, which is an important distinction applied to the analysis in this paper.

In order to deduce additional information about the way networks are structured, *community detection* algorithms are often employed. This refers to the division of the network into groups, or *communities*, according to the pattern of edges in the network such that there are a large number of connections within a group and relatively few connections between communities. A large number of community detection algorithms exist all of which attempt to classify nodes dependent on their relations to other nodes in the network.

3 THE VARIABLE NETWORK

To map the structure of the design tools, a *variable network* is created. The variables added to the network are any parameters which are involved in a calculation in a design tool (length, beam, etc.). The variables used to generate design information will be finite, and hence can be mapped to a network. This network represents variables as nodes. Edges represent functional (mathematical) relationships between variables. These relations are represented as directed edges. A directed edge is drawn from variable A to variable B if and only if A is used to calculate B. That is to say, directed edges point to dependent variables from the variables used to calculate them. An example is shown in Figure 1.

By formulating the network in this way, purely independent variables (inputs) have zero in-degree and non-zero out-degree. Conversely, outputs will contain zero out-degree and non-zero in-degree.

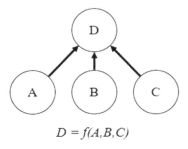

Figure 1. Representative variable network representing a variable D as a function of variables A, B and C.

Nodes with non-zero in and out degrees are intermediate dependent variables. Also note that in this representation, no nodes should have both zero in and out degrees, as this would imply variables exist which have no effect on any other variable. In general, the graph should also be acyclic. This is intuitive, as any directed cycle would result in a circular variable references which cannot occur. This is not to say there are no cycles in the design process, since we know the point-based design approach is iterative. Instead, this variable network is representative of a single loop around the design spiral, or through a single iteration of design information. Each subsequent iteration of a design will affect the independent variables (inputs), which will in turn propagate through the network to alter the dependent variables.

Using this network generation process, the complete variable network can be built by including all variables used in a design tool, by evaluating the equations used to generate the design information. Note that the resulting network does not explicitly represent the mathematical function relating the variables, it simply represents that variables are functionally related. This is advantageous in generating the network, especially in the limit of large networks, which can arise as software and toolkits become more complicated. For small tools, it may be viable to map variables manually, however for larger software packages, the network generation tool could be automated by tracing data references.

The construction of the variable network in this manner yields a number of benefits. Firstly, it is quite straightforward to create. Also, it provides transparency to the way a design tool and design process is structured from an information perspective. As design can be viewed as the act of generating knowledge for decision making over time (Parker 2014), this approach provides a framework that enables quantitative analyses. The framework is flexible enough to enable the generation of significant insight to the design process, independent of the implemented design tools used, by using various network-centric metrics and algorithms. This information can be utilized to better align tools with the selected design approach.

4 HIERARCHICAL NETWORK STRUCTURES

As discussed in Section 2, identifying communities within a network not only provides additional information about the network's structure, but aims to uncover implicit relations within the network by studying the connections of nodes. Traditional community detection approaches are based on modularity maximization which aim to find natural divisions of a network into groups of nodes which have many edges within the group, and few between groups. Many algorithms aim to do so by solely considering the structure of the network, while others determine divisions using the way information flows through the network (Newman 2010). The flow of information, as well as the information itself can reveal a lot about a network's structure and hence community structures. The first algorithm applying information theory to community detection in networks was posed by Rosvall & Bergstrom (2008), and was given the name *InfoMap*.

InfoMap reveals inherent hierarchical structures implicit within networks. The authors apply information theory to a network by conducting a random walk on the network long enough to visit all nodes at least once. Over the course of this random walk, certain nodes are more likely to be visited than others due to the network structure. Then, Huffman codes (Huffman 1952) are used to assign each node a unique binary bit string, the length of which is based on the frequency which a node is visited in the random walk. In this way, a random walk can be quantified by a single binary string based on the nodes encountered. Then, for each possible community division of the network, the task is to find the shortest length of the bit description of this random walk. This will be the most efficient partition of the network from an information-transmission perspective. Making use of Shannon's source coding theorem (Shannon 1948), the average number L of bits per step in the walk is equal to the entropy of the random walk, which is used to derive the map equation, defined as:

$$L = qH(Q) + \sum_g p_g H(P_g) \quad (1)$$

where q is the fraction of time that a random walk spends hopping between groups, p_g is the fraction of time the random walk spends within and exiting group g, Q represents a sequence of entry label nodes (nodes in which a random walk enters a group), and P_g represents the sequence of the rest of the nodes in group g through which the random walk passes. The function $H(x)$ corresponds to Shannon's information entropy and is defined as:

$$H(x) = -\sum_i x_i \log_2 x_i \quad (2)$$

Thus, $H(Q)$ accounts for the entropy of the sequence of groups that a random walk passes through, and $H(P_g)$ is the entropy of the nodes traversed within group g. This method has been shown to effectively partition graphs into communities, and has been extended to determine multiple hierarchical partitions within networks (Rosvall & Bergstrom 2009).

5 CASE STUDY

A representative case study was created to demonstrate the creation and analysis of the variable network. The case study is based on the design activities of University of Michigan (UofM) Naval Architecture and Marine Engineering senior undergraduates in completing their capstone ship design course. Students utilize a number of in-house design tools to conduct a preliminary point-design of a containership, given a set of design requirements. Many decades of students have utilized these tools for their designs, so they are well verified and validated. Studying these point-based tools enables the structure of traditional design tools to be understood by studying the structure of the associated variable networks in the context of an integrated design activity.

The tools considered in this case study include the UofM Cost Prediction, Midship Section design, and Weights I spreadsheets, the UofM Propeller Optimization Program (POP) (based on Parsons (1975) and Oosurveld (1975)) and UofM Powering Prediction Program (PPP) (based on Holtrop & Mennen (1982) and Holtrop (1984)). These tools are used to determine the ship's cost, structure, weight, propeller design, and powering requirements, respectively. While there are additional tools used by students, the tools considered in this case study provided enough variables to demonstrate the method. Larger scale implementations of the method are left for future work.

These preliminary design tools are based on a combination of regression analyses, experimental data, and theory, which are implemented through the form of C++ codes, Excel spreadsheets, and Fortran codes. The differences in implementations of analysis codes are akin to implemented design tools in industry used by disparate design teams. The different analysis methods are included to highlight the flexibility of this approach, as results are not predicated on the implemented analysis tool, but rather focus on the structure of combined variables as a method of yielding a solution.

Each of these tools were dissected to determine the equations and variables used in their analyses. Due to the limited size and numbers of the tools utilized, it was practical to study the structure of each software manually. For each considered tool, each utilized variable was recorded, in addition to the dependencies between variables. Careful attention was paid to ensuring all variables were

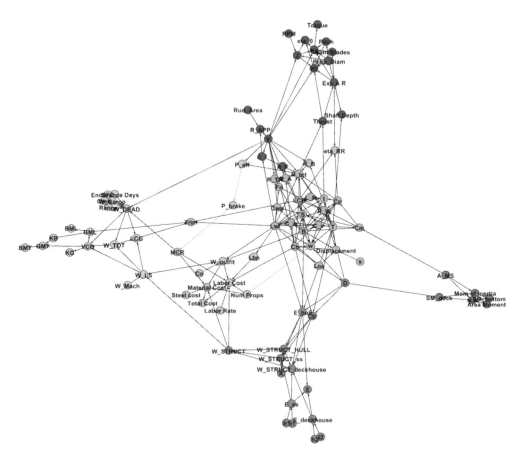

Figure 2. Variable network created using UofM design tools for preliminary containership design. The resulting graph visualization illustrates clear clustering. Nodes are colored according to the Louvain algorithm and represent different identified communities.

accounted for, and accounting for various notations between tools. These relations were used to create the variable network shown in Figure 2 using the process described in Section 3. The created network contains 92 nodes, and 209 edges.

6 RESULTS

6.1 Variable network interpretations

The network shown in Figure 2 provides interesting insights as to the structure of the design analysis tools. The network has been arranged as a force-directed graph, which aims to minimize the number of crossing edges and locates similar nodes closer together, and less similar nodes farther apart. Given this visualization the structure of the network reveals a number of variable clusters. Interestingly, these clusters correspond to their associated design discipline. The Louvain algorithm (Blondel 2008) is applied to identify communities, which are shown in different colors. Also notable is that the resulting variable network is indeed acyclic.

The standard modularity maximization approach confirms the communities apparent through the network visualization. The identified clusters are as follows: dark blue nodes represent powering and propeller design, light blue nodes represent overall weight and stability, green represents cost, red represent structural weight, and grey represents midship structural design. These clusters were expected, as they correspond to the five design tools considered in the case study. Interestingly, a sixth cluster emerges in the center of the graph, which is connected to all other clusters. This does not appear to coincide with any specific tool, and is shown in pink. Many of the variables in this cluster correspond to the hullform shape (L,B,T, Displacement, etc.) and classic naval architecture parameters. Not surprisingly, this cluster also contains a large proportion of the independent variables (nodes with in-degree of zero) as these entail the design requirements.

From this consideration a number of insights can be drawn about the way the design process is structured. First, the size of the communities (in terms of the number of variables within groups) provides insight to both the fidelity and complexity of the discipline's analysis method. Groups with larger numbers of variables are likely to be of higher fidelity, and those with a large number of edges are likely to be more complex tools. Additionally, the community structure suggests that the naval architecture specialty is central to the design activity, as it is the only community with connections to all other design disciplines. It also illustrates communities which are dissimilar—in this case the midship section analysis is not directly tied to the stability calculations, however they are related through structural weight and through the hullform parameters.

The directionality of these links is also important to consider. The naval architecture discipline has a large number of independent variables (variables with in-degree zero), and as a result, the central cluster contains a large number edges pointing to other clusters. This provides insight into the initial flow of information, as traditionally these are variables which must be decided early on in order to progress the design. The directionality of the flows provides context for the sequence in which design information is generated. The paths from independent variables through the network reveal the order in which decisions must be made in order to design the vessel. However, if these variables are fixed (through requirements for example), they will immensely reduce the design flexibility by limiting the number of negotiation pathways through the network. This quantifies a main advantage of SBD—by not fixing these variables in early design stages, information is able to flow much more readily through the network. This will be further explored in Section 7.

6.2 Exploring hierarchical network structures

While the detailed structure of the variable network provides valuable insight into the way design disciplines are connected, and the way the tools used predicate design spaces—additional insights can be gained through an analysis of the hierarchical network structure. The map equation was applied to the representative variable network using the *InfoMap* algorithm created by Edler & Rosvall, and is displayed in Figure 3.

In Figure 3, nodes now represent identified design communities, and edges represent the flow of

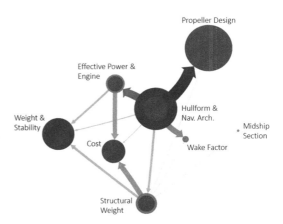

Figure 3. Hierarchical structure of the case study variable network using the *InfoMap* algorithm to identify design disciplines.

information. The node size corresponds to the time the random walk spent in the community, and the size of the edges represent the per step probability of moving between modules. Hence, the edges provide information between how likely information is to be transmitted on any given step of the design process between groups. The ratio of the inner to outer circles in each node represent the ratio of flow within and out of the community, respectively. Note that the node size of communities which are only pointed to (have no outflow of information) represent the inner circles of other nodes.

The communities identified by the map equation are very similar to those identified using the modularity maximization approach. The information-theoretic approach highlights the same broad communities, and elicits two additional communities. One community corresponds to the hullform wake factor, which arises from the high dependence on hullform parameters (as it is estimated using regression data) including Depth (D), which is important to the hull weight estimates. The other additional community arises from *InfoMap* separating the powering and propeller communities, which was merged in the modularity approach.

The edges in Figure 3 also provide valuable information about the flow of information between design communities. While the edges in Figure 2 represent variable dependencies, the edges in Figure 3 represent dependencies on design information. For example, the large flow from the naval architecture module to propeller design module suggests that the propeller discipline requires extensive information about the hull shape before the propeller can be designed. This dependency is intuitive and arises as a result of the point-based design structure, in which there exists a required sequence of analyses to be conducted to generate design information.

The directions of the flows reflect the central role the naval architecture discipline has on the other design areas, as all flows are directed out of the community. This means decisions made by the naval architect will have widespread impacts to all other design groups through the large amount of information flow. It also means that the majority of the negotiated design variables in this design approach will be naval architecture related—pertaining to the hullform. Conversely, the cost and stability communities have zero out-degree, which illustrate their analyses are predicated on decisions made by other groups. This is an interesting result, as stability and cost are disciplines which often emerge as design failures which are only obvious in late design stages (Cavas 2007, Shields 2017b). The emergence of design failures is a result of the number of dependencies on other groups' design decisions and the path by which they are generated as they require information from other groups before they can be calculated. They often are characterized by necessary rework, failure to create a converged design, or through increased design effort (Braha & Bar-Yam, 2007).

Perhaps the most insight to be gained from Figure 3 lies not in considering the existing edges, but those which do not exist. The absence of an edge between two groups which are seemingly closely related suggests a communication pathway does not exist, when it likely should. For example, consider the propeller and powering modules—these are highly interrelated design efforts in practice—engine selection and propeller optimization are highly dependent on one another. However, given the structure of the variable network, they are primarily related through the ship's speed (V), which is an independent variable (with in-degree zero) and resides within the naval architecture module. Given the directions of the flows, the two groups' analyses are both predicated on information transmitted from the naval architecture group. This type of insight can be used to reveal the underlying sequence of the design process, but could also be extended to analyze the effects of fixing requirements (in the form of variables) on the efficiency and rate of information transfer between groups.

7 IMPLICATIONS FOR SET-BASED DESIGN

The variable network outlined in Section 3 and the associated hierarchical representation outlined in Section 4 have presented the framework of studying a design activity. The case study presented and analyzed in Sections 5–6 have presented the types of insights which can be gained from such an analysis. However, the question still remains: what implications does this framework have on structuring a SBD design activity?

The first step in the SBD process, identified by Sobek et al. (1999), is mapping the design spaces. This is conducted to define feasibility, explore tradeoffs, and communicate sets of possibilities. Determining the design spaces from the outset of a project is integral to the success of the design activity. Herein lies many of the key contributions of the variable network approach and hierarchical structure analyses. The key contributions of the framework as they pertain to SBD are summarized below. The proposed framework:

- Maps initial design spaces, and generates information about the design process.
- Utilizes information flows to quantify design spaces and communication pathways between groups to make informed decisions about structuring design disciplines.
- Evaluates the structure of the design tools and processes independent of the implemented medium.
- Enables design tools to be better aligned with the selected design approach.

The variable network lends itself to a number of key insights which enable SBD. Clearly the network enhances the transparency of design tools, and how that fits into the larger design process. Illustrating functional dependencies between variables from the outset of the design activity enables designers to understand the design approach and understand tradeoffs. Additionally, the size of the resulting variable network provides context as to the size, scope, and complexity of the design activity. The network decouples the structure of the design problem from the medium through which it is implemented. This reduces the confusion between a design tool and design method, which is a key role in reducing designer understanding (McKenney 2013). The method aims to better incorporate the decision maker into the design process, and reduce overreliance on design tools, by opening the 'black box'.

The generation of the variable network can be conducted in parallel, which enables different functional groups to work concurrently. By doing so, it also enables functional groups to uncover potentially unknown interdependencies between variables which are crucial considerations from a SBD perspective. These unknown interdependencies are crucial to identify early in the SBD process, to ensure adequate communication pathways are opened for set negotiations. These interdependencies can exist as resident pathogens in the design system or design process, and can lead to emergent failures over time

(Leveson 2004, Perrow 2011). Identifying these interdependencies early on assist in front-loading the design process to identify issues when there is maximal design freedom, which is directly aligned with the SBD mantra.

By analyzing the hierarchical structure of the variable network, inter-variable dependencies can be used to determine inter-group dependencies. This scalability, based on the optimal flow of information, further bridges the gap between the structure of design tools to the design approach. This provides insight as to how design tools can be potentially re-aligned to better enable SBD, by classifying variables into design spaces based on the network structure. For example, variables (nodes) only connected to other nodes within their own group represent *internal design spaces*. Alternatively, variables pointing to nodes in different groups, but only pointed to by nodes within their cluster would be classified as *broadcast spaces*. Finally, *negotiated spaces* arise from nodes with zero in-degree (independent variables) which are those which must be negotiated to generate solutions. These classifications of design spaces would be relatively straightforward to determine from the network based on structure alone and are directly tailored from the design tools and approach being used. This is crucial as novel design activities will have unique and distinct design spaces dependent on both the nature of the design activity and the tools being utilized, and the method incorporates both.

In addition to classifying variables into design spaces, the network can be used to evaluate communication pathways between functional groups using the network's implicit hierarchical structure. Intra-group connections in the variable network, such as those shown in Figure 2, provide context about specific variable negotiations, while connections in the hierarchical structure (Figure 3) provide a higher-level context about the quantity and directionality of information flows between groups. The process of viewing the macro-level information flows between groups enables missing communication pathways to be identified, and enables the variable network to be restructured to solve these issues. The impact of changing the relationship between variables is immediately measurable by observing the way the hierarchical structure of the network changes. Restructuring the variable network to manage communication pathways between groups can be used to tailor design tools to better enable SBD.

The method is extendable to elicit information about potentially problematic areas of a design approach. A number of issues associated with structuring a SBD activity identified by Singer (2003) include: clearly defined design variables, balancing of workloads across design disciplines, and agent information overload. Many of these issues can be addressed using the variable network framework and resulting hierarchical structure. By clearly defining variables in a mathematical sense, the subjectivity of variables is eliminated. Analyzing optimal information flows in the hierarchical variable network provides insight to the balance of design tool utilization, which can be used to better balance workflows. These information flows can also be analyzed from a variable perspective to provide insight into which variables or groups are exhibiting information overload.

Information overload may be identified through information transmission metrics, and by determining variables with edges to a large number of disciplines. Highly connected and congested nodes may suggest that variable refinement is required, which may be able to adjust the tool to be better utilized in the SBD framework. As an example, consider the representative variable networks shown in Figure 4. In the network on the left, total weight (W_{TOT}) is expressed as a function of machinery weight (W_M), outfitting weight (W_O), structural weight (W_{STR}) and deadweight (W_{DWT}). In this case, W_{TOT} has edges to three other design disciplines (*D1*, *D2*, and *D3*). If information overload were an issue at this node, variable refinement may be applied to restructure the communication pathways. The network on the right illustrates variable refinement whereby an additional intermediate variable representing lightship weight (W_{LS}) has been added. Note that both cases would yield the same answer given the same inputs, yet the networks have different structures. In this example, the addition of the W_{LS} variable reduces the information overload in W_{TOT} by enabling negotiations with *D1* through W_{LS}. In this example, this would be a relatively minor change to an implemented tool, yet this restructuring process could present significant challenges in more complicated tools.

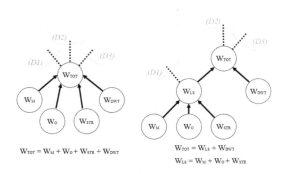

Figure 4. Two variable network structures representing the same function, with different communication pathways for negotiated design spaces.

8 CONCLUSIONS

This paper presents a method for determining the structure of a design approach simply by enumerating the variables considered in a design effort, and by tracking their dependencies through a set of design tools. The structure of the resulting variable network is not dependent on the actual form of the variable functions, nor is it dependent on the medium through which the analysis is conducted. The network provides a number of insights to a design effort. First, community detection can be used to determine functional groups (the design tools) without defining them a-priori, and allows for additional groups to emerge (such as in the case of the naval architecture group) which may not exhibit themselves through other design tools. By considering the hierarchical network structure, the flow of information within and between functional groups can be determined, which not only highlights inter-discipline dependencies, but quantifies the amount of information flow required between groups. This information provides insight to address potential issues earlier in the design process, as well as to identify the applicability of existing design tools to be integrated into the SBD framework.

The presented case study highlights the importance of the role of the naval architect in the design activity. In the traditional point-based approach, the decisions that the naval architect makes early on has ramifications in all other design disciplines, and allows for emergent design failures in groups with no out-degree. If these tools were to be utilized in a SBD framework, the connectivity of the naval architecture discipline to all other design disciplines shifts its role from not only generating design information, but also to acting as an integration manager. Given the structure of the considered variable network, the naval architect would be responsible for both determining variables and enabling the negotiations between a large number of design disciplines. The absence of communication pathways between seemingly related fields may help tailor the approach of the design tools to be more effective in relation to negotiated design spaces and SBD overall.

Significant research remains to be completed in this domain. Further work includes:

- Determining ideal information-flow structures of design disciplines in SBD.
- Studying dynamic network generation over the course of a design to incorporate increasing information fidelity.
- Time dependent analyses of information flows and routings, expanded to consider decisions made and rates of design convergence.
- Expanding the *InfoMap* analysis to allow partial variable memberships across different groups.
- Considering weighted edges between variables.
- Determining the impact of fixed requirements on design flexibility, information flow, and time to design.

While the case study considered a relatively simple ship design, the authors believe the method is scalable to incorporate more complicated designs, and flexible enough to be applied across a wide range of industries and engineering design efforts outside of the naval realm.

ACKNOWLEDGEMENTS

We would like to thank Kelly Cooper from the Office of Naval Research (ONR) for providing support for this project. This work was funded under grant number N00014-15-1-2752. We would also like to thank Michael Sypniewski for his insightful comments and helpful review of the paper.

REFERENCES

Bernstein, Joshua I. (1998). Design Methods in the Aerospace Industry: Looking for Evidence of Set-Based Practices. *Master of Science Thesis, Massachusetts Institute of Technology.*

Blondel, V.D., Guillaume, J.L., Lambiotte, R., & Lefebvre, E. (2008). Fast unfolding of communities in large networks. *Journal of statistical mechanics: theory and experiment, 2008(10), P10008.*

Braha, D., and Y. Bar-Yam (2007), The Statistical Mechanics of Complex Product Development: Empirical and Analytical Results. *Management Science, 53 (7), 1127–1145.*

Cavas, Christopher P. (2007). Is New U.S. Destroyer Unstable? Experts Doubt Radical Hull; Navy Says All Is Well. Retrieved from https://www.defensenews.com/naval/2015/12/03/instability-questions-about-zumwalt-destroyer-are-nothing-new/.

Cuneo, B.J. (2013). Development of a Human Centric Multidisciplinary Design Optimization Method Using Fuzzy Logic Systems and Controllers. *Ph.D. Dissertation. University of Michigan.*

Doerry, N., & Steding, M. (2009). Ship Design Manager Manual. *Naval Sea Systems Command.*

Doerry, N. & Fireman, H. (2009), Fleet Capabilities-Based Assessment. *Naval Engineers Journal, 121: 107–116.*

Eccles, T.J. (2010). Ship Design and Analysis Tool Goals *(COMNAVSEA memo 9000 Ser 05T/015).* Washington, D.C.

Edler, D & Rosvall, M. The MapEquation software package, available online at http://www.mapequation.org.

Evans, J.H. (1959). Basic Design Concepts. *Naval Engineers Journal, Vol. 21, Nov.*

Gray, A.W. (2011). Enhancement of Set-Based Design Practices Via Introduction of Uncertainty Through the Use of Interval Type-2 Modeling and General Type-2 Fuzzy Logic Agent Based Methods. *Ph.D. Dissertation. University of* Michigan.

Holtrop, J. (1984). A statistical re-analysis of resistance and propulsion data. *Int Shipbuild Prog, 31, 272–276.*

Holtrop, J., & Mennen, G.G. (1982). An approximate power prediction method.

Huffman, D.A. (1952). A method for the construction of minimum-redundancy codes. *Proceedings of the IRE, 40(9), 1098–1101.*

Keane, R., Fireman, H. & Billingsley, D. (2006), Leading a Sea Change in Naval Ship Design: Toward Collaborative Product Development. *2005 SNAME Maritime Technology Conference and Expo and Ship Production Symposium, Houston, TX*

Leveson, N. (2004). A New Accident Model for Engineering Safer Systems. *Safety Science (42).*

Liker, J.K., Sobek, D.K., Ward, A.C., & Cristiano, J.J. (1996). Involving Suppliers in Product Development in the United States and Japan: Evidence for Set-Based Concurrent Engineering. *IEEE Transactions on Engineering Management, 43(2), pp. 165–178.*

McKenney, T.A., Buckley, M.E., & Singer, D.J. (2012). The Practical Case for Set-Based Design in Naval Architecture. *Paper presented at the International Marine Design Conference, Glasgow, UK.*

McKenney, T.A. (2013). An Early-Stage Set-Based Design Reduction Decision Support Framework Utilizing Design Space Mapping and a Graph Theoretic Markov Decision Process Formulation. *PhD thesis, University of Michigan.*

Mebane, W.L., Carlson, C.M., Dowd, C., Singer, D.J., & Buckley, M.E. (2011). Set-based design and the ship to shore connector. *Naval Engineers Journal, 123(3), 79–92.*

Newman, M.E.J. (2010). Networks: An Introduction. *Oxford: Oxford University Press.*

Oosterveld, M.W.C., & Oossanen, P.V. (1975). Further computer-analyzed data of the Wageningen B-screw series. *Int Shipbuild Progr 22, 251–62.*

Parker, M.C. & Singer, D.J. (2015). Analyzing the dynamic behavior of marine design tools using network theory. *Ocean Engineering 106, 227–237.*

Parker, M.C. (2014). A Contextual Multipartite Network Approach to Comprehending the Structure of Naval Design. *Ph.D. thesis, University of Michigan.*

Parsons, M.G. (1975). Optimization methods for use in computer-aided ship design *(No. 13).*

Parsons, M.G., Singer, D.J. & Sauter, J.A. (1999), A Hybrid Agent Approach for Set Based Conceptual Design. *10th International Conference on Computer Applications in Shipbuilding (ICCAS), Cambridge, Vol. 2, pp.207–221.*

Perrow, C. (2011). Normal accidents: Living with high risk technologies. *Princeton university press.*

Rosvall, M., & Bergstrom, C.T. (2008). Maps of random walks on complex networks reveal community structure. *Proceedings of the National Academy of Sciences, 105(4), 1118–1123.*

Rosvall, M., Axelsson, D., & Bergstrom, C.T. (2009). The map equation. *Eur. Phys. J. Special Topics, 178, 13–23.*

Sullivan, P.E. (2008). Ship Design and Analysis Tool Goals *(COMNAVSEA memo 9000 Ser 05D/047). Washington, D.C.*

Shannon, C.E. (1948), A Mathematical Theory of Communication. *Bell System Technical Journal, 27: 379–423.*

Shields, C.P.F. & Singer, D.J. (2017a). Naval Design, Knowledge-Based Complexity, and Emergent Design Failures. *Naval Engineers Journal, 129(4), 75–86.*

Shields, C.P.F. (2017b). Investigating Emergent Design Failures Using a Knowledge-Action-Decision Framework. *Ph.D. Thesis. University of Michigan.*

Singer, D.J., Doerry, N., & Buckley, M.E. (2009). What Is Set-Based Design? *Naval Engineers Journal, 121(4), 31–43.*

Singer, D.J. (2003), A Hybrid Agent Approach for Set-Based Conceptual Ship Design through the Use of a Fuzzy Logic Agent to Facilitate Communications and Negotiation. *Ph.D. Dissertation. University of Michigan, 2003.*

Sobek, D (1997). Principles that Shape Product Development Systems: A Toyota-Chrysler Comparison. *PhD thesis, University of Michigan.*

Sobek, D. & Ward, A.C. & Liker, J. (1999). Toyota's Principles of Set-Based Concurrent Engineering. *Sloan Management Review. 40.*

Sypniewski, M.J. & Singer, D.J. (2017). A Framework for Asessing the Effects of Bias in Early-Stage Ship Design. *International Conference on Computer Applications in Shipbuilding (ICCAS). Singapore.*

Ward, A.C., Liker, J.K., Christiano, J.J., & Sobek, D.K. (1995). The Second Toyota Paradox: How Delaying Decisions Can Make Better Cars Faster. *Sloan Management Review, 36(2), pp. 43–61.*

Structural design

Probabilistic assessment of combined loads for trimarans

H.C. Seyffert & A.W. Troesch
Department of Naval Architecture and Marine Engineering, The University of Michigan, Ann Arbor, Michigan, USA

J.T. Knight & D.C. Kring
Navatek Ltd., South Kingstown, Rhode Island, USA

ABSTRACT: Combined loading presents formidable challenges to the design of multihulls. Without the canon of anecdotal knowledge, legacy designs, and test data that benefits monohull design, significant concerns around multihulls remain open-ended questions. In Lloyd's Register "Rules for the Classification of Trimarans," rule load combinations are suggested to ensure that the "hull structure complies … with the [LR] acceptance criteria" (Lloyd's Register 2017). These eight standard load combination cases were constructed using a deterministic Equivalent Design Wave methodology (Blanchard & Ge 2007). The load combinations, though, do not explicitly include probabilistic aspects apart from the underlying LR Rule assumption (20-year return period, or probability of exceedance at 10^{-8}). Given that the small number of load combinations is a "practical attempt to reduce the number of load cases to a reasonable number," it is not clear whether these cases are exhaustive when considering an irregular wave environment, nor is the degree of conservatism apparent. Should a detailed probabilistic assessment of lifetime performance be required, it is imperative that the load combination cases are accurate, sufficiently conservative, and exhaustive. In this paper, load combinations are examined using the Design Loads Generator, Aegir, and a copula model for the joint distribution of ship loads.

1 INTRODUCTION

1.1 *Motivation*

Trimaran design is inherently more complicated than monohull design due to the major (and obvious) challenges posed by multi-hull vessels. In particular, anecdotal knowledge based on decades of design experience does not necessarily apply. The loads experienced by trimarans are not as well defined as for typical monohulls, requiring novel, first-principle approaches to the analysis and design of structures. Additionally, combined loading on cross-deck structures may be significant. The interaction of different loads on these structures must be considered for adequate design performance. This paper considers the assessment of lifetime combined loading on a trimaran in a probabilistic sense through a short-term analysis.

1.2 *Background on combined loading methods*

Many authors have considered the problem of extreme values of combined loads, and a few major methods are reviewed here. Combined loading methods range from purely analytical approaches, to methods that estimate asymptotic distributions, to pure Monte Carlo Simulation based approaches.

Madsen, Krenk, & Lind (1986) consider combined loading in an analytical approach by analyzing processes that may achieve maximum values at different times. Such an approach makes use of Rice's level-crossing formula to approximate the mean out-crossing rate of a parametric curve by a load vector (Rice 1944). The issue with such formulations is that few closed-form solutions exist for out-crossing rates.

Naess & Moan (2013) discuss a similar method, but again note the difficulty in solving for the out-crossing rate of a vector of loads on a failure surface. The authors mention the possibility of simplifying the problem through a piece-wise linearization of the failure threshold, but admit the difficulty still inherent in evaluating such an expression. The authors suggest Monte Carlo-based methods to overcome the analytical challenges.

The First and Second Order Reliability Methods, FORM and SORM, can analyze combined loading by constructing a failure surface with a pre-determined response level and finding the most probable input that leads to the desired response. Jensen (2007) used FORM to determine the most probable wave input leading to the extreme roll of a containership. This method may become computationally intractable though, for high-dimension problems, due to the required linearization schemes. Also, for non-linear systems, it is unclear how useful the most probable input of a lower-order approximate model is. Without ensembles of statistically equivalent inputs which lead to design responses, the statistics of a non-linear response cannot be directly determined, but only inferred.

Ewans & Jonathan (2014) get around the difficulty of finding a joint distribution for multiple loads through the use of asymptotic distributions with max-stability. Maxima from max-stable distributions can be modeled by a generalized extreme value distribution. Therefore, no assumption about the load distributions needs to be made. Heffernan & Tawn (2004) introduced the conditional extreme model, which is based on assuming asymptotic forms of conditional distributions. Combined loads are considered through distributions of input loading based on a single load experiencing a maximum value. Less clear, though, is how to consider situations in which simultaneous moderate values cause interesting system responses.

Copulas have also been considered by many authors as a sound method to analyze linear load combinations, for example: deWaal & van Gelder (2005), Bastian, Pahlow, Hundecha, & Schumann (2009), and Bartoli, Mannini, & Massai (2011). However, copula analysis has not been widely applied to consider combined loading on ships for structural design, problems which present notable complexities.

Finally, full Monte Carlo simulations may be used to determine complex marine system responses to combined loading. However, for long exposures or many loads, this method may not be reasonable to collect converged statistics.

1.3 Design Loads Generator (DLG)

The Design Loads Generator method constructs an ensemble of short Gaussian inputs with a given spectral definition that lead to extreme values of an associated linear response (Alford (2008), Kim (2012)). The DLG makes use of the Gaussian extreme value distribution (see, e.g. Ochi (1990)) to assemble inputs that lead to extreme values that are members of an associated extreme value distribution for a given return period. Using the DLG, it is possible to capture the statistical extreme response of a system response for a given exposure. Kim, Engle, & Troesch (2013) successfully employed the DLG to determine nonlinear long-term combined wave bending and whipping for two hull forms in harsh sea conditions.

The DLG offers an advantage because it makes no assumption on asymptotic distributions. The underlying assumptions of the DLG are that the driving environment is Gaussian and there is a linear transfer function that relates the input-output system. As deep-water ocean waves are generally considered Gaussian and many vessel structural loads can be related to the incoming wave environment by a linear transfer function, these assumptions may be appropriate. With the ensemble of focused, short wave time series, a high fidelity non-linear simulator may then be employed to generate a conditional extreme value distribution.

1.4 Extreme value theory

The DLG makes use of extreme value theory, which is briefly summarized here. The interested reader should consult Ochi (1990) for a derivation of the given results.

The extreme value distribution is the distribution of extreme responses that a system experiences for a given exposure. A set of order statistics from a random sample $(x_1, x_2, \cdots, x_n)$ with size n is defined as $Y_1, Y_2, \cdots, Y_n$. This random sample is drawn from a distribution with the pdf $f_X(x)$. The joint pdf of $Y_1, Y_2, \cdots, Y_n$ is called $g(y_1, y_2, \cdots, y_n)$. The pdf of the largest value Y_n is found by Equation 1. Then, the most probable extreme value, $\bar{y}_n$, is defined as the solution to Equation 2. The most probable extreme value $\bar{y}_n$ is related to the number of samples n in the limit as $n \to \infty$ by Equation 3.

$$g(y_n) = nf_Y(y_n)\{F_Y(y_n)\}^{n-1} \qquad (1)$$

$$\frac{d}{dy_n} g(y_n) = 0 \qquad (2)$$

$$\frac{1}{n} \approx 1 - F_Y(\bar{y}_n) \qquad (3)$$

An issue of using the most probable extreme value as a design value for engineering purposes is that $\bar{y}_n$ has about a 63.2% probability of exceedance. Therefore, a risk parameter α can be defined. Then, there is an extreme value $\hat{y}_n$ that satisfies Equations 4–5.

$$\int_0^{\hat{y}_n} g(y_n) dy_n = \{F(\hat{y}_n)\}^n = 1 - \alpha \qquad (4)$$

$$p(X > \hat{y}_n) = 1 - F_Y(\hat{y}_n) = \frac{\alpha}{n} \qquad (5)$$

This formulation allows the definition of an extreme value associated with an exposure (which can be expressed by the number of cycles or samples n) and a risk parameter α (i.e. with a given probability of exceedance). Risk parameters will be used when combining a vessel exposure in a certain sea state with a design probability of exceedance, like that given by the Lloyd's Register rules. Here, the risk parameter α is used to define the linear system response which has the defined probability of non-exceedance $PNE = 1-\alpha$.

1.5 Copula method

A copula model is also used to analyze the load combination cases offered by LR for this trimaran hull form. A copula is a multivariate *pdf* in which each marginal distribution is continuous and uniformly distributed. This property is satisfied by the *CDF* of any random variable, which is by definition uniformly distributed between [0,1]. Therefore, copulas can be used to describe the correlation between marginal distributions of different random variables. Given a random vector of marginal distributions $(X_1, X_2, \cdots, X_d)$, the *CDF* transform of the vector is $(U_1, U_2, \cdots, U_d) = (F_1(X_1), F_2(X_2), \cdots, F_n(X_d))$, where $F_i(X_i) = p(X_i \leq x)$. The copula is then a function C which satisfies Equation 6. Equation 7 relates the copula back to the marginal *pdf*'s by the inverse *CDF*.

$$C(u_1, u_2, \cdots, u_d) = p(U_1 \leq u_1, U_2 \leq u_2, \cdots, U_d \leq u_d) \quad (6)$$

$$C(u_1, \cdots, u_d) = p(X_1 \leq F_1^{-1}(u_1), \cdots, X_d \leq F_d^{-1}(u_d)) \quad (7)$$

The method used here is similar to the Weibull approach for univariate extreme loads. The marginal time history for a load component i is given as $x_i(t)$. When analyzing trimaran multivariate loading, one approach is to sequentially select the individual load component which is to be maximized (call this load m), and then study the joint distribution of the other load components, conditioned on the load m being a local maximum. In this approach, events (maxima) are indexed by Equation 8, and Equation 9 gives the multivariate load event vector which describes the conditional joint distribution.

$$I = \left\{ t : \frac{d}{dt} x_m(t) = 0, \frac{d^2}{dt^2} x_m(t) < 0 \right\} \quad (8)$$

$$\overline{\xi}_j = \left(x_1(I_j), \cdots, x_d(I_j) \right) \quad (9)$$

$$F_{\overline{\xi}_N}(u_1, \cdots, u_d) = p\left(\xi_{N,1} \leq F_1^{-1}(u_1), \cdots, \xi_{N,d} \leq F_d^{-1}(u_d) \right) \quad (10)$$

A copula is fit to the vector of maxima events $\overline{\xi}_j$, and the distribution of the extreme maxima $F_{\overline{\xi}}$ is given by $F_{\overline{\xi}} = [F_C]^N$ where F_C is the *CDF* of the copula. Then, the copula can be related back to the original vector space by the inverse *CDF*, to calculate the probability of non-exceedance (*PNE*) of the extreme multivariate load $\overline{\xi}_N$. This formulation also gives the surface which is related to an overall *PNE* by Equation 10. The copula method is used to generate distributions of the loads on the trimaran hull, and will be compared with distributions assembled by the DLG and the standard load combinations offered by LR.

2 LLOYD'S REGISTER TRIMARAN RULES

2.1 Combined loading rules

Lloyd's Register (2017) specifies standard load combination cases to be used for trimaran structure design. Eight rule load combination cases are specified as testing cases to determine whether the "longitudinal, transverse and shear strength of the hull structure complies with the acceptance criteria." Seven cases consider physical loading, and the eighth considers roll motion, which is not considered in this paper. The coordinate system used for the LR rules, as well for the following analysis, is given in Figure 1.

M_{WH}/M_{WS} = hogging/sagging vertical wave bending moment
M_{SPH}/M_{SPS} = hogging/sagging splitting bending moment
M_{LT} = longitudinal torsional bending moment
M_H = horizontal bending moment
M_{TT} = transverse torsional bending moment

2.2 Development of rule load combination cases

The rule load combination cases defined by Lloyd's Register are developed using the Equivalent Design Wave (EDW) approach (Blanchard & Ge 2007). This design wave is a regular wave which leads to a response with a given long-term design value. The design value is defined as the largest value a load achieves over the exposure associated with a defined probability of exceedance. The seven cases give specific load combinations to test structural adequacy, where for each case a specific global load on the trimaran is maximized. A load combination factor, or *LCF*, for each load indicates the percentage of the design value that the load experiences for that specific case. The *LCF* for each load M_i is calculated by first considering the *RAO* of the maximized load, M_{max}, within the appropriate heading. The maximum value of that *RAO* is determined, a_{max}, and the corresponding phase, ε_{max}, and encounter frequency, $\omega_{e,max}$, are noted. The time, t_{max}, when M_{max} experiences its maximum value is calculated by Equation 11. The height of the equivalent design wave, h_{max}, is then given by Equation 12.

$$\cos\left(\omega_{e,max} t_{max} + \varepsilon_{max} \right) = \pm 1 \quad (11)$$

$$h_{max} = \frac{R}{a_{max}} \qquad (12)$$

R is the design value for M_{max}. The amplitude, a_i, and phase, ε_i, of each secondary load M_i at $\omega_{e,max}$ should be found. The LCF of each load is then calculated by Equation 13. Note, by construction, the LCF of the maximized load M_{max} is 1, as for M_{max}, $a_i = a_{max}$, $\varepsilon_i = \varepsilon_{max}$, and $h_{max}a_i/R_i = h_{max}a_{max}/R = 1$.

$$LCF_i = \frac{h_{max}a_i \cos(\omega_{e,max}t_{max} + \varepsilon_i)}{R_i} \qquad (13)$$

R_i is the design value for the given secondary load M_i. This design value should reflect the operating profile (speed, heading, exposure), and specifically the heading that leads to the largest response. Seven standard load combination cases using this methodology are defined by LR, shown in Table 1.

The total loading that the trimaran structure must survive is the sum of the static and dynamic loads (within the respective planes). This loading condition is to be applied to a FEA model to determine whether the structure performance is adequate. Note that for load cases in which the LCF is negative (i.e. $LCF_{M_{TT}}$ is negative in Cases 1, 2, and 6, and LCF_{M_H} is negative in Cases 5 and 7), the negative sign indicates the component is "considered reversible." For comparison with the DLG and the copula model, hogging moments are considered positive and sagging moments are considered negative, as with the LR coordinate system. Any LCF which relates to a sagging moment will be shown as a negative value, corresponding to negative sagging moments, whereas the negative "reversible" LCF values given in Table 1 are shown as positive.

It will be interesting to compare these LCF values from LR, constructed using the EDW approach, with distributions found from the DLG and the copula model. It should first be noted that the standard LCF values in Table 1 are not based on the trimaran hull form used for this paper. When using the rule load approach, as is done here, the standard load cases from Table 1 are to be applied to the trimaran hull. The LR rules do note, though, that the vessel response, and corresponding LCF values, may vary for different trimaran hulls- specifically that the fore-aft placement of the side-hulls can have a significant impact. Whether the standard load combination cases offered by LR are universal will be determined.

On another point, the spirit of the EDW and the DLG and copula analysis are the same, as all the methods consider a single maximized load and recover responses of other loads at the same instant. The difference, though, is that the DLG constructs ensembles of irregular wave inputs that lead to this maximum response, which gives a distribution of responses. The other responses are driven by this constructed irregular wave input. The copula model estimates the distributions of responses by the copula-driven sampling of long simulations of the vessel loads. The EDW method

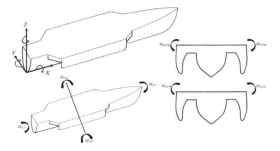

Figure 1. Coordinate system of trimaran hull used by LR and the DLG and copula analyses. Arrows indicate positive directions of the specified loads. The vertical wave bending moment M_{WH}/M_{WS} is found by the wave pressure distribution integrated across the depth of the hull, and M_H is found by the pressure distribution integrated across the breadth of the hull.

Table 1. Load combination factors defined by Lloyd's Register (2017).

Case & Heading	Dynamic Loads						
	M_{WH}	M_{WS}	M_H	M_{SPH}	M_{SPS}	M_{LT}	M_{TT}
1) Head	1.0	0	0	0.3	0	0	−0.2
2) Head	0	1.0	0	0	0.3	0	−0.2
3) Beam	0.1	0	0	1.0	0	0.2	0
4) Beam	0	0.1	0	0	1.0	0.2	0
5) Oblique	0	0	−0.3	0.4	0	1.0	0.3
6) Oblique	0	0	1.0	0.4	0	0	−0.2
7) Oblique	0	0.2	−0.2	0.6	0	0	1.0

uses a regular wave to drive the load response to return a single representative value of those loads at the given maximized instant. Whether the regular EDW can accurately capture the simultaneous response of multiple loads, and whether a single value is even representative of a distribution of responses, will be examined.

Note that for the copula analysis, the load distributions are extrapolated and thus the pressure distribution acting on the hull at the time of maximum load is lost. This contrasts with the DLG method where wave inputs are constructed, which gives these pressure distributions. The EDW method can also generate such distributions, but it is not clear how accurate these distributions will be due to the usage of a regular wave.

2.3 Design value specification

The LR rules reference Standard Wave Data (IACS Rec. No. 34 2001) for the determination of design values. Based on this recommendation, a design value is determined by the 20-year return-period response (corresponding to 10^8 wave encounters) in a North Atlantic environment using the Bretschneider spectrum. It is assumed that all wave headings have an equal probability of occurrence. The Lloyd's Register rules also stipulate that the evaluation of a long-term response be based on a design value with an overall 10^{-8} probability of exceedance. Each load design value is based upon the heading that leads to the largest response of that particular load.

3 METHODOLOGY FOR ANALYSIS

3.1 Probabilistic assessment

The major question that arises from the loading combination cases is what, if any, probabilistic basis exists to justify the given standard cases. Using the EDW approach may not be accurate for an irregular wave environment over a long exposure, and it is unclear how conservative or exhaustive the load combination cases are. As well, are the standard LR load cases applicable for every hull form? These questions are considered here.

3.2 Trimaran specifications

For this paper, the trimaran hull has the lines plan given by Figure 2 and the hull specifications given in Table 2. The linear version of the high-order potential flow code, Aegir, is used to generate transfer functions of multiple loads at the planes indicated in Table 2 with respect to the coordinate frame in Figure 1 (Kring, Milewski, & Fine 2004).

Aegir generates transfer functions of multiple loads at the planes indicated in Table 2, and the relationship between the loads and load planes is given in Table 3. The transfer function non-

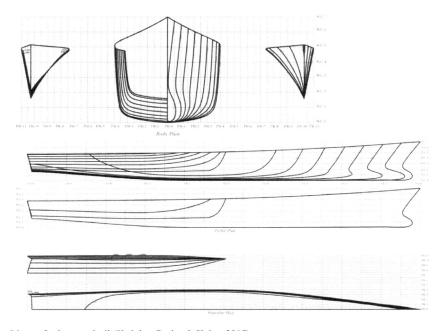

Figure 2. Lines of trimaran hull (Knight, Craig, & Kring 2017).

Table 2. Vessel & Aegir simulation specifications. Note that the [x,y,z] origins of the midship and center hull planes are with respect to the coordinate frame given by Figure 1.

Parameter	Value
Hull overall length (LOA)	110.0 m
Hull waterline length (L)	106.4 m
Total Draft	4.897 m
Beam	30.48 m
Origin of midship load plane [x,y,z]	[45.14, 0, 3.0] m
Origin of centerhull load plane [x,y,z]	[26.75, 5.48, 3.0] m
Water density	1026.06 kg/m^3
Vessel mass	3,301,440 kg
Number of wave frequency components	100
Speed (Froude Number)	12.803 m/s (0.4)

Table 3. Transfer functions of loads from Table 1.

Load	Transfer function
M_{WH}/M_{WS}	Midship vertical bending
M_{SPH}/M_{SPS}	Centerhull vertical bending
M_{LT}	Midship torsional bending
M_H	Midship horizontal bending
M_{TT}	Centerhull torsional bending

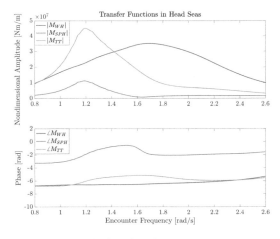

Figure 3. Transfer functions for head seas.

dimensional amplitudes and phases are shown in Figures 3–5. For the vertical wave and splitting bending moments, only the hogging conditions are plotted. The sagging conditions have the same amplitudes, with phases 180° out of phase with the hogging phases.

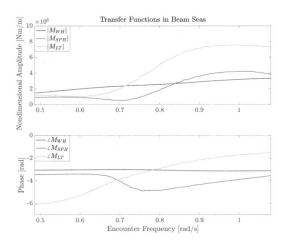

Figure 4. Transfer functions for beam seas.

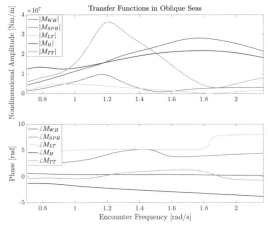

Figure 5. Transfer functions for oblique seas.

Table 4. Operation profile (North Atlantic).

Parameter	Value
Spectrum type	Bretschneider
Significant Wave Height H_s	12.5 m
Zero-Crossing period T_z	9.5 s
Probability of sea state occurrence	4.4e-5
Lifetime of vessel	20 years
Total exposure of vessel in given sea state	7.71 hours
Risk parameter α	2.27e-4
Probability of Non-exceedance PNE = 1 − α	0.9998
Overall probability of exceedance	10^{-8}

An operation profile is assigned to match the LR specifications in Table 4. To satisfy the overall required 10^{-8} probability of exceedance criteria, the risk parameter α is applied, as demonstrated by Equations 4–5. From the spectrum definition and the Standard Wave Data specifications, the probability of the sea state occurrence is defined, leading to a 7.71-hour exposure out of the 20-year lifetime. The design value of each load is then the maximum linear load value in any heading in a 7.71-hour exposure in the given sea state, with a probability of non-exceedance $1 - \alpha = 0.9998$. These criteria lead to the required overall load probability of exceedance 10^{-8}.

It is important to note that this overall probability of exceedance (10^{-8}) defined by LR seems to be based upon "the number of low stress reversals that might occur on a period of 20 years" (Blanchard & Ge (2007)), which is a long-term estimate based on all sea states and operational conditions. A design value may be defined as the most-probable maximum in 7.71 hours (out of the 20-year service life), but over the 20-year service life, there is approximately a 62.3% chance of exceeding this value, necessitating the addition of the risk parameter α. However, it is not clear how to relate a short-term analysis, like that performed here, to the long-term analysis parameters defined by LR. Without applying a risk parameter α, the analysis would be unacceptably non-conservative, while the risk parameter $\alpha = 2.27\text{e-}4$ might be extreme.

Ochi (1981) acknowledges the difficulty of choosing an appropriate α for a short-term analysis to relate to a long-term analysis. Ochi uses the example of a vessel designed using an extreme design value in a given sea state associated with $\alpha = 0.01$. Ochi notes that if the vessel is expected to encounter that sea severity 20 times during its lifetime, α must be divided by 20 to maintain the 99% criteria. Considering extreme midship bending moments of the MARINER-type ship (Russo & Sullivan (1953)) Ochi finds that "design extreme values do not increase substantially with increasing α– value", and concluded that a short-term approach can be adequate for the estimation of extreme values as long as the difference in the number of encounters is considered. This, in some ways, indicates that the choice of α is ambiguous. However, both the DLG and copula methods are valid for any choice of α, which is likely to be set by classification societies as empirical data is collected. For this paper, the risk parameter used will be $\alpha = 2.27\text{e-}4$, as given in Table 4.

The design values $R_{PNE\,=\,0.9998}$ conditioned on heading and associated with a $PNE = 0.9998$ are given in Table 5, along with the heading in which the true (i.e. largest) design value occurs. Note that these headings do not necessarily line up with the standard load cases given in Table 1. For example, M_{SPH}/M_{SPS} is predicted to experience its maximum value in beam seas, by LR Cases 3–4, but the design value for M_{SPH}/M_{SPS} occurs in oblique seas for this hull. This already is an important consideration.

Table 5. Design values for loads in all headings with $PNE = 0.9998$. The values in bold-face are the largest design value seen in any heading- these are $R_{PNE\,=\,0.9998}$. Note that "–" indicates no transfer function was constructed for that heading.

Load	Head seas	Beam seas	Oblique seas
M_{WH}/M_{WS}	**4.311e8 Nm**	3.993e7 Nm	3.090e8 Nm
M_{SPH}/M_{SPS}	9.054e7 Nm	3.551e7 Nm	**9.159e7 Nm**
M_{LT}	–	**7.264e7 Nm**	5.879e7 Nm
M_H	–	–	**2.829e8 Nm**
M_{TT}	**4.224e8 Nm**	–	3.509e8 Nm

Based on the vessel form, the load transfer functions will vary, meaning that the rules may assume a particular load is maximized in a certain heading, when in reality, the design value occurs in a different heading. This indicates that the standard load combinations suggested by LR may not be a "one size fits all" for evaluating different trimaran hull forms.

3.3 Target Extreme Value (TEV)

With the DLG, it is necessary to define the exposure, which dictates the rareness of the distribution of extreme values. The DLG gives an ensemble of Gaussian inputs that lead to extreme values of an associated linear output that are members of the extreme value distribution for a given exposure. This exposure is captured by the target extreme value, or *TEV*. Using the notation of Equations 1–5, the *TEV* for the maximized load, M_{max}, i.e. TEV_{max}, is defined in Equation 14, where σ_{max} is the standard deviation of the load M_{max}. Here, the extreme value, $\hat{y}_n$, associated with the risk parameter, α, is used.

$$TEV_{max} = \frac{\hat{y}_n}{\sigma_{max}} \quad (14)$$

$TEV_{max}\sigma_{max}$ is the design value for the maximized load, M_{max}. The DLG uses TEV_{max} to construct an ensemble of short wave time series that lead to a distribution of responses for the maximized load M_{max} where the $TEV_{max}\sigma_{max}$ event (e.g. a 5σ event) is the most probable. This distribution follows the Gaussian extreme value distribution as given in Equation 1 which satisfies Equations 4–5.

4 ASSESSMENT OF LOADING CASES USING THE DLG AND COPULA MODEL

This section assesses the standard load combination cases offered by the LR rules in a qualitative sense. Based on the ambiguity of the choice of risk parameter α, it is not constructive to compare dimensional load values found in this analysis to those that LR would indicate. Therefore, this section is meant to compare the relative distributions of loads, not specifically the magnitude of the load values.

For each load case in Table 1, the DLG constructs an ensemble of wave inputs that lead to a distribution of responses for the maximized load, M_{max}, about the most probable $TEV_{max}\sigma_{max}$ event. $TEV_{max}\sigma_{max}$ is the design value conditioned on heading, given in Table 5, which corresponds to 7.71 hours in the sea state out of the 20-year exposure in the specific heading, with a probability of exceedance of 10^{-8}, calculated by Equations 4, 5, and 14. These wave input time series are all representative of the operation profile spectral definition, and time series of all loads due to those wave inputs are constructed. At the time of the maximum value of the maximized load, each load is normalized by its respective design value R (Table 5), which gives the load normalized as a LCF.

The copula model assembles distributions of the loads based on the operating profile and risk parameter, and also normalizes these distributions by the design value, leading to an LCF value. Note that for computational reasons, the copula model cannot fully recover the operational profile associated with the 10^{-8} probability of exceedance given by Table 4. The copula model generates distributions of linear responses for the 7.71-hour exposure with $\alpha = 0.015$, $PNE = 1 - \alpha = 0.985$, leading to an overall probability of exceedance 6.6×10^{-7}, versus 10^{-8} for the LR rules and DLG simulations.

For these copula distributions associated with a higher risk parameter α (i.e. lower PNE), the design values, called $R_{PNE} = 0.985$, are proportionally lower, meaning that the copula results normalized as a LCF can be compared with the DLG results and the LCF values given by LR in Table 1. The design values $R_{PNE} = 0.985$ and resulting TEV for the $PNE = 0.985$ case are shown in Table 6. It is of interest to compare the copula distribution to DLG distributions at the same $PNE = 0.985$. DLG distributions can be constructed for $PNE = 0.985$ and normalized by the design values $R_{PNE} = 0.985$ in Table 6 for comparison with the copula distributions with this risk parameter $\alpha = 0.015$. The copula model was used to generate load distributions for Cases 1–4.

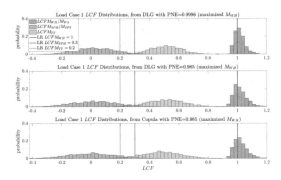

Figure 6. Distribution of LCF values from the DLG for $PNE = 0.9998$, DLG for $PNE = 0.985$, and copula for $PNE = 0.985$ for Case 1, compared with the LCF values reported by LR in Table 1. For Case 2 (not plotted), all distributions are reflected across $LCF = 0$.

4.1 Head seas, cases 1–2

The load distributions for Case 1 are shown in Figure 6. Case 2 distributions look the same for both the DLG and copula methods, except are reflected across $LCF = 0$, since all the transfer functions are linear. The top and middle insets show the DLG distributions for $PNE = 0.9998$ and $PNE = 0.985$, respectively. The copula distributions of all loads for $PNE = 0.985$, normalized by $R_{PNE=0.985}$, are shown in the bottom inset. The vertical lines indicate the rule LCF values indicated by LR for the appropriate load. Some observations are that while $LCF_{M_{WH}}$ follows a typical extreme value pdf, the other loads are more normally distributed. The different PNE values do not seem to have a significant effect on the shape or mean value of the three distributions. For a $PNE = 0.985$, both the DLG and copula give approximately equivalent results. However, the most probable values of the $LCF_{M_{SPH}}$ and $LCF_{M_{TT}}$ are significantly different than LR, both in relative ordering and value.

4.2 Beam seas cases 3–4

The load distributions for Case 3 are shown in Figure 7. Case 4 distributions are reflected across $LCF = 0$. The top and middle insets show the DLG distributions for $PNE = 0.9998$ and $PNE = 0.985$, respectively. The bottom inset gives the copula distributions for $PNE = 0.985$ normalized by $R_{PNE=0.985}$. The vertical lines indicate the LCF values indicated by LR for the appropriate load.

Table 6. Design values $R_{PNE} = 0.985$, resulting TEV, and heading where $R_{PNE} = 0.985$ is observed, using $\alpha = 0.015$ or $PNE = 1 - \alpha = 0.985$.

Load	$R_{PNE} = 0.985$	TEV	Heading of $R_{PNE} = 0.985$
M_{WH}/M_{WS}	3.800e8 Nm	5.11	Head Seas
M_{SPH}/M_{SPS}	7.984e7 Nm	5.06	Oblique Seas
M_{LT}	6.356e7 Nm	5.00	Beam Seas
M_H	2.482e8 Nm	5.10	Oblique Seas
M_{TT}	3.718e8 Nm	5.07	Head Seas

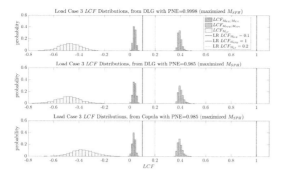

Figure 7. Distribution of *LCF* values from the DLG for *PNE* = 0.9998, DLG for *PNE* = 0.985, and copula for *PNE* = 0.985 for Case 3, compared with the *LCF* values reported by LR in Table 1. For Case 4 (not plotted), all distributions are reflected across *LCF* = 0.

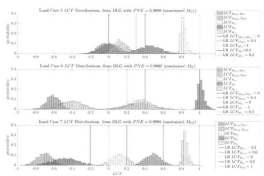

Figure 8. Distribution of *LCF* values from the DLG for *PNE* = 0.9998 for Cases 5–7, compared with the *LCF* values reported by LR in Table 1.

Recall that although the LR maximized load for these cases is the splitting moment M_{SPH}/M_{SPS}, the design value for M_{SPH}/M_{SPS} occurs in oblique seas, not in beam seas. That is why the distribution of $LCF_{M_{SPH}}$ is not centered around *LCF* = 1. This distribution does not achieve the design value. Even though the DLG is finding wave environments which lead to exposure-period-maxima of M_{SPH}/M_{SPS}, these maxima are not as large as they would be if the DLG constructed waves in oblique seas meant to maximize M_{SPH}/M_{SPS}. Similarly, the copula distribution of $LCF_{M_{SPH}/M_{SPS}}$ is not as large as it would be if constructed in oblique seas. The distributions of $LCF_{M_{WH}/M_{WS}}$ from the DLG and copula models are significantly lower than predicted by the LR rules. The distribution of $LCF_{M_{LT}}$ though from the DLG and copula model, is significantly larger than predicted by LR, and of a different sign.

4.3 Oblique seas cases 5–7

In Cases 5–7, the maximized loads are the longitudinal torsional, horizontal, and transverse torsional bending moments, respectively. These distributions are shown in Figure 8. Note that only M_H achieves its design value in oblique seas. That is why the distributions of M_{LT} and M_{TT} are not centered around *LCF* = 1. For Case 5, the distributions of $LCF_{M_{WH}/M_{WS}}$ and $LCF_{M_{SPH}/M_{SPS}}$ are both clustered around *LCF* = 0. For Case 6, the distributions of $LCF_{M_{SPH}/M_{SPS}}$ and $LCF_{M_{LT}}$ both lie around the *LCF* = 0.3, though the $LCF_{M_{SPH}/M_{SPS}}$ distribution has a much higher variance. Note that for none of the cases do the LR *LCF* values really capture the distributions of loads assembled by the DLG.

4.4 LCF from EDW approach for trimaran hull, with headings from Table 1

It has already been noted that the standard *LCF* values recommended by LR may not be fully appropriate, mainly because of the discrepancy between the headings in which LR predicts the load design values will occur and the headings where these design values do occur (compare Tables 1 and 5 for this hull configuration). This difference alone indicates that the trimaran hull in this paper has a frequency response which is significantly different than the hull used to generate the *LCF* values in Table 1. This was confirmed by the notable differences between the distributions of loads generated by the DLG and copula models, and the corresponding *LCF* values suggested by LR.

A natural question is then: would *LCF* values calculated by an EDW approach for this specific hull form accurately reflect the DLG or copula distributions? These *LCF* values can be calculated using the transfer functions shown in Figures 3–5 with Equations 11–13 for *PNE* = 0.9998. Note for Cases 3, 4, 5, and 7, in which the maximized load experiences its design value in a different heading, the EDW height, h_{max}, is found as the wave height which leads to the most probable extreme value generated by the DLG for that maximized load. This is a fair difference, since these loads do not actually achieve their design value for the given heading. This approach is similar to LR's alternative load development approach, except that the headings from Table 1 are preserved to compare the resulting *LCF* values with the distributions from Figures 6–8.

For Cases 1–4, Table 7 gives the *LCF* values given by LR, LCF_{LR}, the values calculated using the EDW method for this trimaran hull, LCF_{EDW}, the most probable *LCF* of the assembled DLG distribution for $PNE = 0.9998$, $LCF_{DLG,PNE\,=\,0.9998}$, and the most probable LCF of the assembled copula distribution for $PNE = 0.985$, $LCF_{copula,PNE\,=\,0.985}$. Table 8 gives LCF_{LR}, LCF_{EDW}, and $LCF_{DLG,PNE\,=\,0.9998}$ for Cases 5–7.

Table 7. Cases 1–4 LCF values given by LR Table 1, LCF_{LR}, LCF calculated for specific trimaran hull using EDW, LCF_{EDW}, the most probable LCF of the assembled DLG distribution for $PNE = 0.9998$, $LCF_{DLG,PNE\,=\,0.9998}$, and the most probable LCF of the assembled copula distribution for $PNE = 0.985$, $LCF_{copula,PNE\,=\,0.985}$.

Load	LCF_{LR}	LCF_{EDW}	LCF_{DLG} $PNE=$ 0.9998	LCF_{copula} $PNE=$ 0.985
Case 1–2, Head seas				
M_{WH}/M_{WS}	±1	±1	±1	±1
M_{SPH}/M_{SPS}	±0.3	∓0.02	0	0
M_{TT}	0.2	±0.10	±0.51	±0.46
Case 3–4, Beam seas				
M_{WH}/M_{WS}	±0.1	±0.05	±0.03	±0.03
M_{SPH}/M_{SPS}	1	±0.39	±0.39	±0.39
M_{LT}	0.2	∓0.47	∓0.49	∓0.46

Table 8. Cases 5–7 LCF values given by LR Table 1: LCF_{LR}, LCF calculated for specific trimaran hull using EDW: LCF_{EDW}, and the most probable LCF of the assembled DLG distribution for $PNE = 0.9998$: $LCF_{DLG,PNE\,=\,0.9998}$.

Load	LCF_{LR}	LCF_{EDW}	LCF_{DLG} $PNE = 0.9998$
Case 5, Oblique seas			
M_{WH}/M_{WS}	0	−0.07	0
M_{SPH}/M_{SPS}	0.4	−0.36	0.04
M_{LT}	1	0.81	0.81
M_H	0.3	0.51	0.46
M_{TT}	0.3	0.17	0.11
Case 6, Oblique seas			
M_{WH}/M_{WS}	0	−0.8	−0.61
M_{SPH}/M_{SPS}	0.4	0.27	0.36
M_{LT}	0	−0.03	0.37
M_H	1	1	1
M_{TT}	0.2	−0.07	−0.37
Case 7, Oblique seas			
M_{WH}/M_{WS}	−0.2	0.37	0.41
M_{SPH}/M_{SPS}	0.6	−0.71	0.01
M_{LT}	0	0.11	0.31
M_H	0.2	−0.44	−0.19
M_{TT}	1	0.83	0.54

For Cases 1–2, the maximized load M_{WH}/M_{WS} does achieve its design value in head seas, so LCF_{EDW} is unity. Since the DLG constructs waves to lead to a distribution of responses around the design value, with the design value being the most probable point, $LCF_{DLG,PNE\,=\,0.9998}$ is also unity. Using the EDW methodology for this hull form predicts that there will be a negligible contribution of M_{SPH}/M_{SPS}. The $LCF_{DLG,PNE\,=\,0.9998}$ and $LCF_{copula,PNE\,=\,0.985}$ for this load are also nearly zero, but the distribution of M_{SPH}/M_{SPS} widely varies about that point, achieving nearly 40% of its design value in both hogging and sagging at its largest. The LCF_{EDW} for M_{TT} predicts a large contribution, but not as large as the values actually observed by the DLG and copula methods.

For Cases 3–4, the maximized load M_{SPH}/M_{SPS} does not achieve its design value in beam seas, so LCF_{EDW} is less than one. By construction, this value lines up with the most probable LCF value constructed by the DLG and copula models. The EDW method closely recovers the most probable point of the M_{WH}/M_{WS} and the M_{LT} distributions assembled by the DLG/copula (i.e. $LCF_{EDW} \approx LCF_{DLG,PNE=0.9998} \approx LCF_{copula,PNE=0.985}$).

For Cases 5–7, only M_H achieves its design value in oblique seas. The LCF_{EDW} values, for the most part, better capture the distributions of responses given by the DLG/ copula models, but are not always accurate, nor do these values allow the wide variance seen in these distributions.

This analysis illuminates the issues associated with an EDW approach, especially coupled with a "one size fits all" approach for standard load combination cases. Calculating standard LCF values with the EDW approach may not give load combination cases that apply to a wide range of hull forms, as seen with this trimaran hull. These cases illuminate the potential pitfall of using a general set of load combinations cases for a variety of multihull designs. Given a different placement of the side-hulls, the interaction between the major loads on the vessel can change drastically. This interaction is somewhat considered by the EDW approach, but since the standard LR load combination cases were not specifically tailored to this trimaran hull, the predicted distribution of loads by LR were often inaccurate.

It is important to consider the headings that the loads actually achieve their design value. Some of the loading combination cases from LR may not truly test the system, because none of the considered loads achieve particularly large values (i.e. Cases 3–4). With the load cases from Table 1, no case has a M_{SPH}/M_{SPS} value larger than about 80% of its design value. Even for cases where the EDW method can accurately predict the most probable LCF value found from the DLG or copula methods (i.e. for

Cases 3–4), using a single *LCF* value does not capture the variance in the observed load distributions.

4.5 Recommendation on load combination cases

Based on the above analysis, it is clear that there may be no universal set of load combination cases to test all trimaran hull forms. However, the spirit of the load combination cases is still applicable. It is important to consider the vessel response to the major global loads, like those given in Table 1. New load combination cases are defined to more realistically describe the combined loading experienced by this trimaran hull. The major driver for the new load cases is that one of the global loads is experiencing its design value, and the simultaneous values of secondary loads are collected. Here, based on the headings where each load experiences its design value, as in Table 5, new cases are defined in Table 9.

Cases 1*, 2*, and 6* remain in the original headings from Table 1, so their distributions are still represented by Figure 6 and the middle inset of Figure 8. For Cases 3*, 4*, 5*, and 7*, the DLG constructs ensembles of wave environments that lead to extreme responses of the associated maximized load in the given heading from Table 9. The distribution of responses is the extreme value distribution for the specified load for a 7.71-hour exposure with the risk parameter $\alpha = 2.27e-4$, leading to a distribution of responses around the most-probable value, which is the design value, $R_{PNE = 0.9998}$. Distributions of the secondary loads, conditioned on the maximized load experiencing its maximum value, are also assembled. Figure 9 shows the distributions for the updated Case 3*. Case 4* (not plotted) has the same distributions reflected across $LCF = 0$. Figure 10 gives the updated Cases 5*, and Figure 11 gives the updated Case 7*. Using the peak value of the DLG distributions for Cases 1*–7* offers updated *LCF* values for the new load combination cases given in Table 10. For Cases 1*–2*, two sub-cases are defined to account for the high variance in M_{SPH}/M_{SPS}.

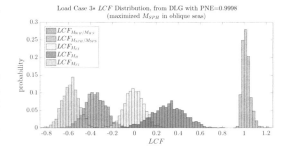

Figure 9. Distribution of *LCF* values from the DLG for Case 3* for maximized M_{SPH} in oblique seas, associated with $PNE = 0.9998$. Case 4* distributions (not plotted) are reflected across $LCF = 0$.

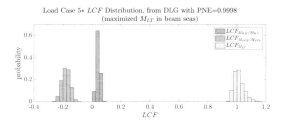

Figure 10. Distribution of *LCF* values from the DLG for Case 5* for maximized M_{LT} in beam seas, associated with $PNE = 0.9998$.

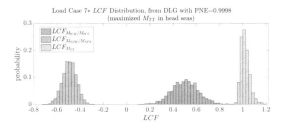

Figure 11. Distribution of *LCF* values from the DLG for Case 7* for maximized M_{TT} in head seas, associated with $PNE = 0.9998$.

Table 9. Recommended headings for new load combination cases 1*–7* for the trimaran hull shown in Figure 2. Note that Cases 3*–5*, and 7*, as defined by Table 1 have a different heading. Cases 1*, 2*, and 6* remain in the same heading from Table 1.

Case	Maximized load	Heading
1*/2*	M_{WH}/M_{WS}	Head Seas
3*/4*	M_{SPH}/M_{SPS}	Oblique Seas
5*	M_{LT}	Beam Seas
6*	M_H	Oblique Seas
7*	M_{TT}	Head Seas

Table 10. *LCF* values defined by DLG distributions for Cases 1*–7*.

Case & heading	Dynamic loads						
	M_{WH}	M_{WS}	M_H	M_{SPH}	M_{SPS}	M_{LT}	M_{TT}
1*a) Head	1.0	0	0	0.2	0	0	+0.51
1*b) Head	1.0	0	0	0	0.2	0	+0.51
2*a) Head	0	1.0	0	0.2	0	0	−0.51
2*b) Head	0	1.0	0	0	0.2	0	−0.51
3*) Oblique	0	0.39	0.33	1.0	0	0	−0.56
4*) Oblique	0.39	0	−0.33	0	1.0	0	0.56
5*) Beam	0	0	0	0	0.19	1.0	0
6*) Oblique	0	0.61	1.0	0.36	0	0.37	−0.37
7*) Head	0.51	0	0	0	0.49	0	1.0

5 SIMULTANEOUS LOAD COMBINATIONS

The above figures compared the *LCF* distributions from the DLG and copula model, the *LCF* values indicated by the LR rules, and *LCF* values calculated by the EDW approach for this specific trimaran hull using the headings from Table 1. In many of the cases, for the trimaran hull considered here, the LR standard load combination cases did not adequately capture the distribution of loads found from the DLG or copula model. Even when updated *LCF* values were found, from either an EDW approach for the specific hull form using the same headings from Table 1, or using the most probable *LCF* from the DLG or copula distributions, reflecting the updated Cases 1*–7* in Table 10, there is still large variance present in the load distributions. It may be instructive to know the simultaneous *LCF* values, and not just distributions of these values.

This can be considered by looking at the load time series constructed by the DLG for each load combination case. Bounding planes exceeding the standard LR *LCF* values for a given case can construct a threshold surface to illuminate when, if at all, a parametric curve of all the loads exceeds all the *LCF* values. For a case that has n non-zero *LCF* values, the threshold surface will consist of $2n$, n–dimensional cubes cornered at each $\left(\pm LCF_{M_1}, \pm LCF_{M_2}, \cdots, \pm LCF_{M_n}\right)$ point. Each cube is unbounded out from its corner origin.

The times series of the n loads considered by a LR standard combination case, represented as an *LCF*, can be plotted as a parametric curve on the n–dimensional failure surface $\left(LCF_{M_1}(t), LCF_{M_2}(t), \cdots, LCF_{M_n}(t)\right)$. Anything outside of this surface represents a threshold crossing, meaning that all *LCF* values have been exceeded. Such an analysis can further show the correlation structure between the loads. As an example, consider the parametric time series for the loads in Cases 1–2.

5.1 Head seas cases 1–2

For Cases 1–2, where the three loads M_{WH}/M_{WS}, M_{SPH}/M_{SPS}, and M_{TT} are considered, the threshold surface is visualized in 3 dimensions: $\left(\pm LCF_{M_{WH}/M_{WS}}, \pm LCF_{M_{SPH}/M_{SPS}}, \pm LCF_{M_{TT}}\right)$. The *LCF* values are those from LR given by Table 1. A threshold exceedance is defined as when all three loads simultaneously exceed the respective *LCF* value. The threshold region consists of 8 unbounded 3-dimensional cubes with an origin corner at (±1, ±0.3, ±0.2). The parametric curve $\left(LCF_{M_{WH}/M_{WS}}(t), LCF_{M_{SPH}/M_{SPS}}(t), LCF_{M_{TT}}(t)\right)$ shows the value of each load *LCF* at a particular instant in time.

Figure 12 shows the parametric curve $\left(LCF_{M_{WH}/M_{WS}}(t), LCF_{M_{SPH}/M_{SPS}}(t), LCF_{M_{TT}}(t)\right)$ along with the threshold regions where exceedances are recorded. Each blue line is a single parametric curve $\left(LCF_{M_{WH}/M_{WS}}(t), LCF_{M_{SPH}/M_{SPS}}(t), LCF_{M_{TT}}(t)\right)$ generated by the DLG. The red stars are times when the curve crosses the threshold surface. The shape of this parametric curve indicates the correlation between the different loads. The ellipsoid is slanted from the positive $LCF_{M_{TT}}$ direction toward the positive $LCF_{M_{WH}/M_{WS}}$ direction. Figures 13, 14, and 15 are projections of Figure 12 on the $(M_{SPH}/M_{SPS}, M_{TT})$, $(M_{WH}/M_{WS}, M_{SPH}/M_{SPS})$ and $(M_{WH}/M_{WS}, M_{TT})$ planes, respectively.

Figure 13 shows that positive M_{TT} values are correlated with negative M_{SPH}/M_{SPS} values, when conditioned on a maximum of M_{WH}/M_{WS}. This would be difficult to determine simply from the distributions in Figure 6, where the load values across the different distributions are not explicitly linked. This behavior is reinforced by the distributions in Figure 11, where maximized M_{TT} values in head seas correspond to a distribution of negative M_{SPH}/M_{SPS} values.

Figure 14 shows that there little correlation between M_{WH}/M_{WS} and M_{SPH}/M_{SPS} based on the near-circular structure of the parametric curve. This is reinforced by the distributions shown in Figures 6, though the distributions in Figure 9 for oblique seas indicate that this is not an exactly symmetric relationship. When M_{WH}/M_{WS} is maximized in head seas (Figure 6), the corresponding $LCF_{M_{SPH}/M_{SPS}}$ values are about uniformly distributed around zero, while when M_{SPH}/M_{SPS} is

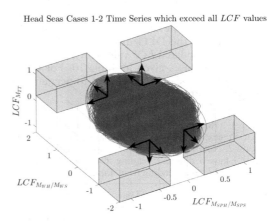

Head Seas Cases 1-2 Time Series which exceed all *LCF* values

Figure 12. Parametric curve of ($LCF_{M_{WH}/M_{WS}}(t)$, $LCF_{M_{SPH}/M_{SPS}}(t)$, $LCF_{M_{TT}}(t)$) and threshold regions where exceedances occur. Any threshold surface region where no exceedances occur is not shown for simplicity.

maximized in oblique seas (Figure 9), the $LCF_{M_{WH}/M_{WS}}$ values take on rather significant negative values.

Figure 15 shows a high positive correlation between M_{WH}/M_{WS} and M_{TT}, as evidenced by Figure 6 in which large positive $LCF_{M_{TT}}$ values are simultaneously recovered for maximized M_{WH}. In the load distributions for M_{TT} in head seas (where M_{TT} experiences its design value, Figure 11), there is also a strong connection between large positive M_{TT} values and M_{WH} values.

5.2 *Relevance & future work*

This sort of analysis illustrates key features of the loads acting on the trimaran hull, specifically how

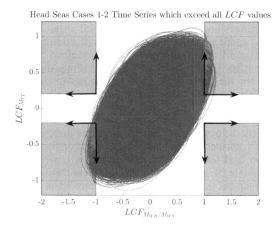

Figure 15. Projection of Figure 12 on the $\left(LCF_{M_{WH}/M_{WS}}, LCF_{M_{TT}} \right)$ plane.

these loads interact in different headings. This can be useful for defining load combination cases with an explicit probabilistic basis. With this approach, it is possible to determine the overall probability of exceeding the load combination case over the vessel's exposure. Clearly the analysis is a useful metric for connecting allowable risk for a vessel design with load combination cases meant to evaluate structural integrity. This is an active area of research, with promising results so far.

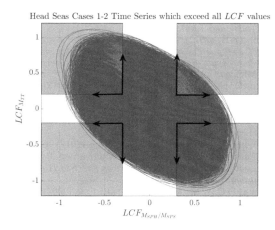

Figure 13. Projection of Figure 12 on the $\left(LCF_{M_{SPH}/M_{SPS}}, LCF_{M_{TT}} \right)$ plane.

6 CONCLUSIONS

This paper examined combined loading on a trimaran hull and compared the standard rule load combination cases offered by Lloyd's Register with distributions from the Design Loads Generator and copula models. Lloyd's Register offered seven standard load combination cases meant to reduce to a reasonable number the amount of possible load combinations to check with structural acceptance criteria. These loads were developed using an EDW methodology by looking at seven global loads. Each load was maximized in the heading which presumably led to its maximum response, and the simultaneous responses of the other global loads was noted, leading to loading combination factors, or *LCF* values, for each load. The standard *LCF* values offered by Lloyd's Register may be applied to any trimaran hull, within the noted geometry bounds, and indicate load values to be applied to a FEA model to check structural compliance.

The rule load combination cases offered by Lloyd's Register were tested for the specific trimaran hull from Figure 2, but were found to be largely

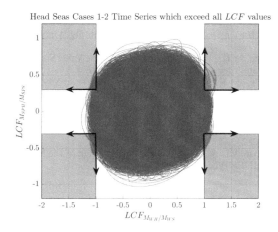

Figure 14. Projection of Figure 12 on the $\left(LCF_{M_{WH}/M_{WS}}, LCF_{M_{SPH}/M_{SPS}} \right)$ plane.

non-applicable for the hull. The major differences occurred because this trimaran hull experiences maximum values of its global loads (the design values) in different headings than that assumed by the LR rules. Even in cases where the headings did line up for the design values of maximized loads, the *LCF* values offered by LR were often misleading compared to the distributions of loads assembled by the DLG and copula models. Using the EDW methodology with the same headings directed by the LR rules to calculate *LCF* values for this specific hull form did not always give a reasonable estimate of the load distributions. The DLG and copula models showed strong agreement in their load distributions.

The DLG and copula distributions were used to offer more appropriate load combination cases, but found that even a single *LCF* value cannot capture the large variance in some load distributions. Regardless, updated load combination cases were recommended. These cases covered in a much more exhaustive sense the potential extremes of the global loads that this trimaran hull may experience. Parametric curves of the loads were constructed by the DLG to show the potential of analyzing simultaneous load combinations in a full probabilistic sense. This area of future work will allow the connection between allowable risk levels with load combination cases which truly evaluate that risk.

ACKNOWLEDGEMENTS

The work in this paper was supported by the Office of Naval Research (ONR) through the Probabilistic Assessment of Design Events for Complex Systems Subject to Stochastic Input project (program manager Kelly Cooper) and with Government support under and awarded by the National Defense Science and Engineering Graduate (NDSEG) Fellowship. Navatek, Ltd. is supported by the Office of Naval Research, PM Kelly Cooper, under Contract Number N00014-17-C-2022.

REFERENCES

Alford, L.K. (2008). *Estimating Extreme Responses using a Non-Uniform Phase Distribution*. Ph. D. thesis, The University of Michigan.

Bartoli, G., C. Mannini, & T. Massai (2011). Quasi-Static Combination of Wind Loads: a Copula-Based Approach. *Journal of Wind Engineering and Industrial Aerodynamics 99*(6–7), 672–681.

Bastian, K., M. Pahlow, Y. Hundecha, & A. Schumann (2009). Probability Analysis of Hydrological Loads for the Design of Flood Control Systems Using Copulas. *Journal of Hydrologic Engineering 15*(5), 360–369.

Blanchard, T. & C. Ge (2007). Rules for the Classification of Trimarans. *Journal of Naval Engineering 44*(1).

de Waal, D.J. & P.H.A.J.M. van Gelder (2005). Modelling of ExtremeWave Heights and Periods Through Copulas. *Extremes 8*(4), 345–356.

Ewans, K. & P. Jonathan (2014). Evaluating Environmental Joint Extremes for the Offshore Industry using the Conditional Extremes Method. *Journal of Marine Systems 130*, 124–130.

Heffernan, J.E. & J.A. Tawn (2004). A Conditional Approach for Multivariate Extreme Values (with Discussion). *Journal of the Royal Statistical Society: Series B (Statistical Methodology) 66*(3), 497–546.

IACS Rec. No. 34 (2001). Standard Wave Data.

Jensen, J.J. (2007). Efficient Estimation of Extreme Non-Linear Roll Motions using the First-Order Reliability Method (FORM). *Journal of Marine Science and Technology 12*(4), 191–202.

Kim, D.-H. (2012). *Design Loads Generator: Estimation of Extreme Environmental Loadings for Ship and Offshore Applications*. Ph. D. thesis, The University of Michigan.

Kim, D.-H., A.H. Engle, & A.W. Troesch (2013). Estimates of Long-Term Combined Wave Bending and Whipping for Two Alternative Hull Forms. *Society of Naval Architects and Marine Engineers- Transactions* (120), 1–20.

Knight, J.T., M. Craig, & D.C. Kring (2017). Copula Functions for Prediction of Multivariate Extreme Loads. *International Conference on Fast Sea Transportation Nantes, France*.

Kring, D.C., W.M. Milewski, & N.E. Fine (2004). Validation of a NURBS-based BEM for Multihull Ship Seakeeping. *25th Symposium on Naval Hydrodynamics St. John's, Newfoundland and Labrador, Canada*.

Lloyd's Register (2017). Rules for the Classification of Trimarans.

Madsen, H.O., S. Krenk, & N.C. Lind (1986). *Methods of Structural Safety*. Englewood Cliffs: Prentice-Hall, Inc.

Naess, A. & T. Moan (2013). *Stochastic Dynamics of Marine Structures*. Cambridge: Cambridge University Press.

Ochi, M.K. (1981). Principles of Extreme Value Statistics and their Application. *Trans. SNAME*.

Ochi, M.K. (1990). *Applied Probability & Stochastic Processes in Engineering & Physical Sciences*. Wiley series in probability and mathematical sciences.

Rice, S.O. (1944, July). Mathematical Analysis of Random Noise. *The Bell System Technical Journal 23*(3), 282–332.

Russo, V.L. & E.K. Sullivan (1953). Design of the MARINERType Ship. *Trans. SNAME 61*.

Trimaran structural design procedure for a large ship

John C. Daidola
Webb Institute, New York, USA

ABSTRACT: The structural design of a large trimaran is presented for a vessel of both high strength steel and titanium. The structural design procedure consists of prescriptive classification society rules as well as verification through a grillage analysis for ultimate strength. The results for both material designs are presented and compared, including midship section scantlings, hull structural weight, material properties and performance in a marine environment. The vessel considered is significantly larger than any which have been built to date and is intended for fast cargo transportation. The structural design procedure has been adopted following comparison to other approaches.

1 INTRODUCTION

The interest in trimaran hull forms of all types has been on the upswing for nearly two decades. Significant hydrodynamic research and testing has occurred primarily focused on resistance, seakeeping and hull loads. At the same time, structural aspects have had less attention. Notably however, since the mid-2000's significant efforts have been expended by and on behalf of the classification society Lloyd's Register (LR) on trimaran structure. These have more recently resulted in the Rules for Classification of Trimarans (LRTR) (LR, 2016) which address both commercial and naval vessels. Also, recently a computer based model for grillage analysis of trimaran hull structures has been developed by NAVATEC (Knight, 2017).

The vessel considered herein is a high speed trimaran cargo carrying ship 264 m (866 ft.) in length, with a design speed of over 35 knots, and constructed of H-36 high strength steel as well as titanium Ti-3 AL-2.5V.

Previous studies (Daidola, Arnold, Imbesi, 2017) have shown that the LRTR and monohull derived methods provide some comparable results for the main hull structure, while the LRTR provides information for the design of the wet deck and cross deck structure. Given the dearth of applications to large trimarans and the resulting low level of experience in the predictions, ultimate strength has been adopted herein as an additional analysis to confirm the adequacy of the structure.

For high speed vessels less hull weight and the surface condition of the underwater hull as it relates to roughness and frictional resistance are particularly important. For these reasons and to support lower maintenance of structure, a titanium material hull structure has been considered in addition to steel, although life cycle cost comparisons were not intended. With its lower stiffness for titanium, the resulting structure also provides an additional ultimate strength analysis data point.

The sections which follow present the principal features of the vessel, the nature and extent of the analyses, and comparison of estimates. Analyses represent aspects of preliminary structural design only.

2 TRIMARAN CHARACTERISTICS

The characteristics of the large high speed trimaran sealift ship from previous studies (Daidola, Arnold, Imbesi, 2017) are presented in Figures 1, 2 and Table 1. The principal characteristics and vessel loading cases of interest are given in Table 1, the inboard and outboard profiles in Figure 1, and transverse sections considered for structural analysis in Figure 2.

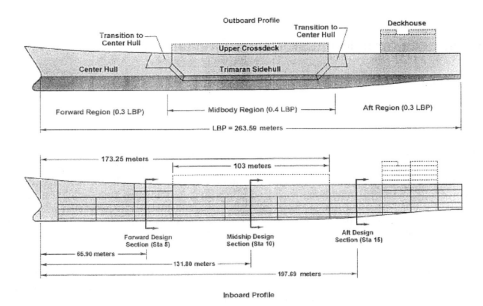

Figure 1. Large sealift trimaran inboard and outboard profiles.

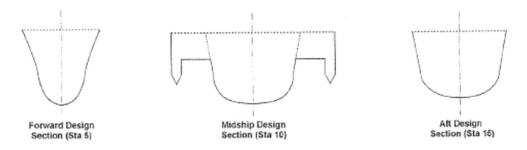

Figure 2. Large sealift trimaran body plan and design sections.

Table 1. Hull characteristics—Large sealift trimaran.

Characteristic	Load Case 1 Light ship + Full fuel	Load Case 2 Full load: Light ship + Full fuel + Cargo
Length Overall	264 m (866.1 ft.)	264 m (866.1 ft.)
Length BP (LBP)	263.6 m (864.8 ft.)	263.6 m (864.8 ft.)
Beam Overall	49.5 m (162.4 ft.)	49.5 m (162.4 ft.)
Beam Main Hull	25.8 m (84.6 ft.)	25.8 m (84.6 ft.)
Beam Side Hulls	3.00 m (9.8 ft.)	3.00 m (9.8 ft.)
Draft	8.07 m (26.5 ft.)	8.91 m (29.2 ft.)
Displacement	27300 Mtons (26800 LT)	31200 Mtons (30700 LT)
Block Coefficient	0.443	0.437
Trim (+ by stern)	1.42 m (4.65 ft.)	0.22 m (0.717 ft.)
LCG (+ fwd. of CG)	−11.7 m (−38.5 ft.)	−10.1 m (−33.2 ft.)
Cross-Structure Wet Deck Height above Waterline	6.13 m (20.1 ft.)	5.29 m (17.4 ft.)

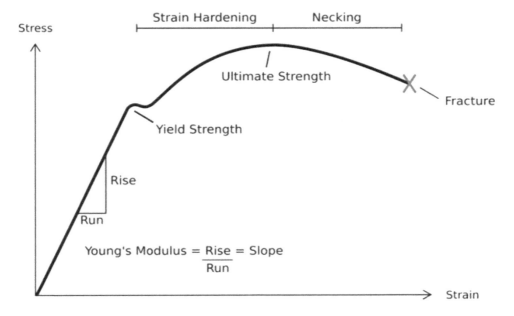

Figure 3. Allowable stress and ultimate strength analysis.

3 STRUCTURAL DESIGN APPROACHES

The LRTR follows the classification society prescriptive formula design approach to an allowable stress normally based on experience and keeping the stresses resulting from the design loads under a given working stress level. The allowable stress, or working stress, is a fraction of the strength properties of the material utilized, typically the yield and ultimate strengths. Most of the world's ships are designed in this manner.

In contrast, the ultimate strength analysis is based on the explicit consideration of the conditions under which the structure is loaded and under which the structure will fail, as well as the material properties. It typically represents the collapse of the structure due to loss of stiffness and strength and which may be related to (Paik, Thayamballi, 2003):

- Loss of equilibrium or static stability of part or the entire structure.
- Gross yielding, rupture or fracture.
- Instability due to buckling and plastic collapse of plating, stiffened panels and support structure.

The essence of the difference between these two approaches can be gleaned from Figure 3. The prescriptive design of structure from classification society rules is most often based on the proportional limit with allowable stress below the yield strength. If buckling is considered as an ultimate strength then a higher load limit above the yield strength will be specified. The actual ultimate strength may yet be higher and represents a truer measure of the structural capacity.

The allowable stress approach maintains a load limit below the yield strength. From a purely elastic buckling approach the design could allow the load to increase to a point between the yield and ultimate strength. If post buckling behavior is considered it is possible to increase the load to the ultimate strength. In any event, if the load causes a displacement beyond the point of ultimate strength, collapse of the structure should be anticipated.

As a result of these relationships and the dearth of experience with the type of vessels considered herein, the approach has been to adopt the LRTR for the hull structural definition of scantlings with the understanding the resultant position in Figure 3 is below the yield strength. In order to provide confidence that the resulting structure under the prescribed loading will in fact not exceed the ultimate strength in Figure 3, a complimentary ultimate strength structural analysis is also carried out.

4 MATERIAL PROPERTIES

For high speed vessels less hull weight and the surface condition of the underwater hull as it relates to roughness, frictional resistance and ultimately powering are important. For these reasons a high strength

Table 2. Properties of selected hull structural material.

Material	High strength steel, HS-36	Titanium, Ti-3Al-2.5v
Yield strength, σ_y	355 MPa (51 Ksi)	483 MPa (70 Ksi)
Ultimate strength, σ_u	490 MPa (71 Ksi)	621 MPa (90 Ksi)
Modulus of elasticity, E	200 GPa (29000 Ksi)	103 GPa (15000 Ksi)
Density	7.9 g/cm^3 (0.284 lbf/in^3)	4.4 g/cm^3 (0.16 lbf/in^3)
Poisson's ratio	0.26	0.30

steel and also a titanium material hull structure has been considered. The titanium, with it resistance to corrosion and erosion, will be beneficial both from powering and vessel maintenance points of consideration. A comparison of life cycle costs of the vessels designed with the two different materials was not intended for this study. With its lower stiffness for titanium, the resulting structure also provides an additional ultimate strength analysis data point.

HS-36 high strength steel was chosen as a commonly utilized material in shipbuilding. Titanium is represented by a family of commercial alloys with a very wide range of strength, unique physical properties compared to conventional metals, and exceptional resistance to corrosion and erosion in seawater and other aqueous chloride media (Schutz, Scaturro, 2010). Selection of titanium alloys for seawater applications is normally predicated upon the expected practical immunity to uniform and/or localized attack in seawater. The titanium alloys traditionally used in marine/seawater service can be welded, machined, and fabricated, with some adjustments, using similar methods and equipment used for stainless steels. Standard mill product forms for these marine titanium grades are commercially available, and typically specified via relevant ASTM product specifications which provide composition and tensile property requirements. The titanium selected for consideration is Ti-3 Al-2.5v which is in the category which offer medium strength, high fracture toughness, and fabrication for welded structure.

The properties for both materials are provided in Table 2.

5 LRTR DESIGN PROCEDURES

The LRTR are configured such that they are to be applied with Complimentary Rules which are either those for the Classification of Ships (commercial), Special Service Craft, or Naval Ships (utilized herein). Furthermore, the ship's rule length is to be between 70 m and 250 m with vessels outside theses limits specially considered. The large high speed trimaran slightly exceeds the maximum length for the LRTR. Notwithstanding this exceedance of limitation, the LRTR was applied in this assessment, but this aspect is a source of uncertainty.

5.1 Longitudinal hull girder strength

Table 3 provides the maximum midship longitudinal bending moments from LRTR formulations. The moments shown in Table 3 are all hogging moments, likely due to the result of the ship's still water bending moment being a hogging moment.

5.2 Local structure

The LRTR provides prescriptive formulations for the required strength of local structure as well as transverse bending moments, wet deck pressures, and allowable stresses. The formulations for plating include a corrosion allowance in absence of specific ship specification requirements and this is applied to the net scantlings determined by LRTR formulas. In the case of titanium, this allowance was set to "0" to account for its resistance to corrosion. Once the local structure is initially determined, the design primary longitudinal and transverse bending moments as well as the wet deck pressures can be applied to confirm the structure is adequate in terms of not exceeding the allowable stresses.

The moment of inertia (MOI) and section modulus (SM) resulting from the local scantling design must be compared to the minimum required LRTR values as well as those required to absorb the maximum primary hull girder bending moments in Table 3 within the allowable stress. As a result, the

Table 3. Maximum longitudinal hull girder bending moments for load Cases 1 and 2.

	Station 5 m* tonne*10^{-3} (ft*LT*10^{-3})	Midship m* tonne*10^{-3} (ft*LT*10^{-3})	Section 15 m* tonne*10^{-3} (ft*LT*10^{-3})
LRTR – Load case 1	168 (543)	355 (1150)	248 (800)
LRTR – Load case 2	155 (500)	314 (1020)	231 (747)

Table 4. Midship MOI and SM—Steel and Titanium.

	MOI m^4 (ft^4)	SM m^3 (ft^3)
Steel	305 (35400)	27.5 (970)
Titanium	300 (34820)	26.9 (949)

local scantlings met all requirements in the midship region without further increase, but required additional increases in structure at Station 5 and Station 15 to fully absorb the bending moments, while significantly exceeding the minimum required values of MOI and SM [158 m^4 (18300 ft^4), 17.4 m^3 (614 ft^3) respectively steel]; [158 m^4 (18300 ft^4), 16.4 m^3 (579 ft^3) respectively titanium], Table 4.

Figure 4 provides the midship section at Station 10. Table 5 provides the plating and frame scantlings for LRTR scantlings shown in red regions in Figure 4. The internal deck structure was originally

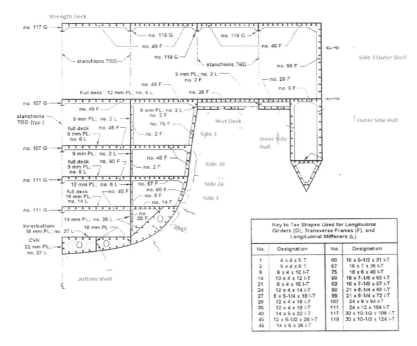

Figure 4. Midship station 10 design scantlings in steel.

Table 5. Midship design scantlings for steel and titanium.

	Steel		Titanium	
	Plating (mm)	Longitudinal Stiffener (in × in × lbf/ft total, lbf/ft net) Designation	Plating (mm)	Longitudinal Stiffener (in × in × lbf/ft total, lbf/ft net) Designation
Description				
Bottom Shell Area	16	10 × 4 × 15 I-T, 11.55	14	10 × 4 × 6.5 I-T, 5.03
Bilge Shell Area	14	12 × 4 × 16.5 I-T, 13.11	14	10 × 4 × 8.45 I-T, 6.6
Side 1 Area	16	10 × 4 × 15 I-T, 11.55	16	10 × 4 6.5 I-T, 5.03
Side 2a Area	16	10 × 4 × 15 I-T, 11.55	16	10 × 4 × 6.5 I-T, 5.03
Side 2b Area	16	8 × 4 × 13 I-T, 9.79	16	8 × 4 × 5.63 I-T, 4.19
Side 3 Main Hull Area	16	8 × 4 × 10 I-T, 10.0	16	5 × 4 × 3.24 T, 3.24
Wet Deck Shell Area	10	12 × 4 × 19 I-T, 14.76	8	12 × 4 × 7.89 I-T, 6.5
Sidehull, Inboard Shell Area	16	10 × 4 × 12 I-T, 8.93	16	8 × 4 × 5.63 I-T, 4.19
Sidehull, Keel Area	16	12 × 4 × 14 I-T, 11.55	16	10 × 4 × 6.5 I-T, 5.03
Sidehull, Outboard Shell Area	16	10 × 4 × 12 I-T, 8.93	16	8 × 4 × 5.63 I-T, 4.19
Side 3 Outer Area	16	8 × 4 × 10 I-T, 7.44	16	5 × 4 × 3.24 T, 3.24
Sheer Strake Area	16	8 × 4 × 10 I-T, 7.44	16	5 × 4 × 3.24 T, 3.24
CVK	14	8 × 5-1/4 × 18 I-T, 11.82	12	8 × 4 × 7.32 I-T, 5.52
Weather Deck	14	8 × 4 × 10 I-T, 7.44	14	6 × 4 × 3.94 T, 3.97

developed under criteria somewhat different from the LRTR, but in both cases the loads are due to cargo rather than the more unique hydrodynamic external loads reflected in the structural scantlings in Table 5 and the requirements for hull girder moment of inertia. For simplification of the analysis, the internal scantlings were assumed identical. In the case of titanium, these internal scantlings were retained despite the greater strength in order to limit deflections resulting from the reduced modulus of elasticity of the titanium.

As can be seen in Table 5, the required plating thickness for titanium is essentially the same for steel, even though it is stronger and the corrosion allowance has been eliminated. At the same time, there is a general decrease in the scantling requirements for stiffeners.

6 GRILLAGE ANALYSIS

As an additional means of analyzing the adequacy of the structure defined by the LRTR the ultimate strength behavior was investigated. The overall failure of ship structure is primarily governed by the buckling and plastic collapse of the stiffened panels in the hull structure, especially in the deck and bottom. These panels, or grillages, are an assembly of plates and support members such as longitudinals and transverses. The interaction between the plate elements and support members with respect to their geometrical and material properties, loading condition and initial imperfections play an important role in their failure.

NAVATEC (Knight, 2017) has developed a code to analyze flat, rectangular grillages and panels following the approach of Paik (Hughes and Paik, 2010) to determine compressive ultimate limit states. The model of the structure assumes flat panels so that the true ship section must be linearized into discrete segments, as shown in Figure 5. The analyses assume simply-supported boundary conditions along all the edges of every grillage. The ultimate limit states are numbered representing different failure modes defined as follows and the lowest stress value among these is assumed to be the ultimate strength of the grillage and the maximum capacity to resist longitudinal hull girder stress:

Mode 0: Panel-type collapse spanning multiple frames, between longitudinal girders. (i.e. Girders have not collapsed, but everything between them has.)
Mode 1: Panel collapse; collapse of the plating and stiffeners together as one unit, where frames and girders are assumed to form simply-supported boundary conditions.
Mode 2: Plating collapse; collapse of the plating between stiffeners.
Mode 3: Beam-column type collapse; buckling failure of a stiffener with its attached width of plating.
Mode 4: Buckling of the stiffener web.
Mode 5: Stiffener tripping; flexural-torsional buckling of the stiffener with its attached width of plating.
Mode 6: Gross yielding.

The panels identified for the midship section of the large high speed trimaran are shown in Figure 6. The results from application of this procedure to the LRTR structure are given in Tables 6 and 7 for steel and titanium respectively for the Load Case I (Lightship + Full Fuel), which

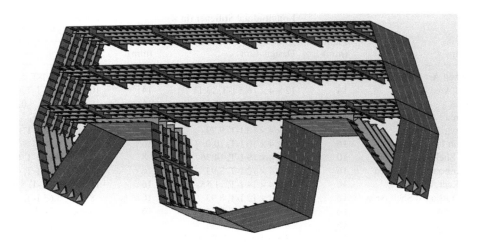

Figure 5. Simplified section module (Knight, 2017).

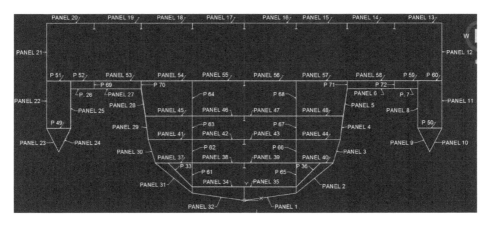

Figure 6. High speed trimaran midship panels.

Table 6. Results of grillage analysis for steel [MPa (x6.88 Ksi)].

Panel	Mode 0	Mode 1	Mode 2	Mode 3	Mode 4	Mode 5	Mode 6
1	271	269	276	239	281	227	278
2	261	259	238	236	251	203	273
3	251	253	256	232	251	218	268
4	250	247	243	228	246	217	264
5	269	280	275	271	307	307	296
6	153	246	187	194	217	136	281
7	265	246	185	194	216	135	281
8	245	245	242	228	245	216	264
9	256	256	252	251	248	221	267
10	258	256	252	251	248	221	267
11	255	242	232	226	236	208	263
12	223	230	240	210	247	219	261
13	247	267	276	261	281	254	268
14	247	265	275	261	280	253	268
15	244	265	270	260	276	251	267
16	229	255	223	255	233	213	262
35	266	268	279	268	289	275	270
36	265	266	226	268	233	210	270
39	250	261	244	261	247	230	264
40	261	268	277	267	275	253	269
43	259	258	167	264	191	165	268
44	265	272	224	271	243	209	275
47	247	258	167	264	191	165	268
48	260	271	223	271	242	208	275
50	261	273	239	273	255	220	276
56	248	257	201	260	215	195	264
57	258	268	262	266	270	243	270
58	255	268	270	268	277	248	271
59	261	269	271	268	277	248	271
60	259	268	270	268	277	248	271
65	260	267	222	272	234	220	272
66	252	255	191	264	206	187	264
67	233	231	157	265	183	167	265
68	185	227	158	265	184	168	265
71	255	259	209	271	236	213	271
72	248	247	197	250	207	195	250

Table 7. Results of grillage analysis for titanium [MPa (x6.88 Ksi)].

Panel	Mode 0	Mode 1	Mode 2	Mode 3	Mode 4	Mode 5	Mode 6
1	288	265	295	293	284	267	293
2	279	262	257	293	262	238	293
3	262	257	275	284	269	255	284
4	254	241	256	275	257	245	275
5	287	291	307	335	369	370	335
6	217	271	177	323	198	170	324
7	293	255	171	318	177	159	319
8	258	239	255	275	260	245	275
9	267	262	268	282	263	251	282
10	269	262	268	282	263	251	282
11	267	233	243	274	249	235	274
12	236	158	252	271	269	251	271
13	255	283	304	287	319	288	287
14	254	280	300	287	317	286	287
15	251	279	295	284	311	282	284
16	228	253	235	275	253	232	275
35	287	291	313	294	331	312	294
36	283	287	246	294	243	227	294
39	261	278	263	280	260	248	280
40	276	290	310	292	298	281	292
43	277	279	181	290	204	179	290
44	285	302	249	305	264	230	305
47	262	279	181	290	204	179	290
48	278	299	247	305	262	228	305
50	278	302	266	308	277	241	308
56	259	273	215	282	229	208	282
57	274	292	292	294	296	268	294
58	264	274	278	276	277	255	276
59	275	292	302	296	304	275	296
60	272	292	301	296	304	274	296
65	290	323	268	335	199	211	335
66	260	268	201	281	216	197	281
67	241	242	166	283	205	178	283
68	195	238	168	283	207	179	283
71	263	272	223	296	266	228	296
72	263	272	223	296	266	228	296

represents a condition of primarily external loading and results in the largest bending moments, which are in hogging, as shown in Table 3.

The LRTR derived allowable stress for hull girder bending of the steel vessel is 244 MPa and for the titanium vessel 259 MPa. These dictate the minimum MOI and SM. Table 4 indicates that for steel the actual MOI and SM of the vessel are significantly greater than the minimum required as a result of local structure requirements. On the basis of actual SM, applying the bending moment in Table 3 would result in a stress 37% less than allowable or 154 MPa (22.4 Ksi). Consequently, this latter quantity is considered the respective highest bending stress expected in the steel vessel. For the titanium vessel the actual SM results in a stress 39% less or 158 MPa (23 Ksi).

Tables 6 and 7 identify the capacities of all structure as determined from the grillage analysis for steel and titanium respectively. Table 6 indicates that for the steel structure the ultimate strength of the grillage is less than the maximum bending stress of 154 MPa (22.4 Ksi) for Panels 6 and 7 of the wet deck where stiffener tripping results in 136 MPa (19.78 Ksi). As this structure is closer to the neutral axis of the hull girder, the actual bending stress will be less. Of course, the structural scantlings can be increased. Table 7 indicates that for the titanium structure all grillages have sufficient ultimate strength.

7 HULL STRUCTURAL WEIGHT

A weight estimate of the resulting ship structure was accomplished utilizing the Ship Work Breakdown Structure (SWBS) (NAVSEA, 1981). The results of the weight estimate for SWBS 100, Hull Structure, is shown in Tables 8 and 9 for steel and titanium respectively. The titanium hull structure is approximately 4290 Mton or 46% lighter than steel. This represents 14% of the ship's full load displacement. Furthermore, 116 Mton of this change is the result of reduced scantlings in the shell portions of the hull or 5% of the weight of these elements.

On a relative basis comparing the difference in steel to that of titanium Clark (Clark, Cusumano, Richardson, 2015) carried out a comparative structural design of a 125 m in length trimaran combatant utilizing a monohull based structural design approach and then separately determined cross and wet deck structural requirements by other approaches. The study considered mild steel and the same titanium adopted herein. The results indicates a 33% reduction in hull girder weight and they attribute bucking considerations to limiting the reduction of scantlings and the attendant weight. Although details of the calculation procedure they utilized are not available the implication is that it considered the above mentioned *Mode* 3: Beam-column type collapse only. The reduction in comparative titanium weight for the large trimaran was significantly greater and buckling was not the controlling factor. There are many variables which could account for the differences but for one the ships are very different in size.

Table 8. SWBS 100 hull structural weight summary for steel.

Total Ship Work Breakdown Structure (SWBS) Weight Summary

SWBS Group	Description	Total weight (Mton)	Percent	VCG (m from baseline)
111	Shell Plating	2312.2	24.52%	10.69
113	Inner Bottom	539.0	5.72%	4.08
116	Longitudinal Shell Frames/Stiffeners	689.8	7.31%	7.37
117	Transverse Shell Frames/Stiffener	612.9	6.50%	10.67
121	Longitudinal Structural Bulkheads	878.4	9.31%	11.76
122	Transverse Structural Bulkheads	403.8	4.28%	13.62
131	14.4 m Level Structure	1198.3	12.71%	16.12
132	11.0 m Level Structure	462.8	4.91%	10.46
133	8.0 m Level Structure	621.9	6.59%	8.88
134	5.0 m Level Structure	518.2	5.49%	4.68
136	22.6 m Level Structure	1193.1	12.65%	22.67
100	General Hull Structure (Summary)	9430.4	100.00%	12.04

Table 9. SWBS 100 hull structural weight summary for titanium.

Total Ship Work Breakdown Structure (SWBS) Weight summary—Titanium

SWBS Group	Description	Total weight (Mton)	Percent	VCG (m from baseline)
111	Shell Plating	1262.5	24.57%	10.69
113	Inner Bottom	300.3	5.84%	4.08
116	Longitudinal Shell Frames/Stiffeners	307.5	5.98%	7.37
117	Transverse Shell Frames/Stiffener	341.5	6.65%	10.67
121	Longitudinal Structural Bulkheads	489.4	9.52%	11.76
122	Transverse Structural Bulkheads	225.0	4.38%	13.62
131	14.4 m Level Structure	667.6	12.99%	16.12
132	11.0 m Level Structure	257.8	5.02%	10.46
133	8.0 m Level Structure	346.5	6.74%	8.88
134	5.0 m Level Structure	288.7	5.62%	4.68
136	22.6 m Level Structure	651.4	12.68%	22.67
100	General Hull Structure (Summary)	5138.2	100.00%	12.09

8 VESSEL COST

A comparison of the cost of the steel and titanium hull variants would require a full life cycle analysis to be fair. As mentioned earlier, a comparison was not intended for this current study on structural design. However, some observations are offered.

The cost of the titanium base metal is in the range of an order of magnitude more than steel, while at the same time the hull structural weight is in the order of one half. There is little experience with large structure fabrication in titanium and it should be expected it will be of greater cost than in steel. As a consequence, the acquisition cost of the titanium vessel will be greater.

On the other hand, the lighter titanium ship will not corrode, saving on maintenance and the hull roughness will be less saving on fuel. These characteristics will be present throughout the vessel's life. Furthermore, a vessel that does not corrode will have a longer life. As a consequence, it should be expected that the operating costs will be lower while carrying more cargo than the steel variant.

If a life cycle cost approach is adopted, then each of the acquisition and operational characteristics need to be considered over the projected life of the vessel. Furthermore, if the vessel under consideration is a naval type, the cost of the hull will be a smaller percentage of the total vessel acquisition cost and consequently the cost of steel versus titanium will be more muted.

9 CONCLUSIONS

The primary conclusions resulting from this study are:

- The application of the LRTR to either high strength steel or titanium produces ship scantlings which are adequate from a grillage analysis validation as well.
- The reduction in vessel structural weight when using titanium is the result of a combination of both the lower density of titanium as well as the smaller scantlings dictated by its higher strength.
- The reduction in structural weight of the vessel studied was 46%, which represents 14% of the vessel's full load displacement.
- The application of titanium to the vessel allows for up to 14% more deadweight, reduced resistance and therefore reduced propulsion power and fuel consumption, as well as reduced hull structural maintenance over its lifetime.

ACKNOWLEDGEMENTS

This work has been in part supported by the Office of Naval Research (ONR) under grant no. N00014-10-1-0652.

REFERENCES

Clark, C., Cusumano, T., Richardson, R. (2015), Structural Design of a Trimaran Small Combatant using DPSS, NSWCCD-CISD-2015/008.

Daidola, J.C., Arnold, A.M., Imbesi, M.J. (2017), Trimaran Preliminary Structural Design Procedures for a Large High Speed Sealift Ship, Naval Engineers Journal, June 2017.

Hughes, O.F., Paik, J.K. (2010), Ship Structural Analysis and Design, SNAME.

Knight, Joshua T. (2017), User Guide for the Hull Module Analyzer, NAVATEK Ltd.

Lloyds Register (2016). Rules for the Classification of Naval Ships.

Lloyds Register (2016). Rules for the Classification of Trimarans.

NAVSEA (1981), Command, Ship Work Breakdown Structure, NAVSEA S9040-AA-IDX-010/SWBS.

Paik, J.K., and Thyamballi, A.K. (2003). Ultimate Limit State Design of Steel-Plated Structures, Chichester, United Kingdom, Wiley.

Rosborough (2004). Ship Hull Characteristics Program (SHCP) User's Manual, NSWCCD.

Schutz, R.W., Scaturro, M.R. (2010), Titanium Alloys for Marine Environments and Seawater Service, Titanium Seawater EFC (Draft).

Analysis of calculation method of hull girder residual strength for cruise ship

Yingchao Pu
China Classification Society Rules Research Institute, Shanghai, China

Guijie Shi
State Key Laboratory of Ocean Engineering, China Strategy Institute of Ocean Equipment Engineering, Shanghai Jiaotong University, Shanghai, China

ABSTRACT: The increase of cruise ship tonnage causes the design load gradually close to hull girder ultimate strength. Therefore, evaluating residuals strength after collision and grounding damage becomes more important. However, at present it is difficult to find relevant research achievements and it is urgent to evaluate residual strength in design stage. In this paper, one cruise ship is selected as the research object with assuming a reasonable Damage extent. The finite element modeling method of whole ship is discussed and nonlinear finite element method is used to analyze the effect of collision and grounding damage on residual strength. And also, the influence of superstructures on the failure mode and failure sequence of hull girder is studied in detail. Then, considering the superstructures effectiveness the incremental iteration method is modified to evaluate cruise ship residual strength. At last, a principled method for determining the effectiveness value after collision and grounding damage is proposed, so as to provide reference for evaluating cruise ship residual strength in the early design stage.

1 INTRODUCTION

In order to meet the rapid increasing demand of cruising market, the cruise ship tonnage has shown an increase trend. According to Clarkson data in January 2017, the total tonnage of large cruise ship of more than 70,000 GT accounts for 40% of the existing fleet. Larger cruise ship leads to the increase of superstructure layers, the plating and stiffeners is allowed to present local nonlinear buckling and plastic deformation, and the hull design load is more close to the hull girder ultimate capacity.

The cruise ship has a large number of openings on the side and deck structures, its side moves a distance from outer side to middle longitudinal section on the lifeboat deck, and also there are many discontinuous bulkheads, pillars and inner walls and so on. Under the combined action of still water and wave moment, the cruise generally presents hogging deformation, but the deformation of superstructures is inconsistent with the main hull. There is horizontal shear stress and vertical tension between the superstructures and main hull. The superstructures cannot fully participate in hull girder longitudinal strength. The effectiveness of superstructures is influenced by many factors, such as side openings, deck openings, side shrinkage, pillar and partial bulkheads [1]. It is difficult to summarize a general formula for calculating superstructure effectiveness, which is still a thorny problem in the structural design of cruise ship.

In history, the cruise ship has occurred a number of marine accidents, and each accident has been highly concerned in maritime industry. In 2012 the Costa Concordia cruise ship grounded in Italy sea [2], in 1999 the Norwegian Dream cruise ship collided with the container ship in British sea, and in 1986 the MS Mikhail Lermontov cruise grounded in New Zealand. According to LR statistics, from 1990 to 2004, more than 48% of casualties on cruise ship was attributed to collision and grounding damage [3]. Collision or grounding damage reduces hull girder ultimate strength, and cabin flooding after structural damage may also increase still water load. In ship design stage, it should be ensured that the ship has enough reserve strength to withstand collision and grounding damage, in order to provide enough time to rescue passengers and ships, and protect life and property safety. The SOLAS convention has a mandatory requirement for the damage stability of passenger ships [4], but the assessment of residual strength is still a blank. The residual strength assessment should ensure that the sum of the static water and wave bending moment after damage does not exceed the ship residual strength.

The research on hull residual capacity is focused on two aspects: the statistical analysis of Damage extent and calculation method of residual capacity.

Based on the historical statistics of collision and grounding accidents, the probability of collision and grounding damage was summarized, and the Damage extent relevant to ship dimensions was proposed [5]. The number of cruise ship fleet is relatively small, and accident data analysis should depend on other ship types. There are three major types of seabed indenters as defined in [6], namely 'rock', 'reef' and 'shoal'. Not only denting but also tearing by grounding damage may eventually trigger collapse of the hull girder by bending or shearing and cause hazardous consequence [7]. After damage, the transverse section of hull structure is asymmetrical, and local member failure leas to translation and rotation of the neutral axis at the same time. The nonlinear finite element method (NFEM) and the incremental iterative method (SMITH) are usually used to calculate residual strength after damage.

The residual strength evaluation criteria can be determined by reliability method, which ensures that the safety level of residual strength is equivalent to the intact state. Based on reliability method, IACS CSR rules puts forward the evaluation criteria for residual strength of bulk carriers and tankers [8]. The rules clearly specify the scope of collision and grounding damage, the wave bending moment only considers 67% of the rule value, and the still water bending moment due to cabin flooding increases by 10%. However, the increase of the still water bending moment corresponding to cabin flooding may exceed 10% for all the conditions of damaged stability calculation [9]. Therefore, the reasonable evaluation method of residual strength is still a research hot point.

At present, the residual strength research is mainly concerned with cargo ships with single deck, such as bulk carriers, tankers, container ships, etc. The residual strength for cruise ship is also very important for safety redundancy. The effect of cruise ship superstructure on residual strength is different from that of the conventional cargo ship. It is necessary to study residual strength of cruise ship. In this paper, one 70,000 GT cruise ship as the research object, NEFM is used to analyze the effect of damage on residual strength with reasonable assumption of collision and grounding Damage extent. The linear elastic analysis by whole ship direct calculation is used to study the effect of superstructure effectiveness on hull girder longitudinal strength. The distribution of superstructure effectiveness is then summarized, which is applied to modify the incremental iteration method for cruise residual strength calculation. The principle selection method of superstructure effectiveness after collision and grounding damage is proposed, which can provide reference for the residual strength evaluation in the design stage of cruise ship.

2 CRUISE SHIPS

2.1 Main dimension

A cruise ship is selected as the research object in this paper. The 3rd deck is set as the main deck. The structures above the 3rd is seemed as superstructures. The hull structures have an obvious side shrinkage on the third deck. The main parameters of the cruise ship are 250 m in length, 32.2 m in width, and 32 m in depth, and 10.6 m height at the main deck.

The mid-section is shown in Figure 1, including structural scantlings of main hull and superstructures. The hull structures are designed based on CCS Rules for Cruise Ships (2017).

Based on general arrangement of the cruise ship, the middle section locates within the scope of the maximum bending moment, and the section also locates boundary edge of the pool on the topmost deck, which destroys the structure continuity in the two superstructure decks. Therefore, the middle section is set as the main section for analyzing residual strength.

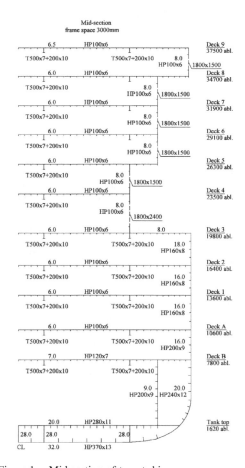

Figure 1. Mid-section of target ship.

Table 1. Damage extent.

Case		Damage extent, m
Collision	Height, h	0.75D
	Depth, d	B/16
Grounding	Height, h	min (B/20 and 2 m)
	Width, b	0.60B

2.2 *Damage extent*

Determining the Damage extent after collision and grounding accidents is a complex job. Since the lack of samples for cruise structural damage, it is difficult to get a reasonable damage assumption by historical data. After comparing rules requirements of classification societies, including ABS, DNV and IACS, the Damage extent is assumed based on CSR relevant requirement. Taking into account of higher freeboard of cruise ship, the collision Damage extent in height direction is measured from downwards 1 m below the upper edge of main deck. The Damage extent is listed in Table 1.

3 NONLINEAR FINITE ELEMENT METHOD

3.1 *FE model*

Nowadays, nonlinear finite element method has been successfully applied in calculating ultimate and residual capacity for ordinary single deck ships by using whole ship model or cargo hold model. However, cruise ship has extremely complicated load transition routes. Accordingly, it is also very hard to analyze ultimate strength and residual strength by nonlinear finite element method. This paper has fabricated three kinds of FE models to do residual strength analysis, including:

1st model: A finite element model simulating the "genuine" whole ship is established in which all structures including longitudinal and transverse bulkheads, pillars, stiffeners and opening, have been considered. However, the non-concern local area of the model may collapse earlier due to stress concentration, which enhanced the difficulty of numerical stability and terminated the calculation process before the collapse of whole hull girder. So this kind of FE model is difficult to analyze residual strength.

2nd model: Another kind of FE model only includes the scope of 3 web frames in the center of ship length, which has greatly reduced the element quantity. The difficulty is reasonably simulating the boundary conditions at model edge. The transverse section will not conform to plane assumption because of partially effectiveness of superstructures on longitudinal strength. The boundary condition is derived from global analysis under longitudinal bending deformation in elastic stage. Nevertheless, the boundary condition could not reveal the nonlinear characteristics in plastic stage. So, this model cannot be used to analyze residual strength, either.

3rd model: The 3rd kind of FE model is fabricated based on simplifying the whole ship. The concerned area at the middle, including 3 web frames length, is modeled by fine meshing, fore and aft part are modeled by coarse meshing, and a transit region between fine meshing and coursing meshing extends to 2 web frame length. The detailed modeling method is as following:

1. Fine meshing locates at middle length. The mesh sized is set as s/6 (s, longitudinal space), as shown in Figure 2. Therefore, there will be enough elements to simulate plating buckling mode between stiffeners.
2. The transition region extends about 3 web frames. Element size gradually changes from s/6 to web frame space. The plate thickness in the transitional region is deliberately increased 0.5 mm, so as to prevent that the collapse area occurs first to the focusing area.
3. Coarse meshing extends backwards and forwards from transition region. The mesh size is set as web frame space. Longitudinal stiffeners are merged at element long edge. The model length is equal to actual ship length 243 m, breadth 32.2 m, and depth to the upmost deck.
4. In the model, the window openings, discontinued longitudinal bulkheads and large openings on decks in superstructure are considered.

Finally, the simplified whole ship model has elements number about 720,000 and nodes number about 580,000. Using this model, the stress and displacement characteristics in central region of the ship could be revealed properly under longitudinal bending moment, while the element number and nodes number are reduced to about half which makes great advantage for the calculating efficiency and improves the numerical convergence compared to the "genuine global model". Besides, in the coarse mesh zones, the elements for web frames

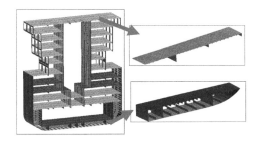

Figure 2. Fine meshing in target region.

are modeled in beam type and the discontinued structures are eliminated which greatly improves the numerical convergence since the failure modes for these structures earlier than the central focusing area are ignored, and a reasonable ultimate and residual capacity could be obtained.

3.2 Loads and boundary conditions

It is assumed that the bilinear material curve is chosen for the material used in the sample ship. The characteristics for the mild material and high tensile strength steel are listed in Table 2.

The loads applied on the ship are mainly longitudinal bending moment and longitudinal shear force. Since the goal in this paper is to analyze the ultimate capacity and residual capacity under longitudinal bending moment acting on ship, the loads and load applying methods should be focused on. In this paper, the longitudinal target bending moment distribution curve is changed to lateral pressure distribution acting on the bottom of ship (bottom plate downwards from the inner bottom) which leads to that the bending moment at middle length reaches the largest value along ship length, so as to model the loading condition under longitudinal bending moment. The boundary conditions are listed in Table 3, mainly constraining the rigid body displacement.

The model applied on load and boundary conditions is shown in Figure 3.

3.3 Initial imperfection

Three types of initial imperfection are to be considered in order to induce the buckling modes in the ultimate and residual capacity direct analysis. The initial geometric imperfections are global deflection of stiffened plate, local deflection of plate and tripping deflection of stiffener webs, and the value could be referred to in [10]. The imperfection deflections applied on stiffened plates are shown in Figure 4.

As per cruise ships, the hogging condition is the main loading condition which leads the compression at bottom structures. Since initial geometrical imperfection give little influence to the extending areas, the initial geometrical imperfection is applied on the bottom, sides and the lowest deck below the neutral axis of transverse section.

The initial imperfections are as following:

Global deflection of stiffened plate:

$$w_{S0} = B_0 \sin\frac{\pi x'}{l} \sin\frac{\pi y'}{S} \qquad (1)$$

Local deflection of plate:

$$w_{P0} = A_0 \sin\frac{m\pi x'}{l} \sin\frac{\pi y'}{s} \qquad (2)$$

Tripping deflection of stiffener webs:

$$w_{T0} = C_0 \frac{z'}{h'_w} \sin\frac{\pi x'}{l} \qquad (3)$$

where l = stiffener length between primary strengthen members; s = stiffener space; S = space

Table 2. Properties for materials.

Items	Symbol	Value
Young's modulus (MPa)	E	206000
Plastic modulus (MPa)	E_T	1000
Passion ratio	N	0.3
Yield strength (MPa)	Reh	Mild steel: 235 High tensile steel: 355

Table 3. Boundary condition.

Position	Tx	Ty	Tz	Rx	Ry	Rz
Bottom intersection of central section and afterward end	–	Fix	–	Fix	–	Fix
Bottom intersection of central section and forward end	Fix	Fix	–	Fix	–	Fix
Nodes of bottom plate at afterward and forward end	–	–	Fix	–	–	–

Figure 3. Simplified whole ship model.

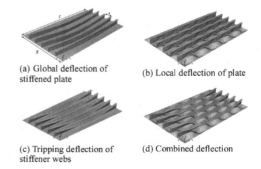

(a) Global deflection of stiffened plate
(b) Local deflection of plate
(c) Tripping deflection of stiffener webs
(d) Combined deflection

Figure 4. Illustration of initial imperfections.

of primary members; h'_w = height of stiffener web; m = buckling mode number of plating, set as the minimum integer for $l/s \leq \sqrt{m(m+1)}$; $A_0 = s/200$; $B_0 = l/1000$; $C_0 = l/1000$.

3.4 Results and analysis

According to the damage assumption in 2.2, the damaged extent of collision and grounding condition between two web frames is set on the finite element model so as to calculate the ultimate capacity of the hull girder. It is shown in Figure 3 that the displacement and stress contour plot of hull girder under collision and grounding condition at the time of collapse of structures. In order to make comparison for intact and damaged conditions, the ultimate strength for hull girder in intact condition was also calculated.

Based on the analysis of NFEM results (see Figure 5) and the failure sequences, the following conclusions could be made:

1. The "simplified whole model" could be applied in the ultimate and residual capacity prediction and a reliable result could be obtained.
2. The superstructures of cruise ships have important influence to the ultimate strength and the upmost deck would buckle first than other structures, which indicates that the superstructure contributes much effort to the ultimate strength. However, the longitudinal stress on the main deck is low and the buckling mode rarely happens.
3. In hogging condition, bottom structures buckle first. Nevertheless, for this ship, the longitudinal plate buckles first than inner bottom plate since the thickness of longitudinal bulkhead plate is much lower.
4. The assumption of extent of collision in 2.2 has little influence to the failure sequences while only reduces the hull capacity a little both in hogging and sagging conditions.
5. The grounding damage assumed in this paper changes the failure modes under hogging condition, and reduces the ultimate capacity by 21%. From this opinion, the assessment of cruise ship residual strength should be focused on the grounding condition.

4 INCREMENTAL-ITERATIVE METHOD (SMITH METHOD)

4.1 Superstructure effectiveness analysis

Since the superstructure is huge and there are many decks on a cruise ship, the plane assumption of hull girder under longitudinal bending moment no longer exists. Many researchers paid lot of efforts to the superstructure effectiveness for ships with huge superstructures, and in [11] a method to calculate the effectiveness was proposed, which is adopted in this paper. Through global finite element analysis, the stress distribution on sections was obtained and effectiveness on each deck was calculated as in Table 5. By analyzing superstructure effectiveness, we could predict that the effectiveness for transverse sections of this ship is about 0.5.

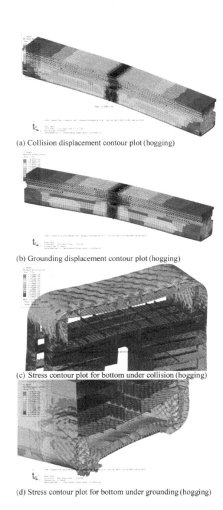

(a) Collision displacement contour plot (hogging)

(b) Grounding displacement contour plot (hogging)

(c) Stress contour plot for bottom under collision (hogging)

(d) Stress contour plot for bottom under grounding (hogging)

Figure 5. Displacement and stress contour of damaged condition (hogging).

Table 4. Hull ultimate capacity results (NFEM, × 10^{12} Nmm).

Cases	Intact ①	Collision ②	Grounding ③	②/①	③/①
Hog	9.50	9.40	7.52	99.1%	79.2%
Sag	4.40	3.70	3.75	83.7%	85.0%

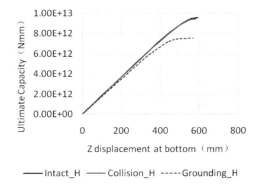

Figure 6. Hull ultimate capacity diagram in hogging.

Table 5. Effectiveness of decks above main deck.

Location	0.4 L	0.5 L	0.7 L
Deck 11	0.49	–	0.04
Deck 10	0.41	–	0.08
Deck 9	0.36	0.38	0.42
Deck 8	0.47	0.62	0.40
Deck 7	0.48	0.52	0.49
Deck 6	0.49	0.43	0.68
Deck 5	0.51	0.38	1.01
Deck 4	0.63	0.66	1.12
Average	0.49	0.50	0.53

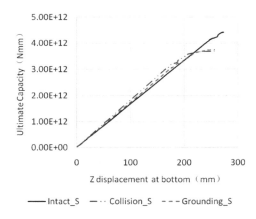

Figure 7. Hull ultimate capacity diagram in sagging.

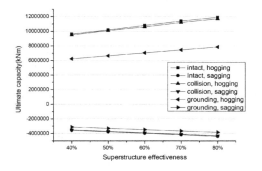

Figure 8. Effect of superstructure effectiveness on ultimate capacity of cruise ship at middle section.

4.2 Appliation of incremental-iterative method

The incremental-iterative method has been widely used in ship hull ultimate capacity calculation and the procedures could be referred in IACS CSR rules. In this paper, the hull girder ultimate capacity of middle section has been calculated using incremental-iterative approach assuming that the effectiveness of superstructure varies from 40% to 80% with the increment 10%. In the preliminary design stage, since the superstructure effectiveness could not be obtained easily for the reasons that the deck openings, side shell openings, and shear lagging effect would contribute to the change of effectiveness distribution, the difference along the width and depth direction are not taken into consideration. That is, in order to make it convenient in the preliminary design stage, the effectiveness of superstructure above main deck is set a unique value.

In Figure 8, it is seen that the ultimate capacity of cruise ships is linearly related to the effectiveness, i.e. the residual capacity of hull girder linearly increased with the increase of effectiveness both in hogging and sagging condition. The hull girder ultimate capacity in different effectiveness is listed in Table 6, showing that the ultimate capacity at the effectiveness 80% is 1.16, 1.16 and 1.18 times of that at the effectiveness 50% respectively under intact, collision and grounding condition. So proper effectiveness estimation is essential in the ultimate capacity assessment for cruise ships.

Further, the ultimate capacity of collision condition is 98% of that of intact condition, which means that collision damage contributes limited influence to the decrease of hull girder capacity both in hogging and sagging condition for cruise ships. The reason is that because the neutral axis is near the main deck where the extent of collision damage is, structures close to main deck contribute little to the longitudinal strength.

However, under grounding condition, the hull girder capacity of middle section is only 65% and 87% compared with intact condition in hogging and sagging condition respectively. Since the hogging condition is the critical loading phenomenon, grounding contributes obvious decrease for the

Table 6. Influence of effectiveness to the hull girder capacity ($\times 10^6$ kNm).

Superstructure Effectiveness	40%	50%	60%	70%	80%
Intact, hogging	9.6	10.2	10.8	11.4	11.9
Intact, sagging	−3.6	−3.8	−4.0	−4.2	−4.4
Collision, hogging	9.5	10.1	10.6	11.2	11.7
Collision, sagging	−3.5	−3.7	−3.9	−4.1	−4.3
Grounding, hogging	6.2	6.7	7.1	7.5	7.9
Grounding, sagging	−3.1	−3.3	−3.5	−3.7	−3.9

hull girder capacity in hogging. For this reason, the grounding condition should be paid careful attention to in the residual strength assessment for cruise ships.

4.3 Modification of incremental-iterative method

Based on previous analysis, it is seen that for cruise ships, the effectiveness of superstructure is a key problem to the hull girder capacity analysis. Based on the NFEM result, proper value for effectiveness could be set for collision and grounding conditions for the sample ship. In Figure 9 and Figure 10, the ratio of the value of incremental-iterative method to that of NFEM is shown at different superstructure effectiveness percentage in hogging, sagging and average conditions.

According to Figure 9 and Figure 10, it is seen that the residual capacity of incremental-iterative method is nearly the same as that of NFEM when the superstructure effectiveness is set 50% and 70% under collision condition and grounding condition respectively. This indicates that for collision and grounding, the superstructure contributes differently to the hull girder residual capacity. Referring to the effectiveness result in 4.1, it is shown that the superstructure effectiveness is almost 50% in intact condition and after collision, e.g. the effectiveness is almost the same. However, in grounding condition, the superstructure effectiveness increases to 70% since that the neutral axis moves up and the superstructures contribute more to the longitudinal strength.

So in the preliminary design stage for cruise ships with side shrinkage above main decks, the superstructure effectiveness might be set as 50% for collision condition while it might be increased to about 70% for grounding conditions. However, if the design stage goes deeper, e.g. more drawings brought out, global finite element analysis done, influences of large openings on decks and side taken into consideration, and pillars load transition effort taken into consideration, the stress distribution of transverse section might be much clearer and more correct effectiveness could

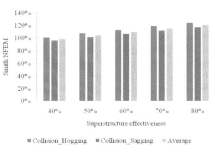

Figure 9. Ratio of Smith method to NFEM (collision).

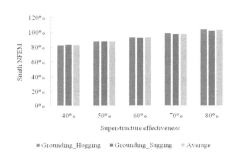

Figure 10. Ratio of Smith method to NFEM (grounding).

be estimated. The hull girder capacity evaluation could be more accurate.

5 CONCLUSIONS

The cruise ship has multi-layer superstructures and many discontinuous structures such as openings, discontinuous bulkheads and pillars, etc. The superstructures have large influences on residual strength after collision and grounding damage. In this paper, the residual strength and failure mode of cruise ship after collision and grounding damage are analyzed in detail by NFEM. The incremental iteration method is modified to evaluate residual strength of cruise ship. At last, a principled selection method of superstructure effectiveness value is put forward. The main conclusions of this paper are listed as follows:

1. By comparing three type of modeling method proposed in this paper, the simplified finite element model can be used to evaluate the residual strength of cruise ship. The target area is modeled in detail by fine meshing, and the non-concerned area is modeled by coarse meshing, which reduces the element number greatly,

avoids the local failure of non-concerned areas and improves the convergence efficiency of nonlinear computation.
2. The superstructures of cruise ship have great influence on hull girder ultimate strength. When reaching the limit state, the longitudinal stress of the main deck is always not high, which is different with conventional cargo ships. In sagging condition, superstructure decks present buckling failure earlier than main deck.
3. The grounding damage assumed in this paper changes the failure mode under hogging condition, and reduces the ultimate capacity by 21%. The assessment of cruise ship residual strength should be focused on the grounding condition.
4. The contribution of superstructure to residual strength of hull girder after collision and grounding damage is different. When calculating the cruise ship residual strength in the early design stage, the superstructure effectiveness after collision damage can be roughly estimated to be 50%, and the value after grounding damage can be increased to 70%.

REFERENCES

[1] Alie, M.Z.M., et al., The Influence of Superstructure on the Longitudinal Ultimate Strength of a RO-RO Ship. 2016.
[2] Casoli, E., et al., Assessment of the impact of salvaging the Costa Concordia wreck on the deep coralligenous habitats. Ecological Indicators, 2017. 80(Supplement C): p. 124–134.
[3] SAFEDOR, FSA for Cruise Ships – Subproject 4.1. 2005.
[4] Vassalos, D., Damage survivability of cruise ships – Evidence and conjecture. Ocean Engineering, 2016. 121(Supplement C): p. 89–97.
[5] Pedersen, P.T. and S. Zhang, Effect of ship structure and size on grounding and collision damage distributions. Ocean Engineering, 2000. 27(11): p. 1161–1179.
[6] Alsos, H.S. and J. Amdahl, On the resistance of tanker bottom structures during stranding. Marine Structures, 2007. 20(4): p. 218–237.
[7] Bin, S., et al., An analytical method to assess the damage and predict the residual strength of a ship in a shoal grounding accident scenario. Journal of Ocean Engineering and Science, 2016. 1(2): p. 167–179.
[8] IACS, Common Structural Rules for Bulk Carriers and Oil Tankers. 2016.
[9] Downes, J., et al., A Method for the Quantitative Assessment of Performance of Alternative Designs in the Accidental Condition. 2007.
[10] IACS, TB Report Structural Redundancy. 2013.
[11] Wang, F.H., et al., Study on hull girder strength calculation methods of large naval ship. Journal of Ship Mechanics, 2005. 9(3): p. 48–59.

Integrated knowledge-based system for containership lashing bridge optimization design

Chuntong Li & Deyu Wang
State Key Laboratory of Ocean Engineering, Shanghai Jiao Tong University, Shanghai, China
Collaborative Innovation Center for Advanced Ship and Deep-Sea Exploration, Shanghai, China

ABSTRACT: An integrated optimization design system for containership lashing bridge based on Knowledge-Based Engineering (KBE) approach has been developed. The professional knowledge, design standards, rules, domain knowledge and expert experience of lashing bridge design are stored in the knowledge base under this proposed knowledge-based system. Based on the basic knowledge base, the system realizes the design, calculation, analysis, and optimization of the lashing bridge structure. A series of advanced template files is established in this paper, including templates for CAD structural graphics, FEM graphics and knowledge. Knowledge templates are comprised of design specifications, design rules, expert experience, constraint rules, expressions, macros, etc. The repetitive modeling work is avoided while the knowledge is being reused and the duration of the engineering design phase is effectively shortened. The knowledge engineering-based optimizing design system can provide support and suggestions for optimizing the structure, effectively avoiding errors and conflicts, and eventually attaining the lightweight structure.

Keywords: Knowledge-Based System (KBS); Optimization; Knowledge; Design automation

1 INTRODUCTION

Lashing bridge is encompassed by a square pipe or a built-up lashing pillar with upper and lower platforms, tightly stabilizing a cargo container on the deck or hatch cover to resist rolling or heeling of the ship. Lashing bridge design is one of the key technologies for the large container ship. It is not only related to the actual number of containers and the flexibility of the container placement but also affects the determination of main dimensions and general layout of the ship. To survive under the competitive environment of world-class shipyards, large shipyards are required to adopt information technology more frequently and rationally (Shin and Han, 1998; Zhou, Qiu and Hua, 2007). Considerable knowledge from different disciplines is involved in the design of lashing bridge. The product design, innovation, and product quality improvement are evidently obstructed as the design is complex in process and dependent on design experience. In the meantime, as also presented from the actual production design, it is urgently required to update the current state is through the method of KBE.

KBE application aims to shorten time and save the cost of new product development, which is primarily accomplished through the automatically repetitive design tasks while the product and process knowledge are being captured, retained and reused (Liese, 2004; La Rocca, 2012). In the product design, coordination between various departments is insufficiently available, or the designer lacking professional knowledge and being inexperienced shall prolong the time for production. As Baxter (2007) notes, 'about 20% of the designer's time is spent searching for and absorbing information', and '40% of all design information requirements have been satisfied by personal stores, even though more suitable information may be available from other sources'. As Park (2002) summarizes, the KBE can help engineers to design and manufacture the ship product completely and rapidly. The KBE framework shall give proper advice to designers, and accordingly the design, the efficiency can be improved (Yong Chen, Ze-Lin Liu and You-Bai Xie; Verhagen, Bermell-Garcia, van Dijk and Curran; 2012).

KBE is regarded as one among the most vigorous branches of artificial intelligence. It principally seeks to incorporate relative design experience, design standards and specifications into design software to design new product (Chapman and Pinfold, 1999; Penoyer, Burnett, Fawcett and Fawcett; 2000; McMahon, Lowe, Culle; 2004). Sanya and Shehab (2014) presentes a novel KBE framework for effectuating platform-independent knowledge-enabled product design systems. Through adopting this method, knowledge reuse shall be promoted, and platform-specific approaches shall be eliminated. In the design domain, KBE is the most frequently adopted method to support customization and automotive design that can shorten the lead time, improve quality and gain more profit (Wei Guo,

Jiafu Wen, Hongyu Shao and Lei Wang, 2015). An intelligent optimization design system of hull structure is developed by Yang (2014) whereby KBE technology is applied. A successful application of KBE method is presented by Jin-ju Cui and De-yu Wang (2013) in ship structural design and optimization, and design cycle can be reduced evidently. Yang (2012) presents a KBE methodology for ship hull structural member design. The present KBE technology provides appropriate suggestions, support and information to achieve reuse and accumulation of knowledge. A KBE system is developed by Corallo (2009) for low-pressure turbine automation.

As established in the literature, 20% of the time is spent on innovative tasks, while the remaining 80% is engaged in repetitive tasks (Verhagen. et al, 2012). The repetitive tasks can be evidently reduced (80% automation) through KBE technology. The crux of rapid design system for Knowledge-based product is to incorporate computer-aided design software with knowledge engineering. The Knowledge-Based System (KBS) is encompassed by considerable knowledge, reasoning, decision-making and design rules. In this paper, Neighborhood Rough Set theory and Support Vector Machine technology are introduced, and the object-oriented knowledge representation method, mixed knowledge reasoning method and integration tool are adopted to design lashing bridge intelligently. The parametric structural model is transformed into the mathematical optimization model by means of KBE technology. The objective of minimizing structural quality shall be fulfilled through establishing knowledge templates and constraint rules. The KBE system is well independent, offering the high level of knowledge reuse in the course of lashing bridge optimization design.

2 KNOWLEDGE-BASED ENGINEERING

KBE is a novel technology taking on an enormous potential for engineering design applications (La Rocca, 2012). Knowledge-Based Engineering pertains to diverse interpretations and relevant definitions (Baxter et al. 2007; Chapman and Pinfold, 2001). A more involved definition is proposed by Chapman & Pinfold (2001) who indicate that 'KBE makes an evolutionary stride forward in Computer-Aided Engineering (CAE) and counts as an engineering method denoting an incorporation of Object-Oriented Programming (OOP), Artificial Intelligence (AI) and Computer-Aided Design (CAD) technologies, contributing to customized or variant design automation solutions'. The application of the KBE approach primarily contains three parts, i.e. knowledge acquisition, knowledge representation, and knowledge reasoning.

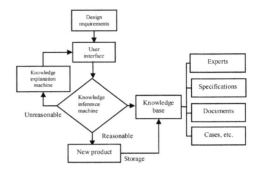

Figure 1. The flow chart of knowledge-based engineering products design.

The object-oriented knowledge representation method is adopted in this paper, with rules, frame, and process abiding by this method to establish a hybrid knowledge representation. This method outstrips the defects taken on by the single knowledge representation method and expresses the knowledge as an object-oriented class structure. In this regard, object-oriented knowledge representation is deemed to be ideal for the design knowledge and experience of lashing bridge. Knowledge reasoning technology is based on the product knowledge, drawing another or some other judgments through certain reasoning strategy. Yet Case-Based Reasoning (CBR) is excessively dependent on the case library, so it is difficult to realize the design innovation. Therefore, CBR, Rule-Based Reasoning (RBR) and Model-Based Reasoning (MBR) are adopted to improve the reasoning accuracy and efficiency. New product design process of KBE is overviewed in Figure 1.

3 ROUGH SETS THEORY AND SUPPORT VECTOR MACHINE

Attribute reduction and feature selection count as important applications of Rough Set theory in knowledge discovery. In the neighborhood rough set reduction calculation, the set of reduction is obtained by judging the significance of condition attributes. Neighborhood rough set theory initially presented by Pawlak (2012) is a quite useful mathematical theory in dealing with imprecise, incomplete and incompatible knowledge. The theory has been successfully implemented in many applications (Yong et al. 2014), e.g., feature selection (Hu et al. 2008), classifier (Hu et al. 2008), and rule learning (Du et al. 2011), etc. In this paper, the concept of spherical neighborhood in topological space and the feature selection algorithm of numerical data are adopted by virtue of neighborhood rough set model (Hu et al. 2008).The algorithm can be applied to numerical attributes directly and does not need to be discretized. In this regard, this algorithm shall eliminate the discretization process in contrast to the classical rough set method, and there is no need to establish neighborhood graphs in the sample space in contrast to the k-step neighborhood model.

$L = \langle U, Q, V, f \rangle$ is an information system, where U is a finite set of all samples $U = \{x_1, x_2, x_3, \ldots, x_n\}$, called a universe. Q is a finite set of sample attributes, which includes sample attribute and decision attribute. $V = \bigcup_{a \in Q} V_a$, where V_a is a domain of the attribute. $f : U \times Q \to V$ is called an information function. Given an N-dimensional space Ω, $\Delta = R^N \times R^N \to R$ where $\Delta(x_i, x_j)$ is a metric function, $x_i \in U$, it usually satisfies:

1. $\Delta(x_i, x_j) \geq 0$, if and only if $x_i = x_j, \Delta(x_i, x_j) = 0$;
2. $\Delta(x_i, x_j) = \Delta(x_j, x_i), \forall x_i, x_j \in R^N$;
3. $\Delta(x_i, x_k) \leq \Delta(x_i, x_j) + \Delta(x_j, x_k), \forall x_i, x_j, x_k \in R^N$;

There are three widely used metric functions for distance calculation.

1. Manhattan distance function: $\Delta(x_i, x_j) = \sum_{k=1}^{n} |f(x_i, a_k) - f(x_j, a_k)|$;
2. Euclidean distance function: $\Delta(x_i, x_j) = \left(\sum_{k=1}^{N} |f(x_i, a_k) - f(x_j, a_k)|^2 \right)^{1/2}$;
3. Chebychev distance function: $\Delta(x_i, x_j) = \left(\sum_{k=1}^{N} |f(x_i, a_k) - f(x_j, a_k)|^p \right)^{1/p}, p = \infty$;

Definition 1. Real space Ω and a finite set U, $U = \{x_1, x_2, x_3, \ldots, x_n\}, \forall x_j \ \delta(x_i)$ is defined as:
$\delta(x_i) = \{x \mid x \in U, \Delta(x_i, x_j) \leq \delta\}, \delta \geq 0$

Definition 2. Given a set of objects U and a neighborhood relation N over U, we call <U, N> is a neighborhood approximation space.

Lower approximations defined as: $\overline{N}X = \{x_i \mid \delta(x_i) \cap X \neq \varnothing, x_i \in U\}$

Upper approximations defined as: $\underline{N}X = \{x_i \mid \delta(x_i) \subseteq X, x_i \in U\}$

$\forall X \subseteq U, \underline{N}X \subseteq X \subseteq \overline{N}X$. The boundary region of X in the approximation space (<U, N>) is defined as:
Boundary: $BN(X) = \overline{N}X - \underline{N}X$, Positive: $Pos(X) = \underline{N}X$, Negative: $Neg(X) = U - \overline{N}X$.

Definition 3. Given a neighborhood decision system $NDT = (U, N, D), (X_1, X_2, \ldots, X_n)$ are the object subsets, $\forall B \subseteq A, \delta_B(x_i)$ is the neighborhood information granule. The upper and lower approximations of decision D with respect to attributes B are defined as:

$\overline{N}_B D = \{\overline{N}_B X_1, \overline{N}_B X_2, \overline{N}_B X_3, \ldots, \overline{N}_B X_N\}$

$\underline{N}_B D = \{\underline{N}_B X_1, \underline{N}_B X_2, \underline{N}_B X_3, \ldots, \underline{N}_B X_N\}$

where $\overline{N}_B X = \{x_i \mid \delta_B(x_i) \cap X \neq \varnothing, x_i \in U\}$, $\underline{N}_B X = \{x_i \mid \delta_B(x_i) \subseteq X, x_i \in U\}$

Boundary: $BN(D) = \overline{N}_B D - \underline{N}_B D$, Positive: $Pos_B(X) = \underline{N}_B X$, Negative: $Neg_B(X) = U - \overline{N}_B X$.

The dependency degree of D to B is defined as:

$k_D = \gamma_B(D) = Card(Pos_B(D))/Card(U)$

Dependence reflects the proportion of samples in the domain that can be accurately classified. Obviously:

1. $0 \leq \gamma_B(D) \leq 1$;
2. If $B_1 \subseteq B_2 \subseteq B_3 \subseteq \ldots \subseteq A$, then $\gamma_{B1}(D) \leq \gamma_{B2}(D) \leq \gamma_{B3}(D) \leq \ldots \leq \gamma_A(D)$.

There are two ways to define the neighborhood: one is the number of objects contained in the neighborhood, such as the classical k-nearest neighbor method, and the other is defined according to the maximum distance from the center of the neighborhood to the boundary. This paper uses the second method. The "k-step neighborhood model" is a neighborhood model built based on the k-Nearest Neighbor (KNN) algorithm. The comparison between Pawlak rough sets and neighborhood rough sets is shown in Figure 2. The lower approximation

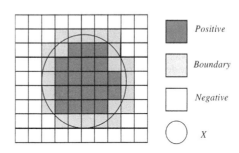

(a) Rough set in discrete feature space.

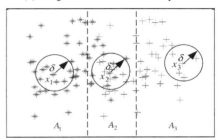

(b) Rough set in numerical feature space.

Figure 2. Comparison of Pawlak rough sets and neighborhood rough sets.

and the upper approximation of the set X are shown in Figure 2 (a). The large rectangle represents the universe, the small square represents the equivalence class of the equivalence relation R and the elliptic graph represents the rough set X. The upper and lower approximations consist of 21 and 44 small squares respectively. As shown in Figure 2 (b), samples x_1 and x_3 belong to the lower approximation of classes * and +, and samples x_2 is the boundary of classes * and +. If $\delta = 0$, then Neighborhood relations degenerate into equivalence relations, and the neighborhood particles are transformed into equivalence classes. We can see that Pawlak rough set is a special case of neighborhood rough sets.

Support Vector Machine (SVM) technology is first proposed by Vapnik. SVM primarily falls into two types, i.e. support vector classification (SVC) and support vector regression (SVR). Through drawing upon the strengths taken on by SVR in reckoning with small samples and prediction, the traditional process is avoided from induction to deduction, and the conduction is realized efficiently reasoned from training sample to prediction sample. The regression results of SVM are closely associated with the selection of parameters. RBF kernel functions shall basically not result in data catastrophes and are more efficient than sigmoid kernel functions (Keerthi et al. 2003). For the $\varepsilon - SVR$ model, the regression precision and generalization ability of SVR model are determined by three parameters: the penalty parameter C, the kernel width σ and the upper limit of error ε. The LIB-SVM (Chang et al. 2011) toolbox developed by Professor Lin Chih-Jen of National Taiwan University was employed to acquire regression prediction models and parameters.

4 BASIC THEORY

4.1 Application of neighborhood rough sets theory

The increase of samples and features not only provides considerable knowledge but also brings plenty of redundant information. This paper summarized the rules of condition attribute and decision attribute, laying the scientific basis for the next knowledge reasoning. A neighborhood rough set model is adopted to reduce attribute. A fast forward algorithm is formulated as follows for attribute reduction (Du et al. 2011):

Algorithm: Fast forward heterogeneous attribute reduction based on neighborhood rough sets (HARNRS).

Input: $NDS = (U, A \cup D, f)$, delta
Output: red.
1: $\varnothing \to red, U \to S$;
2: **while** $S \neq \varnothing$
3: **for** each $a_i \in C - red \varnothing \in POS_i$;
4: **for** each $O_j \in S$ compute $\delta(O_j)$;
5: **if** $\exists X_k \in U/D$ such that $\delta(O_j) \subseteq X_k$
6: $POS_i \cup O_j \to POS_i$
7: **end if**
8: **end for**
9: **end for**
10: find a_k such that $|POS_k| = \max_i |POS_i|$
11: **if** $POS_k \neq \varnothing$
12: $red \cup a_k \to red\ S - POS_k \to S$
13: **else**
14: **exit while**
15: **end if**
16: **end while**
17: **return** red, **end**

There is a parameter delta controlling the size of neighborhoods. The algorithm ensures that the core is not reduced. For more information about neighborhood rough sets and HARNRS algorithm, please refer to (Du et al. 2011). The rules between condition attributes and decision attributes are listed in Table 1 and Table 2.

4.2 Knowledge base

The knowledge base is defined as a collection of expert experience, business specifications, various design rules, cases and other types of knowledge, which are commonly applied in designing structure initially and eventually realizing optimization. The knowledge is sorted and summarized into some rules and problem-solving strategies, and it provides best guidance and recommendations for

Table 1. Lashing bridge structure condition attributes.

Structural design conditions attributes	
C_1	ship length between perpendiculars (mm)
C_2	ship width (mm)
C_3	total number of containers
C_4	the highest number of layer
C_5	container spacing (mm)
C_6	the height of the top of the hatch (mm)
C_7	the distance between hatch (mm)
C_8	distance between bridge and container (mm)
C_9	number of columns

Table 2. Lashing bridge structure decision attributes.

Structural design features attributes	
D_1	lashing bridge length (mm)
D_2	lashing bridge width (mm)
D_3	the height of the first layer (mm)
D_4	the height of the second layer (mm)
D_5	the height of the third layer (mm)
D_6	overall weight(kg)

designer in design. Object-oriented and if-then hybrid knowledge representation method is used to implement knowledge and system integration. Knowledge base design is illustrated in Figure 3. For instance, the wall thickness of the pillar shall be no less than the minimum of the following two values.

$$\begin{cases} t = \dfrac{br}{20l} \\ t = \dfrac{b}{18} \end{cases}$$

where b denotes the panel width of the section, l is the effective length of the pillar, 0.8 times the full length of the pillar, and r indicates the minimum inertia radius.

The design results of the lashing bridge structure are more sensitive to the expert experience and it often requires the evaluation of finite element analysis. To ensure the profile to be overall stabilized, the following requirements must be satisfied.

$$\frac{l}{b} \leq 1000 \sqrt{\frac{1}{\sigma_y}}$$

where l denotes the span of the profile; b represents the width of the small wing; σ_y indicates the material yield limit.

For the Walkway panel, for the ratio of frame spacing to plate thickness $\mu = s/t < 70-80$, regardless of the impact exerted by the middle surface stress, if the ratio $\mu = s/t > 70-80$, the impact exerted by the middle surface stress shall be considered. The walkway board is primarily to satisfy the requirements of securing work, so the thickness of the walkway board set ranges from 6 mm to 10 mm.

Lashing bridge structure should provide a safe working environment for securing. Some additional ergonomic and safety-related requirements are extracted from 'Code of Safe Practice for Cargo Stowage and Securing' (CSS Code), 'Guidelines for the Preparation of the Cargo Securing Manual', 'International Convention for the Safety of Life at Sea', abstracted into design rules and inspection rules. The design results are ensured to be rational. For example, the width of working areas shall not be less than 750 mm. The horizontal distance from the lashing securing points to the containers shall not exceed 1100 mm, and not be less than 220 mm for lashing bridges and 130 mm for other positions.

4.3 Standard template library

Standard template library is encompassed by the parametric standard parts library, parametric assembly library and parametric finite element model library. Lashing bridges are comprised of some standard components, thus it is of necessity to establish a standard library of parameterized model components and store the information of the standard components in the knowledge base. Through introducing the design table and macro programs to establish standard parts library, characteristic parameters of parts are extracted and stored in design tables. The structure of design table for angle bars is shown in Figure 4. Relations

Table 3. Allowable criteria for yielding.

Direct stress (N/mm²)	Shear stress (N/mm²)	Von Mises stress (N/mm²)
0.88 σ_o = 314.2	0.40 σ_o = 142	0.90 σ_o = 319.5

Where the main material of lashing bridge structure denotes AH36 high strength steel, the yield limit is $\sigma_o = 355 N/mm^2$.

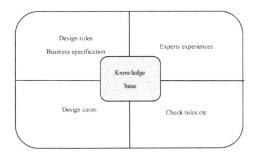

Figure 3. Structure of knowledge base.

Figure 4. The application of angle bar design table.

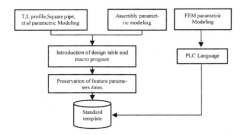

Figure 5. The design flowchart of standard template library.

between standard parts are established on the basis of standard tables and formulas check rules. New structural parts can be adjusted in parameters, and designers can update the standard parts library whenever necessary (Yang et al. 2012). The process of the standard template library construct is illustrated in Figure 5.

The system of constraint relations between components is acquired through analyzing geometrical features, static structures and dynamic features of the products and components. The hierarchical module of the lashing bridge is further compartmentalized as a binary tree structure. A complete top-down design is realized through receiving the upper layer information and driving the lower layer to generate an adaptive component model. In considering the structural characteristics of lashing bridge or components, a group of main parameters derived from the functional parameters is employed as driving parameters, and other secondary parameters are bound by them. Accordingly, the model can be overall changed rapidly as driven by the primary parameters, meeting the customers' individual requirements of customized products or components. The overall assembly structure is controlled whereby WAVE technology and associated replication geometry method are applied, ensuring the parameter association of the whole assembly and parts. UG/WAVE technology can be used to establish the relevant relations of parameters between different parts, realizing the relevant replication of parts (Li et al. 2016). The parametric FEM library is constructed by PCL language, which is beneficial for finite element analysis and optimization. The parametric finite element model was constructed by PCL language, the corresponding *.bdf file was generated during optimization iteration, and then the result of finite element analysis was obtained by Nastran software.

5 CASE STUDY

5.1 Design process explanations

The intelligent design process of the lashing bridge based on Knowledge Engineering is shown as follows. The first step refers to knowledge collection, collation and mining, and building knowledge base, selecting the design variable values in line with the knowledge base. The second step is to choose the proper design process to establish the system of constraint relations among components, and build a parametric model to realize the top-down design process. The third step is to obtain the design variable values of the lashing bridge through the hybrid knowledge reasoning method. CBR is established following the previous design examples in the case library, RBR is based on established design and inspection rules, and MBR is an inference model built on SVM. The integration of knowledge and design process is achieved through the object-oriented knowledge representation method. The final step is to check the design variables with the knowledge check rules to ensure the reasonability of the structure. As the test results indicate, the system will give relevant design evaluation and recommendations for designers, making the final design results meet the requirements of the user, norms, and expert experience.

5.2 Application of KBE of design

Three kinds of knowledge reasoning methods incorporated with hybrid reasoning are adopted. The knowledge reasoning mechanism is elucidated in Figure 6. The system carries out CBR according to the customer's needs and carries on the similarity analysis to the reasoning result. If the similarity requirements are satisfied, the similar instance is indicated to exist in the instance database, and the scheme can be output directly. If the similarity fails to satisfy the similarity requirements, it implies that there is no similar instance. The prediction model based on support vector machine regression theory is the basis of MBR reasoning. The

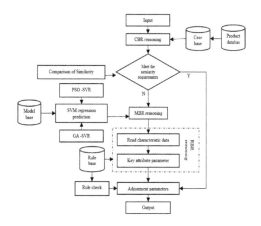

Figure 6. The flow chart of lashing bridge intelligent design.

overall parameters of the structure are obtained by the MBR reasoning. Then the parameters are corrected following RBR reasoning. The parameters are modified by expert experience, design specification and other knowledge. The data generated in the reasoning process is allowed to be manually modified by the designer. So, we had two paths to obtain the original dimensions of the lashing bridge. The first was based on CBR (Case based reasoning), MBR (Model based reasoning) and RBR (Rule based reasoning) to obtain the overall size of the structural framework and part of the structural dimensions. The second path was based on the design knowledge and experience to build relationships and formulas, as well as the structure of the relationship between the various components to obtain a specific detailed design dimensions. Manual intervention is an important part of the entire design process. Eventually, the initial design parameters of the structure are obtained.

For example, by entering the parameters as shown in Figure 7, the CAD model of 4250TEU, 10000TEU and 20000TEU containership lashing bridge structure is given by the system as shown in Figure 8. The check results of the structural design rationality are shown in Figure 9. The initial design parameters as shown in Table 3.

A Parametric Finite Element Model library for lashing bridge is established by PCL Language. The finite element model can be easily established through the structural parameters of the lashing bridge acquired in the foregoing reasoning. Take the 4250TEU lashing bridge as an example, finite element model is established, and the mesh size is about 100 mm × 100 mm. In particular, the CCS 〈China Classification Society〉 specification states that the value of lashing force on each lashing bar is 175 KN in the absence of specific lashing force data at the beginning of the design. The Finite Element analysis results in MSC/Patran is presented in Figure 10. Results of strength analysis: the maximum resultant stress is 221 N/mm^2, and the maximum shearing stress of is 118n/mm^2. As the analysis results bespeak, the maximum resultant stress and maximum shearing stress of the main structural members of the lashing bridge are less than the allowable values, thus the strength of the lashing bridge meets the requirements.

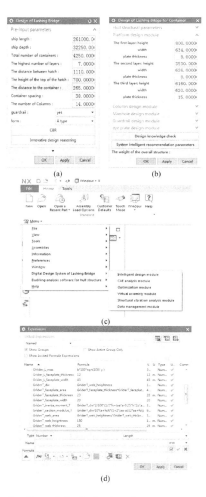

Figure 7. Part of the design interfaces.

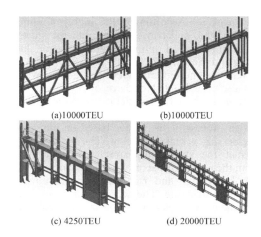

Figure 8. Part of lashing bridge structures.

Figure 9. Knowledge-based inspection window.

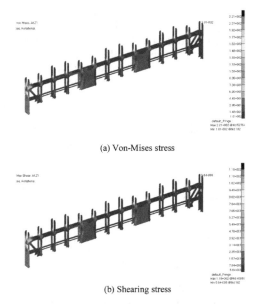

(a) Von-Mises stress

(b) Shearing stress

Figure 10. Finite element analysis results (4250TEU).

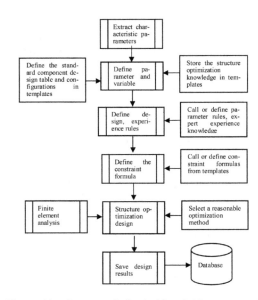

Figure 11. Process of the lashing bridge structural optimization.

5.3 *Optimization process explanation*

In the lashing bridge structure design, diverse design requirements and constraints are involved, and numerous design requirements and constraints are involved. Accordingly, the highly nonlinear, multi-modal, and convergence problems are triggered in structural optimization (Guo et al. 2007; Yang and Wang 2012).

KBE technology can not only reuse knowledge but also provide reliable advice, offer technical support, and reduce the conflict between the information and improve work efficiency. The optimization design process of the bridge based on knowledge engineering is clarified below. The first step is to extract the structural characteristic parameters from the results of the intelligent design and establish knowledge and rule templates required to optimize structure. The second step is to define parameter variables and constraints rules. The parametric structural model is transformed into a mathematical optimization model. The final step is to select the proper design process to support the optimization and eventually save the design result. The specific workflow is shown in Figure 11.

5.4 *Structural optimization problem*

The 4250TEU lashing bridge structure primarily contains lashing pillar (square pipe), bracing (square pipe), shear walls, L bar, T bar and plate structure. Among them, the lashing pillar, bracing, shear wall, walkway board, the support plate of Walkway board besides, and the L bar. Girder webs and panels are thickness sheets with national standard, and the configuration parameters for these panels are selected as design variables. The web thickness and faceplate thickness corresponding to the design variables are selected from a standard plate library. Each set of design parameters in the design table corresponds to the configuration parameters of the standard structure. The knowledge of design specifications, expert experience, and structural stability provides structural constraints. Lashing bridge structure optimization design constraints include specifications for plating thickness, section modulus and buckling theory for local stability of profiles. In the optimization process, the constraint relationship (including the wall thickness constraint, the constraint of the T-profile manufacturability dimension parameter, the structural local stability requirement, etc.) between the parameters is established based on the knowledge base. The value range of some design variables is determined by design table. Optimization calculate was mainly achieved by MATLAB software. Visual C++ is adopted to encapsulate the mathematical optimization data, the finite element data, database and design table to achieve the goal optimization. Finite element analysis is integrated in the optimization process. The results of structural strength obtained from each finite element calculation during iterative optimization are used as constraints. The optimization objective is to reduce the weight of the structure.

Optimal design results are listed in Table 3. The model of 4250TEU lashing bridge is transformed into mathematical optimization model by KBE technique and optimized through adopting the

Table 3. 4250TEU lashing bridge structure optimization and primary result comparison.

Variables and targets	Original	Optimized
Vertical bracing panel thickness	200 × 200 × 14.2 × 14.2 mm	200 × 200 × 12 × 12 mm
Diagonal bracing panel thickness	200 × 200 × 14.2 × 14.2 mm	200 × 200 × 12 × 12 mm
Shear wall	12 mm	10 mm
Walkway board	10 mm	8 mm
The support plate of Walkway board	15 mm	12 mm
L bar	125 × 75 × 10 × 10 mm	125 × 75 × 8 × 8 mm
Von-Mises stress	221 N/mm²	259 N/mm²
Shearing stress	118 N/mm²	135 N/mm²
Mass	31.867t	28.565t

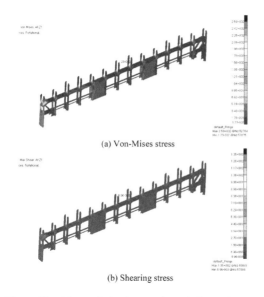

(a) Von-Mises stress

(b) Shearing stress

Figure 12. The optimized strength analysis results.

Multi-island Genetic Algorithm. The total weight of the lashing bridge structure after optimization reaches 31.867t, a reduction of 10.36% in the structural weight. The optimized strength analysis results are shown in Figure 12. In this design case, the proposed method performs well in design efficiency and final design results. It can be concluded that the proposed approach shows great potential, and it can be applied to similar and even more complex design problems.

6 CONCLUSIONS

In this paper, a knowledge-based system, adopting the artificial intelligent technology, is presented for intelligently designing lashing bridge. Several vital technologies of intelligent design are incorporated into the intelligent design system. The design features of the lashing bridge are selected by the neighborhood rough set theory. The design knowledge base and the standard template library are established. In the intelligent design, CBR, RBR and MBR reasoning methods are incorporated with each other to complete the basic structure design, ensuring the structure to be well-performed and reasonable. The optimization constraints and constraint rules are set according to the knowledge base. Based on the KBE technology, the mathematical optimization model is constructed, and the weight of the structure is reduced.

The proposed approach shows great potential, and the KBE technology is conducive to optimizing the design of lashing bridge. The objective of the lightweight structure is fulfilled through adopting KBE method. Knowledge base and standard part library are established to reuse knowledge and designers' dependence on knowledge and experience is lowered. A quick and convenient design idea for lashing bridge designers is provided, improving the automation and intelligent level of the design process. The KBE technology liberates designers from heavy and repetitive works so that they can put more efforts into creative works.

ACKNOWLEDGMENTS

This present paper is supported by both the project of High-tech Ship Research Projects Sponsored by MIIT (NO [2016]25) and The Chinese Government Key Research Project KSHIP-II project (Knowledge-based Ship Design Hyper-Integrated Platform) No 201335.

REFERENCES

Baxter D, Gao J, Case K, Harding J, Young B, Cochrane S, Dani S. 2007. An engineering design knowledge reuse methodology using process modelling. *Research in Engineering Design* 18(1): 37–48.

C.B. Chapman and M. Pinfold. 2001. The application of a knowledge based engineering approach to the rapid design and analysis of an automotive structure *Advances in Engineering Software* 32: 903–912.

Chang C C, Lin C J. 2011. LIBSVM: a library for support vector machines. *ACM Transactions on Intelligent Systems and Technology* 2 (3): 27.

Chapman CB and Pinfold M.1999.Design engineering—a need to rethink the solution using knowledge based engineering. *Knowledge-Based System* 12 (5–6): 257–267.

Corallo, A., Laubacher, R., Margherita, A., Turrisi, G. 2009. Enhancing product development through knowledge-based engineering (KBE) A case study in the aerospace industry. *Journal of Manufacturing Technology Management* 20(8): 1070–1083.

Cui J J and Wang D Y. 2013. Application of knowledge-based engineering in ship structural design and optimization. *Ocean Engineering*. 72: 124–139.

D. Baxter, J. Gao, K. Case, J. Harding, B. Young, S. Cochrane, S. Dani. 2007. An engineering design knowledge reuse methodology using process modelling. *Research in Engineering Design* 18: 37–48.

Du Y, Hu Q, Zhu P, Ma P. 2011. Rule learning for classification based on neighborhood covering reduction. *Information Sciences* 181(24): 5457–5467.

Gianfranco La Rocca. 2012. Knowledge based engineering: Between AI and CAD. Review of a language based technology to support engineering design. *Advanced Engineering Informatics* 26: 159–179.

Guo XD, Yu MH, Guan YF, Hu JJ, Zhu SJ. 2007. Optimization of large crude oil carrier midship section structure based on the relative difference quotient algorithm. *Journal of Ship Mechanics* 11(4):615–621.

Hezhen Yang, Jinfeng Chen, Qijin Lu, Ning Ma. 2014. Application of knowledge-based engineering for ship optimisation design. *Ships and Offshore Structures* 9(1): 64–73.

Hu Q, Liu J, Yu D. 2008. Mixed feature selection based on granulation and approximation. *Knowledge-Based System* 21 (4): 294–304.

Hu Q, Yu D, Liu J, Wu C. 2008. Neighborhood rough set based heterogeneous feature subset selection. *Information Sciences* 178(18): 3577–3594.

Hu Q, Yu D, Xie Z. 2008. Neighborhood classifiers. *Expert System Application* 34(2): 866–876.

I.O. Sanya and E.M. Shehab. 2014. An ontology framework for developing platform-independent knowledge-based engineering systems in the aerospace industry. *International Journal of Production Research* 52 (20): 6192–6215.

Keerthi S S, Lin C J. 2003. Asymptotic behaviors of support vector machines with Gaussian kernel. *Neural computation* 15 (7): 1667–1689.

La Rocca, G. 2012. Knowledge based engineering: Between AI and CAD. Review of a language based technology to support engineering design. *Advanced Engineering Informatics* 2(26): 159–179.

Li L, Tang H, Guo S, Huang L,Xu L. 2016. Design and implementation of an integral design CAD system for plastic profile extrusion die. *The International Journal of Advanced Manufacturing Technology*; 89: 543–559.

McMahon C, Lowe A, Culley S. 2004. Knowledge management in engineering design: personalization and codification. *Journal of Engineering Design* 15 (4): 307–325.

Park J H and Storch R L. 2002. Overview of ship–design expert systems. *Expert Systems* 19 (3): 136–141.

Pawlak Z. 2012. Rough sets: Theoretical aspects of reasoning about data. Springer Science & Business Media.

Penoyer JA, Burnett G, Fawcett DJ, Liou SY. 2000. Knowledge based product life cycle systems: principles of integration of KBE and C3P. *Computer-Aided Design* 32(5–6):311–320.

Shin YJ and Han SH. 1998. Data enhancement for sharing of ship design models. *Computer-Aided Design* 30(12): 931–41.

Verhagen W J, Bermell–Garcia P, van Dijk R E, Curran R. 2012. A critical review of Knowledge-Based Engineering: An identification of research challenges. *Advanced Engineering Informatics* 26(1): 5–15.

Wei Guo, Jiafu Wen, Hongyu Shao and Lei Wang. 2015. Implementation of knowledge-based engineering methodology in hydraulic generator design. *Advances in Mechanical Engineering* 7(5): 1–13.

Yang H Z, Chen J F, Ma N, Wang D Y. 2012. Implementation of knowledge-based engineering methodology in ship structural design. *Comput-Aided Design* 44(3): 196–202.

Yang HZ, Wang AJ. 2012. Fatigue reliability based design optimization of bending stiffener. *Journal of ship Research* 56(2):120–128.

Y Chen, ZL Liu, YB Xie. 2012. A knowledge-based framework for creative conceptual design of multi-disciplinary systems. *Computer-Aided Design* 44(2): 146–153.

Yong L, Wenliang H, Yunliang J, Zhiyong Z. 2014. Quick attribute reduct algorithm for neighborhood rough set model. *Information Sciences* 271: 65–81.

Zhou XH, Qiu Y, Hua G. 2007. A feasible approach to the integration of CAD and CAPP. *Computer-Aided Design* 39(4): 324–38.

Enhanced structural design and operation of search and rescue craft

F. Prini, R.W. Birmingham, S. Benson & R.S. Dow
Marine, Offshore and Subsea Technology Group, School of Engineering, Newcastle University, Newcastle upon Tyne, UK

P.J. Sheppard & H.J. Phillips
Royal National Lifeboat Institution, Poole, UK

M.C. Johnson, J. Mediavilla Varas[†] & S. Hirdaris*
Lloyd's Register, Southampton, UK

ABSTRACT: This paper is a conceptual overview of the approach adopted to enhance the structural design and operation of high-speed craft through a set of design curves. It outlines how to construct a structural limit curve as a function of speed and sea state severity. The curve contributes to defining the 'operational envelope' of the craft, which informs design teams and crews on the limits to speed for the safe operation in waves. The construction of a structural limit curve requires an understanding of the loads that the structure sustains in operation and how it responds. This is achieved through a combination of numerical hydrodynamic and structural simulations, towing tank tests and sea trials. The study, which is conducted on the Severn Class lifeboat of the Royal National Lifeboat Institution (RNLI), has wider applicability in the design of future lifeboats and small high-speed craft. Explicit results, not presented here, are referenced throughout the paper.

1 INTRODUCTION

Small high-speed craft normally operate at a wide range of speeds and in different sea conditions. There are nevertheless restrictions to speed when operating in waves. Installed propulsive power, crew endurance, equipment functionality and structural integrity are all factors that set boundaries to the safe operation of the vessel defining its 'operational envelope'.

In calm seas the maximum speed is typically limited by the installed propulsive power. As the sea becomes more severe it is often the crew comfort, the on-board equipment and the structural strength that become of concern. As suggested by (Riley and Marshall 2013), the maximum allowable speed can be limited by either accelerations, which have a negative effect on personnel and equipment, or by large loads that lead to excessive structural stresses. Which of these operating limits is reached first partially depends on the size and weight of the craft and on the seat vibration damping systems installed (Cripps, Cain, et al. 2004).

This paper presents a conceptual overview of the method adopted to construct a structural limit curve as a function of speed and sea state severity. This curve contributes to defining the 'operational envelope' of the craft, which informs design teams and crews of the speed limits for the safe operation in waves. The paper does not present explicit results, which are referenced throughout the paper.

2 RATIONALE

The Royal National Lifeboat Institution (RNLI) operates a number of lifeboat classes, all designed with a specified service life. The Severn Class (Figure 1 and Table 1), consisting of a fleet of 45 vessels, first entered service in 1995.

These lifeboats would now be approaching the end of their original operational life, but due to their exceptional in-service performance the RNLI has started a life extension programme to extend the operational life of the fleet to 50 years (Roberton 2015).

[†]Contributed to this work whilst at Lloyd's Register. Now at Materials Innovation Institute, Delft, NL.
*Contributed to this work whilst at Lloyd's Register. Now Visiting Research Scholar at Maritime Technology Group, Department of Mechanical Engineering, Aalto University, Espoo, FI.

Figure 1. RNLI Severn Class lifeboat.

Table 1. Severn Class main particulars.

Length overall	LOA	17.00	m
Length waterline	LWL	15.50	m
Beam overall	BOA	5.62	m
Draught	T	1.37	m
Displacement (full load)	Δ	42000	kg
Speed	V	25	kn

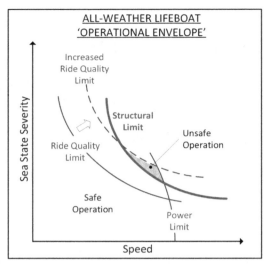

Figure 2. Conceptual representation of the operational envelope of an all-weather lifeboat. (Adapted from Riley & Marshall (2013)).

For the Severn Class, as for the other lifeboats designed and operated by the RNLI, crew's endurance has traditionally been one of the main limitations to speed during operation in rough seas. Excessive motions and accelerations can reduce the crew's operational effectiveness and raise the risk of personal injury. In those situations the coxswain in command of the vessel is aware of the shocks experienced and will usually adapt the speed, and/or the heading, with respect to the prevailing weather and sea conditions. The structural limit of the lifeboat is assumed to be far beyond the crew safety limit since structural failures have proved to be extremely rare, although do occasionally occur (Phillips et al. 2009).

Many lifeboats in the Severn fleet have already been fitted with more modern replacement engines. New technologies to improve the ride quality are also available that, if adopted, would reduce the crew's exposure to accelerations and provide the possibility to operate at higher speeds. Whilst this is beneficial for the response to emergency call-outs, it also implies the potential for operating closer to the structural limit of the vessel. Furthermore, improved seat damping system, and thus the ride quality, could limit the ability of the coxswain to appreciate the loads being sustained by the structure. This has the potential to push the operation of the vessel close to or even beyond the structural limit. Such a scenario is graphically illustrated in Figure 2, an adaptation from the original plot presented by Riley and Marshall (2013).

The Severn life extension programme highlighted the need to gain a better understanding of the structural strength of the RNLI's all-weather lifeboats. This could only be achieved by predicting with confidence the loads sustained by the structure during operation. A study undertaken by Newcastle University, the RNLI and Lloyd's Register set out to investigate the loads and the consequential structural response. Work is ongoing to produce a set of structural response curves that will provide the RNLI's design, maintenance, training and operational teams with an enhanced insight into the operational envelope of the Severn.

3 SEAKEEPING LOAD PREDICTION METHODS

3.1 Numerical methods

Hydrostatic loads, due to self-weight and buoyancy forces, can be determined with an acceptable degree of accuracy (Phelps 1997). Differently, there is less guidance for the prediction of hydrodynamic loads on a small high-speed craft. There is also no standard approach to the inclusion and treatment of hydroelastic effects.

With regard to the latter the American Bureau of Shipping (2011) suggests that for small craft only in some cases is there significant interaction between loads and response, for which a fully hydroelastic approach should be adopted. Examples of this are springing of multi-hulls and the dynamic response of panels to slamming impacts. Also designs optimised for low structural weight that result in a higher flexibility of the hull could require a hydroelastic approach.

In the majority of cases, and especially for small stiff hulls, rigid body motions are dominant and therefore wave loads are not significantly influenced by dynamic effects. The prediction of hydrodynamic loads can be carried out by treating wave and dynamic loads independently. Wave loads are computed on the assumption that the body is rigid. Highly dynamic effects, such as those induced by slamming, are computed separately and superimposed.

The majority of seakeeping predictions are carried out within the framework of potential flow theory, assuming an incompressible, inviscid and irrotational fluid. The problem can be further simplified if the ship-wave system is idealised as linear, as first suggested by St. Denis & Pierson (1953). This is usually the case for conventional displacement vessels travelling in light or moderate seaways, for which motion and wave amplitudes are small. The sea surface can be modelled as a linear superposition of regular sinusoidal waves of all frequencies. Motions and wave-induced loads can then be studied in elementary regular waves and the principle of superposition applied to determine the overall response in an irregular seaway. Under these assumptions seakeeping predictions can be made even more efficient by solving the problem in the frequency domain. Several linear potential flow methods such as 2D strip theories and 3D panel methods have been developed and are incorporated into commercially available software packages.

In rough seas and at high speed the potential flow approximations cannot be made as the ship-wave interaction becomes nonlinear. A number of approaches have been developed of which a review is given in Prini et al. (2015).

Semi-empirical methods are usually used for practical design and are implemented in the scantling rules of most Classification Societies (Det Norske Veritas 2012, American Bureau of Shipping 2014, Lloyd's Register 2014). The assumption underpinning these methods is that transient non-uniform pressures can be modelled as 'equivalent' quasi-static uniform pressures that, if applied to the structural component, will produce the same maximum deflection and peak stress as those produced by the actual loading (Heller & Jasper 1960).

Theoretical approaches to investigate water impact pressures have been developed since the 1920s (Von Kármán 1929, Wagner 1932), later implemented and extended by others (Stavovy and Chuang 1976, Zhao and Faltinsen 1993, Zhao et al. 1997). Advanced methods based on solving the Reynolds Averaged Navier-Stokes (RANS) and Euler equations have been applied to a range of problems and their use is also being investigated for seakeeping predictions on the Severn Class (Aktas et al. 2017).

3.2 Experimental methods

Experimental measurements of seakeeping responses of a vessel are an alternative to numerical methods and provide a way of validating their predictions.

Most of the tests at model scale are conducted in towing tanks or wave basins, which provide ease of taking measurements and good control of the wave environment (Lloyd 1989). Regular and irregular wave patterns can be generated and repeated during subsequent runs. Seakeeping motion tests have been carried out extensively in the past and the choice of the measurement apparatus often depends on the facility's practice, the test objectives and on whether the model is towed by a carriage or equipped with its own propulsion system.

External pressures are usually measured with pressure transducers. Since they provide point-measurements, the complete pressure field on the hull bottom can only be reconstructed from arrays of transducers, as proposed by Rosén (2005). The use of slamming patches to measure the hydrodynamic force exerted on a cut-out of a bottom panel has also been investigated (Manganelli et al. 2003).

Seakeeping experiments are also conducted to measure hull girder load effects. Ideally a 'hydro-structural' scaled model would be used to measure loads at any longitudinal position, however, due to the practical complexities in satisfying the structural similarity at model scale, the use of a segmented model is most common (ITTC 2011). The segmentation consists in cutting the hull shell into a number of segments so that the hull does not provide any continuous structural support. The hull girder strength is given by an internal backbone structure. Load measurements are taken at the segmentation cuts by means of strain gauges on the backbone beam or load cells connecting the segments. Two types of segmented model exist depending on the stiffness of the connecting structure: rigid or elastic. A rigid segmented model has a much higher stiffness than the actual vessel and greater natural frequency than the wave encounter frequency (ITTC 2011). Wave loads can be measured and compared with numerical computations. However, dynamic and impact load effects, such as whipping and springing, require the stiffness of the hull girder to be appropriately reproduced at scale. For these loads an elastic segmented model that represents the rigidity of the prototype hull should be used.

In spite of the numerous advantages of model testing it should be recognised that scaling is problematic. The towing force only resembles the thrust of an actual propulsion system and the wave environment tends to lack the confused nature of the sea.

To investigate the seakeeping of a vessel in real operational conditions, sea trials are necessary. They are nevertheless expensive and time consuming, which is why they are not carried out on a regular basis. If conducted for design purposes they also require a prototype vessel to be built first.

An example of full-scale tests conducted on an instrumented small high-speed naval craft (LWL = 9.5 m, V = +40kn) was presented by Rosén & Garme (1999). At a larger scale, other examples are the sea trials conducted by the US Navy with a wave-piercer catamaran (Jacobi et al. 2014) and those conducted with the research vessel Triton (Grassman and Hildstrom 2003, Renilson et al. 2004), a joint program between the United Kingdom and the United States to assess the trimaran hull form for implementation in future warship designs.

3.3 *The RNLI load curve*

Due to the extreme conditions in which the RNLI's lifeboats operate and their challenging structural requirements, the methods adopted for other high-speed craft presented limitations (Cripps et al. 2005). Consequently a load prediction method developed in-house has been used by the RNLI design team for some time (Cripps, Phillips, et al. 2004, Cripps et al. 2005).

This approach treats the design loads in terms of equivalent static ultimate pressures. The maximum ultimate pressure for the design of a new lifeboat is determined as a function of load displacement and operational speed. This pressure value is modified according to the longitudinal position along the hull and for the topsides, which carry only a percentage of the pressure applied over the hull bottom. For each panel a pressure value is then found, which can be applied uniformly over the whole of the respective panel.

4 A DIRECT CALCULATION METHOD

Semi-empirical methods, as implemented by most Classification Societies, and the RNLI's load prediction method have been successfully employed for design purposes. However, because of their nature, they cannot be used for direct calculations of the structural response to the numerous load cases of different combinations of speed, heading and sea.

RANS and Euler equation solvers are increasingly more popular and have successfully been used for single case studies. Yet two aspects limit their use in seakeeping. Firstly, if the non-linear behaviour is dominant, simulations should be carried out in the time domain as nonlinearities are history-dependent. The principle of superposition, applied by linear codes, does not hold anymore. Each combination of heading, speed, displacement, regular and irregular wave pattern has to be investigated. The second drawback is the significant computational resource still required for these methods. This, together with the large number of scenarios to investigate, made these codes unsuitable for this study.

It was concluded that a single reliable method to predict the whole loading scenario during operation in waves was not available. As a result a systematic approach combining different methods was adopted, as shown in Figure 3. This consists of a numerical hydrodynamic model based on linear potential theory (CFD Model), a global finite element model for the computation of the structural response (FE Model), and experimental tests at both model scale (Small-Scale Tests) and full scale (Full-Scale Tests).

The principle underlying this approach is that rigid body motions are dominant and wave loads are not significantly affected by high frequency dynamic effects induced by slamming, which can be accounted for independently. Loads of different natures that act on the structure at the same time (hydrostatic, wave and slamming-induced) are calculated separately and superimposed in the generation of a load case for the structural analysis (Figure 3). Hydrostatic and wave loads are predicted numerically with a CFD model.

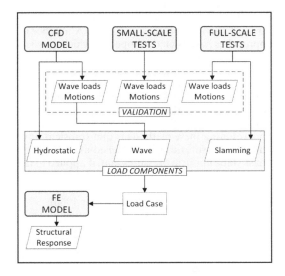

Figure 3. Direct calculation approach. Overview of the tasks.

Slamming-induced loads are predicted based on experimental data from sea trials.

Because of the different operational modes of the Severn, from displacement to planing, it was still necessary to validate the wave loads predicted with the seakeeping simulation model. This was done by comparing the results from the seakeeping simulations against experimental data. This process is represented in Figure 3 as 'validation'. A detailed explanation of the tasks undertaken and of the methods adopted to compare the results and assess the accuracy of the wave loads predicted numerically is given in the following sections.

5 NUMERICAL MODEL

A global finite element model of the Severn Class was developed with the marine design software MAESTRO. Details of the first model under construction are given in Prini *et al.* (2015). This model was further updated and refined to improve its accuracy. The refined model is shown in Figure 4.

The entire vessel with its main structural components were represented through a combination of shell, beam and rod elements. The laminate properties were embedded into each element as layered orthotropic for shells; uniform orthotropic for beams; and isotropic for rods.

The structural mass was automatically computed by the software, based on the elements and the material properties used. Other masses were represented, according to their nature, as: volume masses, scaled-up structural mass, point masses and large solid masses whose centre of gravity lies at a distance from the supporting nodes. The computed centre of gravity position was checked against the target value calculated from an analysis of the inclining test data of the fleet.

MAESTRO integrates hydrostatic, hydrodynamic and structural analysis through a hydrostatic balance tool, a potential flow solver MAESTRO-Wave and a linear finite element solver (Ma *et al.* 2012, Zhao *et al.* 2013, MAESTRO Version 11.5.0 2017).

5.1 *Hydrostatic and hydrodynamic analysis*

Static equilibrium on the waterline is reached by applying hydrostatic balance and inertia relief methods embedded into MAESTRO (Ma *et al.* 2012, Zhao *et al.* 2013, MAESTRO Version 11.5.0 2017). The first provides equilibrium in heave, pitch and roll by iteratively adjusting draught, trim and heel of the vessel. The latter adjusts additional accelerations to reach equilibrium in surge, sway and yaw. Static equilibrium balance is used in two circumstances. Before running any hydrodynamic analysis, equilibrium on the still waterline is sought to define the attitude of the vessel and the wetted elements to be used by the hydrodynamic solver. The second purpose of the static equilibrium balance is to compute the hydrostatic pressures that will form part of the load case for the structural analysis. For this purpose, the vessel is re-balanced on a sinusoidal wave rather than on the still waterline.

The computational tool MAESTRO-Wave was used to predict the hydrostatic and wave loads of the vessel. Figure 5 shows the steps undertaken and the output from the analysis. Two linear potential flow codes, based on the theory proposed by Salvesen *et al.* (1970), were used: at speeds up to Froude numbers of 0.4 (10 knots for the Severn Class) a 2D strip theory using a zero-speed Green function was used; at speeds above

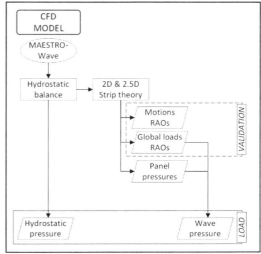

Figure 5. Direct calculation approach. Numerical simulation model for the prediction of hydrostatic and wave loads.

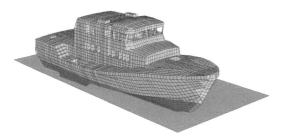

Figure 4. Finite Element (FE) global model of the Severn Class lifeboat.

Froude numbers of 0.4 a 2.5D strip theory using a Rankine Source method and a forward speed correction term in the free surface computation was employed. The equations of motions are formulated based on the structural mesh (Zhao et al. 2013). This overcomes the challenge of transferring the pressure mapping from the hydrodynamic model to the corresponding structural model.

The simulations were run in the frequency domain, at a range of speeds, headings and wavelengths. Panel pressures, motions and hull girder load RAOs (Response Amplitude Operators) were calculated. It should be noted that in this paper the convention is followed whereby the Response Amplitude Operator is the same as its transfer function and used in its unsquared form.

The accuracy of the predictions was assessed by comparing motions and global loads RAOs against those measured experimentally. This is shown in Figure 5 as 'validation'.

5.2 *Structural analysis*

The MAESTRO linear finite element solver is used to run the structural analysis. The sequence of the tasks undertaken is shown in Figure 6. One of the first steps is the definition of a load case, which consists of all those loads that act on the structure simultaneously. What is often of interest, especially for design purposes, is the response to loads that are experienced in severe waves, or sea states, that characterise the extreme environmental and operating conditions of the craft (American Bureau of Shipping 2011). The limits of numerical methods, as discussed in the previous sections, make these loads challenging to estimate and experimental data cannot always be produced for each scenario. A common approach is to define an equivalent wave (often referred to as equivalent 'extreme' or 'design' wave) through its amplitude and frequency. The loads that would be experienced in such a wave, for a given vessel's loading condition, heading and speed, are then predicted through scaling and extrapolation of loads known for other scenarios.

For the present study, the load components forming the load cases are: hydrostatic, wave and slamming loads. This last term includes the slamming load effects on the hull girder (whipping) and the slamming impact pressures acting at a local level on the bottom and bow panels.

Once the extreme wave is defined, the hydrostatic pressure is computed through the static equilibrium balance method explained earlier.

Wave loads obtained from the hydrodynamic simulations are known in terms of their RAOs, hence for regular waves of unit amplitude. These loads, and their related pressures, are scaled linearly according to the amplitude of the extreme wave. This process is at the basis of linear theory and is related to the very notion of transfer function (St. Denis and Pierson 1953).

It is also necessary to define the extreme value of the global and local load effects induced by slamming. A measure of these loads was obtained from tests conducted at full scale. The information consists of a number of short-term data sets. Extreme values can therefore be calculated by extrapolation of measured data. The nonlinear nature of slamming also implies that linear extrapolation by means of a transfer function may not be possible. Instead suitable statistics can be applied, as explained for example by Ochi (1981), Hughes (1983) and Clarke (1986). Once the extreme value of the response is found, it must be applied as a load to the structural model. Two different methods were adopted.

The whipping response caused by slamming can be thought of as an addition to the vertical bending moment induced by waves. The extreme value of both these components is now known, so the dynamic response can be accounted for through a scaling factor applied to the linear response. Wave-induced pressures are scaled up so that the magnitude of the resulting bending moment includes both the wave and the slamming terms.

Slamming loads reacted upon by the local structure are modelled as additional pressures. Although these pressures would be transient in nature, they are applied as equivalent static uniform pressures

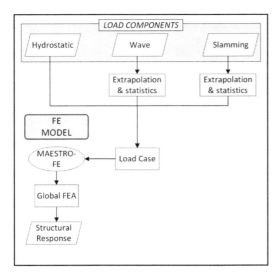

Figure 6. Direct calculation approach. Finite Element (FE) model for the prediction of the global structural response.

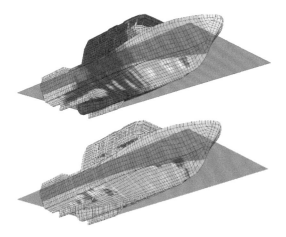

Figure 7. Example of the output from the structural analysis, showing displacements (top) and maximum longitudinal stresses (bottom).

to the structural panels of the bottom shell. The procedure adopted to recover a static uniform pressure from experimental measurements will be outlined later. For now it suffices to consider that the measurements were taken on a number of panels along the length, from near amidships to the bow. From these measurements it is possible to calculate a maximum equivalent static pressure value and a longitudinal distribution factor to take into account that the slamming pressure varies in magnitude along the hull length and that not the entire hull may be subject to slamming impacts. From these it is possible to construct a static pressure field to apply to the structural model.

Once the load case is defined the inertia relief method is applied to restrain the model. This supersedes the application of rigid restraints to prevent unlimited rigid body motions, which often results in unrealistic deformation patterns and artificial stress concentrations.

The finite element analysis is performed as linear elastic, with stresses, strains and displacements output for evaluation. An example is shown in Figure 7.

6 SMALL-SCALE TESTS

Motions and hull girder loads were also predicted through seakeeping experiments conducted in a towing tank. Figure 8 outlines the experimental process.

Two scale models of the Severn lifeboat were tested: a 'solid' and a 'segmented' model. The first is a conventional model for measuring rigid body motions. The latter is a rigid segmented model that, in addition to body motions, allowed the measurement of hull girder loads at three segmentation cuts (Figure 9). The backbone structure holding the hull segments together consisted of three aluminium beams of square hollow section instrumented with strain gauges. The strain gauge layout was devised to measure vertical and horizontal bending moments and vertical shear force. The beams were calibrated with a test rig through 3 and 4 point bending tests.

The segmentation and sealing of the hull shell and the presence of the backbone structure introduced further complexities in the model design, building and testing process. Furthermore, it was necessary to ensure that the two models showed

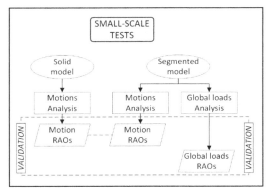

Figure 8. Direct calculation approach. Small-scale tests with a 'solid' and a 'segmented' model for the prediction of rigid body motions and global wave loads.

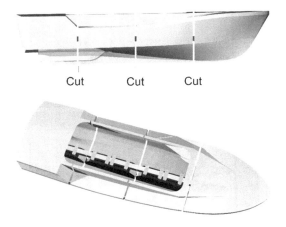

Figure 9. Segmented model of the Severn Class lifeboat with three segmentation cuts. The hull segments are held together by an internal backbone beam. Load measurements are taken at the segmentations by means of strain gauges on the beam.

445

similar seakeeping characteristics. The segmented model was therefore built with a solid hull shell. Preliminary tests in calm water and in waves were run and results compared with those from the solid model. The hull shell was later segmented and the same tests run again. Motion results, from the first and second set of tests were compared to ensure that the segmentation had not altered the seakeeping behaviour of the model.

Details of the two models, the facility, the test apparatus and setup are given in (Prini *et al.* 2016). Two groups of seakeeping tests were completed: at forward speed in head waves and at zero speed with different headings.

Tests at forward speed were conducted with the model attached to a standard free-to-heave-and-pitch dynamometer. Motion data was therefore collected for heave and pitch only and load data for vertical bending moment and vertical shear force.

Tests at zero speed were performed with the model positioned at the centre of the tank. A set of mooring lines constrained the model in yaw, surge and sway. Motion data was collected through an optical tracking system for heave, pitch and roll. Hull girder loads were measured for vertical and horizontal bending moments and vertical shear force. Figure 10 shows the setups for the two groups of tests.

The experiments were conducted in regular waves generated by a wavemaker. For each speed, or heading, a range of wavelengths was tested and the wave elevation measured. This allowed reconstruction, through a peak-to-trough analysis, of

Figure 10. Experiments setup. Model towed by a carriage for tests at forward speed (top). Model constrained by mooring lines for tests at zero speed (bottom).

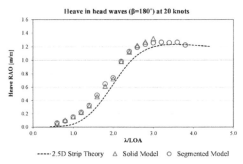

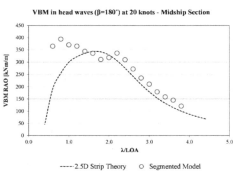

Figure 11. Comparison between numerical and small-scale test data. Heave RAO in head waves at 20 knots (top). Amidships vertical bending moment in head waves at 20 knots (bottom). The magnitude of the response is plotted per metre of wave height against the ratio wavelength/ship length (λ/LOA).

the RAOs of the responses from their respective time histories.

The motions and load RAOs were then compared against those obtained from the seakeeping simulations. Figure 11 shows an example for heave and amidships vertical bending moment at 20 knots in head waves. Results are presented in terms of RAOs. The magnitude of the response is plotted per metre of wave height against the ratio wavelength/ship length. The comparison between the small-scale test data and the numerical results is represented in Figure 8 as 'validation'.

7 FULL-SCALE TESTS

Tests at full scale were conducted on an instrumented Severn Class lifeboat to determine her seakeeping behaviour and the wave and slamming loads experienced in real operational conditions. Details of the trials procedure, instrumentation layout and data collected can be found in Prini *et al.* (2018).

The tests took place in the North Sea offshore from Tynemouth (UK) and consisted of 11 trials conducted at speeds ranging from 5 to 25 knots and in different sea states, with a significant wave height from 0.3 to 4.6 metres. The route followed during each of the trial was a 'star pattern' devised to include headings from head to following seas, both port and starboard, at 45-degree increments. The length of each leg of the star was set to allow a minimum of 100 wave encounters to occur, based on the expected mean period of the sea spectrum. An example of star pattern and actual route followed during one of the trials is shown in Figure 12. Details of different types of trial trajectories can be found in Johnson (2004).

The lifeboat had been fitted with 1 triaxial accelerometer, 1 triaxial rate gyro, 58 linear strain gauges and 2 thermocouples for the temperature compensation of the strain signals. The sensors were positioned and oriented to measure: accelerations and angular velocities at the centre of gravity of the vessel; vertical and horizontal bending strains of the hull girder at five longitudinal locations; local panel deflection due to pressure loads, slamming and green water on six bottom panels, two bow panels and two panels on the fore deck.

All the sensors were wired into one data acquisition unit and their signals sampled at different frequencies depending on their nature: acceleration and strain signals recording local pressure loads were sampled at 2048 Hz; angular velocities and strain signals measuring hull girder loads at 256 Hz.

A measure of the sea state was also necessary to correlate motions and structural response to the wave environment. In addition to visual observations, a directional Waverider buoy was used. The buoy was deployed central to the trial area and left free to float for the whole duration of each trial. Forward (North), transverse (West) and vertical (heave) displacements were generated through a GPS-based motion sensor at 1.28 Hz. Additional wave data was also obtained from two wave buoys moored approximately 10 nautical miles N and 23 nautical miles ESE from Tynemouth, operated by the Channel Coastal Observatory (CCO) and the Centre for Environment, Fisheries and Aquaculture Science (Cefas) in the UK.

The analysis of data from sea trials is typically more challenging than for tests conducted in a controlled environment: the vessel's response to a random sea state is irregular; signals are often affected by noise and drift; and quasi-static and dynamic effects are superimposed. The data post-processing was based on the principle that these components tend to occur at different and distinct frequencies: drift is a very low frequency component or even a slowly moving trend; wave-induced responses are related to the wave encounter frequency; and slamming-induced responses tend to occur at higher frequencies. It was therefore possible to isolate the response of interest by applying the appropriate frequency-based filters. An example of how the response of a panel due to slamming is isolated by applying a high-pass filter to the raw signal is shown in Figure 13.

Another aspect driving the data analysis was the type of data describing the sea state. A wave buoy provides an effective measure of the wave environment in the trial area, but this measure is still relative to the buoy's location. It is impossible to relate

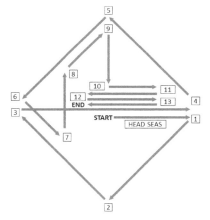

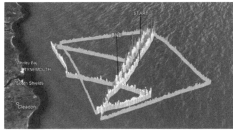

Figure 12. Example of star pattern followed during the trials. Legs sequence and direction (top). Actual route followed with NNE waves (bottom). Accelerations are shown superimposed.

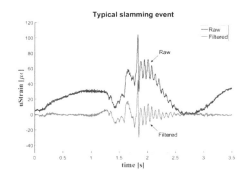

Figure 13. Strain recorded at the centre of hull bottom panel during a slamming event. A high-pass filter is applied to the raw signal to isolate the strain due to slamming.

a particular event in the response to the wave that has generated it. Figure 14 outlines the analysis of the data obtained from the full-scale tests.

The RAOs of motions and hull girder loads were computed through a spectral analysis. The response spectrum and the wave encounter spectrum were first calculated. The transfer functions were found from the ratio of the spectral ordinates at each encounter frequency defining the wave encounter spectrum. This analysis was conducted for headings from head to beam seas for which no negative encounter frequencies occur. The RAOs were compared against those obtained numerically with a loading condition representative of the full load departure of the lifeboat used for the trials. This step is shown in Figure 14 as 'validation' and an example of RAO plot with both numerical and full-scale data is shown in Figure 15.

8 NONLINEAR AND DYNAMIC ASPECTS

8.1 Numerical model

Strip theory introduces several simplifications to solve the ship-wave interaction problem for which a detailed explanation is given by many authors including Salvesen *et al.* (1970), Hughes (1983) and Lloyd (1989). The main simplification is that the underwater part of the hull is approximated by a number of prismatic segments, or strips. Forces are calculated independently for each strip using two-dimensional flow theory and the vessel's response is obtained by integration over the various segments (Hughes 1983).

Moreover, because the analysis is based on potential theory, fluid viscosity is neglected. This implies that hydrodynamic lift is not present and that the wetted surface does not change with speed. It also implies that motion damping can only be attributed to wave radiation (Lloyd 1989). This is generally adequate for most motions with the exception of roll, for which viscous damping is important. As a consequence potential flow solvers tend to underestimate roll damping. A correction factor, in terms of a critical roll damping ratio, was therefore defined for the hydrodynamic analysis.

In order to linearise the problem, further simplifications are introduced, which imply that: the hull is wall-sided; and the amplitudes of waves and motions are small (Hughes 1983). The assumption of 'wall-sidedness', in particular, means that linear strip theories predict the same value of wave-induced bending moment for both sagging and hogging.

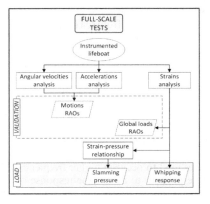

Figure 14. Direct calculation approach. Full-scale tests on an instrumented lifeboat for the calculation of rigid body motions, global wave loads and slamming-induced load effects: whipping of the hull girder and local slamming pressures.

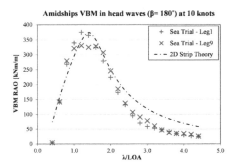

Figure 15. Comparison between numerical and full-scale test data. Vertical bending moment near amidships in head waves at 10 knots. The magnitude of the response is plotted per metre of wave height against the ratio wavelength/ship length (λ/LOA). Sea trial data is presented for two head sea legs carried out during the same trial.

8.2 Small-scale tests

One of the advantages of conducting tests in regular waves is that it is possible to observe, from the time histories of the responses, occurrence of some nonlinearities. Nonlinear motion responses could be captured with either of the two models tested, whilst load nonlinearities are only observable with the segmented model. However, these are limited to some aspects of the response at a global level only.

The differences between sagging and hogging bending moments arising from the hull shape and from the hydrodynamic differences between the entry and the exit of the hull at the waterplane are captured. The hog-to-sag ratio could be calculated from a peak-to-mean and trough-to-mean value analysis.

Slamming-induced dynamic responses, such as whipping, require the hull girder stiffness to be reproduced at model scale. Since the primary strength of the segmented model was provided by a 'rigid' backbone beam, assessment of these load effects was not possible at model scale. Nonlinearities due to irregular seas or scaling effects were also neglected.

8.3 Full-scale tests

Work is being conducted to calculate the effect of slamming at a global and local level. These are whipping of the hull girder and the local response of the hull bottom panels, the bow and the deck reacting to the applied pressures (Figure 14). The nonlinearity and the highly dynamic nature of these loads makes it impossible to solve the problem in the frequency domain, hence a time-domain analysis has to be performed to find the peak values of the response. Suitable statistical approaches and probability distributions, such as the Gumbel, Weibull or Generalized Extreme Value, can be applied to linearise and extrapolate the load magnitude to find extreme values with a given probability of exceedance (or return period). An example of how a Gumbel distribution can be fitted to measured data to predict the magnitude of extreme values with a given return period is shown in Figure 16.

As detailed in the previous sections, the whipping response can be accounted for through a scaling factor applied to the linear response. The magnitude of the resulting bending moment will include both the wave and the slamming terms.

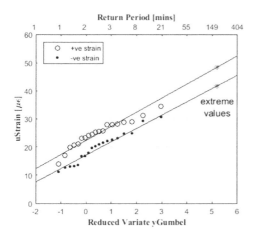

Figure 16. Example of Gumbel plot to predict extreme slamming strains on a panel based on observed slamming events. The extreme values are computed for a return period of 3 hours based on 19 minutes of data recording. The peak positive and negative strain values recorded over every minute are shown.

Slamming loads reacted upon by the local structure are instead modelled as additional pressures. To achieve this, one more step is necessary, which is the conversion from a strain value to a pressure load that can be applied to the structural model.

A slamming impact is typically characterised by a pressure front that travels rapidly. Attempting to recover a dynamic pressure field from strain readings and applying it dynamically to the finite element model was not practical for this study. Instead an equivalent static uniform pressure can be found. This is the pressure that, if applied to the whole panel, produces at its centre the same strains as those produced by the actual pressure field. This correlation can be found from a local finite element model of the panel under consideration through a linear static structural analysis.

9 THE STRUCTURAL RESPONSE CURVE

The previous sections concerned the prediction of the loads caused by a given extreme wave and the computation of the vessel's response. It is left to define what is an 'extreme wave' and to define an approach to assess the results of the structural analysis to identify a 'structural limit'.

An extreme wave is taken to be a wave that causes an extreme value of a ship's response, such as vertical bending moment or vertical acceleration. In order to relate a wave to a ship's response, it is often assumed that the response is linearly proportional to the wave amplitude, even when extrapolated to higher sea states, and hence that the highest seaway will produce the largest response (Lewis 1988). It is then possible, from an examination of the response in elementary regular waves (or RAOs), to find the combination of heading and wave frequency that causes that response to reach its maximum. With the wave frequency found, then the amplitude of the highest wave that the vessel is likely to encounter over a period of time should be estimated. Given the stochastic nature of the sea and the many possible operating conditions of a vessel, this procedure is inevitably related to the concept of probability.

Only within a limited period of time, typically 1 to 4 hours, can the sea be considered to remain nearly uniform and statistically stationary (Lewis 1988). Hence, idealised wave spectra (ITTC 2002) are commonly employed to formulate short-term descriptions of the sea. For long-term predictions the use of wave scatter diagrams is most common. These are constructed based on visual observations and/or measured data and are therefore relative to particular sea areas. Atlases of this type were published for example by Hogben and Lumb (1967), Bales et al. (1981) and Hogben et al. (1986). A more recent collection of wind and wave frequency distributions for sites around the British Isles was edited by the Southampton Oceanography Centre (2001). From these descriptions of the sea state it is possible to find the probability of a wave of being exceeded by a higher wave over a given period of time.

Set the return period, an equivalent regular wave (described by its heading, frequency and amplitude), which simulates the magnitude and location of an extreme value of a ship's response, can be determined (American Bureau of Shipping 2011).

The purpose of this study was to determine a structural limit curve as a function of speed and sea state severity. The procedure to achieve this is presented here for a return period of three hours (Figure 17). Speed on the abscissa can be conveniently expressed in terms of speed-over-ground in knots. A suitable description of the sea can be obtained from an idealised spectrum together with a measure of wave height and period. The significant wave height is used on the ordinate.

In order to work with two axes only, a number of parameters must be fixed. These are: loading condition, heading, spectral shape and associated period. Multiple plots can be created for different combinations of these parameters, or one can be created for the most severe. Within any plot, the structural response curve is constructed from specific points with coordinates given by speed and significant wave height. Each point on the plot represents the structural response of the vessel to a load case consisting of the extreme loads that are likely to be experienced in that combination of speed and significant wave height. The extreme loads are the most probable maximum loads, computed through the definition of an equivalent extreme wave, expected to occur once within the return period.

The response of the structure to the loads imparted by an extreme wave can then be assessed. Depending on the probability level used to determine the extreme wave, and on the operational profile of the vessel, the designer can be satisfied with different levels of adequacy. For design loads a linear dynamic response is often sought. For investigations of the ultimate strength to a wave representative of survival condition, the designer could accept damage of local structural members, as long as the overall integrity of the ship is not compromised. Ultimately more than one structural response curve can be produced, according to the chosen level of adequacy of the structure and/or the deemed urgency of a particular operation. For example, a curve could be created for a non-urgent, standard passage or transit and an alternative curve for urgent or time-critical services. The latter would require clear operating procedures and training as it means the boat would be working closer to or may even cross the structural limit bringing with it the increased likelihood of some structural damage to the secondary structure (e.g. stiffeners).

Because of the statistical nature of the approach and of the dependence on the return period, structural response lines should be read as lines with an associated risk of structural failures rather than hard lines with exact numbers.

The same procedure outlined here can be adopted for predicting the extreme lifetime loads and the associated structural response. This can be done by using a long-term description of the sea combined with information on the operational profile of the vessel to account for the time spent at each combination of speed and sea state.

10 CONCLUSIONS

The structural design of small high-speed craft has traditionally relied on semi-empirical methods. Whilst successfully employed for design purposes, these methods are less suitable for the direct calculation of seakeeping loads. A study has been undertaken to investigate the loads sustained by the RNLI's lifeboats during operation and the consequential structural response.

Numerical simulations, towing tank experiments and sea trials were conducted to predict the major loads sustained when in operation. The numerical model was validated through comparison against experimental data.

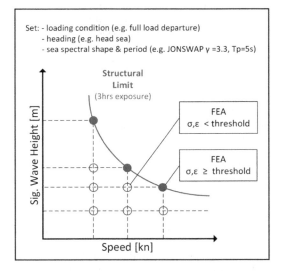

Figure 17. Direct calculation procedure to construct a structural response curve as a function of speed and sea state severity.

Hydrostatic, wave and slamming loads are accounted for through a combination of seakeeping simulations and experimental data collected during the sea trials. The loads experienced during operation in rough seas are computed through an 'equivalent regular extreme wave' approach. Wave and slamming loads are linearised and scaled in the generation of a load case that represents the extreme loads experienced in given operating conditions.

The response of the structure to a load case is studied with a global finite element model of the vessel. A structural limit curve (Figure 2), as a function of speed and sea state severity, can then be defined from an analysis of the response of the structure to a range of load cases. Once the work is completed structural response curves will be constructed for different stress (or strain) thresholds and it will be possible to comment on their accuracy. Theories to determine the possible modes of failure of composite structures could also be employed.

11 OUTCOMES

Work is now being conducted by the RNLI to obtain empirical confirmation of the limits to speed in waves imposed by the installed propulsive power and crew endurance—the 'Ride Quality' and 'Power Limit' curves of Figure 2. Together with the work on the structural limit described in detail in this paper, the outcome will make it possible to gain a better insight into the operational envelope of the Severn Class.

This new knowledge will inform the design teams on areas of possible improvements in view of the life extension programme. The outcome will also have wider applicability to the design of future all-weather lifeboat classes and small high-speed craft in general.

This study also sets the basis for the development of a structural monitoring system to support the operation of search and rescue craft. By informing the crews of the loads being sustained by the structure, it is possible to optimise the onboard comfort whilst minimizing the risk of structural damage.

Ultimately, it is expected that this approach to the design and operation of lifeboats will result in improved performance, better response to emergency call-outs and increased safety for the onboard crews.

ACKNOWLEDGEMENTS

The project has been funded by the Royal National Lifeboat Institution and supported by Lloyd's Register. This paper is an overview of a PhD Thesis, which will be submitted in 2018, supervised by Prof Robert S. Dow, Prof Richard W. Birmingham and Dr Simon Benson.

The authors would like to thank the RNLI crews of Tynemouth Lifeboat Station and the Newcastle University technicians' team who respectively made the sea trials and the towing tank tests possible.

REFERENCES

Aktas, B., Prini, F., and Benson, S., 2017. Full-Scale Unsteady RANSE CFD Seakeeping Simulations of a High-Speed Craft. Presented at the 20th Numerical Towing Tank Symposium, Wageningen, The Netherlands, 16–21.

American Bureau of Shipping, 2011. Guidance Notes on Structural Direct Analysis for High-Speed Craft.

American Bureau of Shipping, 2014. Rules for Building and Classing High-Speed Craft.

Bales, S.L., Lee, W.T., and Voelker, J.M., 1981. *Standardized Wave and Wind Environments for NATO Operational Areas*. David W. Taylor Naval Ship R&D Center, No. DTNSRDC/SPD-0919-01.

Clarke, J.D., 1986. Wave Loading in Warships. Presented at the Advances in Marine Structures Conference, Dunfermline, UK: Elsevier Applied Science, 1–25.

Cripps, R.M., Cain, C., Phillips, H.J., Rees, S., and Richards, D., 2004. Development of a New Crew Seat for All Weather Lifeboats. Presented at the SURV 6 Conference: Surveillance, Pilot & Rescue Craft, London, UK: The Royal Institution of Naval Architects, 69–75.

Cripps, R.M., Phillips, H.J., and Cain, C., 2004. Development of Integrated Design Procedures for Lifeboats. Presented at the SURV 6 Conference: Surveillance, Pilot & Rescue Craft, London, UK: The Royal Institution of Naval Architects, 9–15.

Cripps, R.M., Phillips, H.J., and Cain, C.F., 2005. Development of Integrated Design Procedures for Lifeboats. *International Journal of Small Craft Technology*.

Det Norske Veritas, 2012. Rules for Classification of High Speed, Light Craft and Naval Surface Craft.

Grassman, M.J. and Hildstrom, G.A., 2003. *Structural Trials of the RV Triton - a Status Update and Quick-Look Report*.

Heller, S.R. and Jasper, N.H., 1960. On The Structural Design of Planing Craft. *Quarterly Transactions of The Royal Institution of Naval Architects*, 49–65.

Hogben, N., Dacunha, N.M.C., and Olliver, G.F., 1986. *Global Wave Statistics*. British Maritime Technology Limited.

Hogben, N. and Lumb, F.E., 1967. *Ocean Wave Statistics*. London, UK: Her Majesty's Stationery Office.

Hughes, O.F., 1983. *Ship Structural Design: a Rationally-Based, Computer-Aided, Optimization Approach*. Wiley-Interscience.

ITTC, 2002. *The Specialist Committee on Waves. Final Report and Recommendations to the 23rd ITTC*. Venice.

ITTC, 2011. *Global Loads Seakeeping Procedure*. International Towing Tank Conference, No. 7.5-02-07-02.6.

Jacobi, G., Thomas, G., Davis, M.R., and Davidson, G., 2014. An Insight into the Slamming Behaviour of Large High-Speed Catamarans through Full-Scale Measurements. *Journal of Marine Science and Technology*, 19 (1), 15–32.

Johnson, M.C., 2004. Improvements in the Conduct and Interpretation of Ship Seakeeping Trials. PhD Thesis. University of Southampton, Southampton, UK.

Lewis, E.V., 1988. *Principles of Naval Architecture*. The Society of Naval Architects and Marine Engineers.

Lloyd, A.R.J.M., 1989. *Seakeeping: Ship Behaviour in Rough Weather*. Ellis Horwood.

Lloyd's Register, 2014. Rules and Regulations for the Classification of Special Service Craft.

Ma, M., Zhao, C., and Danese, N., 2012. A Method of Applying Linear Seakeeping Panel Pressure to Full Ship Structural Models. Presented at the 11th International Conference on Computer Applications and Information Technology in the Maritime Industries, Liege, Belgium, 50–61.

MAESTRO Version 11.5.0, 2017. MAESTRO User Manual.

Manganelli, P., Wagemakers, B., and Wilson, P.A., 2003. Investigation of Slamming Loads Using Slam Patches on a Scale Model of an Open60' Class Yacht. *The International Journal of Small Craft Technology*, 145 (B1), 47–62.

Ochi, M.K., 1981. Principles of Extreme Value Statistics and their Application. Presented at the Extreme Load Response Symposium, Arlington, VA, USA: The Society of Naval Architects and Marine Engineers, 15–30.

Phelps, B.P., 1997. *Determination of Wave Loads for Ship Structural Analysis*. Melbourne, AU: Aeronautical and Maritime Research Laboratory, No. DSTO-RR-0116.

Phillips, H.J., Sheppard, P.J., Venning, G., Austen, S.J., and Houchen, S., 2009. Theoretical and Practical Aspects of Conducting a Major Composite Repair. Presented at the SURV 7 Conference: Surveillance, Pilot & Rescue Craft, Poole. UK: The Royal Institution of Naval Architects, 117–124.

Prini, F., Benson, S., Birmingham, R., Dow, R.S., Phillips, H.J., Sheppard, P.J., and Mediavilla Varas, J., 2015. Seakeeping Analysis of a High-Speed Search and Rescue Craft by Linear Potential Theory. Presented at the International Conference on Lightweight Design of Marine Structures, Glasgow, UK, 87–96.

Prini, F., Benson, S., Birmingham, R.W., Phillips, H.J., Sheppard, P.J., Mediavilla Varas, J., Johnson, M.C., Hirdaris, S., and Dow, R.S., 2018. Full-Scale Seakeeping Trials of an All-Weather Lifeboat. Presented at the SURV 9 Conference: Surveillance, Pilot & Rescue Craft, London, UK: The Royal Institution of Naval Architects.

Prini, F., Birmingham, R.W., Benson, S., Phillips, H.J., Sheppard, P.J., Mediavilla Varas, J., Johnson, M., and Dow, R.S., 2016. Motions and Loads of a High-Speed Craft in Regular Waves: Prediction and Analysis. Presented at the 24th International HISWA Symposium on Yacht Design and Yacht Construction, Amsterdam, NL, 1–14.

Renilson, M., Scrace, R., Johnson, M., and Richardson, C., 2004. Trials to Measure the Hydrodynamic Performance of RV Triton. Presented at the International Conference on Design & Operation of Trimaran Ships, London, UK: The Royal Institution of Naval Architects, 5–18.

Riley, M.R. and Marshall, J., 2013. *Empirical Equations for Developing Ride Severity Envelopes for Planing Craft Less Than 55 Feet in Length*. Combatant Craft Division, Code 83, NAVSEA, Report No. NSWCCD-83-TM-2013/36.

Roberton, D.M.V., 2015. Residual Life Assessment of Composite Structures: With Application to All Weather Lifeboats. PhD Thesis. University of Southampton, Southampton, UK.

Rosén, A., 2005. Impact Pressure Distribution Reconstruction from Discrete Point Measurements. *International Shipbuilding Progress*, 52 (1), 91–107.

Rosén, A. and Garme, K., 1999. Slamming Studies on High-Speed Planing Craft through Full-Scale Trials and Simulations. Presented at the 5th International Conference on Fast Sea Transportation, Seattle, Washington, USA.

Salvesen, N., Tuck, E.O., and Faltinsen, O., 1970. Ship Motions and Sea Loads. Presented at the SNAME Annual Meeting, New York, NY, USA: The Society of Naval Architects and Marine Engineers, 1–30.

Southampton Oceanography Centre, 2001. *Wind and Wave Frequency Distributions for Sites around the British Isles*. Southampton, UK: Health and Safety Executive.

St. Denis, M. and Pierson, W.J., 1953. On the Motions of Ships in Confused Seas. Presented at the SNAME Annual Meeting, New York, NY, USA: The Society of Naval Architects and Marine Engineers, 280–357.

Stavovy, A.B. and Chuang, S.L., 1976. Analytical Determination of Slamming Pressures for High-Speed Vehicles in Waves. *Journal of Ship Research*, 20 (4).

Von Kármán, T., 1929. *The Impact on Seaplane Floats during Landing*. Washington, DC, USA: National Advisory Committee for Aeronautics, Technical Note No. 321.

Wagner, H., 1932. *Loading of Seaplanes*. Washington, DC, USA: National Advisory Committee for Aeronautics, Technical Memorandum No. 622.

Zhao, C., Ma, M., and Hughes, O., 2013. Applying Strip Theory Based Linear Seakeeping Loads to 3D Full Ship Finite Element Models. Presented at the 32nd International Conference on Ocean, Offshore and Arctic Engineering, Nantes, France: The American Society of Mechanical Engineers.

Zhao, R. and Faltinsen, O., 1993. Water Entry of Two-Dimensional Bodies. *Journal of Fluid Mechanics*, 246, 593–612.

Zhao, R., Faltinsen, O., and Aarsnes, J., 1997. Water Entry of Arbitrary Two-Dimensional Sections with and without Flow Separation. Presented at the 21st Symposium on Naval Hydrodynamics, Washington, DC, USA: National Academy Press, 408–423.

The anti-shock design of broadside structure based on the stress wave theory

Zi-fei Meng, Ji-cai Lang, Si-bo Xu, Chen Feng & Ping-ping Wang
College of Shipbuilding Engineering, Harbin Engineering University, Harbin, China

ABSTRACT: The anti-shock design of broadside structure in large ships has attracted great attention in the field of naval architecture and ocean engineering. The impact of broadside structure subjected to underwater explosion is a complex Fluid-Structure Interaction (FSI) problem. In this paper, the Coupled Eulerian-Lagrangian (CEL) method is proposed to solve this problem. Firstly, the CEL method is verified by a free-field underwater explosion model. Afterwards, the relationship between incidence, reflection and transmission of stress wave in a section-varying bar was derived by the stress wave theory. Moreover the propagation characteristics of stress waves in the side plate structure under underwater explosion loads were discussed based on CEL method. We found that section-varying structure can reduce the intensity of stress wave. This study provided solid basis for the design of impact resistance of ship side structures.

1 INTRODUCTION

The anti-shock design of ship has been paid great attention by researchers all over the world, in order to defend torpedo, missile and other weapons attack. Recently, many methods have been utilized such as applying multi-layer protective structure at the side and using high strength steel material to improve the anti-explosion and anti-shock ability of warships (Kang et al. 1996, Paik et al. 1998, Paik 1999 and Wu 2010). However, there are still many shortages in these measures. For instance, it could increase the weight of structure or bring trouble to construction. Therefore, it is calling for efficient ways to improve the anti-explosion ability of warships. In this paper, the stress wave theory and the Coupled Eulerian-Lagrangian (CEL) method were used to investigate the propagation characteristics of stress wave in the side plate structure subjected to underwater explosion. A new way to improve the impact resistance of the side plate was purposed, providing solid basis for the new design of the side structure in the future.

2 THE STRESS WAVE THEORY IN A SECTION-VARYING BAR

The relationship between incidence, reflection and transmission of elastic wave in a section-varying bar is analyzed in this section.

The longitudinal motion of a bar is studied in the coordinate system, as shown in Figure 1, where A_0 is the cross-section area, ρ_0 is the density and F is the force.

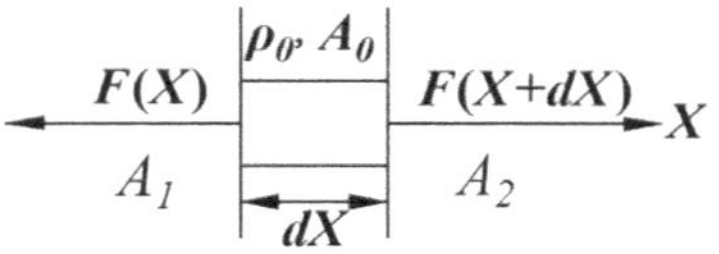

Figure 1. The sketch of a uniform cross-section bar.

We make two assumptions before deriving the equations:

Firstly, the cross-section remains constant when the bar is deformed.

Secondly, the stress and strain obey Hooke's law.

The stress is positive in tension and the velocity of particle is positive in X axis in a later article.

According to the impulse theorem, there are

$$[F(X+dX)-F(X)]dt = Fdt = \rho_0 A_0 dXv \quad (1)$$

Then

$$\frac{F}{A_0} = \sigma = \rho_0 \frac{dX}{dt}v = \rho_0 Cv \quad (2)$$

where v is the particle velocity, σ is the stress and C is the wave velocity.

Subsequently, let the elastic wave propagates from one section A_1 to another section A_2. Supposing σ_I, σ_R and σ_T are the stress in the bar under the action of the incident, reflected and transmission wave respectively. Accordingly, v_I, v_R and v_T are the particle velocity in the bar under the action of the incident, reflected and transmission wave respectively. According to the continuity condition and Newton's third law, the particle velocity/stress should be equal after reflection and transmission at the position of section-varying (WANG 2005). Then

$$v_I + v_R = v_T \tag{3}$$

$$A_1(\sigma_I + \sigma_R) = A_2 \sigma_T \tag{4}$$

According to Equations (2) and (3):

$$\frac{\sigma_I}{(\rho_0 C)_1} - \frac{\sigma_R}{(\rho_0 C)_1} = \frac{\sigma_T}{(\rho_0 C)_2} \tag{5}$$

According to Equations (4) and (5):

$$\begin{cases} \sigma_R = \dfrac{A_2 - A_1}{A_1 + A_2} \sigma_I \\ v_R = -\dfrac{A_2 - A_1}{A_1 + A_2} v_I \end{cases} \tag{6}$$

$$\begin{cases} \sigma_T = \dfrac{2A_1}{(A_1 + A_2)} \sigma_I \\ v_T = \dfrac{2A_1}{(A_1 + A_2)} v_I \end{cases} \tag{7}$$

Therefore, if the stress σ_I is obtained, the stress σ_R and the stress σ_T can be calculated by Equations (6) and (7).

If the stress wave propagates from a large cross-section to a small section ($A_1 > A_2$), the intensity of the transmission wave will be greater than that of the incident wave. Accordingly, if the stress wave propagates from a small cross-section to a large section ($A_2 > A_1$), the intensity of the transmission wave will be smaller than that of the incident wave according to Equations (6) and (7).

3 NUMERICAL METHOD

3.1 Introduction of CEL method

It is well-known that two kinds of numerical methods based on mesh are Lagrangian method and Eulerian method which have been widely used in the field of numerical simulation. Lagrangian method is always used to describe the motion of structure while the Eulerian method can deal with large deformation problems of fluid. The CEL method is a combination of Lagrangian method and Eulerian method. When the material is moving, its physical quantities are mapped to the Eulerian meshes. The Eulerian meshes like "monitoring points" that track the real-time flux of the fluid. In Lagrangian meshes, the material sticks to the mesh nodes, and the meshes are deformed with the movement of the material. The mass, momentum and energy are transferred with the movement of the mesh elements. In CEL method, the structure is calculated by the Lagrangian method and the fluid is calculated by the Eulerian method. At the interface between the fluid and the structure, Lagrangian elements are subjected to the pressure from Eulerian materials, while Eulerian materials are subjected to the geometric constraint imposed by Lagrangian elements. The information exchange will be realized by mapping between two sets of meshes. According to the above principles, the CEL method has some obvious advantages in calculating large deformation of objects, such as collision, hydrodynamics and interaction between fluid and structure (Li et al. 2014).

3.2 Governing equation

The governing equations for the Lagrangian method are as follows (Benson & Okazawa 2004):
The conservation of mass equation.

$$\frac{D\rho}{Dt} + \rho \nabla \cdot \mathbf{v} = 0 \tag{8}$$

The conservation of momentum equation.

$$\rho \frac{D\mathbf{v}}{Dt} = \nabla \cdot \boldsymbol{\sigma} + \rho \mathbf{b} \tag{9}$$

The conservation of energy equation.

$$\frac{De}{Dt} = \boldsymbol{\sigma} : (\nabla \otimes \mathbf{v}) \tag{10}$$

where $\rho, \mathbf{v}, \boldsymbol{\sigma}, \mathbf{b}, e$ are the density, velocity, stress, mass force and internal energy respectively.

The governing equations for the Eulerian method are as follows (Benson & Okazawa 2004):
The conservation of mass equation.

$$\frac{\partial \rho}{\partial t} + \nabla \cdot (\rho \mathbf{v}) = 0 \tag{11}$$

The conservation of momentum equation.

$$\frac{\partial \rho \mathbf{v}}{\partial t} + \nabla \cdot (\rho \mathbf{v} \otimes \mathbf{v}) = \nabla \cdot \boldsymbol{\sigma} + \rho \mathbf{b} \tag{12}$$

The conservation of energy equation.

$$\frac{\partial e}{\partial t} + \nabla \cdot (e\mathbf{v}) = \boldsymbol{\sigma} : (\nabla \otimes \mathbf{v}) \tag{13}$$

where $\rho, \mathbf{v}, \boldsymbol{\sigma}, \mathbf{b}, e$ are the density, velocity, stress, mass force and internal energy respectively.

3.3 Eulerian volume fraction method

The distributions of materials are calculated to describe the deformation of fluid in each increment. The initial distribution of materials is calculated by "the volume fraction tool" in ABAQUS, which carries out Boolean operations between the Eulerian body and the reference body. For instance, the tool maps the T profile as a reference body to the Eulerian body and then calculates the Eulerian volume fractions, as shown in Figure 2. The gray part represents the water in the Eulerian grids while the number represents the volume fraction of the water in the Eulerian element. Accordingly, the white part represents the T profile whose material is assigned in the Lagrangian body. The motion of the fluid is observed by the EVF tab in post-processing.

3.4 Penalty contact method

In the CEL method, the contact between Eulerian domain and Lagrangian domain is discretized by general contact algorithm, which is based on the penalty contact method. The penalty contact method is not much stringent compared to the kinematic contact method. Seeds are created on the Lagrangian element edges and faces while anchor points are created on the Eulerian material surface. The penalty method approximates hard pressure–overclosure behavior. This method allows small penetration of the Eulerian material into the Lagrangian domain. The contact force F_p enforced between seeds and anchor points is proportional to the penetration distance d (Qiu et al. 2011).

$$F_p = kd \tag{14}$$

The factor k is the penalty stiffness which depends on the Lagrangian and Eulerian material properties.

3.5 Numerical validation

The free-field underwater explosion model is applied to verify the CEL method by comparing the simulation results with the empirical formula results. The size of Eulerian (water) domain is $0.60 \times 0.01 \times 0.01$ m³ inside of which the TNT is placed in the center.

The charge of spherical TNT is 0.055 g, the density is $\rho = 1600$ kg/m³, and the detonation velocity is $D = 6930$ m/s. The state equation of JWL is adopted for explosive products (Simulia 2016). The relevant parameters in the equation are listed in Table 1. The mesh size convergence tests are carried out to determine the acceptable mesh size. In the free-field model, the mesh size for Euler is 0.5 mm, and the Eulerian zone contains 480,000 elements.

The density of water is $\rho = 1000$ kg/m³. The equation of state in U_S-U_P form is applied and the sound velocity of water is $C_0 = 1480$ m/s.

In order to reduce the effect of the boundary on the shock wave, the boundary of the Eulerian domain is set as the non-reflecting boundary. When the explosion is carried out in the free field, the peak pressure of the shock wave can be obtained by the empirical formula given by Cole and Zamyshlyayev (Cole 1948 and Zamyshlyayev 1973). The peak pressure of shock wave at different distances calculated by CEL is plotted in Figure 3, which is compared with empirical formula results, where R is the detonation distance, r is the initial radius of the charge. The pressure time-history curves calculated by CEL and empirical formula at $R/r = 6$ are shown in Figure 4.

Table 1. The relevant parameters of JWL equation.

Variable	Value
A/GPa	371.2
B/GPa	3.21
R_1	4.15
R_2	0.95
W	0.3
$e/J \cdot kg^{-1}$	4.29×10^6

Figure 2. The sketch of Eulerian volume fraction.

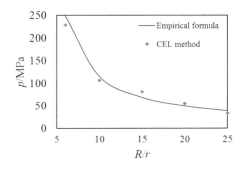

Figure 3. Peak pressure contrast at different distances.

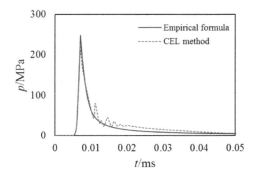

Figure 4. Pressure time-history curves obtained by CEL and empirical formula at $R/r = 6$.

It can be seen from Figures 3 and 4 that the results calculated by the CEL method are similar to those from empirical formula. Therefore, the impact load of shock wave generated by underwater explosion can be well simulated by CEL method.

4 PROPAGATION CHARACTERISTICS OF STRESS WAVES IN BROADSIDE PLATE SUBJECTED TO UNDERWATER EXPLOSION

4.1 *Model establishment*

Furthermore, we apply the CEL method to study the propagation characteristics of elastic wave in the broadside plate subjected to shock wave produced by underwater explosion. It will involve in fluid-structure interaction which is so complex that the analytical solution can not be obtained by the stress wave theory.

The computational model established is shown in Figure 6. The size of Eulerian domain is $2.0 \times 2.0 \times 1.6$ m³, in which the depth of water is 1.5 m, and the relevant material parameters of water keep in line with the section 3.5. The non-reflecting boundary condition is also set around the Eulerian domain. The radius of the spherical explosive charge is 0.0195 m. The center of detonation is 1.0 m away from the left of water domain, 0.49 m away from the bottom of the structure. And the TNT material is consistent with the section 3.5. The size of side plate structure is $1.0 \times 1.0 \times 0.8$ m³, which is immersed in water and its thickness is 12 mm. The material of this structure is steel. #1-#10 are the measuring points. The measuring points of #1, #2, #3 and #4 are located in the root of T profile, #5 and #6 are located in the top of the side plate structure, #7 and #8 are located in the middle of the web of the first T profile, and #9 and #10 are located in the middle of the web of the second T profile, as shown in Figure 6 (b). The distance between T profiles is 0.2 m, and the geometric drawing of the T profile is plotted in Figure 6 (c). In this model, T profiles are modeled by the Lagrangian grids. Accordingly, the water and TNT are modeled by the Eulerian grids. The mesh size convergence tests are carried out, as shown in Figure 5, where U_{max} is the maximum of displacement at #1 point. Results of three mesh sizes differ several millimeters. The suitable mesh size in this model is adopted according to Figure 5. The mesh size for Euler is 10 mm, and the Eulerian zone contains 6,400,000 elements. The mesh size for Lagrange is also 10 mm, and the Lagrangian zone contains 51,800 elements.

4.2 *Propagation characteristics of stress waves from broadside plate to the T profile Web*

The spectrum analysis method is used to investigate the propagation characteristics of stress wave from the side plate structure to the web of T profile.

The acceleration time-history data of #1, #2, #7 and #8 are extracted and the corresponding spectral curves are plotted by fast Fourier transform (FFT), as shown in Figure 7. It can be seen that the intensity of the stress wave in the higher frequency range [500, 2000] HZ from the outboard to the web of the first T profile decreases greatly, while the intensity of the stress wave in [0, 500] HZ has not obviously change. It shows that most of the high frequency signals will be filtered out when the stress wave passes through the first section from small to large.

The spectrum curves of #3, #4, #9 and #10 are plotted in Figure 8. It can be seen that the intensity of the stress wave in [0, 500] HZ from the plate to the web of the second T profile decreases greatly, especially in Figure 8 (a) while in other frequency bands the intensity of stress wave in the web is slightly different from that in the side plate. It shows that most of the low frequency signals will be filtered out when the stress wave passes through the form of small to large cross-section again.

4.3 *Propagation characteristics of stress waves in broadside plate*

The acceleration time-history data of #1-#6 are extracted, and the corresponding spectral curves

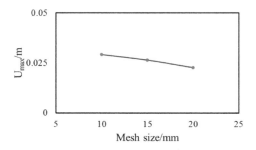

Figure 5. The mesh size convergence tests.

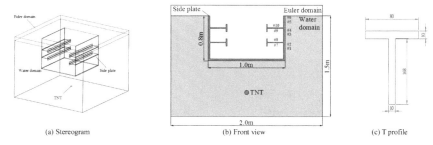

Figure 6. The simulation model of side plate under underwater explosion.

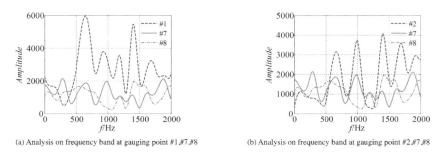

Figure 7. Analysis on frequency band at various gauging points.

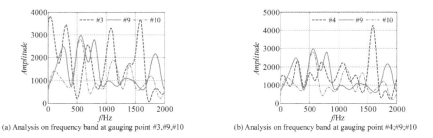

Figure 8. Analysis on frequency band at various gauging points.

are obtained by fast Fourier transform (FFT), as shown in Figure 9.

It can be observed from the diagrams that the amplitude of #3 and #4 in [500, 2000] HZ is smaller than that of #1 and #2. It indicates that there is a large attenuation trend when the stress wave in this frequency range passing through the first T profile. While the amplitude of #3 and #4 in [0, 500] HZ is larger than that in #1 and #2, it means that the intensity of the stress wave in this frequency range is slightly enhanced after passing through the first T profile. In a word, most of the high frequency signals is filtered out while the low frequency signals is enhanced when the stress wave passes through the first T profile. The former is caused by the fact that the high frequency signal of the stress wave will be filtered out when it passes through the first section from small to large. The latter is caused by the principle that the intensity of the stress wave increases from large section to small section, as described in chapter 2.

From Figures 9 (c) and (d) we can find that the amplitude of #5 and #6 is smaller than that of #3 and #4 in [0, 500] HZ, which shows that the intensity of the stress wave in this frequency range decreases after passing through the second T profile. However, the amplitude of #5 and #6 is slightly smaller than that of #3 and #4 in [500, 2000] HZ. It indicates that the low frequency signals of the stress wave will be filtered out after passing through the same T profile again. It is consistent with the conclusion in the section 4.2.

As presented in Figure 9, the amplitude of the stress wave in the side plate reduced from 5000-6000 (#1-#2) to 1000–2000 (#5-#6), indicating the ability to resist impact can be greatly improved by setting two T profiles in the side plate.

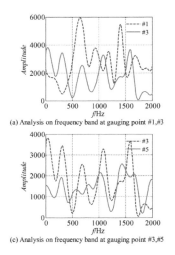
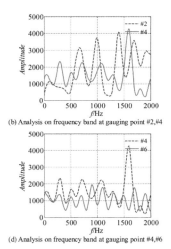

Figure 9. Analysis on frequency band at various gauging points.

5 CONCLUSIONS

Based on the stress wave theory and CEL method, the propagation characteristics of stress wave in side plate structure subjected to underwater explosion are studied in this paper. The following conclusions are obtained:

1. When the stress wave propagates from a large cross-section to a small section, the intensity of the transmission wave will be greater than that of the incident wave.
2. Most of the high frequency signals of the stress wave will be filtered out when the cross-section is changed from small to large. However, when the section changes again, it will filter out most of the low frequency signals.
3. For the side plate with two T profiles, the intensity of the stress wave in the high frequency range decreases while the low frequency increases slightly after passing through the first T profile. When the stress wave passes through the second T profile, the intensity of the high frequency stress wave changes little while the intensity of the low frequency signal decreases.

According to the above conclusions, we find a new method to improve the impact resistance of ship side structures. We can apply section-varying structure such as profile at the side plate to reduce the intensity of stress wave produced by shock wave of underwater explosion. In addition, if the ratio of side plate thickness and profile web height is adjusted properly, the intensity of stress wave propagating to the hull would be reduced accordingly. It is of certain significance for studying the effect of section-varying structure on vibration and shock resistance.

REFERENCES

Benson, D.J. & Okazawa, S. 2004. Contact in a multi-material Eulerian finite element formulation. *Computer Methods in Applied Mechanics & Engineering* 193(39–41):4277–4298.

Cole, R.H. 1948. Underwater explosions. *New Jersey: Princeton University Press* 100–127. USA.

Kang, J.M. Kim, J.Y. Lee, K.J. et al. 1996. Effect of design variables on the behavior of VLCC structure in collision. *Proceeding of the 13th MSC Users conference*: 1–12. Japan.

LI Gang, AI Sen, TANG Xiao-han. 2014. The Application and Study on the Ground Separation of a Large-Scale Payload Fairing Based on CEL Algorithm in Abaqus. *Chinese Conference on Computational Mechanics and QIAN Ling-xi prize for Computational Mechanics in 2014.*

Paik, J.K. 1998. Experimental investigation of structural crashworthiness in ship collision and grounding using double skinned structural models. *Pusan National University, Department of Naval Architecture and ocean Engineering.*

Paik, J.K. Chung, J.Y. Pedersen, P.T. et al. 1999. On rational design of double hull tanker structures against collision. *The society of naval architectures and marine engineers annual meeting preprints.*

Qiu G, Henke, S. Grabe, J. 2011. Application of a Coupled Eulerian–Lagrangian approach on geomechanical problems involving large deformations. *Computers & Geotechnics* 38(1):30–39.

Simulia D. 2016. Abaqus Version 2016 Documentation. *USA: Dassault Systems Simulia Corp.*

WANG Li-li. 2005. Foundation of stress waves. *National Defense Industry Press.*

WU Yong-bin. 2010. Research on Design and Simulation Technology of ship Anti-shock Protection structure. *Harbin Engineering University.*

Zamyshlyayev, B.V. 1973. Dynamic loads in underwater explosion. *AD-757183.*

Multiobjective ship structural optimization using surrogate models of an oil tanker crashworthiness

P. Prebeg, J. Andric, S. Rudan & L. Jambrecic
Faculty of Mechanical Engineering and Naval Architecture, University of Zagreb, Zagreb, Croatia

ABSTRACT: One of the possible solutions for reduction of consequences of tanker ship accidents in Adriatic Sea or any other sensitive marine environment, is to investigate possible improvements in tanker structural design. The paper proposes tanker structural design methodology that leads to improved structural safety with multiobjective optimization using hull girder ultimate strength and ship crashworthiness quality measures as objectives in addition to the traditionally used weight or cost quality measures. Implementation of the methodology to the presented test case of Aframax class tanker structural design shows the current status of the undergoing research and illustrates benefits but also an added level of complexity to the concept design phase.

1 INTRODUCTION

Overall objective of a standard oil tanker structural design process is to simultaneously increase the ship-owner's profit and reduce shipyard production cost, while satisfying all rules prescribed by IACS Harmonized Common Structural Rules for Bulkers and Oil Tankers—CSR BC & OT (IACS 2017). The goal of ongoing research partially covered by this paper, is to investigate possibilities of introducing additional structural safety measures as additional objectives. The focus is on the measures capable of identification of hull structural integrity (ship crashworthiness, hull girder ultimate strength). Selected design parameters, having significant effect on design solution, have to be identified and discussed with the stake-holders in concept (CDP) and preliminary (PDP) design phases, where the most far-reaching decisions are made.

Depending on the level of details modeled and the extent of the structural model (partial model to full ship model), nonlinear finite element analysis of a single variant could take from an hour to several days. Even the simplest possible model are usually too demanding for direct usage in structural optimization during preliminary design phase and especially during a concept design phase. The possible solution is a creation of an appropriate surrogate models that could replace demanding nonlinear numerical models in structural optimization. The main purpose of the work presented in this paper is to evaluate proposed methodology that incorporates multiobjective optimization with objectives that reflects a level of structural safety. In the presented case study crashworthiness surrogate model is trained using LS Dyna implementation of nonlinear finite elements method (NLFEM) with explicit solver, while the ultimate strength is calculated directly in optimization using Bureau Veritas Mars 2000 with implemented Smith progressive collapse analysis method for hull girder ultimate strength calculation (Andrić et al. 2014).

Since an inclusion of surrogate models in optimization process requires execution of analysis runs that are necessary to build those surrogates, special considerations are necessary to reduce that number to acceptable level, while maintaining a level of accuracy acceptable for optimization. An increase of control factors leads to the nonlinear increase in number of computer simulations, in this case ship NLFEM calculations, to obtain the same level of surrogate model accuracy. This is the reason why it is necessary to reduce the number of control parameters for crashworthiness surrogate model to only those that have significant effect. This work also suggests the approach for the reduction of the design variants that are not rational from the viewpoint of the traditional oil tankers structural design practice. In that sense, selection of the appropriate design experiments plan includes the Stage I CSR BC & OT prescribed rule calculation in order to filter out unrational designs.

Proposed methodology includes multiobjective optimization with minimization of weight, maximization of crashworthiness and maximization of hull girder ultimate strength, using CSR BC & OT Stage I for calculation of structural adequacy.

2 DESIGN SYNTHESIS TECHNIQUES—BACKGROUND

Although the methods/tools available for design synthesis like optimization methods, surrogate modeling, are wildly used today in naval architecture and ship structural design, there is still need to cover some of them that are very important and yet not well understood in practice.

2.1 *Surrogate modeling*

Surrogate/approximation/metamodeling, is the key to surrogate assisted optimization. It can be stated that surrogate modelling actually evolves from classical Design of Experiments (DOE) theory, in which polynomial functions are used as response surfaces, or surrogate models. One of the most cited handbooks with detail overview of DOE for classical (physical) experiments is (Montgomery 2001), while the overview of surrogate modeling for deterministic computer experiments (DACE – design and analysis of computer experiments) can be found in e.g. (Fang et al. 2006), (Simpson et al. 2001).

The main difference between "classical" and computer experiments is nonexistence of random error for deterministic computer experiments, which according to (Sacks et al. 1989) leads to the conclusion that surrogate model adequacy is determined solely by systematic bias and that the classical notions of experimental blocking, replication and randomization are irrelevant. In depth review of surrogate modeling for computer based engineering design can be found in (Simpson et al. 2001) and (Wang and Shan 2007).

Steps necessary for generation of surrogate models are:

- planning of experiments or sampling,
- execution of simulations with original analysis methods,
- generation of the surrogate model
- validation of the surrogate model adequacy.

After selection of the appropriate experimental design according to the appropriate criteria, see e.g. (Goel et al. 2008), and when the necessary computer runs are performed, the next step is to choose a surrogate model type and corresponding fitting method. Many alternative models and methods exist, and there is no clear answer which is better. The selection of appropriate surrogate model depends mostly on characteristic of physical phenomenon that is approximated. Some of widely used surrogate models in engineering include: Response Surfaces – RS, Radial basis functions – RBF, Artificial Neural Network – ANN, Support Vector Machine – SVM, Multivariate Adaptive Regression Splines – MARS.

Generally, the value of a target criteria response y at some location x can be written as:

$$y(\mathbf{x}) = \hat{y}(\mathbf{x}) + \varepsilon_b \quad (1)$$

where $\hat{y}(\mathbf{x})$ is surrogate model of response y, while ε_b is a surrogate model error or bias. As already stated, one of the characteristics of deterministic computer experiments is nonexistence of random error ε_r, and that is the reason why it is not included in expression (1).

In this research, RS method will be used as surrogate modelling method, so it's basic theoretical background is given in the following subchapter.

2.1.1 *Response Surfaces (RS)*

Probably the most widely used surrogate modeling method is response surfaces (RS) that approximates criteria functions using low order polynomials, mostly simple linear and quadratic or some specific polynomials like orthogonal Legendres polynomials.

General matrix formulation can be written as:

$$\hat{y}_{RS} = \mathbf{B}^T \boldsymbol{\beta} \quad (2)$$

where $\mathbf{B}$ is a k-tuple of a used polynomial functions, while $\boldsymbol{\beta}$ is a k-touple of unknown corresponding coefficients. If a mostly linear polynomial is used, $\mathbf{B}$ and $\boldsymbol{\beta}$ are:

$$\mathbf{B}^T = \{1 \quad x_1 \ldots \quad x_i \ldots \quad x_k\} \quad (3)$$

$$\boldsymbol{\beta}^T = \{\beta_0 \quad \beta_1 \ldots \quad \beta_i \ldots \quad \beta_k\} \quad (4)$$

The unknown coefficients $\boldsymbol{\beta}$ are usually determined using least square regression analysis by fitting the response surface approximation into existing data:

$$\boldsymbol{\beta} = \left(\mathbf{B}_{1-n} \mathbf{B}_{1-n}^T\right)^{-1} \mathbf{B}_{1-n} \mathbf{y}_{1-n} \quad (5)$$

where $\mathbf{y}_{1-n}$ is n-tuple of n known response values, while $\mathbf{B}_{1-n}$ is $k \times n$ matrix with the calculated values of selected basis functions at locations $1-n$.

RS popularity for modeling of a deterministic computer experiments, besides its good characteristics for certain type of problems, is due to the fact that surrogate modeling itself evolves from classical Design of Experiments theory where RS was used for the description of physical phenomena (Montgomery 2001). Discussion of the statistical pitfalls associated with the application of RS to deterministic computer experiments can be found in (Sacks et al. 1989) and (Simpson et al. 2001). Some of the applications in engineering are given in e.g. (Arai and Shimizu 2001), (Prebeg et al. 2014) and (Goel et al. 2007).

2.2 Multi-objective optimization and nondominated designs

Although the problems in engineering are always constrained, optimization methods for unconstrained problem are still relevant because it is possible to use some general approaches for incorporation of constraints, e.g. penalization of objective function with relative unsatisfaction of constraints, see e.g. (Coello Coello et al. 2007). Those methods are also interesting due to the fact that significant research activity of mathematical society exist in that area.

The problems in engineering are usually consisted of several design objectives, but since solving of multi-objective problem incorporates a significant amount of additional complexity, it is common practice in today's engineering to reduce the problem to the most important objective or to some complex objective that incorporates several objectives into one objective function. Whenever possible, the optimization problem should be solved as multi-objective problem since determination of nondominated solutions or Pareto frontier provides significantly better insight into a design problem then a single solution provided by a solution of single objective optimization problem. This is usually possible in the product development concept design phase, since mathematical models in that phase are usually simple, while the number of design variables is not too high.

One of the easiest ways to illustrate terms related to the concept of nondominance is to start with the description of relations between one solution and all other obtained solutions in example of unconstrained optimization problem with two objectives that needs to be minimized.

By comparing solution A (marked ● in Figure 1a) with all other obtained solutions it is possible to state:

- Solution A is better with respect to objective 1 from all solutions that are on the right side of it (marked ◐)
- Solution A is better with respect to objective 2 from all solutions that are above it (marked ⊜)
- Solution A is better with respect to all objectives from the solutions that are on the upper right side of it (marked ⊕). It can be stated that those solutions are dominated by solution A. Those solutions are in the solution's A dominance cone.

When described comparison is carried out for all of the obtained solutions, each solution will define its dominance cone. Only those solutions that are not inside of other solutions dominance cone are interesting for decision making. Those solutions are called Pareto solutions, nondominated solutions or all together Pareto frontier (marked ● in Fig-

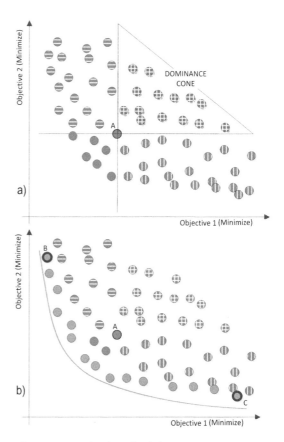

Figure 1. Nondominated solutions.

ure 1b). They are called nondominated because none of the solutions dominate on them, or in another words they are not in the dominance cone of any of the obtained solutions. Blue line in Figure 1 illustrates theoretical Pareto frontier. It is important to notice that the solution B is the best with respect to the Objective 1, while the solution C is the best with respect with Objective 2. Single objective optimization that would treat Objective 1 or Objective 2 as the objective would return as a solution only the corresponding project, and it would not be able to provide the designer with the information on the tradeoffs between those two objectives.

As an alternative to the simplified graphical definition of the concept of nondominance, one of the next definitions of nondominance can be used:

- Solution is non-dominated if it is better than any other obtained solution in at least one objective.
- Solution is nondominated if there exists no feasible vector **x** which would decrease some objective without causing a simultaneous increase in at least one other objective.

3 PROPOSED METHODOLOGY FOR THE TANKER STRUCTURAL DESIGN WITH AN IMPROVED STRUCTURAL SAFETY

The presented methodology is a an extension of a methodology for ship structural design given in e.g. (Zanic 2013) and (Zanic et al. 2013). For the detailed description of the proposed tanker structural design procedure, different type of diagram blocks will be used as shown in Figure 2. The figure contains four different types of blocks where each block is identified with several items: Analysis block (3 items), Optimization block (6 items), Design of experiments—DOE block (5 items) and Decision making block (2 items). The items in the blocks are used to specify the following:

- $\mathbf{x}$ – variables
- $\mathbf{d}$ – descriptors
- $\mathbf{a}$ – attributes
- $\mathbf{y}$ – objectives
- $\mathbf{g}$ – constraints
- AM – applied analysis modules
- SM – applied synthesis modules
- Res – obtained results

The procedure presented in Figure 2 implies that the design process needs to start and end with an extensive structural analysis as given by the Blocks 1a and 5, which is the standard design practice in most of the design offices and shipyards. In addition to that, this design procedure prescribes several additional blocks, and includes multiobjective optimization with weight and several structural safety measures as objectives. The first three block levels are part of CDP, while the last is PDP. The proposed improvement, with respect to the standard industrial practice, is done in CDP where the decisions about specific topology and geometry variants selection have to be made because those decisions influences structural safety in a great deal. Some additional improvements could also be done in PDP, but it is a just a small portion of impact that decisions in CDP have.

As given above, Block 1a is used to make standard 2D cross section analysis to check feasibility of the investigated variant. An additional effort, with respect to the standard industrial practice, should be made to generate a reasonable number of topology/geometry (T/G) variants based on the selection of possible topology $\mathbf{d}^T$ and geometry descriptors $\mathbf{d}^G$ (e.g. web frame spacing, number of side stringers, double side width, etc.). Each T/G variant should be manually dimensioned by changing scantling of structural members with a goal of producing a feasible or near feasible solution with smallest structural weight. This type of redimensioning is still

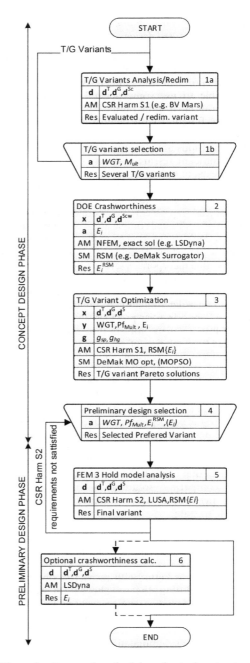

Figure 2. Proposed methodology for tanker structural design with improved structural safety.

standard industrial practice, however it is usually done for only one (the prototype) T/G variant.

Block 1b is used to select only those T/G variants that are feasible and realistic candidates for the final design. In other words, it should be use to exclude design variants that have unacceptable levels of weight or unacceptable levels some other relevant attributes. In that way the next block (computationally demanding crashworthiness analysis) will deal with only those variants that are of practical importance.

The purpose of the Block 2 is generation of response surface model of internal energy E_i absorbed by collision or grounding as a measure of crashworthiness. The set of control parameters/variables includes selected topology $\mathbf{d}^T$ and geometry descriptors $\mathbf{d}^G$ and the subset of scantling variables $\mathbf{d}^{Scw}$ that have a great effect on the measure of crashworthiness (e.g. thickness of the side shell).

Block 3 is used to perform multiobjective optimization with weight, and structural safety measures (crashworthiness, hull girder ultimate strength) as objectives on each of the selected T/G variants. The optimization has to be done with constraints/requirements prescribed by CSR BC & OT using all scantling descriptors as design variables and all topo geo variables that can be automatically changed in the model. Alternatively, if it is not possible to use some or all T/G variables in optimization, the solution sequence should generate a separate optimizations for each (or some relevant subset) T/G variant that are generated by recombination of possible levels of those T/G variables. All nondominated designs obtained for each T/G variant should enter the final pool of designs as a candidates for the selection in the next block.

Block 4, where the preliminary design variant is selected from the set of nondominated solutions, is the last block in CDP, but it could also be seen as a first block in PDP. As a criteria for decision making, weight, internal energy (calculated by RSM), and probability of hull girder ultimate strength failure are proposed to be used.

The three hold FEM model is used in Block 5 to verify accuracy of the CDP model, and to dimension parts of the structure such as transverse bulkheads, web frames and double bottom that cannot be adequately dimensioned with the models used in CDP. If the PDP variant turns to be infeasible according to CSR BC & OT, a try to repair it manually should be done by changing the scantling variables that are not included in as control parameters/variables in DOE plan in Block 1, which means to change $\mathbf{d}^{S\text{-}Scw}$ set of scantling descriptors as indicated in diagram. In that way crashworthiness measure will remain on the same value. If it is not possible to find feasible solution, it is necessary to return to the Block 4 as indicated in the diagram. Return to Block 4 is also suggested if it is possible to find a feasible solution, but with a relatively large changes in structural scantlings, especially if $\mathbf{d}^{S\text{-}Scw}$ set of scantling descriptors needs to be changed and/or if it is necessary to change scantlings that influence the most the hull girder ultimate strength (e.g. deck thickness).

4 IMPLEMENTATION OF THE PROPOSED METHODOLOGY FOR CDP

As already stated, in design practice an application of certain methodology depends on availability of certain analysis and synthesis tools/software. The used implementation of proposed methodology is shown in Figure 3 and uses design support frameworks DeMak (Octopus Designer), an in house software developed at UNIZAG FSB, where synthesis tools are needed.

The CSR BC & OT based models uses Mars 2000 application for generation of the models and calculations, while newly developed module MAGIC (Mesh generator for ship crashworthiness in CDP) in an in-house software View3D is used for the automatic preparation of LSDyna analysis models for crashworthiness. Optionally, LSPrepost or FEMAP can be used for the LSDyna model generation. The reason why the new modules for automatic FEM model preparation is made, although initially it was planned to use FEMAP, is because after some while it became apparent that it would be a to huge task to prepare relevant number of T/G variants (50–100) manually. It also raises a risk of generating an unintentional modeling error to undesired level. Additional identified problem was that manually created models that have used available meshing algorithm in FEMAP resulted with a relatively different meshes, which is unwanted since that adds an additional "noise" in the RSM model. LSDyna solver is used for crashworthiness analysis.

View3D MAGIC is developed to enable automatic FEM mesh generation for simplified crashworthiness structural models. Wherever it is possible, elements are quadratic (100 × 100 mm), Triangles are used only when necessary (less than 1% of elements are triangular). It is possible to parametrically change:

- Topology parameters like number of web frames, number of side stringers, side stiffener spacing.
- Geometry parameters like double side width, height of web/breadth of flange of longitudinal and transfer stiffeners.
- Scantlings like outer shell thickness, inner hull thickness, side stringer thickness, and off course web/flange thickness of longitudinal and transfer stiffeners.

The usage of module is limited to simple structures as is the tanker structure around ship midlength, so it is convenient to use it to model a central part of tanker structure between two

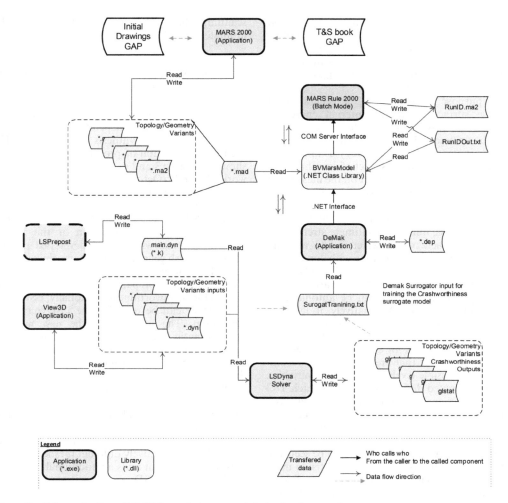

Figure 3. Schematic view of DeSS for tanker structural design implementation.

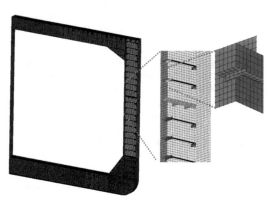

Figure 4. Tanker mesh detail generated by View3D MAGIC.

transverse bulkheads. Figure 4 show a detail of the generated mesh.

5 APPLICATION OF PROPOSED METHODOLOGY TO AFRAMAX TANKER STRUCTURAL DESIGN

A generic Aframax class oil tanker is selected for a test case for application of the proposed methodology. Basic ship characteristics are given in Table 1.

5.1 *Design block 1*

As explained in Section 4, Bureau Veritas Mars 2000 Application is used for modelling and calculations of structural adequacy of 2D cross section

Table 1. Aframax class oil tanker principle dimensions.

LOA	247.24 m
B	42 m
D	21 m
Scantling draught	15.6 m
Displacement	133000 t
Max. service speed	15.9 kn
Ship center of gravity by length	5.599 m
Ship center of gravity height	12.050 m
Max. still water VBM in hogging	2745862 kNm
Max. still water VBM in sagging	2079009 kNm

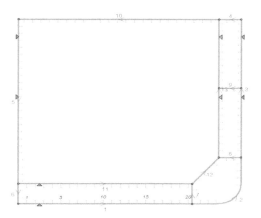

Figure 6. Midship section of the tanker prototype variant.

Figure 5. Mars 2000 snapshot—test case basic ship data.

Table 2. Investigated section variants and T/G variables.

No	Name	DSW mm	WFS mm	NSTR
1	D20-3.4-1	2000	3400	1
2	D21-3.4-1	2100	3400	1
3	D22-3.4-1	2200	3400	1
4	D20-3.4-2	2000	3400	2
5	D21-3.4-2	2100	3400	2
6	D22-3.4-2	2200	3400	2
7	D20-3.4-3	2000	3400	3
8	D21-3.4-3	2100	3400	3
9	D22-3.4-3	2200	3400	3
10	D20-3.9-1	2000	3900	1
11	D21-3.9-1	2100	3900	1
12	D22-3.9-1	2200	3900	1
13	D20-3.9-2	2000	3900	2
14	D21-3.9-2	2100	3900	2
15	D22-3.9-2	2200	3900	2
16	D20-3.9-3	2000	3900	3
17	D21-3.9-3	2100	3900	3
18	D22-3.9-3	2200	3900	3
19	D20-4.5-1	2000	4500	1
20	D21-4.5-1	2100	4500	1
21	D22-4.5-1	2200	4500	1
22	D20-4.5-2	2000	4500	2
23	D21-4.5-2	2100	4500	2
24	D22-4.5-2	2200	4500	2
25	D20-4.5-3	2000	4500	3
26	D21-4.5-3	2100	4500	3
27	D22-4.5-3	2200	4500	3

according to CSR BC & OT & OT Stage 1 rules. Figure 5 show basic ship data enteres in Mars 2000.

The initial, prototype version of midship section (Figure 6) is based on a midship section structure of a similar ship that was available to authors. Scantlings have been changed manually in order to obtain near feasible solution.

Based on the general ship design solution data/ constraints and expected impact to the structural safety (especially collision crashworthiness) one topology (number of side stringers – NSTR) and two geometry variables (double side width – DSW, web frame spacing – WFS) with appropriate three levels for each of them were selected to generate additional T/G variants.

Table 2 present characteristics of all 27 T/G variants obtained by the recombination of previously given variables and their levels.

Figure 7 show three sections with different number of side stringers.

After redimensioning of all T/G variants it has been identified all of them are feasible rational structural design variants with respect to the all used criteria, so it was decided to use all of them for the DOE and optimization in the next two design blocks.

5.2 Design block 2 crashworhiness RSM model

Since collision in Adriatic Sea is of the particular interest to the authors, the collision parameters were based on the possible collision scenarios in

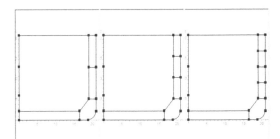

Figure 7. Section variants with different number of side stringers.

Figure 8. Maritime traffic in the Adriatic (http://atlas.shape-ipaproject.eu/).

Adriatic. Adriatic Sea is a semi-enclosed narrow sea stretching from north-west to south-east mainly between two countries, Italy and Croatia. Due to shape of the Sea two most important traffic routes are north-west to south-east or longitudinal routes and west to east, or transversal routes, Figure 8. Commonly large merchant ships are sailing over longitudinal routes (thick red lines) to bring the cargo to large northern harbours in Koper, Rijeka, Trieste, Venice etc. while ferries and leisure ships are sailing over transversal routes (grey lines), connecting large cities on both west and east cost of Adriatic. Due to the nature of such trafficking an orthogonal collision of a tanker and a ferry was assumed to be a reasonable collision scenario for Adriatic. Figure 8. also shows level of Environmental vulnerability to oil spills where colour ranges from blue that identifies areas with a low vulnerability, to red that identifies very high vulnerability.

In order to study consequences of specified collision scenario a calculation model in commercial software package LS-Dyna was set. It consists of two ships in concern:

- A struck ship, being an (designed) Aframax class tanker
- A striking ship, being a typical international ferry of the Adriatic Sea.

The main striking ship particulars are listed in Table 3. Due to the complexity of the problem, the both ship models are reduced in order to enable the study of all most important physical aspects of their collision and yet at the same time to enable the reasonably fast calculation. More elaborated description on the the preparation of models and evaluation of the particular RSM model can be found in preliminary study given in (Prebeg et al. 2016).

The preliminary study presented in (Prebeg et al. 2016) includes two control parameters: one geometry parameter (double side width) and one scantling parameter (thickness of the side shell). Both parameters were tested on three levels.

In order to prepare surrogate models of struck ship crashworthiness, first it is necessary to select the relevant measures of crashworthiness. Internal energy absorbed by the structure during collision is usually used as a crashworthiness criteria (see e.g. (Klanac et al. 2009), (Ehlers 2010)). Usually, maximal internal energy is used, however, for practical purposes, maximal internal energy is substituted with Internal energy absorbed during the first 1.2 seconds $E_{it=1.2}$.

Table 3. Striking ship (ferry).

Length over all	121.83 m
Ship mass	4730 t
Ship with cargo mass (assumed)	6889 t
Draft aft	5.25 m
Draft fore	5.30 m
Middle draft	5.28 m
Ship center of gravity height	8.38 m
Ship center of gravity length	61.08 m

Table 4. Preliminary study control parameters.

Parameters	Level 1 mm	Level 2 mm	Level 3 mm
Double side width (DSW)	2000	2100	2200
Side shell thickness (ts)	13.5	15.5	17.5

In order to study influence of all effects, including removal of experiments, full factorial design is used, which result with the total of 9 experiments (see Table 5). Since full quadratic RS model for two parameters have 6 unknowns, that requires 6 experiments for their determination, the remaining experiments are used for the evaluation of model error.

Figure 9 shows resulting surrogate model in 3D plot together with the numerical experiments used for the generation (marked with spheres). Based on the surrogate model evaluation, which is given in (Prebeg et al. 2016), it can be concluded that the accuracy of an internal energy surrogate model with respect to the used control parameters is more than adequate for use in optimization.

5.3 Design block 3

As described in Section 3, this design block is used to perform multiobjective optimization with weight, and structural safety measures (crashworthiness, hull girder ultimate strength), while the constraints/requirements are prescribed by CSR BC & OT. Design Tools implementation, which is particularly demanding due to this block especially, is given in Section 4. The summarized test case design problem definition is given in the sequel, while more info are given in (Jambrecic 2017).

5.3.1 Design variables

The total set of design variables in Block 3 includes topology / geometry variables, as given in Block 1 Table 2, together with midship section scantlings.

Plate thickness variables minimum, maximum values (lower and upper bounds) and step are given in Table 6. It also shows panel names that are used in Mars 2000 and their aliases used in DeMak in parallel. Position of panels on cross section are presented in Figure 10.

Table 7 gives definition of stiffener scantling variables characteristics. The same characteristics are used for all stiffener groups for simplicity. Position of stiffener groups on cross section are presented in Figure 11.

The reader should note that breadth between stiffeners is not used as a variable at this point since automatic modification demands implementation of rather complex logic of changes in Mars 2000 that is not yet implemented, but it is an undergoing effort.

5.3.2 Design constraints

As explained earlier, tanker structure should satisfy constraints/requirements that are prescribed

Table 5. Experiment results.

Model	ts, mm	DSW, mm	tbreach, s	Ei, breach, mJ	Ei, t = 1.2, mJ
A20	13.5	2000	0.3158	3.26 E + 10	1.267 E + 11
B20	15.5	2000	0.3774	4.55 E + 10	1.297 E + 11
C20	17.5	2000	0.3750	4.77 E + 10	1.308 E + 11
A21	13.5	2100	0.3922	4.57 E + 10	1.245 E + 11
B21	15.5	2100	0.3513	4.01 E + 10	1.265 E + 11
C21	17.5	2100	0.3871	4.76 E + 10	1.277 E + 11
A22	13.5	2200	0.3807	4.18 E + 10	1.253 E + 11
B22	15.5	2200	0.4129	5.09 E + 10	1.269 E + 11
C22	17.5	2200	0.4237	5.50 E + 10	1.275 E + 11

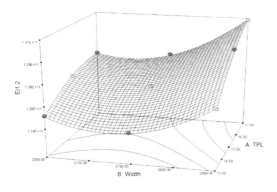

Figure 9. Surrogate model 3D plot.

Table 6. Panels thickness of plating variables (TPL).

No	MARS 2000	DeMak	Min. mm	Max. mm	Step mm
1	BTM	Pan1_Str1	18	25	0.5
		Pan1_Str2	16	25	0.5
2	BILGE	Pan2_Str1	16.5	20	0.5
3	SIDE SHELL	Pan3_Str1	16.5	20	0.5
		Pan3_Str2	16.5	20	0.5
		Pan3_Str3	16.5	20	0.5
4	DECK STRINGER	Pan4_Str1	13.5	20	0.5
5	BULKHEAD	Pan5_Str1	14	20	0.5
		Pan5_Str2	14	20	0.5
		Pan5_Str3	14.5	20	0.5
6	KEELSON	Pan6_Str1	15	20	0.5
7	DBTM GIRDER	Pan7_Str1	13.5	20	0.5
8	SIDE STRINGER	Pan8_Str1	12.5	18	0.5
9	SIDE STRINGER_1	Pan9_Str1	12.5	18	0.5
10	DECK	Pan10_Str1	14	20	0.5
11	INNER BTM	Pan11_Str1	17	22	0.5
		Pan11_Str2	17	22	0.5
12	HOOPER	Pan12_Str1	15	20	0.5
13	INNER HULL	Pan13_Str1	14	20	0.5
		Pan13_Str2	14	20	0.5
		Pan13_Str3	14	20	0.5

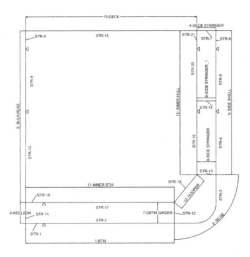

Figure 10. Position of panels on midship section.

Table 7. Stiffener scantling variables characteristics.

Name	Acronym	Min.	Max	Step
Height of web	HW	200	500	10
Thickness of web	TW	9	15	0.5
Breadth of flange	BF	140	300	5
Thickness of flange	TF	16	30	0.5

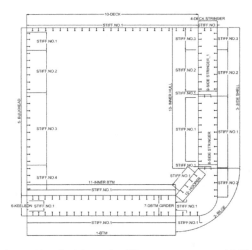

Figure 11. Position of stiffener groups on midship section.

by CSR BC & OT. BV Mars 2000 is used for the calculation of adequacy parameters.

Figure 12 and Figure 13 panel/strake and stiffener related checks that are done by Mars 2000 according to CSR BC & OT Rules.

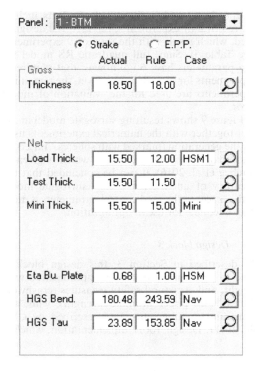

Figure 12. Mars 2000 snapshot—panel related constraints.

In addition to panel and stiffener related constraints, Mars 2000 also calculates hull girder ultimate strength. Only intact sagging condition have been used as a constraint since preliminary tests have shown that this condition is always critical for any investigated variant. The reader should note that the integration of Mars 2000 calculation results is done using the output file produced by Mars 2000 executed in batch mode, which is an unconventional way to get the Rules' Scantlings and not officially advised by Bureau Veritas.

Normalized adequacy parameter g was used in optimization as actual constraints for each of the given requirements was used. In general, it is calculated as follows:

$$g = \frac{C - D}{C + D} \quad (6)$$

where C is structural Capability, while D is structural Demand.

5.3.3 Design objectives

As given in Section 3, Figure 2, the proposed set of objectives includes structural weight (or cost) and additional safety measures hull girder ultimate strength and crashworthiness measure. At the time of

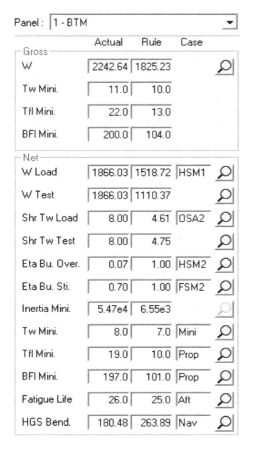

Figure 13. Mars 2000 snapshot—stiffener related constraints.

Structural weight of longitudinal structure is calculated automatically in Mars2000, but since design variables includes web frame spacing that can change number of web frames, it is necessary to include a mass of transverse structure to enable realistic ordering of variants with respect to structural weight. A simplified approach was used in which central tank structure without bulkheads was use as a reference structure, while each transverse frame have the same structural mass regardless of the number of frames in tank. As shown in (Jambrecic 2017) calculated mass of the referent web frame is 9.39 t.

5.3.4 *Optimisation results*

Previously defined multiobjective optimization problem was solved using MOPSO Hypercube multiobjective algorithm that is implemented in an in-house design support environment DeMak. Pareto front containing 32 non-dominated projects was generated, as shown on Figure 15.

5.4 *Design block 4 – design selection*

Block 4, where the preliminary design variant is selected from the set of nondominated solutions,

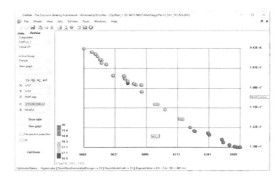

Figure 15. Nondominated designs—Design block 3 results.

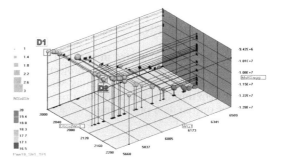

Figure 16. Nondominated designs in 3D with marked preferred designs.

Figure 14. Mars 2000 ultimate strength.

finishing this paper surrogate model for crashworthiness was not yet prepared for the total scope of variables used in presented test case so it was not used as an objective. Optimization is carried out with the next two objectives: minimization of weight and maximization of hull girder ultimate strength in sagging.

is the last block in CDP, As a criteria for decision making, weight, and hull girder ultimate strength failure in sagging is used in the test case. Figure 16 shows generated Pareto designs, which is used as a set of potential candidates for selection of preferred design, in a chart where the geometrical variable DSW is on axis X, Weight is on axis Y and ultimate moment in sagg is on Z. Color shows range of deck thickness of plating, while size of the spheres is proportional with the number of side stringers.

Since this paper presents current status of the research, the formal selection is not made at this point. Designs D1 and D2 are marked on the figure, more as an interesting design variants for comparison then as highest graded preferred designs. Design D1 is a design with the lowest structural weight, while D2 is a bit heavier but also with a relatively high hull girder ultimate strength moment in sagging. D1 double side width is 2000 mm with 1 side stringer, while the double side width of D2 is 2100 mm with 2 side stringers. Panel buckling criteria is critical for most of the panels, while fatigue criteria is critical for most of the stiffeners. Neither of hull girder ultimate strength criteria are critical for any panel or stiffener for any of designs, including D1. It is interesting to mention that initial designs of all 27 T/G variants was filtered out during optimization process as dominated by some other solutions, however further quantification of gains with respect to initial designs has not been made.

6 CONCLUSIONS

The work presented in the paper proposes tanker structural design methodology that leads to improved structural safety using hull girder ultimate strength and ship crashworthiness quality measures. Implementation of the methodology to the presented test case of Aframax class tanker structural design shows the current status of the undergoing research and illustrates benefits but also an added level of complexness to the concept design phase.

ACKNOWLEDGMENT

This work has been supported in part by Croatian Science Foundation under the project 8658. Thanks are due to the Bureau Veritas Mars 2000 team for help during preparation of a module for automatic output data processing.

REFERENCES

Andrić, J., Kitarović, S. and Bičak, M. 2014. IACS incremental-iterative method in progressive collapse analysis of various hull girder structures. *Brodogradnja: Teorija i praksa brodogradnje i pomorske tehnike,* 65(1), 65–77.

Arai, M. and Shimizu, T., 2001. OPTIMIZATION OF THE DESIGN OF SHIP STRUCTURES USING RESPONSE SURFACE METHODOLOGY. *Practical Design of Ships and Other Floating Structures. Proceedings of the Eighth International Symposium on Practical Design of Ships and Other Floating Strucutres.* Shanghai, China: Elsevier, 331–339.

Coello Coello, C.A., Lamont, G.B. and Van Veldhuizen, D.A., 2007. *Evolutionary algorithms for solving multi-objective problems.* 2nd ed. New York: Springer.

Ehlers, S. 2010. A procedure to optimize ship side structures for crashworthiness. *Proceedings of the Institution of Mechanical Engineers, Part M: Journal of Engineering for the Maritime Environment,* 224(1), 1–11.

Fang, K.-T., Li, R. and Sudjianto, A., 2006. *Design and modeling for computer experiments.* Boca Raton, FL: Chapman & Hall/CRC.

Goel, T., et al. 2008. Pitfalls of using a single criterion for selecting experimental designs. *International Journal for Numerical Methods in Engineering,* 75, 127–155.

Goel, T., et al. 2007. Response surface approximation of Pareto optimal front in multi-objective optimization. *Computer Methods in Applied Mechanics and Engineering,* 196(4–6), 879–893.

IACS, 2017. Common Structural Rules for Bulk Carriers and Oil Tankers (CSR BC & OT). International Association of Classification Societies.

Jambrecic, L., 2017. *Muli-objective structural optimization of an oil tanker.* (Master's thesis). University of Zagreb. (In Croatian).

Klanac, A., Ehlers, S. and Jelovica, J. 2009. Optimization of crashworthy marine structures. *Marine Structures,* 22(4), 670–690.

Montgomery, D.C., 2001. *Design and analysis of experiments.* 5th ed. New York; Chichester: Wiley.

Prebeg, P., Rudan, S. and Andric, J., *Evaluation of Surrogate Models of Internal Energy Absorbed by Oil Tanker Structure during Collision.* ed. 15th International Conference on Computer Applications and Information Technology in the Maritime Industries (COMPIT), 2016 Lecce, Italy.

Prebeg, P., Zanic, V. and Vazic, B. 2014. Application of a surrogate modeling to the ship structural design. *Ocean Engineering,* 84, 259–272.

Sacks, J., et al. 1989. Design and Analysis of Computer Experiments. *Statistical Science,* 4(4), 409–423.

Simpson, T.W., et al. 2001. Metamodels for Computer-based Engineering Design: Survey and recommendations. *Engineering with Computers,* 17(2), 129–150.

Wang, G.G. and Shan, S. 2007. Review of Metamodeling Techniques in Support of Engineering Design Optimization. *Journal of Mechanical Design,* 129(4), 370–380.

Zanic, V. 2013. Methods and concepts for the multi-criteria synthesis of ship structures. *Ships and Offshore Structures,* 8(3–4), 225–244.

Zanic, V., Andric, J. and Prebeg, P. 2013. Design synthesis of complex ship structures. *Ships and Offshore Structures,* 8(3–4), 383–403.

Improved ultimate strength prediction for plating under lateral pressure

M.V. Smith, C. Szlatenyi, C. Field & J.T. Knight
Navatek, Ltd., South Kingstown, Rhode Island, USA

ABSTRACT: Semi-analytical membrane stress-based methods for estimating plate ultimate strength offer important benefits to early stage ship design and structural optimization of large stiffened plate structures. Using these methods, it is feasible to rapidly design an efficient and safe structure for new and unique vessels from first principles. Membrane stress methods predict plate collapse by directly solving the governing equations of plate bending under large deflection. These methods are lower fidelity than finite element methods, but with significantly lower setup and computational expense. Combined longitudinal compressive/tension, transverse compressive/tension, edge shear, and lateral pressure loads can all be considered. However, the existing method does not accurately capture the effect of lateral pressure loads. This paper presents an adjustment to the membrane stress method that improves the accuracy of plate collapse prediction for plates subject to combined loads including lateral pressure. The lateral pressure load is treated differently depending on the relative magnitudes of the added deflection due to lateral pressure in the single and critical half-wave number shape. This approach is compared with results from the finite element model program FEMAP with a NASTRAN solver. Accuracy of the membrane stress-based ultimate strength prediction is significantly improved, at minimal computational expense, making it well-suited to set-based ship design and trade space exploration where many variants of a structural design must be analyzed in a short time.

1 INTRODUCTION

Designers are continually developing new arrangements and shapes for ships and offshore structures. Increasingly large design spaces are being explored, often with automated optimization techniques, in the early stage of design (Doerry, et al., 2015). This requires the quick and efficient analysis of many distinctive structures. The results of such exploration are often novel ideas that are at the fringes or completely outside previous designs. Traditional ship and structure shapes are more easily analyzed because they have the benefit of many generations and iterations of real-world structures that can be used as data for regression analyses. However, first-principles approaches, which solve the governing physics, as op-posed to interpolating or extrapolating from previous experience, must be used for these new designs. Finite element analysis (FEA) is an accurate and reliable first-principles approach that is industry standard for structural analysis. However, the user setup and computational cost of FEA can be prohibitive, particularly for early-stage design or exploration of many different concepts. Certain structural information may not even be available in the early stages of design. Consequently, simplified representations of the structure are often analyzed, leaving much of the power of FEA unused. Semi-analytical membrane stress-based methods are a lower-fidelity first-principles approach that can bridge the gap between the high cost, high generality of FEA and the low cost, low generality of other empirical approaches. Thus, membrane-stress methods are increasingly being applied to investigate large unknown design spaces for early-stage design and optimization of stiffened plate structures. This paper focuses on improvements to plate ultimate strength predictions from a membrane-stress method (Paik, Thayamballi, & Kim, 2001; Paik, Thayamballi, & Lee, 2004), and its implications on early-stage ship design and modern design practices such as set-based design.

The remainder of this paper reviews the ultimate strength formulation for the plating between stiffeners using membrane stress-based methods. These methods are explained in detail in <u>Ship Structural Analysis and Design</u> (Hughes & Paik, 2010) and closely paraphrased here. An adjustment to these methods is then presented that increases the accuracy of plate ultimate strength estimation for those plates subject to combined loads including lateral pressure. The paper concludes with a comparison of the method with nonlinear FEA for a selection of steel plates, and the implications for early-stage design are discussed.

2 STRUCTURAL IDEALIZATION

2.1 Geometry and loading conditions

The scope of this paper includes formulations for only the ultimate strength of the plate elements between stiffeners. Figure 1 depicts a plating element subject to combined loads. The plating dimensions are represented by a, b, and t defined as the plate length in x-direction, plate width in the y-direction, and plate thickness; respectively. Compressive or tensile stresses can be applied in the x-direction, σ_{xav}, and y-direction σ_{yav}. Edge shear is defined by τ_{av} and lateral pressure by p.

The edge constraints for the plate make several assumptions based on the plate being one in an arrangement of many plates within a grillage, and therefore interact with one another along the stiffener and frame boundary. The surrounding stiffeners have much higher bending rigidities than the plating, so the out-of-plane bending at the edges is small. Under compression it is common for one plate deflect out of plane and the adjacent plate to deflect out of plane in the opposite direction. In this case the rotational restraint at the edges must be small. Thus, the plate boundary conditions are assumed to be simply supported, and that the edges remain straight to make a conservative estimation.

2.2 Initial imperfections

Grillages are fabricated by welding. The heat of the welding process expands and then contracts material at the welding location. This creates both stresses in the plate and out-of-plane deflections of the plate.

Figure 2 shows the simplified residual stress field in the plate due to welding (Paik, Thayamballi, & Kim, Advanced Ultimate Strength

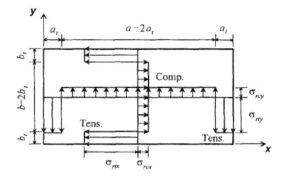

Figure 2. Residual stress field of welded plate. Source: Paik, Thayamballi, & Kim, Advanced Ultimate Strength Formulations for Ship Plating Under Combined Biaxial Compression/Tension, Edge Shear, and Lateral Pressure Loads, 2001.

Formulations for Ship Plating Under Combined Biaxial Compression/Tension, Edge Shear, and Lateral Pressure Loads, 2001). Tensile stresses develop along the edge of the plate at the welding locations. Countering compressive stresses develop on the interior of the plate ensuring static equilibrium. The tensile and compressive residual stresses denoted are σ_{rt} and σ_{rc}, respectively. The length and width of the tensile zone are denoted a_t and b_t.

The resulting initial out-of-plane deflection, w_0, is well represented by the summation of sinusoidal half waves in Equation 1, where m is the half-wave number in the longitudinal direction, and n is the half-wave number in the transverse direction. M and N are the maximum half-wave numbers in the longitudinal and transverse direction.

$$w_0 = \sum_{m=1}^{M}\sum_{n=1}^{N} A_{0mn} \sin\frac{m\pi x}{a} \sin\frac{n\pi y}{b} \qquad (1)$$

3 ULTIMATE STRENGTH ESTIMATION BY MEMBRANE-STRESS METHOD

The formulations in the following sections are closely paraphrased from (Hughes & Paik, 2010) and (Paik, Thayamballi, & Kim, Advanced Ultimate Strength Formulations for Ship Plating Under Combined Biaxial Compression/Tension, Edge Shear, and Lateral Pressure Loads, 2001).

3.1 Governing equations

Equations 2 and 3 are the equilibrium and compatibility equations for plate bending, where F is the stress function and w is the added deflection due to applied load.

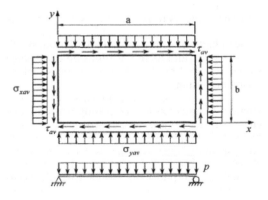

Figure 1. Simply-supported plating subject to combined loads. Source: Paik, Thayamballi, & Kim, Advanced Ultimate Strength Formulations for Ship Plating Under Combined Biaxial Compression/Tension, Edge Shear, and Lateral Pressure Loads, 2001.

$$D\left(\frac{\partial^4 w}{\partial x^4} + 2\frac{\partial^4 w}{\partial x^2 \partial y^2} + \frac{\partial^4 w}{\partial y^4}\right)$$
$$-t\left[\frac{\partial^2 F}{\partial y^2}\frac{\partial^2(w+w_0)}{\partial x^2} - 2\frac{\partial^2 F}{\partial x \partial y}\frac{\partial^2(w+w_0)}{\partial x \partial y}\right. \quad (2)$$
$$\left.+\frac{\partial^2 F}{\partial x^2}\frac{\partial^2(w+w_0)}{\partial y^2} + \frac{p}{t}\right] = 0$$

$$\frac{\partial^4 F}{\partial x^4} + 2\frac{\partial^4 F}{\partial x^2 \partial y^2} + \frac{\partial^4 F}{\partial y^4}$$
$$-E\left[\left(\frac{\partial^2 w}{\partial x \partial y}\right)^2 - \frac{\partial^2 w}{\partial x^2}\frac{\partial^2 w}{\partial y^2}\right. \quad (3)$$
$$\left.+ 2\frac{\partial^2 w_0}{\partial x \partial y}\frac{\partial^2 w}{\partial x \partial y} - \frac{\partial^2 w_0}{\partial x^2}\frac{\partial^2 w}{\partial y^2} - \frac{\partial^2 w}{\partial x^2}\frac{\partial^2 w_0}{\partial y^2}\right] = 0$$

The membrane stresses inside the plate are given by equations 4, 5, and 6.

$$\sigma_x = \frac{\partial^2 F}{\partial y^2} - \frac{Ez}{1-\nu^2}\left[\frac{\partial^2 w}{\partial x^2} + \nu\frac{\partial^2 w}{\partial y^2}\right] \quad (4)$$

$$\sigma_y = \frac{\partial^2 F}{\partial x^2} - \frac{Ez}{1-\nu^2}\left[\frac{\partial^2 w}{\partial y^2} + \nu\frac{\partial^2 w}{\partial x^2}\right] \quad (5)$$

$$\tau = \tau_{xy} = \frac{\partial^2 F}{\partial x \partial y} - \frac{Ez}{2(1+\nu)}\frac{\partial^2 w}{\partial x \partial y} \quad (6)$$

3.2 Membrane stress-based method

The membrane stress-based method, also known as the edge-orientated plastic hinge approach, considers the membrane stress distribution inside the plate. Figure 3 shows the membrane stress field in a square plate subjected to uniaxial compression. Before buckling, the membrane stress is uniform along the loaded edge and zero on the unloaded edge. After buckling, as the plate begins to deflect, the membrane stress in the edges becomes non-uniform. If the unloaded edges are free to move in plane then no stresses will develop in that direction. If the unloaded edges are constrained then a non-uniform stress field will develop.

As the applied load increases, the deflection will increase, eventually causing the membrane stress to exceed the material's yield stress on the outer edges of the plate. The plate will not collapse while it is possible to redistribute the stress field so that the plate edges remain straight. Plate collapse will thus occur when the stress at the boundary exceeds the yield stress of the material. The von Mises yield criteria (Eq. 7) is used to combine the membrane stresses calculated in the x-direction, y-direction, and shear. The plate is assumed to collapse when

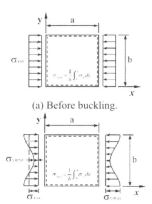

(a) Before buckling.

(b) After buckling, the unloaded edges move freely in plane.

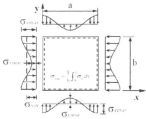

(c) After buckling, the unloaded edges remain straight.

Figure 3. Membrane stress field during plate collapse. Source (Paik, Thayamballi, & Kim, Advanced Ultimate Strength Formulations for Ship Plating Under Combined Biaxial Compression/Tension, Edge Shear, and Lateral Pressure Loads, 2001).

the von Mises stress exceeds yield anywhere on the boundary.

$$\sqrt{\sigma_x^2 - \sigma_x\sigma_y + \sigma_y^2 + 3\tau^2} = \sigma_Y \quad (7)$$

3.3 Membrane stress calculation

The nonlinear governing differential equations can be solved to determine the membrane stresses as follows. Plates subjected to biaxial compression and lateral pressure loads are considered.

Equation 9 is the initial deflection and Equation 10 is the added deflection. In the equations A_{omn} and A_{mn} are the amplitudes of the deflections. The values of m and n are the half-wave numbers in buckling in the x- and y-directions, respectively. The buckling half-wave number is assumed be 1 in the short direction of the plate. So, for a plate with $a/b \geq 1, n = 1$, and m is taken to be a range of integers satisfying the following inequalities, Equation 8, up to two times the aspect ratio:

1. When $\sigma_{xav} < 0$ and $\sigma_{yav} < 0$ (both compressive): $\frac{(m^2/a^2+1/b^2)^2}{m^2/a^2+\sigma_{yav}/\sigma_{xav}b^2} \leq \frac{[(m+1)^2/a^2+1/b^2]^2}{(m+1)^2/a^2+c/b^2}$
2. When $\sigma_{xav} > 0$ (tensile): $m = 1$
3. When $\sigma_{xav} < 0$ and

$$\sigma_{yav} \geq 0 : \frac{a}{b} \leq \sqrt{m(m+1)} \qquad (8)$$

$$w_0 = A_{0mn} \sin\left(\frac{m\pi x}{a}\right) \sin\left(\frac{b\pi y}{b}\right) \qquad (9)$$

$$w = A_{mn} \sin\left(\frac{m\pi x}{a}\right) \sin\left(\frac{b\pi y}{b}\right) \qquad (10)$$

The stress function, F, can be defined as in Equation 11.

$$\begin{aligned}F = & \left(\sigma_{xav} + \sigma_{rx}\right)\frac{y^2}{2} + \left(\sigma_{yav} + \sigma_{ry}\right)\frac{x^2}{2} \\ & + \frac{EA_{mn}(A_{mn} + A_{0mn})}{32} \\ & \left(\frac{n^2 a^2}{m^2 b^2}\cos\frac{2m\pi x}{a} + \frac{m^2 b^2}{n^2 a^2}\cos\frac{2n\pi y}{b}\right)\end{aligned} \qquad (11)$$

The Galerkin method is then applied to solve for the unknown amplitude of deflection. Equation 12 is the resulting third order equation.

$$\begin{aligned}&C_1 A_{mn}^3 + C_2 A_{mn}^2 + C_3 A_{mn} + C_4 = 0 \\ &C_1 = \frac{\pi^2 E}{16}\left(\frac{m^4 b}{a^3} + \frac{n^4 a}{b^3}\right) \\ &C_2 = \frac{3\pi^2 E A_{0mn}}{16}\left(\frac{m^4 b}{a^3} + \frac{n^4 a}{b^3}\right) \\ &C_3 = \frac{\pi^2 E A_{0mn}^2}{8}\left(\frac{m^4 b}{a^3} + \frac{n^4 a}{b^3}\right) \\ &\quad + \frac{m^2 b}{a}\left(\sigma_{xav} + \sigma_{rex}\right) + \frac{n^2 a}{b}\left(\sigma_{yav} + \sigma_{rey}\right) \\ &\quad + \frac{\pi^2 D m^2 n^2}{tab}\left(\frac{mb}{na} + \frac{na}{mb}\right)^2 \\ &C_4 = A_{0m}\left[\frac{m^2 b}{a}\left(\sigma_{xav} + \sigma_{rex}\right) + \frac{n^2 a}{b}\left(\sigma_{yav} + \sigma_{rey}\right)\right] \\ &\quad - \frac{16ab}{\pi^4 t}p \\ &\sigma_{rex} = \sigma_{rcx} + \frac{2}{b}\left(\sigma_{rtx} - \sigma_{rcx}\right)\left(b_t - \frac{b}{2n\pi}\sin\frac{2n\pi b_t}{b}\right) \\ &\sigma_{rey} = \sigma_{rcy} + \frac{2}{a}\left(\sigma_{rty} - \sigma_{rcy}\right)\left(a_t - \frac{a}{2\pi}\sin\frac{2\pi a_t}{a}\right)\end{aligned} \qquad (12)$$

Implicit in this derivation is that the total deflection is the superposition of two sinusoidal components. The component due to lateral pressure has a shape with $m = n = 1$, whereas the component due to axial compression has a shape with m given by Equation 8. The quality of this assumption will be evaluated later. The membrane stresses in the plate are calculated by taking the second partial derivatives in the x- and y-directions of the forcing function (Eqs. 13 and 14).

$$\begin{aligned}\sigma_x = & \frac{\partial^2 F}{\partial y^2} = \sigma_{xav} + \sigma_{rtx} \\ & - \frac{m^2 \pi^2 E A_{mn}(A_{mn} + 2A_{0mn})}{8a^2}\cos\frac{2n\pi x}{b}\end{aligned} \qquad (13)$$

$$\begin{aligned}\sigma_y = & \frac{\partial^2 F}{\partial x^2} = \sigma_{yav} + \sigma_{rty} \\ & - \frac{n^2 \pi^2 E A_{mn}(A_{mn} + 2A_{0mn})}{8b^2}\cos\frac{2m\pi y}{a}\end{aligned} \qquad (14)$$

For plate ultimate strength calculations membrane stresses need only to be calculated along the edges of plates, so the calculations for membrane shear stress have been omitted. The edge conditions chosen ensure that there is no shear stress on the plate edges.

4 COMPARISON WITH FINITE ELEMENT

Ultimate limit strength estimation results from the membrane-stress method of calculation are compared with results from finite element analysis (FEA). The FEA program FEMAP was used with a NASTRAN solver.

The material model used is elastic perfectly plastic with a Young's modulus of 206 GPa and yield stress of 315 MPa. The geometry takes advantage of symmetry across the xz-plane. Symmetry across the yz-plane is not used, as it is not applicable when a pressure load (m = 1) is superposed with a buckled shape containing an even number of half waves. The half plate was modeled as a 0.15 × 1.2 m rectangle. Convergence tests resulted in a reasonable mesh size of 22 × 176 elements. CQUAD4 elements are used as they are most appropriate for this thin plate analysis. Constraints along the short edges are [TX, TZ, RX, RZ] and along the long edge is [TY, TZ, RY, RZ]. These are to represent the simply supported and straight boundary conditions. The symmetry condition [TY, RX, RY] is used along the long edge on the line of symmetry.

Before the ultimate limit state analysis, the flat plate is fixed and deformed in the z-direction according to the initial deflection function (Eq. 9). In this function, m is chosen equal to the plate aspect ratio with $n = 1$. Nodes are then translated by the results of this deformation and the analysis results discarded to eliminate stresses that develop from this deformation step.

A two-step analysis is run using a nonlinear static solver to determine the ultimate limit state. First, a uniform pressure load is applied (when present) over 100 time steps. Next, a displacement is imposed on an edge over 100 time steps. This applied strain method is preferred to determine plate ultimate limit state as it enables post ultimate stresses to be determined, unlike an applied stress method. For each displacement time step, nodal constraint forces are recorded along the edge opposite the applied displacement. The average stress is then computed and a stress-strain or load-shortening curve for the plate can be created. The ultimate limit stress is the maximum stress the plate can support.

4.1 Axial compression

The membrane stress method of ultimate limit strength estimation shows good agreement with estimations from FEA for plates under longitudinal and transverse compression. Table 1 gives the plate and material data. Figure 4 plots the comparison of membrane stress ultimate strength estimation with FEA for a steel plate of aspect ratio four under longitudinal compression, and Figure 5 under transverse compression. Agreement was similarly good for other aspect ratios. A range of plate slenderness ratios from one to five was chosen because this range occurs frequently in marine structures. The slenderness ratio was varied by only changing the plate thickness. Equation 15 is the formula for plate slenderness, β.

$$\beta = \frac{b}{t}\sqrt{\frac{\sigma_Y}{E}} \quad (15)$$

4.2 Combined axial and lateral pressure loads

Plates under both axial compression and a range of lateral pressures were analyzed. These plate loading combinations are important in marine design because structures under load often experience lateral pressure loading from hydrostatics and slamming. The lateral pressure is applied as a constant

Table 1. Plate and material data.

a	1200 mm
b	300 mm
σ_Y	3.15 E + 08 N/m²
E	2.06 E + 11 N/m²
w_{0pl}	0.0015 mm

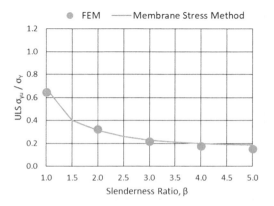

Figure 5. Plating ultimate strength comparison subject to transverse axial compression.

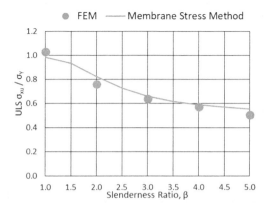

Figure 4. Plating ultimate strength comparison subject to longitudinal axial compression.

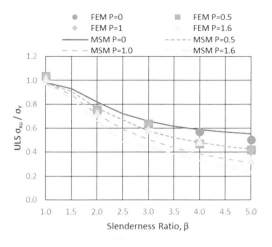

Figure 6. Ultimate strength of plating subject to combined longitudinal axial compression and lateral pressure.

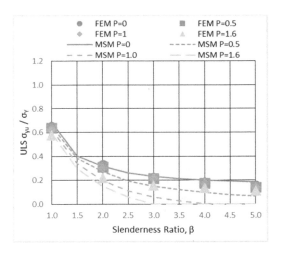

Figure 7. Ultimate strength of plating subject to transverse axial compression and lateral pressure.

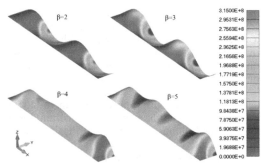

Figure 8. Plate deformation at ULS under longitudinal axial compression and 0.1 MPa lateral pressure.

force in the out-of-plane direction, uniformly spread over the plate. Figures 6 and Figure 7 compare the membrane-stress method calculation for ultimate limit strength with FEA for plates under longitudinal and transverse compression with lateral pressure.

5 IMPROVEMENTS FOR COMBINED AXIAL AND LATERAL PRESSURE LOADS

The membrane stress resulting from combined biaxial compression and lateral pressure loads on plates is poorly approximated for large values of lateral pressure by equations 13 and 14. Further investigation using FEA revealed that the deflected shape function assumed in the membrane stress method does not accurately model the deflections occurring in some slender plates subject to combined axial compression and lateral pressure loads. Figure 8 shows exaggerated images of the deformation of a plate subject to both longitudinal compression and lateral pressure loads. The plates are colored according to the von Mises stress on the top face, red corresponding to areas of stress exceeding the material yield.

The membrane-stress method assumes that the plate deflection has a sinusoidal shape. For plates with low slenderness ($\beta = 1-2$), this is a good assumption. However, at higher slenderness ratios this does not appear to be a good assumption. In these cases, the total added deflection is not well-approximated by the superposition of two sinusoids, as assumed by Equation 12. Instead, the deformation is dominated by an $m = n = 1$ component. The following is a method for better approximating the two different deflection shapes and determining which shape will dominate.

5.1 Prediction of the deflection shape

The membrane stress approach can be improved by first predicting which deflection shape is going to dominate, given the magnitude of the lateral pressure loads and geometry of the plate. To predict which deflection shape will dominate, the added deflection is first calculated subject to the action of lateral pressure alone. Equation 12 is used with $\sigma_{xav} = 0$, $\sigma_{yav} = 0$, $m = n = 1$, and $p = $ lateral pressure load. This deflection will be known as A_{pl}. The initial deflection, A_0, is set equal to zero in this first analysis because there are no significant weld-induced initial imperfections with this mode shape. Then, Equation 12 is used a second time to calculate the deflection due to axial compression alone, setting m according to Equation 8. This deflection will be known as A_{pm}. If $A_{pl} > A_{pm}$, then the ultimate strength of the plate is recalculated using a single half-wave component ($m = 1$), otherwise the calculation remains the same as before. One could consider this to be an attempt to predict 'snap-through' of the plating prior to the membrane-stress analysis.

5.2 FEA comparison

Figures 9 and Figure 10 show the improved comparison for ultimate strength calculation using the updated membrane-stress method accounting for changes in deflection shape due to lateral pressure. With increasing lateral pressure there is a weakening of plate, as previously seen. The magnitude of the ultimate strength from the membrane stress method is now much closer to the magnitude calculated with FEA. The largest slenderness ratio tested showed the most improvement.

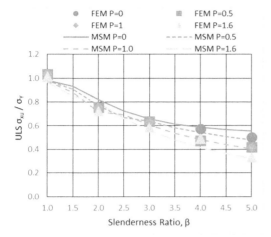

Figure 9. Ultimate strength of plating subject to combined longitudinal axial compression and lateral pressure.

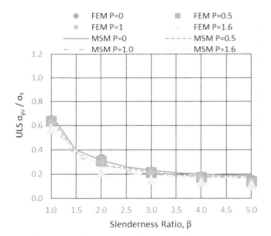

Figure 10. Ultimate strength of plating subject to combined transverse axial compression and lateral pressure.

6 CONCLUSIONS

This advancement in ultimate strength prediction using membrane-stress methods shows that there are still improvements that can be made to the semi-analytic methods. Future work can expand upon this advancement by the expanding the plate shapes, material properties, and load combinations to find further improvements. One area of particular importance is the applicability to the design of aluminum plated structures. Similar membrane-stress based methods, as well as regression models, have been used for the rapid analysis of aluminum structures in the past, with particular focus on the impact of the heat affected zone (Paik & Duran, 2004; Paik, van der Veen, Duran, & Collette, 2005). However, further research is needed into possible coupling between the heat affected zone and lateral pressure loading within the membrane stress formulations presented in this paper.

Computational efficiency can be a deciding factor when choosing between analysis methods in early-stage ship design. This is because modern set-based design practices typically involve the generation of thousands, or perhaps millions, of variations on the ship design. While surrogate modeling techniques are often used to approximate the solution from a computationally-expensive analysis across a wider design space, it is preferable to perform the analysis directly whenever possible. The comparisons presented in this paper demonstrate that a first-principles estimate of plate ultimate strength is possible using membrane stress-based approaches, and that their accuracy is comparable to nonlinear finite element analysis, at least for steel plates. With plating being one of the fundamental building blocks of ship structural strength, such semi-analytical formulations could have a significant impact in design settings where many variations need to be analyzed directly. Computational performance may vary, but preliminary experience suggests that the semi-analytical approach used in this paper is 1500x faster than the nonlinear finite element. The exact computational savings cannot be calculated at this time, and is sensitive to the level of discretization and choice of solver in the finite element model. However, this semi-analytic membrane stress formulation is a promising alternative between finite element analysis and other empirical techniques.

ACKNOWLEDGEMENTS

This work is supported by the Office of Naval Research, PM Kelly Cooper, under Contract Number N00014-15-C-5134 and N00014-17-C-2022.

REFERENCES

Doerry, N., Gerner, M., MacKenna, A., Pearce, F., Bassler, C., Hannapel, S., & McCauley, P. (2015). Concept Exploration Methods for the Small Surface Combattant. *World Maritime Technology Conference.* Providence.

Hughes, O.F., & Paik, J.K. (2010). *Ship Structural Analysis and Design.* Jersey City: The Society of Naval Architects and Marine Engineers.

Knight, J.T. (2017). Rapid, early-stage ultimate limit state structural design for multihulls. *6th International Conference on Marine Structures.* Lisbon, Portugal.

Knight, J., & Piro, D. (2017, October 20). Theory Manual: Early Stage Structural Design of Multihull Vessels. South Kingstown, RI, United States of America.

Paik, J.K., & Duran, A. (2004). Ultimate Strength of Aluminum Plates and Stiffened Panels for Marine Applications. *Marine Technology*, 108–121.

Paik, J.K., Thayamballi, A.K., & Kim, J.B. (2001). Advanced Ultimate Strength Formulations for Ship Plating Under Combined Biaxial Compression/Tension, Edge Shear, and Lateral Pressure Loads. *Marine Technology*, 9–28.

Paik, J.K., Thayamballi, A.K., & Lee, J.M. (2004). Effect of Initial Deflection Shape on the Ultimate Strength Behavior of Welded Steel Plates Under Biaxial Compressive Loads. *Journal of Ship Research*, 45–60.

Paik, J.K., van der Veen, S., Duran, A., & Collette, M. (2005). Ultimate compressive strength design methods of aluminum welded stiffened panel structures for aerospace, marine and land-based applications: A benchmark study. *Thin-Walled Structures*, 1550–1566.

Experimental reproduction of ship accidents in 1:100 scale

M.A.G. Calle & P. Kujala
Marine Technology Research Group, Department of Mechanical Engineering, Aalto University, Espoo, Finland

R.E. Oshiro & M. Alves
Group of Solid Mechanics and Impact Structure, Department of mechatronics and Mechanical Systems, Polytechnic School of the University of Sao Paulo, Sao Paulo, SP, Brazil

ABSTRACT: Scaled experiments are important in the crashworthiness analysis of marine structures since real-scale tests are difficult and expensive to be carried out. However, scaled tests in marine structures considered up to a maximum scale reduction of ten times, still resulting in large-size marine structures tests. In this work, three ship accident scenarios were experimentally modeled in 1:100 reduced scale: frontal collision of an oil tanker with bulbous bow against a rigid structure, ship grounding event caused by a sharp rock and the collision between two identical ships. In general, the resulting kinematic responses and collapse modes observed in the tested marine structures shows a good agreement with that observed in real-scale structures. Additionally, resulting average forces from the 1:100 scaled experiments showed to be within the same order of magnitude with that obtained from real-scale empirical approaches and large-scale experiments after bringing them all to actual-scale by similarity laws.

1 INTRODUCTION

Naval accidents involving collision and grounding are strongly associated with structural damage and they are responsible for over half of all ship losses in worldwide, particularly in areas associated with intense ship traffic. Most of the more catastrophic naval oil spill accidents occurs due to collision and grounding. Besides permanent damage in ship structures, these accidents can potentially lead to various consequences such as oil spill, degradation of the marine environment, explosions, human losses, blocking of ships traffic among others.

For these reasons, there is a continuous interest in improving the ship structures crashworthiness when subjected to collision accident. Through the years, different lines of action were developed to address it, such as simplified analytical formulations, experimental tests and Finite Element (FE) modeling. Nowadays, FE modeling are largely used to model ship collision events. However, some relevant variables to describe ship accident scenarios are more difficult to include numerically than experimentally, for instance, ship cargo (and if it is liquid, granular or solid), influence of surrounding water, harsh environment conditions, pre-damaged structures, explosion among other aspects. Furthermore, FE modeling depends on experimental tests to validate its methods and assumptions.

Performing real-scale tests in marine structures are very expensive. For this reason, there are few literature about collision tests in actual-scale marine structures (Vredeveldt & Wevers, 1992). Actually, most experiments used a dimensional reduction up to ten times (Hagiwara et al., 1983; Amdahl & Kavlie, 1992; Rodd & McCampbell, 1994; Kuroiwa, 1996; Villavicencio et al., 2014). Miniature marine structures using reduction scales higher than 10 times are not normally used to evaluate the ship crashworthiness, unless when making major geometrical simplifications (Tabri et al., 2008). The manufacture of large quantity of complex components in a real ship structure is one of the main difficulties to overcome when reducing drastically the dimensional scale.

However, other strong limitation is the validity of the structural responses when using significant reduction scales of the experiments. Results from collision experiments using reductions scales higher than ten times could potentially not reflect the represented real-scale case as stated by diverse authors (Booth et al., 1983; Drazetic et al., 1994; Schleyer et al., 2004).

In this work, three ship accident scenarios were experimentally reproduced in 1:100 reduced scale using metallic thin-walled structures to evaluate its capability to reproduce structural collapse modes, kinematic and reaction forces. These responses are also compared against predicting empirical formulations and large-scale test responses after bringing them all to real-scale using similarity laws.

2 REDUCED SCALE EXPERIMENTS

A Suezmax oil tanker with 159,900 DWT was chosen as a reference real-scale structure for the reduced-scale experimental modeling. A reduction factor of $\beta = 1:100 = 0.01$ was used for the models.

To manufacture the miniature ship structures, a SAE1008 carbon steel sheet with 0.25 mm thickness and 189 MPa yield stress was used given its mechanical compatibility with actual ship-building plates (Grade A steel).

All metallic parts were manufactured using laser cutting technique, fit together by tight slots and welded by laser in a planned/balanced sequence to avoid geometry distortions induced by heat.

Three main experiments in reduced scale were considered in this work: the frontal collision of the oil tanker against a rigid structure, a grounding accident where the bottom structure of the oil tanker is torn by a sharp rock and a perpendicular collision between two oil tankers in high seas.

All these cases bring as a result permanent structural damage in determined regions of the ship structures. For this reason, it was only conceived for the experiments as metallic structures the structural areas plastically affected by the impact. All tests were performed in dry conditions, i.e., not considering surrounding water in the experimental arrangement.

Most of the data acquisition in these tests was carried out using non-contact photogrammetric techniques. All test events were recorded using a black/white high-speed camera at a rate of 20,000 frames/s. From these recordings, the structure motion was measured via post-processing of digital images. From motion, velocity, acceleration and force data were also evaluated. A color high-speed camera was also used to evaluate qualitative aspects at lower rates around 4,500 frames/s. Doppler laser sensors also supported the tests in order to measure the velocity of the ship structures.

2.1 Head-on ship collision

An accidental frontal collision scenario of an oil tanker against a rigid wall structure, like a bridge abutment or a reinforced floodgate, was here reproduced experimentally in a reduced 1:100 scale. In this experiment, only the ship bulbous bow was manufactured as a metallic thin-walled structure, analogous to actual-size structures. The rest of the ship body was made using glass-fiber and wood and the total ship length was 2.54 m (Fig. 1).

The geometrical simplification of the metallic bow structure was mandatory due to technical constraints. The external of the bulbous bow structure was conceived as the composition of a pyramidal base, a truncated cone and a flat deck. Five main transversal frames, four longitudinal frames and a keel compose the internal bow stiffening. Secondary stiffening structures and small detailed connections were disregarded.

The total mass of the entire ship was set in 63.81 kg. A horizontal slider was used to restrict the ship's movement to the horizontal direction. A vertical plane anvil acts as the target rigid wall structure (Fig. 1). The ship was pushed by a cable

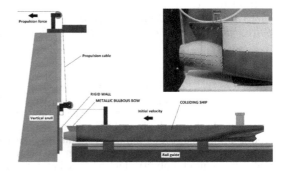

Figure 1. Frontal ship collision test scheme in 1:100 reduced scale and detail of bulbous bow.

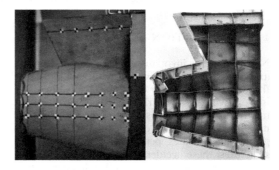

Figure 2. Collapse mode of the bulbous bow: internal and external side views.

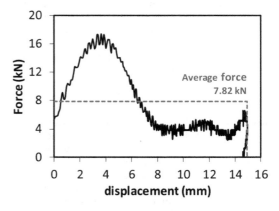

Figure 3. Force-displacement response from miniature frontal ship collision test.

connected to a falling mass in a drop weight rig machine, so achieving an impact velocity of 1.91 m/s. The impact test took about 25 ms.

The main collapse mode observed in the bulbous bow structure was the folding and buckling mechanism of the bulb (Fig. 2). A complete inward folding of the first transversal frame section resulted in a permanent bulb shortening of about 10.3 mm. Minor material rupture was detected in the welded intersection joints as observed in Figure 2.

The force-displacement response shows an initial peak force related with the fold generation (Fig. 3). These force peaks are also observed in crushing test of large-scale bow bulbs (Endo et al., 2002; Yamada & Endo, 2007). The crushing process stops before the generation of a second structural fold.

2.2 Ship rounding

An accidental scenario of ship grounding in which the ship bottom collides against a rock obstacle was reproduced experimentally in reduce 1:100 scale.

In this experiment, only the ship bottom structure was built as a thin-walled metallic structure made of 0.25 thick mild steel sheets. Its geometry consists in a double plate panel structure of $210 \times 230 \times 23$ mm, Figure 4, with four horizontal and four vertical internal web frames uniformly distributed (Calle et al., 2017). Additional secondary stiffening structures and cutouts were disregarded.

The rest of the ship was simplified as a rigid platform with ship-equivalent mass mounted in a horizontal slider in order to constraint its movement in one direction (Fig. 4).

The ship bottom structure was attached in the rigid platform at an angle of 19.8°, Figure 4, in order to mimic the relative horizontal + vertical movements against the seabed obstacle bringing as a consequence a combined structural tearing + penetration damages respectively. The rigid obstacle was represented by a sharp indenter with cylindrical nose placed in the end of the slider.

The total mass of the rigid platform and the ship bottom structure was set in 51.04 kg.

The velocity propulsion of the rigid platform was also induced by the falling mass of a drop weight rig machine so leading to an impact test velocity of 2.47 m/s. The grounding test lasted about 100 ms.

The predominant collapse mode observed in the ship bottom structure was a continuous-progressive longitudinal torn fissure of the external and internal plates with a negligible reduction of the plate thickness (Fig. 5). It can be also observed in Figure 5 a sequential rupture of three internal transversal reinforcements. The remaining cut off material was accumulatively folded and pushed by the indenter (Fig. 5). The resulting torn cracks in both external and internal plates did not show symmetric paths.

Some material rupture in the welding joints was detected, particularly when crushing the last internal transversal frame. Negligible structural influence or damage was detected in the surrounding regions of the fissure.

A gradual increase of the reaction force in the indenter and some peaks were observed in the force-displacement response as shown in Figure 6. The gradual increase of the force is related with the

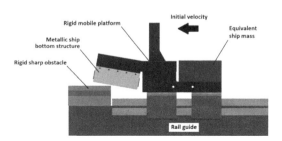

Figure 4. Ship grounding test scheme in 1:100 reduced scale.

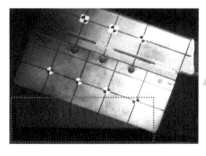  

Figure 5. Collapse mode of ship bottom structure after grounding test: external and internal side views (indenter in dotted line) and low view of torn fissure.

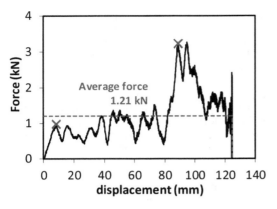

Figure 6. Force-displacement response of ship grounding test.

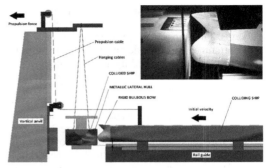

Figure 7. Ship-to-ship collision test arrangement in 1:100 reduced scale.

gradual penetration of the indenter in the external and internal plates of the ship bottom structure. Besides, the force peaks resulted from the sequential rupture of the internal transversal web frames and the external/internal plates (pointed out with an X in Figure 6) as corroborated by test video analysis.

Analogous intermittent force peaks can be also observed in the reaction force measured from large-scale ship grounding tests as reviewed by Zhang (1999).

2.3 *Ship-o-ship collision*

One of the most critical ship collision scenarios in which an oil tanker ship collides perpendicularly the middle hull of an identical ship was reproduced experimentally in reduce 1:100 scale. This scenario was reproduced considering two identical oil tanker ships.

For this experiment, only a section of the target hull structure of the collided ship was built as a thin-walled metallic structure made of 0.25 thick mild steel sheets. In the same way as in the ship grounding experiment, the hull structure also consists in a double plate panel structure ($210 \times 230 \times 23$ mm) with four horizontal and four vertical internal web frames. On the other hand, the bulbous bow of the colliding ship was built as a wooden solid-rigid structure. The rest of the bodies of the two oil tanker structures were manufactured using a fiberglass shell and internal wooden reinforcement structures.

In order to produce the linear movement of the colliding ship, a horizontal slider was employed as a linear guide (alike the miniature head-on ship collision test). The collided ship was set in a standing still position by hanging it by flexible steel cables, Figure 7, so allowing the free motion in the horizontal plane to recreate the friction absence when floating in high seas.

The mass of the colliding ship was 123.3 kg, yet the mass of the collided ship was about 40% heavier in order to replicate the hydrodynamic resistance of the surrounding water to the lateral ship movement induced during the impact. Ship masses were set by distributing sandbags as internal cargo along the bodies of both ships. The mass centers of both ships were aligned with the collision axis in order to obtain pure linear movements of both ships in the collision axis direction with no rotation.

The initial velocity of the colliding ship was set in 4.72 m/s while the collided ship stood still. The final linear velocity of both ships in the collision axis resulted in 1.68 m/s with no rotation. This test took about 45 ms.

The rigid bow penetrated the target lateral hull structure by pushing both external and internal plates. It resulted in a general symmetric collapse mode as seen in Figure 8. Other minor collapse modes can be such as membrane deformation in the external plate, buckling of internal web frames and plate folding in various regions of the structure (Fig. 8). It can be also noticed small cracks in several areas of the structure, generally beside welded joints and intersections.

The absence of edge constraints in the external plate of the hull structure allowed its general folding collapse mode instead a localized plate rupture due to bow penetration as commonly observed in real-scale ship collision accidents.

Figure 9 shows the velocity histories for the colliding and collided ships from the miniature ship-to-ship collision test. Given that not all velocity history could be acquired for the miniature collided ship, this curve was partially extrapolated (dotted line) according to the qualitative video analysis.

Some remarkable kinematic characteristics are clearly observed: a progressive velocity decrease of the colliding ship and velocity increase of

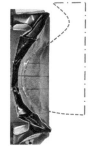

Figure 8. Collapse mode of the lateral ship hull after/during ship-to-ship collision test: external frontal, internal side and external inclined views.

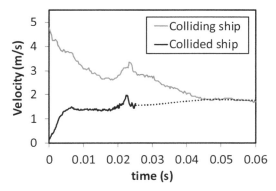

Figure 9. Experimental kinematic response of the ship-to-ship collision test.

llided ship along the entire collision event, a transitory velocity increase of both ships caused by the internal movement/contact of their cargos against the internal ship structure (bulkheads), and a final joint velocity of both ships after crushing.

It is worth mentioning that the internal motion of the cargo produces consequently lower contact force levels and larger crushing displacements than when considering rigid ship bodies as verified numerically by Calle et al. (2017).

3 SIMILARITY

Similarity or similitude is the technique to reproduce a structure behavior in a different scale. Sometimes, it is convenient to replace a prototype (real-scale structure) by a model (scaled structure) for diverse reasons. The correspondence between the structural response of a prototype and its model is given by the scaling factors. These factors are generated by dimensional analysis assuming that both prototype and model have identical mechanical and physical properties.

However, some works had reported lack of similarity in diverse metallic structures subjected to impact tests (Booth et al., 1983; Drazetic et al., 1994; Schleyer et al., 2004). A new set of scaling factors were proposed by Oshiro & Alves (2004, 2009), the so-called VSG-m similarity laws, to specifically scale structures subjected to impact. These VSG-m laws amend this lack of similarity between prototype and model responses caused by the strain rate sensitivity of the materials and by slight differences in their flow stresses' level. Table 1 presents the scaling factors from the traditional similarity approach (MLT) and from the new VSG-m approach.

β is the dimensional scale factor, $\beta = (L)_m/(L)_p$, q is the parameters from the Norton constitutive formulation for strain rate sensitivity of the material given by $\sigma_d = \sigma_0 [\dot{\varepsilon}/\dot{\varepsilon}_0]^q$, and β_{σ_0} is a factor that relates the yield stresses of the model and prototype materials when they are different in the form $\beta_{\sigma_0} = (\sigma_0)_m / (\sigma_0)_p \neq 1$.

3.1 Head-on ship collision

As commented previously, a Suezmax oil tanker with 159,900 DWT was selected as a reference real-scale structure. The reference ship was assumed to be made of a common shipbuilding material, DH-36 steel plate with 25 mm thickness and 332 MPa yield stress (Choung et al., 2013).

The test parameters from the miniature head-on ship collision experiment were converted to their real-scale equivalents, Table 2, using the MLT and the VSG-m scaling laws presented in Table 1. The VSG-m scaled variables were calculated considering $\beta_{\sigma_0} = 189$ MPa/332 MPa $= 0.57$ and the parameter $q = 0.08$ as evaluated by Oshiro et al. (2017).

Hence, for instance, according to the VSG-m similarity laws, the ship velocity for the real-scale head-on ship collision event represented by the miniature experiment carried out in this work

Table 1. Scaling factors.

Variable	Symbol	Scaling factors	
		MLT	VSG-m
Length	β	β	β
Velocity	β_V	1	$\beta_{\sigma_0}^{1/(2-q)} \cdot \beta^{q/(q-2)}$
Mass	β_m	β^3	β^3
Force	β_F	β^2	$\beta_{\sigma_0}^{2/(2-q)} \cdot \beta^{4(1-q)/(2-q)}$
Time	β_t	β	$\beta_{\sigma_0}^{1/(q-2)} \cdot \beta^{2/(2-q)}$
Energy	β_E	β^3	$\beta_{\sigma_0}^{2/(2-q)} \cdot \beta^{(6-5q)/(2-q)}$

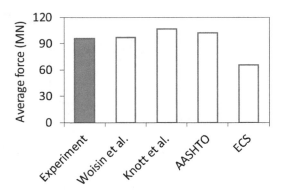

Figure 10. Comparison between average forces obtained from the miniature head-on ship collision experiment and diverse empirical approaches.

Table 2. Scaled real-scale variables from miniature head-on ship collision test.

Variables	Miniature experiment	Real-scale equivalence	
		MLT	VSG-m
Displacement	14.9 mm	1.49 m	1.49 m
Velocity	1.91 m/s	1.91 m/s	2.11 m/s
Mass	63.81 kg	63,810 tons	63,810 tons
Average force	7.82 kN	78.2 MN	95.85 MN
Time	0.02 s	2.0 s	1.81 s
Energy	190.6 J	190.6 MJ	233.5 MJ

Table 3. Scaled real-scale variables from miniature ship grounding test.

Variables	Miniature experiment	Real-scale equivalence (MLT)
Displacement	125 mm	12.5 m
Velocity	2.47 m/s	2.47 m/s
Mass	51.04 kg	51,040 tons
Average force	1.21 kN	12.1 MN
Time	0.097 s	9.7 s
Energy	150.8 J	150.8 MJ

should be 2.11 m/s instead 1.91 m/s as suggested by traditional similarity laws. This increase is caused mainly due to the difference in flow stress levels in the materials for prototype and model.

So, considering the VSG-m similarity approach, the first reduced scale test corresponds to an actual-scale head-on ship collision against a flat rigid structure, such as a pier or bridge abutment, at a velocity of 2.11 m/s (4.1 knots). This ship has a total mass of 63,810 tons, which represents a 40.5% loaded 159,900 DWT ship. The impact test provoked a partial bulb crushing of 1.49 m displacement with an average force of 95.85 MN. The total time spent to crush the bulbous bow structure was about 1.81 s.

The average force evaluated from the miniature head-on ship collision experiment (Fig. 3) is within the same order of magnitude of that evaluated for real-scale head-on ship collision accidents using diverse empirical approaches given by ship safety standards (AASHTO, 1991; ECS, 2002) and ship collision researches (Woisin & Gerlach, 1970; Knott & Bonyun, 1983) as seen in Figure 10. The empirical formulations for these empirical approaches are listed in Appendix A.

3.2 Ship grounding

The second miniature test represents to an actual ship grounding accident in which the ship bottom structure is torn by a sharp rock in the seabed at a total velocity of 2.47 m/s (4.8 knots). The miniature ship mass is equivalent to a Suezmax oil tanker with a total mass of 51,040 tons. The test parameters for its equivalent real-scale ship grounding experiment are obtained using the MLT similarity laws and are listed in Table 3.

The average horizontal reaction force measured from the 1:100 scaled ship grounding experiment was 1.21 kN considering a total grounding displacement of 125 mm. Therefore, using the MLT similarity laws, it corresponds in real-scale to an average force of 12.1 MN and a maximum displacement of 12.5 m.

Next, in order to have an idea about the level of forces in real ship grounding events, the scaled average force is compared against that obtained from large-scale ship grounding experiments. These experiments involves the four large-scale grounding experiments performed by Naval Surface warfare Center US (NSWC) in 1:5 reduced scale (Rodd,

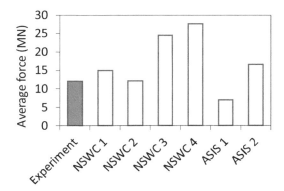

Figure 11. Comparison between average forces obtained from miniature and large-scale ship grounding experiments.

1997) and the two performed by Association of Structural Improvement of Shipbuilding Industry (ASIS) in 1:4 reduced scale (Wang et al., 1997) as reviewed by Zhang (1999).

To make comparable both miniature and large-scale experimental responses, traditional similarity laws were used in both cases to bring them all to actual-scale. The large-scale experiments' data and scaling factors used to bring them to the same scale are detailed in Appendixes B.

The average forces obtained from the miniature and large-scale ship grounding tests are compared in Figure 11. These results seem to be within the same level of forces albeit different test parameters (ship bottom structure design, maximum penetration, plate thickness, material type, number of bulkheads among others) were considered in each one of these experiments.

4 CONCLUSIONS

The aim of this work was to evaluate the feasibility of reproducing experimentally ship collision accidents in a 1:100 reduce scale.

The focus of these experiments was not only to replicate the kinematic response of the ship bodies during the collision, but also the collapse modes of the damaged structures.

In this sense, three noteworthy ship accident events were reproduced in miniature: a head-on ship collision, a ship grounding and a collision between two ships.

Initially, strong limitations were found during the manufacture of the miniature ship structures. The major restriction was the technical unfeasibility to replicate all details of real naval structures due to the restricted access for welding in the interior of small-sized structures and complexity. In order to overcome these issues, a major simplification of the ship structure geometry and a planned assembly procedure were necessary.

In the case of the head-on ship collision and ship grounding tests, all miniature experiments showed complex collapse modes of the ship structures that correspond with damaged structures observed in equivalent large-scale experiments. The same qualitative correspondence can be observed when comparing the reaction forces.

On the other hand, when analyzing the miniature ship-to-ship collision results, the collapse mode resulted to be different from that observed in real-scale accidents because of the different mechanical constraints used for the external hull plate. However, it could be verified in the experiment the strong influence of the cargo sloshing in the global kinematic of the event.

In general, both MLT and VSG-m approaches showed to be suitable to bring miniature tests' responses to their real-scale equivalents so showing, for instance, average forces compatible with that estimated for large-scale experiments by empirical approaches.

Experimental reduced scale modeling showed to be an appropriate method to evaluate the structural response of marine structures when subjected to collision accidents. It can be a suitable technique for improving the structural ship design when subjecting to collision. Furthermore, the main potential advantage in using miniature experiments is to easily introduce complex parameters into the analysis such as structural misalignments, welded joints, surrounding water, diverse types of internal cargo, additional ship components, capsizing risk among others.

However, in order to reproduce the behavior of large-scale structures with suitable precision, it is necessary to carry out more miniature tests to validate the accuracy of this new technique that combines miniature experiments with application of purpose-designed scaling laws.

ACKNOWLEDGEMENTS

This work has been jointly supported by Sao Paulo State Research Foundation (FAPESP, process no. 2011/11733-3), National Council for Scientific and Technological Development (CNPq, process no. 501978/2014-3) and Brazilian Research Funding Agency (FINEP, process no. 01.10.0511.00).

REFERENCES

AASHTO (American Association of State Highway and Transportation Officials), 1991. Guide specification and commentary for vessel collision design and highway bridges. Washington DC, USA.

Amdahl, J., Kavlie, D., 1992. Experimental and numerical simulation of double hull stranding, DNV-MIT Workshop on Mechanics of Ship Collision and Grounding, Oslo.

Booth, E., Collier, D., Miles, J., 1983. Impact scalability of plated steel structures. In: Jones N, Wierzbicki T. Structural Crashworthiness. London: Butterworks; 136–174.

Calle, M.A.G., Oshiro, R.E., Alves, M., 2017. Ship collision and grounding: Scaled experiments and numerical analysis. International Journal of Impact Engineering; 103:195–210.

Choung, J., Nam, W., Lee, J.Y., 2013. Dynamic hardening behaviors of various marine structural steels considering dependencies on strain rate and temperature. Marine Structures; 32:49–67.

Drazetic, P., Ravalard, Y., Dacheux, F., Marguet, B., 1994. Applying non-direct similitude technique to the dynamic bending collapse of rectangular section tubes. International Journal of Impact Engineering; 15:797–814.

ECS (European Committee for Standardization), 2002. Eurocode I—Actions on Structures. Parts 1–7: General actions-Accidental actions due to impact and explosions, 3rd draft. Brussels, Belgium.

Endo, H., Yamada, Y., Kitamura, O., Suzuki, K., 2002. Model test on the collapse strength of the buffer bow structures, Marine Structures, 15:365–381.

Hagiwara, K., Takanabe, H., Kawano, H.A., 1983. Proposed method of predicting ship collision damage. International Journal of Impact Engineering; 1(3):257–279.

Knott, M.A., Bonyun, D., 1983. Ship collision against the sunshine skyway bridge. IABSE Colloquium on Ship Collision with Bridges and Offshore Structures, Copenhagen, Denmark.

Kuroiwa, T., 1996. Numerical Simulation of Actual Collision and Grounding Experiments, International Conference on Design and Methodologies for Collision and Grounding Protection of Ships, USA.

Oshiro, R.E., Alves, M., 2004. Scaling impacted structures. Archive of Applied Mechanics; 74:130–145.

Oshiro, R.E., Alves, M., 2009. Scaling of structures subject to impact loads when using a power law constitutive equation. Int J of Solids and Structures; 46:3412–3421.

Oshiro, R.E., Calle, M.A.G., Mazzariol, L.M., Alves, M., 2017. Experimental study of collision in scaled naval structures. International Journal of Impact Engineering; 110:149–161.

Rodd, J.L., 1997. Frame design effects in the rupture of oil tanks during grounding accidents. International Conference on Advances in Marine Structures III, Dera, Rosyth, UK.

Schleyer, G.K., Hsu, S.S., White, M.D., 2004. Scaling of pulse loaded mild steel plates with different edge restraint. International Journal of Mechanical Sciences; 46:1267–1287.

Tabri, K., Määttänen, J., Ranta, J., 2008. Model-scale experiments of symmetric ship collisions. Journal of Marine Science and Technology; 13:71–84.

Villavicencio, R., Liu, B., Guedes, S.C., 2014. Experimental and numerical analysis of a tanker side panel laterally punched by a knife-edge indenter. Marine Structures; 37:173–202.

Vredeveldt, A.W., Wevers, L.J., 1992. Full-scale ship collision tests, First Joint Conference on marine Safety and Environment/Ship production, Netherlands.

Wang, G., Ohtsubo, H., Liu, D., 1997. A simple method for predicting the grounding strength of ships. Journal of Ship Research; 41(3):241–247.

Woisin, G., Gerlach, W., 1970. On the estimation of forces developed in collisions between ships and offshore lighthouses, IALA Conference, Stockholm, Sweden.

Yamada, Y., Endo, H., 2007. Experimental and numerical study on the collapse strength of the bulbous bow structure in oblique collision, Marine Technology; 45(1):42–53.

Zhang, S., 1999. The Mechanics of Ship Collisions. PhD Thesis, Department of Naval Architecture and Offshore Engineering, Technical University of Denmark.

APPENDIX

A. Evaluation of average forces obtained for head-on ship collision accident.

Reference	Average force formulation	Force (MN)
Miniature experiment	$\bar{F} = \left(1/\delta \int F d\delta\right)/\beta_F \cdot 10^{-6}$	95.85
Woisin & Gerlach	$\bar{F} = V^{2/3} L_{pp}^2 / 1100$	96.78
Knott & Bonyun	$\bar{F} = 0.88\sqrt{DWT}\left(\dfrac{V}{8}\right)^{2/3}\left(\dfrac{D_{act}}{D_{max}}\right)^{1/3}$	106.46
AASHTO	$\bar{F} = 0.122 V \sqrt{DWT}$	102.45
ECS	$\bar{F} = V\sqrt{K(M \cdot 10^3)} \cdot 10^{-6}$	65.42

where $\bar{F}$ is the average force in MN, F is the instantaneous force, δ is the bulb crushing displacement, V is the ship velocity (2.11 m/s), L_{pp} is the ship length (254.2 m), DWT is the ship deadweight tonnage in tons (159,500 tons), D_{act}/D_{max} is the ratio between the vessel displacement at impact time and when fully loaded (40.5%), K is the equivalent stiffness of the ship (15 MN/m for sea-going vessels) and M is the ship mass in tons (63,810 tons).

B.1. Force scaling factors (β_F) for ship grounding experiments.

Reference	Scale	Scaling factor for force
Miniature tests	1:100	10,000
NSWC tests	1:5	25
ASIS tests	1:4	16

B.2. Evaluation of real-scale average forces from ship grounding experiments.

Reference	Energy (MJ)	Displacement (m)	Average force (MN)	Real-scale average force (MN)
Miniature experiment	0.00015	0.125	0.0012	12.10
NSWC-Test 1	3.25	5.46	0.595	14.88
NSWC-Test 2	2.65	5.46	0.485	12.13
NSWC-Test 3	5.34	5.46	0.978	24.45
NSWC-Test 4	6.03	5.46	1.104	27.61
ASIS-Test 1	2.20	5.00	0.440	7.04
ASIS-Test 2	3.10	3.00	1.033	16.53

Hydrodynamic design

Experimental validation of numerical drag prediction of novel spray deflector design

C. Wielgosz, M. Fürth, R. Datla & U. Chung
Stevens Institute of Technology, Hoboken, NJ, USA

A. Rosén
KTH Royal Institute of Technology, Stockholm, Sweden

J. Danielsson
Petestep AB, Stockholm, Sweden

ABSTRACT: Due to their specific hull form, planing craft are vessels that can overcome wave resistance and reach high operational speeds by transforming the hydrodynamic pressure, unfavourable in terms of resistance, into lift. The different resistance components acting on a planing hull have been studied and modelled in the past. By empirical studies, it is known that a noticeable percentage of the total resistance is caused by the whisker spray resistance, hence, devices to decrease its contribution are of great interest. The most common spray deflection devices applied are spray rails. An innovative spray deflection concept, called deflectors, has been introduced and evaluated using numerical simulations. The present paper considers a systematic towing tank experiment for studying the deflectors and validating the numerical study. Results show that the validation of the numerical test case is not easily obtained. Suggestions for improved approaches for experimental investigations of spray deflectors are discussed.

1 INTRODUCTION

Planing craft are characterized by a V-shaped hull form and a transom stern, as illustrated in Figure 1. Their straight buttocks, in conjunction with transom stern and trim angle, make the craft act as a "lifting surface". This allows the craft to rise most of the submerged hull over the water surface and thus diminishing the wetted area and the associated frictional drag as well as the wave making drag, (Savitsky 1985). Additionally, the V-shape and the stern transom favour an effective flow separation at the stern and sides, causing the separation drag to be kept at its minimum.

A high speed craft underway will acquire an equilibrium position that is characterized by an equilibrium trim angle and immersion. This condition will generate three distinct flow areas on the bottom of the hull known as pressure area, spray area and dry frontal area, as showed in Figure 2.

The pressure area, located behind the stagnation line, carries the loads. In this area, the flow is mostly longitudinal with some transverse flow toward the chines. The flow at the stagnation line follows the direction of the stagnation line. The spray area, located between the stagnation line and the spray edge, contributes to the total drag but it

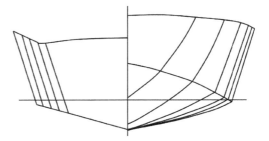

Figure 1. Example of planing craft hull form.

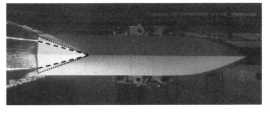

Figure 2. Underwater picture showing the three distinct flow areas of the bottom of a planing craft.

is not considered load-carrying. The direction of the fluid flow in this area is almost a reflection of the incoming flow velocity direction with respect to the stagnation line. The remaining area of the hull is outside the water and in contact with the air. The different flow directions are schematized in Figure 3.

In the original Savitsky method planing craft resistance was characterized by the pressure drag (D_p) and viscous drag (D_v) present in the pressure area (Savitsky 1964). The importance of the viscous drag contribution on the total resistance due to whisker spray was known and studied by Clement (1964). Via towing tank tests, Clement showed the possibility to achieve a reduction in total resistance deflecting the whisker spray laterally by means of spray rails. Later, the contribution of the whisker spray resistance was theoretically modelled, leading to an updated prediction method that includes whisker spray and air resistance, Savitsky et al. (2007). Hence, the predicted total resistance (R_{TOT}) of a planing hull is composed of the hydrodynamic resistance (R_{hd}) (present in the pressure area and the spray area), pressure drag (D_p), viscous drag (D_v) and whisker spray resistance (R_{ws}), and the aerodynamic resistance (R_a) (present in the dry frontal area) as showed schematically in Figure 4.

In addition to the studies on the added resistance caused by the whisker spray, Payne (1982) considered the energy present in the spray sheet. Applying spray rails, which redirect the spray laterally, this energy would be wasted. A new concept to redirect the spray, called deflectors, has been developed by the Swedish company Petestep AB, a schematic representation is presented in Figure 5.

This novel concept redirects the spray flow backwards and downwards, with the aim to recover the energy contained in the spray sheet and generate a reduced wetted area on the bottom, thus causing a diminished resistance. Computational Fluid Dynamic (CFD) studies by Olin et al. (2017) and experimental full scale tests run by the company, show promising results on decreased resistance. The focus of this paper is to represent the numerical setup in an experimental environment, validate the results of said numerical study by comparing experimental values of bare hull and deflectors hull resistance, gain a better understanding of the physical phenomenon undergoing when utilizing the deflectors and study the possibility to scale such phenomenon.

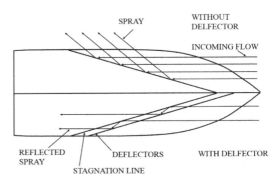

Figure 5. Novel concept of spray deflectors designed by Petestep AB (Olin et al. 2017).

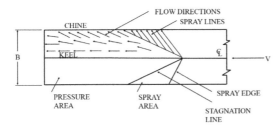

Figure 3. Schematization of the fluid flow directions in the different bottom areas of a planing craft (Savitsky 1964).

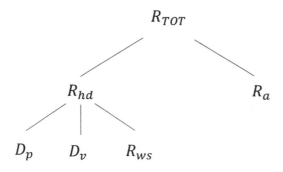

Figure 4. Schematization of the total resistance and its components..

2 EXPERIMENTAL SETUP

The tests are carried out at the Davidson Laboratory, Stevens Institute of Technology, Hoboken, NJ.

2.1 Scaling techniques

The hull evaluated by Olin et al. (2017) is a prismatic hull with its main characteristics showed in Table 1. Since the simulations use a captive method, the chosen physical model needs to be prismatic, with the same deadrise angle and the same equilibrium running conditions. The selected model, denominated M20, is a wooden model, showed in Figure 6, with the main characteristics, overall length L, beam B, height H, deadrise angle β and wetted beam B_{wetted}, shown in Table 1. The model is prismatic, with a slightly higher deadrise angle because the model was readily available in the Davidson Laboratory and it is assumed that the difference in deadrise

Table 1. Main characteristics of full scale boat and tested model M20.

Main characteristics	Full scale	Model M20	Scale
L [m]	7.0	1.26	1:6
B [m]	2.56	0.23	1:14
H [m]	–	0.15	–
β [deg]	18.6	20	–
B_{wetted} [m]	1.563	0.1563	1:10

Table 2. Values kept consistent to achieve similitude between full-scale and model scale.

Quantity	CFD Boat	Model M20
τ [deg]	3.80	3.80
λ [-]	0.848	0.848
C_v [-]	5.25	5.25
V [kn]	40.0	12.6
B_{wetted} [m]	1.563	0.1563

Figure 6. The wooden model, denominated M20, used in the towing tests.

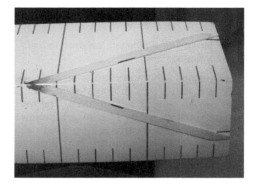

Figure 7. The add-on deflectors applied on the hull of the M20 model, view from above.

Figure 8. The add-on deflectors applied on the hull of the M20 model, view from behind to highlight their tapered design.

angle can be negligible. Moreover, since the full scale and model boat have different hull proportions, the model is slender, it is decided to scale the phenomenon by means of the wetted beam B_{wetted}. The chosen scale is 1:10 so that the condition of dry chines, present in full scale, is achieved.

To have the same equilibrium conditions, in numerical simulations and model testing, is fundamental to allow comparison between the two different methodologies. The similitude is achieved by keeping the trim angle τ, the wetted-length to beam ratio λ and the speed coefficient $C_v = V/(g \cdot B_{wetted})^{1/2}$, a Froude number defined as in the formula, consistent with the CFD simulations. The phenomenon is scaled from full scale to model scale by means of wetted pressure area, obtained by the scaled wetted beam and constant values of τ, λ, and velocity, obtained by applying Froude's law of similitude as suggested by the International Towing Tank Conference 2002. The different values are showed in Table 2.

In addition, the model is fixed in heave and pitch to reproduce the captive method used by Olin et al. (2017). Further, the Longitudinal Centre of Gravity (LCG) position required to reach the equilibrium condition warrant the model to be fixed in heave and trim. In this specific experimental setting and running condition, the LCG position would be instable and not reproducible. In full scale, this LCG position can be obtained by aerodynamic and propulsion effects.

To obtain the comparison between the two model scale configurations, bare hull and deflectors, the deflectors needed to be scaled and produced. The characteristics and position of the deflectors, function of the spray sheet thickness and the location of the stagnation line, are found by scaling the deflectors designed for the numerical simulations with a factor of 1:10. The scaled deflectors are produced in-house from extruded PVC and applied on the hull by means of double sided tape, they are also tapered to the hull by usage of modelling clay as showed in Figure 7. In full scale, the deflectors are built-in the hull, they are not adds-on the hull bottom surface. In Figure 8 it is possible to perceive the design of the deflectors, they have their maximum height at the stern and taper down to zero at their meeting point.

Attention is given to the possibility of scaling spray effects in model scale. When the spray leaves the chines, it has a different behaviour in model scale, it appears as a blister, a thin sheet, rather than droplets as in full scale, but the dimensions of the spray do not vary (Axt 1947). In addition, the beam of the chosen model is consistent with the recommendations given by ITTC (2002) regarding model sizes for spray testing.

2.2 Experimental equipment, calibration and recording system

The Davidson Laboratory hosts a high speed towing tank that is 95.4 m long, 5.00 m wide and 1.98 m (variable up to 2.29 m) deep. The towing apparatus is composed by a lightweight, monorail supported, unmanned carriage that houses signal condition and other equipment. It is possible to obtain a constant towing speed up to 18.3 m/s (~66 km/h) with speed control of 0.003 m/s.

For resistance tests, the International Towing Tank Conference (2002) suggests measuring lift, drag, trim and sinkage while testing. To record lift and drag acting on the model, the Davidson Laboratory linear three-component balance is used. The balance is equipped with three Linear Variable Differential Transformer (LVDT) sensors, measuring drag and forward and aft lift. From forward and aft lift the values for pure lift and pitching moment are obtained. To record heave and trim the Davidson Laboratory heave wheel, LVDT based, and the Schaevitz type linear servo inclinometer are used.

All the instrumentation is statically calibrated and in Table 3 the maximum difference between applied and measured forces for the three-component balance are shown.

The recording system of every signal is as follows; the signal is recorded over a span of 30 m where the running speed is considered constant, the transducer signal is converted and recoded by the 16-bit analog to digital converter and data recorder. The discrete raw signal is averaged by the inhouse DAP51c program and the output value is used to compute the quantity measured by the sensor.

Table 3. Maximum difference between applied and measured forces.

Drag [N]	Lift [N]	PitchM [Nm]
± 0.18	± 0.22	± 0.14

Table 4. Speed matrix for M20 towing tests.

M20 speed matrix		Full scale speed
[kn]	[m/s]	[kn]
11.9	6.1	38
12.4	6.4	39
12.6	**6.5**	**40**
13.0	6.7	41
13.4	6.9	42

2.3 Experimental procedure

The ITTC (2002) suggests collecting at least 15 data points for a resistance test. The Conference also suggests choosing a speed range which extend at least 5% below and above the speeds for which reliable data are required. In this study, only the speed of 40 knots is of interest, 12.6 knots in model scale, used in the numerical simulations.

To follow the recommended guidelines, the model is towed three times for the speed matrix presented in Table 4, for both configurations bare hull and deflectors. The bare hull configuration is run two times for increasing speeds, sequences of runs denominated "Bare hull 1" and "Bare hull 3" in the results, and once for decreasing speeds, sequence denominated "Bare hull 2". Deflectors configuration is run twice for decreasing speeds, "Deflectors 1" and "Deflectors 3" sequences, and once for increasing speeds, "Deflectors 2" sequence.

The instrumentation is zeroed at the beginning of the tests for deflector configuration and again zeroed before the bare hull configuration is tested. Between each run an adequate interval of time is taken to let the water in the tank reach again calm state. Those actions are taken to comply with other recommendations given by the ITTC (2002).

3 RESULTS AND ANALYSIS

3.1 Experimental results

Lift, resistance, pitch moment, trim and heave are recorded for the two different configurations, bare hull and deflectors, for the aforementioned speed matrix (from 6.1 m/s to 6.9 m/s) for three test sequences. The recorded values for bare hull resistance and lift are showed in Figure 9 and Figure 10 and for deflectors resistance and lift in Figure 11 and Figure 12. The values for pitch moment are not presented since not relevant for this study.

The bare hull results show good repeatability with negligible differences between the measurements for the three different series.

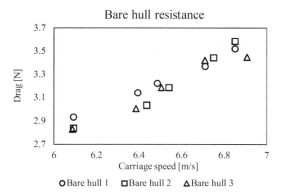

Figure 9. Resistance results for bare hull configuration.

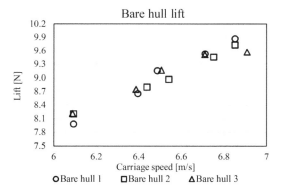

Figure 10. Lift results for bare hull configuration.

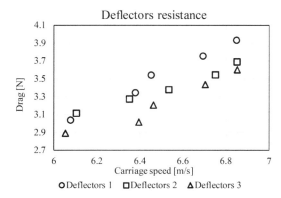

Figure 11. Resistance results for deflectors configuration.

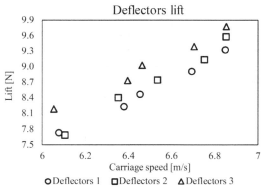

Figure 12. Lift results for deflectors configuration.

magnitude of 0.57 N, around 6.5% of the lift measurement for the same speed, as showed in Table 5.

The causes of the low repeatability of the tests are:

- The small size of the model resulting in small forces that cannot be correctly measured. The possible difference between drag values, due to the calibration of the balance, recall Table 3, can be ± 0.18 N, a value that it is not negligible and of the same order of magnitude as the recorded quantities.
- The dimensions of the model coupled with the condition of the dry chines make the model skim on the water causing noticeable vibrations of the model while running, vibrations that can interfere with the recording.
- The vibrations are amplified by having the model clamped in heave and trim, since the possibility of damping is reduced.

For clarity the average values of lift and drag, for both configurations, are presented in Table 6.

Moreover, the values for sinkage and trim angle are constantly monitored, they should be constant since the model is clamped. The maximum difference in heave and trim values for the two configurations are reported in Table 7.

3.2 *Theoretical analysis of the different drag components*

A theoretical analysis of the physical model bare hull resistance components for the design speed of 12.6 knots is carried out to compare with the experimental results. To verify if the contribution of the resistance components in full scale is comparable with their contribution in model scale, the same analysis is completed for the full scale bare hull configuration. This aspect it is interesting since in the two different scales the flow states of

The deflectors results are however rather scattered with a maximum difference in drag magnitude of 0.33 N, roughly 10% of the lowest drag measure for that speed; and a difference in lift

Table 5. Drag and lift for deflectors configuration at the running speed of 6.5 m/s.

Runs' set	Drag [N]	Lift [N]
Set 1	3.54	8.46
Set 2	3.37	8.73
Set 3	3.21	9.03

Table 6. Average values of drag and lift for bare hull and deflectors configurations.

Speed [m/s]	Bare hull averaged results		Deflectors averaged results	
	Drag [N]	Lift [N]	Drag [N]	Lift [N]
6.1	2.86	8.13	3.01	7.86
6.4	3.06	8.73	3.21	8.45
6.5	**3.20**	**9.09**	**3.38**	**8.74**
6.7	3.41	9.50	3.58	9.14
6.9	3.52	9.75	3.74	9.56

Table 7. Maximum differences of heave and trim values during tests.

Maximum difference in:	Bare hull	Deflectors
Heave [cm]	0.140	0.109
Trim angle [deg]	0.088	0.085

Table 8. Summary of the contribution of each resistance component for bare hull configuration in model and full scale determined using Savitsky's method (1964, 2007).

Resistance	Full scale	% of tot	Model M20	% of tot
D_p [N]	879.3	32%	0.9408	24%
D_v [N]	1115	41%	2.2121	58%
R_{hd} [N]	1994.3	73%	3.1529	82%
R_{ws} [N]	465.5	17%	0.3995	10%
R_a [N]	271.3	10%	0.2998	8%
R_{TOT} [N]	2731.1	100%	3.8522	100%

the whisker spray area are different, in model scale laminar flow is predominant, in full scale, transitional flow is present. The different components of the resistance are determined using Savitsky's method (1964, 2007) and the obtained values are presented in Table 8.

Looking at the results from the theoretical analysis it is noticeable that the theoretical prediction for the total resistance of the model overestimates the actual resistance found by experimental testing. The theoretical total resistance is 3.85 N while the experimental resistance, on average at the design speed of 6.5 m/s, is 3.20 N, as showed in Table 6. Analysing the different components of the total resistance, the experimental resistance can almost be fully predicted by the hydrodynamic resistance component of the theoretical model.

Focusing on the contribution of each resistance component in full scale and model scale, as showed in Table 8, it can be observed that:

- Full scale hydrodynamic resistance (R_{hd}) contributes to 73% of the total resistance, of which 41% it is caused by the viscous resistance (D_v) and 32% by the pressure resistance (D_p).
- Model scale hydrodynamic resistance (R_{hd}) has a contribution of 82% to the total resistance, of which 57% is produced by the viscous resistance (D_v) and just 24% by the pressure resistance (D_p).
- The hydrodynamic resistance has a greater contribution to the total resistance in model scale and it is mainly generated by viscous drag. This difference it is caused by the different Reynolds numbers; in model scale the Reynolds number is 8.68×10^5, for the design speed of 12.6 knots, while in full scale the Reynolds number is 2.65×10^7 for the corresponding design speed of 40 knots. This discrepancy in Reynolds numbers generates different values for the frictional coefficient for fully turbulent flow $C_f = 0.075/(\log_{10}(Re)-2)^2$ that leads to this consistent difference of contribution by pressure and viscous resistances in full and model scale.
- Full scale whisker spray resistance (R_{ws}) has a contribution of 17% to the total resistance, while in model scale its contribution is only 10%. This significant difference is justified by the different state of the flow in the two different scales, laminar flow in model scale and transitional flow in full scale.
- Air resistance (R_a) contribution is similar in both scales, there is a 10% contribution in full scale and an 8% contribution in model scale. Air resistance is not affected by scaling effects, at least for these design speeds and theoretical model (Savitsky 2007).

The main differences between resistance contributions are present in the viscous drag and whisker spray drag. These two drag components are functions of the Reynolds number, hence of the flow state present in the wetted area. For this setup, it seems that is not possible to scale the phenomenon down to model scale. The greater contribution of the viscous drag in model scale, compared to full scale, decreases noticeably the contribution of the whisker spray, a central aspect in this validation. It is important to specify that the theoretical formula to compute the whisker spray resistance is the state of the art but not a standardized methodology (Savitsky 2007). Its validation does not consider the difference between the Reynolds numbers of the whisker spray for the two different scales, assuming the same friction coefficient in the spray area.

4 CONCLUSIONS AND RECOMMENDATIONS

4.1 Conclusions

The validation of a numerical study, Olin et al. (2017), of a novel spray deflectors design is carried out by systematic towing tank experiments. The full scale model is scaled to model scale and two different hull configurations, bare hull and deflectors, are tested in three sequences of the chosen speed matrix.

Unfortunately, from the lack of repeatability in the experimental results it is clear that the size of the tested model is too small and that the set up applied in the numerical simulations is not feasible for physical evaluations. The reduced size of the model generates forces of small magnitude that can be affected by the three component balance calibration error. Moreover, its size coupled with the dry chines condition generate high vibrations that are amplified by the model being clamped in heave and trim. Hence, for this specific setup and running condition no benefit is gained by model scale experimental testing. If modifications to the test setup do not lead to consistent results, further full scale experimental tests are required.

The theoretical results show that the difference between full scale and model scale contribution to the total resistance by the whisker spray resistance are noticeably different, with a 10% contribution in model scale and a 17% contribution in full scale. The difference is caused by the different flow state in the whisker spray area. Thus, further studies into scaling effects of whisker spray resistance is recommended.

4.2 Recommendations

If similar tests are to be performed, it is recommended to use a larger model and to change the testing condition. A bigger model would:

- Increase the magnitude of the forces recorded.
- Have a bigger wetted hull bottom area that could reduce the vibrations.
- Possibly lead to a state of the flow in the whisker spray area closer to transitional flow, increasing the whisker spray resistance contribution to the model scale total resistance.

If increasing the model size does not achieve a better repeatability, the suggestion would be to change running condition to obtain:

- A more feasible position of LCG so that the model will achieve the equilibrium heave and trim angle required during the test by free run, decreasing all the vibrations due to the clamped model.
- A wetted chines configuration that will generate a bigger wetted area, leading to a more stable running condition and thus decreasing the vibrations in the system.

After facing these challenges, the planned continuation for the project is to build a model ex novo and to choose a running condition that will allow the model to be free in heave and trim. With such a running condition it will be also possible to test the model in waves and to study the behaviour of the deflectors in rough seas.

REFERENCES

Axt W.C. 1947. The effect of scale, surface tension and acceleration on the main spray characteristics of geometrically similar flying-boat hull models. *Davidson laboratory technical note no. 59*.

Clement E.P. 1964. Reduction of planning boat resistance by deflection of whisker spray. *Technical report, David Taylor model basin reports* AD0454407.

International Towing Tank Conference 2002. Recommended procedures and guidelines: Testing and extrapolation methods, High speed marine vehicles, Resistance test. 7.5-02-05-01.

Olin L. et al. 2017. Numerical modelling of spray sheet deflection on planning hulls. *Journal of Engineering for the Maritime Environment*. Vol. 231, Issue 4, pp. 811–817.

Payne P.R. 1982. The spray volume shed by an uncambered planning hull in steady planning. *Ocean Engineering* Vol. 9. No. 4, pp. 373–384.

Savitsky D 1964. Hydrodynamic design of planning hulls. *Mar Technol*. 1(1): 71–95.

Savitsky D. 1985. Planing Craft. *Naval Engineering Journal* 97:113–141.

Savitsky D. et al. 2007. Inclusion of whisker spray drag in performance prediction method for high-speed planing hulls. *Mar Technol*. 44(1): 35–56.

Experimental and numerical study of sloshing and swirling behaviors in partially loaded membrane LNG tanks

M. Arai & T. Yoshida
Yokohama National University, Yokohama, Japan

H. Ando
Monohakobi Technology Institute, Tokyo, Japan

ABSTRACT: It is well known that partial loading in the membrane LNG tanks may lead to a violent sloshing phenomenon and it can cause structural damages to the tanks. In this study, in order to understand the basic characters of sloshing, we carried out model experiments and measured the liquid global forces, sloshing pressures and liquid motion in the model tanks. In the model experiments, a liquid rotating motion in the tank (swirling) was observed in some test cases. Swirling phenomenon has not been discussed in detail in membrane tanks, however, in some cases we observed that the liquid motion became very violent. We, therefore, studied membrane tanks with different tank-length-to-tank-breadth ratio and examined the conditions of the occurrence of swirling. We also examined the sloshing loads in irregular waves and investigated the relationship between sloshing loads and ship's motion response spectra. A practical formula that predicts the sloshing load was obtained. The results obtained from this study are considered to be useful for not only the designers but also for the operators of LNG carriers.

1 INTRODUCTION

Due to the growing demand for natural gas in the world, numbers and size of Liquefied Natural Gas (LNG) Carriers are increasing. Also there appeared new topics related to the LNG cargo transportation such as the application of LNG carriers to the shuttle transportation of the produced natural gas from the Floating LNG (FLNG), adoption of LNG as the fuel for ships, etc. In these new transportation modes, partial filling in the LNG tank is inevitable. However, violent sloshing phenomena may occur in the partially loaded membrane LNG tanks, and it can cause structural damages to the tank. In this study, in order to understand the characters of sloshing, we carried out model experiments and measured the liquid global forces, sloshing pressures and liquid motion in the model tanks. In the model experiments, a liquid rotating motion in the tank, i.e., swirling, was observed in some test cases. Swirling phenomenon has been studied by some researchers but those researches were basically for rectangular tanks. For example, Faltinsen et al. 2005 utilized multi-modal approach and analyzed three-dimensional liquid motion in a square-base tank. Vaziri et al. 2015 used a fully non-linear potential theory and studied sloshing in four rectangular tanks with different length-to-breadth ratio. In Chen et al. 2010, a time dependent Finite Difference Method was used to simulate liquid sloshing in a rectangular tank. However, swirling phenomenon has not been discussed in detail for the prismatic tanks which have the shape of membrane tanks of LNG carriers. Also above mentioned studies were carried out with regular excitation. We, therefore, studied membrane tanks with different tank-length-to-tank-breadth ratio and examined the conditions of the occurrence of swirling. We also studied the sloshing loads in irregular waves and investigated the relationship between sloshing loads and ship's motion response spectra. A practical formula that estimates the sloshing load was obtained, and we applied it to predict the sloshing load in the tanks of an LNG carrier which navigates in the North Pacific Ocean. The results obtained from this study are considered to be useful for not only to the designers but also for the operators of LNG carriers.

2 MODEL EXPERIMENTS

2.1 Experimental setup

A series of model experiments were carried out at Monohakobi Technology Institute (MTI) Yokohama Laboratory, as shown in Figure 1. A 1:40 scale tank mounted on a moving table was used (Figure 1, left, Tank-L). The Tank-L is a scale model of a membrane tank with length × breadth × depth dimensions of 971 × 952 × 689 mm. Ten Kyowa

PS-05KD pressure gauges were placed along the walls (see Figure 2). The moving table was excited with regular and irregular sway motions. A smaller model tank of 1:68.75 scale was also used to measure the transverse and longitudinal forces acting on the tank (Figure 1, right, Tank-S) by using a two-directional load cell (see Figure 3). The internal shapes of the two tanks were exactly the same, and only the tank size was different.

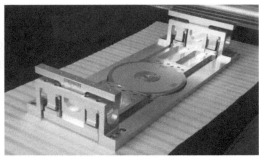

Figure 3. Load cell.

Figure 1. Model tank mounted on a moving table.

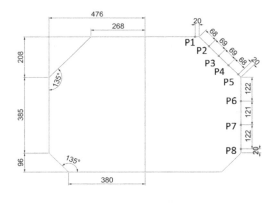

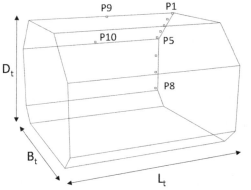

Figure 2. Model tank dimensions and pressure gauge locations (Tank-L, units in mm).

2.2 Regular excitation experiments

Sway excitation was applied to the large tank in the tank's lateral direction at its sloshing natural frequency while changing the filling level to 30, 50, 70, 80 and 90%.

Simple linear theory may be used for estimating the sloshing natural frequency of a tank having a rectangular cross sectional shape. The natural frequency of the tank when assuming a rectangular cross section is given by the following equation:

$$f_n = \frac{1}{2\pi}\sqrt{\frac{g\pi}{b}\tanh\frac{\pi h}{b}} \quad (1)$$

where, h is the liquid height in the tank, b is the tank width, and g is the gravitational acceleration.

The actual membrane LNG tank has an octagonal cross section with chamfers where the tank width is narrow at top and bottom parts of the tank. Therefore, when the static liquid surface is in the vicinity of the tank ceiling and the tank bottom, the width of the free surface which is narrowed by the chamfers (i.e., $b=B'$) is used as the value of b in the equation (1) (see Figure 4).

As shown in Figure 4, the predicted natural frequencies of the chamfered tank by equation (1) agree well with the experimental and numerical values obtained from the preliminary studies carried out by Yokohama National University.

2.3 Irregular excitation experiments

Several ISSC-type wave spectra were prepared for the irregular excitation tests. Then the irregular excitation time series were created using the sway motion response amplitude operator (RAO) of an LNG carrier. Table 1 shows the tested irregular excitation conditions, where H_s is the significant wave height and T_{mean} is the mean wave period.

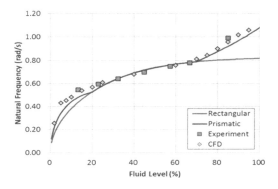

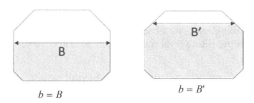

$b = B$ $b = B'$

Figure 4. Estimation of 1st mode natural frequency.

Table 1. Irregular test conditions.

Case	Filling level [%]	H_S [m]	T_{mean} [s]	Duration [s]
1	50	0.072 (4.95)	1.510 (12.52)	1200 (9950)
2	50	0.144 (9.9)	1.510 (12.52)	1200 (9950)
3	50	0.08572 (5.89)	1.152 (9.55)	1200 (9950)
4	50	0.144 (9.9)	1.124 (9.32)	1200 (9950)

Note: Values in parentheses indicate actual scale ones.

3 NUMERICAL METHOD

The numerical method used in the analysis of sloshing was based on the finite difference technique developed by the authors (Arai et al. 2002, 2006a, 2006b) for partial load conditions. The numerical method is outlined below.

The flow field in the tank is assumed to be three-dimensional, and the liquid is assumed to be incom-pressible and inviscid. The governing equations in the Cartesian coordinate system (x, y, z) fixed to the tank are the mass continuity equation:

$$\frac{\partial u}{\partial x} + \frac{\partial v}{\partial y} + \frac{\partial w}{\partial z} = 0 \qquad (2)$$

and Euler's equations of motion:

$$\frac{\partial u}{\partial t} + u\frac{\partial u}{\partial x} + v\frac{\partial u}{\partial y} + w\frac{\partial u}{\partial z} = -\frac{1}{\rho}\cdot\frac{\partial p}{\partial x} + f_x \qquad (3)$$

$$\frac{\partial v}{\partial t} + u\frac{\partial v}{\partial x} + v\frac{\partial v}{\partial y} + w\frac{\partial v}{\partial z} = -\frac{1}{\rho}\cdot\frac{\partial p}{\partial y} + f_y \qquad (4)$$

$$\frac{\partial w}{\partial t} + u\frac{\partial w}{\partial x} + v\frac{\partial w}{\partial y} + w\frac{\partial w}{\partial z} = -\frac{1}{\rho}\cdot\frac{\partial p}{\partial z} + f_z \qquad (5)$$

in which the variables are defined as follows:
u, v, w: components of the velocity in x, y and z directions, respectively,
ρ: liquid density,
p: pressure,
f_x, f_y, f_z: external forces in x, y and z directions, respectively.

Change of the free surface of the liquid over time was calculated by the following equation:

$$\frac{\partial H}{\partial t} + u\frac{\partial H}{\partial x} + v\frac{\partial H}{\partial y} = w \qquad (6)$$

where $H = H(t,x,y)$ is the vertical position of the free surface. This simplified expression of the free surface cannot treat the extremely local deformation of the free surface such as spray. However, the present method has the excellent feature of numerically stable computation and can also simulate the violent flow in the tank, including the impact of the liquid on the tank ceiling.

4 COMPARISON AND DISCUSSION OF MODEL TEST AND NUMERICAL RESULTS

4.1 Sloshing pressure

Figure 5 compares the measured and computed pressure time histories of 50% filling case. Regular excitation was given and the pressure at P5 (lower part of the chamfer (see Figure 2) and at P8 (lower part of the side wall) were shown.

4.2 Hydrodynamic force on tank wall and effect of tank size

Figure 6 shows an example of the computed force time histories. In Figure 6, the magnitude of the force is non-dimensionally expressed using: liquid density (ρ), gravitational acceleration (g), tank length (L_t), tank breadth (B_t) and liquid height (h). Regular sway motion was applied to the tank. F_y is the computed lateral force component, and F_x is the longitudinal force component. As an initial condition of the numerical simulation, the free surface was given a small inclination (i.e., 1 degree) to generate the swirling motion. This small initial disturbance was necessary since, without the disturbance, it was difficult to generate the swirling

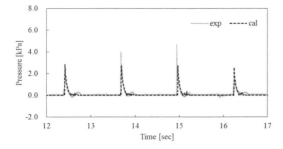

(a) P5

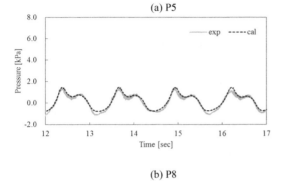

(b) P8

Figure 5. Pressure histories.

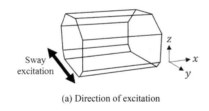

(a) Direction of excitation

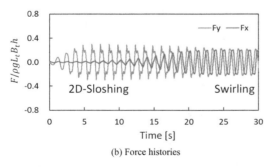

(b) Force histories

Figure 6. Direction of excitation and force histories.

motion in the numerical simulation. At the beginning of the simulation, only F_y was generated as shown in Figure 6, but after a period of time, swirling motion started, and the amplitude of F_x increased to be almost the same as that of F_y. From this figure, it can be said that the ratio F_x/F_y can be used as an index with which to judge the occurrence of swirling.

As presented in Figure 6, the occurrence of swirling was easily distinguished by the emergence of the longitudinal force (F_x). In this study, the tank breadth (B_t = 952 mm) was kept constant and the tank length was changed for several cases, and the computed results were examined. The tanks were excited with a regular sway excitation with a 20-mm amplitude at the natural sloshing period. Figure 7 summarizes the computed F_x/F_y ratio for tanks with different L_t/B_t ratios for two conditions, 30% and 50% filling levels, respectively. It can be clearly seen that the swirling occurred with high intensity when the L_t/B_t ratio was near 1.0. For both cases, swirling occurred in the range $0.9 < L_t/B_t < 1.10$. As for the filling level, we observed that swirling occurred when the filling level was between 30% and 70%.

4.3 *Comparison between experimental results and numerical results of irregular excitation experiments*

In order to study the sloshing response in irregular seaways, we excited the model tank (Tank-S) on a

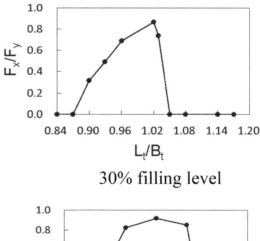

30% filling level

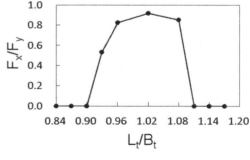

50% filling level

Figure 7. Computed F_x/F_y for different L_t/B_t ratios.

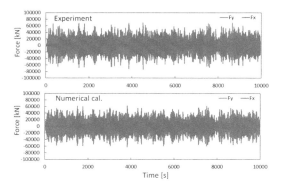

Figure 8. Comparison of measured and computed force histories (50% filling, irregular sway motion, $H_{1/3}$ = 5.89 m, T_{mean} = 9.55 s on actual ship scale).

moving table with irregular sway motions. The time histories of the tank's sway motion were prepared using the typical wave spectrum data as shown in Table 1 and the ship's sway motion RAO (Response Amplitude Operator). Examples of the measured and computed time histories of F_x and F_y are shown in Figure 8. The force and time scales in Figure 8 are converted from the model scale to the actual ship scale using Froude's law.

Although, the same time history of the tank motion was used in the model experiment and the numerical simulation, the measured and computed force histories did not perfectly match each other. One possible reason for this is the difference in the rotating direction, i.e., clockwise or anti-clockwise, of the liquid motion in the tank. As observed in the model experiments with regular excitation, a small disturbance in the flow can trigger the swirling and determine the direction of the rotating motion. Incidentally, either clockwise or anti-clockwise motion may start. The swirling motion couples with the sloshing motion, and the accumulated liquid motion response changes the liquid motion time history afterward. Therefore, it was very difficult to numerically simulate the time histories of the force under irregular excitation. However, in some parts of the force time histories, very similar patterns were reproduced by our numerical computation. One example of such a case is shown in Figure 9 and Figure 10. A part of the time histories presented in Figure 8 is shown with an enlarged time scale in Figure 9 (F_y) and Figure 10 (F_x). As shown in Figure 9, the computed F_y, the lateral force, agrees well with the measured one in this time period (Figure 9(a)). On the other hand, the computed F_x, the longitudinal force, did not match the measured one (Figure 10(b)). As shown in Figure 10(c), however, if we reverse the sign of the force it agrees

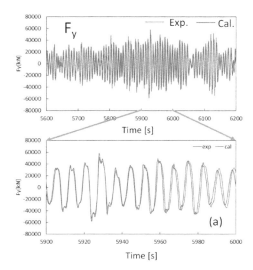

Figure 9. Comparison of measured and computed force histories of F_y (part of Figure 8 is enlarged in time, irregular sway motion, $H_{1/3}$ = 5.89 m, T_{mean} = 9.55 s on actual ship scale).

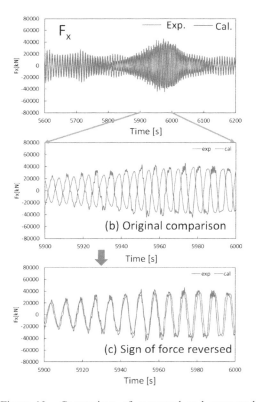

Figure 10. Comparison of measured and computed force histories of F_x (part of Figure 8 is enlarged in time, irregular sway motion, $H_{1/3}$ = 5.89 m, T_{mean} = 9.55 s on actual ship scale).

quite well with the measured data. This means that the rotating direction of the numerical simulation (Figure 10(b)) was opposite that of the measured one.

4.4 Effect of L_t/B_t and size of actual tanks

Sloshing responses in irregular excitation for the tank with different L_t/B_t were studied. The same irregular excitation used for the case shown in Figure 8 was applied. Figure 11 presents the results and again it can be seen that the swirling occurred with high intensity when the L_t/B_t ratio was near 1.0. For these cases, swirling occurred in the range $0.9 < L_t/B_t < 1.10$.

In order to examine the L_t/B_t ratio of the actual LNG tanks, we collected the tank size information of 14 constructed or designed membrane-type LNG carriers, and 6 of them had tanks that met the above-mentioned criterion (i.e., $0.9 < L_t/B_t < 1.10$). Table 2 presents the tank dimensions of those ships. If the liquid cargo is partially loaded in the tanks with the tank length to tank breadth ratio near 1.0, swirling is expected to occur.

5 SLOSHING LOAD ESTIMATION IN ACTUAL SEAS

As shown in Section 4.3, measured F_y histories in irregular waves can be reasonably reproduced by the numerical method. Figure 12 presents the return period of the peak value of F_y where zero-up cross method was applied to obtain the peak value of F_y. In Figure 12, numerical values by two dimensional (2D) and three dimensional (3D) computations are compared with experimental ones. In this study, 2D computation means that the mesh division in the longitudinal direction of the

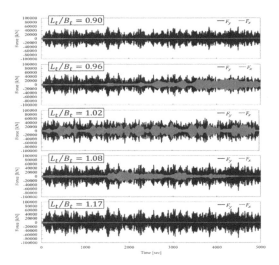

Figure 11. Force histories by irregular excitation for different L_t / B_t ratios.

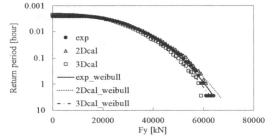

Figure 12. Return period (50% filling, $H_{1/3} = 5.89$ m, $T_{mean} = 9.55$ s on actual ship scale).

Table 2. Tank L_t/B_t ratios of actual or designed ships.

Ship	NO. 4 Tank			No. 3 Tank			No. 2 Tank			No. 1 Tank		
	L_t (m)	B_t (m)	L_t/B_t	L_t (m)	B_t (m)	L_t/B_t	L_t (m)	B_t (m)	L_t/B_t	L_t (m)	B_t (m)	L_t/B_t
1	46.05	42.65	1.080	46.05	42.65	*1.080	46.05	42.65	*1.080	31.09	36.53	0.851
2	46.05	40.31	1.142	46.05	40.31	1.142	46.05	40.31	1.142	35.50	36.80	*0.965
3	49.60	49.90	*0.994	49.60	49.90	*0.994	49.60	49.90	*0.994	39.87	40.57	*0.983
4	47.07	41.63	1.131	47.07	41.63	1.131	47.07	41.63	1.131	33.81	32.11	*1.053
5	38.28	37.81	*1.012	43.58	37.81	1.153	43.89	37.81	1.161	Wedge Shape		
6	40.00	37.81	*1.058	44.75	37.81	1.184	44.75	37.81	1.183	31.45	33.75	*0.932

*Indicates the tank with $0.90 < L_t/B_t < 1.1$.

tank is not considered and 2D flow in the tank's transverse section is solved. Therefore, swirling phenomenon is not reproduced by the 2D computation. Figure 12 shows a certain value of peak F_y emerges once in a certain "return period". By fitting the distribution function of the Weibull distribution to the measured and computed values, we estimated the maximum peak value of F_y in three hours. Since there is a tendency that the F_y in the swirling phase is a little smaller than that in the sloshing phase, the results of 2D computation becomes slightly larger than the measured and 3D computation values. However, the differences between them are not significant, and we can conclude that the measured responses of F_y by irregular excitation can be reproduced by both 2D and 3D computations.

In order to obtain the relationship between the tank excitation and the generated sloshing load, the numbers of the peak values in the time histories of F_y was counted by zero-up cross method. Results of the counted peak numbers are shown in Table 3.

As shown in Table 3, the number of peaks obtained by the F_y histories do not agree with the number of excitations (i.e., the number of encounter waves). On the other hand, they are close to the number of peaks estimated by the natural period of sloshing motion of the tank (T_N). It seems that the liquid sloshing numbers in the tank by the irregular tank excitation is not determined by the tank excitation but is determined by the natural period of sloshing. This may lead to the possibility that the component of the irregular excitation which

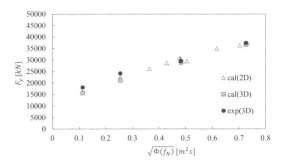

Figure 14. Correlation between F_y and $\sqrt{\Phi(f_N)}$.

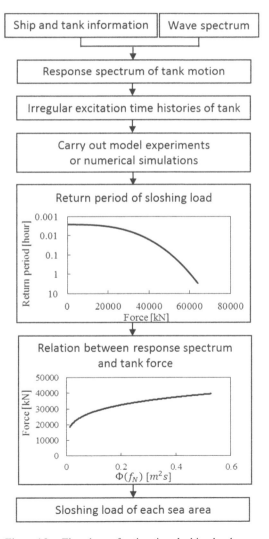

Figure 15. Flowchart of estimating sloshing load.

Table 3. Number of sloshing.

Cash	Number of peak observed	Number of excitation	(Duration of exp.)/T_N	3 hour maximum [kN]
1	1539	1047	1284	39209
2	1492	684		52376
3	1394	906		64099
4	1365	1099		81096

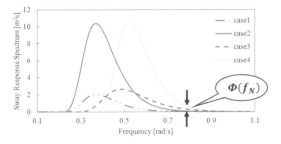

Figure 13. Sway response spectra of small tank (full scale).

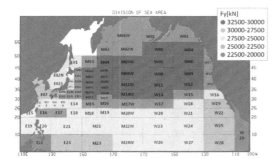

Figure 16. Distribution of F_y in the North Pacific Ocean in winter.

has the sloshing natural period dominates the phenomenon. Then let us focus on the magnitude of the tank sway motion spectrum at the sloshing natural frequency (see $\Phi(f_N)$ of Figure 13).

Figure 14 indicates the relation between the maximum F_y values in three hours and the $\sqrt{\Phi(f_N)}$. Figure 14 suggests that there is a strong relation between $\sqrt{\Phi(f_N)}$ and F_y. By applying the least squares method to the data shown in Figure 14, we obtained a formula that can be used to estimate the global sloshing load to the tank's side wall as:

$$F_y = 3.3 \times 10^4 \sqrt{\Phi(f_N)} + 1.4 \times 10^4 \qquad (7)$$

Note that the equation (7) is adjusted to be used for the actual scale LNG tank considering the density of the liquefied natural gas.

Figure 15 shows the flow to estimate the sloshing load in a certain sea area. As an example of sloshing load estimation based on the flow presented in Figure 15, 3-hour maximum value of the global sloshing load exerted on the tank side wall was obtained for each of the sea area in the North Pacific Ocean in winter (see Figure 16). Although we need to further advance the method to include all of the six ship motion components other than only sway motion, we can grasp the general tendency of the level of sloshing severity in the ocean by Figure 16. For example, the area between Japan and the US in the north part of the Pacific Ocean is the harsh area from a sloshing point of view.

6 CONCLUSIONS

In this study, the basic characters of sloshing in membrane LNG tanks were studied. From this study, we obtained the following results:

- Swirling occurs if the membrane tanks have the tank length to tank breadth ratio between 0.9–1.1 and filling level between 30%-70%. There exist a considerable number of LNG carriers which have such tank length to tank breadth ratio.
- Although swirling load is in general smaller than the sloshing one, it should be noted that the load distribution pattern of swirling differs from that of sloshing.
- Global tank load can be evaluated reasonably by our numerical method.
- We proposed a flow to estimate the sloshing load using the wave spectrum and the RAO of the ship.
- By using the proposed method, global sloshing load exerted on the tank side wall was obtained for each of the sea area in the North Pacific Ocean in winter.

REFERENCES

Arai, M., Cheng, L.Y., Kumano, A., Miyamoto, T. 2002. A technique for stable numerical computation of hydrodynamic impact pressure in sloshing simulation, *Journal of Society of Naval Architects of Japan*, Vol.191, 299–307.

Arai, M., Makiyama, H.S., Cheng, L.Y., Kumano, A., Ando, T. 2006. Numerical and experimental study of 3-D sloshing in tanks of LNG carriers, *Proceedings of OMAE 2006*, Paper No. OMAE 2006-92235.

Arai, M., Makiyama, H.S., Cheng, L.Y., Kumano, A., Ando, T., Imakita, A. 2006. Numerical analysis of 3-D sloshing in tanks of membrane-type LNG carriers, *ICSOT, Design, Construction & Operation of Natural Gas Carriers & Offshore Systems*, 201–209.

Chen B.F., Wu C.H. 2010. Effects of excitation angle and coupled heave-surge-sway motion on fluid sloshing in a three-dimensional tank, *Journal of Marine Science and Technology*, Vol. 16, 22–50.

Faltinsen O.M., Rognebakke O.F., Timokha, A.N. 2005. Classification of three-dimensional nonlinear sloshing in a square-base tank with finite depth, *Journal of Fluids and Structures*, Vol. 20, 81–103.

Karuka, G.M., Arai, M., Ando, H. 2017. Sloshing and swirling in partially loaded prismatic chamfered tanks, *Proceedings of OMAE 2017*, Paper No. OMAE 2017-61562.

Vaziri N., Chern M.J., Borthwick A.G.L. 2015. Effects of base aspect ratio on transient resonant fluid sloshing in a rectangular tank: a numerical study, *Ocean Engineering*, Vol. 105, 112–124.

A numerical trim methodology study for the Kriso container ship with bulbous bow form variation

M. Maasch, E. Shivachev, A.H. Day & O. Turan

Department of Naval Architecture, Ocean and Marine Engineering, University of Strathclyde, Glasgow, UK

ABSTRACT: The application of Computational Fluid Dynamics (CFD) is the fastest developing area in marine fluid dynamics as an alternative to Experimental Fluid Dynamics (EFD). While EFD employs well-established methods for predicting a ship's performance, CFD is still challenged to reach a reliable level of accuracy in a reasonable amount of time. In the present study, this issue is addressed in the context of trim optimization by exploring the combination of time-inexpensive potential flow simulations with high-fidelity Unsteady Reynolds-Averaged Navier-Stokes (URANS) simulations. This approach allowed covering a broad fore body design space by running a large number of potential flow simulations while at the same time important flow effects due to viscous forces were included by running URANS simulations over the full speed range for a small set of simulations. The KCS baseline design results were validated against an experimental towing tank dataset ensuring a valid CFD setup and thus demonstrating its competitiveness to EFD.

1 INTRODUCTION

Fluctuating fuel prices and the newly adopted mandatory measures by the International Maritime Organization (IMO) to reduce emissions have been driving the shipping industry to become more efficient. Energy efficiency is becoming an integral part of ship design. Ship hull forms are traditionally designed to perform best for one operating condition (design speed and design draft); however, cargo ships often operate in off-design conditions.

One of the methods to improve the hydrodynamic performance of ships when sailing at a speed different to the design speed or in adverse loading conditions is to operate the ship at a trim angle. This allows bringing certain ship hull geometry features, such as the bulbous bow, the stern bulb or the transom back into the design position (in reference to the design conditions). The potential of further improving the energy efficiency of ships when operating in trimmed conditions could be investigated by optimising those hull parts.

Changing the bulbous bow shape in order to adapt its design to the adverse operating conditions is a challenging task, as one needs to make sure that the new geometry also works in the design conditions. The new design should result in a compromise that works better on average over a realistic set of loading conditions than the original hull. Ideally, the hull geometry change should also be restricted to a small part of the ship so that the new design can be applied as a retrofit option. Ship hull optimization is a complex and important aspect of ship design. The available scope of ship design optimization largely extended with the use of numerical tools, both for Computer Aided Design (CAD) and Computational Fluid Dynamics (CFD). Various marine software packages include some functionality to alter the shape of the numerical hull surface representation such as points-based modification of standard surface models (e.g. NURBS surfaces) or parametric geometry objects. The overall goal of ship geometry optimization is to improve the operational performance, often targeting on a low fuel consumption. An indication of an improved performance can be given by the ship hull resistance reduction. Ship resistance was traditionally predicted by towing tank experiments only. However, with the rapid developments in computer technology, numerical ship hull design became widely used and nowadays assists or even replaces experimental towing tests. Still, numerical results are compared to the experimental data if available for validation purposes.

Both topics, numerical trim optimization and numerical bulbous bow optimization, independently, have already been investigated within various studies using the KCS and other ship geometries. Filip et al (2014) presented a bulbous bow retrofit analysis for the KCS container ship under slow steaming conditions using Reynolds-averaged Navier-Stokes Equations (RANSE) simulations for a small number of design variants, certainly limited by the extensive simulation time. Wagner et al

(2014) carried out a scenario-based optimization of the KCS bulbous bow for four different operating conditions at various speeds and drafts using a potential flow solver. This allowed creating a high number of design variants; however, Wagner et al further suggested to include sophisticated RANSE simulations for better results accuracy and to validate the potential solver results. Vroegrijk et al (2015) performed a full-scale bulbous bow optimization on a container ship by using a combination of potential flow and RANSE simulations. For the ship at different drafts and speeds, the results showed that the potential flow simulations were not able to accurately predict the performance trend between different bulbous bow variants. Hence, Vroegrijk et al concluded that only RANSE simulations should be used within in a bulbous bow optimization study. FORCE Technology (Reichel et al., 2014) performed an extensive series of experimental trim model tests for different ship types. This study concluded that the change in trim mostly affected the wave making resistance component of the total resistance. As mentioned before, by trimming the ship, the bulbous bow and other energy saving geometry features are rotated back into an ideal operational point. This conclusion allows suggesting that a potential flow solver could be well suited for a trim study, as it can quickly estimate the wave making resistance.

Following the above review of previous studies, this paper presents a trim optimization study in combination with a bulbous bow retrofit for both design and adverse operating conditions using the benchmark Kriso Container Ship (KCS).

The performance of the KCS was calculated using potential flow simulations as well as URANS simulations. Furthermore, simulation results were validated by experimental tests.

2 TRIM STUDY METHODOLOGY

Based on the Kriso Container Ship (KCS) benchmark hull, the setup of this trim study consisted of three stages. At stage 1, the bulbous bow of the KCS hull geometry was partially parametrised within the parametric CAD modeller of the software CAESES by FRIENDSHIP Systems. At stage 2, the original KCS and 39 KCS bulbous designs were simulated in a virtual towing tank using the marine flow code Shipflow (SHF) by Flowtech AB for steady ship hydrodynamics. At stage 3, based on the wave making resistance at seven trim angles and three operational speeds, predicted by the potential flow code, one of the best design candidates was chosen to be simulated using the RANSE code STAR-CCM+ by SIEMENS. Furthermore, the original KCS was also simulated using the RANSE code to perform a comparison to the experimental results. Figure 1 shows the structure for the proposed methodology of this trim study.

The shown strategy fulfilled two purposes. First, the most beneficial trim angle for three operational speeds could be derived for the original KCS and validated by experimental tests. Second, it could be estimated how much the KCS performance would benefit from a bulbous bow retrofit for different speeds. Therefore, the total resistance difference $\Delta R_{TM}^{v,\theta}$ of each design to the original KCS was calculated at a given speed and trim angle (Eq. 1).

$$\Delta R_{TM}^{v\theta} = R_{TM,Original}^{v\theta} - R_{TM,Modified}^{v\theta} \quad (1)$$

The operational profile was defined by weighting factors t_{vi} (Eq. 2), representing the time spent in each speed.

$$\sum_{0}^{i} t_{vi} = 1 \quad (2)$$

Finally, the total resistance improvement $\Delta R_{TM}^{Profile}$ for a specific operational profile (Eq. 3) could be derived.

$$\Delta R_{TM}^{Profile} = \Delta R_{TM}^{v1} t_{v1} + \Delta R_{TM}^{v2} t_{v2} + \Delta R_{TM}^{v3} t_{v3} \quad (3)$$

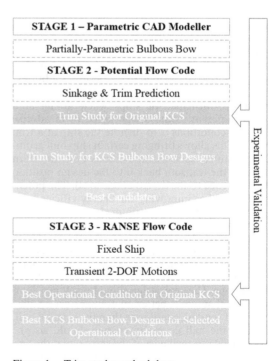

Figure 1. Trim study methodology.

2.1 Parametric computer aided design of the KCS

The partially parametric KCS geometry model is based on the publicly available IGES file (http://www.simman2008.dk/KCS/kcs_geometry.htm) with the specifications given in Table 1. In order to allow a feasible variation of the KCS bulbous bow shape, a partially parametric modelling approach based on the deformation of image objects of the KCS hull was chosen. Although not as flexible as the fully parametric modelling approach, the partially parametric modelling approach is well suited for the modification of local geometry features of an already existing geometry. To make this study an example of a potential bulbous bow retrofit the area of shape modification was restricted to the bulb and a small part of the underwater fore ship. Figure 2 illustrates the KCS in starboard view. The area of modification is highlighted in golden colour.

Table 1 lists the KCS model properties for full scale and model scale. The model scale geometry of the original KCS hull was built for experimental testing in the Kelvin Hydrodynamics Laboratory at the University of Strathclyde.

Three geometry shift functions were applied to the bulbous bow, which allowed changing its length (dx), width (dy) and tip height (dz). Figure 3 shows an example of each shift function independently by comparing the original KCS fore ship (grey colour) with the modified shape (golden colour). Care was taken that the geometry modification had no effect on the fair transition of the bulbous bow into the fore ship shape. The geometry setup was

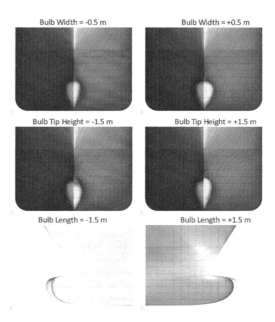

Figure 3. Geometry shifts of the KCS bulbous bow.

then coupled with the variation algorithm SOBOL (available in CAESES) that quasi-randomly created 40 variants within the chosen boundaries.

2.2 Scope of operational conditions

The present trim study comprises of 40 bulbous bow design variants (including the original KCS design), simulated in seven different trim angles for three different speeds. The trim of a floating ship is defined as the difference in forward draft T_f and aft draft T_a, and can be expressed as distance t in unit meter (Eq. 4) or as angle θ in unit degree (Eq. 5).

$$t = T_f - T_a \qquad (4)$$

$$tan(\theta) = \frac{T_f - T_a}{L_{pp}} \qquad (5)$$

The displacement of the KCS was kept constant throughout the whole study. For specifying the range of trim angles, only those cases were considered that would allow an operation in self-propulsion conditions with the propulsor sufficiently submerged. At a constant displacement, higher bow-down trim angles would cause the propulsor to get closer to the free surface, resulting in a loss in operating performance. Hence, the propeller tip clearance, i.e. the distance from the top propeller blade tip in top position to the undisturbed water level in relation to the propeller diameter, was

Table 1. KCS hull properties.

Dimensions	Full scale	Model scale
Scale	1	75
LPP (m)	230	3.0667
BWL (m)	32.2	0.4293
D (m)	19	0.2533
T (m)	10.8	0.144
Displacement (m³)	52030	0.1203
S w/o rudder (m²)	9530	1.651
CB	0.651	0.651
CM	0.985	0.985

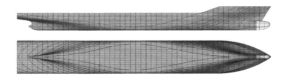

Figure 2. KCS hull geometry (starboard and bottom-up view).

monitored. As a threshold, a propeller tip clearance of 15% of propeller diameter was set which yielded a critical bow down trim angle of $\theta = 0.75°$ (see Figure 4).

The bow-up trim angle range is often not limited by the propeller tip clearance as the propulsor moves away from the free surface. Hence, the trim angle range was kept variable. During the computation of KCS in bow-up conditions the analysis yielded an increase in total resistance which led to the decision to restrict the bow-up trim angle range to −0.75° as no further improvement was expected. Thus, the final trim angle range was set to ±0.75°.

The three chosen operational speeds represented the KCS slow-steaming speed of 18 knots (v_1), a medium speed of 21 knots (v_2) and the KCS design speed of 24 knots (v_3), here given as full-scale speeds. Defined by the above stated limits the set of performed simulations consisted of 21 operational conditions for 40 KCS designs. The created results pool served as basis to derive a performance trend of the KCS for different operating profiles. The below points summarise the scope of the present trim study:

- 40 different hull variants were created (including the original KCS)
- Those variants were used to perform potential flow code simulations
- At three speeds at $Fn = 0.195, 0.227, 0.269$
- At seven trim angles from $\theta = -0.75°$ to $0.75°$ in steps of $\theta = 0.25°$
- The geometry of the original KCS and the best bulbous bow design candidate was used to perform URANS simulations
- Numerical results for the original KCS geometry were validated against experimental data

2.3 *Experimental data*

Prior to the presented numerical trim study, experimental tests were performed for the KCS model for various operational conditions. The tests were carried out in the Kelvin Hydrodynamics Laboratory of the University of Strathclyde. The experimental setup and the results were presented by Shivachev (2017). For this study, the results were further post-processed by calculating the non-dimensional resistance coefficients for a corrected water temperature of 15°C following procedures proposed by the ITTC (ITTC 7.5-02-02-01). For the measured fresh water temperature that defined the water density ρ_M and kinematic viscosity v_M, the monitored total resistance force of the KCS model R_{TM} at a carriage speed v_M was converted to its non-dimensional total resistance coefficient C_{TM} considering the hydrostatic wetted surface S_M (Eq. 6).

$$C_{TM} = \frac{R_{TM}}{0.5 \rho_M S_M v_M^2} \quad (6)$$

The frictional resistance coefficient $c_{FM,ITTC}$ was calculated by the ITTC-1957 frictional correlation line (Eq. 8) for the model Reynolds number Re_M (Eq. 7), considering the hydrostatic water line length $L_{M,WL}$.

$$Re_M = \frac{v_M L_{M,WL}}{v_M} \quad (7)$$

$$C_{FM,ITTC} = \frac{0.075}{(\log Re_M - 2)^2} \quad (8)$$

The wetted surface (see Eq. 6) and the waterline length of the KCS (see Eq. 7) varied for each calculated trim angle as shown in Figure 5.

The residuary resistance coefficient C_{RM}, which was assumed to equate to the wave making resistance coefficient $c_{RM} = c_W$ (Eq. 9), was defined as difference of the total resistance coefficient and the frictional resistance coefficient. The form factor k was determined by the performed Prohaska model tests at level trim to $(1+k) = 1.0118$.

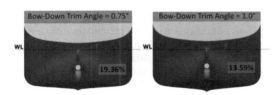

Figure 4. KCS propeller tip clearance (as percentage of propeller diameter).

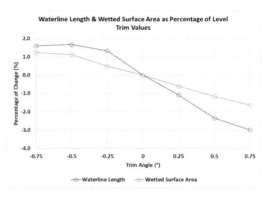

Figure 5. Variation of waterline length and wetted surface due to trim.

Table 2. Experimental total resistance at 15°C.

Trim (°)	Total resistance (N) at 15°C		
	Fn = 0.195	Fn = 0.227	Fn = 0.269
−0.6	4.444	5.856	7.759
−0.25	4.232	5.651	7.659
0	4.166	5.607	7.512
0.25	4.114	5.478	7.427
0.6	4.142	5.568	7.632

$$c_W = c_{TM} - c_{FM,ITTC}(1+k) \qquad (9)$$

In order to correct the viscous effects of the measured results to a water temperature of 15°C, the frictional resistance coefficient $c_{FM,ITTC}^{15°C}$ (Eq. 11) was re-calculated by considering the model Reynolds Number $Re_M^{15°C}$ (Eq. 10) for the kinematic viscosity $\nu_M^{15°C}$ at 15°C.

$$Re_M^{15°C} = \frac{v_M L_{M,WL}}{\nu_M^{15°C}} \qquad (10)$$

$$c_{FM,ITTC}^{15°C} = \frac{0.075}{\left(\log Re_M^{15°C} - 2\right)^2} \qquad (11)$$

This yielded the corrected total resistance coefficient $c_{TM}^{15°C}$ (Eq. 12) and finally the corrected total resistance $R_{TM}^{15°C}$ at 15°C (Eq. 13).

$$c_{TM}^{15°C} = c_{FM,ITTC}^{15°C}(1+k) + c_W \qquad (12)$$

$$R_{TM}^{15°C} = 0.5 \, c_{TM}^{15°C} \rho_M S_M v_M^2 \qquad (13)$$

The temperature corrected total resistance of the experiments for the measured trim angles at the three speeds is given in Table 2.

The total resistance was used to validate both the potential flow and the URANS simulations.

2.4 Numerical simulation setups

The potential code simulations (stage 2) were performed using a Rankine panel code to calculate the wave making resistance coefficient c_W from transverse wave cuts on the disturbed free surface (Flowtech International AB, 2017). A panel mesh study was performed for the original KCS hull geometry in level trim conditions for three speeds. Table 3 shows the resolution of each part mesh for a coarse mesh, a medium mesh and a fine mesh setup.

Figure 6 illustrates the panel mesh setup for the KCS hull parts and the free surface.

The results for the wave making resistance coefficient c_W, computed on three meshes of different

Table 3. Panel mesh resolution for mesh study.

Mesh	KCS hull					
	Total	Aft	Boss	Main	Bulb	Free surface
Coarse	**3603**	70	25	792	70	2646
Medium	**8996**	192	40	1995	204	6605
Fine	**13904**	300	50	3024	315	10215

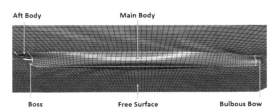

Figure 6. Panel mesh visualisation (coarse).

density, were used to calculate the total resistance (Eq. 12) which was then compared to the experimental measurements. Although the total resistance predicted by the panel mesh code converged towards the experimental results, the error of around 5% was considered too high. The SHF wave cut method gave an accurate trend prediction for the wave making resistance coefficient compared to the experiments, but it failed to calculate the coefficient within the correct range of magnitude. The observed magnitude offset is a known effect of the wave cut method (Flowtech User Support, 2017). Given that the same mesh setup, i.e. free surface panel mesh resolution for the wave cut calculations, was used throughout all simulations, the c_W-magnitude offset was considered to be similar for each calculated simulation result.

Since a rather accurate prediction of the impact of a bulbous variation on the total ship resistance was an important target of this study, the coefficient was corrected in magnitude by a value of $c_{W,Correction} = 3.5 \cdot 10^{-4}$ for each calculated variant. The result of the correction for the original KCS geometry at slow steaming speed is shown in Figure 7. The applied correction allowed recalculating the results for the mesh study, which led to a difference of under 1% compared to the experiments for the fine panel mesh.

The convergence trend of the mesh study for three speeds is shown in Figure 8. The fine panel mesh setup was chosen for the trim study. In total, 840 potential flow simulations were carried out, covering the original KCS geometry and 39 bulbous bow designs.

The RANSE code simulations (stage 3) were used to compute the total resistance of the KCS

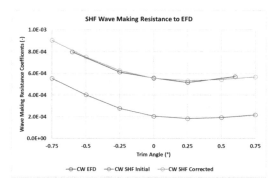

Figure 7. Wave making resistance coefficient correction.

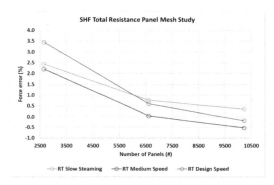

Figure 8. Percentage error of total resistance for three speeds predicted on three panel mesh resolutions.

directly. The numerical mesh and the solver setup summarised below were already discussed in Shivachev (2017) and remained unchanged.

- Model scale at 1/75 to original size of the KCS
- Hexahedral mesh of around 0.6 million cells in a cuboid domain
- Half domain setup with the centre plane (symmetry plane) along the ship keel line
- Local mesh refinements around the KCS hull and on the free surface to provide a higher resolution of local flow phenomena
- All-y+ treatment with a target of y+ > 30
- RANSE solver with a k-ε turbulence model
- Implicit unsteady 1st-order time model with a time step of $\Delta t = 0.01 L_{M,WL} / v_M$, following the recommendations of the ITTC (2011)
- Volume of Fluid model to capture the free surface interface between two immiscible fluids, i.e. water and air
- 2-DOF dynamic ship motions were resolved using the Dynamic Fluid Body Interaction (DFBI) module

The original KCS geometry was simulated at design speed over the chosen trim angle range with and without an active DFBI model, i.e. with a fixed hull and the hull free for sinkage and trim. Therefore, the simulations were started with a fixed ship. After the convergence of the total resistance, the DFBI motion model was activated. Again, the simulation results, i.e. the total resistance and the dynamic sinkage and trim motions, were allowed to converge. The same approach was followed for the best KCS bulbous bow design, which was simulated for all three speeds over the chosen trim angle range in order to allow for a performance trend analysis. In total, 28 CFD simulations were carried out.

3 NUMERICAL RESULTS

As outlined in Section 2, the numerical results pool, consisting of 868 simulations, was first assessed regarding the performance of the original KCS geometry (Section 3.1). Therefore, the numerical results were compared against the experimental measurements for each speed and trim angle, thus proving a valid numerical setup. The results of the potential flow simulations (SHF), i.e. wave profiles, the wave making resistance and the total resistance, were analysed in order to explain the differences in performance over the trim angle range. Finally, the RANSE simulation results were compared to the potential flow code results and to the experimental measurements.

Section 3.2 presents the results pool assessment for the KCS bulbous bow designs in comparison to the original KCS geometry for each trim angle. By assessing the impact of the bulbous design change on the wave making resistance, the minimization of energy losses associated with the generation of waves could be estimated. Then, the same approach was chosen to evaluate the total resistance reduction. Finally, the best KCS bulbous design was simulated using the RANSE code to check the validity of the potential flow simulations and to include the impact of viscous forces.

3.1 Original KCS geometry

The results assessment for the original KCS geometry was performed to derive the most beneficial trim angle and to validate the numerical results with the experimental data. Therefore, the first part of the results pool evaluation focused on the total resistance variation. Figure 9 shows a direct comparison of the numerical and the experimental results over the trim angle range for three speeds. It can be noted that the trend and the magnitude of the numerical results agrees well with the experiments. For both data sets a bow-down trim of $\theta = 0.25°$ seems to be the most beneficial condition.

The slight difference in the results originated from the corrected wave making resistance prediction of the panel code as the frictional resistance component, based on the ITTC-1957 frictional correlation line (Eq. 8), was equal for both the numerical and the experimental results.

Figure 10 shows the free surface elevation at design speed over the trim range along the KCS hull trough the domain. The KCS forepeak is positioned at x/Lpp = 0, the aft peak at x/Lpp = 1. For the bow-down trim conditions, the KCS generates a slightly higher bow wave. The stern wave profiles at the aft peak are higher for the bow-down trim as well; however, the disturbances faded more quickly downstream compared to the bow-up trim condition profiles. Hence, the prediction of shallower waves for bow-down conditions resulted in a reduced wave-making resistance calculated by the wave cut method.

The same trend can be observed in Figure 11, which shows the free surface elevation at $y/Lpp = 0.1$ through the computational domain. For steady-state flow simulations, those wave cuts can be interpreted as flow pattern traveled downstream from the hull.

Whereas the bow wave has settled for all trim conditions while traveling downstream, the wave profile in the wake of the ship at the position $x/Lpp \approx 1.5$ remains pronounced which results from the hull stern wave (observed in Figure 10 at $x/Lpp \approx 1.4$. This effect is also shown in Figure 12, which illustrates a comparison of the free surface for $\theta = 0.75°$ (RHS) and $\theta = 0.25°$ (LHS).

The above figure also shows a pronounced bow wave for the bow-down trim, which suggests that an adapted bulbous bow could in fact further improve this operational condition.

Figure 13 compares the wave making resistance at each trim angle against the level trim predictions. For bow-up trim conditions, the wave making resistance shows a large increase of up to 65% above level trim. For $\theta = 0.25°$ bow-down trim, the wave making resistance finds its lowest value, which again agrees well with the experimental findings.

Figure 14 shows the impact of the trim variation on the total resistance. Due to the influence of the waterline length and the wetted surface on the frictional resistance component (see Figure 5), the total resistance shows minor trend differences for bow-up trim compared to the wave making resistance.

Due to the low fraction of the wave making resistance on the total resistance, for the original KCS between 5% for the slow steaming speed up to 11% for the design speed, larger improvements in the wave making resistance only resulted in small improvements in the total resistance. In agree-

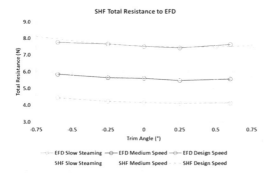

Figure 9. Total resistance (SHF) validation.

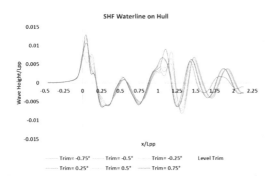

Figure 10. Waterline elevation on the KCS hull.

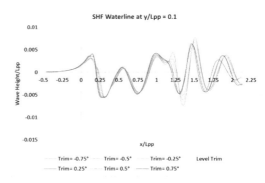

Figure 11. Waterline elevation on KCS hull.

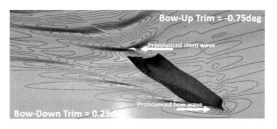

Figure 12. Free surface contour plots comparison.

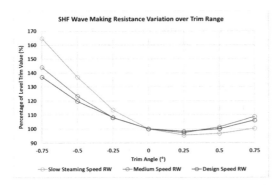

Figure 13. Wave making resistance comparison to level trim value over trim range.

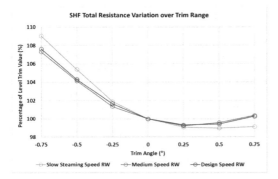

Figure 14. Total resistance comparison to level trim value over trim range.

ment with the experiments, a maximum resistance reduction of 1% could be achieved by trimming the original KCS hull $\theta = 0.25°$ to bow for medium and design speed. For the slow steaming speed, the resistance reduction kept nearly constant for all bow-down trim angles. Finally, the RANSE simulation results for the total resistance of the original KCS were in good agreement with the experimental measurements.

Figure 15 compares the numerical results, i.e. the potential flow calculations and the RANSE calculations with the KCS fixed and free for sinkage and trim, with the experiments. Due to the wave making resistance coefficient correction applied to the potential flow results, the total resistance agrees well with the experiments. The RANSE computations with an active DFBI model predicted the total resistance with an error of under 3% compared to the experiments. The fixed KCS simulation results, however, showed an error of around 8.5%. This suggests, that the effect of the dynamic sinkage and trim plays an important role, which should be included in such trim studies. Further, it can be

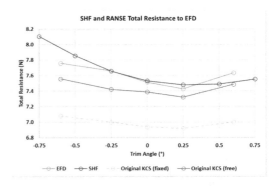

Figure 15. Total resistance comparison for numerical and experimental results.

noted that the RANSE trend prediction is more accurate compared to the potential flow results.

3.2 KCS bulbous bow designs

For the KCS bulbous bow variation, the results evaluation was again based on the wave resistance and the total resistance. Performing calculations for a high number of different KCS designs allowed deriving an improved bulbous bow design, which led to the minimization of energy losses associated with the generation of waves.

Figure 16 shows the impact of an adapted bulbous bow on the wave making resistance by plotting the resistance reduction for each speed at each trim angle as percentage improvement (right axis). The results show that the bulbous bow variation has a larger impact for the KCS operating in bow-up trim conditions. The largest improvement of around 2.7% was achieved for a bow-up trim angle of $\theta = -0.5°$ for the slow steaming speed. This indicates that even though the immersed transom of the KCS dominates the generation of waves, a redesign of the bulbous bow can still have a significant impact on the wave making resistance.

As outlined above, the fraction of the wave making resistance on the total resistance was around 5% for the slow steaming speed, thus the impact of the bulbous variation on the total resistance was small. The possible improvements for the total resistance are shown in Figure 17.

Similar to the improvements in the wave making resistance, the largest reduction of total resistance could be achieved for bow-up trim angles. For the optimum trim angle of $\theta = 0.25°$ only a minor reduction of around $\Delta R_{TM} = 0.14\%$ was found at slow steaming speed.

Due to the low variation of the results, it was not expected that the weight-based results pool analysis would show larger improvements for different operational profiles. Table 4 presents the

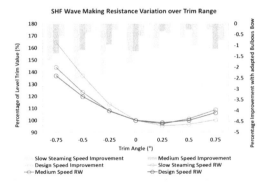

Figure 16. Wave making resistance comparison including possible improvement by adapted bulbous bow.

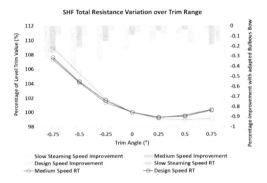

Figure 17. Total resistance comparison including possible improvement by adapted bulbous bow.

Table 4. Best design candidates for different operational profiles.

Profile #	Speed seights			KCS Design	@ θ (°)	$\Delta R_{TM}^{Profile}$ (%)
	t_{v1}	t_{v2}	t_{v3}			
1	2/3	1/6	1/6	D33	0.25	1.14
2	1/6	2/3	1/6	D10	0.25	0.82
3	1/6	1/6	2/3	D24	0.25	0.73
4	1/3	1/3	1/3	D24	0.25	0.75

combined improvements in total resistance of the trim optimization and the bulbous retrofit for four operational profiles of varying speeds.

The table lists the speed weights, the KCS design number (out of 40) for which the improved resistance was calculated, the ideal trim angle and the combined total resistance improvement. Profile #1 represents the KCS operating in slow steaming conditions 66% of its time at a speed of 18 knots. This profile type returns the largest improvements of $\Delta R_{TM} = 1.14\%$ compared to level trim operation for Design 33. For the other profiles, the total savings were below 1%.

Figure 18 highlights the importance of including a dynamic motion model in the RANSE simulations. Whereas the total resistance prediction for the fixed ship simulations did not agree with the experiments and the potential code computations, the results accuracy increased when including 2-DOF motions. The minor performance increase (here shown at design speed) predicted by the potential solver is not reflected for all trim angles in the RANSE simulations. This could be due to the initial error of the RANSE simulations of around 3%.

In order to analyze the best bulbous bow designs (see Table 4) from a geometric point of view, Table 5 presents the change of the bulbs in length, width and tip height compared to the original KCS design.

All bulbous designs show a reduction in bulb tip height when operated at a trim angle of $\theta = 0.25°$. KCS Design 33 has an increased bulb length of $dx = 0.891m$.

As can be seen in Figure 19, the presence of the extended bulb generates a slightly longer and shallower bow wave which then results in an improved fore shoulder wave pattern. Further, the emerging transom, due to the forward trim, produced a slightly shallower wake field.

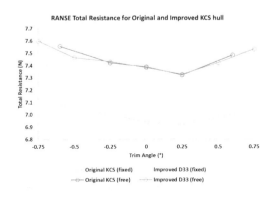

Figure 18. Total resistance comparison including possible improvement by adapted bulbous bow.

Table 5. KCS bulbous bow design specifications.

Design #	Bulb length dx (m)	Bulb width dy (m)	Bulb tip height dz (m)
Original	0	0	0
D33	0.891	−0.484	−0.422
D10	−0.188	0.063	−1.313
D24	0.469	0.156	−1.406

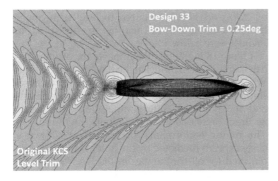

Figure 19. Comparison of KCS design 33 and KCS original design.

4 CONCLUSION

This paper presented a trim methodology study by combining experimental towing tank testing, potential flow and RANSE flow simulations. The nominal performance of the Kriso Container Ship (KCS) was investigated including a numerical bulbous bow shape variation and its influence in combination with the trim study to find an optimal fore ship design for adverse operating conditions. Thus, it was shown that the combined effort of improving the vessel's floating position (trim) with the simultaneous retrofit of local geometry features, such as the bulbous bow, improved the ship's performance.

In total, 868 numerical simulations were performed for 40 KCS designs with a varying bulbous bow design. The potential flow solver was used to compute the wave making resistance of the KCS for different speeds and trim angles and has proven to make accurate predictions on a fine panel mesh. Due to its short run time the solver was used for a large number of simulations in order to create a sufficient results pool. The solver was able to reproduce the effect of different trim angles on the wave making resistance and was successfully validated by experimental results.

The RANSE simulations were run to predict the total resistance of the KCS without making use of empirical formulations, i.e. the ITTC-1957 frictional correlation line. Results have shown that the RANSE simulations follow the total resistance trend, predicted by the experimental tests, more accurately. However, the RANSE results also show a larger error compared to the corrected potential flow results, thus it remains difficult to judge the predictions made for the bulbous bow variation study. Further, the RANSE simulations showed that the results accuracy was highly dependent on the quasi steady-state ship motions, i.e. dynamic sinkage and trim.

The assessment of the results pool was performed by introducing a weighting method that allowed extracting an improved KCS design for certain operational conditions. Results showed that a bow-down trim of $\theta = 0.25°$ was the most beneficial condition for all three investigated speeds. The total resistance of the original KCS could be reduced by $\Delta R_{TM} = 1.0\%$, which was also validated by the experimental test series. The bulbous variation study yielded only small improvements in operational performance of up to $\Delta R_{TM}^{Profile} = 1.14\%$, for a slow steaming operational profile for the adapted KCS design 33.

ACKNOWLEDGEMENT

The authors would like to thank Flowtech International AB for the free access to Shipflow for this study. The numerical RANSE results were obtained using the EPSRC funded ARCHIE-WeSt High Performance Computer (www.archie-west.ac.uk). EPSRC grant no. EP/K000586/1.

REFERENCES

Filip, G. et al., 2013. Bulbous Bow Retrofit of a Container Ship Using an Open-Source Computational Fluid Dynamics (CFD) Toolbox. Transactions—Society of Naval Architects and Marine Engineers 122, 1(V), pp. 244–262.

Flowtech International Ab, 2017. "Shipflow 6.3 User's Manual".

Flowtech User Support, 13 December 2017. Re: KCS bulbous bow study. Type to support@flowtech.se.

ITTC Practical Guidelines for Ship CFD Application, 2011, 7.5 – 03 – 02 – 03, p. 11.

Reichel, M., Minchev, a. & Larsen, N.L., 2014. Trim Optimization—Theory and Practice. TransNav, the International Journal on Marine Navigation and Safety of Sea Transportation, 8(3), pp. 387–392. Available at: http://www.transnav.eu/Article_Trim_Optimisation_-_Theory_and_Reichel,31,521.html.

Shivachev, E., Khorasanchi, M. & Day, A.H., 2017. Trim Influence on Kriso Container Ship (KCS); An Experimental and Numerical Study In Proceedings of the ASME 2017 36th International Conference on Ocean, Offshore and Arctic Engineering. Trondheim, pp. 1–7.

SIEMENS, 2017. "User guide STAR-CCM+ Version 12.02.009".

Vroegrijk, E., Whitworth, S. & Caldas, A., 2015. Validation of Bulbous Bow Optimisation by Viscous CFD.

Wagner, J., Binkowski, E. & Bronsart, R., 2014. Scenario based optimization of a container vessel with respect to its projected operating conditions. International Journal of Naval Architecture and Ocean Engineering, 6 (2), pp. 496–506. Available at: http://dx.doi.org/10.2478/IJNAOE-2013-0195.

Hull form hydrodynamic design using a discrete adjoint optimization method

P. He, G. Filip, J.R.R.A. Martins & K.J. Maki
University of Michigan, Ann Arbor, MI, USA

ABSTRACT: Computational fluid dynamics based optimization is becoming increasing popular in hull form hydrodynamic designs; however, existing optimization studies commonly utilize gradient-free methods which can only handle a handful of design variables. Ship hull geometry is necessarily described by a large number of design variables since the hydrodynamic drag is strongly influenced by small changes in the hull shape. To break this limit, in this paper we utilize a gradient-based optimization approach along with the adjoint method for efficient gradient computation. To demonstrate its power, we optimize the stern region of the KVLCC2 tanker using more than 100 design variables subject to proper geometric constraints (volume and thickness) for practical shape. We also evaluate the impact of propellers on the optimization results. We neglect the free-surface since the Froude number is low and the design changes are in the stern region. Overall, we obtain 2.9% and 1.3% hydrodynamic drag reduction for the cases with and without the propeller, respectively.

1 INTRODUCTION

Minimizing hydrodynamic drag is one of the major tasks in the ship design process as it determines the economic viability of a design. To reduce the drag, traditional hull designs heavily rely on the designers' experience. The process typically involves manual iterations of performance evaluation and geometry modification. With the advance in computing techniques, now we can automate the hull design process using the computational fluid dynamics (CFD) method along with an optimization algorithm (for example Campana et al. 2006). A recent study of turbomachinery by Puente et al. (2017) demonstrated that the CFD-based design optimization approach can significantly reduce the length of the design period, while achieving satisfactory quality compared with a sophisticated human-supervised design tool. The unique challenges of marine operation include the role ship waves, hull-propeller interaction, dynamic stability, seakeeping, and maneuvering, yet a wide range of optimization techniques have been developed and successfully employed to improve ship designs. In this paper, we focus on a specific optimization technique that has unique power to treat problems with arbitrary number of design variables. This is certainly the case when one would like to describe a ship hull form.

There are two main classes of design optimization methods: the gradient-free and gradient-based methods. The gradient-free method was dominant in the hull form optimization studies in the past decade (Luo & Lan 2017, Zhang et al. 2017, Kim & Yang 2010, Tahara et al. 2008). The computational cost of a gradient-free optimization algorithm (e.g., Artificial Neural Network, Genetic Algorithm, and Design of Experiment) scales exponentially with respect to the number of design variables (Martins & Hwang 2016). Due to this limitation, the aforementioned optimization studies only parameterized the hull geometry with a handful of design variables, ranging from 4 to 8. However, the hydrodynamic drag of a ship hull can be strongly influenced by small changes in geometry and it is preferred to describe the hull geometry using a large number of design variables. This will be seen in the optimized shapes in this paper that are difficult to parameterize apriori. To break this limit, the gradient-based method can be used. Instead of generating a large number of CFD samples, the gradient-based method starts from the baseline design and utilizes the gradient (derivative) information to find the most promising direction in the design variable space for improvement. To efficiently compute the derivatives, one can utilize the adjoint method whose computational cost is independent of the number of design variables (Jameson 1988).

Given the above advantage, the coupled gradient-based optimization and adjoint derivative computation method has been widely used in aircraft, automobile, and turbomachinery design optimization studies (Jameson et al. 1998, Nielsen

& Anderson 1999, Mader & Martins 2013, Kenway et al. 2014, Othmer 2014, Lyu et al. 2015, Xu et al. 2015). The adjoint method has also been used in the marine context (Martinelli & Jameson 2007), although their method has not seen wide spread adoption. This is likely due to the challenges of dealing with the naturally unsteady nonlinear interaction of ship waves, and the computational challenges associated with wave breaking. There is another important reason why the adjoint method lags behind in terms of adoption by the hull form design community. We believe it is because the implementation of adjoint derivative computation requires the access to the CFD source code and a dedicated effort for code development; however, current hull form optimization studies commonly rely on commercial CFD solvers whose source codes are not publically available.

Fortunately, we developed an efficient adjoint optimization framework based on an open-source CFD solver (OpenFOAM). Our optimization framework has been successfully applied to constrained aerodynamic optimizations for aircraft and automobiles (He et al. 2017). In this paper, we extend this framework for ship hull form optimization. To demonstrate its power, we conduct hydrodynamic shape optimization for the stern region of the KVLCC2 tanker. We parameterize the hull shape using more than 100 design variables and impose proper geometric constraints (volume and thickness) for practical shape. Moreover, we evaluate the impact of propellers on the optimization results. The free-surface is neglected and a double-body boundary condition is imposed on the calm-water plane.

It is import to note that there are two different methods for formulating the adjoint of a flow solver: continuous and discrete. The continuous approach derives the adjoint formulation from the Navier-Stokes (N-S) equations and then discretizes for numerical solution. In contrast, the discrete approach starts from the discretized N-S equations and differentiates the discretized equations to get the adjoint terms. Although these two approaches handle adjoint formulation in different ways, they both converge to same answer for a sufficiently refined mesh (Nadarajah & Jameson 2000). We opt to use the discrete approach since the adjoint derivative is consistent with the flow solutions. Moreover, it is easier to derive formulation for new boundary condition and objective functions.

The rest of the paper is organized as follows: In Section 2, we introduce the optimization framework and its components. The hull shape optimization results for KVLCC2 is shown in Section 3. Last, the conclusion remarks are summarized in Section 4.

2 METHOD

Our design optimization framework implements a discrete adjoint for computing the total derivative df/dx where f is the function of interest (which for optimization will be the objective and the constraint functions, e.g., hydrodynamic drag), and x represents the design variables that control the geometric shape via Free-Form Deformation (FFD) control point movements. It consists of multiple components written in C++ and Python. In this section, we introduce the overall adjoint optimization framework along with a brief description of its modules. The detailed theoretical background for these modules has been elaborated in He et al. (2017).

2.1 *Adjoint optimization framework*

The modules and data flow for the optimization framework are shown in Figure 1. We use the extended design structure matrix (XDSM) standard developed by Lambe & Martins (2012). The diagonal entries are the modules in the optimization process, while the off-diagonal entries are the data. Each module takes data input from the vertical direction and output data in the horizontal direction. The thick gray lines and the thin black lines denote the data and process flow, respectively. The numbers in the entries are their execution order.

The framework consists of two major layers: OpenFOAM and Python, and they interact through input and output (I/O) files. The OpenFOAM layer consists of a flow solver (simpleFoam), an adjoint solver (discreateAdjointSolver), and a graph coloring solver (coloringSolver). The flow solver is based on the standard OpenFOAM solver simpleFoam for steady incompressible turbulent flow. The adjoint solver computes the total derivative df/dx based on the flow solution generated by simpleFoam. The mesh deformation derivative matrix (dx_v/dx where x_v contains the volume mesh coordinates) is needed when computing the total derivative, and it is provided by the Python layer. To accelerate the partial derivative computation, we developed a parallel graph coloring algorithm whose details were elaborated in He et al. (2017).

The Python layer is a high-level interface that takes the user input, as well as the total derivatives computed by the OpenFOAM layer, and calls multiple external modules to perform constrained optimization. To be more specific, these external modules include:

pyGeo: a surface geometry parameterization module using the FFD approach (Kenway et al. 2010). This approach embeds the geometry into a volume that can be manipulated by moving points at the surface of that volume (the FFD points).

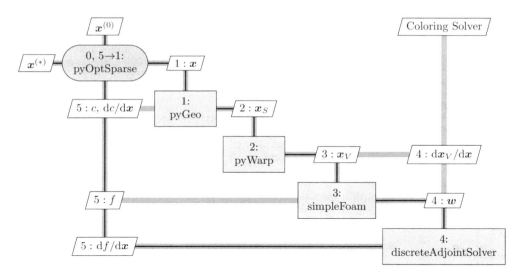

Figure 1. Extended design structure matrix (Lambe & Martins 2012) for the discrete adjoint framework for constrained shape optimization problems. x: design variables; $x^{(0)}$ baseline design variables; $x^{(*)}$: optimized design variables; x_S: design surface coordinates. x_V: volume mesh coordinates; w: state variables; c: geometric constraints; f: objective and constraint functions.

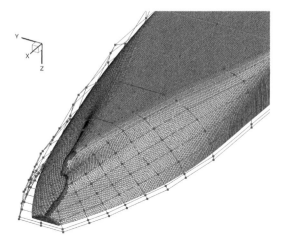

Figure 2. Surface mesh and FFD setup.

The FFD volume is a tri-variate B-spline volume such that the gradient of any point inside the volume can be easily computed. An example of FFD control points is shown in Figure 2. pyGeo also compute the geometric constraints (c) and their derivatives (dc/dx).

pyWarp: a volume mesh deformation module using the analytical inverse distance algorithm (Luke et al. 2012). The advantage of this approach is that it is highly flexible and can be applied to both structured and unstructured meshes. In addition, compared with the radial basis function based method, this approach does a better job of preserving mesh orthogonality in the boundary layer.

pyOptSparse: an open source, object-oriented Python module, extended from pyOpt (Perez et al. 2012), for formulating and solving constrained nonlinear optimization problems. pyOptSparse provides a high-level API for defining the design variables, and the objective and constraints functions. It also provides interfaces for several optimization packages, including some open source packages. In this study, we choose SNOPT (Gill et al. 2002) as the optimizer. SNOPT uses the sequential quadratic programming (SQP) algorithm for the optimization, which solves the nonlinear equations resulting from the Karush-Kuhn-Tucker (KKT) optimality conditions using the quasi-Newton method.

2.2 Total derivative computation using discrete adjoint approach

As mentioned above, to perform gradient-based design optimization, we need to compute the total derivative df/dx. Note that f depends not only on the design variables, but also on the state variables that are determined through the solution of governing N-S equations. Thus:

$$f = f(x, w) \quad (1)$$

where x is the vector of design variables with length n_x and w is the vector of state variables with length n_w.

Applying the chain rule for the total derivative, we obtain:

$$\frac{df}{dx} = \frac{\partial f}{\partial x} + \frac{\partial f}{\partial w}\frac{dw}{dx} \quad (2)$$

A naive computation of dw/dx via finite differences would require solving the governing N-S equations n_x times, which can be computationally expensive for a large number of design variables. We can avoid this issue by making use of the fact that the derivatives of the residuals with respect to the design variables must be zero for the governing equations to remain feasible with respect to variations in the design variables. Thus, applying the chain rule to the residuals, we can write:

$$\frac{dR}{dx} = \frac{\partial R}{\partial x} + \frac{\partial R}{\partial w}\frac{dw}{dx} = 0 \quad (3)$$

where R is the vector of flow residuals. Substituting Equation (3) into (2) and canceling out the dw/dx term, we get:

$$\frac{df}{dx} = \frac{\partial f}{\partial x} - \frac{\partial f}{\partial w}\left(\frac{\partial R}{\partial w}\right)^{-1}\frac{\partial R}{\partial x} \quad (4)$$

Considering the combination of the $\partial R/\partial w$ and $\partial f/\partial w$ terms in Equation (4), we can solve a linear equation:

$$\frac{\partial R^T}{\partial w}\psi = \frac{\partial f^T}{\partial w} \quad (5)$$

to obtain the adjoint vector ψ. Then, this adjoint vector is substituted into Equation (4) to compute the total derivative:

$$\frac{df}{dx} = \frac{\partial f}{\partial x} - \psi^T \frac{\partial R}{\partial x} \quad (6)$$

Since the design variable x does not explicitly appear in Equation (6), we only need to solve Equation (5) once for each function of interest, and thus the computational cost is (almost) independent of the number of design variables (n_x). This is an advantage for three-dimensional shape optimization problems, since the number of functions of interest is usually less than 10 but the number of design variables can be a few hundred.

A successful implementation of adjoint-based derivative computation requires an efficient and accurate computation for the partial derivatives ($\partial R/\partial w$, $\partial R/\partial x$, $\partial f/\partial w$, and $\partial f/\partial x$) in Equations (5) and (6). In this paper, we use the finite difference approach to calculate the partial derivatives accelerated by a parallel coloring algorithm. Our adjoint computation implementation achieved a reasonably good speed and scalability with up to 10 million cells and 1000 CPU cores, as shown in our previous work (He et al. 2017).

3 RESULTS AND DISCUSSION

3.1 Solver verification and mesh independence

It is known in the literature that the CFD simulation results can be sensitive to the numerical setup of a flow solver and the mesh size. Therefore, in this section, we conduct a solver verification and mesh independence study. The objective is to ensure that the simulation results are physically reasonable and the mesh size is sufficient to capture key flow structures for shape optimization.

As mentioned above, we focus on optimizing the stern region of the KVLCC2 hull. The flow simulations are conducted using the OpenFOAM standard solver simpleFoam. The simulation domain size is $5L_{pp}$, $2L_{pp}$ and L_{pp} in the x, y, and z directions, respectively, with L_{pp} being the length between perpendiculars. We generate unstructured hex-mesh with 3, 6, and 20 million cells using the OpenFOAM built-in snappyHexMesh utility. The y^+ is about 1 for all the three meshes. The surface mesh for the KVLCC2 hull is shown in Figure 2. The Reynolds number is selected to be 2×10^6. This lower value enables us to conduct high-fidelity large-eddy simulation validation runs for our optimizations, and the results will be reported in our future work.

A second order linear upwind scheme is used to differentiate the divergence terms, whereas for the diffusion terms, the central differential scheme is adopted. The Spalart-Allmaras turbulence model is utilized for all the simulations. Since the Froude number for KVLCC2 is low, we ignore the wave-making resistance in the drag calculation.

To mimic the impact of propellers, we apply the actuator disk theory and add a source term in the momentum equation. This source term acts as adding a streamwise (thrust) and circumferential (torque) body force in a cylindrical space occupied by the propeller. The total thrust is set to be equal to the hydrodynamic drag of the hull. The corresponding torque is computed following Hoekstra (2006).

Now we compare the simulated velocity contours and profiles at the propeller plane with the experimental data reported in Lee et al. (2003), as shown in Figure 3 and Figure 4. The simulation results agree reasonably well with the experimental data for all the three grid sizes. The simulated total resistance coefficient C_T (4.637×10^{-3}) is larger than the experimental value (4.110×10^{-3}) reported in Kim et al. (2001). The difference is partially attributed to the fact that our simulations are conducted at a lower Reynolds number than Kim et al. (2001); our simulated C_T is 4.345×10^{-3} at the same Reynolds number. We also find that, by increasing the mesh size from 3 to 20 million, there is no significant change in drag and velocity contours and profiles. We conclude that 3 million cells are sufficient to capture the key flow feature for this case, coinciding with that reported in Larsson et al. (2014). Therefore, in the following, we choose the 3 million cell setup for optimization.

3.2 *Hull shape optimization results*

Next we conduct two hull shape optimizations using our discrete adjoint framework. The optimization setup is summarized in Table 1.

Here we setup 108 FFD control points to manipulate the stern shape, as shown in Figure 2.

To ensure a symmetric hull shape in the y direction, we set 54 linear constraints to link the design variable changes between $y > 0$ (y^+) and $y < 0$ (y^-). Moreover, to keep a flat side face of the hull, we limit the y direction thickness at any location to be equal to or smaller than the original maximal thickness of the hull. We also setup a smallest thickness constraint to ensure there is sufficient space to install the propeller shaft. To keep the block coefficient and the water line, we impose a volume constraint to ensure that the hull volume remains unchanged during the optimization. In total, we have set 116 geometric constraints. It is important to highlight that imposing proper geometric constraints is critical to ensure a practical hull shape.

We first consider the case without a propeller. Figure 5 shows the convergence history of C_T and optimality. The optimality is a measure of how well the KKT optimality conditions are satisfied, and

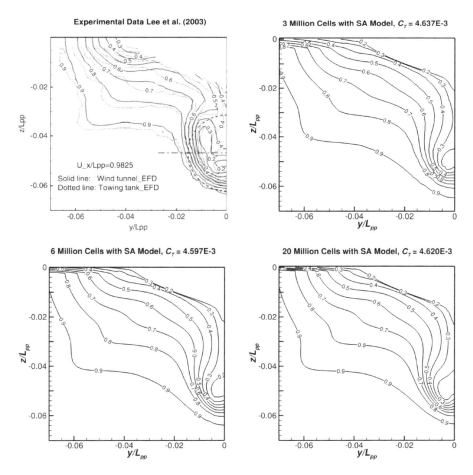

Figure 3. The simulated velocity contours at the propeller plane ($x/L_{pp} = 0.9825$) agree reasonably well with the experimental data reported in Lee et al. (2003).

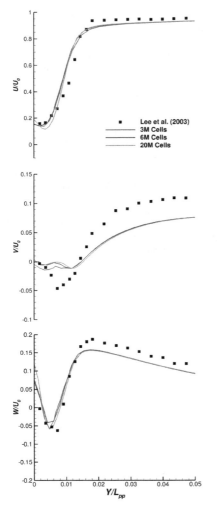

Figure 4. The simulated velocity profiles at the propeller plane ($x/L_{pp} = 0.9825$, $z/L_{pp} = -0.05075$) agree reasonably well with the experimental data reported in Lee et al. (2003).

Table 1. KVLCC2 hull shape optimization setup.

	Variable	Description	No.
Minimize	C_T	Resistance coefficient	
with	y	FFD y-coordinate	108
respect to	Δy^{y+}	Symmetric hull shape	54
subject to	$= -\Delta y^{y-}$	in the y direction	
	$V = V_{baseline}$	Volume constraint	1
	$t \leq t_{side}$	Thickness constraint to keep hull side face flat	36
	$t \geq t_{shaft}$	Thickness constraint for shaft installation	25
	$|\Delta y| < 1$ m	Design variable bounds	
		Total constraints	116

therefore it quantifies how close the design is to an optimum. C_T converges reasonably well in 35 optimization iterations, and we obtain 1.3% of drag reduction. Moreover, the optimality drops about two orders of magnitude which indicates that the shape is close to an optimum. Considering that the KVLCC2 hull was designed by well experienced designers and its shape has been continuously optimized in the past forty years, 1.3% of drag reduction is still a noticeable achievement. In terms of speed, the optimization took 39 hours using 128 CPU cores (Intel Xeon E5-2680 v3 at 2.5 GHz).

Now we compare the baseline and optimized hull shapes, as shown in Figure 6. We observe that the drag reduction is obtained by decreasing the y direction thickness in the lower region of the hull. Due to the volume constraint, the thickness increase in the upper hull. A detailed flow analyses and hydrodynamic drag breakdown are shown in Figure 7. We notice that the drag reduction is mostly achieved by adjusting the pressure distribution (decreasing the pressure drag component $C_{T\,pressure}$), while the viscous drag ($C_{T\,viscous}$) remains same between the baseline and optimized hull shapes. We also observe the center of the stern bilge vortex moves upwards at the propeller plane in the optimized case. Note that the changes are not easy to parameterize before the optimization was conducted.

The above optimization is based on the case without a propeller. It is reasonable to expect that adding the propeller will change the pressure distribution in the stern region and therefore influence the optimization results. Figure 8 shows the convergence history of C_T and optimality for the case with the propeller. Again, we observe a reasonably good

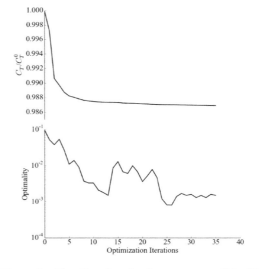

Figure 5. The C_T and optimality converge well in 35 iterations for the case without propeller.

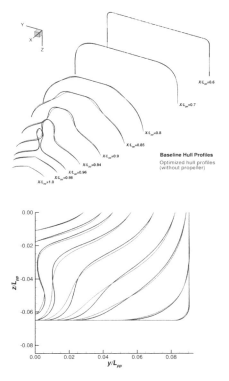

convergence. In addition, we achieve a higher drag reduction (2.9%), compared with the case without the propeller. The optimization took 32 hours.

The comparison between the baseline and optimized shapes with and without the propeller is shown in Figure 9. We find that adding the propeller significantly changes the optimized

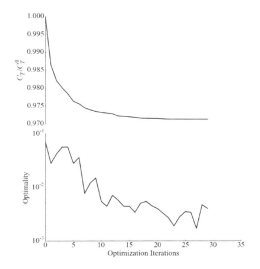

Figure 6. Comparison between baseline and optimized shapes.

Figure 8. The C_T and optimality converge well in 29 iterations for the case with propeller.

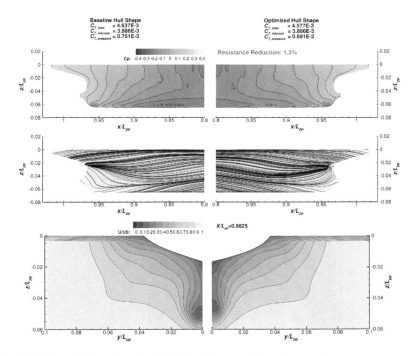

Figure 7. The hydrodynamic drag reduction is 1.3% for the case without the propeller.

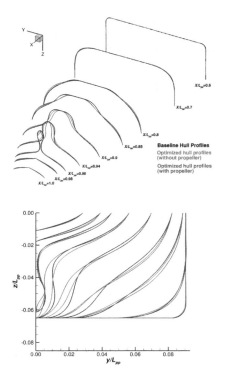

shape. Most noticeably, the thickness in the lower part of the hull further reduces, especially near the propeller shaft. The detailed drag reduction breakdown, as shown in Figure 10, indicates that adding the propeller doubles the pressure drag component, compared with the no propeller case. Again, all the drag reduction is obtained by modifying the pressure distribution and decreasing the pressure drag.

4 CONCLUSION

In this paper, we utilize our newly developed discrete adjoint framework to optimize the stern region of the KVLCC2 hull. The most beneficial feature of using the adjoint method is that its computational cost is independent of the number design variables. Taking this advantage, we parameterize the hull shape using 108 design variables, which allows us to fully capture the impact of small changes on the hydrodynamic drag. Moreover, we impose proper volume and thickness constraints to ensure a practical optimized shape. We conduct two optimizations with and without adding the propeller to evaluate its impact.

Figure 9. Comparison between baseline and optimized hull shapes with and without propeller.

The optimization results show that the drag reduction is mostly achieved by manipulating the pressure distribution of the hull; the pressure drag component reduces while the viscous drag remains

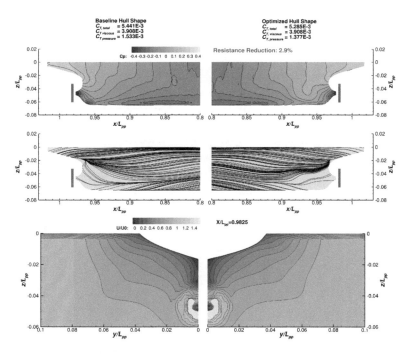

Figure 10. The hydrodynamic drag reduction increases to 2.9% when adding the propeller.

same during the optimizations. For the case without the propeller, we obtain 1.3% of drag reduction. After adding the propeller, the pressure drag increases, and 2.9% of drag reduction is achieved. The two optimizations converge in less than 40 hours.

Overall, we have demonstrated the success of applying the adjoint method for hull design optimization. Since this method converges quickly and does not constrained by the number of design variable, it has the potential of supplementing the existing industrial hull shape design process. Moreover, the optimization framework is naturally extensible to additional disciplines, including scantling design and manufacturing costs.

ACKNOWLEDGMENTS

The computations were performed in the Extreme Science and Engineering Discovery Environment (XSEDE), which is supported by National Science Foundation grant number ACI-1548562, as well as the Flux HPC cluster at the University of Michigan Center of Advanced Computing.

REFERENCES

Campana, E.F., Peri, D., Tahara, Y. & Stern F. 2006. Shape optimization in ship hydrodynamics using computational fluid dynamics. *Computer Methods in Applied Mechanics and Engineering.* 196: 634–651.

Gill, P.E., Murray, W & Saunders, M.A. 2002. SNOPT: An SQP algorithm for large-scale constrained optimization, *SIAM Journal of Optimization*, 12(4): 979–1006.

He, P., Mader, C.A., Maki, K.J. & Martins, J.R.R.A. 2017. An aerodynamic design optimization framework using a discrete adjoint approach with OpenFOAM. *Computers & Fluids*. (conditionally accepted)

Hoekstra, M. 2006. A RANS-based analysis tool for ducted propeller systems in open water condition. *International Shipbuilding Progress* 53: 205–227.

Jameson, A. 1988. Aerodynamic design via control theory, *Journal of Scientific Computing* 3(3): 233–260.

Jameson, A. Martinelli, L., & Pierce N.A. 1998. Optimum aerodynamic design using the Navier-Stokes equations, *Theoretical and Computational Fluid Dynamics* 10(1–4): 213–237.

Kenway, G.K.W., Kennedy, G.J. & Martins, J.R.R.A. 2010. A CAD-free approach to high-fidelity aerostructural optimization, in: Proceedings of the 13th AIAA/ISSMO Multidisciplinary Analysis Optimization Conference, Fort Worth, TX.

Kenway, G.K.W., Kennedy, G.J. & Martins, J.R.R.A. 2014. Scalable parallel approach for high-fidelity steady-state aeroelastic analysis and derivative computations, *AIAA Journal*, 52(5): 935–951.

Kim, W.J., Van, S. H & Kim, D.H. 2001. Measurement of flows around modern commercial ship models. *Experiments in Fluids*. 31: 567–578.

Kim, H. & Yang, C. 2010. A new surface modification approach for CFD-based hull form optimization. Engineering Journal of Hydrodynamics, Ser. B 22(5): 520–525.

Lambe, A.B. & Martins, J.R.R.A. 2012. Extensions to the design structure matrix for the description of multidisciplinary design, analysis, and optimization processes, *Structural and Multidisciplinary Optimization,* 46(2): 273–284.

Lars Larsson, Michel Visonneau, and Frederick Stern 2014. Numerical Ship Hydrodynamics. Springer.

Lee, S., Kim, H. & Kim, W. 2003. Wind Tunnel Tests on Flow Characteristics of the KRISO 3, 600 TEU Containering and 300 K VLCC Double-Deck Ship Models. *Journal of Ship Research* 47(1): 24–38.

Luke, E. Collins, E. & Blades, E. 2012. A fast mesh deformation method using explicit interpolation, *Journal of Computational Physics*, 231(2): 586–601.

Luo, W. & Lan, L. 2017. Design Optimization of the Lines of the Bulbous Bow of a Hull Based on Parametric Modeling and Computational Fluid Dynamics Calculation. *Mathematical and Computational Applications* 22(4): 1–12.

Lyu, Z., Kenway, G.K.W. & Martins, J.R.R.A. 2015. Aerodynamic Shape Optimization Investigations of the Common Research Model Wing Benchmark. *AIAA Journal*, 53(4): 968–985

Mader, C.A. & Martins, J.R.R.A. 2013. Stability-constrained aerodynamic shape optimization of flying wings, *Journal of Aircraft*. 50(5): 1431–1449.

Martins, J.R.R.A. & Hwang J.T. 2016. Multidisciplinary Design Optimization of Aircraft Configurations—Part 1: A modular coupled adjoint approach. Lecture series, Von Karman Institute for Fluid Dynamics, Sint–Genesius–Rode, Belgium.

Nadarajah, S. & Jameson, A. 2000. A comparison of the continuous and discrete adjoint approach to automatic aerodynamic optimization, in: Proceedings of the 38th AIAA Aerospace Sciences Meeting and Exhibit, Reno, NV.

Nielsen, E.J. & Anderson, W.K. 1999. Aerodynamic design optimization on unstructured meshes using the Navier-Stokes equations, *AIAA Journal* 37(11).

Othmer, C. 2014. Adjoint methods for car aerodynamics, *Journal of Mathematics in Industry*. 4(1): 1–23.

Perez, R.E., Jansen, P.W. & Martins, J.R.R.A. 2012. pyOpt: A Python-based object-oriented framework for nonlinear constrained optimization, *Structural and Multidisciplinary Optimization* 45(1): 101–118.

Puente, R., Corral, R. & Rarra, J. 2017. Comparison between aerodynamic designs obtained by human driven and automatic procedures. *Aerospace Science and Technology*. (in press)

Tahara, Y., Peri, D., Campana, E.F. & Stern, F. 2008. Computational fluid dynamics-based multiobjective optimization of a surface combatant using a global optimization method. *Journal of Marine Science and Technology* 13(2): 95–116.

Xu, S., Radford, D., Meyer, M. & Muller, J.D. 2015. Stabilisation of discrete steady adjoint solvers, *Journal of Computational Physics* 299: 175–195.

Zhang, S., Zhang, B., Tezdogan, T., Xu, L. & Lai, Y. 2017. Computational fluid dynamics-based hull form optimization using approximation method. *Engineering Applications of Computational Fluid Mechanics* 12(1): 74–88.

Potential effect of 2nd generation intact stability criteria on future ship design process

Yaohua Zhou, Yanhong Hu & Gaofeng Zhang
Shanghai Rules and Research Institute, China Classification Society, Shanghai, China

ABSTRACT: By the end of sub-committee on Ship Design and Construction session 4 (SDC 4) of International Maritime Organization (IMO), the development of 2nd Generation Intact Stability Criteria (SGISC) has achieved remarkable progress. This new criteria will be a mandatory requirement and replace the current 2008 Intact Stability Code after approval. Different from the hydrostatic criteria currently adopted in ship design process, the stability assessment of SGISC is based on hydrodynamic theory and covers four completely new failure modes of dynamic stability accidents. Thus, its application will exert remarkable influence on the future design process. In this work, sample calculations of Level 1 and Level 2 criteria with various types of ships are presented. Based on the vulnerability check calculations, the potential effect of SGISC on design process is analyzed, including relationship between vulnerability and ship type, as well as the effect of several ship parameters. Also, the effect of estimation method of roll moment of inertia and the potential application of CFD method are analyzed. The experience obtained from the development of SGISC will be helpful in reducing the designers' difficulty when applying it, optimizing future design process, and improving the safety level of new design.

1 INTRODUCTION

As mandatory requirements of the Administrations, the Intact Stability Criteria established basic safety level for full scale ships. The technical standards currently in force are established in 2008 IS Code (IMO 2008), which is developed based on hydrostatic theory, dead ship assumption and experience obtained from accidents. According to the operation records, the ships based on this code have full safety level in most cases. However, the accidents of parametric roll and excessive acceleration (France, Levadou, Treakle, et al. 2001 and Germany 2012) indicate that the dynamic stability failure modes could still danger the safety of current ship designs. Based on this background, the IMO launched the development of SGISC, with an aim to raise the safety level by covering more failure modes.

Besides the dead ship failure model covered in the traditional hydrostatic criteria, four new failure modes are also considered in the SGISC, including parametric roll, pure loss of stability, excessive acceleration and surf-riding/ broaching. The frame work of SGISC includes Level 1/2 vulnerability criteria and direct stability assessment. The former utilizes simplified numerical method or formulas to check the vulnerability of ship designs, and the latter adopts "art of state" methods to simulate the motion responses and safety level of ship designs.

With regard to the mechanism, numerical simulation and experimental study, a great deal of research has been carried out by the academia and Administrations (Neves, Belenky, Kat, Spyrou, Umeda 2011; Fossen, Nijmeijer 2012; Japan 2011, 2012, 2013, 2014 and 2015). Based on these achievements, the development of Level 1 and 2 vulnerability criteria has been basically completed.

As one of the technical research team work for Chinese delegation in IMO, China Classification Society has participated in the relevant research work of SGISC in the past few years. In this work, the experience obtained from these researches is presented. Due to the limitation of complexity of the direct stability assessment, Level 1 and 2 criteria will be most widely used in the future design procedure. Thus, the sample calculations of Level 1/2 criteria for various types of ships are conducted, and potential effect of SGISC on design process is analyzed, including vulnerability of different types of ships, effect of estimation method of roll moment of inertia and the potential application of CFD approach.

2 FAILURE MODELS AND VULNERABILITY CRITERIA

2.1 *Parametric Roll (PR)*

Parametric roll is an amplification of roll motions caused by periodic variation of transverse stability in waves[1].

The Level 1 criterion of PR includes two checks, which judge vulnerability by calculating the variations of metacentric height in waves and weighted vulnerability index C_1 (IMO 2015).

$$\frac{\Delta GM}{GM_c} \leq R_{PR} \quad (1)$$

$$C_1 = \sum_{i=1}^{N} W_i C_i < R_{PR0} \quad (2)$$

where ΔGM is amplitude of the variation of the metacentric height; GM_c is metacentric height in calm water; R_{PR} is standard value of 1st check; W_i is the weighting factor for wave; C_i is index for each specific wave and R_{PR0} is standard value of criteria.

The Level 2 criterion estimates the vulnerability index C_2 by calculating roll amplitudes of PR with solving 1 D.O.F roll motion equation.

$$\ddot{\phi} + 2\alpha\dot{\phi} + \gamma\dot{\phi}^3 + \omega_\phi^2 \frac{GZ}{GM} = 0$$
$$GZ = GM\left(\phi + \sum_{k=1}^{n} l_{2k+1}\phi^{2k+1}\right) + GZ_w \quad (3)$$

$$C_2 = \sum_{i=1}^{N} W_i C_i < R_{PR0} \quad (4)$$

where GZ_w is the restoring variation due to waves; ω_ϕ is natural roll circular frequency; ω_e is encounter frequency; The l_{2k+1} coefficients shall be determined with a least square fit to the GZ curve in calm water; α and γ are non-dimensional damping coefficients; ϕ is roll angle; GM is metacentric height in calm water and $k! = 1 \cdot 2 \cdots k$.

If equation (1), (2) or (4) is satisfied, the loading condition is considered to be non-vulnerable to Level 1 criterion 1st/2nd check or Level 2 criterion.

2.2 Pure Loss of Stability (PLS)

Pure loss of stability is a phenomenon that, due to significant change of submerged hull, the transverse stability is dramatically reduced and danger the ship's safety.

The Level 1 and 2 criteria judge the vulnerability by calculating the GM values and characteristics of GZ curves in waves (IMO 2015).

$$GM_{\min} > R_{PLA} \quad (5)$$

$$\max(CR_1, CR_2) > R_{PL0} \quad \begin{aligned} CR_1 &= \sum_{i=1}^{N} W_i C1_i \\ CR_2 &= \sum_{i=1}^{N} W_i C2_i \end{aligned} \quad (6)$$

$$C1_i = \begin{cases} 1 & \phi_V < R_{PL1} \\ 0 & otherwise \end{cases} \quad C2_i = \begin{cases} 1 & \phi_s > R_{PL2a} \\ 0 & otherwise \end{cases} \quad (7)$$

where $GM_{\min}$ is minimum value of metacentric height; R_{PLA}, R_{PL0} is standard value of criteria; $C1_i, C2_i$ is indexes of criterion1 and 2 (IMO 2015); ϕ_V is angle of vanishing stability; $R_{PL1} = 30$ deg; ϕ_s is heel angle due to centrifugal heeling lever of turning and $R_{PL2a} = 15/25$ degress.

If equation (5) or (6) is satisfied, the loading condition is considered to be non-vulnerable to Level 1 criterion or Level 2 criterion.

2.3 Excessive Acceleration (EA)

Excessive acceleration is a phenomenon that large accelerations are caused by roll motions due to synchronous resonance of ship in irregular waves and danger the crews' safety.

The Level 1 criterion judges the vulnerability by checking acceleration of a specific sea condition [1].

$$\phi k_L (g + 4\pi^2 h/T^2) < R_{EA1} \quad (8)$$

where ϕ is characteristic roll amplitude; k_L is factor taking into account simultaneous action of roll, yaw and pitch motions; g is gravitational acceleration; h is height above the roll axis of the location where passengers or crew may be present; T is roll period and R_{EA1} is standard value.

The Level 2 criterion judges the vulnerability by checking weighted vulnerability index C of acceleration in irregular waves, which is by calculating the response of ship in unit amplitude wave.

$$I_{xx}\ddot{\phi} + B\dot{\phi} + C_\phi \phi = M_{FK} \quad (9)$$

$$C = \sum_{i=1}^{N} W_i C_i < R_{EA2} \quad (10)$$

where I_{xx} is roll moment of inertia; B is equivalent roll damping; C_ϕ is restoring moment; M_{FK} is heeling moment of wave; R_{EA2} is standard value, which is still undecided by IMO and is taken as 0.0281 for criteria check (IMO 2016).

If equation (8) or (10) is satisfied, the loading condition is considered to be non-vulnerable to Level 1 criterion or Level 2 criterion.

2.4 Surf-Riding/Broaching (SR/B)

Broaching (a shortening of "broaching-to") is a violent uncontrollable turn that occurs despite maximum steering efforts to maintain course. Broaching is usually preceded by surf-riding which occurs when a wave, approaching from the stern, "captures" a ship and accelerates the ship to the speed of the wave (IMO 2016).

The Level 1 criterion judges the vulnerability by checking service speed of ship in calm water (IMO. 2015).

$$L > 200m, Fn \leq 0.3 \tag{11}$$

where L is length of ship and Fn is Foude number of service speed.

The Level 2 criterion judges the vulnerability by estimating vulnerability index C, which is calculated according critical ship speed by solving 1 D.O.F motion equation.

$$T_e(u_{cr}; n_{cr}) - R(u_{cr}) = 0 \tag{12}$$

$$C < R_{SR} \tag{13}$$

where $T_e(u_{cr}; n_{cr})$ is the thrust delivered by the ship's propulsor(s) in calm water; $R(u_{cr})$ is the calm water resistance of ship at the ship speed of u_{cr}; R_{SR} is standard value and n_{cr} is the commanded number of revolutions of propulsor(s) corresponding to the threshold of surf-riding.

If equation (11) or (13) is satisfied, the loading condition is considered to be non-vulnerable to Level 1 criterion or Level 2 criterion.

2.5 Dead Ship (DS)

Dead ship assumes that a ship has lost its power and has turned into beam seas, where it is rolling under the action of waves as well as heeling and drifting under the action of wind. Drift-related heel is a result of action of a pair of forces: wind aerodynamic force and hydrodynamic reaction caused by transverse motion of the ship (IMO 2016).

The Level 1 criterion adopts the "weather criteria" (IMO 2008), and made small modification with wave steepness factor (IMO 2015). The Level 2 criterion judges the vulnerability by estimating a long-term probability vulnerability index C, which is calculated according short-term roll motion characteristics by solving 1D.O.F motion equation. In this roll equation, actions of beam waves, mean wind and gusty wind are taken into account. If equation (15) is satisfied, the loading condition is considered to be non-vulnerable to Level 2 criterion.

$$\ddot{x} + 2 \cdot \mu_e(\sigma_{\dot{x}}) \cdot \dot{x} + \omega_{0,e}^2(\phi_s) \cdot x = \omega_0^2 \cdot m(t) \tag{14}$$

$$C < R_{Ds0} \tag{15}$$

where x is the roll angle with respect to the static heeling angle due to the action of mean wind; $\mu_e(\sigma_{\dot{x}})$ is the equivalent linear roll damping coefficient depending on standard deviation of the roll velocity $\sigma_{\dot{x}}$; $\omega_{0,e}(\phi_s)$ is an equivalent roll natural frequency; ω_0 is the upright roll natural frequency; $m(t)$ is the time dependent roll moment due to the action of waves and gustiness and R_{Ds0} is standard value.

3 SAMPLE CALCULATIONS AND VULNERABILITY ANALYSIS

3.1 Sample ships

In order to analyze the vulnerability of current ship designs and the potential effect of SGISC on

Table 1. Principal dimensions of sample ships.

Ship	Length m	FN	Breadth m	Draught m	Ship	Length m	FN	Breadth m	Draught m
Container ship1	87.3	0.218	15.20	3.4–4.5	Ro-Ro ship1	217.0	0.222	32.26	8.1–10.4
Container ship2	130.0	0.188	21.00	4.6–7.9	Ro-Ro passenger ship1	108.85	0.276	21.00	4.9–5.0
Container ship3	139.6	0.262	23.25	5.0–8.3	Ro-Ro passenger ship2	127.5	0.276	23.40	6.5
Container ship4	167.0	0.257	27.60	6.2–10.4	Ro-Ro passenger ship3	151.0	0.237	24.00	4.6–6.0
Container ship5	251.88	0.250	32.20	6.1–12.5	Ro-Ro passenger ship4	164.0	0.257	28.00	6.3
Container ship6	242.0	0.191	37.30	6.3–13.4	Public service vessel1	70.4	0.386	10.20	3.8–4.0
Container ship7	262.0	0.254	40.00	11.5	Public service vessel2	85.0	0.446	11.14	3.1–3.3
Container ship8	284.16	0.248	32.20	7.0–13.4	Public service vessel3	86.0	0.312	15.20	4.7–5.3
Container ship9	265.8	0.259	40.30	13.8–14.0	Public service vessel4	151.2	0.382	21.80	5.9–7.3
Container ship10	320.0	0.234	42.80	8.3–14.6	OSV1	64.2	0.277	17.20	5.2–6.1
Container ship11	319.0	0.248	48.60	7.9–14.7	OSV2	66.0	0.302	16.80	4.8–6.3
Bulk carrier1	115.0	0.195	22.00	5.1–8.2	Oil tanker1	320.0	0.148	60.00	9.9–22.6
Bulk carrier2	187.0	0.171	32.26	6.1–11.2	LNG1	92.0	0.227	16.40	4.3–4.8
Bulk carrier3	217.0	0.167	32.26	14.3	LNG2	126.0	0.164	21.50	3.1–4.1
Bulk carrier4	222.0	0.155	32.26	6.1–14.4	Salvage vessel1	88.0	0.313	15.20	5.3–6.3
Bulk carrier5	285.0	0.146	46.00	9.3–18.1	Salvage vessel 2	98.5	0.358	16.20	5.4–5.8
Bulk carrier6	314.6	0.139	52.50	11.1–17.6	Salvage vessel 3	112.7	0.313	16.00	5.4–6.0
Bulk carrier7	321.0	0.138	57.00	9.3–21.6	Tugboat1	80.0	0.301	20.00	4.6–7.2
Research ship1	88.8	0.256	17.20	5.5–6.0	Tugboat2	47.9	0.299	14.00	4.3–4.8

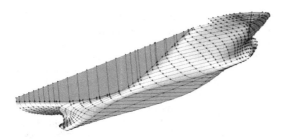

Figure 1. 3D model of container ship 10.

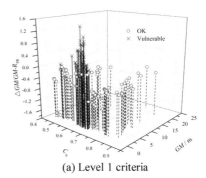

(a) Level 1 criteria

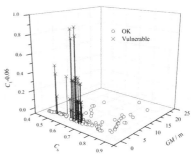

(b) 1st check of Level 2 criteria

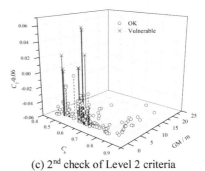

(c) 2nd check of Level 2 criteria

Figure 2. Results of PR criteria.

design procedure, 38 full scale ships with various types are checked. All these designs are approved by Administrations according to the current weather criterion, and are under commercial service. Table 1 shows the principal dimensions of the sample ships. Figure 1 shows the 3D model of Container ship 10.

3.2 Vulnerability analysis

A total of 147 loading conditions of sample ships are checked with Level 1 and Level 2 criteria of PR, PLS and EA. For DS and SR/B criteria, only Level 1 criteria are applied.

For DS criteria, due to the reduction of wave steepness factor on the basis of mandatory weather criteria, all loading conditions satisfy the Level 1 criteria. With regard to SR/B criteria, public service vessel, OSV, salvage vessel and tugboat may be judged as vulnerable due to their small lengths and high Froude numbers of service speed. These ships' $L = 66{\sim}152$ m, and $Fn = 0.301{\sim}0.446$. However, in the absence of necessary model test data, it is unable to further check Level 2 criteria for these vulnerable ships.

Figures 2–4 show the results of Level 1/2 criteria of PR, PLS and EA.

According to Figures 2 and 3, with the reduction of GM values, the vulnerability of PR and PLS dramatically increased. 6.8% of the loading conditions are judged as vulnerable to PR, and 31.3% are vulnerable to PLS.

For PR criteria, container ships, public vessels and Ro-Ro ships may be judged as vulnerable, whose $L = 86{\sim}320$ m and $C_b = 0.49{\sim}0.69$.

For PLS criteria, container ships, OSVs, public vessels, tugboats and Ro-Ro ships may be vulnerable, whose $C_b = 0.6{\sim}0.7$. However, as shown in Figure 3, it is found that the results of Level 2 criteria are inconsistent with Level 1 criteria. This is the so called "inconsistency problem", which is still under discussion by IMO and needs further validation.

For EA criteria, as shown in Figure 4, the vulnerability also increased with the increasing of GM values. Since only 2.72% loading conditions are judged as vulnerable to Level 1 criteria, and all of them satisfy the Level 2 criteria, the requirements could be regarded as easier to meet by design means. The container ships and Ro-Ro ships may be vulnerable, whose $L = 217{\sim}329$ m and $C_b = 0.58{\sim}0.63$.

According to the sample calculations, several types of ships may be vulnerable to failure models such as PR, PLS, EA and SR/B. After the SGISC be adopted as mandatory technical requirements by the Administrations, the current designs will

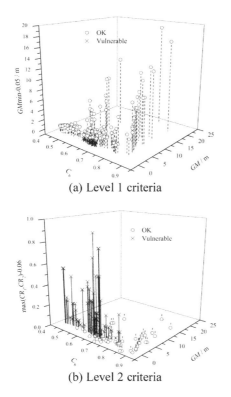

Figure 3. Results of PLS criteria.

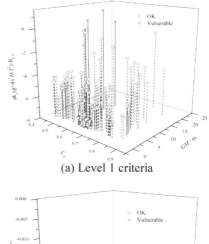

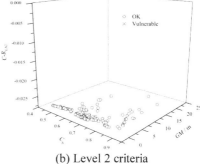

Figure 4. Results of EA criteria.

need to be adjusted and new designs to be developed accordingly.

In general, good designs for PR and PLS need to reserve sufficient transverse stability in wave, by keeping enough *GM* values or reducing the variations of *GZ* curves in wave for loading conditions. This usually means optimizing hull lines or setting upper limit for the height of gravity center (KG). On the contrary, setting lower limit for KG of ballast loading condition can lead to satisfaction of EA. Increasing roll damping coefficients will also be helpful in reducing the vulnerability of PR and EA, but it should be realized by passive measure such as modify the hull lines or enlarge the bilge keel. All these measures need more time spent on the optimization of hull lines and loading conditions. Since the design of hull lines needs to take a variety of influencing factors into account, this procedure will be more challenging and lead to more human resources consumption.

On the other hand, the calculation of SR/B Level 2 criteria needs characteristic data of ship hull speed and resistance. This means scaled model tests may be necessary. Therefore, the design iterative procedure will be more complicated and expensive.

4 EFFECT OF ESTIMATION METHOD OF ROLL MOMENT OF INERTIA AND POTENTIAL APPLICATION OF CFD APPROACH

4.1 *Effect of estimation method of roll moment of inertia*

In order to conduct motion simulations for Level 2 criteria or direct stability assessment, it is necessary to calculate the roll moment of inertia, which is usually ignored in the traditional design procedures. In the absence of detailed design data, IMO has provided several simplified formulas for estimation (IMO 2016) from the practical point of view. However, as a key issue of estimating natural roll period, this parameter is important for the calculation of vulnerability index of PR, DS and EA failure models. Error of roll moment of inertia may lead to completely different response characteristics.

Figure 5 shows an example of comparison of PR motion responses of 4 different estimation methods (Zhou, Ma, Gu 2013). The numerical simulations adopt 3 D.O.F model (Zhou, Ma, Lu, Gu 2015), including roll, heave and pitch.

According to these results, it is helpful to accurately reflect the motion response characteristics by ensuring the estimation accuracy of roll moment of inertia. This means that the estimation error may lead to deviation of vulnerability level assessment. Considering the inevitable errors of estimation formula, one of the potential solutions for accurate computation is to utilize FEM model (as shown in Figure 6), which should be based on detailed design plans. On the other hand, it should also be noted that the calculated roll moment of inertia is difficult to validate by means of full scale test of natural roll period. Therefore, it is still necessary to carry out further research on feasible test method.

4.2 Potential application of CFD approach

Different from weather criteria which are based on the traditional hydrostatic theory, the SGISC adopted hydrodynamic approaches in the Level 2 criteria and direct stability assessment. For most situations, these hydrodynamic approaches are based on potential flow theory. These numerical simulations need to estimate input parameters of ship such as roll damping, thrust and resistance, which are unable or difficult to estimate by simulations. Traditionally, scaled model test is the most reliable method to obtain these data. IMO suggests following the procedures in MSC.1/Circ.1200 or procedures approved by the Administrations. Alternatively, roll damping could also be estimated by simplified Ikeda's method, which is developed based on experience obtained from model tests. However, there is no similar simplified formula suggested for thrust and resistance by IMO.

With the advances of CFD approach in recent years (ITTC 2014), the viscous flow simulations could be regarded as design aids, especially for optimizing analysis of energy saving and resistance reduction. Researches show that the CFD approach could achieve more accurate numerical simulations for nonlinear phenomenon and reflect the effect of optimal designs. This provides a new solution for the direct calculation of parameters such as roll damping, thrust and resistance.

Figure 7 shows an example of simulations of roll damping by CFD (Zhou 2015 and 2016). Figure 8 shows an example of simulations of semi-submerged propeller.

The comparisons between simulations and model tests show that the CFD approach has a certain degree of practical value, and can be applied to ship design. As a flexible tool, the CFD approach could be utilized for optimum design and predict the performance before scaled model tests, which will greatly reduce the number of design iterations and save cost. Without doubt, the mesh generation and parameter settings of CFD approach need to be fully validated by scaled model tests and approved by the Administrations before its application in design process of actual ship.

Furthermore, it is even possible that the CFD approach will be put into use in the direct stability assessment of failure models in the near future. Figure 9 shows an example of direct simulations of parametric roll of Container ship 10. Although the CFD approach may be more time-consuming than potential flow method, it has incomparable advantages in the simulation of strong nonlinear motion response.

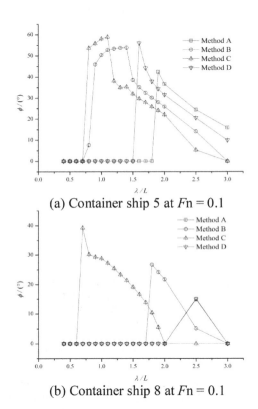

Figure 5. Comparison of PR motion response.

Figure 6. FEM model of Container ship 8.

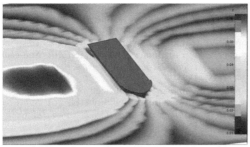

(a) Free decay of a 3100TEU container ship ($Fn=0$)

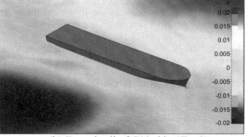

(b) Forced roll of C11 ship ($Fn=0$)

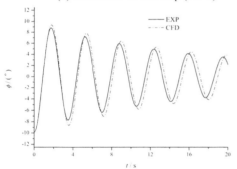

(c) Time history of free decay (3100TEU container ship)

Figure 7. Simulations of roll damping by CFD.

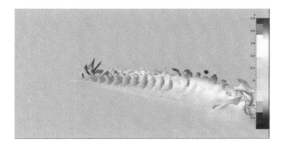

Figure 8. Simulations of semi-submerged propeller by CFD.

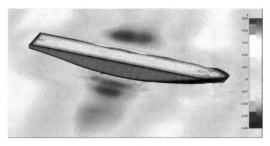

(a)

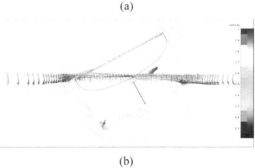

(b)

Figure 9. Simulations of parametric roll by CFD (Container ship 10).

5 CONCLUSIONS

Based on vulnerability analysis of Level 1 and Level 2 criteria for full scale ships and the experience of developing SGISC, the potential effect of SGISC on the design process has been analyzed in this work, including vulnerability of different types of ships, effect of estimation method of roll moment of inertia and the potential application of CFD approach. With the above research, the following conclusions are drawn:

The current designs of container ships, public vessels, OSVs, tugboats, salvage vessel and Ro-Ro ships may be vulnerable to Level 1 or Level 2 criteria of SGISC. *GM* may have significant effect on the vulnerability of PR, PLS and EA, and the accuracy of roll moment of inertia should be ensured as much as possible. Furthermore, the CFD approach may be more widely used in the future design process, especially for the vulnerability check of SGISC and subsequent optimal iterations.

REFERENCES

Fossen, T.L., Nijmeijer, H. 2012. *Parametric Resonance in Dynamical Systems*. Dordrecht:Springer.
France, W.N., Levadou, M., Treakle, T.W., et al. 2001. An Investigation of Head-Sea Parametric Rolling and its

Influence on Container Lashing Systems, *In SNAME Annual Meeting 2001 Presentation*: 1–24.

Germany. 2012. Further background information on the proposal by Germany with regard to the new generation intact stability criteria (INF.10). *Stability and Load lines and on Fishing vessels safety (SLF) 54th session*: 1–74.

IMO. 2008. *International code on intact stability, 2008.* London: International Maritime Organization.

IMO. 2015. Development of second generation intact stability criteria-Report of the working group (Part 1) (WP.4/Annex 1~3). *Ship Design and Construction (SDC) 2nd Session, IMO*: 11–29.

IMO. 2016. Finalization of second generation intact stability criteria-Report of the working group (Part 1) (WP.5/Annex 1~7). *Ship Design and Construction (SDC) 3rd Session, IMO*: 13–119.

ITTC. 2014. Specialist Committee on CFD in Marine Hydrodynamics-Final report and Recommendations to the 27th ITTC. *Proceedings of 27th International towing tank conference*, Copenhagen: 1–53.

Japan. 2011. Information collected by the Correspondence Group on Intact Stability (INF.12). *Sub-Committee on Stability and Load lines and on Fishing vessels safety 54th session, IMO*: 1–147.

Japan. 2012. Information collected by the Correspondence Group on Intact Stability (INF.15). *Sub-Committee on Stability and Load lines and on Fishing vessels safety 55th session, IMO*: 1–211.

Japan. 2013. Information collected by the Correspondence Group on Intact Stability regarding the second generation intact stability criteria development (INF.8). *Sub-Committee on Ship Design and Construction 1st session, IMO*: 1–132.

Japan. 2014. Information collected by the Correspondence Group on Intact Stability (INF.10), *Sub-Committee on Ship Design and Construction 2nd session, IMO*: 1–198.

Japan. 2015. Information collected by the Correspondence Group on Intact Stability (INF.10). *Sub-Committee on Ship Design and Construction 3rd session, IMO*: 1–164.

Neves, M.A., Belenky, V.L., Kat, J.O., Spyrou, K. Umeda, N. 2011. *Contemporary Ideas on Ship Stability and Capsizing in Waves.* Dordrecht:Springer.

Zhou, Y., Ma, N., Gu, X. 2013. Effect of calculation methods of roll moment of inertia on sensitivity prediction of parametric rolling for container ships. *Shipbuilding of China* 54(3):11–20.

Zhou, Y., Ma, N., Lu, J., Gu, M. 2016. A study on hybrid prediction method for ship parametric rolling. *Journal of Hydrodynamics* 28(4): 617–628.

Zhou, Y., Ma, N., Lu, J., Gu, X. 2015. A Study on Applicability of CFD Approach for Predicting Ship Parametric Rolling. *Proc. of the 12th Int'l. Conf. on the Stability of Ships and Ocean Vehicles*, Glasgow: 465–474.

Zhou, Y., Ma, N., Shi, X., Zhang, C. 2015. Direct calculation method of roll damping based on three-dimensional CFD approach. *Journal of Hydrodynamics* 27(2): 176–186.

Operational profile based evaluation method for ship resistance at seas

P.Y. Feng
Shanghai Key Laboratory of Ship Engineering, Shanghai, China

S.M. Fan
Science and Technology Laboratory of Water-Jet Propulsion, Shanghai, China

Y.S. Wu & X.Q. Xiong
Marine Design and Research Institute of China (MARIC), Shanghai, China

ABSTRACT: This paper proposes an evaluation method for the ship resistance at seas, taking into account both the calm water resistance and the wave added resistance of the ship. In the light of the Product Lifecycle Management (PLM) concept, a long-term performance prediction model based on the ship's operational profile is proposed, which systematically takes into account the influencing factors such as the ship's speed profile, the navigation sea areas and the long-term wave statistics. Through this model, a more comprehensive and more robust comparison between different hull form designs can be achieved from the long-term operational perspective. To demonstrate the method, two cruise ship hull forms are evaluated in order to determine the preferred design in terms of smaller total resistance at seas. One design adopts a bulbous bow while the other one adopts the vertical bow concept. The calm water towing test of the two hull models are performed in the towing tank of MARIC, and the added resistance in regular waves are evaluated numerically. Using the proposed evaluation model, the resistance performance of the two designs in waves are analyzed in detail. It is shown that the long-term mean wave added resistance of the two hull forms are very close to each other, which are approximately 20% of their calm water resistances. Finally, the comparison result shows that the total resistance of the vertical bow design is 2.7% lower than that of the bulbous bow design.

1 INTRODUCTION

The Energy Efficiency Design Index (EEDI) was made mandatory for new ships at MEPC 62 with the adoption of amendments to MARPOL Annex VI (Resolution MEPC.203(62) 2011). The EEDI requires a minimum energy efficiency level per capacity mile for different ship type and size segments, which was the first legally binding climate change treaty to be adopted since the Kyoto Protocol. As EEDI is going to enter Phase II with an aim of lowering EEDI baseline by up to 20%, it is now of urgent need to develop and explore possible methods to improve the ship's energy efficiency.

Minimization of the ship's resistance has long been the goal of ship designers and is no doubt one of the effective ways to improve ship efficiency. Reduction of the ship's resistance means less fuel consumption during operation and can consequently lead to less green-house-gas emission. Therefore, hull form optimization and hydrodynamic design innovation for minimum resistance has attracted a lot of attention in the ship design field.

The development of hull form optimization for minimum resistance is developing at an incredibly fast pace, from model scale to full scale, from local optimization to fully parametrical optimization (Yang et al. 2016). In preliminary applications, only the wave-making resistance is considered based on potential theory (Suzuki et al. 1999, Zhang 2012, Goren et al. 2017). With the technological advancement of Computational Fluid Dynamics (CFD) and computing powers, fully viscous simulations have been made possible which can compute the ship's total resistance in calm water (Li et al. 2014, Park et al. 2015).

Nowadays, it is almost unanimously agreed in the ship design field that hull form design or optimization focusing on only one specific condition (i.e. one draft with one speed) is seldom effective and may lead to deteriorated performance during actual operations. The common practice to address this fallacy is to optimize the ship's resistance for a series of drafts and speeds, hence the operational profile based evaluation of the ship's resistance. The operational profile should be developed closely together with the ship owner to ensure that the selected operational profile reflects the planned operation of the ship. Temple et al. (2012) studied the multi-objective hull form optimization to compare build cost and lifetime fuel consumption. Wagner et al. (2014) studied the scenario based optimization of a container vessel with respect to its projected operating conditions. However, these studies only consider the ship's resistance in calm water, thus still missing one key factor of the operational profile, i.e. the waves.

In the presence of waves, ships will experience an additional resistance known as the added resistance. According to statistics, there is slightly more than 11% probability worldwide that an open sea ship will encounter waves with height less than 0.5 m (Bolbot et al. 2016). In other words, ship's resistance in waves is a better reflection of her economic competitiveness than that in calm water. Aiming at the reduction of wave added resistance, the Axe-bow and Leadge-bow concepts are proposed and investigated by several authors (Hirota et al. 2005, Hwang et al. 2013). However, only the influence of bow shape on wave added resistance is discussed in these studies, so there is no comprehensive analysis on how they will affect the ship's actual performance during operation. Therefore, it is hard to conclude whether these novel bow shapes are superior to the original bulbous bow design. Referring to the state-of-the-art studies, Yu et al. (2017) carried out bow hull-form optimization in waves of a 66,000 DWT bulk carrier, where multi-objective functions are applied to minimize the wave-making resistance in calm water and added resistance in regular head wave at one specific frequency. This is a better way to evaluate the ship's resistance performance at sea, but one regular wave frequency is far from enough.

In this study, an evaluation method for the ship resistance at seas is proposed, taking into account both the calm water resistance and the wave added resistance of the ship. In the light of the Product Lifecycle Management (PLM) concept, a long-term performance prediction model based on the ship's operational profile is established, which systematically takes into account the influencing factors such as the ship's speed profile, the navigation sea areas and the long-term wave statistics. Through this model, a more comprehensive and more robust comparison between different hull form designs can be achieved from the long-term operational perspective.

To demonstrate the method, two cruise ship hull forms are evaluated in order to determine the preferred de-sign in terms of smaller total resistance at seas. One design adopts a bulbous bow while the other one adopts the vertical bow concept. The calm water towing test of the two hull models are performed in the towing tank of MARIC, and the added resistance in regular waves are evaluated numerically. Using the proposed evaluation model, the resistance performance of the two designs in waves are analyzed in detail according to their actual operational profile and the long-term wave statistics of their operational sea areas.

2 RESISTANCE EVALUATION MODEL

In order to evaluate the resistance performance of different hull form designs considering the ship's actual operational profile, three important factors need to be taken into account, namely, the speed profile, the draft/trim profile and the environmental conditions. The speed profile and the draft/trim profile can be highly correlated and are often presented in pair. This is especially the case for containerships. For cruise ships, the drafts usually don't change very much during the operation, but the ship may operate at several speeds under different circumstances.

On the other hand, the real long-term resistance performance of the ship at actual seas cannot be represented by that in regular waves or under certain short-term sea conditions. To put it in simple words, one hull form design may have minimum wave added resistance under one sea state but suffer from large wave added resistance under other sea states. This is because wave added resistance is a function of wave height and wave period, so all possible wave conditions that the ship might encounter should be considered.

Taking the above factors into account, the following ship resistance evaluation model is proposed, as illustrated in Figure 1, which involves four layers of evaluation.

In layer 1, the short-term mean wave added resistance R_{AW}^{MS} is estimated by spectral analysis based on the principle of linear superposition according to Equation 1:

$$R_{AW}^{MS}\left(H_S,T_Z;V_j,d_i\right) = 2\int_0^\infty R_{AW}\left(\omega;V_j,d_i\right) \cdot \frac{S_\zeta(\omega,H_S,T_Z)}{\zeta_a^2} d\omega \quad (1)$$

where R_{AW} is the ship's mean wave added resistance in regular waves; S_ζ is the energy spectrum to describe a short-term sea state, with H_S the significant wave height and T_Z the mean zero-crossing wave period; ω is the regular wave frequency; ζ_a is the amplitude of the regular wave. Notice that the wave added resistance is proportional to the square of wave amplitude. R_{AW}^{MS} corresponding to all relevant ship speeds V_j and drafts d_i need to be calculated.

In layer 2, the short-term mean wave added resistance of each sea state is weighted against the

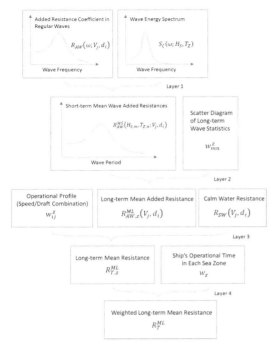

Figure 1. Evaluation model of ship resistance at seas.

occurrence percentage of that particular sea state denoted by w_{mn}^z, so as to estimate the long-term mean wave added resistance $R_{AW,z}^{ML}$ as follows:

$$R_{AW,z}^{ML}(V_j,d_i) = \sum_{m=1}^{N_H}\sum_{n=1}^{N_T} w_{mn}^z \cdot R_{AW}^{MS}(H_{S,m},T_{Z,n};V_j,d_i) \quad (2)$$

In layer 3, the ship's total resistances corresponding to each speed and draft combination in the operational profile are calculated as the summation of calm water resistance R_{SW} and the long-term mean wave added resistance $R_{AW,z}^{ML}$, and are further weighted against the percentage of that particular operational condition denoted by w_{ij}^z:

$$R_{T,z}^{ML} = \sum_{i=1}^{N_d}\sum_{j=1}^{N_V} w_{ij}^z \cdot \left[R_{SW}(V_j,d_i) + R_{AW,z}^{ML}(V_j,d_i)\right] \quad (3)$$

Finally, in layer 4, the weighted long-term mean resistance of the ship across a total of N_Z operational sea area is obtained as follows:

$$R_T^{ML} = \sum_{z=1}^{N_Z} w_z \cdot R_{T,z}^{ML} \quad (4)$$

where the weight w_z is determined by the ship's operation time in each sea area.

This long-term estimation model captures most of the relevant influencing factors on ship's resistance during actual operation at sea, hence different designs become directly comparable in terms of the overall performance both in calm water and in waves.

3 HULL DESIGN AND OPERATIONAL PROFILE

Figure 2 shows the two cruise ship hull designs to be compared in this study. Design A on the top adopts the bulbous bow while Design B at the bottom adopts the vertical bow.

The main dimensions of the two designs are listed in Table 1. Due to the vertical bow, the length of Design B is larger than that of Design A.

This cruise ship targets the Asian market and is meant to serve on routes among major ports such as Shanghai, Busan, Nagasaki and so on, as shown in Figure 3. The operational region is further divided into three sea zones, according to the characteristics of the long-term wave statistical data. We assume that the cruise ship will spend most of her time in zone Z01 and same amount of time in zones Z02 and Z03, hence the values of w_z are $w_1 = 0.6$, $w_2 = 0.2$ and $w_3 = 0.2$.

The operational profile of the cruise ship is listed in Table 2. Since the draft of the cruise ship is not likely to change very much during the operation, there is only one draft to be considered in the operational profile. On the other hand, there

(a) Design A: Bulbous Bow

(b) Design B: Vertical Bow

Figure 2. Cruise ship hull designs.

Table 1. Main dimensions of the two designs.

Name	Design A	Design B
Length Perpendicular (m)	239.0	252.0
Waterline Length (m)	248.7	261.7
Breadth (m)	32.25	32.25
Fore Draft (m)	7.4	7.4
Aft Draft (m)	7.4	7.4
Block Coefficient	0.657	0.623
Displacement Volume (m^3)	37474	37467
Wetted Surface Area (m^2)	8833.9	8968.1
Longitudinal Center of Buoyancy (m)	−7.9	−11.3
Vertical Center of Gravity (m)	14.65	14.65
Radius of Pith Inertia (m)	66.1	66.1

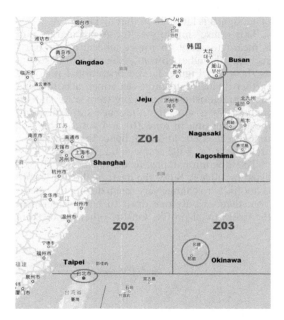

Figure 3. Potential operational sea area of the cruise ship.

Table 2. Main dimensions of the two designs.

Condition	Design speed	Cruise speed	Slow steaming
Speed (kn)	22.0	19.5	14.5
Draft (m)	7.4	7.4	7.4
Percentage (%)	10	60	30

are three speeds included in the operational profile, i.e. the design speed, the cruise speed and the slow steaming speed. According to this profile, this cruise ship will operate at the 19.5 kn cruise speed for the majority of the time (60%); next is the 14.5kn slow steaming condition, which takes up 30% of the operational profile; the least frequent condition is, ironically, the 22kn design speed (10%). This again demonstrates the importance of optimize the ship's performance based on her operational profile rather than just the design condition. We assume that this profile holds for the ship in all three sea zones, hence the values of w_{ij}^z are the same for $z = 1,2,3$.

4 DESIGN COMPARISON

Based on the operational profile described above, we proceed to the comparison of the two designs in order to find the better one with lower expected long-term mean resistance R_T^{ML}.

Firstly, the calm water resistances of the two designs are compared as shown in Table 3. As can be seen, for the three speeds in the operational profile, the calm water resistances of the vertical bow design are all smaller than those of the bulbous bow design, especially under the slow steaming condition. Taking the operation profile into account, the weighted reduction is 4%. This result seems to be counter-intuitive because the bulbous bow design is generally believed to be effective at reducing the wave resistance, thus should be superior to the vertical bow design. However, in the current case, the length of Design B is noticeably longer than that of Design A, which is beneficial for the overall resistance reduction. Besides, this bulbous bow design is compromised to achieve lower resistance across the whole speed range, therefore it is not the optimal design solution for one certain speed.

Next, the wave added resistances of the two designs are evaluated. The mean wave added resistances in regular waves are calculated by NEWDRIFT+, which is an in-house program developed by National Technical University of Athens – Ship Design Laboratory (NTUA – SDL). NEWDRIFT+ is a 6DOF, 3D panel code for the seakeeping and wave induced loads analysis of

Table 3. Calm water resistance comparison.

Speed (kn)	Design	R_{SW} (kN)	Difference (%)
14.5	A	456.9	−9.3
	B	414.4	
19.5	A	789.8	−2.4
	B	770.8	
22	A	1060.4	−4.6
	B	1011.7	
Weighted	A	716.99	−4.0
	B	687.97	

ships, which allows the prediction of the added resistance of ships advancing with certain speed in waves. The theoretical background and the validation of the added resistance module are elaborated in reference (Liu et al. 2011).

Figure 4 shows the comparison of the mean wave added resistance coefficients of the two designs under the speeds according to the operational profile. As can be seen, the coefficients of the two designs follow the same trend and the speed influence on wave added resistance is not strong. However, it can be observed that the coefficients of Design A are smaller in short wave ranges than those of Design B, while the coefficients of Design B around the peak region are smaller than those of Design A. This can be explained by the fact that the vertical bow design helps reduce the ship motions in long waves, thus suffering from lower wave added resistance when ship motions are the mean cause of added resistance. On the other hand, wave added resistance in short waves is mainly caused by wave reflection/diffraction, which is closely related to the ship's bow shape. In this sense, the bow shape of Design A is superior to that of Design B.

If we further look at the wave spectrum of two typical sea states that ship is highly likely to encounter during operation at seas, as shown in Figure 4. It can be seen that the wave energy mainly concentrates in short wave components. Therefore, the wave added resistance responses in short waves are more important than those in long waves for the current design.

This is demonstrated in Figure 5, which shows the comparison of short-term mean wave added resistances per square wave height under various wave periods. The ITTC two-parameter spectrum is adopted in this study for the short-term prediction of wave added resistance:

$$S_\zeta(\omega, H_S, T_Z) = \frac{124 H_S^2}{T_Z^4 \omega^5} \exp\left\{-\frac{496}{T_Z^4 \omega^4}\right\} \qquad (5)$$

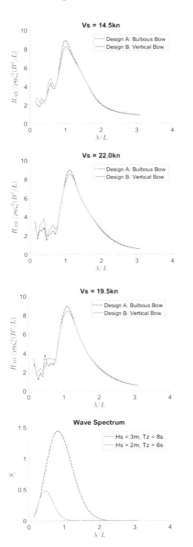

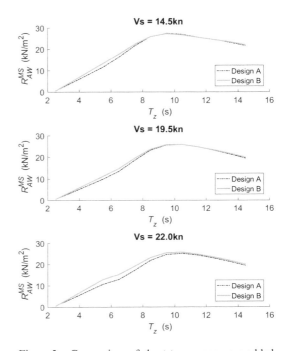

Figure 4. Mean wave added resistance coefficients and typical wave spectrum.

Figure 5. Comparison of short-term mean wave added resistances.

As can be seen, although the wave added resistance responses of Design B around peak region are smaller, the short-term mean values are slightly larger than those of Design A. However, we have to admit that the differences are very small and even negligible in terms of engineering significance.

Finally, the expected long-term mean wave added resistance $R_{AW,z}^{ML}$ is obtained by taking into account the occurrence probability of all combinations of short-term sea states (H_S, T_Z) according to the accumulated long-term wave statistics. Figure 6 shows the contours of occurrence percentage of the sea states. In zone Z01, the most likely sea states that the ship may encounter are wave heights between 1 m and 2 m, with wave periods around 7 s and smaller than 4 s. The sea states in zones Z02 and Z03 are similar and the most likely sea states that the ship may encounter are wave heights around 1.5 m with wave periods around 4 s.

The long-term prediction results of each sea zone and each ship speed are shown in Table 4. According to the result, the expected long-term mean wave added resistances of Design A are slightly smaller than those of Design B. To be more specific, the final weight value R_{AW}^{ML} of Design A is 156.4 kN, which is 3.3% smaller than that of Design B (161.5 kN).

Table 5 summarizes the resistance data of the two designs based on our proposed model. From the result, it is demonstrated that for low speed (14.5kn), the wave added resistance is approximately

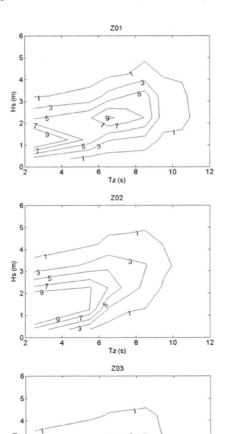

Figure 6. Contours of sea state occurrence percentage (%).

Table 4. Long-term prediction of mean wave added resistance.

	Design A: Bulbous bow			
	Mean wave added resistance (kN)			Weighted average R_{AW}^{ML} (kN)
Sea zone	$V_1 =$ 14.5 kn	$V_2 =$ 19.5 kn	$V_3 =$ 22.0 kn	
Z01	169.5	152.4	146.5	157.0
Z02	170.4	153.6	147.9	158.1
Z03	164.4	148.6	143.4	152.8
All	168.7	151.9	146.2	156.4
	Design B: Vertical bow			
	Mean wave added resistance (kN)			Weighted average R_{AW}^{ML} (kN)
Sea zone	$V_1 =$ 14.5 kn	$V_2 =$ 19.5 kn	$V_3 =$ 22.0 kn	
Z01	173.3	157.5	157.6	162.2
Z02	174.2	158.5	158.8	163.2
Z03	167.7	153.2	153.5	157.6
All	172.4	156.8	157.0	161.5

Table 5. Comparison of resistance performance.

Speed (kn)	Design	R_{AW}^{ML}/R_{SW} (%)	R_T^{ML} (kN)	Difference (%)
14.5	A	36.9	625.6	−6.2
	B	41.6	586.8	
19.5	A	19.2	941.7	−1.5
	B	20.3	927.6	
22	A	13.8	1206.6	−3.1
	B	15.5	1168.7	
Weighted	A	21.8	873.3	−2.7
	B	23.5	849.5	

40% of calm water resistance, which is quite high. As the speed increases to 22kn, this ratio drops to around 15%. The weighted wave added resistance according to the operation profile is approximately 20% of calm water resistance. This implies that the priority should be to optimize the hull form for lower calm water resistance and it is rational to put five times more weights on the calm water resistance over wave added resistance when setting up the optimization, if a single objective optimization framework is adopted.

The final weighted total resistance according to the operation profile is 873.3 kN for Design A and 849.5 kN for Design B. This corresponds to a 2.7% reduction of total resistance of Design B compared to Design A. Recall the comparison result from Table 3, the reduction of calm water resistance of Design B over Design A is 4%. This implies that Design A has smaller wave added resistance during operation at sea, compared to Design B.

5 CONCLUSIONS

The main contribution of this study is the proposal of an evaluation method for the ship resistance at seas, taking into account both the calm water resistance and the wave added resistance of the ship. This long-term performance prediction model based on the ship's operational profile systematically takes into account the influencing factors such as the ship's speed profile, the navigation sea areas and the long-term wave statistics. Through this model, a more comprehensive and more robust comparison between different hull form designs can be achieved from the long-term operational perspective. And the prediction result can provide valuable insight for hull form optimization.

Using the proposed model, two cruise ship hull forms are compared in terms of the total resistance at seas. To conclude, the comparison result shows that the total resistance of the vertical bow design is 2.7% lower than that of the bulbous bow design, and the long-term expected wave added resistance is approximately 20% of calm water resistance. It should be pointed out that neither the two hull forms are the final optimized ones, and subsequent hull form optimization using the proposed resistance estimation model will be carried out in the future.

ACKNOWLEDGEMENTS

This study is supported by the High-tech Ship Research Project funded by the Ministry of Industry and Information Technology of China (No. [2016]25) and the National Natural Science Foundation of China (No. 51709240).

REFERENCES

Bolbot, V., Papanikolaou, A.2016. Optimization of ship's bow form for the added resistance in waves. *International Conference on Marine Technology*, Dhaka, Bangladesh.

Goren, O., Calisal, S., Danisman, D. 2017. Mathematical programming basis for ship resistance reduction through the optimization of design waterline. *Journal of Marine Science and Technology* 22: 772–783.

Hirota, K., Matsumoto, K., Takagishi, K., Yamasaki, K., Orihara, H., Yoshida, H. 2005. Development of bow shape to reduce the added resistance due to waves and verification of full scale measurement. *First International Conference on Marine Research and Transportation (ICMRT)*, Ischia, Italy.

Hwang, S., Kim, J., LEE, Y., AHN, H., VAN, S., KIM, K. 2013. Experimental study on the effect of bow hull forms to added resistance in regular head waves. *12th International Symposium on Practical Design of Ships and Other Floating Structures (PRADS)*, Changwon, Korea.

Janson, C., Larsson, L. 1997. A Method for the Optimization of Ship Hulls from a Resistance Point of View. *Twenty-First Symposium on Naval Hydrodynamics*, Trondheim, Norway.

Li, S., Zhao, F., Ni, Q. 2014. Bow and stern shape integrated optimization for a full ship by a simulation-based design technique. *Journal of Ship Research* 58(2): 83–96.

Liu, S., Papanikolaou, A., Zaraphonitis, G. 2011. Prediction of added resistance of ships in waves. *Ocean Engineering* 38: 641–650.

Park, J., Choi, J., Chun, H. 2015. Hull-form optimization of KSUEZMAX to enhance resistance performance. *International Journal of Naval Architecture and Ocean Engineering* 7: 100–114.

Resolution MEPC.203(62). 2011. Amendments to MARPOL Annex VI on Regulations for the Prevention of Air Pollution from Ships by Inclusion of New Regulations on Energy Efficiency for Ships. *International Maritime Organization*, MEPC 62/24/Add.1, London.

Suzuki, K., Iokamori, N. 1999. Studies on minimization of wave making resistance based on Rankine source method. *Journal of Society of Naval Architects of Japan* 185: 9–19.

Temple, D., Collette, D. 2012. Multi-objective hull form optimization to compare build cost and lifetime fuel consumption. *International Marine Design Conference*, Glasgow, Scotland, 2012.

Wagner, J., Binkowski, E., Bronsart, R. 2014. Scenario based optimization of a container vessel with respect to its projected operating conditions. *International Journal of Naval Architecture and Ocean Engineering*, 6: 496–506.

Yang, C. & Huang, F. 2016. An overview of simulation-based hydrodynamic design of ship hull forms. *Journal of Hydrodynamics, Ser. B* 28(6): 947–960.

Yu, J., Lee, C., Lee, I., Choi, J. 2017. Bow hull-form optimization in waves of a 66,000 DWT bulk carrier. *International Journal of Naval Architecture and Ocean Engineering* 9: 499–508.

Zhang, B. 2012. Shape Optimization of Bow Bulbs with Minimum Wave-Making Resistance Based on Rankine Source Method. *Journal of Shanghai Jiaotong University*, 17(1): 65–69.

On the importance of service conditions and safety in ship design

R. Grin, J. Bandas, V. Ferrari, S. Rapuc & B. Abeil
Maritime Research Institute Netherlands, Wageningen, The Netherlands

ABSTRACT: A good ship design is a balanced design, in which the performance is not limited (or outperformed) by one specific characteristic. Today's state of the art design tools help the designer to make this trade-off, for instance by means of multi-objective optimisation techniques. However these techniques require clear objectives, and underlying calculation approaches should be accurate, otherwise a suboptimal solution is found in the best case (or worse, no solution is found at all). The design objectives are complicated by including real operational conditions, such as 'off-design' speed and loading condition, wind, waves, and confined water. In this 'design for service', these objectives are often not readily available and calculation approaches are not clearly defined. This paper provides some ideas and guidelines about how to approach such a problem applied to a 120 m cruise vessel.

1 INTRODUCTION

In the present paper, a balanced design approach is described which takes into account the service conditions of a vessel. Objectives and calculation approaches were defined for a 120 m expedition type cruise vessel in operational conditions. As a first step, a matrix was created with operational issues and the accompanying hydrodynamics (see Appendix A).

The operational issues were subdivided into safety aspects, onboard activities and port calls. For each of the operational aspects a listing is made of specific issues that might hamper the operation. Note that the 'onboard activities' aspect is specific for passenger ships. However such a matrix can be suited for any ship by replacing it with the ship's mission(s) or more generally its earning capacity. In addition future versions of the matrix should also include the cost side to account for e.g. fuel consumption.

The hydrodynamics axis is also subdivided into three groups:

- Linear ship behaviour like roll, accelerations and shipping green water
- Non-linear ship behaviour (like parametric roll and slamming-induced vibrations)
- Controllability (like course keeping in waves, broaching risk and crabbing)

Appendix A provides a more complete listing of all hydrodynamic issues within each group. For all of the issues, suitable calculation approaches are provided (tools) and objectives (criteria) are suggested.

The contents of this paper follow the same hydrodynamic subdivision as above. After introducing the ship and adopted operational profile in section 2, the linear ship behaviour is discussed in section 3. Traditionally this is the focus of operability studies, as tools and criteria are generally proven and readily available in the industry.

In section 4, non-linear ship behaviour is discussed. It starts with an evaluation of the IMO vulnerability criteria. As soon as these criteria come into force, they will be the starting point for ship designers to evaluate seakeeping risks. As parametric roll is a known issue for cruise vessels it is discussed in more detail and results of time domain seakeeping calculations are shown. The chapter is concluded with a straightforward evaluation of the risk of bow flare slamming.

The last section discusses controllability issues. It starts with the course keeping capability in wind and waves. As cruise ships have a large wind area this might be quite challenging. Results of a time domain seakeeping-manoeuvring tool are presented. The IMO Safe Return To Port regulations are briefly discussed, but the emphasis of the second part of the chapter is on harbour operations (crabbing and tender operations), which are a key element in cruise ship operations.

2 OPERATIONAL PROFILE

2.1 *Cruise ship design*

A typical expedition type cruise vessel design has been taken as reference ship (see Table 1). This twin-shaft vessel is equipped with diesel-direct propulsion, two rudders, one stern thruster and two bow

Table 1. Main dimensions of cruise vessel.

Quantity	Units	Value
Length perpendiculars	[m]	120.0
Breadth	[m]	17.0
Draught	[m]	4.0
Displacement	[tonnes]	5,800
Metacentric height	[m]	1.7
Natural period of roll	[s]	14.3
Design speed	[kn]	18

thrusters. For roll reduction, the ship has bilge keels and one pair of 6 m² retractable high aspect ratio fin stabilisers.

Some of the main passenger facilities include a restaurant, small theatre, lounge and pool. Most of the public areas are located one deck below the top deck, at 18 m above the ship's baseline.

The lounge, pool and tender area are selected to illustrate the operability. The lounge is chosen because of its forward location at two-thirds of the ship length. At this location, the vertical motions are relatively large due to pitching. The outdoor pool is selected because of its aftward location, only 7 m forward of the aft perpendicular. This is one of the most sheltered locations outside while in transit. The tender area is located at the stern. For bigger cruise vessels, the tender areas are typically located in the midships area to allow creating leeway over a large length. However, for the size of cruise vessel being studied, this is often not possible because of limited space. Therefore the tender area is located at the stern.

2.2 Adopted operational profile

Expedition type cruise vessels are deployed for cruising in remote areas, such as the Arctic, British Columbia, the Amazon, Antarctica or the Pacific islands. These cruises visit smaller ports, often with few to no facilities and often anchor and use tenders to go to shore.

For this paper, a typical operation is considered with cruises in Antarctica and the Pacific islands in the summer of the Southern hemisphere. When heading to the Northern hemisphere, one or two longer repositioning cruises are done. While in the Northern hemisphere, the vessel will mainly sail the Arctic until summer ends. Because of this sailing profile, the encountered weather conditions are relatively harsh. This is clearly shown in the wave scatter diagram in Figure 1. This scatter diagram is based on hindcast data along the sailing route for a period of 10 years, provided by the European Centre for Medium-Range Weather Forecasts (ECMWF).

Even in summer, there is a 0.1% chance of encountering wave heights between 8 and 9 m and

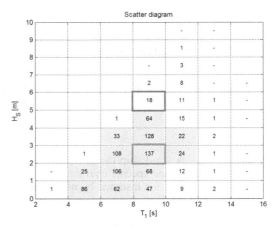

Figure 1. Wave scatter diagram for operational profile.

a mean wave period between 10 and 12 s. This may seem like a small chance, but it means the vessel could encounter these conditions for an average of 4 hours each year if no operational measures like rerouteing are taken.

Two wave conditions are selected to illustrate the operability. The first is a Typical Operational Condition (TOC) with a significant wave height (H_s) of 2.5 m and a mean wave period (T_1) of 9 s. This condition occurs 13.7% of the time. The second condition occurs 1.8% of the time and is denoted TAC (Typical Adverse weather Condition).

Note that all wave conditions should be addressed for a full operability assessment or when operational guidance is given to the ship's crew.

3 LINEAR SHIP BEHAVIOUR

3.1 Classical criteria

The study by the Nordic Co-operative Organization for Applied Research (NORDFORSK, 1987) summarised criteria for tolerable ship behaviour and these are widely used in the industry. These criteria are used for reference in this study.

Nordforsk gives a criterion for the vertical accelerations (AZ) of 0.02 g root mean square (RMS). This criterion is valid for older people and passengers on a cruise liner. It is stated that this criterion is 'close to the lower threshold below which vomiting is unlikely to take place'. Criteria for roll and transverse accelerations (AY) are also provided, though it is mentioned that these criteria may not be as reliable as the AZ criterion. The criteria for RMS roll and RMS AY are 2 deg and 0.03 g, respectively.

The NORDFORSK study supplies criteria for roll, AY and AZ individually. However, these criteria do not account for combined motions which may lead to higher discomfort (see section 3.2).

In addition, the NORDFORSK study suggests criteria for shipping green water (5% of the wave encounters) and slamming (around 3% for a 120 m ship). Note that there are no separate criteria provided for passenger ships, so these limits apply for all merchant ships.

The slamming criterion in NORDFORSK is the classical Ochi criterion, which refers to bottom slamming and assumes a slam when the fore ship comes out of the water and its re-entry velocity is higher than a certain threshold. This approach is questionable for passenger ships as these typically have no flat bottom in the bow sections (forward of ordinate 14 the reference ship has V-shaped sections). Therefore the Ochi slamming criterion cannot be accurately used to assess bow slamming, even though it will certainly occur (see section 4.4).

The well-known Motion Sickness Incidence, abbreviated as MSI (Colwell, 1989), accounts for the human sensitivity to vertical accelerations and habituation (if you are exposed to ship motions for a longer time, you get used to them). A commonly applied criterion for passengers is 10% MSI within the first hours of exposure (ISO 2631/3–1985). Within this paper, the 'first hours' are interpreted as 4 hours, which is in line with the criteria provided in NATO STANAG 4154 (NATO STANAG, 2000). A more recent comfort indicator, which accounts for vertical and horizontal accelerations, is given in section 3.2.

Figure 2 shows the operational limits for TOC assuming active fin stabilisers (computations on the basis of the linear strip-theory tool SHIPMO). In head and bow quartering seas, the MSI criterion at the lounge is exceeded at speeds above 10 kn. Vertical accelerations are exceeded for headings above 75 deg from the stern at all speeds. Transverse accelerations are exceeded in a small region close to beam seas. This occurs due to the combined sway and yaw, rather than the roll motions. Shipping of green water over the bow is unlikely in this condition.

Based on these Nordforsk criteria, passenger comfort would only be acceptable in stern quartering and following seas, leaving little opportunity for the ship's crew to reduce ship motions when the ship's track requires her to sail in a 'wrong' direction. This situation is even more difficult for TAC. However, the picture becomes more acceptable when using 'combined criteria'.

3.2 Combined criteria

Recent developments in the EU Compass project (Turan, 2006) resulted in the Motion Illness Rating (MIR), which does account for the combined effect of AY and AZ on seasickness. In addition, the MIR accounts for the human sensitivity to certain frequencies of motion, as well as gender, age and acclimatisation to ship motions. The MIR ranges from 0 to 100, where 0 means 'I feel alright' and 100 means 'I feel absolutely dreadful'. The MIR used in the present work assumes the following 'typical' cruise vessel population: 50% male, 50% female and an average age of 45. In addition, it is assumed that the passengers are not yet adapted to ship motions, which makes them more susceptible to seasickness (exposure time one hour). A MIR criterion of 10 is chosen, which means 'fairly comfortable' (Dallinga and Bos, 2010).

While seasickness plays an important role in comfort, so does the ability to perform activities without disruption. Disruptions relate to the ability to keep one's balance while standing, walking, climbing stairs and dancing. These activities are indicated by the 'effective gravity angle' (EGA), which is the angle between the horizontal accelerations and the sum of the vertical acceleration and gravity. A high EGA cause for instance standing persons to look for support, tipping or sliding of glasses but is also an indicator for sloshing in a pool. In the study of Dallinga and Bos, an RMS EGA of 2.0 deg is adopted for smaller cruise vessels and RoPax vessels.

In addition to the MIR and EGA, a green water criterion is considered as well, based on a criterion of once every 3 hours. This is stricter than the NORDFORSK criterion, which results in an average of one event per 3 minutes. This seems more appropriate for cruise vessels, as passengers have

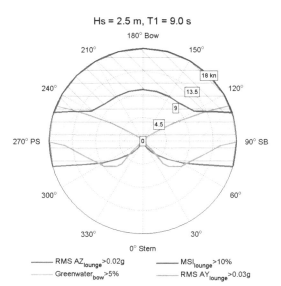

Figure 2. Operational limits in TOC when classical criteria are applied.

clear sight of the foredeck and might be concerned seeing large quantities of water on deck.

Figures 3 and 4 show the operational limits for TOC and TAC, respectively, when the combined criteria are applied. These show that the conditions in the region between beam and head seas are still the most demanding. There is a risk of shipping green water in the TAC. This can be avoided by changing heading or reducing speed, though doing so may increase the risk of parametric roll (see section 4.2).

3.3 Comfort at passenger areas on deck

For this expedition cruise vessel, the weather decks are important as they give a clear view of the surroundings (e.g. wildlife spotting) and the main pool is located outside as well. These locations will be one of the first passenger facilities that will be closed in adverse weather. Reasons for doing so include high wind speeds, large ship motions and potentially icing and rain.

A simple assessment of the wind comfort at the pool areas is provided here, though CFD or wind tunnel tests would be more appropriate to derive local shielding effects. For this study, a wind-angle-dependent shielding is assumed where 90% of the undisturbed relative wind velocity occurs in following to beam winds and reduces to 30% in head winds. A first suggestion for the limiting relative wind speed is 10 m/s (high 5 Beaufort).

Figure 5 shows the resulting speeds and headings in TAC (true wind 19 m/s) where the limiting wind speed is exceeded and discomfort due to wind can be expected. Note that in TOC (true wind 6 m/s), the relative wind never exceeds 10 m/s.

If excessive sloshing occurs inside the pool, it will be closed and, if necessary, emptied (note that emptying the pool is usually a precaution if adverse weather is expected). Therefore, identifying in which conditions sloshing occurs is important.

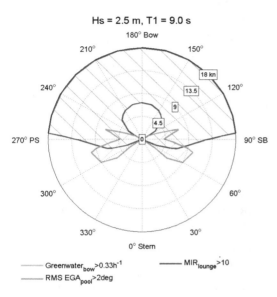

Figure 3. Operational limits in TOC when combined criteria are applied.

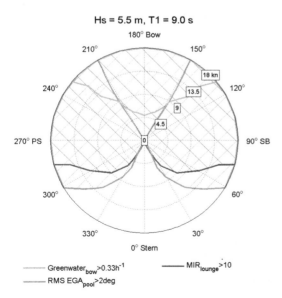

Figure 4. Operational limits in TAC when combined criteria are applied.

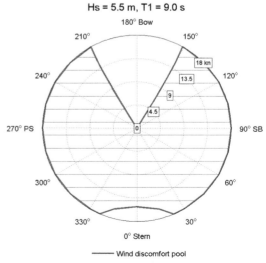

Figure 5. Wind discomfort at the pool area in TAC.

Assuming the pool is located at 18 m above the keel at station 2 and has a simple rectangular shape (5.5 m long, 3.5 m wide, 1.5 m deep), then the natural period of the main sloshing modes can be determined by calculating the phase velocity (c) using the dispersion relation. In the equations below, the natural period can be obtained by assuming the wave number k and iterating until ω equals ω_e.

$$\omega^2 = kg \tanh(kh) \quad (1)$$

$$c = \frac{\omega}{k} \quad (2)$$

$$T_e = \frac{2l}{c} \text{ or } \omega_e = \frac{\pi c}{l} \quad (3)$$

In this example, the natural periods are 3.2 s for longitudinal and 2.3 s for transverse sloshing. Ship motions at these periods are likely to generate potentially dangerous splashing, leading to mitigating actions. In addition, 'semi-static' spilling may occur at long pitch or roll periods.

A typical simplified approach makes use of the Response Amplitude Operator (RAO) that gives the transfer function of the water motions in the pool per degree EGA in longitudinal and transverse direction. Based on these RAOs, the swimming pool response in irregular seas can be calculated as a function of heading and speed. The operational limits of the pool then result from combining this response with a criterion on the allowable spilling (e.g. N times per hour). However, depending on pool geometry, multiple sloshing modes might exist. Therefore this approach gives a first indication only. For more complicated swimming pool geometries, CFD or model tests on a moving platform are more appropriate to derive the limits.

4 NON-LINEAR SHIP BEHAVIOUR

4.1 *IMO vulnerability criteria*

The framework of the International Maritime Organization (IMO) Second Generation Intact Stability Criteria (SGISC) is based on three types of assessment procedures:

- Level 1 (L1): vulnerability indication based on a relatively simple rule-of-thumb.
- Level 2 (L2): stochastic assessment based on scatter diagrams of short-term sea states in the environments under consideration to calculate a probability index of the vulnerability.
- Level 3 (L3): Direct Stability Assessment (DSA), which uses simulations of the vessel to estimate the number of stability failure events. This approach would be the most accurate type of assessment available, as the relevant physics themselves would be calculated or measured. The drawback would be the time required to generate an adequate statistical sample of stability failure events.

Discussion in the SGISC working group on L1 and L2 is close to finalised, but documentation on the L3 DSA is still in the starting phase. On the other hand, the robustness and consistency of the L2 assessments as applied to different ship types are still being studied, with the proposal by the IMO working group to use DSA to explore the quality of these. Table 2 contains a summary of the documented L1 and L2 assessment methods (IMO, 2015 and 2016).

L1 and L2 assessments for the studied cruise vessel were carried out, using an in-house developed Python implementation. Vessel hydrostatics were calculated using the freeware hydrostatics package ConvertHullForm. Roll damping and RAOs of accelerations and motions were calculated using linear strip-theory (SHIPMO). As roll damping is difficult to accurately predict numerically, it was tuned based on model test results of a similar sized cruise ship. Resistance and propulsion characteristics were obtained from curve fitting a speed/power prediction and calibrated with model test results as well.

Table 3 shows that according the IMO vulnerability assessment the risks of loss of stability,

Table 2. IMO vulnerability criteria guidelines SDC2/WP.4, (IMO, 2015) and SDC3/WP.5 (IMO, 2016).

Failure mode	Level 1	Level 2	Explanatory
Pure loss of stability	SDC2/WP.4, annex 1	SDC2/WP.4, annex 1	SDC3/WP.5, annex 3
Parametric roll	SDC2/WP.4, annex 2	SDC2/WP.4, annex 2	SDC3/WP.5, annex 4
Surf-riding broaching	SDC2/WP.4, annex 3	–	SDC3/WP.5, annex 5
Dead-ship condition	SDC3/WP.5, annex 1	SDC3/WP.5, annex 1	SDC3/WP.5, annex 6
Excessive acceleration	SDC3/WP.5, annex 2	SDC3/WP.5, annex 2	SDC3/WP.5, annex 7

Table 3. IMO vulnerability for 120 m cruise vessel.

Vulnerability mode		Fails?
Pure loss of stability	Level 1	No
	Level 2	No
Parametric roll	Level 1	No
	Level 2	No
Dead ship	Level 1	Yes
	Level 2	Yes
Excessive acceleration	Level 1	Yes
	Level 2	No
Surf-riding/broaching	Level 1	Yes
	Level 2	Yes

parametric roll and excessive accelerations are acceptable whereas the risks in dead ship condition and the risk of broaching are unacceptable. See sections 4.2 to 4.4 for a further discussion of the results of the first three vulnerability modes.

Table 3 also shows that for the 120 m cruise ship, L1 is equal to or more conservative than L2 assessment (which should indeed be the case). Note that there are several publications (e.g. Tompuri et al, 2017 or Schrøter et al, 2017) where L2 fails and L1 is passed. As almost no published data is available presenting the results of an L3 assessment, it is actually not known if L2 is over-conservative or L1 is under-conservative. This is also recognised by the IMO working group, which is calling for a further calibration of L1 and L2 by means of DSA.

4.2 Parametric roll

As a first rule of thumb, one could use the following indicators for the risk of parametric roll:

– Wave encounter frequency close to twice the natural roll frequency of the vessel.
– Wave length similar to the ship length. The definition of 'similar to' is not strictly defined, but is considered to fall within 0.8 and 2.0 times the ship length.

When both indicators are true and when sailing in head or following seas, the risk of parametric roll should be studied in more detail. While useful as a first indication, both indicators disregard the amount of roll damping and magnitude of the stability variations in waves, which both depend on the hull shape and presence of appendages. For example, a pronounced flare at the bow and stern increases the risk of parametric rolling. It also does not give any indication of the threshold wave height required to trigger large rolling motions. Level 1 and 2 assessments take these aspects into account in a limited way.

The L1 assessment is based exclusively on hydrostatic considerations with the ship in regular waves and the wave height parameter proportional to the ship length by a factor of 0.0167. This means that for smaller ships, a lower wave height should be used. For the assessment of the 120 m cruise vessel, this results in a wave height of 2 m (amplitude of 1 m). The resulting conclusion is that the ship with retracted stabilisers is considered not sensitive to parametric roll. However, a wave height of more than 2 m is quite common (in the present operational profile in 58% of the time).

The L2 assessment takes into account the occurrence of a particular combination of height and period in a wave scatter diagram. Depending on the risk of the ship rolling more than 25 degrees in each condition, it is added to a weighted summation.

While the IMO guidelines also allow simpler calculation methods, for the purposes of this paper, the risk is obtained from a time domain simulation using the 6-DOF time domain code FREDYN (Gkikas and Van Walree, 2014). An example of the results is shown in Figure 6.

Based on the two indicators mentioned at the beginning of this section, parametric rolling could only occur in head waves for the 120 m cruise ship. After applying the weighting factors, the L2 assessment concludes that this ship fulfils the parametric roll criteria and therefore this is considered not to be a major concern.

However, while a threshold of 25 deg for the maximum roll might be tolerable for a container vessel at full load from an acceleration point of view, it is extremely high for a cruise vessel where such angles will certainly lead to passenger injuries. In light of the recommendations published by Belenky (2006), a threshold amplitude on the basis of tolerable accelerations, depending on ship type, seems warranted. Note that, as per section 4.3, the criterion for Stability Loss also depend on ship type (15 deg for passenger ships, 25 deg for others).

In addition to the above, the risk of occurrence of parametric roll should be examined for three ship speeds, 50%, 86% and 100% of the service speed, without accounting for involuntary speed loss in waves. As a result, speeds lower than half the service speed are left out of the IMO assessment. However parametric roll is expected to occur frequently below 50% of the service speed and it is impossible for the vessel to maintain the service speed in higher sea states.

Figure 7 shows the result of the L3 DSA that includes the L2 results at 8 kn (red dashed line), as well as additional FREDYN simulations at 3 kn (orange dashed line). This speed is associated with the maximum sustained speed in severe storm conditions (or when sailing on one shaft in case of

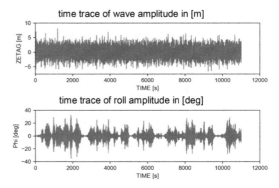

Figure 6. FREDYN calculation of parametric roll, head waves, H_s = 5.5 m, T_2 = 7.5 s V_s = 8 kn.

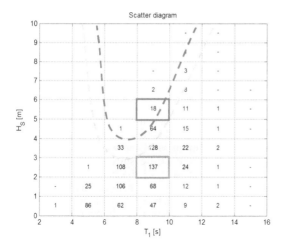

Figure 7. Wave conditions in which parametric roll occur (max. roll above 25 deg), red dashed line denotes 8 kn speed and orange dashed line 3 kn.

malfunction of one of the main engines). Above the dashed lines the max roll angle exceeds 25 deg.

It is worth noting that a number of conditions where parametric roll can be anticipated, fall within the scatter diagram. However, this is not detected after applying the L2 weighting factors. This becomes especially concerning after considering the lower speed of 3 kn, which leads to even more occurrences. In addition, a maximum roll angle above 25 deg it is extremely high for a cruise vessel where such angles will certainly lead to passenger injuries.

As parametric rolling is a serious safety concern, even small chances of occurrence should not be neglected. It is therefore the opinion of the authors that a DSA of parametric roll for realistic wave and sailing conditions is highly recommended, if the indicators at the beginning of this section show that there is a risk.

4.3 Stability loss

Similarly to parametric roll, the risk of loss of stability was examined through a two-tier assessment. The L1 assessment requires a minimum GM value of 0.05 m in waves and includes free-surface effects. This GM value is calculated in two ways: a simple hydrostatic calculation for vessels with a large bow flare or widening stern (fulfilling the condition given in 2.10.2.2. SDC (IMO, 2015) or a more complex calculation for all other ships. The second option applies to the 120 m cruise ship and places the static vessel in a regular wave with a height proportional to the ship length by a factor of 0.0334, leading to H = 4 m. This resulted in a minimum GM of 0.36 m, excluding free-surface correction. Therefore the vessel will comply with the L1 assessment as long as the free-surface correction does not exceed 0.31 m.

The L2 assessment follows the same scheme as that for parametric roll based on a summation of weight coefficients from a complete wave scatter diagram. However, in this case the additional criteria to add these coefficients to the summation are:

- the angle of vanishing stability should remain higher than 30 deg
- or the angle of heel under the action of a heeling lever, the magnitude of which depends on, amongst others, wind speed and wave height, remains higher than a fixed value, set depending on the type of vessel (15 deg for passenger ships, 25 deg for others).

The conclusion of this assessment is that the vessel does not show any likelihood to experience stability loss in waves. The angle of vanishing stability is noted to be hardly sensitive to the selected wave condition.

4.4 Weather criterion

Table 3 showed that the 120 m cruise vessel does not comply with the L1 and L2 vulnerability criteria for dead ship condition. These criteria are (partially) based on the IMO code on intact stability (IMO, 2006a and 2008) includes a severe wind and rolling criterion (weather criterion). The objective of this criterion is to verify that the free drifting dead ship can survive in beam seas where the roll motions must not exceed the down-flooding angle.

A more direct approach is possible in irregular seas, including the effect of wind (IMO, 2010a). This approach can be followed with time domain seakeeping codes like FREDYN, or by means of model tests. This was not performed within the scope of this paper.

4.5 Slamming

Slamming at the bow is one of the primary reasons for the ship's crew to slow down. Underlying reasons for this decision could be passenger discomfort and avoiding sliding and tipping of furniture or even structural damage. It makes sense to assume that when passenger comfort is satisfactory, damage to ship structure and risk of sliding is negligible.

There are no well-defined criteria for slamming pressures and whipping accelerations. The Vibration Dose Value (VDV) can be used to judge the measured transient vibrations. For comfort, a 30-minute VDV criterion of $1.0 \text{ m/s}^{1.75}$ was adopted (Dallinga and Bos, 2010). These VDV values are

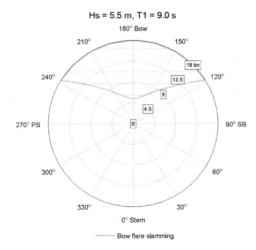

Figure 8. Risk of bow flare slamming in TAC.

estimated with a simple empirical model based on model test results of a 173 m RoPax ferry with three different bows with varying flare angle (Grin, Dallinga and Boelen, 2016). This model estimates the 30-minute VDV on basis of wave steepness, vessel speed, relative wave velocity and bow flare.

Figure 8 shows the resulting operational limits for TAC (discomfort due to bow flare slamming is not expected for TOC). Head and bow quartering seas are shown as most demanding, in which cases passenger comfort could be improved by reducing speed or changing heading. Note that this plot gives a first indication only. For a proper assessment of discomfort due to slamming-induced vibrations, dedicated model tests and finite element computations are needed to determine the behaviour of the ship's structure.

Other slamming-related issues could be stern slamming when sailing at low speed, at anchor or while at the quay. The risk of impact loads is especially high for hull features like anchor pockets, (partially overhanging) lifeboats, balconies and large glass structures at observation decks.

5 SHIP CONTROLLABILITY

5.1 Manoeuvring in waves

When investigating the ship operability, it is important to also include course controllability. A captain will always try to avoid conditions where the ship cannot maintain course or where the rudder needs to steer to its limit to do so. Of course, such issues can be addressed through free-sailing model tests in quartering waves as typical time domain simulation codes are not able to combine the sea-keeping and manoeuvring response of the ship together with an accurate estimate of the wave drift forces. This is why, within the Cooperative Research Ships (CRS, http://www.crships.org), the ManWav time domain simulation tool was developed for manoeuvring in waves. This tool focuses on low-frequency wind and wave forces in combination with the ship's powering and manoeuvring performance to simulate low frequency track keeping, sustained speed and manoeuvring performances of the ship in wind and wave conditions.

For the present cruise ship, ManWav simulations were performed for the two sea states, TOC and TAC, assuming a constant torque of 85% engine power. To analyse the simulations, two criteria were used to assess the ship operability:

– The rudder angle should be less than 25 deg; and
– The track deviation should be less than 1.0 L_{pp}.

Based on the obtained sustained speed and these two criteria, the operability plot presented in Figure 9 is obtained. From this figure, it is seen that in TOC the ship is able to maintain a minimum speed of 12 kn and the criteria are never exceeded. However, in TAC, the mean sustained speed goes down to 8 kn and the defined criteria are exceeded in both bow and stern quartering seas. As one may expect, the most challenging headings for course keeping are stern quartering seas.

While the ManWav tool allows propeller and rudder ventilation to be accounted for, this was not included within the scope of the current study. In some cases, propeller ventilation may lead to significant extra loss of thrust. In addition, it is noted that autopilot settings have an impact on the ship's

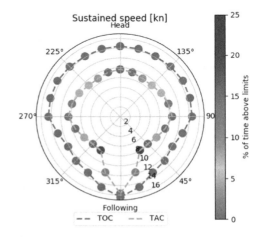

Figure 9. Polar plot presenting for both sea states TOC and TAC the ship sustained speed and % of time above the course keeping limits as function of the wind/wave heading.

capability to keep track, but since these were low-frequency simulations, this effect is considered to be limited.

Last but not least, when discussing the ship manoeuvring performance in waves, one should investigate the risk of broaching. Typically, one can expect that a 120 m cruise ship could run into broaching in long stern quartering seas (as the IMO L1 and L2 vulnerability criteria predicted).

5.2 Safe return to port

Since mid-2010, IMO has required a 'Safe Return To Port' (SRTP) assessment for all passenger ships (IMO, 2006b and 2010b). The main goal of SRTP is to ensure a ship can sail safely to the nearest port under her own propulsion after for instance a fire incident. The worst-case scenario considering the power generation and propulsion configuration should be accounted for. Although not mandatory, a minimum speed of 6 kn while heading straight into Beaufort 8 wind and corresponding sea conditions is recommended, taking into account the intended area of operation. It is not specified which H_s/T_p combination condition to choose for this condition and if reaching a speed of at least 6 kn in head seas means the vessel would be safe in all conditions. As shown in section 5.1, quartering seas can be more demanding in certain conditions.

A check on this was not performed within the scope if this paper.

5.3 Harbour operations

5.3.1 Crabbing

When designing a passenger ship, it is important to take the low-speed manoeuvrability into account. Auxiliary propulsors such as bow and stern thrusters must be chosen such that the ship will be able to manoeuvre in port without assistance and possibly in adverse environmental conditions. This aspect needs to be assessed already early in the design stage, because the presence of thruster tunnels in the hull has an influence on the ship resistance. For this purpose, some typical scenarios need to be defined and analysed in order to verify the adequacy of the auxiliary thrusters.

One of the most critical of these scenarios is the crabbing manoeuvre, where the ship approaches or moves away parallel to a quay. The vessel then uses all available propulsors to generate a transverse force to balance the external loads. As initial verification, a simplified assessment can be performed based on the following assumptions: considering only wind as external force and operating in deep, unrestricted water. The first hypothesis can be considered realistic, since many harbours are sheltered from strong currents and high waves.

However, the second assumption would not be correct in reality, because quay and shallow water have a significant influence on ship manoeuvrability.

The forces generated by the ship's propulsion are calculated (Quadvlieg and Toxopeus, 1998) as well as the wind forces. By comparing the ship and wind forces, considering a static equilibrium, the wind speed limit is found. Above such speed, the ship would not be able to maintain position. By repeating this procedure for different wind directions, a polar plot is obtained showing the crabbing capability.

Such a plot indicating the crabbing capability for the 120 m cruise ship is shown in Figure 10. This is based on deep unrestricted water, two bow thrusters and one stern thruster and considering two different power levels: 300 kW and 450 kW/thruster. Figure 10 shows that with 300 kW, the ship is not able to sustain winds up to 30 kn in all directions. However, this does become possible with 450 kW. Depending on the expected wind in the foreseen destinations of the ship, an adequate power of the tunnel thrusters can be chosen.

By estimating the hydrodynamic transverse drag force at 90°, it is also possible to evaluate the transverse speed of the ship using the same static formulation of equilibrium between forces. This is another important parameter for defining the low-speed manoeuvrability and is relatively easy to evaluate during sea trials.

5.3.2 Port manoeuvring

A separate, but complementary, approach for verifying the ship's low-speed manoeuvrability in the design stage is to simulate the standard IMO manoeuvres, consisting of zig-zags and turning circles, at low speed. This allows evaluation of the typical manoeuvring characteristics, such as rate of turn, drift angle and initial turning ability, with the

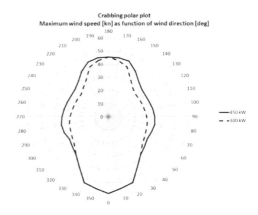

Figure 10. Crabbing polar plot showing the maximum wind speed that the manoeuvring ship can sustain as function of the wind direction.

ship sailing slowly. This makes it possible to verify if the ship can safely manoeuvre in confined spaces.

5.3.3 *Tender operations*

A cruise vessel will often dock in harbour, which makes it easy for passengers to go ashore. However, this is not possible for all destinations. This applies especially to remote areas frequented by an expedition type cruise vessel. In that case, the ship will anchor or drift in place and the passengers will be ferried ashore using ship tenders (customised lifeboats), shore tenders or RHIBs (rigid-hulled inflatable boats). This is often combined with some limited form of dynamic positioning to keep the ship in place or to maintain a fixed heading to create a lee for safe tender operations and/or to avoid dragging anchor due to fishtailing (large sway and yaw motions behind the anchor).

If boats are used to bring passengers ashore, the excessive motions between the boat and tender platform can make passenger transfer slow and/or dangerous, while slamming against the platform might also hamper the operation. As a first indication of the operational limits, single-body calculations were done at zero speed to calculate the relative water motion at the position of the tender platform. These calculations were performed with PRECAL, the linear panel code developed within the CRS. This assumes that the tender boat follows the local wave elevation behind the cruise vessel, including diffraction and radiated waves, but without interaction effects and line and fender loads. For a more complete assessment, multi-body computations are required.

Figure 11 shows the estimated operational limits at 9 s wave period, for tender operations at the stern of the 120 m cruise vessel. These limits are based on the assumption that passenger transfer is no longer possible when relative motions exceed 0.5 m amplitude in more than 1% of the wave encounters.

6 CONCLUSIONS

Continuing the development of the 'design for service' concept, this paper evaluates the performance of a 120 m expedition type cruise vessel using the latest numerical tools and methods available at MARIN to cover a balanced and practical variety of operational aspects. The many aspects of 'design for service' discussed in the present paper are summarised in Appendix A.

For the 120 m cruise vessel, two wave conditions were selected on the basis of the operational profile:

– A typical operational condition (TOC) at 2.5 m wave height and 6 m/s wind
– A typical adverse weather condition (TAC) at 5.5 m waves and 19 m/s wind.

For TOC and TAC a selection of hydrodynamic issues have been evaluated. It is important to note that for a real operability assessment all wave conditions and all aspects that might hamper the operation should be accounted for.

From this limited assessment it is shown that in TOC, seasickness limits the onboard activities most. Non-linear and controllability issues are not expected in TOC, except for tender operations which cannot be performed above approximately 1 m waves.

In TAC the situation is more critical. In particular the risk of parametric roll is large and the vessel has controllability issues in stern quartering seas. Note that the risk of parametric roll was evaluated assuming retracted fin stabilisers; with active stabilisers and speeds above 8 kn the risk is negligible in TAC. In addition crabbing is just possible with the two 450 kW bow thrusters and impossible with the 300 kW ones.

For the next design iteration the designer should pay attention to sloshing in the pool and fishtailing behind the anchor in TOC. These were not evaluated in this round, but are deemed limiting. In addition one could consider the relocation of the important passenger locations to lower the seasickness levels. In TAC the design would benefit from an increase in roll damping to lower the risk of severe rolling. Furthermore the controllability in stern quartering waves and broaching risk is a point of concern.

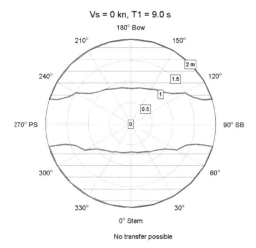

Figure 11. Operational limits for tender operations at the stern.

For the moment there is no combined seakeeping and manoeuvring tool that is capable of capturing all hydrodynamic issues in a reliable and efficient manner. In addition the criteria (objectives) are not well defined for many of them. However it is felt that the operability matrix gives guidance to the different aspects that might hamper the operation and make the hydrodynamic issues more specific. It gives a broad overview which makes it a powerful starting point to define critical issues in the various design stages.

It is deemed not yet feasible to optimise for the full matrix, but one or a few of these critical issues can be used in multi-objective optimisation techniques to improve the design. Note that the 'optimised' design must be cross-checked with the complete matrix again, to avoid unbalanced designs.

With the ever increasing computational power and further improvement in tools it should become possible to include the complete matrix in these optimisation techniques, making it possible to really 'design for service'.

REFERENCES

Belenky, V. 2006. Numerical procedures and practical experience of assessment of parametric roll. *STAB2006*, Brazil.

Colwell, J. 1989. Human Factors in the Naval Environment: A Review of Motion Sickness and Biodynamic Problems. *Technical Memorandum 89/220*, Canada.

Crepier, P., Veldhuis, C.H.J. and Starke, A.R. 2013. Design for Service: Effects of Loading Conditions on Powering Performances. *18th Symposium on Practical Design of Ships and Other Floating Structures PRADS 2013*, Changwon City, Korea.

Dallinga, R. and Bos, J. 2010. Cruise Ship Seakeeping and Passenger Comfort, *HPAS 2010*. Glasgow, UK

Dallinga, R., Loeff, G., and Thill, C. 2003. Design for Service in Ferry-Design. *3rd International Conference on High Performance Marine Vessels 2003*. Shanghai, China.

Dallinga, R., van Daalen, E., Grin, R. and Willemstein, T. 2004. Scenario Simulations in Design for Service. *9th Symposium on Practical Design of Ships and Other Floating Structures 2004*, Luebeck-Travemuende, Germany.

Gkikas G. and van Walree F. 2014. Development of a Time Domain Boundary Element Method for the Seakeeping Behavior of a Cruise Ship under Combined Strong Wind and Extreme Irregular Beam Seas. *Proceedings of the ASME 2014 33rd International Conference on Ocean, Offshore and Arctic Engineering OMEA2014–23376*. San Francisco, US.

Grin R., Dallinga R. and Boelen P. 2016. Voyage Simulation Techniques as Design Tool for Ferries. *RINA Conference on the Design and Operation of Passenger Ships*. London, UK.

IMO. 2006a. Interim Guidelines for Alternative Assessment of the Weather Criterion. *International Maritime Organization MSC.1 / Circ.1200*. London, UK.

IMO. 2006b. Adoption of Amendments to the International Convention for the Safety of Life at Sea, *International Maritime Organization MSC.216(82)*. London, UK.

IMO. 2008. Adoption of the International Code on Intact Stability, 2008', *International Maritime Organization MSC 83/28/Add.2 Annex 13*, London, UK.

IMO. 2010a. A procedure for determining a GM limit curve based on an alternative model test and numerical simulations. *International Maritime Organization SLF 53 / Inf.3*. London, UK.

IMO. 2010b. Interim Explanatory Notes for the Assessment of Passenger Ship Systems' Capabilities after a Fire or Flooding Casualty *International Maritime Organization MSC.1/Circ.1369*. London, UK.

IMO. 2015. Development of second generation intact stability criteria, Development of amendments to part B of the 2008 IS code on towing, lifting, and anchor handling operations, Report of the working group (Part 1). *International Maritime Organization SDC 2/WP.4*. London, UK.

IMO. 2016. Finalization of second generation intact stability criteria, Amendments to part B of the 2008 IS code on towing, lifting, and anchor handling operations, Report of the working group (Part 1). *International Maritime Organization SDC 3/WP.5*. London, UK.

ISO. 1985. Evaluation of human exposure to whole-body vibration—Part 3: Evaluation of exposure to whole-body z-axis vertical vibration in the frequency range 0,1 to 0,63 Hz. *International Organization for Standardization*. Geneva, Switzerland.

NORDFORSK. 1987. Assessment of ship performance in a seaway. *NORDFORSK, Nordic Co-operative Organization for Applied Research*. Copenhagen, Denmark.

NATO STANAG. 2000. Standardization Agreement 4154 – Common procedures for seakeeping in the ship design process. *North Atlantic Treaty Organization*. Brussels, Belgium.

Schrøter, C.,Lützen, M., Erichsen, H., Jensen, J.J. et al. 2017. Sample applications of the Second Generation Intact Stability Criteria—robustness and consistency analysis, *16th Int. Ship Stab. Workshop*. Belgrade, Serbia.

Tompuri, M., Ruponen, P. and Lindroth, D. 2017. On the consistency of the Level 1 and 2 vulnerability criteria in the Second Generation Intact Stability. *16th Int. Ship Stab. Workshop*. Belgrade, Serbia.

Turan O. COMPASS. 2006. A rational approach for reduction of motion sickness & improvement of passenger comfort and safety in sea transportation. *EU COMPASS Project G3RD-CT-2002-00809*.

Quadvlieg, F and Toxopeus, S. 1998. Prediction of crabbing at early design stage. *7th Conference on Practical Design of Ships and Other Floating Structures PRADS 1998*, the Hague, the Netherlands.

APPENDIX A

Table A1: Tools and criteria.

		safety aspects					onboard activities							port calls					Tool used in paper	Recommended tool	Applied criterion
		collision avoidance	extreme ship motions	structural damage	helicopter operations	closure of weather decks	closure of pool	restaurant/bar	fitness/sports	swimming pool/wellness	theatre/cinema	scenic view	sleeping/relaxing	slow steaming	(un)berthing	anchoring	station keeping	schedule keeping (dis/embarking)			
linear behaviour	seasickness					x	x	x	x	x	x	x							2	2	MIR < 10
	ability to move around		x			x	x	x	x	x	x								2	2	RMS EGA<2 deg
	sliding/tipping of furniture		x				x	x	x	x									-	2	EGA
	sloshing in pool		x		x					x									2, 9, 10	2, 9, 10	relative motion
	local wind speed				x	x	x	x	x										3	9, 10	Uwind < 10 m/s
	green water	x	x									x							2	2, 9	1 per 3 h
non-lin. behaviour	stability loss in waves		x																1	7	<15 deg roll
	parametric roll		x																1	7	<25 deg roll
	safe return to port	x																	-	2, 5	Vs > 6kn
	dead ship condition		x																1	7, 9	<25 deg roll
	impact loads			x															-	9	local load
	slamming induced vibrations		x							x									4	9	VDV<1 m/s$^{1.75}$
controllability	propeller/rudder ventilation	x													x			x	-	5, 7, 9	thrust loss
	track keeping	x	x											x					5	5, 7, 9	<25 deg rudder
	broaching	x																	1	7, 9	yaw
	sustained speed	x	x											x				x	5	2, 5	speed
	crabbing														x				6	6, 9	position offset
	ship-ship interaction													x	x		x		2	7, 9	relative motion
	fishtailing behind anchor															x			-	7, 9	position offset
	(port) manoeuvring	x	x												x				-	8, 9	IMO MSC 137(76)
	dynamic positioning		x													x	x		-	7, 9	position offset

1 IMO level 1 and 2 vulnerability criteria (Marin: Python code)
2 Linear Strip-theory (Marin: Shipmo) or linear panel code (Marin: Fatima, CRS: Precal)
3 Approximate shielding coefficient
4 Empirical method
5 Low frequency time domain seakeeping-manoeuvring tool (CRS: ManWav)
6 Mean force manoeuvring model (CRS: Cognac)
7 Nonlinear time domain seakeeping code (Marin: Panship, AnySim, CRN: Fredyn)
8 Time domain manoeuvring model (Marin: Sursim, CRS: MPP)
9 Model tests (seakeeping, manoeuvring or wind tunnel)
10 CFD (Marin: Comflow, ReFresco)

First principle applications to docking sequences

C. Weltzien
Hamburg University of Technology, Hamburg, Germany

ABSTRACT: Nowadays many shipyards have their own floating docks. In view of all the challenging projects these days, many yards have to produce designs which are close to or even beyond the structural limits of their docking facilities. This paper presents a first principle based method to calculate the key numeric values to extend the scope of application of floating docks and platforms.

The method calculates the hydrostatic stable equilibrium of the interacting bodies for selected floating situations. The subsequent structural calculations with non-linear wood layers result in the block force distribution. Moreover, the resulting deflection lines of ship and dock are considered in the hydrostatic calculations. The modified buoyancy distribution is then included in the structural calculations.

Therefore, the described method provides a useful tool to minimize local and global stresses and deformations of the interacting bodies during the docking procedure by fast optimization of the block system arrangement and the ballasting sequences.

NOMENCLATURE

A_{wl}	Area of waterline
dFD	Mass difference of dock
dF_S	Mass difference of ship
$dM\varphi_D$	Heel moment of dock
$dM\varphi_S$	Heel moment of ship
$dM\vartheta_D$	Trim moment of dock
$dM\vartheta_S$	Trim moment of ship
E	Elastic modulus
f_r	Relaxation factor
I	Area moment of inertia
$I_{\eta S}$	Area moment of inertia of waterline around the transversal axis through its centre
$I_{\xi S}$	Area moment of inertia of waterline around the longitudinal axis through its centre
$(\vec{p}_{off})_D$	Distance vector between aft perpendicular of dock and point of contact
$(\vec{p}_{off})_S$	Distance vector between aft perpendicular of ship and point of contact
$q(x)$	Line load
$\vec{r}_{off}$	Distance vector between aft perpendiculars of ship and dock
ζ_B	Vertical coordinate of centre of buoyancy in global system
ζ_G	Vertical coordinate of centre of mass in global system
V	Displaced volume
$\vec{wl}_{1,x}$	First row of waterline matrix
$\vec{wl}_{2,x}$	Second row of waterline matrix
$w^{IIII}(x)$	Fourth derivative of bending line
$w_{act}(x)$	Bending line calculated by the last iteration step
$w_{new}(x)$	Bending line during next iteration step
$w_{old}(x)$	Bending line during last iteration step

1 INTRODUCTION

Only the most experienced professionals operate the floating structures of ship yards. By incorrect operation the ship as well as the dock itself can be severely damaged. The dock masters must be sure that their dock provides the necessary capacity and structural strength to bear the respective construction. In most cases the dock masters have carried out similar applications before. The question that arises nowadays on ship yards is how to handle new types of ships. The constructions become more complex and heavier with a higher centre of gravity. In some cases the structures have larger dimensions as well. This paper presents the calculations of the docking sequences of different applications by a first principle based method.

At a ship yard a new ship type has to be docked for repair. This ship has larger dimensions and a higher total weight than the previous ships. As long as a sufficient load capacity and the structural limits of the yard's dock are not guaranteed, the risk by docking the new ship type is too high. This paper provides a first estimation on the feasibility of the new project by recording measurements of a performed docking process and comparing the related calculations to the new one.

At the Pella Sietas ship yard a steel section is built on a floating platform. For the docking process, the platform has to be tugged into deeper water. This paper shows a possible floating process by taking into account hydrostatic and structural calculations. Thereby the ballasting of the floating platform at different time steps is presented in order to avoid stability failure and ensure both global and local stresses of the interacting bodies are minimized.

2 MULTIBODY HYDROSTATIC

2.1 Hydrostatic stable equilibrium

The method presented calculates the hydrostatic stable equilibrium of the interacting bodies for selected floating situations. Three types of interaction are possible:

1. Ship and dock have no contact
2. Ship and dock have point contact
3. The whole ship keel rests on keel blocks

In the following the hydrostatic equilibrium is calculated for the three different cases.

In the first case the hydrostatic equilibrium of each body is calculated separately. Hydrostatic equilibrium is given when there is an equilibrium of forces and moments on the floating body, as explained in the lecture script (Krüger, S. 2008):

$$dF = 0 \tag{1}$$

$$dM\vartheta = 0 \tag{2}$$

$$dM\varphi = 0 \tag{3}$$

In the second case ship and dock have point contact. During a whole docking process this situation occurs twice. Firstly while lifting, when trim of dock and ship differ slightly just before their first contact. Secondly while floating, when the ship refloats while the dock underneath has trim.

The method calculates the position of contact by solving the following system of equations:

$$dF_S + dF_D = 0 \tag{4}$$

$$dM\vartheta_S - |dF_S| \cdot (\vec{wl}_{1,x})_S \cdot (\vec{p}_{off})_S = 0 \tag{5}$$

$$dM\vartheta_D - |dF_S| \cdot (\vec{wl}_{1,x})_D \cdot (\vec{p}_{off})_D = 0 \tag{6}$$

$$dM\varphi_S - |dF_S| \cdot (\vec{wl}_{2,x})_S \cdot (\vec{p}_{off})_S = 0 \tag{7}$$

$$dM\varphi_D - |dF_S| \cdot (\vec{wl}_{2,x})_D \cdot (\vec{p}_{off})_D = 0 \tag{8}$$

The value of contact force corresponds to the mass difference of ship respectively of dock.

The third case describes the situation when the whole ship keel has contact to the keel blocks. In this case ship and dock have the same trim. Under the assumption that bilge blocks are applied, dock and ship have the same heel as well. The equilibrium of the whole system is then calculated by solving the following system of equations:

$$dF_S + dF_D = 0 \tag{9}$$

$$dM\vartheta_S + dM\vartheta_D - |dF_S| \cdot \vec{wl}_{1,x} \cdot \vec{r}_{off} = 0 \tag{10}$$

$$dM\varphi_S + dM\varphi_D - |dF_S| \cdot \vec{wl}_{2,x} \cdot \vec{r}_{off} = 0 \tag{11}$$

Under the following conditions the calculated equilibrium is a stable equilibrium (Krüger, S. 2008):

$$A_{wl} > 0 \tag{12}$$

$$\frac{I_{\eta S}}{V} + \varsigma_B - \varsigma_G > 0 \tag{13}$$

$$\frac{\frac{I_{\xi S}}{V} + \varsigma_B - \varsigma_G - \left(\frac{I_{\eta S}}{V}\right)^2}{\frac{I_{\eta S}}{V} + \varsigma_B - \varsigma_G} > 0 \tag{14}$$

2.2 Lever arm curves and hydrostatic tables

The lever arm curve of a floating system represents its hydrostatic stability. During the docking process it must be ensured that ship and dock will not capsize. For this reason the lever arm curve of the whole system is calculated at each floating position with free trim. Moreover the hydrostatic tables of the whole system can be displayed. The critical situations during the floating process can be identified by means of these outputs.

3 STRUCTURAL INTEGRITY

3.1 Block forces

The calculation of the block force distribution is based on the deformation method and is originally implemented by (Greulich, M. 2013).

In longitudinal direction ship and dock are divided into Timoshenko beams of equal length. These elements provide the stiffness of dock and ship regarding inertia and shear forces. On the nodes in between, loads are applied and deflections are calculated. The keel blocks, on which the ship rests during docking, are modelled as nonlinear spring elements. The block stiffness factors are based on the spring characteristics of Kunow, who made different investigations during launching in the 1980s (Kunow, 1974).

In detail the calculation of the block force distribution is explained in previous publication (Dankowski, H. & Weltzien, C. 2017).

3.2 Bending line

The bending line results from the fourth integration of the line load (Lehmann, E. 2005):

$$E \cdot I \cdot w^{IV}(x) = -q(x) \quad (15)$$

For the dock and the ship the particular loads of buoyancy, weight and block forces are added up to a resulting load distribution. By integration and division by elastic modulus and second moment of area the bending lines of dock and ship are calculated.

4 INTERACTION OF HYDROSTATIC AND STRUCUTRAL INTEGRITY

4.1 Influence of the bending line to the buoyancy distribution

The buoyancy distribution of a floating body changes with its deflection. To consider this effect in the hydrostatic calculations the frames are adjusted to the calculated bending line. The modified hull form of a bent floating dock is shown in Figure 1.

In the case of low stiffness of the floating body the buoyancy distribution overreact to the changing deflection. For this reason a relaxation of the bending line is implemented. The bending line that is applied on the body during the calculation is calculated by the following formula:

$$w_{new}(x) = f_r \cdot w_{act}(x) + (1 - f_r) \cdot w_{old}(x) \quad (16)$$

The iteration process is terminated when the difference between the bending lines of two sequential steps is less than a given residuum. The effect of bending in the buoyancy distribution is shown in Figure 2.

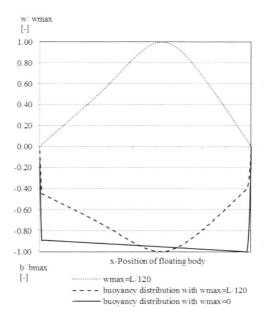

Figure 2. Different buoyancy distributions of a floating body due to its bending line.

Figure 1. Hullform of a floating body without bending and with bending.

5 FIRST APPLICATION

As described in the introduction, a new ship type has to be docked in a yard's own floating dock for repair. Lack of experience with this larger project points out the necessity of a reliable calculation method. To evaluate the presented method, a performed docking process in this yard's own floating dock has been measured. The recorded data is compared to the calculation results. As the results are realistic, the calculation method gives a good estimate on the feasibility of the new project.

5.1 Measurements of the performed docking process

The relevant values with regard to the structural reliability of the dock are the deformations of the keel blocks as well as the deflection of the dock. The deformations of the keel blocks result in the block force distribution that directly affects the dock. The dock deflection depends on its structural stiffness.

The measurements have been recorded in three different passes. The first pass has been performed directly before docking and provides the input values for the calculations. The second pass has been recorded during docking. These measurements will be directly compared to the calculation results. The last pass has been done directly after docking. It indicates plastic deformations of the keel block woods and gives a hint on their spring characteristics. In each pass the height of the keel blocks from pontoon deck to the top of the wood layers, as well as the deflection of the dock, have been measured. The blocks heights have been measured with a thin wooden board and a laser device, the dock deflection with a laser device and by reading off the draughts of the pontoon. Figure 3 shows the recorded data of the first pass.

Each keel block consists of a concrete block with a height of 2m and three wood layers on top. By subtracting the concrete block heights of the total block heights it can be noticed that the wood layers in the middle of the block system are higher than the outer wood layers. The reason is the high initial deflection of the dock as shown in Figure 3. It is a construction related deflection and the different heights of wood balance this offset to get a plane contact surface to the ship keel. The keel block heights during all three passes are shown in Figure 4. The dock deflection varies by max. 2cm.

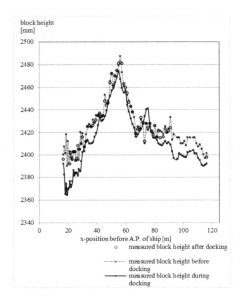

Figure 4. Measured total keel block heights before, during and after docking.

During the docking process all wood layers are impressed due to the ship's weight. The deflections of the outer keel blocks are higher than the deflections of the middle ones. After docking, the woods of the outer keel blocks are still deflected. They are plastically deformed.

To determine the displacement of the dock at each pass, the freeboard of the pontoon deck at six different locations has been measured by using a folding ruler. The fillings of the water ballast tanks have been measured by sensors. In the first and third pass the draughts of the ship has been read off as well to estimate the ship's load distribution during docking.

5.2 *Calculations of the performed docking process*

The construction documents of ship and dock provide information about their hull form, their lightship weight and cross-section value distributions. To determine the loading condition during the measurements the ballast water tanks of the dock are filled as shown by the sensors and the cranes are located in the same positions. In the case that the calculated draughts of the dock and the ship differ slightly from the measured ones, compensating wedges are applied on the dock and the ship.

Figure 5 shows the weight distribution during docking.

For the calculations the average deflection is applied as initial deflection on the dock. The initial

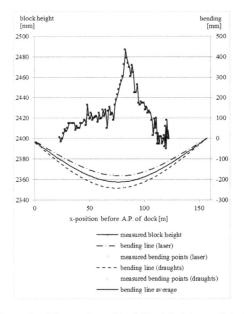

Figure 3. Measured total keel block heights and dock deflection before docking.

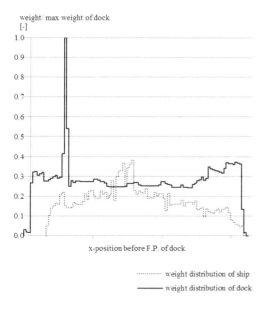

Figure 5. Weight distribution during docking.

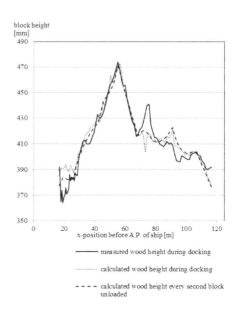

Figure 6. Measured and calculated deformation of wood layers during docking.

heights of wood layers of the keel blocks are set to the measured heights. These initial values are shown in Figure 3.

5.3 Comparison of measurements and calculations

The calculated deflection of the dock during the docking process varies less than 2 cm. This value corresponds to the measured one.

The deformation of the wood layers during the docking process as result of calculations and measurements are shown in Figure 6. In the case that each keel block is loaded, the calculated deformations are too low. During the measurements it was noticed that nearly every second keel block was unloaded due to the slightly varying keel block heights. For this reason a further calculation has been done that takes into account this notice. These results correspond much better to the measurements. It is notable that at the outer keel blocks of the stern the measured deflections are much higher than the calculated ones, while at the bow it is the other way round. One explanation is that due to the backward trim of the ship the rigid body motion impresses the rearward keel blocks additionally and lightens the forward ones. In overall the calculations and the measurements have a good agreement.

As the results show, the presented method is a reliable first principle tool to consider docking processes.

5.4 Calculation results of the prospective docking process in comparison to the performed docking process

The new ship type is 23% heavier than the previous ships. For this reason the buoyancy of the dock will be increased by reducing the remaining ballast water to a minimum level that pumps can obtain and that longitudinal strength is sufficient. In addition steel blocks will be used instead of the presented concrete blocks. Besides their lower weight, the steel blocks have the additional advantage that they are much more stable than the tippy concrete blocks. The new ship type has larger dimensions, but it still fits into the given dock.

Therefore the hydrostatic stability as well as the geometric feasibility is ensured. In the following it is shown that the structural reliability of the dock is given as well. The main modifications of the keel block system for the new ship type are described in the following.

Firstly the keel block system is modified. Due to the stable steel blocks the blocks can be positioned in a homogenous distribution. Due to the tippy concrete blocks, during the previous docking processes the middle keel blocks had twice the distance than the outer ones. The different block distributions are shown in Figure 7. In the performed case the outer keel blocks had twice the stiffness than the inner ones and the load peak was appropriate high. With the homogenous block system the load can be better distributed to the keel blocks and the load peak will be reduced.

Secondly the height of the wood layers of all keel blocks is increased. The softer blocks distribute the load better over the whole block system.

Advantages of the new ship type regarding docking are a wider flat keel and less distance between flat keel and ship endings. This leads to a greater contact area between blocks and the ship's keel so the block system is softer and additionally it yields shorter overhangs that reduce the load peaks at the endings. The block force distribution of the previous and the prospective docking is shown in Figure 8.

The calculated block force distribution shows that though the second ship is heavier, the load peaks are smaller. The shorter overhangs, especially in the bow area, the wider flat keel and the homogenous block distribution lead to this effect. As the given dock withstands the first ship type without damages, it can be concluded that the new ship type can be safely docked as well.

6 SECOND APPLICATION

As described in the introduction, a steel section is built on a floating platform at the Pella Sietas ship yard. For the docking process the platform has to be tugged into deeper water. In the following the calculated operation of this project is presented. The docking process will be performed in the future.

6.1 The initial situation

During the towing operation the freeboard to the pontoon deck should be large enough in order to prevent the deck from flooding by possible waves in the harbour. Therefore and for less resistance the trim should be limited. In addition the platform should have a low deflection. The initial situation is shown in Figure 9.

6.2 The lowering process

The platform has to immerse until the buoyancy of the section corresponds to its weight. At this moment the section can be tugged off the platform. The whole floating process is a challenge due to the sensitivity of the whole system. As the platform has no side boxes, is has very low stiffness. For this reason it directly reacts to the filling of ballast water tanks. Only a controlled ballast water sequence can avoid high deflections and

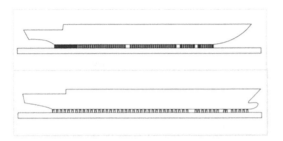

Figure 7. Keel block distribution of the prospective ship. The hullform is modified for publishing.

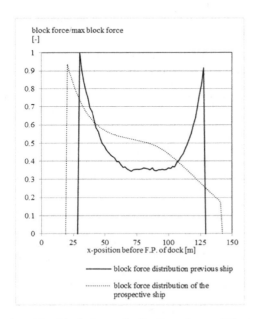

Figure 8. Block force distribution of two different ships.

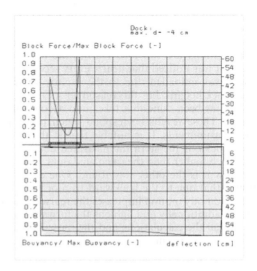

Figure 9. Initial situation for the towing operation.

therefore high local stresses that may cause damage to the section and to the platform. Moreover the hydrostatic stability of the whole system can decrease rapidly for example if the pontoon deck is flooded and the waterline's area moment of inertia is extremely reduced. Therefore the trim must be greater than 1m otherwise the risk of capsizing around the transverse axis is too high.

In the following the proposed floating process is described by four floating situations. In situation 4 the section is 40 percent floated. The floating situations are shown in Figure 10, the related draughts and trims are shown in Figure 11.

The platform has 36 ballast water tanks in total. These tanks are ordered into three rows along the width of the platform. The starboard and portside tanks are filled simultaneously, the middle tanks separately. Once the tanks are completely filled, the next tank will be filled. In this

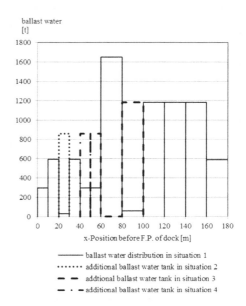

Figure 12. Ballast water distribution in 4 floating situations.

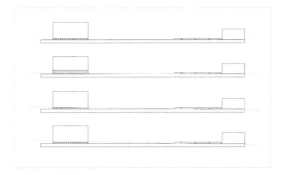

Figure 10. 4 floating situations, bending of dock is not shown.

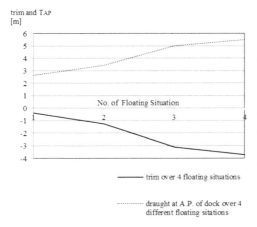

Figure 11. Trim and draught of the platform over 4 floating situations.

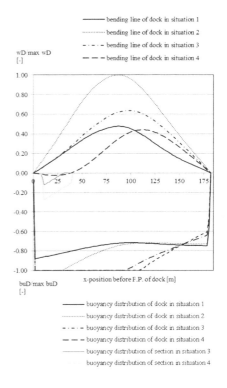

Figure 13. Bending line of the dock and buoyancy distribution of ship and dock in 4 floating situations.

way the free surfaces are reduced. The ballast water distribution in the four floating situations is shown in Figure 12.

In situation 1–2 the rearward middle ballast water tanks are filled to increase the backwards trim. Due to the greater mass at the stern of the platform, the bending of the platform increases, as shown in Figure 13. For the floating process the resulting hogging case is of advantage to a certain extent: The section becomes partly floated whilst the platform's area moment of inertia reduces only gently. In situation 2 the defined maximum bending of the platform is reached. For this reason the ballast water tanks in the middle of the platform is filled. In situation 3 the bending is reduced and the section is partly immersed. In the next step the ballast water tanks in front of the section are filled so the trim increases and the section is 40% floated. In the last step the middle tanks underneath and in front of the section will be filled to float the section completely.

7 CONCLUSIONS AND OUTLOOK

The paper demonstrates that the presented calculation method is an efficient tool to optimize the docking process. Different aspects, such as the block force distribution, the bending line of ship and dock, and the ballast water tank fillings, are considered. The conducted measurements before, during and after the docking process on a ship yard demonstrate that the calculated values are realistic. The comparison of two different docking processes with different block systems demonstrates the importance of a well-considered keel block set-up to minimize the local stresses by a sufficient load distribution over all keel blocks. In this context the height of the different wood layers on a keel block has a significant influence on the load distribution as they affect the block's nonlinear spring characteristics.

The floating process of a platform with very low stiffness illustrates, that the bending line has a significant influence to the floating position. As the platform reacts extremely sensitive to the variation of the ballast water tank fillings, it is very important to perform a well-considered filling sequence to have sufficient structural strength and hydrostatic stability. For this reason it is of advantage to have a useful tool that considers both, the hydrostatic stability and structural strength, and takes into account the interaction of both aspects. The presented calculation method is such a tool.

In the future it is planned to include the righting levers of the keel blocks into the lever arm curves. This extension enables the recalculation of occurred capsizing accidents on docks and may prevent further accidents.

ACKNOWLEDGMENTS

This investigation has been carried out within the research project MOPS, a project to develop new numerical methods for offshore and polar ships. Special thanks are given to the Federal Ministry of Economics and Technology (BMWi) for funding and supporting this research project.

I would also like to thank the Pella Sietas ship yard for providing this challenging application and the Marinearsenal Wilhelmshaven for their great support.

REFERENCES

Dankowski, H. & Dilger, H. 2013. Investigation of the Mighty Servant 3 Accident by a Progressive Flooding Method [Konferenz]//Proceedings of the ASME 2013 32nd International Conference on Ocean, Offshore and Arctic Engineering.

Dankowski, H. & Weltzien, C. 2017. Calculation of the Hydrostatic and Structural Integrity of Docking Sequences [Konferenz]//Proceedings of the ASME 2017 36th International Conference on Ocean, Offshore and Arctic Engineering.

Greulich, M. 2013. Numerische Analyse eines Schiffes in einem Schwimmdock [Bericht]: Project Thesis/ Hamburg University of Technology, Insitute of Ship Design and Ship Safety.

Krüger, S. 2008. Hydrostatik von Schiffen [Vorlesungsskript]: Schriftenreihe Schiffbau.

Kunow, R. 1974. Messung der Kantenpressung bei Stapelläufen [Bericht]: Tech. rep./Research Center for German Shipbuilding.

Lehmann, E. 2005. Grundzüge des Schiffbaus [Buch]. – [s.l.]: Eike Lehmann, 2005. – Bd. 1.

Ship mooring design based on flexible multibody dynamics

H.W. Lee, M.I. Roh & S.H. Ham
Seoul National University, Seoul, Republic of Korea

ABSTRACT: The mooring system is a system for maintaining a floating object such as a ship in a fixed position. For the design of the mooring system, the dynamic, coupled analysis of the ship and the mooring system should be performed in advance. Traditionally, the FEM (Finite Element Method) has been adopted to model the mooring system connected to a single body such as the ship. However, it is not applicable for a multibody system such as two ships in tandem offloading. In this study, mooring lines are modeled by one-dimensional flexible body which is composed of several nodes and elements. Also, the flexible multibody dynamics is used to solve the equations of motion of the multibody system (ships, mooring lines, etc.) without any iteration. Furthermore, the interactions between the system such as collision and friction are considered. For the collision, the non-interpenetration constraints are adopted, and the holding power of an anchor is also taken into account. Finally, the dynamic, coupled analysis of the mooring system connected to an LNG FPSO and an LNG carrier in tandem offloading is performed. As a result, we obtained the optimized design of the mooring system which is satisfied with the guidelines for the mooring design such as OCIMF (Oil Companies International Marine Forum).

1 INTRODUCTION

In ship and ocean engineering field, the dynamic analysis of floating multibody systems such as floating cranes or offshore structures have been widely performed. To predict the dynamic motion and the loads on the floating bodies, the mooring analysis is very important. However, the existing studies simply modeled the mooring line connected to the multibody system using linear springs or analytic solutions. However, these simple models cannot consider the elastic stiffness and the weight of the mooring line together, and are not suitable to consider the dynamic effect. Therefore, the FEM (Finite Element Method) based mooring line model is generally adopted for accurate simulation.

There are numerous works on the FEM based mooring line model (Kim et al. 2010, Kim et al. 2013, Ku et al. 2016). However, in traditional works, they were only adopted to the single floating body, not multibody system. Therefore, in this study, we suggested FEM based mooring line model applicable to the multibody system. Then, we simulated various applications including tandem offloading of LNG FPSO and LNG carrier to find the optimal design of the mooring line.

2 THEORETICAL BACKGROUNDS

2.1 Flexible multibody dynamics

The multibody system is a system which consists of multiple bodies connected to each other with joints. The multibody dynamics is a discipline to express the motion of the multibody system using the mechanical constraints. The general form of the Discrete Euler-Lagrange equation of the multibody system is expressed as follows.

$$\begin{bmatrix} \mathbf{M} & -\mathbf{G}_k^T \\ \mathbf{G}_k & \varepsilon \end{bmatrix} \begin{bmatrix} \mathbf{v}_{k+1} \\ h\boldsymbol{\lambda}_k \end{bmatrix} = \begin{bmatrix} \mathbf{M}\mathbf{v}_k - h\mathbf{F}(\mathbf{q}_k, \mathbf{q}_{k-1}) \\ -\mathbf{g}(\mathbf{q}_k)/h + \Gamma\mathbf{G}_k\mathbf{v}_k \end{bmatrix} \quad (1)$$

The bodies that consists the multibody system can be divided as rigid and flexible body. The floating body with mooring system can be regarded as a multibody system with the rigid floating body connected to the flexible bodies, which are mooring lines, with ball joints.

The mooring line can be modeled as multiple flexible beam elements connected to each other. We used the Absolute Nodal Coordinate Formulation (ANCF), which is suitable to consider the large deformation. In ANCF, the beam can be expressed by the displacement node vectors and the slope vectors at nodes. It is expressed by shape function $\mathbf{S}$ and nodal coordinate $\mathbf{q}$. For the shape function, the cubic Hermite shape function is used.

$$\mathbf{r}(x) = \mathbf{S}\mathbf{q} = \begin{bmatrix} s_1\mathbf{I} & s_2\mathbf{I} & s_3\mathbf{I} & s_4\mathbf{I} \end{bmatrix} \begin{bmatrix} \mathbf{r}_0 \\ \mathbf{r}'_0 \\ \mathbf{r}_l \\ \mathbf{r}'_l \end{bmatrix}, (0 \leq x \leq l) \quad (2)$$

In multibody system, the bodies are connected by one or more joints. The kinematic joints can be expressed as constraint equations in mutibody dynamics. The constraint equations of the ball joint between the rigid and flexible bodies are as follows.

$$C_q^{(ball)} = \frac{\partial C^{(tr,3)}}{\partial q} = \left[0\ I\ 0\ 0 - I\ \widetilde{R(\theta_G)r_B} \right] \quad (3)$$

For the verification, the displacement of the flexible cantilever beam is compared with the analytic solution. The result is shown in Table 1 with different beam property and the number of the elements.

2.2 Interaction with the seabed

The existing researches modeled the contact of the mooring line with the seabed using linear springs. However, it is not appropriate for the element using 3rd order shape function as the buckling can occur to the elements on the seabed. For this, we added one more constraint called slope constraint, which makes the slope of the mooring line be parallel to the seabed when contact occurs. Therefore, we suggested the non-interpenetration constraints to consider the contact with the seabed, and slope constraints to prevent the buckling.

Each constraint can be derived using the kinematic description of the mooring line elements. For the contact point P of the flexible mooring line A and the body fixed point C on the seabed B, the non-interpenetration constraint can be derived as follows. $\mathbf{n}$ is the normal vector of the seabed, and $\mathbf{S}$ is the shape function.

$$\begin{aligned} g^1 &= {}^E\mathbf{n} \cdot {}^E\mathbf{r}_{P/C} \\ &= {}^E\mathbf{n} \cdot \left({}^E\mathbf{r}_{P,flexible} - {}^E\mathbf{r}_C \right) \\ &= {}^E\mathbf{n} \cdot \left\{ \mathbf{S}(\xi)\mathbf{q}_A - \left({}^E\mathbf{r}_B + {}^E\mathbf{r}_{C/B} \right) \right\} \\ &= {}^E\mathbf{R}_B{}^B\mathbf{n} \cdot \left\{ \mathbf{S}(\xi)\mathbf{q}_A - \left({}^E\mathbf{r}_B + {}^E\mathbf{R}_B{}^B\mathbf{r}_C \right) \right\} \end{aligned} \quad (4)$$

Likewise, if the slope vector of the point P is expressed as $\mathbf{v}_p$, the slope constraint in case of the collision can be obtained as

$$\begin{aligned} g^2 &= {}^E\mathbf{v}_P \cdot {}^E\mathbf{n} \\ &= \mathbf{S}_\xi \mathbf{q} \cdot {}^E\mathbf{n} \end{aligned} \\ \left({}^E\mathbf{v}_P = \frac{d\mathbf{S}}{d\xi}\mathbf{q} = \mathbf{S}_\xi \begin{bmatrix} {}^E\mathbf{r}_{N_3} \\ {}^E\mathbf{v}_{N_3} \\ {}^E\mathbf{r}_{N_4} \\ {}^E\mathbf{v}_{N_4} \end{bmatrix} \right) \quad (5)$$

Meanwhile, in the real world, the mooring line can penetrate inside the seabed due to its weight. Generally, this can be considered using seabed stiffness, which is the spring coefficient of the traditional linear spring method. In the same manner, the constraint method can also consider the seabed stiffness using the regularization term ε, which determines how strongly the constraints should be satisfied. Therefore, the non-interpenetration constraint can be violated according to the value of ε. The ε is the inverse number of the spring coefficient.

Moreover, to consider the friction between the mooring line and the seabed, we adopted the friction model suggested by Choi & Yoo (2012) shown in Figure 1. With this model, the stick between the mooring line and the seabed can be realized.

Figure 3 shows a simple example of the moving barge connected to a mooring line. The holding power of the anchor is not considered, which means

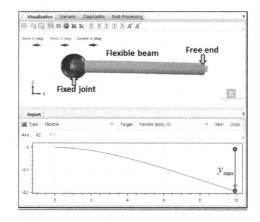

Figure 1. Verification of the flexible cantilever beam.

Table 1. Comparison with analytic solution.

Beam property						Cantilever (Analytic)	Cantilever (Simulation)		
E	Density	L	D	I	w	y_{max}	y_{max}		
[GPa]	[ton/m³]	[m]	[m]	[m⁴]	[ton/m]	[mm]	1 EA	3 EA	5 EA
10	1	10	1	0.0491	0.7854	19.620	19.620	19.620	19.620
100	1	10	1	0.0491	0.7854	1.962	1.962	1.962	1.962
210	7.85	10	1	0.0491	6.1654	7.334	7.334	7.334	7.334

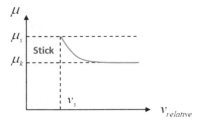

$$\mu(v_r) = a^{-(v_r+b)} + \mu_k \quad |v_r| > v_s$$

$$\mu(x_r, v_r) = \frac{k_{stick} x_r + c_{stick} v_r}{N}, \quad |v_r| \le v_s$$

$$b = -v_s - \log_a(\mu_s - \mu_k)$$

Figure 2. Friction coefficient-relative velocity relation of the stick friction model.

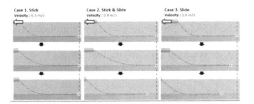

Figure 3. An example of moving barge connected to a mooring line.

that the end of the mooring line on the seabed is assumed to be free. The static and sliding friction coefficients are given to be 0.98 and 0.74, which are the recommended values between the chain and the sand from Taylor and Valent (1984). As the barge moves, the frictional force between the mooring line and the seabed is exerted in opposite direction. In case 1, the frictional force prevents barge from moving. As the velocity of the barge become higher, the sliding occurs in case 2. With higher velocity, case 3, the mooring line slides as the barge moves.

2.3 External loads

The current force plays an important role on the behavior of the mooring system. To calculate the current force exerted on the mooring line, the Morison equation is used. u and v is the velocity of the current and the mooring line each.

$$F = \rho V \dot{u} + \rho C_a V(\dot{u} - \dot{v}) + \frac{1}{2}\rho C_d A(u-v)|u-v| \quad (6)$$

The current velocity is not constant according to the water depth (Fig. 2).

Therefore, with the given data at the surface, the current velocity can be obtained using following

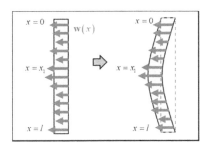

Figure 4. The current force exerted on the mooring line.

equation, where z is height which is negative in downward, and d is the water depth which is positive.

$$u(z) = u(0) \cdot \left(\frac{z+d}{d}\right)^{1/7} \quad (7)$$

Lastly, the calculated current force should be integrated along the mooring line. Then, the final form of the exerted current force can be obtained as

$$\begin{aligned}
\mathbf{Q}_{added} &+ \mathbf{Q}_{damping} \\
&= -\rho C_a \frac{1}{4}\pi d^2 l \left(\int_0^1 \mathbf{S}^T \mathbf{S} d\xi\right) \ddot{\mathbf{q}}(t) , \\
&+ \left(\int_0^1 \mathbf{S}^T d\xi\right)\frac{1}{2}\rho C_d d \cdot (\mathbf{u}-\mathbf{v})^2 l
\end{aligned} \quad (8)$$

where d is the diameter of the mooring line, and l is the length of the element.

3 APPLICATIONS

3.1 Developed program

For the dynamic simulation of the mooring system, the program named 'SyMAP' (SyDLab's Multiphysics Analysis Program) is developed. Figure below shows the configuration of SyMAP. It is developed in C# programming language and Windows Presentation Foundation (WPF). With this program, the mooring line can be generated easily, and the tension, displacement and the bending moment of the mooring line can be obtained (Ham et al. 2016, Ham et al. 2017a, Ham et al. 2017b, Ham et al. 2017c).

3.2 Tandem offloading of LNG FPSO and LNG carrier

For the application, the tandem offloading simulation of LNG FPSO and LNG carrier is performed. The FPSO and the LNG carrier are connected with the catenary line. The figure below shows an example of the mooring system of the FPSO from MM-Marine (Fig. 5).

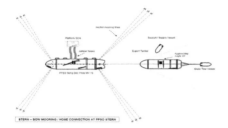

Figure 5. Mooring system of FPSO.

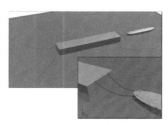

Figure 6. The configuration of the tandem offloading simulation of the LNG FPSO and LNG carrier.

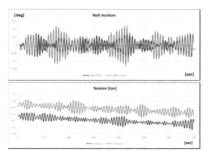

Figure 7. The motion of the LNG FPSO and carrier and the tension of the catenary line.

With reference of this data, we performed the tandem offloading simulation using SyMAP. The mooring line is simplified to be 4, and the FPSO is connected to the LNG carrier with 2 catenary lines. For the environmental condition, the JONSWAP spectrum is used for the irregular wave, and the significant height and the period are set to be 1.65 m, and 5.1 sec respectively. The configuration of the system is shown in Figure 6.

As a result, the behavior of the LNG FPSO and carrier was obtained, and the motion and the tension of the mooring lines could be calculated. Figure 7 shows the roll motion of the LNG FPSO and carrier, and the tension of the catenary line connecting them.

For the optimal design of the mooring line, we compared several different mooring design, with aspect of the behavior of the system, and mooring tension. Finally, we checked if the optimized design of the mooring system satisfies the guidelines for mooring design.

4 CONCLUSIONS

In this study, we suggested the FEM based mooring line model which is applicable to the multibody system. The suggested model comprises flexible beam elements connected to each other, and to barge and the seabed with ball joints. To consider the interaction between the mooring line and the seabed, the non-interpenetration and slope constraints are suggested. Then, the friction model which can realize stick is adopted. The external forces such as current force is considered using Morison equation. Finally, we developed the dynamic simulation program for the mooring analysis. Then, the tandem offloading of the LNG FPSO and carrier is simulated. As a result, we found the optimized design of the mooring system which satisfies the guidelines.

REFERENCES

Ham, S.H., Roh, M. Il, Kim, J.S. 2017c. Numerical analysis of wreck removal based on multibody system dynamics. *Journal of Marine Science and Technology* 1–15.

Ham, S.H., Roh, M. Il, Lee, H. 2016. Simulation of load lifting with equalizers used in shipyards. *Automation in Construction* 61: 98–111.

Ham, S.H., Roh, M.I., Hong, J.W. 2017b. Dynamic effect of a flexible riser in a fully connected semisubmersible drilling rig using the absolute nodal coordinate formulation. *Journal of Offshore Mechanics and Arctic Engineering* 139(5): 1–10.

Ham, S.H., Roh, M.I., Lee, H., Hong, J.W., Lee, H.R. 2017a. Development and validation of a simulation-based safety evaluation program for a mega floating crane, *Advances in Engineering Software* 112: 101–116.

Kim, B.W., Sung, H.G., Hong, S.Y., Jung, H.J. 2010. Finite element nonlinear analysis for catenary structure considering elastic deformation, *Computer Modeling in Engineering & Science* 63(1): 29–45.

Kim, B.W., Sung, H.H., Kim, J.H., Hong, S.Y. 2013. Comparison of linear spring and nonlinear FEM methods in dynamic coupled analysis of floating structure and mooring system, *Journal of Fluids and Structures* 42: 205–227.

Ku, N. & Cha, J.H. 2016. A study on moored floating body using non-linear FEM analysis, *Applied Ocean Research* 2(2): 85–92.

Tang, H.J., Huang, C.C., Chen, W.M. 2011. Dynamics of dual pontoon floating structure for cage aquaculture in a two-dimensional numerical wave tank, *Journal of Fluids and Structures* 27: 918–936.

Ship concept design

Managing complexity in concept design development of cruise-exploration ships

Ali Ebrahimi, Per Olaf Brett & Jose Jorge Garcia
Ulstein International AS, Ulsteinvik, Norway
Department of Marine Technology, Norwegian University of Science and Technology, Trondheim, Norway

ABSTRACT: Technology innovation, dynamic market conditions, changing customer requirements are trends and influences that ship design and shipbuilding companies cannot escape from. This situation represents an increase in the complexity of the conceptualization, designing and building of ships, which often is cited as a reason for system development difficulty and or failure in the delivery of new buildings—time, price and or quality wise. To improve the effectiveness of the design and construction processes of vessel new buildings, complexity should be understood, and handled in an effective way. This paper discusses the conceptual design complexity in cruise ship design and presents a practical approach to handle such complexity in early design phases. We propose the application of multi variate data analysis and multi criteria decision making approaches as an effective way to handle the conceptual design decision making complexity of cruise ships. Several practical design cases/problems are presented and discussed in this paper to demonstrate the issue at hand.

1 INTRODUCTION

Research looking into integrated technical, operational and commercial challenges in the cruise sector is scarce, and partly fragmented. This is result of a because, lack of a unified theme and a coherent theoretical framework cruise shipping research, and related tourism research in general. Research into cruise based tourism has focused on a variety of issues such as customers' experiences and their satisfaction (Brida, *et al.*, 2010), customer loyalties, port experiences and their motivations (Petrick and Durko, 2015) in addition to general management issues (e.g. Gibson, 2006). Recent scientific and public discourses also highlight potential negative environmental impacts of cruise tourism in various regions of the world, such as the Antarctica. Other concerns relate to the image of the cruise industry, is residents' attitudes towards the industry as well as safety and risk issues (Klein, 2005), which may have become more salient because of recent accidents and the importance of safety in this segment.

Nevertheless, commercial and environmental issues are addressed in the literature, and technical and operational aspects of the cruise vessel design as well. Application of light weight material, (Noury *et al.*, 2002), improved fuel/energy efficiency (Górski *et al.* 2013, Wartsila, 2015) and structural weight minimization and weight optimization (Žanic, 2014) have been some of these areas of interests for researchers.

This paper discusses different complexity drivers and complexity aspects in the conceptual design of cruise-exploration ships. Five main aspects of complexity from a system engineering standpoint are reviewed, including: structural, behavioral, contextual, temporal and perceptual aspects. Further, we propose a link between the challenges of the conceptual design of cruise-exploration ships a each of these five complexity factors. It is discussed and argued how different complexity elements within conceptual design of cruise ships influence the three vessel design perspectives: A: Design for Efficiency, B: Design for Effectiveness, as well as C: Design for Efficacy. Some of the factors studied include: vessel economics measures considering price per passenger rates, vessel capital expenses and costs (CAPEX), operational expenses and costs (OPEX) and specific voyage related costs (VOYEX) beside other ship performance measure of merits in terms of safety, operability, comfort and luxury level, survivability in intact and damage condition and environmental footprint.

2 CONCEPTUAL SHIP DESIGN AND DECISION MAKING COMPLEXITY

Ship conceptual design is a complex endeavor requiring the successful coordination of many disciplines, not only limited to technical, operational and commercial aspects (Papanikolaou, 2010a, Ulstein and Brett, 2015). Such a process is coupled with the selection of the better solution or solutions out of many feasible ones on the basis of a set of criteria within certain given and forced upon constraints. (Papanikolaou, 2010b, Ulstein and Brett 2012, Ebrahimi, et al., 2015).

In the early phases of ship conceptual design, it is difficult for the designers and other decision makers to comprehend the business situation from a holistic perspective. It results from the lack of experience and broader knowledge. The consequences of changes and alterations on the solution space boundary conditions, on overall performance of the ship design solution, is not very well developed in the ship design industry (Brett et al., 2012). This complexity affects the decision-making process in different phases of the product design process. In such a complex decision-making environment, relatively small changes in design are typically associated with very large consequences in the resulting performance yield outcomes. For example, according to practical design experiences at Ulstein, it is common to see revisions and changes on the preliminary expectations during the conceptual design. Addition/reduction on the number of passengers, improve/reduce vessel luxury level or region of operation are some examples. Each of these single changes might lead to substantial consequences on vessel's technical, operational and commercial performance. For instance, adding number of passengers requires larger hotel area, potentially an additional deck, which may lead to stability issues, requiring eventually a larger platform. Larger the platform and a higher passenger capacity result on a higher CAPEX and OPEX. Meanwhile, it can benefit the guests by lowering the fee per night of the travel. On the other hand, operational performance and the operability of the vessel also are influenced by varying the platform size or region of operation. Therefore, in such circumstances, making an appropriate trade-off among different expectations, wishes and the objectives is an essential issue in the early design phase. Decisions taken at this stage, considerably affect also the downstream phases: manufacture, assembly, operation, maintenance, and disposal of the ship, in terms of cost, quality and function performed by the vessel. It is a common practice in this cases the use of alternative lightweight materials like aluminum or carbon fiber in the construction of 2–3 top decks in accommodation. In addition to technical impacts on the design features like weight and stability, by making such decision, vessel economy, construction strategy and vessel disposal are directly influenced. More the possible alternatives in the decision-making with higher internal interaction, increases the decision-making complexity. Especially, when the consequences in vessel's performance of making different decisions are vague in the early design phases. To handle such complexity, it is proposed to approach conceptual design decision making, under different paradigms (Figure 1).

Decision-making includes a vessel design logic, integrates value creation functionality and capability to perform and develop knowledge into a decision model that accurately and concisely portrays the real problem at hand (Skinner, 2001). On the other hand, classic ship design, consists: different rules, regulations and several design guides/practices that influences the decision-making parameters and simultaneously increases the complexity of the process. Systems thinking used in the model encourages to explore inter-relationships (context and connections), design perspectives and boundaries.

To handle the complexity in decision-making process of conceptual ship design, very often, numerous smaller decisions must be made before a complex decision can eventually arrive at its final conclusion. Each decision may directly or indirectly have a bearing on other subsequent decisions. For example, in case of exploration cruise, making proper decision in the vessel area of operation, target customers and amenities, both exploration or onboard is essential in early phases. These early step decisions will input directly to the next step decisions such as number of passengers and luxury level. These decisions will result on a minimum the size for the vessel, number of crew, vessel layout and relevant amenities installed onboard the vessel. It is seen in many design cases, deficiency of clear holistic picture on the interactions of these decisions and their influence on final design solution leads to the design solutions which are ill-positioned in the market place.

Rhodes and Ross (2010) propose five essential aspects for the engineering of complex systems. This decomposition embraces the 1) structural: related to the form of system components and their interrelation—ships, 2) behavioral: related to performance, operations, and reactions to stimuli,

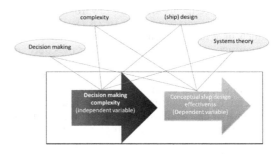

Figure 1. Decision making complexity handling paradigms.

3) contextual: related to circumstances in which the systems exist, 4) temporal: Related to dimensions and properties of the systems overtime like market uncertainties and 5) perceptual aspects related to stakeholder preferences, perceptions, and cognitive biases to evaluate the system complexity.

3 CRUISE AND CRUISE-EXPLORATION CONCEPTUAL DESIGN ASPECTS

3.1 Passenger ship market and categories

Generally, a passenger ship is a merchant ship whose primary function is to carry passengers on the sea, that typically returns to the same port at which it embarked from after its voyage is completed. Over the past few years, passenger line routes have seen diversification as well as global expansion. There is now a greater choice of worldwide passenger destinations and the number of transatlantic routes is rising as well. In response to the substantial growth tourism market and passenger lines expansion globally, the number of contracted vessels in passenger ship segment has also had substantial increment within last four years by the record of, 660 new build contracts in cruise, cruise-exploration and Ropax-ferry vessel categories between 2013 and September 2017. (IHS, 2017 – Figure 3). Valid perception of the market dynamics, reduces the temporal complexity of decision making process by mitigating incorporated uncertainties. It is essential in the macro decision-making levels to have proper overview of contracting activities and fleet growth to set new ordering strategies in a correct place and time.

Figure 3 demonstrates cyclic fleet growth with positive trend in cruise exploration market within last 50 years. Such analysis for different categories inside the segment supports the decisions of expanding fleet in the right category for decision makers. However, in addition to market dynamics, proper comprehension of the characteristics,

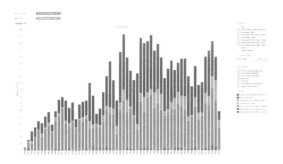

Figure 2. Cruise and Ropax annual contracting activities (Ref).

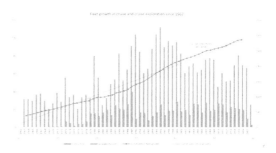

Figure 3. Cruise and exploration-cruise fleet growth (IHS 2017).

similarities and discrepancies of different vessel categories in the passenger ship segment is an important parameter in early decision-making phase. It is observed how misunderstanding of these criteria's leads to the new order decisions by the customers with different expertise and experiences than the chosen vessel category they want to play a role in. That type of wrong decisions dramatically influences the competitive advantages and market reputation of the operating cruise company in this context.

It is recognized by the authors of this paper, that commercial, operational and technical aspects of design, deviates among different categories of passenger vessels.

Generally, passenger ships are categorized in 7 groups including:

A. Liner type cruise operation: B1: Coastal Cruise: B2: Cruise Ferry (Ropax). C1: Exploration/ Expedition Cruises, C2: Arctic Cruises, D: Mega Cruises and E River cruises where the main focus of this paper is the exploration/ expedition cruises.

Exploration/Expedition Cruises are typically characterized as a vessel related to leisure activities. This type of cruise is focused on providing opportunity for passengers to explore as more as possible destinations within travel frame work. Therefore, normally short transit distance and many port of calls with longer stays at ports are typical criteria's in this cruise type segment. Vessels are normally small in the range of 80–150 m length with high to ultra-luxury hotel to accommodate normally between 50–750 passengers. Speed is not typically an important design criterion for exploration type cruise ships due to flexible schedule. Providing sufficient space onboard the vessel for exploration tools like kayaks, zodiac, paddleboards, glass-bottom boats, snorkeling gear, underwater submersibles and etc, is essential design feature in this segment.

3.2 Cruise-exploration ship concept design complexity

Traditional cruise ship design is characterized by a series of tradeoffs that are often made without thorough consideration as to their overall impact to the design as a holistic system (Zanic, 2014). Typically, with the expectation of ship size, number of guests, interior luxury, vessel speed and the operational region the most important parameters such as initial acquisition cost, profitability, operational cost range, operational profile and allocation of revenue making spaces are not included in the design decision making process.

Design particulars are dependent on the objective opinion of the costumer/user the requirements. These requirements are often ambiguous and typically change over time. (Zink, 2000). Therefore, understanding the simultaneous impact of the requirements, design variables, like vessel size, luxury level, environmental foot print and fuel type, targeting market, operational region, costumer experiences, and emerging technologies during the concept development stage is critically important and to some extent illusive in many cases.

Figure 4, demonstrates simplified causal map between design perspectives and complexity aspects of cruise-exploration conceptual design. Design parameters, vessel particulars and design objectives are (intermediate variable or independent) dependent and independent variables which formulate the concept based on their internal causalities. In this model all functional related parameters of the vessel including vessel hull form/size and related marine systems create the structural complexity of the system while vessel stability and environmental footprint are the main composers contributing to the behavioral complexity. Different operational regions are considered as different contexts where the vessel interacts with the environment. Meanwhile, operability and comfort criteria are the drivers of contextual complexity, which can vary in different contextual circumstances. Motion sickness and motion induced indexes are the two specific comfort criteria in the cruise ships which can dramatically influence the passenger's experiences. Therefore, coming up to the right combination of main dimensions and systems like active roll reduction system and fin stabilizers are very effective here. Operability in the swimming pools or the restaurant is another parameter which influences the early design decision. In a vessel with 150 m length, operating in 3 m Hs based on operation limiting criteria (Nordforsk, 1987) created by vessel accelerations, operability in restaurant will be almost 7% higher when restaurant is located around the longitudinal center of bouncy compared to moving it 25% of the length towards both vessel ends. It is clearly seen how decisions on vessel layout as a part of structural complexity is directly influenced by the operational performance of the ship.

Business sustainability issues over the time, changes in market situation and technological advancement within ship lifecycle period are those who contribute in temporal complexity. Such parameters naturally contain level of risk and uncertainty which requires to be handled in appropriate way in conceptual design stage. Vessel luxury

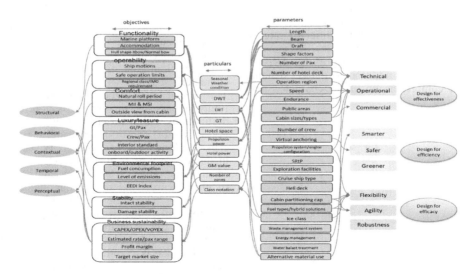

Figure 4. Causal map between design perspectives and system complexity aspects. Adapted from Ulstein cruise concept design practices. model is simplified by elimination of some casual line.

level, travel experience and price paid by passengers for travel are part of perceptual parameters from passenger point of view. On the other hand, vessel commercial parameters, external shape and hotel layout from customers perspective, are the parameters which escalates the perceptual complexity for designer and decision maker in the early stage.

The amount and diversity of parameters contributing in the model and their internal interactions are a representation of the level of complexity in such conceptual design decision-making process.

For example, as displayed in Figure 4, different parameters contribute to shape the feeling of leisure and luxury among passengers, by leveraging tangible and intangible elements. The area allocated per passenger, both public and private spaces, is one of the factors which largest influence on the luxury perception and is defined by Gt/Pax or a space ratio. This is a tangible factor, defined based on room sizes or level of the crowdedness in public areas which stimulates the guests/passenger's emotions towards their luxury feeling. On the other hand, the level of service and attention given to the passengers is another criterion contribution to the overall luxury perception. This factor is measured as the number of crew members (including marine crew but also hotel personnel and exploration guides) per passenger, and is also called service ratio (Crew/Pax). as Although it is an intangible element, it influences the feeling of happiness and leisure in the costumers.

The style selected for the hotel area, or the type of material utilized for the hotel outfitting also conveys the luxury feeling. While for instant, in Ropax vessels to meet the pre-defined schedule has higher priority for passengers than luxury feeling, which entirely changes design strategy in the early conceptual phases. Based on these factors vessel luxury level can vary from modest to ultra-luxury solution (Figure 5).

4 MULTI VARIATE DATA ANALYSIS AND MULTI CRITERIA DECISION MAKING

We argue that an effective decision-making framework for handing the initial complexity in initial cruise-exploration ship design should include multivariate data analysis (MDA) and multi criteria decision making (MCDM) approaches. Through review of three design perspectives by the use of multi criteria decision analysis and multi variate data analysis, a uniform philosophy is created to articulate a framework that would enable decision makers and designers to more accurately assess and evaluate the implications and the consequences of the changes in design alternatives in overall performance of the alternative design solution.

4.1 Multivariate data analyses on cruise-exploration ship data

Multivariate data analysis refers to any statistical technique used to analyze data that arises from more than one variable. This essentially models reality where each situation, product, or decision involves more than a single variable. Ulstein applies different multivariate data analysis techniques based on the knowledge discovery from data base including i) Data base integration; ii) Data cleaning; iii) Clustering analysis; iv) Multivariate regression analysis; v) Parametric Equation development and parametric study of clusters; and vi) Data validity and sensitivity analysis. (Ebrahimi, Brett et al 2015). Such multi—variate approaches are used in conceptual design development of cruise ships (Figures 6, 7 and 8) to support the decisions based on facts and figures coming out of the, real market data analysis.

Figure 5 shows two-dimensional density-based clustering analysis on luxury elements of exploration cruise segment. Applying such a data analytical approach it is seen vessels are grouped in 5 clusters based on their service and space ratio simultaneously.

segments are partly overlapping and represent seamless steps in vessel luxury, based on size and

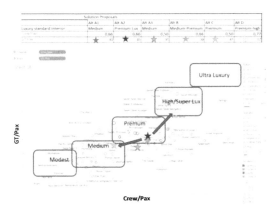

Figure 5. Cruise ship luxury level contracting.

Figure 6. Price/(Loa*B*D) vs number of passengers.

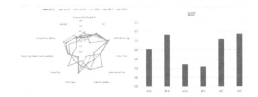

Figure 7. GOOF index comparison.

service ratios. It is seen in the medium luxury group GT/Pax is between 0,2–0,4 and crew/pax ratio is between 0,35–0,55, while these values improve to 0,6–0,8 for space and 0,65–0,85 for service ratio respectively in super luxury vessel group, which has direct impact on CAPEX, OPEX, VOYEX and price per passenger per night of the vessels. Similar clustering approach is applicable on vessel sizes based on dimensional parameters or vessel cost units over GT or L*B*D of the ship.

Such information caters better grasp of the market vessels related to requested design criteria to narrow down both design and solution space in early design phases. that handle better the decision-making complexity by having more knowhows of design solution position-performance in real market place. Plotting different developed solutions on the graphs demonstrates, wither the design solution well or ill positioned in the market place based on dimensioning criteria which leads to more early phase design decisions.

4.2 Multi criteria decision making on cruise-exploration concept design decision making complexity handling

Multiple-criteria evaluation problems consist of a finite number of alternative solutions explicitly known in the beginning of the solution process. Each alternative is represented by its multiple criteria performance indicator. The problem may be defined as finding the good solution by available alternatives.

The intention is to focus on how to select among possible solutions and differentiate/select better among different perspectives and their influence on design decision-making. Giving more priority to each of these design perspectives will lead to alternative design solutions with emphasis on different particulars and attributes. It is essential that such overall priorities are well explained and understood within its conceptual design solution space to handle better the aspects of complexity.

Utilizing a multiple criteria decision-making model leads to the identification the better compromised set of solutions among relevant benchmarks/preferences and pre-selected potential design solutions. Typically, in a cruise exploration design process, performance indicators such as cost, construction quality, luxury and comfort, operability, safety, fuel economy, maneuvering, sea keeping behavior, price per passenger, operational area and environmental foot print of the vessel are the main criteria, which are considered critically in an early design stage. Hence, designers should be able to make an early process trade-off among these capabilities and capacities. Higher the vessel safety by more stability criteria in GM value leads to stiffer vessel with lower comfort for passengers. Luxury and cost is another trad off which needs to be handled in a proper way. The influence of selecting arctic area as operational region, to increase revenue with inferior environmental impacts beside adding more complexities on structural and behavioral aspects. Polar code requirements on more steel weight, propulsion system, material type and the cost consequences within vessel lifecycle, all are some of these contradictory objectives, which makes conceptual design decision making and ranking process among the solutions more complex.

Goodness-of-fit GOOF Index developed by Ulstein (Ulstein and Brett, 2012, 2015), and Ulstein Benchmarking model (Ulstein and Brett, 2015), (Ebrahimi, et al., 2015) are the effective approaches to handle ranking problem among alternative peer solutions when different contradictory design objectives influences the decision-making process. In this model different design criterion of solutions are measured towards the preliminary expectations in both qualitative and quantitative manner and vessels are scored and ranked accordingly to come up with the 'better' fitted solution, among peers in solution space, then selected solution is compared with market peers in the benchmarking model to demonstrate the competitiveness of the final solution in the market place.

5 CASE STUDY

Case study part of the paper depicts the design of exploration cruise to be operated in Norden Europe area with following expectation criteria from costumer. (Table 1)

To approach this design problem, 6 alternative design solutions are generated (Table 2) in

Table 1. Design expectation list.

Number of Pax	Number of crew	Max speed	Ice class	GT
250–280	160–180	17	1B	Around 10000

Table 2. Alternative design solutions.

	Solution proposals					
	Alt A1	Alt A2	Alt A3	Alt B	Alt C	Alt D
Luxury standard Interior	Medium	Premium Lux	Medium	Medium-Premium	Premium	Premium-high
Ice class notation	Ice B1	Ice B1	Ice B1	Ice B1	Ice B1	Ice B1
GT	10400	11700	9100	10250	10750	14900
LOA	130	130	130	120	130	150
LBP	120	120	120	110	120	140
Beam	19,5	19,5	19,5	20,0	19,5	21,3
L/B	6,15	6,15	6,15	5,50	6,15	6,57
Depth to main deck	7,3	7,3	7,3	7,3	7,3	7,8
Max. Draft	5,5	5,5	5,5	5,5	5,5	5,7
Design Draft	5,3	5,3	5,3	5,3	5,3	5,5
Passengen	260	260	250	260	260	260
Crew	172	172	125	172	130	200
Crew/Pax	0,66	0,66	0,50	0,66	0,50	0,77
Gust Cabin arrangement (Qty*area)						
Luxury	6*70	6*70	5*50	6*70	6*70	6*70
balcony large cabins	20*32	20*32	20*32	20*32	20*32	20*32
Window cabins	104*25	104*25	100*26	104*25	104*25	104*25
Public area/Pax m^2	8,0	13,0	6,0	8,0	12,0	20,00
Crew cabin Arrangement	12*single 80 * 2 men	12*single 80 * 2 men	12*single 80 * 2 men	12*single 80 * 2 men	10*single 60 * 2 men	20*single 90 * 2 men
Dwt	1,420	1,230	1,550	1,350	1,320	1,600
DWT/Pax	5,5	4,7	6,2	5,2	5,1	6,2
Engine _KW_Total	10,000	10,500	9,700	10,500	10,150	12,300
Propulsion Power	6,800	6,800	6,800	7,320	6,800	7,450
GT/Pax	40	45	36	39	41	57
Max. Speed (knots)	18	18	18	18	18	18
LWT	6200	6400	6100	5900	6300	8500
St weight	3150	3260	3100	2980	3220	4350
Number of accommodation deck excluding top deck/bridge	5	5	5	5	5	5
MSI Operability in 5.1 Hs 30% from LCG Norwegian Sea	98,0%	98,0%	98,0%	98,0%	98,0%	98,0%
Vertical Acceleration Operability 0.05*g limiting 30% from LCG	79,0%	79,0%	79,0%	78,0%	79,0%	83,0%
GM	1,58	1,38	1,73	1,93	1,45	1,55
Natural Roll Period sec	11,8	12,62	11,3	11	12,21	13
SFC Ton/day	20	21,1	19,5	20,3	20,2	25,3
Spped * GT/Power	18,72	20,06	16,89	17,57	19,06	21,80
Price/Pax night break even point USD @ 75%	510	570	440	505	480	700
USD/GT	9,807	9,231	9,341	9,122	9,023	9,396
Opex/Day	53600	59700	51700	53400	47500	67000
USD_Newbuiling_ Price adjusted	94,500,000	108,000,000	85,000,000	93,500,000	97,000,000	140,000,000
USD/L*B*D	5,107	5,836	4,593	5,337	5,242	5,618
USD Price/pax	363,462	415,385	340,000	359,615	373,077	538,462

Table 3. GOOF Index ranking table.

Ranking parameters for GOOF index		Alternative Solutions						Range		Scale	
		Alt A1	Alt A2	Alt A3	Alt B	Alt C	Alt D	min	max		
Natural Roll Period Tr	O	2,7	3,5	2,3	2,0	3,1	3,8	9,9	14,3	1	5
FC ton/day	T	4,0	3,6	4,2	3,9	4,0	2,0	17,55	27,83	5	1
Vertical Acc Op	O	4,9	4,9	4,9	4,8	4,9	5,3	0,35	0,8	1	5
crew/Pax	T	3,1	3,1	1,8	3,1	1,8	4,0	0,4	0,9	1	5
public area/Pax	T	1,6	2,8	1,1	1,6	2,6	4,5	5,4	22	1	5
Interior quality	T	2,3	3,7	1,0	2,3	2,3	3,7	2	5	1	5
OPEX/day	C	3,6	2,8	3,8	3,6	4,4	1,9	42750	73700	5	1
Price/Pax	C	2,2	2,9	1,5	2,2	1,9	4,3	396	770	1	5
Targeting market sustainabilit	C	3,2	2,9	3,5	3,2	3,3	2,3	100	1000	5	1
Vessel Price MUDS	C	4,1	3,4	4,6	4,1	3,9	1,7	76500000	154000000	5	1
USD/GT	C	3,3	3,0	2,8	3,2	3,4	2,7	8121	10335,57	5	1
GOOF		35,1	36,5	31,5	34,1	35,6	36,1				

conceptual phase, to fulfill expectation criteria by varying main dimensions, cabin sizes, number of crew and vessel luxury level. Among the solutions, solution D is larger solution well-fitting with premium-high luxury where alternative A3 resembles a modest-medium luxury solution in the requested size. This variation among the solution is to depict the influence of different design technical and operational aspects on the commercial factors.

It is seen how implementing such changes in design increases vessel price from 85MUSD for alternative A3, to 140MUSD for alternative D. The cost consequence of such substantial CAPEX increment for passengers is almost extra 300USD per day to be onboard alternative D, while they will benefit from higher luxury and larger privet and public spaces onboard the vessel. Solutions are represented in different market data analytics graphs to figure out ill positioned solutions based on market place information. Figure 5 displays the GT/Pax and Crew/Pax level of solutions compared to market peers as luxury perception measure.

Figure 6 depicts price/unit of size (L*B*D) with number of passengers onboard. It is seen all solutions are belonging to cluster 4, while unit price of the vessel size well fits with market vessels in the design size range. Several other graphs such as power vs GT, Price vs size and number of passengers are prepared for better demonstration of developed solutions towards market. Further to such analysis it is required to select better fitted solution to owner expectation by GOOF index.

To be able to rank the solutions some technical operational and commercial features are selected and each alternative solutions are scored based on the values in scale between 1–5. In the ranking features Fuel consumption (FC), Crew/Pax, interior quality and public area/pax are technical features which are controlled in design phase. However, these parameters influence the commercial perspective literally Natural roll period and Vertical accelerations are operational indicators which directly influence the POB comfort and Price/Pax is calculated for all cases based on break-even point to fulfil OPEX, CAPEX and VOYEX with 75% utilization rate. Targeting market sustainability is an indicator based on price/pax night. Higher the fees score in this feature is less which shows the population target people who can afford the cost is less which increases the risk of less vessel utilization annually (Table 3).

Vessels are scored in different features differently. While alternative D is the most expensive solution and scores least in price, but this solution has higher revenue making capability as well due to higher luxury level of the services and spaces onboard the ship. Alt A3 which is smaller solution with medium luxury level does not seem very promising solution in this context to fulfil owner requirement. As a result of study Alt A2, and Alt D seems the most viable solutions with higher goodness of fit to owner expectations (Figure 7). Among these two alternative solutions, the one fitted better to target price of the decision maker is the solution to be selected.

6 CONCLUSION

Different complexity drivers and complexity aspects in the conceptual design phase of exploration/expedition cruise ships argued in this paper. State-of-the-art multicriteria decision-making approaches and multi variate data analytic techniques including clustering approaches are proposed and discussed as an efficient tool in this paper to handle the complexity in the early conceptual phase of this vessel segment. Five main aspects of complexity from system engineering point of view, including: structural, behavioural, contextual, temporal and perceptual aspects are discussed, linking challenges of the conceptual design of cruise ships to each of the aspects. It is argued and displayed in casual maps, how different complexity aspects within conceptual design of cruise ships, influence the three Ulstein developed vessel design perspectives: Design for Efficiency, Design for Effectiveness, as well as Design for Efficacy in conceptual design phase. Causal mapping between complexity aspects and design perspectives argued within case studies in this paper and approaches to handle and manage the complexity in an efficient way is argued. Such handling of complexity in conceptual design is reviewed and discussed with respect to its advantages and improvements in overall design performance yield in contrast to market solutions within discussed case study. Handling complexity in early design phase and the concept development of 260 Pax cruise ship as a case study is discussed in this paper. Disparate design objectives and mission requirements, including vessel luxury level, cruising routs, vessel speed, different onboard/onshore entertainment approaches, capacities besides potential revenue making capability, operating costs and capital costs are discussed for different possible scenarios, relating to different complexity aspects and design perspectives.

REFERENCES

Brida, J.G., et al., (2010). *Cruise visitors' intention to return as land tourists and recommend a visited destination.*

Callon, W., et al., (2017), How does decision complexity affect shared decision making? An analysis of patient-provider antiretroviral initiation dialogue. *Patient Education and Counseling 100, 919–926.*

Caprace, J.D., and Rigo, P., (2011) *Ship complexity assessment at the concept design stage,* Journal of Marine Scienc and Technology, 16,1, 68–75.

Ebrahimi, A. et al., (2015), Better decision making to improve robustness of OCV designs, *Int'l Maritime Design Conference, vol. 3, pp. 1–13.*

Gaspar, H.M., et al., (2012), Handling complexity aspects in conceptual ship design. *Int'l Maritime Design Conference, Glasgow, UK.*

Gibson, P., (2006). Cruise operations *Burlington, MA: Elsevier (Butterworth-Heinemann).*

Górski, W., Teresa A.-G., and Burciu, Z. 2013. "The Influence of Ship Operational Parameters on Fuel Consumption." *Ship Design and Research Centre, Gdynia.*

Holland, J.H., (2005), Studying complex adaptive systems. *Jrl Syst Sci & Complexity, 19: 1–8.*

IHS Fairplay, (2017), IHS Maritime World Register of Ships Q3 2017.

Klein, R. (2005) Cruise Ship Squeeze: The New Pirates of the Seven Seas, *New Society Publisher, Canada.*

Levander, K., (2006), System Based Ship Design. Norwegian *University of Science and Technology, Trondheim, Norway.*

Liew. A.S., (2005), Complex decision making processes: their modelling and support. *Proceedings of the 38th Hawaii International Conference on System Sciences.*

Magee, C.L..and De weck, O.L., (2004), Complex system classification, *Proceedings, 14th Annual International Symposium of the International Council on Systems Engineering.*

Maritime University, Gdynia, Morska.

Nordforsk, (1987).

Noury. P., et al. (2002), Lightweight construction for advanced shipbuilding—recent development. *Competetive ship building ETSIN, Madrid.*

Papanikolaou, A., (2010a), Holistic Ship Design Optimization Merchant and Naval Ships. *Computer-Aided Design Volume 42, Issue 11, November.*

Papanikolaou, A., (2010b), An integrated method for design of Ro-Ro Passenger vessels. *Naval architecture, ocean and marine engineering, January P 24–37.*

Petrick, J.F., and Durko, A.M., (2015), Segmenting luxury cruise tourists based on their motivations. *Tourism in Marine Environments.*

Rhodes, D.H., and Ross, A.M., (2010), Five aspects of engineering complex systems: emerging constructs and methods, *Proceedings, 4th Annual IEEE Systems Conference, April, San Diego.*

Skinner, (2001) Introduction to Decision Analysis *ISBN-13: 978-0964793866.*

Ulstein, T and Brett, P.O., (2012) Critical systems thinking in ship design approaches *11th International Maritime Design Conference, 2012.*

Ulstein, T and Brett, P.O., (2015) "What is a better ship? – It all depends...," *IMDC 2015, vol. 1, pp. 49–69.*

Wartisla. Improving energy efficiency in the cruise and ferry industry. *Industry white paper.*

Zanic, V et al (2000) Structural design methodology for large passenger and roro/passenger ships. *In 7th Int. Marine Design Conf. (IMDC).*

Zanic, V., (2014) Design synthesis of complex ship structures *Ships and Offshore Structures, 8:3-4, 383-403, DOI: 10.1080/17445302.2013.783455.*

Concept design considerations for the next generation of mega-ships

K.M. Tsitsilonis, F. Stefanidis, C. Mavrelos, A. Gad & M. Timmerman
Department of Naval Architecture, Ocean and Marine Engineering, University of Strathclyde, Glasgow, Scotland, UK

D. Vassalos
Maritime Safety Research Centre, Department of Naval Architecture, Ocean and Marine Engineering, University of Strathclyde, Glasgow, Scotland, UK

P.D. Kaklis
Department of Naval Architecture, Ocean and Marine Engineering, University of Strathclyde, Glasgow, Scotland, UK

ABSTRACT: The cruise industry represents one of the fastest growing shipping industries, supported by a 62% demand increase for the last decade, and an order-book in excess of $25bn. The sustained growth in demand is complemented by increasing vessel size, which represent investments in the order of billions. Megaships serve a specific market segment, with capacities in excess of 6,000 guests and crew. Increased capacities though lead to an increasing number of people at risk, hence safety considerations are key to these developments. By considering the current trends of the industry, a techno-economic feasibility for a 12,000 passenger mega-liner, with a significantly increased safety level, is investigated. The increased size of the cruise ship enables enhanced on-board guest experience, whilst ensuring an unparalleled safety level.

1 INTRODUCTION

Cruise passengers have increased, in a global scale, by 33% in the last 10 years and are expected to reach approximately 30 million by 2020 (CLIA, 2016), with capacity utilisation of cruise ships being consistently over 100% (FCCA, 2016). This indicates that the demand for cruising continuously outstrips supply, increased by 62% in the last 10 years (CMW, 2017). Furthermore, considering that only 4% of the global population has experienced a cruise, the cruise market represents a large unexploited potential (Dowling, 2017).

Consequently, to address the rapid growth in demand, the size of the average cruise ship in the world fleet has increased by 30% in the past decade, with passengers and crew numbers ranging upwards of 6,000 people (ISL, 2017). Amongst this rapid growth of the cruise market (Dowling, 2017; Cassidy, 2016), cruise ship size is considered as a quintessential parameter in achieving economies of scale (Dowling, 2017; Chang et al. 2017; Hemmen & Antonini, 2016) and addressing the needs of the mass market (Georgsdottir & Oskarsson, 2017).

As a result of the above, guest experience on board the vessel is not only the catalyst for the decision to cruise, but most importantly the link between the cruise company customer and its profitability (Dowling, 2017). Thus, multiple studies that have been conducted, concluding that guest experience has to be tailored to their cultural norms and expectations (Elliot & Choi, 2011; Baker, 2014; Hung & Petrick, 2013). Such a concept is well understood, however, it is confronted as a barrier to enter a new geographical market rather than a competitive advantage in the micro-economic long run. This implies that vessels lack the ability to quickly and efficiently transform so as to exploit emerging market segments (Sun et al., 2014); this way committing investments of billions of dollars (Cruise Critic, 2017), in short term market trends.

However, as the number of passengers and crew increases, safety has become an issue of paramount importance (Mileski et al. 2014; Papanikolaou et al. 2013). Therefore, safety should be addressed in arrangements that act as an integral part of the design phase, rather than preventative measures applied to constrain the design (IMO, 2011). The benefits of introducing safety through the aforementioned process, becomes apparent in the public and safety perception of passengers (Lee & Arcodia, 2017), cost associated with accidents (Eliopoulou et al. 2016) and most importantly in dynamically quantifying and managing potential loss of life (Neves et al. 2011).

Having identified the above trends and needs in the design of current cruise ships, a concept techno-economic feasibility study is conducted in order to design: a) a cruise ship with 12,000 passenger capacity; b) and increased level of safety via the use of first-principles tools within a Risk-Based design context.

2 BUSINESS CASE

2.1 *PESTLE feasiblity*

PESTLE represents the six factors that affect the feasibility of the design as dictated by the achievement of the three aforementioned objectives. These factors are: Political, Economic, Social, Technological, Legal and Environmental (PESTLE). Political influences resulting from Donald Trump's election and possible future impacts of Brexit are reflected as changes in the immigration policies of these countries, which contribute significantly to the cruise ship industry, this way possibly discouraging potential future travellers (Enelow, 2016; Leppert, 2016). On the contrary, China's five-year economic development plan promotes the cruise industry, improving the prospects on this growing market segment on the far east (Hennelly, 2017).

Considering the economic factors, it is evident that the cruise ship industry is in an advantageous position for the short term. In specific, the early bookings of the majority of passengers provide cruise companies with interest-free loans, whilst once on board, the lack of substitute goods implies larger profit margins for on-board spending activities (Dowling, 2017). In addition, demand in the market is strong, as projections have RCCL and Carnival Cruises experience increase in passengers by 30–40% respectively, from 2014 by the year 2022 (Dowling, 2017) and to account for that the industry has secured an order book of $25bn for the next 5 years (ISL, 2017). However, due to the high price elasticity of demand, the industry is faced with unsustainable profits in the long term, as ticket prices, which comprise 70% of the total, revenue have been in a continuously decreasing trend for the past decade (Dowling, 2017).

Within a social context, recent accidents have had a negative impact on the reputation of cruise industries, demanding a safety-focused environment and raising concerns of the current safety level of large cruise ships (Lee & Arcodia, 2017; Martin, 2014). Furthermore, the market is tending more towards a mass market, especially in Asia (Dowling, 2017; Sun et al. 2014).

Technology has had an overall positive influence on the design as well as the operation of the vessel.

Technological advances have continuously reduced the footprint of the industry and increased operational and business efficiency (CIN, 2014; Chang et al. 2017).

Legal factors that improve the perception of safety of passengers are the 'passenger bill of rights' have also been adopted after the Costa Concordia accident (CLIA, 2018). Finally, with regards to the environment, the cruise ship industry has faced criticism in the past (Dragović et al. 2015), however, in combination with technology considerable progress has been made in reducing the industry's carbon footprint (Cruise Critic, 2017).

2.2 *Market segmentation*

The cruise market can be segmented on a geographic, demographic and psychographic basis.

Geographically, despite the potential demand growth in China, the largest market segments belong to the Caribbean and Mediterranean, which are still holding a prominent position as both destinations and main embarkation points. In specific, since 2007 the two aforementioned cruise markets represent over 37.3% and 18.7% of the global cruise ship capacity, respectively (CLIA 2014, 2015). Thus, by examining the demand and seasonality (Rodrigue & Notteboom, 2012), infrastructure of local economies and ports (CTO, 2017), and nautical accessibility in each port (Shipping Guides Ltd., 2017), the itineraries selected are: **a)** Western Mediterranean from Barcelona; **b)** Western Caribbean from Miami; **c)** Eastern Caribbean from Miami and **d)** One-way cruise itinerary from Barcelona to Miami (and return), the ports of which are listed in Table A-1 in the Appendix.

Based on the selected itineraries, by using a service speed of 20 kn, considering the time spent at each port (Cruise Web, 2018; Cruise Mapper, 2018) and calculating the distance between destinations, the operating profile of the vessel is obtained as illustrated in (Fig. 1). Overall, 21 cruises in Europe,

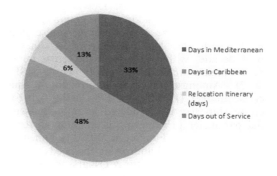

Figure 1. Ship operating profile.

13 in the East Caribbean, 13 in the West Caribbean and 2 from Miami to Barcelona and return (i.e. the relocation itinerary) are scheduled per annum.

Demographically, within the context of satisfying the mass market, North Americans and Europeans in the ages from 30–39 (Generation Y) to 50–59 (Generation X) and expected incomes of 60–69 K and 100–199 K per year are targeted, as these represent the largest demographic market share (CLIA 2014, 2015).

The psychographic segmentation focuses on the demands and expectations of the aforementioned target groups. More specifically, the Generation Y demands and expectations include (Elliot & Choi, 2011): **a)** solitude & isolation; **b)** break from the day-to-day environment; **c)** enriched perspective of life; **d)** relaxation; and **e)** not get physically pampered. On the other hand, the Generation X seeks to: **a)** renew personal connections; **b)** keep family ties; **c)** stimulate intellectually; **d)** relax in luxury. Using this information, a set of on-board activities were devised which are tailored to the needs of each market segment that can be found in Table A-3 in the Appendix.

2.3 Financing and business plan

Initially, the building cost of the designed vessel is estimated to be approximately $2.23bn using regression analysis on a database of 50 vessels (Jeremy, 2016; Kavussanos & Visvikis, 2016). To fund such a project, the private funds can be obtained from a typical 50–50 joint venture in the form of an Special Purpose Vehicle (SPV) (Coopers, 2011) of a reputable cruise ship company of high credit rating, and a private equity firm. To further spread the financing risk, the bank financing can be provided from a bank syndicate of typical shipping banks, with a risk guarantee of an Export Credit Agency (ECA). As the size of the ship will be the primary factor in determining its building cite, a reputable cruise ship building shipyard with a large enough dry-dock is STX France (STX France, 2017). Therefore, the ECA employed would be the French COFACE (COFACE, 2018), that will provide the inter-creditor agreement and risk guarantee to the bank syndicate and export the credit to STX France. In addition, the ECA loan agreement will be set up between the SPV and the bank syndicate.

The business plan is devised on the basis of significantly increasing the share of revenues that is derived from on-board activities in order to decrease dependency on the continuously decreasing ticket prices as mentioned above. Thus, the first step to a new business model is to improve the guests' willingness to spend on board by compensating with significant reduction on the nominal ticket price. By considering the financial statements of the three largest cruise ship companies by market share (RCCL, 2016; Carnival Corporation, 2016; Norwegian Cruiseline Holdings Ltd., 2016), the nominal ticket price of $1,153 per passenger is reduced by 25%.

Having decreased the ticket price, on-board spending can be used to compensate. On-board spending is comprised primarily by on-board activities, however, as shopping has the potential of becoming an integrated experience in cruising (Dennis, 2005), galleries can be introduced on board. A conservative number of $3,000 of sales per m^2 and 2 m^2 of shopping space per capita were assumed (CapitaLand Mall Trust, 2016; GGP Retail Real Estate, 2016). The galleries can be designed to service 3,000 passengers per cruise, and the cruise ship operator can obtain 20% of the revenue in concessions. By constructing a probabilistic model of the targeted market segments (Craciun et al., n.d.) and using as input typical values of the willingness to pay as a percentage of the nominal ticket price (CMW, 2017), the total passenger spending in on-board related activities is estimated.

The costs are subdivided into payroll, fuel, food & beverages and other costs. The two latter cost components were taken as percentages of typical cost components found in the financial statements of the largest companies in the market. The fuel is estimated by considering the correlation of the fuel costs with the fuel price, and a prediction was made using a price of $60 per barrel (Calcuttawala, 2017). The payroll costs are determined by splitting the crew into 6 payroll denominations with the minimum wage dictated by the flag of Panama and using the same probabilistic model structure as also adopted for the revenue

Table 1. Cruise ship profit and passenger spending.

Cruise Ship	Designed USD	Typical USD
Nominal Ticket Price	865	1,153
Nominal Ticket Price	1,454	1,431
Total Profit per cruise	9.728 Mn	2.803 Mn

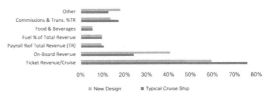

Figure 2. Cost & Revenue analysis comparison.

estimation above. The results of the cost & revenue analysis compared to that of a typical cruise ship are illustrated in (Fig. 2).

Considering the profit per cruise, the total number of cruises per year derived in Section 2.2, the ECA loan repayment terms (OECD, n.d.), linear depreciation and dry-docking costs, as well as a typical rate of return of 10%, the net present value of the project turns positive after the first 10 years of operation.

3 SHIP DESIGN

3.1 Ship design process

The design process is treated as an optimisation problem, starting with the determination of the design constraints, the search of initial values, the listing of the design variables and finally the selection of appropriate measures of convergence. The most prominent constraints of the set of main dimensions are derived firstly from the shipyard's dock dimensions, which are: L ≤ 450 m, B ≤ 70 m (STX France, 2017). Secondly, the port accessibility is considered: T ≤ 9.5 m for 87% accessibility based on current port drafts for the itinerary destinations (Shipping Guides Ltd., 2017). The initial values of length, beam, depth to main deck and gross tonnage are extrapolated from a database containing the largest cruise ships to date in terms of passenger capacity (RINA, n.d.). The area calculations used as guidelines the results from the statistical analysis above and are based on the requirements of 12,000 passenger capacity and equivalent passenger to accommodation area ratio as in the parent vessels. Empirical formulae for the calculation of the hydrostatics and weight estimation are then considered, the results of which are used to examine feasible solutions by determining the positive initial stability and estimated displacement convergence with the software in which the hull is modelled.

3.2 Deck plans

In order to ensure passenger comfort and accommodate 12,000 passengers, the cruise ship design is treated a top-down process beginning with the deck plans. In the absence of other information, two cruise ships of different passenger capacities, Anthem of the Seas and Harmony of the Seas, were used to establish a relationship between the area ratios allocated for staterooms and entertainment per deck (Fig. 3). The results of this analysis showed that the staterooms area should occupy about 30% percent of the total area, whereas the remaining 70% is allocated for entertainment purposes, stairways, elevators and void spaces. In addition, by considering also the design constraints

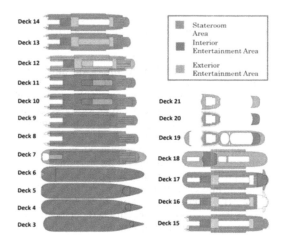

Figure 3. Designed deck plans of the new design.

Table 2. Deck area allocation.

Item	HoS*	AoS**	Design***
Passenger Capacity	5,497	4,182	12,000
Length [m]	362.1	348.0	425.5
Breadth [m]	47.5	41.4	28.3
Height**** [m]	55.0	52.0	59.7
m² of deck area per passenger	33.0	34.6	28.3
Stateroom % of total Area	31.5%	31.5%	31.5%
Entertainment % of Total Area	68.5%	72.5%	70%

*Harmony of the Seas, ***Investigated Design, **Anthem of the Seas, ****Height to top deck.

and the results of the regression analysis, the main dimensions were derived from the allocation of the total area per deck as listed on Table 2:

3.3 Hydrostatics and intact stability

The form coefficients and hydrostatic values (apart from the vertical centre of gravity (VCG) and initial stability (GM)) were calculated by considering the relevant formulae found in the literature (Papanikolaou, 2014; Schneekluth & Bertram, 1998; Friis & Andersen, 2000; Watson, 1998), based on which the hull is modelled using the initial hull design software (Bentley, n.d.) Tables 3 and 4. The estimated hydrostatics were subsequently validated by Maxsurf's software analysis. For the calculation of the vertical centre of gravity (VCG) and initial stability (GM), to ensure passenger comfort in rolling motions, the rolling period of the new design was set the same as that of the parent vessel, from

Table 3. Form coefficients-values.

Item	C_B	C_M	C_P	C_{WP}
Value	0.704	0.911	0.774	0.787

Table 4. Designed vessel hydrostatic values.

	Designed	Parent
VCB [m]	5.2	4.3
BM [m]	30.8	19.4
KM [m]	36.0	24.1
T_{roll} [s]	13.8	13.8
GM [m]	8.0	4.2
VCG [m]	27.9	19.8
GZ_{max} [m]	2.5@25°	1.3@25°

Table 5. Weight estimation.

	Designed	Parent
(1) Steel [t]	67,104	31,217
Hull Outfitting [t]	8,510	3,753
Internal Outfitting [t]	49,425	21,146
Machinery [t]	13,555	8,562
(2) Lightship [t]	145,154	64,679
Deadweight [t]	28,397	14,003
Displacement [t]	176,481	81,743
Gross Tonnage [gt]	376,634	168,666
DWT/Displacement [%]	17.8	17.1
(1)/(2) [%]	46.2	48.3

which the desired metacentric height is calculated (Papanikolaou, 2014):

$$GM = \left(\frac{2 \times C_{Troll} \times B}{T_{roll,parent}} \right)^2 \quad (1)$$

where, B is the designed vessel's beam, $T_{roll,parent}$ is the parent vessel rolling period and C_{Troll} is the rolling period coefficient:

$$C_{Troll} = 0.373 + 0.023 \times \frac{B}{T} - 0.043 \times \frac{L_{WL}}{100} \quad (2)$$

where, T is the draft and L_{WL} is the waterline length of the designed vessel.

With the above hydrostatic values, the following criteria are satisfied; HSC Code 2000 Ch.2 Part B: Passenger craft, IMO A.749(18) Code of Intact Stability and IMO SOLAS, II-1/8.

The greatly improved stability of the vessel in terms of high GM is a result of the smaller L/B ratio in comparison to the parent vessel, the trade-offs of which are evaluated in the following sections.

3.4 Weight estimation

Weight estimation is applied first on the parent vessel and the difference ratio between its actual displacement and the estimated weight allowed the definition of a correction factor that is used to calibrate the displacement estimation of the new design. The lightship of the vessel is considered to include the following weight components: Hull Steel, hull outfitting, interior outfitting and machinery and their weights were estimated considering the relevant formulas found in the literature (Papanikolaou, 2014; Schneekluth & Bertram, 1998; Friis & Andersen, 2000; Watson, 1998; Lamb, 2003). Considering the deadweight, the weight of freshwater, stores and provisions, and black and grey water, these are scaled from the parent ship based on the number of passengers. Finally, the estimation of the fuel and lube oil weight is scaled based on estimated power requirements and the Admiralty Coefficient (Papanikolaou, 2014). The draft of the designed vessel is adjusted such that the geometric displacement is greater than the estimated displacement by 5% as a safety margin. The results of the weight estimation are listed in Table 5 below:

The values of the last two percentage ratios increase the credibility of the weight estimation since they are very close to the expected range of percentages (23% DWT/Displacement and 52% for Steel weight/Lightship) (Papanikolaou, 2014).

3.5 Initial cost estimation and cost risk

An effective cost estimation framework is a critical decision support tool in the design of the vessel. Unfortunately, methodologies that could provide cost estimating relationships require shipbuilding data that are not easily accessible. Hence, the cost prediction is utilised in order to derive a relationship of the cost rate of change in relation to the governing design parameters of length and beam. The cost risk is statistically estimated to include any uncertainties (Lamb 2003). The procedure included the estimation of unit costs per weight group of Steel, Outfitting and Machinery considering the labour, material and overhead expenses (Michalski 2004). The total cost of the new design, calibrated by the fractional difference between the estimated and actual shipbuilding cost of the parent vessel, was calculated:

3.6 Parametric Optimisation

The effect of low L/B ratio was examined under: **a)** the scope of stability **b)** the resistance **c)** the resulting

Table 6. Predicted cost & cost risk.

Item	Total cost	Cost risk
(Bn $)	2.037	± 0.189

Table 7. Total power estimation.

	HoS	Designed
Total Area [m²]	160,163	340,000
Hotel Load [MW]	28.2	66.9
Propulsion Power [MW]	60@25 kn	102@20 kn

Output/Constrain	Mod.1	BoS	Mod.2	Mod.3	Mod.4	Mod.5	Mod.6
Capital Cost [Bn $ ±0.189]	2.003	2.037	2.051	2.107	2.238	2.316	2.383
Resistance @ 20kn [kW]	91,703	85,165	84,206	78,613	71,872	69,952	68,541
Stability (GM) [m]	16.7	13.6	13.0	9.9	4.7	2.6	0.6
Draft [m]	9.62	9.70	9.78	9.81	9.89	9.92	10.09
Feasible (F)/Not Feasible (NF)	F	F	NF	NF	NF	NF	NF

Figure 4. Parametric optimisation output and constraints,.

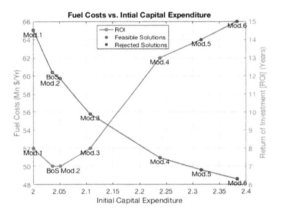

Figure 5. Fuel OPEX vs CAPEX and ROI.

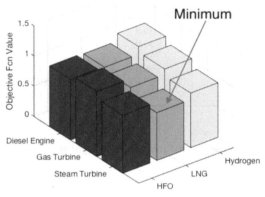

Figure 6. Multi-objective optimisation for machinery selection: Objective function.

sion power), **d)** fuel price of 357$/tonne. The results of the analysis are depicted in the figure below.

It is evident that the CAPEX is growing with the increase of L/B ratio whereas the fuel OPEX decreases. However, the selection of low L/B ratio is justified as the fuel cost decreases at a lesser rate in proportion than the increase of CAPEX. As a consequence, the selected L/B ratio for the designed vessel provides the minimum ROI.

CAPEX **d)** the fuel OPEX and **e)** return of investment (ROI). In all cases where designs with a different L/B ratio were tested, the LxB was kept constant to maintain the required deck area. The feasibility of the designs with the varying L/B was determined by the draught and GM constraints (T < 9.7 m & GM > 4.35 m) as those arise from the accessibility to ports and passenger comfort. The MATLAB code for the weight estimation was updated to include the cost prediction model and used in all the parametric runs. The results of the different designs can be found in the figure below whereas the different their principal dimensions can be found in the Appendix:

Having determined the feasibility of the designs based on the constraints, the ROI in the case of different L/B ratios was investigated. For the calculation of fuel OPEX the following assumptions were made: **a)** SFOC: 190 g/kWh, **b)** propulsion efficiency is 90% **c)** 46 seven-day cruises per annum with 2 out of the 7 days staying at port (no propul-

3.7 Power estimation—propulsion configuration

Preliminary estimation of the power requirements for hotel services and propulsion was performed. The hotel power of the new design was derived by assuming 10% transmission losses and the same nominal power requirement per sq. meter (0.176 kW/m²) of one of the largest cruise ships to date; Harmony of the Seas (HoS).

The propulsion system consists of two-20.4 MW Azipod units and two shaft driven propellers delivering 30.4 MW, arranged in two Contra—Rotating Propeller (CRP) configurations. The hydrodynamic efficiency of the CRP configuration was assumed to have a power reduction of 5% per unit (ABB, 2010; ABB 2011; Levander, 2007) compared to a single screw arrangement. The vessels' power was calculated by Wyman's method and 15% sea margin, 3% of shaft losses and 90% of engine load were accounted for.

3.8 *Machinery selection*

The machinery selection process was treated as a multi-objective optimization problem. Thus, combinations of three different types of fuels and engines were examined under the scope of their Lifecycle cost (LCC), power to weight ratio and *fuel risk*. In specific, the *fuel risk* is quantified by the use of a risk matrix evaluating the **a)** the likelihood of future regulatory compliance, **b)** environmental impact and **c)** human health upon exposure. The objective function examined the combinations of HFO, LNG and Hydrogen as a fuel in diesel engines, gas and steam turbines using the Diesel running on HFO as a baseline. The objective function values are illustrated on the figure below:

The steam power plant using LNG as fuel appears to be the optimum solution for on-board power production. The reason being that the weight factor in the objective function was modelled to favor the fuel type with the lower risk since compliance with the emission regulations was considered a priority. Secondly, the steam engine appears to have the lowest LCC which is the second more important factor of the objective function.

3.9 *Design overview*

Starting with the business case, the revenue distribution between ticket and on-board earnings shifted from 70%-30% to 60%-40% in favor of the on-board revenue. This new business model comes as a response to the constant reduction of ticket prices in an effort of compensating this loss via the on-board generated profit. The advantages of having a large ship become apparent in this case due to the abundance of space offered for on-board revenue generating activities.

Trying to accommodate for these design goals whilst ensuring that the building cost and will remain low lead to a new consideration of a lower L/B ratio. This design selection was justified both technically (improved stability) and economically (minimum ROI) despite some initial concerns regarding the increase of resistance and therefore of OPEX.

Finally, compliance with present and future emission rules, increased power requirements, safety and reliability considerations lead to an alternative propulsion configuration (twin CRP propulsion) and a steam power plant using LNG burning boilers (minimum value of the multi-objective optimisation problem of machinery selection).

4 FUTURE PLANS

Having completed the first design iteration, which delivered a cruise ship with acceptable intact stability characteristics that can accommodate the increased passenger capacity, without compromising the revenue-generating entertainment area, the next steps will be driven by the risk-based design concept.

More specifically, the most prominent hazards for our design will be examined by implementing the Hazard Identification (HAZID) technique on a database containing passenger vessels. Based on literature review it is expected that collision, contact and grounding will be the most prominent hazards, followed by fire (Papanikolaou, 2013). Consequently, the primarily focus of the study will be to perform damage stability and evacuation analysis enabling calculation of risk in terms of Potential Loss of Life, which will be made the basis for design improvements to render the ensuing risk tolerable.

Active (e.g anti-collision system) and passive RCOs will be considered and, in particular, the Damage Stability Recovery System (DSRS) (Vassalos et al. 2016) and referring to a highly expandable foam that can acts as an active measure of mitigating the flooding risk in a compartment as well as passive measures such as crashworthiness.

REFERENCES

ABB, 2010. The CRP Azipod Propulsion Concept—The most Economic way from crane to crane.

ABB, 2011. System Project Guide for Passenger Vessels. ABB Oy, Marine, Helsinki.

Baker, D.M., 2014. Exploring cruise passenger's demographics, experience and satisfaction with cruising the Western Caribbean. International Journal of Tourism & Hospitality Reviews, 1(1), pp. 33–43.

Bentley, n.d. MAXSURF: Initial Hull Design Software. [Online]. Available at: https://www.bentley.com/en/products/product-line/offshore-structural-analysis-software/maxsurf [Accessed 2018].

Calcuttawala, Z., 2017. EIA Boosts World Oil Demand Forecast For 2018 By 100,000 Bpd. [Online] Available at: https://oilprice.com/Latest-Energy-News/World-News.html [Accessed Jan. 2018].

CapitaLand Mall Trust, 2016. Annual Report 2016. [Online] Available at: http://cmt.listedcompany.com/misc/capitamall_ar2016/index.html [Accessed Dec. 2017].

Caribbean, R., 2018. Symphony Of The Seas—Itinerary Schedule |Current Position | CruiseMapper. [Online] Available at: http://www.cruisemapper.com/ships/Symphony-Of-The-Seas-1730 [Accessed 2018].

Carnival Corporation & plc, 2016. Financial Information. [Online] Available at: http://phx.corporate-ir.net/phoenix.zhtml?c = 140690&p = irol-reportsannual [Accessed Jan. 2018].

Cassidy, S., O'Hagan, K., Keith, A. & Geier, F., 2016. Cruise Ship Industry. s.l.:Cornell University: Presentation.

Chang, Y.-T., Lee, S. & Park, H., 2017. Efficiency analysis of major cruise lines, Tourism Management. Tourism Management, Volume 58, pp. 78–88.

CIN News, C.I., 2014. Air Lubrication System to Drive Savings for Quantum. [Online] Available at: https://

www.cruiseindustrynews.com/cruise-news/11491-air-lubrication-system-to-drive-savings-for-quantum.html [Accessed Jan. 2018].

CLIA (Cruise Lines International Association Inc)., 2016. Cruise Industry Outlook. Washington: s.n.

CLIA, 2018. Cruise Industry Passenger Bill of Rights [Online] Available at: https://www.cruising.org/about-the-industry/regulatory/industry-policies/other/pbor [Accessed Jan. 2018].

CLIA 2014, 2015. North American Cruise Market Profile January.

CMW (Cruise Market Watch), n.d. Financial breakdown of the typical cruiser. [Online] Available at: http://www.cruisemarketwatch.com/home/financial-breakdown-of-typical-cruiser/ [Accessed Nov. 2017].

Compagnie Française d'Assurance pour le Commerce Extérieur (COFACE), 2018. [Online] Available at: http://www.coface.com/[Accessed Jan. 2018].

Coopers, Pricewaterhouse Coopers, 2011. The next chapter Creating an understanding of Special Purpose Vehicles. [Online] Available at: www.pwc.com/financialregulation [Accessed 2018].

Craciun, M.-D., Chis, V. & Bala, C., n.d.. Methods of discretising continuous variables within the framework of Baysean Networks. [Online] Available at: https://www.emis.de/journals/AUA/pdf/56_389_paper32-ictami2011.pdf. [Accessed Oct. 2017].

Cruise Critic, 2017. Green Cruising. [Online] Available at: https://www.cruisecritic.co.uk/articles.cfm?ID = 528 [Accessed Jan. 2018].

Cruise Critic, 2017. Upcoming Cruise Ship Refurbishmens. [Online] Available at: www.cruisecritic.com [Accessed 2018].

Cruise Market Watch, 2017. Cruise Market Watch—Growth of the Ocean Liner Industry. [Online] Available at: http://www.cruisemarketwatch.com/growth/ [Accessed 2018].

Cruiseweb.com, 2018. Royal Caribbean Symphony Of The Seas 7 Night Eastern Caribbean Holiday. [Online] Available at: https://cruiseweb.com/cruise-lines/royal-caribbean-international/ship-symphony-of-the-seas/7-night-eastern-caribbean-holiday-cruise-departs-miami-florida [Accessed 4 January 2018].

CTO (Caribbean Tourism Organisation), 2017. Annual Reviews & Prospects. [Online]. Available at: http://www.onecaribbean.org/statistics/annual-reviews-prospects/ [Accessed Jan. 2018].

Dennis, C., 2005. Marketing Segmentation for Shopping Centres. In: Objects of Desire. London: Palgrave Macmillan.

Dowling, R., 2017. Cruise Ship Tourism. s.l.: Center for Agriculture and Biosciences International.

Dragović, B. et al., 2015. Ship emissions and their externalities in cruise ports. Transportation Research Part D: Transport and Environment.

Eliopoulou, E., Papanikolaou, A. & Voulgarelis, M., 2016. Statistical analysis of ship accidents and review of safety level. Safety Science, Volume 85, pp. 282–292.

Elliot, S. & Choi, H.C., 2011. Motivational Considerations of the New Generations of Cruising. Hospitality and Tourism Management, Volume 18, pp. 41–47.

Enelow, S., 2016. President Trump: The Travel Industry Reacts With Caution, Not Enthusiasm. [Online] Available at: https://skift.com/2016/11/10/president-trump-the-travel-industry-reacts-with-caution-not-enthusiasm/ [Accessed Dec. 2017].

FCCA (Florida—Caribbean Cruise Association), 2016. Cruise Industry Overview 2016 – State of the Cruise Industry. Florida: s.n.

Friis, A.M. and Poul Andersen, 2000. Ship Design Part 1 and Part 2. s.l.: Dept. of Naval Architecture and Offshore Eng. Technical University of Denmark..

Georgsdottir, I. & Oskarsson, G., 2017. Segmentation and targeting in the cruise industry: an insight from practitioners serving passengers at the point of destination. The Business and Management Review, April. 8(4).

GGP Retail Real Estate, 2016. Annual Report 2016. [Online] Available at: http://investor.ggp.com/financials/annual-reports-proxy [Accessed Dec. 2017].

Hemmen, R.V. & Antonini, P.E.K., 2016. The Delightful Frustration of Cruise Ship Power Plant Design. s.l.:Martin Ottaway: Marine Consultants, Engineers, Surveyors, Naval Architects & Appraisers.

Hennelly, W., 2017. It's full speed ahead for the cruise ship business in China. [Online] Available at: http://usa.chinadaily.com.cn/opinion/2017-03/02/content_28414943.htm [Accessed Dec. 2017].

Hung, K. & Petrick, J., 2013. Why do you cruise? Exploring the motivations for taking cruise holidays, and the construction of a cruising motivation scale. Tourism Management, Volume 32, pp. 386–393.

ISL (Institute of Shipping Economics & Logistics), 2017. Shipping Statistics and Market review 2017 Volume 61 – No. 8. s.l.:s.n.

International Maritime Organisation, 2011. Generic Guidelines for the Development of the Goal Based Standards. London: MSC.1/Circ.1394.

Jan P. Michalski, 2004. 'Parametric method of preliminary prediction of the ship building costs'. Polish Maritime Research.

Jeremy, C., 2016. Average cost of a cruise ship. [Online] Available at: http://themusterstation.com/cruise-ship-cost-to-build/ [Accessed 2018].

Kavussanos, M.G. & Visvikis, I.D., 2016. The International Handbook of Shipping. s.l.:Macmillan Publishers, Springer.

Kim, J.-S., Roh, M.-I. & Hama, S.-H., 2017. A method for intermediate flooding and sinking simulation of a damaged floater in time domain Author links open overlay panel. Journal of Computational Design and Engineering, 4(1), pp. 1–13.

Lamb, Thomas, 2003. Ship Design and Construction. s.l.: SNAME.

Lee, T.H. & Arcodia, C., 2017. Risk perceptions on cruise ships among young people: Concepts, approaches and directions. International Journal of Hospitality Management, Volume 69, pp. 102–112.

Leppert, J., 2016. Forecasting Brexit's Impact on Cruising. [Online] Available at: http://www.travelpulse.com/news/cruise/forecasting-brexits-impact-on-cruising.html [Accessed Dec. 2017].

Martin, H., 2014. Cruise Industry still facing sinking public perception, survey shows. [Online] Available at: http://articles.latimes.com [Accessed Dec. 2017].

Mileski, J.P., Wang, G. & L.L. Beacham, 2014. Understanding the causes of recent cruise ship mishaps and

disasters. Research in Transportation Business & Management, Volume 13, pp. 65–70.
Misra, S.C., n.d. Design Principles of Ships and Marine Structures. In: Ergonomics in Layout Design. s.l.:CRC Press: Taylor & Francis Group.
Neves, M.A.S. et al., 2011. Contemporary Ideas on Ship Stability and Capsizing in Waves. London: Springer.
Norwegian Cruiseline Holdings Ltd., 2016. Annual Reports. [Online] Available at: http://www.nclhltdinvestor.com/financial-information/annual-reports [Accessed Dec. 2017].
OECD (Organisation for Economic Co-operation and Development), n.d. Arrangement on Officially Supported Export Credits, TAD/PG(2014). [Online] [Accessed Nov. 2017].
Oskar Levander, 2007. New Concepts in Ferry Propulsion. Wartsila, Technical Journal.
Papanikolaou, A. e. a., 2013. GOALDS-Goal Based Damage Ship Stability and Safety Standards.. Accident Analysis & Prevention, Pergamon (www.sciencedirect.com/science/article/pii/S0001457513001413).
Papanikolaou, Apostolos, 2014. Ship Design: Methodologies of Preliminary Design. s.l.:Springer.
RCCL Royal Caribbean Cruises Ltd., 2016. Financial Reports [Online] Available at: http://www.rclcorporate.com/investors/financial-information/financial-reports/#year-in-review [Accessed Jan. 2018].
RINA (The Royal Institute of Naval Architects), n.d. Significant Ships, Issue 1990–2016.
RINA (The Royal Institute of Naval Architects), n.d. Significant Ships, Issue 1990–2016.
Rodrigue, J.-P. & Notteboom, T., 2012. The Geography of Cruise Shipping: Itineraries, Capacity Deployment and Ports of Call. Taipei (Taiwan), s.n.
Schneekluth H., and Volker Bertram, 1998. Ship Design for Efficiency and Economy. s.l.:Springer. Shipping Guides Ltd., 2017. Guide to Port Entry 2017–2018. s.l.:s.n.
STX France, 2017. STX France Services. [Online] Available at: http://stxfrance.fr/en/bu-services-2/ [Accessed Jan. 2018].
Sun, X., Feng, X. & Gauri, D.K., 2014. The cruise industry in China: Efforts, progress and challenges. International Journal of Hospitality Management, Volume 42, pp. 71–84.
Vassalos, Dracos et al., 2016. An alternative system for damage stability enhancement.
Watson, David G.M., 1998. Practical Ship Design. Amsterdam: Elsevier.

APPENDIX

Table A1. Ports for each itinerary.

Itinerary	Ports
W. Mediterranean from Miami	Barcelona → Palma de Mallorca → Marseille → La Spezia → Civitavecchia → Naples → Barcelona
E. Caribbean from Miami	Port of Miami → Port of Philipsburg → Port of Charlotte Amalie → Port of Nassau → Port of Miami
W. Caribbean from Miami	Port of Miami → Port of Roatan → Port of Costa Maya → Cozumel → Port of Nassau → Port of Miami
One-way Barcelona to Miami (and return)	Barcelona → Malaga → Miami

Table A2. Qualities of targeted market segments.

ID. No.	Generation Y	Generation X
1	Break from d-d Environment	Renew personal connections
2	Isolation	Keep Family Ties
3	Enriched perspective of life	Stimulate Intellectuality
4	Relax	Not Get Physically Pampered
5	Physical Action/ No Pampering	

Table A3. Selected on-board activities.

Activities & Entertainment	Target Group	Qualities of Targ. Group
Skydiving	Y	1,5
Bungee Jumping	Y	1,5
Privet Health Classes	Y	2,3
Spa	Y	2,4
North Star	X&Y	X-1,Y-1
Amusement Park	X	2
Art Museum	Y	2,3
Waterworks	Y	2
Lazy River	Y	2,4
Adventure Park (Kayaking)	Y	1,5
IMAX Open Cinema	All segments	X-2,Y-4
Wine/Martini Tasting	Y	1,3
Ship Tour	Y	1,3
4D Theatre	Y	1
DJ Classes	Y	1
Live Music Concert	Y	1,5
TED Talks	X&Y	X-2,Y-3
Art Classes/ Seminars	X	1,3
Escape Room	Y	1,3,5
Bowling	X	2
Sky Dinning	All segments	X-1,Y-1
Driving Range	X	1
Theatre Shows	X	3

	Mod.1	BoS*	Mod.2	Mod.3	Mod.4	Mod.5	Mod.6
L/B Ratio	6.1	6.6	6.7	7.4	8.8	9.6	10.3
Length [m]	400	415.5	420	440	480	500	520
Beam [m]	65.4	63.0	62.3	59.5	54.5	52.34	50.3
Disp. [t]	176,721	177,967	179,079	180,403	181,624	182,997	184,062
^{1}Draft [m]	9.62	9.70	9.78	9.81	9.89	9.92	10.09
Est. Cost [Bn$±0.189]	2.003	2.037	2.051	2.107	2.238	2.316	2.383

Figure A-1. L/B model designs.

Optimization attempt of the cargo and passenger spaces onboard a ferry

P. Szymański
Gdańsk University of Technology, Gdańsk, Poland

G. Mazerski & T. Hinz
Deltamarin sp. z o.o., Gdańsk, Poland
Aalto University, Espoo, Finland

ABSTRACT: Optimization mechanism was developed that aimed at designing a ferry that would combine low total cost of ownership with high level of passenger safety. The layouts of truck lanes on car deck and various passenger spaces on upper decks were automatically modified to reduce the building and fuel costs, ensure efficient evacuation while controlling stability performance. Genetic algorithm was utilised and various lessons from previous publications were used to obtain a group of designs for final investigation by the designers. The method will be also able to integrate some comfort features into the optimization algorithm.

1 INTRODUCTION

1.1 Background

Ferry market in East Asia comprises mostly of old and very old second-hand ferries that were not designed for the current market and routes. There have been many ferry accidents in this area of the world, but the problem preventing newbuilding orders is the lack of financing. This led us to attempt to design a ferry that will be very safe and affordable at the same time. It is assumed that such design would encourage the ship-owners (possibly with support from local maritime administrations) to invest in new vessels.

1.2 Aim

In order to combine high safety level with reasonably low investment and operational cost an optimization procedure is proposed. Such process will include a number of simulations and calculations to improve the design and select the best parameters for the intended vessel. To allow such procedure to be executed smoothly, possibly more than once to check various sensitivities, a tool was created that is able to solve this problem and can be used repeatedly for almost any input. To ensure high efficiency of the optimization, the genetic algorithm was selected.

2 OPTIMISATION PROCESS

2.1 Basic ideas

The multi-objective optimization problem of M objectives and J constraints can be formalized with:

$$\min_{x \in X} \{ f_1(x),\ldots,f_M(x) \,|\, g_j(x) \geq 0, j \in [1,J] \} \quad (1)$$

where we search for design alternatives x in the total design space X confined within variable bounds. Goal is to find such x that minimizes the objectives f(x) while satisfying all the imposed constraints g(x). If constraints are satisfied, design is called feasible and belongs to a feasible set Ω:

$$\Omega = \{ x \in X \,|\, g_j(x) \geq 0, j \in [1,J] \} \quad (2)$$

The solution of Equation 1 is a Pareto optimal alternative x^* which is non-dominated by other feasible alternatives, i.e. there is no alternative better than x^* in the objective space Y (whose feasible part is denoted with Y^Ω). Such alternative represents then a rational choice and it belongs to a set of Pareto optima $\widehat{\Omega}$ called also the Pareto frontier, defined as:

$$\widehat{\Omega} = \{ x \in \Omega \,|\, \pm x^k, f(x^k) < f(x), \forall x^k \in X \setminus x \} \quad (3)$$

2.2 Genetic algorithm

A genetic algorithm for multi-objective optimization is a model of machine learning which derives its behavior from a metaphor of the processes of evolution in nature. This is done by the creation within a machine of a population of individuals represented by chromosomes, in essence a set of character strings that are analogous to the base-4 chromosomes that we see in our own DNA. Elements or chunks of elements are swapped between individuals as if by sexual combination and reproduction (crossover), others are changed at random (mutation). New generations appear from clones of the current population, in proportion to their fitness: a single objective function of the parameters that returns a numerical value, to distinguish between good and bad solutions. Fitness is then used to apply selection pressure to the population in a 'Darwinist' fashion (survival of the fittest).

GAs differ from more normal optimization and search procedures in four ways:

- GAs work with a coding of the parameter set, not the parameters themselves.
- GAs search from a population of points, not a single point.
- Gas use payoff (objective function) information, not derivatives or other auxiliary knowledge.
- GAs use probabilistic transition rules, not deterministic rules.

Genetic algorithms require the natural parameter set of the optimization problem to be coded as a finite-length string (analogous to chromosomes in biological systems) containing characters, features or detectors (analogous to genes), taken from some finite-length alphabet. Usually, the binary alphabet that consists of only 0 and 1 is taken. Each feature takes on different values (alleles) and may be located at different positions (loci). The total package of strings is called a structure or population (or genotype in biological systems).

A typical structure of the genetic algorithm includes the following steps:

- (start of first generation, executed once)
- Translation of variables into binary string
- Initializing (usually random) population of individuals (usually 10–20 strings)
- Evaluation of fitness of all individual strings
- (set of repeatable actions starts here)
- Selection of sub-population for offspring reproduction (parents)
- Recombination of genes of selected parents
- Random mutation in mated population
- Evaluation of fitness of all individual strings
- (repeat until pre-defined number of generations or fitness is achieved)

This process has proved to lead to optimum solutions in large varieties of applications, although a number of limitations as well as more advanced modifications of GA exists.

This chapter provides a short overview to facilitate general understanding of the process. Therefore the reader is advised to study the basis of Genetic Algorithms (GA) using the publications referenced at the end of this paper (Dawn 1995, Goldberg 1989, Heitkoetter and Beasley 1994).

2.3 Tools

The optimization tool is based on four programs. These are Microsoft Excel with SolveXL add-on and Rhinoceros with Grasshopper. The remaining data exchange takes place through scripts in VBA and C#.

The role of Microsoft Excel can be defined as the input data interface and final data collection. In the MS Excel spreadsheet, the user enters all constant and variable parameters of the optimization. The spreadsheet calculates all the information needed to create a 3D model and random room arrangements are generated. In addition to generating data for further calculations, final information from 3D solids and optimization results are collected in MS Excel.

The SolveXL – an add – on to MS Excel – is performing the genetic optimization procedure (as described in chapter 2.2). It provides seamless connection with Excel thus allowing long and complex optimizations. Multiple options for setting optimization parameters are available in this program.

Rhinoceros is a display tool. It is a typical CAD program for creating 2D and 3D objects. The tool does not have a computational role, because the created objects are only displayed on the desktop, and are stored in the temporary memory of the computer.

Grasshopper has a very important role as it imports data from MS Excel, creates hullforms and rooms and finally gives geometrical output for optimization process (distances, areas etc.). It is also possible to utilize other programming languages to perform commands or multi-step mathematical operations.

3 APPLICATION EXAMPLE

3.1 Model description

The process described in the previous chapter was applied to a task of searching the optimum combination of ship main dimensions and arrangement of passenger spaces in a design of a modern ro-pax ferry vessel.

Ship design assumptions used for optimization purposes are as below:

- Number of vehicles – 53 trailers
- Passengers – 600 people
- Crew – 70 people
- 4 types of cabins: VIP, Royal, First class, Economy
- 8 types of public spaces for passengers:
  o Public consumption
  o Galley
  o Mid Lounge
  o Shop
  o Toilet
  o Main hall
  o SPA
  o Pub Lounge

In the optimization process, it was assumed that some passengers are in cabins and some are in public spaces.

The optimization tool has the ability to arrange rooms only on two decks. This is a significant difference compared to the original arrangement shown above, which consists of three decks. However such simplification should not have any noticeable effect when checking the feasibility of the method proposed here. In addition, the arrangement of communication and evacuation points is similar in both cases.

3.2 Parameters

Several types of constant and variable parameters are used in the optimization process. They will be described below. Constants can be either fixed for single optimization or permanent for the entire optimization tools.

3.2.1 Objective function

Optimization has two objective functions. The first, which contains only the KMt parameter (distance from the keel to transverse metacenter point) needs to be maximized. It is calculated based on the main dimensions of the vessel and hull geometry. Since the number of truck lanes across the ship can vary so will the KMt.

The second function is a minimizing function. As there is more than one parameter that should be minimized, so this function is a combination of 4 result functions described below. The combination process is producing a simple a dimensionless, non-weighted sum of distances and costs described below. The parameters are following:

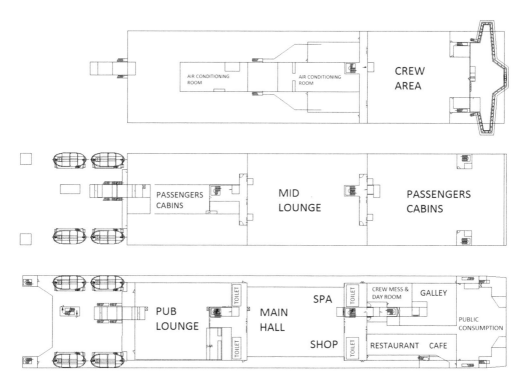

Figure 1. Layotut of decks in original ship design.

- safety of evacuation – average distance from center of each room to life-saving equipment (Mass evacuation system or lifeboat) is multiplied by number of people that are assigned to each room
- functionality of passenger communication onboard – average distance from center of each room to the nearest staircase is multiplied by number of people that are assigned to each room. The methodology behind for safety of evacuation and functionality of passenger communication onboard was developed in earlier publications (Mazerski et al 2014).
- Construction cost – based on ship main dimensions, estimated mass of the ship, area of passenger spaces, propulsion power. The construction cost list was created using a spreadsheet. It was made by parameter method. Data was collected from many ferry designs that were built in a cooperating shipyard.
- Operating cost – amount of fuel consumed annually has been multiplied by the current price of oil. Fuel consumption is based on hull resistance from Holtrop Mennen method (Holtrop 1984, Holtrop and Mennen 1982). It varies with the varying number of truck lanes across the ship. Total cost for 15 year of operations is estimated in this way

3.2.2 Variables

It is assumed that the parameters that are fixed for the specific optimization process, but which can be varied to reflect specific Client requirements can be treated as constants and are therefore described in the following chapter.

There is a set of 23 variables that control the whole optimization process. The first two determine the radius of bilge. This is because the dimensions of the hull are variable. The next two values control the length and position of parallel mid-body.

The next variables determine the order of rooms on the decks. The others decide about the division of passenger cabins into several smaller groups. The variable controlling the width of the car deck determines the main dimensions of the ship. It was assumed that there can be 6, 7 or 8 truck lines across the ferry.

There is also the variable that describes the draft of the vessel. It is coupled with the modification of the height and geometry of the hull. Both draft and geometrical parameters are affecting the hull resistance.

3.2.3 Constants

There is a number of constant used in this optimization process. These can be determined in a way that allows adjusting the process to specific Client requirements. However, in one optimization process they are to be kept constant.

The constants include information about passenger spaces (type, number and size) as well as the parameters of the trucks/cars. In the example presented in this paper there are 15 rooms on two decks and one car deck. The space allocated for each type of vehicle is following:

- for trailers, length – 17.5 m, width – 3.2 m.
- for passenger cars, length – 5 m, width – 2.3 m.

Information about the area of each room is given by the designer before the optimization start. The exceptions are cabins for passengers because it is possible to divide them into smaller segments so that they can be located separately on the decks. The rooms can be allocated over the whole width, or occupy the space from centerline alley to the edge of the deck.

For functional and safety criteria, a network of communication and evacuation points is needed. There are three groups of these:

- center casing – its location depends on the number of truck lines.
- other communication points – these are fore and aft staircases located the side of the vessel (see Figure 1).
- evacuation points – exits to lifeboats and two location of MES (Mass Evacuation System) points.

The coordinates of these points are used to determine the values of the first two factors of the minimizing objective function (see chapter 3.2.1).

3.3 Results

One optimization cycle was carried out. The following parameters for the genetic algorithm were set in the SolveXL software:

- Population size – 13
- Number of generations – 170
- Crossover type – Uniform Random with probability: 1
- Mutation type – Simple by Gene
- Probability of mutation (used variable mutation) – 0.14, after 50 generations, 0.1.

Figure 2 presents the feasible results—each point shows an individual solution with two objective functions plotted on vertical (maximizing function) and horizontal (minimizing function) axis.

The SolveXL program records every 20 generations. The above figure shows all saved generations. The best solutions were obtained by representatives of the last populations. Below, we present two selected solutions from the Pareto front (Result 1 and Result 2). They represent interesting results and fulfill the objective function very well.

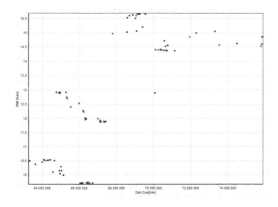

Figure 2. Feasible result plotted against objective functions.

Figure 3. Arrangement from Result 1.

3.3.1 Result 1
Result 1 produced following sequence of rooms:

1. Galley
2. Toilet
3. Pub Lounge
4. Cabins VIP/Royal/First 2
5. SPA
6. Cabins VIP/Royal/First 1
7. Shop
8. Cabins Economy 1
9. Mid Lounge
10. Crew Space
11. Cabins Economy 3
12. Public Consumption
13. Cabins Economy 2
14. Main Hall

Figure 3 illustrates, in a very simplified way, the arrangement from Result 1.

It is a solution with seven load lines across the width of the ship and eight vehicles positioned on these lines along the ship's centerline. In this way, the main dimensions are as follows:

- L_{BP} – 160.25 m
- B – 23.60 m
- H – 12.52 m
- T – 6.00 m

The parameters controlling the midship body and the radius of bilge are as follows:

- Radius of bilge z (% of T) – 54.89
- Radius of bilge y (% of B/2) – 45.05
- Midship body (frame +1, +%) – 99.58
- Midship body (frame –1, –%) – (–3.57)

Below, we present the values of the objective functions that were calculated for the above input data:

Minimizing function – 64 352 337

- Distance to stairs – 6 601 [meters * persons]
- Distance to evac – 18 736 [meters * persons]
- Construction cost – 36 319 000 $
- Cost for ownership [15 years] – 28 008 000 $

Maximizing function – 12.88

3.3.2 Result 2
Result 2 produced following sequence of rooms:

1. Galley
2. SPA
3. Pub Lounge
4. Crew Space
5. Public Consumption
6. Shop
7. Cabins VIP/Royal/First
8. Cabins Economy
9. Main Hall
10. Toilet
11. Mid Lounge

Figure 4 illustrates, in a very simplified way, the arrangement from the table.

It is a solution with eight load lines across the width of the ship and seven vehicles positioned on these lines along the ship's centerline. In this way, the main dimensions are as follows:

- L_{BP} – 139.47 m
- B – 26.80 m

Figure 4. Arrangement from Result 2.

- H – 12.52 m
- T – 6.00 m

The parameters controlling the midship body and the radius of bilge are as follows:

- Radius of bilge z (% of T) – 53.51
- Radius of bilge y (% of B/2) – 84.62
- Midship body (frame +1, +%) – 99.58
- Midship body (frame –1, –%) – (–3.57)

Below, we present the values of the objective functions that were calculated for the above input data:
Minimizing function – 67 963 405

- Distance to stairs – 10 593 [meters * persons]
- Distance to evac – 15 812 [meters * persons]
- Construction cost – 37 319 000 $
- Cost for ownership [15 years] – 30 618 000 $

Maximizing function – 14.975

4 CONCLUSIONS

The work presented in this paper is building on the earlier developments and shows that optimization of passenger spaces can be coupled with the variations in ship's main dimensions. Various functional requirements can be used in addition or instead of the ones used in the example presented above. The relative importance of these parameters can be easily adjusted by using proper weight coefficients in the minimizing function. Provided that proper adjustments are made to this method it can even be used to address the challenge of selecting the optimum fleet of ships for a given route (by varying the number of vessels and their sizes).

Such possibilities show that we were able to provide future ferry designers with a simple but robust tool that will help them quickly generate and evaluate a wide variety of ship arrangements at the concept development stage.

5 FURTHER DEVELOPMENT

In previous chapters, we presented the application of the method for optimization of the stability, safety of passengers, functionality of passengers' spaces and ship's costs. Some fine-tuning of the method is planned in order to allow to use weight factors in the optimization functions. In a longer perspective we plan to include other ship characteristics in the optimization process, like comfort of passengers and the crew and safety of cargo. These aspects depend on ship motions as well as vibration and noise onboard. We plan to include these three features during further work on the ship design optimization methods.

REFERENCES

Dawn, T., 1995, 'Nature Shows the Way to Discover Better Answers'. Scientific Computing World 6 (1995), 23–27.
Dudziak J.: 2008, Teoria Okrętu. Fundacja Promocji Przemysłu Okrętowego i Gospodarki Morskiej, Wyd. 2 popr. i uzup.
Goldberg, D.E., 1989, Genetic Algorithms in Search, Optimization & Machine Learning. Reading: Addison-Wesley.
Heitkoetter, J. Beasley D., 1994, The Hitch-Hiker's Guide to Evolutionary Computation: A List of Frequently Asked Questions (FAQ). USENET: comp.ai.genetic. Available via anonymous FTP from ftp://rtfm.mit.edu/pub/usenet/news.answers/ai-faq/genetic/.
Holtrop, J. 1984, 'A statistical re-analysis of resistance and propulsion data', International Shipbuilding Progress 31 (1984), 272–276.
Holtrop, J., Mennen, G. 1982. 'An Approximate Power Prediction Method', International Shipbuilding Progress 29 (1982), 160–177.
Mazerski, G., Zagórski, P., Hinz, T. 2014, 'Practical Limitation to Design Optimisation of a Small Ferry', COMPIT 2014.
Salminen J, Furustam J.,2009 Introduction to Optimization. Napa Ltd.

Application of a goal based approach for the optimization of contemporary ship designs

O. Lorkowski, K. Wöckner-Kluwe, J. Langheinrich, R. Nagel & H. Billerbeck
Flensburger Schiffbaugesellschaft MBH & CO., KG, Flensburg, Germany

S. Krüger
Institute of Ship Design and Ship Safety, Hamburg University of Technology, Hamburg, Germany

ABSTRACT: Within the last decades, goal based approaches have gained more and more popularity within the maritime industry. In contrast to prescriptive design approaches, goal based design approaches provide sufficient flexibility to evaluate the attained level of safety for novel ship designs, improve a ship's level of safety beyond current rules and regulations, optimise a ship's operability within the design constraints. While the application of goal based standards on safety relevant topics has been investigated in various research projects and is widely supported by authorities and classifications societies, its' application on a ship's operability has not been addressed in detail so far. This optimisation is an important task since the operational expenses (OPEX) and thus life cycle costs are directly related to a ship's operability. Within the German research project MOPS, funded by the Federal Ministry of Economics and Technology, a design tool has been developed to investigate safety and operability related tasks in the context of the holistic design process. In this respect, ship motions are calculated by means of strip theory and incorporated into a multidimensional probability analysis. The results are presented on the example of a recent case study, taken from FSG's data base.

1 INTRODUCTION

Contemporary ship designs are subject to complex design goals, which are set up by international organizations, governments and/or the owner to ensure a certain level of safety, minimize their impact on the environment and maximize their economy for a given business case. These design goals require innovative design solutions, which cannot be assessed within a conventional prescriptive framework, since the impact of specific design parameters is not covered with the required level of detail.

For this reason, the application of goal based methods have gained more and more popularity. They are already state of the art in the area of ship construction or ship safety (IMO 2017) (Hoppe 2015) and their application has been further developed by ship builders e.g. to improve the attained level of safety beyond the prescriptive requirements (Kluwe 2010) (Stoye et al. 2008) (Tellkamp 2015).

The application in fields of the ship's economy and ship operation has barely been explored so far. In this respect, goal based methods may be applied to minimize the ship's operation costs e.g. fuel oil consumption, maximize its' reliability in operation or to demonstrate compliance with emission goals set up by local governments within the tendering process of new ship routes. In this context, goal based approaches allow for a more specific definition of the design goal such as

- The ship's yearly CO_2 emission for a given route an operation profile instead of a required EEDI,
- a maximum delay probability for the ship schedule and the intended area of operation rather than an overall Sea Margin,
- an operability of 90% in the desired area of operation rather than overall absolute limits for the ship's motion in seaway or
- a minimum power consumption on the route for a given schedule rather than a predefined design speed.

The direct definition of such design goals allows the ship designer to apply appropriate state of the art calculation methods (Hoppe 2015) for the assessment of the design rather than following prescribed formulae which might be based e.g. on empirical regression curves. In this context, the design can be adopted in more detail to the design goal and design alternatives can be compared with respect to the long-term impact on the required performance. Consequently, the ship owner obtains a more suitable design for his business case if goal based design methods are applied.

At the same time, goal based approaches require more input data and sometimes a higher computational effort which need to be balanced in the early design stages of the ship.

In the following sections, a design tool is introduced based on RoPax ferry case study. This tool has been developed within the German research Project "MOPS", supported by the "Federal Ministry of Economics and Technology" in Germany.

2 GOAL

For the RoPax ferry example, a speed optimization shall be investigated with the aim to minimize the fuel consumption and emissions for an example route and voyage profile. The basic question behind the speed optimization reads then as follows:

Can the ship speed be adjusted on the respective route segments to minimize the overall power consumption?

In this respect, a sensitivity analysis has been conducted to investigate the influence of the corresponding input data on the results. In particular, the weather conditions may vary throughout the year so that the ship may experience different power demands per season. For this reason, the case study comprises two representative weather scenarios: The yearly 40% quantile of wind and waves which is more likely to be observed in the summer season and the yearly 90% quantile which is more representative for the winter season.

In general, this task can be allocated to goal based design methods where the design goal, in this case the speed optimization, is linked to a certain probability of occurrence such as the limiting probability values for the weather conditions. The following section provides a brief overview about the general scheme of the applied goal based design approach.

3 APPROACH

Figure 1 shows the principal structure of the design tool. At the beginning, the goal needs to be expressed in terms of random variables, which define the long-term operation of the vessel. For the given case study, the random variables are

- speed profile (to be optimized)
- draught/load case
- ship's heading
- wind speed/wind direction
- wave height, wave period, wave direction
- ship's resistance/delivered power

The probability density function of the delivered power P_D can be expressed as a function of

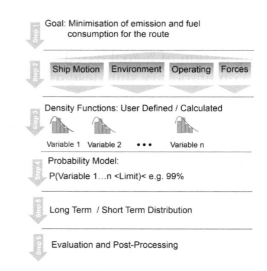

Figure 1. Structure of the goal based approach.

the other random variables via the following probability model:

$$f(P_D^{k,s,h,i}) = p^k(\Delta) \cdot p^s(v) \cdot p^h(\mu) \cdot p^i(v_w, H_s, T_p, \mu_i) \quad (1)$$

where $p^k(\Delta)$ = probability density of the load case, $p^s(v)$ = probability density of the ship speed, $p^s(\mu)$ = probability density of the heading, $p^i(v_w, H_s, T_p, \mu_i)$ = joint probability density function of wind speed, significant wave height peak period and encounter angle. Please note the exponent notation, which is used here to indicate the index of each component.

The integral of equation (1) gives the probability function of the delivered power, which describes the probability that P_D will be smaller or equal to a limiting value P^*_D. This integral reads for numerical/discrete values as:

$$F(P_D \leq P_D^*) = \sum_k \sum_s \sum_h \sum_i p^k(\Delta) \cdot p^s(v) \cdot p^h(\mu) \cdot \ldots$$
$$p^i(v_w, H_s, T_p, \mu_i) \cdot d\Delta \cdot dv \cdot d\mu \cdot dv_w \cdot dH_s \cdot dT_p \cdot d\mu_i \quad (2)$$

In the case that the ship is operated at only one speed and one heading, equation (2) simplifies as

$$F(P_D \leq P_D^*) = \sum_s \sum_i \cdot p^s(v) \cdot \ldots$$
$$p^i(v_w, H_s, T_p, \mu_i) \cdot dv \cdot dv_w \cdot dH_s \cdot dT_p \cdot d\mu_i \quad (3)$$

For the given case study, the probability level of the environmental conditions is explicitly given so that the component

$$\sum_i p^i(v_w, H_s, T_p, \mu_i) \, dv_w \cdot dH_s \cdot dT_p \cdot d\mu_i$$

amounts to 90% and 40% respectively. This defines in turn the absolute values for v_w, H_s, T_p, μ_i, which will be derived in the following chapters.

Each value of $f(P_D^{k,h,s,i})$ and $F(P_D^{k,h,s,i})$ is associated with a certain value of P_D, which is a function of the displacement, speed, encounter angle and the corresponding resistance:

$$P_D^{k,s,h,i} = \frac{R_T^{k,s,h,i} \cdot v^s}{\eta_D^{k,s,h,i}} \tag{4}$$

Herein, $R_T^{k,s,h,i}$ is the ship's resistance including added resistance components due to the environmental conditions and operation, $\eta_D^{k,s,h,i}$ is the propulsive efficiency for the respective combination of ship speed and resistance.

4 OPERATING PROFILE

The ship is assumed to be operated on a weekly schedule with five different transit times (compare Figure 2). From the distance and transit times, average speed values can be derived which will serve as the benchmark for the optimization.

5 ENVIRONMENTAL CONDITIONS

Figure 3 shows the route and respective environmental conditions. Wind and wave data have been derived from a hindcast model, which covers 40 years of hourly sea states and wind speeds with a spatial grid resolution of 0.3° (Pineau & Leballeur 2017). In this respect, a "WaveWatch3" model

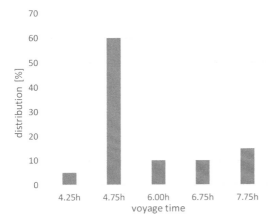

Figure 2. Voyage profile of the ship.

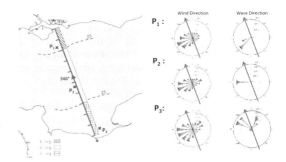

Figure 3. Route and environmental conditions.

Table 1. Summary of environmental conditions.

	Leg 1 40%/90%	Leg 2 40%/90%	Leg 3 40%/90%
v_w [kn]	11/22	14/26	11/22
H_s [m]	0.9/2.4	0.9/2.4	0.5/1.5
T_p [s]	6/9	6/9	5/7
water dpt. [m]	25	>45	25

had been applied to reconstruct the sea states from recorded wind data (Boudiere, et al. 2013).

The route is subdivided into three legs, where the environmental conditions are assumed to be constant. Leg 1 and Leg 3 are characterized by shallow water where the average water depth amounts to 25 m. Leg 2 is a deep-water segment, which comprises together with Leg 1 the largest sea states.

The mean wind and wave direction is approximately 90°, relative to the ship's heading.

The values of the joint probability of wind speed, wave height and wave period may be described by so called "environmental contours" (Winterstein et al. 1993) which comprise various combinations of wind speed, wave height and wave period for a given level of probability. Table 1 summarises the largest wave heights for the 90% and 40% quantiles of the expected sea states and wind speeds:

Coming back to the basic question about the speed optimization, this may now be further detailed as follows:

Can the overall power consumption be minimized by speeding up e.g. on the deep-water segment and slowing down on the shallow water segments?

This question is related to the overall magnitudes of the respective resistance components, which are discussed and described in the next section.

6 RESISTANCE COMPONENTS

The estimation of added resistance components has been addressed in recent research projects and conference papers (Valanto & Hong 2015) (Valanto 2014) (Söding & Shigunov 2013) and is widely discussed for example in the context of a more accurate definition of the sea margin e.g. in (Shigunov 2014). The underlying challenge is to identify suitable approaches for the estimation of the major resistance components.

Figure 4 illustrates graphically the order of magnitude for the major resistance components on the example of the RoPax ferry:

For the given example, the total resistance sums up as:

$$R_{total} = R_{calm} + R_{shallow} + R_a + R_F + R_{Rudder} \quad (5)$$

With R_{calm} as the calm water resistance, $R_{shallow}$ as the shallow water resistance, R_a as the wave resistance, R_F as the fin stabilizer resistance, R_{Rudder} as the rudder resistance. The calm water resistance is of the order of magnitude $1*10^3$ kN, added resistance due to wind, wave shallow water and stabilizer fins are in the order $0.5-1*10^2$ kN, rudder resistance due to course keeping is of $1*10^1$ kN and thus the smallest component for the given case study.

These resistance components need to be superimposed for each route segment to account for the "time-dependent" occurrence during operation.

The following subsections provide a brief overview of the approaches, which have been applied to for the estimation of the respective resistance component.

6.1 Wind

The added resistance due to wind has been derived from wind tunnel tests, which comprise wind load coefficients for an average ocean wind profile:

$$C_x = \frac{R_w}{0.5\rho_w \cdot v_{w10}^2 \cdot A_f} \quad (6)$$

where R_w is the added resistance due to wind, ρ is the density of air, A_f is the front lateral area of the ship and v_{w10} is the wind speed at 10 m reference height.

6.2 Wave

The mean added resistance due to waves has been estimated from 2nd order wave drift forces by applying a strip theory according to (Augener 2016) and including a correction for short waves according to (Faltinsen 1990):

$$\overline{R_a} = \sum_{c=1}^{N} \sum_{b=1}^{M} \overline{\left(\frac{R(\omega_c, \beta_b)}{\zeta_a^2}\right)} \cdot 2 \cdot S_\zeta(\omega_c, \beta_b) d\omega_c d\beta_b \quad (7)$$

Herein, R is the mean wave drift force in the frequency domain for each wave frequency component ω_c and wave direction β_b, normalized by the wave amplitude ζ_a^2 and S_ζ as the wave energy spectrum. In this context, a JONSWAP spectrum with a $\cos^2$ spreading function has been assumed for the area of operation.

The wave drift forces have been validated within the research project DyPos by a model test campaign and show a satisfactory prognosis accuracy for ship-like structures (Augener 2016).

6.3 Shallow water

The added resistance in shallow water has been estimated according to (Schlichting 1934), which corresponds well with model test results for a moderate restriction in water depths. This approach covers

- the additional resistance due to waves produced by the vessel at given speed,
- the additional resistance due to increase of potential flow around the hull in shallow water (blockage effect of the hull)

by applying a relationship between the deep water ship speed and an equivalent shallow water ship speed over ground which is based on the similarity of the generated wave pattern.

6.4 Stabilizer fins

The resistance due to the stabilizer fins has been estimated from drag coefficients, provided by the manufacturer. These drag coefficients are based on model test results and numerical CFD RANS calculations:

$$C_D(\alpha) = \frac{R_F(\alpha)}{0.5\rho_w \cdot v^2 \cdot A} \quad (8)$$

With α as the hydrodynamic angle of attack of the stabilizer fin, v as the ship speed and A as the projected fin area. The hydrodynamic angle of attack has been estimated from time dependent ship motion calculation with RDE-Rolls (Söding et al. 2013) (Kröger 1987) (Petey 1988) for the respective ship speeds and sea states, taking also the control algorithm of the stabilizers fins into account. This sea-keeping tool has been validated

Figure 4. Resistance components with respective order of magnitude(light blue = total resistance, dark blue = resistance components).

and further developed in various research projects and has been proven to predict the roll-motion of the vessel with a high level of accuracy (Billerbeck et al. 2006).

6.5 *Rudder*

The rudder resistance due to course keeping has been estimated from FSG's model test data base, taking also the slip stream of the propeller for different propeller revolutions into account. The corresponding resistance component is calculated similar to equation (7). The rudder angles have been estimated from full-scale measurements (Stoye 2012) for the respective weather conditions. Nevertheless, the overall magnitude of this resistance component is quite small for the investigated ferry case study.

7 ACCURACY OF THE DELIVERED POWER PREDICTION

The accuracy of the delivered power prediction is determined by the accuracy of the predicted resistance and the propulsive efficiency (compare equation 4). These two components can hardly be verified individually in full scale, since only overall power values are obtained e.g. by sea trial measurements. Nevertheless, an attempt is given to quantify a principal order of magnitude, based on model test experience.

For the given case study, the biggest resistance component is the calm water resistance. This component can be prognosed by use model tests or RANS CFD calculations with a high level of accuracy (typically in the lower single digit percent range). The same holds for the stabilizer fin, wind and rudder resistance so that the biggest share of the total resistance is predicted with a high level of accuracy.

The prediction accuracy of the shallow water resistance and wave resistance is a bit smaller than for the calm water resistance but they show a very reliable prediction of principal trends. Since their share on the total resistance is comparatively small for the given example, this prediction accuracy has only a minor impact on the overall result.

The propulsive efficiency of ships, exposed to added resistance such as waves, is many influenced by the open water efficiency of the propellers (Valanto & Hong 2015) (Valanto 2014). This efficiency component can also be predicted by RANS CFD calculations (Stoye 2012) or model tests with a high level of accuracy.

Summing up the findings listed above, a plausible order of magnitude for the prognosis accuracy is assumed to be in the range of ± 4%.

8 OPTIMIZATION ALGORITHM

The optimization has been carried out by use of a tangent search algorithm, which is based on (Hilleary 1966). This algorithm allows to minimize a non-linear objective function (in this case the overall power consumption) in the context of constraint functions (e.g. the lower and upper bound for the ship speed, required transit time due to the voyage profile).

9 RESULTS

Figures 5 and 6 summarize the optimized speeds over the free sailing distance and for the voyage profile example.

From Figures 5 and 6, the followings trends can be observed: For smaller transit times such

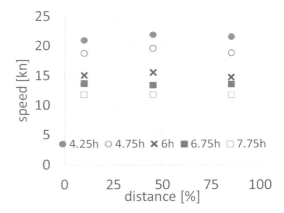

Figure 5. Speed Optimization over the free sailing distance for the 90% quantile weather conditions.

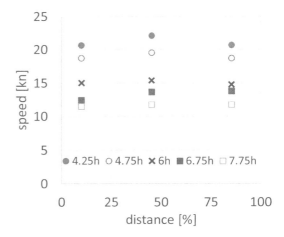

Figure 6. Speed optimization over the free sailing distance for the 40% quantile weather conditions.

as 4.25h and 4.75h larger speeds on leg 2 yield to a reduced power consumption since the shallow water effect causes the mayor difference between the resistance curves of the legs. Consequently, this trend can be observed for both weather conditions (90% and 40% quantile). The optimum speed per transit time varies in the range of 1 knot, which could be considered for an operator guidance.

For the larger transit times, the shallow water effect is less pronounced so that the resistance curves on the legs differ only by the contribution of wind and waves. Since the weather conditions are in principal the same range for the full free sailing distance, no clear tendency can be overserved with respect to an optimum speed profile. Nevertheless, a slight sensitivity with respect to the weather conditions can be observed. For this reason, the optimum speeds for the 90% quantile differ from the optimum speed of the 40% quantile. The optimum speed varies in a range of 1–3 tenth of a knot or even less e.g. for the transit time 7.75 h.

Figures 7 and 8 show a comparison of the delivered power (P_D) consumption from the constant speed profile (= free sailing distance divided by transit time) and the optimized speed profile.

The results are in line with the tendencies observed from Figures 5 and 6: The shorter the transit times, the larger is the potential power saving from the speed optimization. The largest power savings are in the range of 2% compared to the constant speed profile.

These power savings can result in less fuel consumption and may therefore have an impact on the overall life cycle costs/emissions of the ferry. This depends on the frequency of the respective transit time, which is accounted for in the followings life cycle cost analysis.

10 IMPACT ON LIFE CYCLE COSTS

The impact of the speed optimization on the life cycle costs and emissions has been evaluated by use of a Life-Cycle-Performance-Analysis-Tool (LCPA-Tool). The LCPA-Tool has been developed and applied in the EU-funded research projects JOULES and BEST (JOULES 2017) (BESST 2018).

The principal structure of the LCPA-tool is indicated in Figure 9.

The LCPA tool allows for a holistic environmental and economic assessment of ship designs. Most relevant environmental KPIs like Cumulative Energy Demand, Global Warming Potential, Acidification Potential, Eutrophication Potential and Aerosol Formation Potential are considered. From an economic perspective, relevant input parameter like fuel costs, investment costs, discount rates and even external cost can be considered and modified as appropriate.

Figure 10 shows the predicted fuel costs per week for the investigated case study, derived from the LCPA. The results are normalized by the "constant speed" values of the 90% quantile weather condition and are based on a project guide for a state of the art dual fuel engine. In this context, a gas operation has been assumed for the vessel.

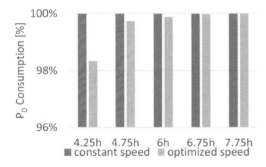

Figure 7. Power consumption comparison for the required transit times and the 90% quantile weather conditions.

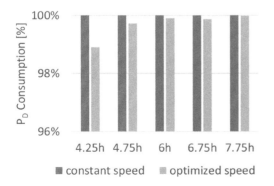

Figure 8. Power consumption comparison for the required transit times and the 40% quantile weather condition.

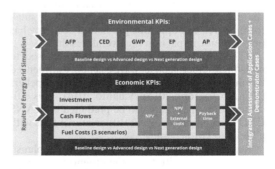

Figure 9. Principal structure of the LCPA-tool.

From Figure 10 follows, that the impact of the speed optimization on the overall fuel costs is quite small. This is mainly due to the fact, that the shorter transit times, which yield to larger power savings, are less frequent in the weekly schedule. In addition, the current fuel price of gas amounts to 40€ per MWh which is quite small. This fact decreases additionally the impact of the speed optimization on the overall weekly fuel costs. Consequently, larger fuel cost savings could be obtained if the impact of the shallow water was more pronounced (e.g. when the ship is more frequently operated at larger speeds) or the price for gas increases. The latter might be possible in the near future since the demand for gas fuel may increase when more gas driven ships are in operation.

Figure 11 shows the predicted emissions per week:

The changes in the emitted emissions are proportional to the results of the overall energy consumption since an almost constant engine load can be achieved for the propulsion plant of the ship.

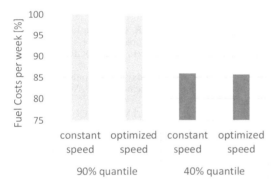

Figure 10. Comparison of the predicted fuel costs, normalized by the constant speed results (90% quantile).

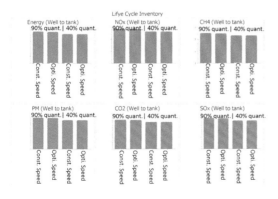

Figure 11. Emissions per week.

11 CONLUSION AND OUTLOOK

In the present paper, a goal based design method has been introduced to analyze the impact of design parameters such as ship speed, environmental conditions and voyage profile on the overall life cycle costs and emissions for a Ro-Pax ferry example.

The approach has well demonstrated the strong holistic character of such design methods, which requires an efficient handling of large data. In this respect, the development of efficient and fast interfaces between the corresponding modules have been found to be of key importance to reduce calculation time and to analyze the impact of certain design parameters on the respective goal. In addition, the corresponding modules of the goal based design method need to comprise an appropriate balance between calculation effort, accuracy and required input data to be suitable for parameter studies in the early design stage of the vessel. In the view of the authors, this might be one of the mayor challenges within the development process of these methods.

For the given Ro-Pax ferry example, a hybrid approach of time-domain and frequency domain sea-keeping calculation methods has been applied to cover the mayor resistance components with a sufficient level of accuracy.

Summing up the above listed findings, it can be concluded that the goal based design method is more complex and demanding in terms of the overall calculation effort.

On the other hand, this design method offers new degrees of freedom for the ship designer and provides a more accurate prediction of the performance of the vessel under realistic operating conditions. For the given Ro-Pax case study, design questions like the size of the propulsion plant could be investigated in the context of a speed optimization or for an overall weather margin. Furthermore, the results could be used to predict the yearly fuel or energy consumption. This can be in particular of importance for gas driven ships since this fuel is not available on short notice. In addition, ships may be designed with respect to a yearly emission target, which might be set up by domestic governments. By applying goal based design approaches, this emission target could be reached more effective than by a corresponding EEDI regulation, since the intended operational profile is covered in more detail.

For these reasons, the authors conclude that goal based design methods offer a significant potential to adopt a ship design with respect to its' intended operation and to quantify the impact to certain design measures on the overall life cycle costs of the ship.

REFERENCES

Augener. P. 2016. Computation of Wave Drift Forces for Dynamic Positioning within the Early Design Stage. Dissertation. Hamburg: Schriftenreihe Schiffbau der Technischen Universität Hamburg-Harburg.

BESST. 2018. EU funded research project: *Breakthrough in European Ship and Shipbuilding Technologies* (BESST). URL: https://trimis.ec.europa.eu/project/breakthrough-european-ship-and-shipbuilding-technologies. Accessed on: 28th February 2018.

Billerbeck, H., Gosch, T., Sommer, H., Tellkamp, J., Werner, K. 2006. Beurteilung der Schiffsicherheit im schweren Seegang, Final Report of the research project SinSee.

Boudiere, E., Maisondieu, C., Ardhuin, F., Accensim, M., Pineau-Guillou, L., Lepesqueur, J. 2013. A suitable metocean hindcast database for the design of Marine energy converters. International Journal of Marine Energy, Volumes 3–4, pp 40–52.

DNV. 2012. Structural Design of Offshore Ships, Offshore Standard DNV-OS-C102.

Faltinsen, O.M. 1990. Sea Loads on Ships and Offshore Structures. Cambridge: Cambridge University Press.

Hilleary, R.R. 1966. The Tangent Search Method of Constrained Minimization; U.S. Naval Postgraduate School. Report/Research Paper No. 59. Monterey, California, USA.

Hoppe, H. 2005. Goal-Based Standards – A new approach to international regulation of ship construction. *WMU Journal of Maritime Affairs* Volume 4 Issue 2: pp 169–180.

International Maritime Organisation (IMO). IMO Goal-based standards. URL: http://www.imo.org/en/OurWork/safety/safetytopics/pages/goal-basedstandards.aspx. Accessed on: 26th September 2017.

JOULES 2017. EU funded research project: *Joint Operation for Ultra Low Emission Shipping* (JOULES). URL: http://www.joules-project.eu/Joules/index.xhtml. Accessed on: 26th September 2017.

Kluwe, F. 2010. Development of a Minimum Stability Criterion to Prevent Large Amplitude Roll Motions in Followings Seas. Dissertation. Hamburg University of Technology, Germany: Hamburg.

Kröger, P. 1987. Simulation der Rollbewegung von Schiffen im Seegang. Dissertation. Germany: Hamburg.

Papanikolaou, A. 2009. Risk-Based Ship Design, Springer-Verlag Berlin Heidelberg.

Papanikolaou, A., Mains, C., Rusaas, S., Szalek, R., Tsakalakis, N., Vassalos, D., Zaraphonitis, G. 2010. GOALDS – Goal Based Damage Stability. *Proceedings of the 11th International Ship Stability Workshop.* pp 46–57.

Petey, F. 1988. Ermittlung der Kentersicherheit lecker Schiffe im Seegang. *Schiffstechnik.* Volume 35. pp 155–172.

Pineau, H. & Leballeur, L. 2017. Metocean statistics. Study Report. Actimar.

Schlichting, O. 1934. Schiffswiderstand auf beschränkter Wassertiefe. JSTG. Volume 35.

Shigunov, V. & Bertram, V. 2014. Prediction of added Power in Seaway by Numerical Simulation. Proceedings of the 9th Int. Conf. on High-Performance Marine Vehicles HIPER. Greece: Athens.

Stoye, S. 2012. Hydrodynamische Berechnungen für Propulsions- und Manövrierorgane im Off-Design, Final Report of the Research Project: OFF-Design.

Stoye, T., Bruns, A., Billerbeck, H., Braathen, A. 2008. Evaluation of Slamming Forces on a Bow of a RoPax-Vessel. Conference Paper, 103. Hauptversammlung der Schiffbautechnische Gesellschaft e.V. Germany: Hamburg.

Söding, H. & Shigunov, V. 2015. Added resistance of ships in waves. *Ship Technology Research > Schiffstechniik*, Volume 62 Issue 1: pp 2–13.

Söding, H., Shigunov, V., Zorn, T., Soukup, P. 2013. Method *rolls* for Simulating Roll Motions of Ships. *Ship Technology Research > Schiffstechnik,* Volume 60 Issue 2: pp 70–84.

Tellkamp, J. 2015 SAFEDOR SP6.4 The 13th Passenger Design Concept. Deliverable 6.4.1 of the European Funded Research Project: SAFEDOR, Germany: Flensburg.

Valanto. P. 2014. Widerstands- und Propulsionsmessungen an einem Kreuzfahrtschiffsmodell. *Schiff & Hafen* Issue 9: pp 50–56.

Valanto. P., Hong, Y. 2015. Experimental Investigation on Ship Wave Added Resistance in Regular Head, Oblique, Beam, and Following Waves. *Proceedings of the Twenty-fifth (2015) International Ocean and Polar Engineering Conference.* Pp 19–26.

Winterstein, S., Ude, T.C., Cornell, C.A., Bjerager, P., Haver, S. 1993. Environmental Parameters for Extreme Response: Inverse FORM with omission Sensitivity, Proceedings of the ICOSSAR-93, Innsbruck.

Zaraphonitis, G., Lee, B.S., Olufsen, O. 2012 "Deliverable 1.2, Final Report", of the European Funded Research Project: GOALDS.

Parametric design and holistic optimisation of post-panamax containerships

A. Priftis & O. Turan
University of Strathclyde, Glasgow, UK

E. Boulougouris
Maritime Safety Research Centre, University of Strathclyde, Glasgow, UK

ABSTRACT: The continuous endeavour of the shipping industry for economic growth has led the shipbuilding industry to explore new designs for ships. Additionally, the introduction of new regulations by the International Maritime Organisation frequently triggers changes in the ship design process. In this respect, proper use of computer-aided ship design systems extends the design space, while generating competitive solutions in short lead time. This paper focuses on multi-objective optimisation of containerships. The developed methodology is demonstrated by the conceptual design and optimisation of a post-panamax containership. The methodology includes a parametric model of the ship's external and internal geometry and the development and calculation of all required properties for compliance with the design constraints and verification of the key performance indicators. Moreover, compliance with the second generation intact stability criteria and structural rules regarding the vessels' midship section was checked. Genetic algorithms were used for the solution of this optimisation problem.

1 INTRODUCTION

1.1 *Container shipping industry*

A number of changes have been observed in the containership industry in the past few years. Although the global containerised trade has been constantly growing since 1996 (UNCTAD, 2016), shipping companies have modified their operation plans, so as to keep expanding their profits. One of the reasons led to these changes is the fluctuation of the fuel price. Since 2008, prices have dropped and nowadays heavy fuel oil (HFO) costs somewhere around 379 $/t. Marine diesel oil (MDO) has been following a similar course and can be found at prices of around 606.5 $/t (Ship & Bunker, 2017). However, the introduction of emission control areas (ECAs) has affected the fuel type ships use. Use of low sulphur fuel is now required in certain parts of the world. In addition, the recent landmark decision by the International Maritime Organisation (IMO) Marine Environment Protection Committee to implement a global sulphur cap of 0.5% m/m (mass/mass) from 1 January 2020 has introduced a step change to the framework of designing and operating ships (IMO, 2016).

The recent improvements in technology and engineering have made the introduction of ultra large container vessels possible. In order to make ocean liners more attractive by lowering the freight rates, ships with notably higher container capacity have been introduced in the containership industry. This trend is known as cascading. These orders consist primarily of very large containerships. The continued influx of such large vessels into the market has led to a large number of vessels being cascaded onto trade lines that historically have been served by smaller vessels (Köpke et al., 2014). In addition, the recent opening of the new Panama Canal locks means that the post-panamax containerships can be utilised in more transport routes, including the trans-Panama services (van Marle, 2016).

An aspect often not taken into account when designing containerships and its importance is closely associated with the size of the vessel is the port efficiency. Although container carriers do not spend considerable amount of time in ports, port efficiency plays a significant role in containership design. The less time a ship spends at port, the more time is available for cruising at sea, which means that vessels can operate in lower speeds and consequently reduce their fuel consumption. Usually, the transport efficiency is optimised by focusing on the schedule of the ships visiting a specific port (Kurt et al., 2015). However, in our case the optimisation focuses on the ship itself and a simplified approach was used, namely monitoring the ratio of the above to below deck containers' number. The larger the ratio, the faster the loading and unloading of containers; thus, the time spent by ships in port is reduced.

1.2 International regulatory framework

Recent developments in the international maritime regulations are going to greatly affect future ship designs and particularly containerships. One major development is the introduction of the EEDI (IMO, 2012a, c, b). The EEDI relates the toxic gas emissions of a ship to her transportation work and is in fact an indicator of a vessel's energy efficiency. The requirement for new ships started with some baseline values in 2013, which will be being lowered successively in three steps until 2025.

New rules have been recently developed regarding the control and management of ships' ballast water and sediments and are being applied to all ships as of September 2017 (IMO, 2004). Although various systems and technologies aiming at the minimisation of the transfer of organisms through ballast water to different ecosystems are currently available, their installation on board ships increases their capital and operating costs. Therefore, research has been focusing lately at solutions to reduce the amount of required ballast water. This problem is more severe for containerships, which inherently carry more ballast water, even at the design load condition, for which the ratio of the containers carried on deck to those carried under deck should be maximised. Thus, design solutions for modern containerships that consider zero or minimal water ballast capacities are very appealing to the ship owners.

As far as safety regulations are concerned, a new generation of intact stability criteria is currently being developed by the IMO (IMO, 2015). The introduction of ships with newly developed characteristic and operation modes has challenged the assumption that the current criteria are sufficient to prove their stability. Hence, the new criteria will be performance-based and will address five modes of stability failure; parametric roll, pure loss of stability, excessive acceleration, stability under dead ship condition and surf-riding/broaching (Peters et al., 2011). As far as containerships are concerned, parametric roll is considered to be one of the most important modes of stability failure (Spyrou, 2005). Pure loss of stability failure mode should also be examined, as the considerable flare found in the aft and fore parts of a containership's hull results in significant changes in the waterplane area as the ship sails through waves. These changes may result in a large roll angle or even capsize (Peters et al., 2011). Likewise, excessive acceleration failure mode should be checked in containerships' case, due to high deckhouses found in such kind of ships. Hence, the draft criteria of level 1 and 2 for excessive acceleration, pure loss of stability and parametric roll failure modes according to SDC 2/WP.4 and SDC 3/WP.5 (IMO, 2015) are applied as part of the optimisation process in this study.

2 PARAMETRIC CAD DESIGN

In recent years, several researchers have presented significant computer-aided design (CAD) methodologies dealing with ship design process and inherently its optimization (Brown & Salcedo, 2003, Tahara, 2015, Yang et al., 2012). A common characteristic of most of the earlier presented works is that they are dealing with specific aspects of ship design or with new system approaches to the design process. On the other hand, the present study deals with an holistic optimisation of a post-panamax, containership, focusing on optimisation of the ship's arrangements, while considering all side effects on ship design, operation and economy (Fig. 1) (Priftis,

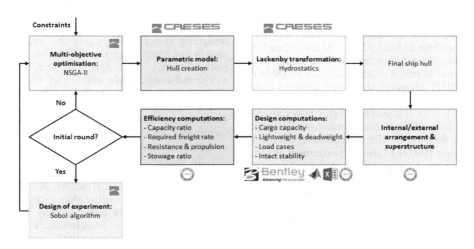

Figure 1. Design optimisation procedure.

2015). Holism is interpreted as a multi-objective optimisation of ship design and is based on the main idea that a system, along with its properties, should be viewed and optimised as a whole and not as a collection of parts (Papanikolaou, 2010). In that respect, efforts are currently being made in the framework of the European Union funded HOLISHIP project (HOLISHIP, 2016). According to the project's approach, a proposed model follows modern computer-aided engineering (CAE) procedures and integrates techno-economic databases, calculation and optimisation modules and software tools along with a complete virtual model which allows the virtual testing before the building phase of a new vessel.

2.1 Geometric model

The geometric model is produced within CAESES® (Friendship Systems, 2017), and consists of four main parts; the main frame, the aft body, the fore body, and the main deck (Fig. 2). Once the initial hull is defined, a Lackenby transformation is applied (Lackenby, 1950). It starts with a hydrostatic and sectional area curve calculation. These are used as input to the Lackenby transformation. By adjusting the prismatic coefficient (C_p) and the longitudinal centre of buoyancy (LCB), the final hull geometry is produced. This process allows shifting sections aft and fore, while fairness optimised B-Splines are utilised (Abt & Harries, 2007). Once the final shape of the hull is created in CAESES®, an IGES file, which contains all the geometry-related information, is generated and imported to NAPA® (NAPA, 2017) to continue with the rest of the design process. A list of in-house developed macros is run to generate the complete ship model (Fig. 3). In particular, macros which generate complex surfaces for the internal cargo arrangement definition, along with simpler ones for the oil, fresh water and water ballast tanks, are run first. The former set of surfaces take the hull surface, double bottom and double side distances into consideration in order to take advantage of as much space available as possible while defining the internal cargo arrangement.

Next, the container arrangement both below and above the main deck is defined. The surfaces generated during the previous stage are used as limits for the arrangement below the main deck. As far as the arrangement above the main deck is concerned, the number of deckhouse decks is related to the maximum number of tiers above the main deck. A plane defined in the previous stage is used to check the compliance with the minimum visibility line regulations when defining the container stacks above the main deck.

2.2 Computations

Using the previously generated planes as limits, the required tanks for the consumables and the water ballast tanks are defined in NAPA®. These tanks along with the cargo holds constitute the completed ship model which is used to create the examined load case. However, before defining the latter, two computations must be run first using custom NAPA® macros; the total resistance and propulsion and the lightweight and deadweight analysis.

The total resistance and the main engine's power are estimated according to the Holtrop and Mennen method guidelines (Holtrop & Mennen, 1978). The calculated value for the main engine's power then is increased by 20% to include a sea margin as well as the impact of hull fouling, representing the common practice in the shipping industry (MAN Diesel & Turbo, 2011). At this stage, the ship's operational speed is defined and set to 21 knots.

The lightweight is divided into three categories; the steel weight, outfitting weight, and the machinery weight. The first weight group is computed using the Schneekluth and Müller-Köster methods. Outfitting and machinery weights are calculated using existing formulas, taking as input several parameters, such as the main dimensions of the ship, as well as the main engine's power (Papanikolaou, 2014). Even though the methods utilised for this step are semi-empirical approaches, and thus, an approximation of the exact values, effort is put to get the most accurate results. In this context, the formulae are calibrated using a similar, existing containership, for which detailed lightship breakdown is available. This allows the calculation of correction factors that improve the final outcome of the model's lightship estimation. Thus, first, all required calculations for the reference ship are performed in Microsoft Excel® (Microsoft, 2010) and a custom macro is developed in NAPA®, including the methods used in the first step, to determine

Figure 2. Modelled aft and fore body created in CAESES®.

Figure 3. Ship model created in NAPA®.

the model's lightship characteristics, including the correction factors in the model's lightship computation.

Afterwards, custom macros responsible for the deadweight analysis are run in NAPA®, generating the necessary values for the determination of the loading cases examined. An operational profile is set up at this stage, so as to reckon the amount of consumables carried on board (Table 1).

Next step is to define the load case examined in this optimisation study. Taking into account the recent regulations regarding the control and management of ships' ballast water and sediments, a zero-ballast "full load departure" condition is defined. The rationale for creating such a load case is to examine how the containership's cargo arrangement should be defined, so that the ship is capable of transporting as many containers as possible while its water ballast tanks are empty. A custom NAPA® macro is run for the load case definition, during which containers are being loaded while intact stability criteria are being monitored to ensure compliance with international regulations. Homogeneous TEU weight is assumed to be equal to 10 tons. The assessment of the initial and large angle stability of the vessel is undertaken for common type loading conditions in accordance with the IMO A.749/A.167 intact stability criteria.

Afterwards, the attained and required energy efficiency design index (EEDI) values are calculated. The determination of EEDI is based on a rather complicated looking (but indeed simple) formula, while it is required that the calculated value is below a reference line set by the IMO regulation for the specific ship type and size (IMO, 2012a, c, b). Using custom NAPA® macros, both the attained and the required values are computed.

Finally, the objectives of this optimisation's study are computed in NAPA®. Taking the load case's results into account, two ratios are defined; the capacity and stowage ratio. The former represents the number of loaded containers to the total container capacity. The latter represents the number of containers that can be loaded above the main deck to the number of containers that can be loaded below the main deck. This ratio represents the ship's port efficiency factor.

The required freight rate is calculated using custom NAPA® macros. This value indicates the minimum rate that evens the properly discounted ship's expenses. The main formula used to calculate the RFR is the following (Watson, 1998):

$$\text{RFR} = \sum_{i}^{N} \left[\frac{\text{PW}(\text{Operating cost}) + \text{PW}\left(\begin{array}{c}\text{Ship acquisition}\\ \text{cost}\end{array}\right)}{(\text{Round trips}) \times (\text{TEUs})} \right]$$

(1)

PW represents the present worth of the respective cost. The overall cost is divided into two categories; the operating cost and the ship acquisition cost. The former is mainly based on the running costs of the ship (e.g. fuel and maintenance costs). As far as the ship acquisition cost is concerned, several data are used as input, including the steel mass of the vessel, cost of steel, discount rate and operation time (Soultanias, 2014).

Once all the above computations are completed, a text file is automatically generated by NAPA®, after calling the required commands. This text file is then read by CAESES® to continue with the second-generation intact stability criteria level 1 and 2 checks. The latter are performed using custom made programmes (or "features" as they are called in CAESES®). Level 1 checks are meant to be simple and conservative, in order to quickly detect any vulnerability to each of the three failure modes. Level 2 checks are more complex, thus less conservative, taking into account more design-related details in order to determine whether the ship is vulnerable to either of the examined failure modes. For each failure mode, several features are developed within CAESES®, connecting various external software to quickly evaluate certain parameters required for these particular computations. Maxsurf® Stability (Bentley Systems, 2014) is run to produce the necessary metacentric height (GM) values for various wave conditions, while Matlab® (Mathworks, 2014) is used to calculate the roll amplitude, where complex equations must be solved.

A set of structural rules is being checked for the midship section of the vessel within CAESES®. Lloyd's Register guidelines are used in this case (Lloyd's Register, 2014). Based on the generated internal geometry, a setup of the midship section is created. The strakes and the stiffeners are taken into account in that respect. A set of input is required to execute the necessary calculations for the structural analysis, which derives from computations that take place at an earlier stage. Section modulus, moment of inertia and stresses are checked in order for the midship section to comply with the aforementioned rules. At first, the

Table 1. Operational profile.

Operational speed (knots)	21
Roundtrip route distance (nm)	13567
Number of ports	18
Average time at port (h)	13.17

minimum requirements for the strake thickness and stiffener section modulus are computed. Then, these results are used as a baseline for the construction of the midship section. Afterwards an internal optimisation takes place in order to calculate the right thicknesses and type of stiffeners so that the requirements are met. At this stage, only the compliance with the structural rules is being checked, however, at a later stage the aim is to connect the steel weight estimation with the midship section definition, based on this step.

3 DESIGN EXPLORATION

Before proceeding to the formal optimisation round, a design of experiment (DoE) is conducted first. This process allows the examination of the design space and the response of several parameters to the change of the model's main characteristics. The algorithm utilised is the Sobol algorithm, a quasi-random sequence which secures the overall coverage of the design space, while overlapping of previous set of sequences is avoided (Mohd Azmin & Stobart, 2015). Through the DoE, the investigation of the feasibility boundaries is ultimately achieved, allowing the detection of the trends of the design variables (Table 2) with regard to the optimisation objectives. In our case, the design engine is assigned to create 250 variants of the initial model. No objectives are set yet, since only the feasibility boundaries are investigated. However, several parameters are evaluated through this process.

The design variables used in this study are presented in Table 2. They consist of TEU arrangement elements, such as the number of bays and rows, specific hull dimensions, such as the double bottom, as well as the variation of the C_p and LCB values. Since the main dimensions of containerships are highly dependent on the container arrangement, the main dimensions of the model derive from these design variables. For instance,

Table 2. Design variables.

Design variable	Min. value	Max. value
Number of bays in hold aft of deckhouse	4	5
Bays	17	20
Y-extent of bilge (m)	4	6
Z-extent of bilge (m)	4	6
Double bottom (m)	2	3
δC_P	−0.075	0.075
δLCB	−0.075	0.075
Rows	15	20
Tiers in hold	8	10
Tiers on deck	6	8

Table 3. Design constraints.

Constraint	Value
Excessive acceleration criteria	= 1
Parametric roll criteria	= 1
Pure loss of stability criteria	= 1
Structural analysis criteria	= 1
"Attained/required" EEDI ratio	≤ 1
Load case draught	≥ 0.925·T_{Design}
Maximum TEU capacity	≤ 8,000
Minimum TEU capacity	≥ 7,000

the breadth of the hull is calculated by taking the number of rows and the beam of each container into account.

Moreover, the constraints are set (Table 3), so as to have a clear view of which of the subsequent variants violate criteria that must be met. It is worth mentioning that the TEU capacity of the model is not constrained during the DoE phase, thus the maximum and minimum number of TEU capacity of the variants is limited to the 7,000–8,000 area only during the formal optimisation round.

4 MULTI-OBJECTIVE OPTIMISATION

The last step to complete the procedure is to set up the formal optimisation round. To achieve that, the non-dominated sorting genetic algorithm II (NSGA-II) is utilised (Deb et al., 2002). In particular, 75 generations are created, having a population size of 10, each. This results in a total of 750 produced variants. The design variables' range remains the same, as the design space proved to be well defined. Contrary to the previous phase, apart from the evaluation of the model's various parameters, four objectives are defined:

- Maximisation of the capacity ratio
- Minimisation of the RFR
- Minimisation of the overall ship resistance
- Maximisation of the stowage ratio

The results of a multi-disciplinary optimisation procedure define the Pareto front of the non-dominated designs. As the decision maker needs to select one design, Multi Attribute Decision Making (MADM) is applied. Several case scenarios are created, so as to determine the optimal of the top solutions to the problem. In this study, three distinctive scenarios are defined, where the significance of each objective is acknowledged differently by assigning specific "weights" following the utility functions technique of decision making theory (Table 4) (Sen & Yang, 1998). In scenario 1, all four

Table 4. Case scenarios.

Objective	Scenario 1	Scenario 2	Scenario 3
Capacity ratio	25%	35%	15%
RFR	25%	35%	35%
Ship resistance	25%	15%	15%
Stowage ratio	25%	15%	35%

objectives are considered to be equally important; hence each one is assigned a weight at saturation of 25%. On the other hand, in scenario 2, the capacity ratio and the RFR are chosen to be more significant for the decision maker (designer, operator) by assigning to them a weight of 35% each, whereas the rest are assigned a weight of 15% each. Similarly, in scenario 3, the stowage ratio and the RFR are assigned a weight of 35% each, while the rest take up 15% of the total, each. After obtaining the results of the formal optimisation run, the data are normalised according to the scenarios. Next, the normalised data are ranked to find the optimal variant of our model. The maximum score that can be achieved after this process for each design, in each case scenario, is 1, whereas the lowest is 0. In all cases, a specific variant dominates in every scenario.

5 DISCUSSION OF RESULTS

5.1 *Base model*

Before proceeding to the actual results, some essential information about the base model is presented, in order to have a clear perspective of the initial hull (Tables 5–6).

5.2 *Design of experiment*

The DoE phase enables the exploration of the huge design space, which is impossible in traditional ship design procedures.

As far as the relation between the number of bays and the RFR is concerned, it is evident that as the number of bays increases, the maximum RFR between the possible variant slightly increases as well. Containerships featuring 17 or 18 bays achieve the lowest possible RFR values (Fig. 4). In Fig. 5, the relation between the total resistance and the double bottom distance is illustrated. A slight increase in the total resistance can be identified, as the double bottom distance rises. It should be noted that the valid designs are described by low resistance values, compared to the invalid ones. Moreover, as far as the relation between the stowage ratio and the number of tiers in hold is concerned, a steep decrease in the ratio can be observed, as the

Table 5. Base model design variable values.

Design variable	Base model
Number of bays in hold aft of deckhouse	4
Bays	18
Y-extent of bilge (m)	5.500
Z-extent of bilge (m)	5.500
Double bottom (m)	2.000
δC_P	0.000
δLCB	0.000
Rows	18
Tiers in hold	9
Tiers on deck	6

Table 6. Base model design objective values.

Objective	Base model
Capacity ratio	1.0000
RFR ($/TEU)	396.43
Ship resistance (KN)	2173
Stowage ratio	0.9423

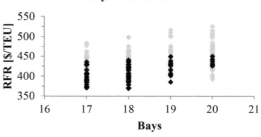

Figure 4. Number of bays vs. RFR.

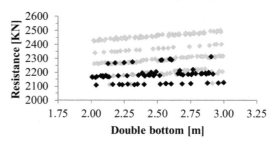

Figure 5. Double bottom vs. Resistance.

number of tiers below the main deck increases. The amount of cargo space in cargo holds increases as more tiers become available below the main deck, hence the number of TEUs stored below the main deck gets higher, leading to a low stowage ratio. In addition, it is worth mentioning that no valid design with ten tiers in hold was produced during the DoE (Fig. 6). Finally, comparing the number of tiers above the main deck to the capacity ratio, a decrease in the achieved value of the latter can be spotted as the number of tiers gets higher. In all possible scenarios the maximum number for the capacity ratio, i.e. 1.0000, was achieved among the design variants (Fig. 7).

5.3 Multi-objective optimisation

Following the NSGA-II run and the evaluation of the results, an improved design, named Des0492, is identified. Des0492 ranked first in every examined scenario. Below, some principal information of the optimised design can be found (Figs. 8, 9, Tables 7, 8).

Table 7. Des0492 design variable values.

Design variable	Des0492
Number of bays in hold aft of deckhouse	5
Bays	18
Y-extent of bilge (m)	5.989
Z-extent of bilge (m)	5.621
Double bottom (m)	2.782
δC_P	−0.035
δLCB	0.044
Rows	16
Tiers in hold	8
Tiers on deck	8

Table 8. Des0492 design objective values.

Objective	Des0492
Capacity ratio	0.9869
RFR ($/TEU)	365.24
Ship resistance (KN)	2122
Stowage ratio	1.5203

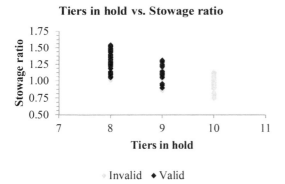

Figure 6. Number of tiers in hold vs. Stowage ratio.

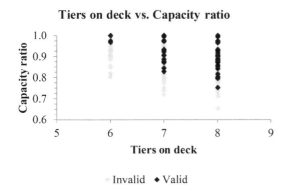

Figure 7. Number of tiers on deck vs. Capacity ratio.

Figure 8. Des0492 hull.

Figure 9. Des0492 ship model.

A set of graphs containing the relation between the optimisation objectives is presented below. The Pareto front is demonstrated by a solid black line in each case.

As far as the values of the capacity ratio and the resistance are concerned, a slight increase in the resistance is observed as the capacity ratio rises. Most of the produced variants feature a low resistance value overall – between 2100 and 2150 KN – (Fig. 10). The relation between the RFR and the resistance is demonstrated in Fig. 11. Des0492 managed to outperform every other variant in terms of the achieved RFR value, while it features one of the lowest resistance values.

In Fig. 12 the resistance is compared with the stowage ratio. In general, a favourable relation is

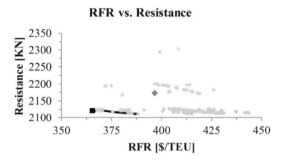

Figure 10. Capacity ratio vs. Resistance.

Figure 11. RFR vs. Resistance.

observed, as the stowage ratio rises when the resistance is decreased. Two groups of variants are visible in the graph, one where the designs have achieved a resistance value of around 2125 KN and a stowage value ranging between 1.25 and 1.55, as well as one where the designs have achieved a resistance value of around 2185 KN and a stowage value ranging between 1.40 and 1.50. Finally, the relation between the stowage and the capacity ratios can be found in Fig. 13. The optimal point is described by a capacity ratio of one and a stowage ratio of around 1.55. Des0492 lies on the Pareto front and is close to that point, featuring both high capacity and stowage ratios.

Below, the baseline model and Des0492 are compared, showing the differences in the objective values (Table 9).

As far as the main dimensions are concerned, the improved design features the same number of bays, while the number of rows is decreased by two, compared to the baseline model. The number of tiers below the main deck is decreased by one and two extra tiers above the main deck are carried in the improved design. The extra tiers found in Des0492 offer the advantage of an increased stowage ratio, as well as a reduced RFR, due to the higher total number of TEUs carried on board. Furthermore, the double bottom distance and the bilge radius are higher in Des0492's case. Overall, Des0492 manages to outperform the original design in every objective but the capacity ratio. Nevertheless, the difference in the latter between the two designs is minimal. In addition, the "attained/required" EEDI ratio for Des0492 for the current state of the rules is equal to 0.79, providing a safety margin from the maximum allowed value set by regulations. A notable improvement can be observed in the port efficiency factor, where an increase of 61.34% is achieved.

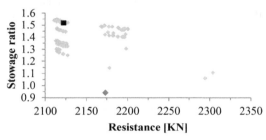

Figure 12. Resistance vs. Stowage ratio.

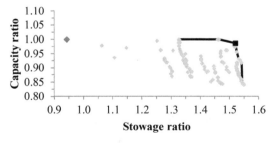

Figure 13. Stowage ratio vs. Capacity ratio.

Table 9. Baseline design vs. Des0492.

Data	Baseline	Des0492	Difference
Capacity ratio	1.0000	0.9869	−1.31%
RFR ($/TEU)	396.43	365.24	−7.87%
Ship resistance (KN)	2173	2122	−2.35%
Stowage ratio	0.9423	1.5203	+61.34%

6 SUMMARY AND CONCLUDING REMARKS

Through the work presented in this paper, the advantages of the utilisation of modern design optimisation in the shipbuilding industry have been demonstrated. By incorporating this type of parametric optimisation process in the early stages of ship design, a much-improved design can be produced, providing numerous benefits to a potential builder and end user (ship owner). Furthermore, it is demonstrated that using modern CAD/CAE systems, it is possible to explore the huge design space with little effort, while generating excellent/ partly innovative results within very short lead times. The presented CAD/CAE setup allows the integration of more advanced tools for the improved modelling of e.g. ship's hydrodynamics. The optimisation can include other areas of ship design as main objectives, allowing naval architects to achieve a greater degree of holism in the design process (Papanikolaou, 2010).

It is evident that the relation of the design process with statutory regulations should be included in the optimisation process as well, as new rules are introduced every year. The present study incorporated new tools for the newly developed second generation criteria for excessive acceleration, pure loss of stability and parametric roll failure modes. The results indicate how the model should be designed to pass certain criteria to comply with international regulations, while it becomes clear that specific design parameters affect the above. The approach presented in this study can be also applied to other containership sizes (Koutroukis et al., 2013), or to other ship types, by using a different CAD hull model and adjusting the design variables and internal compartmentation. More phases of the ship's life cycle can be integrated to future studies, resulting in more comprehensive holistic ship design investigations (Papanikolaou, 2010).

ACKNOWLEDGEMENTS

This work was partially funded by the H2020 European Union project "HOLISHIP-Holistic Optimisation of Ship Design and Operation for Life Cycle" (contract 689074).

REFERENCES

Abt, C. & Harries, S. 2007. Hull variation and improvement using the generalised Lackenby method of the FRIENDSHIP-Framework. The Naval Architect, 166–167.
Bentley Systems 2014. Maxsurf Stability. Windows Version 20 ed.
Brown, A. & Salcedo, J. 2003. Multiple-objective optimization in naval ship design. Naval Engineers Journal, 115, 49–62.
Deb, K., Pratap, A., Agarwal, S. & Meyarivan, T. 2002. A fast and elitist multiobjective genetic algorithm: NSGA-II. IEEE Transactions on Evolutionary Computation, 6, 182–197.
Friendship Systems 2017. CAESES. 4.2.1 ed.
HOLISHIP 2016. Holistic optimisation of ship design and operation for life cycle. EU.
Holtrop, J. & Mennen, G.G.J. 1978. An approximate power prediction method. International Shipbuilding Progress, 25, 166–170.
IMO 2004. BWM - international convention for the control and management of ships' ballast water and sediments. In: INTERNATIONAL MARITIME ORGANISATION (ed.). London, United Kingdom.
IMO 2012a. Consideration of the energy efficiency design index for new ships - minimum propulsion power to maintain the maneuverability in adverse conditions. In: INTERNATIONAL MARITIME ORGANISATION (ed.). London, United Kingdom.
IMO 2012b. Guidelines for calculation of reference lines for use with the energy efficiency design index (EEDI). In: INTERNATIONAL MARITIME ORGANISATION (ed.). London, United Kingdom.
IMO 2012c. Guidelines on the method of calculation of the attained energy efficiency design index (EEDI) for new ships. In: INTERNATIONAL MARITIME ORGANISATION (ed.). London, United Kingdom.
IMO 2015. Development of second generation intact stability criteria. In: INTERNATIONAL MARITIME ORGANISATION (ed.). London, United Kingdom.
IMO 2016. Marine environment protection committee. In: INTERNATIONAL MARITIME ORGANISATION (ed.). London, United Kingdom.
Koutroukis, G., Papanikolaou, A., Nikolopoulos, L., Sames, P. & Köpke, M. 2013. Multi-objective optimization of container ship design. 15th International Maritime Association of the Mediterranean. A Coruña, Spain: Taylor & Francis Group (CRC).
Kurt, I., Aymelek, M., Boulougouris, E. & Turan, O. 2015. A container transport network analysis study on the offshore port system case of west North America coast. International Association of Maritime Economists. Kuala Lumpur, Malaysia.
Köpke, M., Papanikolaou, A., Harries, S., Nikolopoulos, L. & Sames, P. 2014. CONTiOPT - Holistic optimisation of a high efficiency and low emission containership. Transport Research Arena 2014. Paris, France.
Lackenby, H. 1950. On the systematic geometrical variation of ship forms. Transactions of RINA, 92, 289–316.
Lloyd's Register 2014. Rules and regulations for the classification of ships. In: LLOYD'S REGISTER (ed.).
MAN Diesel & Turbo 2011. Basic principles of propulsion. Copenhagen, Denmark.
Mathworks 2014. MATLAB. R2014a ed.
Microsoft 2010. Microsoft Excel.
Mohd Azmin, F. & Stobart, R. 2015. Benefiting from Sobol sequences experiment design type for model-based calibration. SAE Technical Papers, 1.
NAPA 2017. NAPA.
Papanikolaou, A. 2010. Holistic ship design optimization. Computer-Aided Design, 42, 1028–1044.

Papanikolaou, A. 2014. Ship design: methodologies of preliminary design, Netherlands, Springer.

Peters, W., Belenky, V., Bassler, C., Spyrou, K.J., Umeda, N., Bulian, G. & Altmayer, B. 2011. The second generation intact stability criteria: an overview of development. Annual Meeting of the Society of Naval Architects and Marine Engineers. Houston, Texas.

Priftis, A. 2015. Parametric design and multi-objective optimization of a 6,500 TEU container ship. Diploma, National Technical University of Athens.

Sen, P. & Yang, J.B. 1998. Multiple criteria decision support in engineering design, London, United Kingdom, Springer.

Ship & Bunker. 2017. Ship & bunker [Online]. Available: http://shipandbunker.com/prices [Accessed November 2017].

Soultanias, I. 2014. Parametric ship design and holistic design optimisation of a 9 K TEU container carrier. Diploma, National Technical University of Athens.

Spyrou, K.J. 2005. Design criteria for parametric rolling. Oceanic Engineering International, 9, 11–27.

Tahara, Y. 2015. CFD-based hull form / appendage optimization by using deterministic and stochastic optimization theory. 12th International Marine Design Conference. Tokyo, Japan.

UNCTAD 2016. Review of maritime transport. Geneva, Switzerland.

van Marle, G. 2016. Intra-Asia beckons for panamax ships. Container Shipping and Trade. Enfield, United Kingdom: Riviera Maritime Media Ltd.

Watson, D.G.M. 1998. Practical ship design, United Kingdom, Elsevier.

Yang, H., Chen, J., Lu, Q. & Ma, N. 2012. Application of knowledge-based engineering for ship optimisation design. Ships and Offshore Structures, 9, 64–73.

Optimization method for the arrangements of LNG FPSO considering stability, safety, operability, and maintainability

S.H. Lee, M.I. Roh, S.M. Lee & K.S. Kim
Seoul National University, Seoul, Republic of Korea

ABSTRACT: The LNG FPSO is a kind of FPSO (Floating, Production, Storage, and Offloading unit) with an LNG plant, including all ancillary facilities. There are many parts of the process on the topside. In contrast with commercial vessels such as LNG ship and tanker, the design for LNG FPSO requires many considerations such as international standards, and owners' requirements, maintenance philosophy and so on. The arrangement of every LNG facility on limited deck space has to be optimized while maintaining the safety and the operability required for LNG production. And it is necessary to optimize the integration of the topsides with the hull and marine systems considering the stability. In this study, the optimization method for LNG FPSO was proposed to obtain principle dimensions, the arrangement of hull tanks, and the layout of topside modules and equipment considering stability, safety, operability, and maintainability. The proposed method can be used to find the optimal layout of LNG FPSO which satisfies requirements for both topside and hull. The result of this study showed that the proposed method could be used for developing the optimal layout of initial design stage for LNG FPSO and improve work efficiency of the design process.

1 INTRODUCTION

The LNG FPSO is a kind of FPSO (floating, production, storage and offshore) facility with an LNG plant, including all ancillary facilities. The various parts of the process are located topside and distributed as modules. Every element of LNG facility needs to fit into an area less than one quarter the size of a land base terminal while maintaining safety and the flexibility required by LNG production. The design for LNG FPSO requires many considerations such as international codes and standards, owners' requirements, operation and maintenance philosophy, and so on.

A lot of studies for the layout design method have been conducted by many scholars (Lee et al. 2001, Patisatizis et al. 2002, Chung et al. 2011, Park et al. 2011, Mazerski 2012, Ku et al. 2014, Jeong et al. 2015, Kim et al. 2017). In traditional works, they optimize the design of hull or topside of the target, but not both. Since the design of hull and design of topside are interrelated, an arrangement design that considers both at the same time is required. Therefore, this study proposes the integrated method for optimal layout design of LNG FPSO. The proposed method includes simultaneous optimization of principle dimensions, hull tanks, topside modules, and equipment, considering safety, economics, and stability.

2 CONFIGURATION OF INTEGRATED METHOD FOR LAYOUT DESIGN OF LNG FPSO

This study consists of four major items as shown in Figure 1. The first item is the template model defined as a data structure that represents and store information. The template model represents the hierarchical structure of a component of LNG FPSO. The second item is the expert system. In this study, the rule-based expert system is used to reflect the knowledge and international regulations. The rule-based expert system consists of the

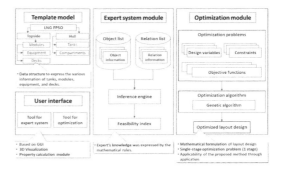

Figure 1. Configuration of integrated method for layout design of LNG FPSO.

knowledge base, database, and inference engine. The third item is the optimization module. In the optimization module, the entire layout design of the LNG FPSO is formulated as the optimal problem. In this study, multi-purpose optimization is performed, and the optimized arrangement result can be obtained. The fourth item is the user interface that enters the rules and makes it possible to perform optimization problems.

3 FORMULATION OF AN OPTIMAL ARRANGEMENT PROBLEM

In this study, the optimization method for LNG FPSO was proposed to obtain principle dimensions, the arrangement of hull tanks, and the layout of topside modules and equipment considering stability, safety, operability, and maintainability. The formulation of the optimization problem for the arrangement design of LNG FPSO is shown in Figure 2.

There are many input data for the layout design of LNG FPSO. The information of hull tank and module, equipment, deck, piping connection and cost parameters is required to optimize the layout. Adjacency coefficient between modules and tanks groups are required too.

In this study, there are many design variables because the tank, module, and equipment arrangement are performed simultaneously with the determination of principal dimensions of LNG FPSO. First, the design variables for determining principle dimensions of the LNG FPSO through tank sizing are the length, width, and height of each cargo tank. The width and height of the cargo tanks storing LNG, LPG, Condensate, and Process liquid are the same, and the length is determined according to the required capacity. Design variable for tanks arrangement is the arrangement sequence of the tanks groups. The locations of each tanks group and it can be represented as an array of the tanks groups "id." After optimization, the array is converted to the arrangement of tanks groups. Design variable for module layout is the arrangement sequence of the module. The locations of each module can be represented as an array of the module "id" (encoding). After optimization, the array is converted to the module arrangement (decoding). Design variable for equipment layout is the location of equipment in each module. The locations of each equipment can be represented as real variables.

In this study, 11 objective functions related to safety, economics, and stability are reflected. A weight factor is assigned to each objective function according to the characteristics and requirements of the LNG FPSO. All objective functions are mathematically formulated. Hull structure weight index, pipe length between topside modules and tanks, pipe length between topside modules, installation area cost, piping cost and ventilation cost have to be minimized considering the economic aspect. Transverse weight distribution index is used to minimize heel due to the weight difference between modules on the port side and starboard side. Adjacency index of tanks group, adjacency index of modules and damage cost considering physical explosion has to be minimized considering the safety aspect. Feasibility index from the expert system has to be maximized.

When arranging the equipment, each equipment should be arranged at once. And equipment should not be overlapped. Passages around the perimeter of the deck and spaces around the equipment should be considered for operability. These aspects are formulated as constraints. In this study, non-overlapping constraints, boundary constraints and intact stability constraints are used to formulate the optimization problem.

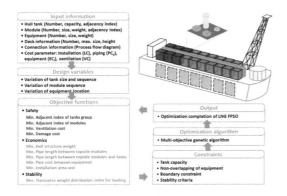

Figure 2. Formulation of an optimal arrangement problem for optimization module.

4 APPLICATION

In this study, the prototype program is developed to apply the proposed method which consists of the template model, the expert system module, the optimization module, and the user interface. The prototype program was developed using C# language and WPF (Windows Presentation Foundation, http://msdn.microsoft.com/) in .Net 4.0 environment.

To verify the applicability of the proposed method and prototype program, they were applied to the layout design of the LNG FPSO. Simultaneous optimization of principle dimensions, the layout of hull tanks, topside modules and equipment was performed for this example.

Multi-objective optimization was performed, and the optimal solution was derived by assigning weight factors to each objective function. The case study was conducted to investigate the influence of the weight factors for each objective function. The basis case was defined based on the reference project, and four case studies were conducted according to the weight factor of the objective function.

Table 1 summarizes the weight factors of each case study. First, the case 1 focused on the safety, and the weight factor was assigned to four objective functions related to safety. In the case 2, the weight factors were given to the objective functions related to economics. In the case 3, the weight factor was given intensively to the objective function related to stability. In the case 4, weight factors were given uniformly to all objective functions. And the same weight factors were assigned to objective function for feasibility index in each case.

Table 2 summarizes the optimization results. First, the results of case 1 were shown that all the safety-related objective functions were improved. In case 2, economic objective functions were improved. Case 3, in which the objective function is concentrated only on stability, is shown that the value of transverse weight distribution index was greatly improved. In case 4, in which the weight factor is uniformly assigned to all objective functions, 8 items out of 10 objective functions were improved.

The optimization result of layout of hull tanks of case 4 is shown in Figure 3 as a representative result. The hull structure weight index decreased due to a significant decrease in length and a decrease in breadth. The arrangement sequence of the tanks groups in the hull was changed. Adjacency index of tanks group was improved, because the LNG tanks group and LNG tanks group, which store the cryogenic liquid, are respectively separated from the AFT part and the FWD part. And LNG tanks group was moved to forward near liquefaction modules. The optimization result of layout of topside modules of case 4 is shown in Figure 4.

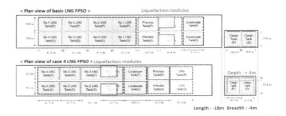

Figure 3. Optimization result of the layout for hull tanks in case 4.

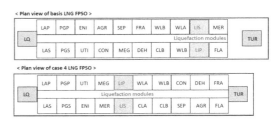

Figure 4. Optimization result of the layout for topside modules in case 4.

Table 1. The weight factors of each case study.

Objective function	Weight factor			
	Case 1	Case 2	Case 3	Case 4
Safety	50	0	0	20
Economics	0	50	0	25
Stability	0	0	50	5
Expert system	50	50	50	50

Table 2. Summary of optimization results.

Description	Basis	Case 1	Case 2	Case 3	Case 4
Length [m]	417	419 (+0.48%)	398 (−4.56%)	402 (−3.60%)	399 (−4.32%)
Breadth [m]	68	68 (0%)	66 (−2.94%)	64 (−5.88%)	64 (−5.88%)
Depth [m]	35.5	35.5 (0%)	39.5 (+11.27%)	40 (+12.68%)	39.5 (+11.27%)
Safety	−0.437	−0.478 (+9.38%)	−0.443 (+1.37%)	−0.401 (−8.24%)	−0.497 (13.73%)
Economics	2.664	2.973 (+11.60%)	2.479 (−6.94%)	2.621 (−1.61%)	2.307 (−13.40%)
Stability	0.54	0.129 (−76.11%)	0.437 (−19.07%)	0.18 (−66.67%)	0.952 (+76.30%)
Expert system	–	2100	2100	2100	2100

5 CONCLUSIONS

In this study, the integrated design framework for LNG FPSO is proposed to obtain the optimal hull sizing, tank arrangement in the hull, module layout in the topside, and equipment layout in the topside modules that satisfy many requirements regarding safety, economic efficiency, operability, and stability. The proposed framework consists of four components. First, the expert system is applied to computerize expert's knowledge and experience systematically. Second, the optimization method is used to yield a better design by formulating the design problem as an optimization problem with a single stage. Third, an arrangement template model is used to store the arrangement data of offshore platform. Fourth, the user interface is developed to the integrated design of the offshore platform by executing the proposed framework. To evaluate the applicability of the proposed framework, a prototype program is developed.

This method was applied to the example of an LNG FPSO to verify the method, and the result shows that the proposed framework could be used for finding a better arrangement and improve the work efficiency of the design process.

REFERENCES

Chung, B.Y., Kim, S.Y., Shin, S.C., Koo, Y.H. 2011. Optimization of compartments layout of submarine pressure hull with the knowledge-based system, *International Journal of Naval Architecture and Ocean Engineering* 3(4): 254–262.

Jeong, S.Y., Roh, M.I., Shin, H.K. 2015. Multi-floor layout model for topside of floating offshore plant using the optimization technique, *Journal of the Society of Naval Architects of Korea* 52(1): 77–87.

Kim, S.K., Roh, M.I., Kim, K.S. 2017. Arrangement method of offshore topside based on an expert system and optimization technique, *Journal of Offshore Mechanics and Arctic Engineering* 139(2): 1–19.

Ku, N.K., Hwang, J.H., Lee, J.C., Roh M.I., Lee, K.Y. 2014. Optimal module layout for a genetic offshore LNG liquefaction process of LNG-FPSO, *Ships and Offshore Structures* 9(3): 311–332.

Lee, K.Y., Roh, M.I., Chi, S.H. 2001. Multidisciplinary design optimization of mechanical systems using collaborative optimization approach, *International Journal of Vehicle Design* 25(4): 353–368.

Mazerski, G. 2012. Optimization of FPSO's main dimensions using genetic algorithm, *ASME 2012 31st International Conference on Ocean, Offshore and Arctic Engineering* 1: 601–610.

Park, K.T., Koo, J., Shim, D.I., Lee, C.J., Yoon, E.S. 2011. Optimal multi-floor layout with consideration of safety distance based on mathematical programming and modified consequence analysis, *Korean J. Chem. Eng.* 28(4): 1009–1018.

Patsiatizis, D.I., Papageorgiou, L.G. 2002. Optimal multi-floor process plant layout, *Computers and Chemical Engineering* 26: 575–583.

Container ship stowage plan using steepest ascent hill climbing, genetic, and simulated annealing algorithms

M.A. Yurtseven, E. Boulougouris & O. Turan
Department of Naval Architecture, Ocean and Marine Engineering (NAOME), The University of Strathclyde, Glasgow, UK

ABSTRACT: Container ship stowage plan is a vital subject on reducing additional cost from shifting and the vessel's turn-around time in maritime ports. However, making proper stowage plan is not straightforward since this subject is literally known as a NP-hard problem. Several studies have been conducted to solve this problem despite it has not been dealt with completely. Hence, the principal objective of this study is to find optimal container stowage plan for container vessel calling at multiple ports.

The most common way of finding the optimal solution is to consult algorithms for container vessel stowage problem. In this study, Steepest Ascent Hill Climbing, Genetic and Simulated Annealing algorithms are implemented considering the same size and two different types of containers (refrigerated and standard containers). After obtaining the numerical test results from three algorithms implementation, their performances for the problem solution are compared and evaluated.

1 INTRODUCTION

Seaborne trade is of the main component in the international trading since it holds the over 80 per cent and more than 70 per cent of world merchandise trade by volume and value, respectively (UNCTAD, 2017).

In the history of maritime industry development, containerization has become a key instrument in liner shipping. The most of the seaborne containerized shipping in the world is operated within liner shipping. This liner shipping is performed via the means of variety capacity of special designed seagoing vessels that can carry over tens of thousands of containers on regular predefined schedules between ports (Tierney, 2015).

For the last few decades, the size of container vessels has continuously increased exceeding over 21,000 Twenty Foot Equivalent Units (TEUs). With the increase in the vessels capacities, it enables the reduction in the vessel running costs. On the other hand, the size growth leads to some serious challenges in vessel operations.

Container Ship Stowage Planning (CSSP) in shipping industry is also known as Master Bay Plan Problem. It can be counted as one of the serious challenges needed to deal with. Because it is literally seen as a NP-hard problem (non-deterministic polynomial-time). Hence, generating well-conceived stowage plan for container vessel is not straightforward and is performed by human planners.

Container vessels visit more than one port carrying thousands of containers in various sizes during their voyages. Before sailing, it is needed to be done the seaworthiness of the vessels meeting stability and the other restrictions properly.

In terms of container stowage on a vessel, the last-in first-out method is followed. Throughout the multiport voyage, several containers on the vessel are temporarily unloaded and reloaded at subsequent ports. This is called shifting which causes the increase in vessel turnaround time and port expenses.

The last few decades, there has been an increasing amount of literature studies on this subject. Different methodologies have been applied to find optimal solution for stowage plan. Up to now, number of studies have used different algorithms. to deal with the CSSP problem.

Avriel & Penn (1993) suggest a 0–1 binary linear programming formulation that aims to minimize the number of shiftings from stowage planning. For the problem solution, the vessel stability is not considered. Besides, the number of port calls and the number of containers to be shipped are determined before sailing starts. The GAMS software is used for the model implementation.

Another stowage plan problem is presented by Martins et al. (2009) on small vessel running in short sea shipping line. In this study, two different approaches, Microsoft Excel Solver and Genetic Algorithm (GA), are used

Liang et al. (2016) present a new optimization approach named Social Network-based Swarm Optimization Algorithm (SNSO) to deal with the slot planning problem in container vessel bays.

Cohen et al. (2017) divide the CSSP into two phases: master bay planning phase and slot planning phase. In the master bay planning phase, container distribution to each bay is solved. In the slot planning phase, a slot is arranged for each container. The authors implement a GA approach solve the CSSP in this study.

In Nikos's thesis (2017), GA approach is adopted to produce a feasible CSSP strategy. By doing that, it is aimed to obtain more accurate result and low computational time. In the problem solution, some constraints are considered such as, the moment of stability between form bow and stern sections, the maximum weight of stack, heel righting moment and also full loaded vessel condition during the voyage.

The aim of this study is to find the optimal CSSP solution on board container vessel. Herein, three different algorithms are deployed by using MatLab software in order to assign containers into slots on a container vessel.

The overall structure of the paper is formed from four sections. The second section explains the problem statement. The third section gives briefly three different algorithms to be deployed in the problem. The final section compares the results and the performance obtained from the algorithms and concludes with authors' final remarks.

2 PROBLEM STATEMENT

In this section, we present some representative examples with variables and constraints then describe the methodology utilized for the problem solution.

The main goal of the study is to reduce the unnecessary container-moves throughout the vessel sailing to decrease the total cost.

For this problem, we have generated a new formula which enables us to calculate the cost value from the number of whole moves occurred at every port. The formula is as follows,

$$C = \sum_{j=1}^{Ncol} \sum_{k=1}^{Nrow} 2k \left(abs \left(sign \left(\prod_{i=1}^{Nrow-k} \left(x(k,j) - x(Nrow-i+1,j) \right) \right) \right) \right)$$

where;
C = cost of total containers movement,
$Ncol$ = the number of column
$Nrow$ = the number of rows
$k = Row^{th}$ number in series of row from top to downward
$x_{(k,j)}$ & $x_{(Nrow-i+1,j)}$ = the destination port number

$$x_{(k,j)} - x_{(Nrow-i+1,j)} = \begin{cases} 0, & x_{(k,j)} = x_{(Nrow-i+1,j)} \\ 1, & x_{(k,j)} \neq x_{(Nrow-i+1,j)} \end{cases}$$

Example 1: Table 1 presents the number of destination ports, the number of refrigerated, unrefrigerated and total containers headed to each port. It is also assumed that refrigerated containers (reefers) can be placed only at the last two bottom rows due to the location of power plugs in the vessel.

For this study, reefers are shown as double-digit number, where the second number in it shows the destination port (e.g. 12 is a reefer going to P2). One-digit number shows an unrefrigerated container showing its destination port (e.g. 2 is an unrefrigerated container going to P2).

It is possible to distribute the containers in the slots as shown in Table 2 and no shifting will be required through the whole journey. Hence the cost function will be optimum and will be equal to 32.

It is possible to consider Table 3 as another solution. However, there are compulsory re-handling in this solution, which leads to additional cost. The total cost will be 40 in this solution.

Table 1. Container distribution for each port with reefers.

	Destination Ports (P = 4)			
	P1	P2	P3	P4
Number of Refrigerated Containers	2	2	2	2
Number of Unrefrigerated Containers	2	2	2	2
Total Number of Containers	4	4	4	4

Table 2. Optimum stowage plan for Example 1 (Total cost is 32).

1	2	3	4
1	2	3	4
11	12	13	14
11	12	13	14

Table 3. Possible stowage plan (Total cost is 40).

1	2	3	4
2	1	3	4
11	12	13	14
11	12	13	14

Similarly, if the one of the refrigerated containers heading to port 1 and port 2 are swapped in Table 2, the updated cost function value will be 44 due to the increase in compulsory shifting.

The reason of the cost difference among three tables above is the sequence of the containers into the slots.

To solve this problem manually is straightforward due to the small number of containers. However, the real life examples will include bigger numbers, which will increase the complexity of the problem. Hence Steepest Ascent Hill Climbing, Genetic, Simulated Annealing algorithms techniques are implemented to find the optimal solution.

This problem model is built under the following assumptions;

- The number of containers and their destination port are known before the voyage starts.
- The bay layout is known, including rows and columns. The whole vessel is imagined to be formed as one bay.
- Each discharging container must be replaced by other container, not travelling to the first four ports, in the same slot at its destination port. That means, container vessel must be operated full capacity during throughout its journey
- For any container, loading and discharging movements cost 2 units.
- Slots can be occupied by only one container.
- Two types (Refrigerated and Standard) and same size (40' long) containers can be loaded and unloaded.
- Refrigerated containers cannot be assigned in non-reefer slots.
- Stability, hatch cover and containers' weight are ignored.

For our problem, a container vessel consists of a single bay with fifteen columns and ten rows (15 × 10), hence 150 containers (variables). As a constraint, refrigerated containers (reefers) can be placed only across the last 4 bottom rows where power plugs exist. Also, the number of vessel calling ports (P) are 5.

Every slot from S1 to S150 has to be filled and the slots assigned for refrigerated containers must be filled only by refrigerated containers.

The data in Table 5 was generated and distributed randomly to the 5 destination ports.

Table 4. Possible stowage plan (Total cost is 44).

1	2	3	4
1	2	3	4
11	12	13	14
12	11	13	14

Table 5. Container distribution according to the ports.

	Destinations Ports (P = 5)				
	P1	P2	P3	P4	P5
Number of Refrigerated Containers	11	13	7	15	14
Number of Unrefrigerated Containers	12	18	30	11	19
Total Number of Containers	23	31	37	26	33

3 THE OPTIMIZATION METHODS FOR THE PROBLEM SOLUTION

Optimization is a process of finding most optimal solution of the problem under given constraints. For the problem in this study, three different algorithms, Genetic, Simulated Annealing, Steepest Ascent Hill Climbing Algorithm are proposed to minimize the number of shiftings.

3.1 Steepest ascent hill climbing algorithm

Steepest Ascent Hill Climbing (SAHC), our first optimization method in this problem, begins by creating a random initial point. Then, new solutions are generated by moving in a random way within the current solution's neighbourhood. In other words, SAHC analyses different moves in the neighbourhood of a single node (Arriaga & Valenzuela-Rendón, 2012). The moves continue until no better solution is found. On the other hand, premature convergence can happen during the climbing and causes the algorithm process termination while it runs. In other words, the algorithm reaches an impasse easily on a local optimum and gets stuck (Weis, 2009).

As shown in Table 7, the different options were employed in SAHC. In Table 6, the number of iteration increases, hence the best solution is obtained whereas the time increases.

As seen in Figure 2, the best cost function is calculated as 370 in 113.5 minutes by SAHC method. Table 7 shows an optimum solution after SAHC method implementation.

3.2 Problem solution using genetic algorithm

Genetic algorithms as a successful computational method in optimization of mathematical complex problems are biologically inspired by the basic principles of natural selection and evolution.

In the point of theoretical steps of Genetic algorithms;

1. A random initial population (chromosomes) is generated Calculate the fitness function of the generated population

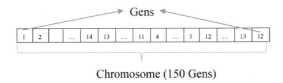

Chromosome (150 Gens)

2. Compare the fitness function with the existing criteria in the problem
   a. If criteria are met, stop it
   b. If criteria are not met, go next step
3. Choose elite member looking the best fitness value
4. Produce offsprings applying crossover and then mutation
5. Replace the new generation with the current generation
6. Go to step 2 and 3 respectively

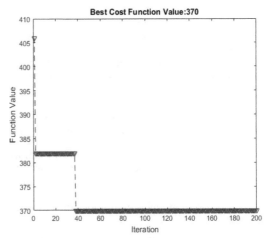

Figure 2. Simulation result of the SAHC method. Total processing time is 113.5 minute and best cost function value is 370.

Table 7. Outline of container distribution in a bay after using SAHC.

3	3	5	1	3	5	4	1	5	5	3	2	3	2	3
4	3	5	1	3	2	4	1	5	5	3	2	3	2	3
4	3	5	1	3	2	4	1	5	5	3	2	3	2	3
4	3	5	2	3	2	4	1	5	5	3	2	3	2	3
4	3	5	1	3	2	4	1	5	5	3	2	3	2	3
4	3	5	1	3	2	4	1	5	5	3	2	1	2	3
14	15	15	12	14	12	14	11	15	15	13	12	11	12	13
14	15	15	11	14	12	14	11	15	15	13	12	11	12	14
14	13	15	11	14	12	14	11	15	15	13	12	11	12	14
14	13	15	11	14	12	14	11	15	15	13	12	11	12	14

Figure 1. SAHC algorithm flow chart.

Table 6. Results from the utilization of different options in SAHC.

Variable	Iteration	Best cost value	Time (Min)
150	10	398	5.2
150	50	374	26.4
150	100	374	54.6
150	200	370	113.5
150	300	372	169.0
150	400	378	216.8
150	500	374	281.1
150	1000	370	526.1
150	2000	370	1063.6

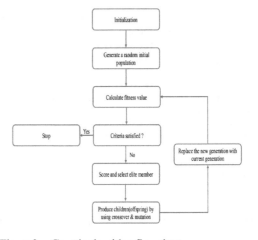

Figure 3. Genetic algorithm flow chart.

Table 8. Results from the use of different options in genetic algorithm.

Variable	Iteration	Population size	Elite count	Best cost value	Time (Min)
150	1000	500	25	404	1.8
150	1000	1000	50	396	3.1
150	2000	500	25	406	4.4
150	2000	1000	50	416	7
150	3000	500	25	402	7.4
150	3000	1000	50	398	10.9
150	5000	500	25	390	15.3
150	5000	1000	50	420	21.5
150	10000	500	25	406	48.2
150	10000	1000	50	378	60.5

Table 9. Outline of container distribution in a bay after using GA.

5	3	3	3	2	3	2	3	3	1	5	3	4	3	2
5	5	3	3	2	3	2	4	3	1	5	3	4	3	2
5	5	3	3	2	3	2	4	3	1	5	1	4	3	2
5	5	3	3	2	3	1	4	3	1	5	1	4	3	2
5	5	2	3	2	3	2	4	5	1	5	1	4	3	2
5	5	2	3	2	3	1	4	5	1	5	1	4	3	2
15	15	12	15	12	14	12	14	15	11	15	11	14	13	12
14	15	12	13	12	14	11	14	15	11	15	11	14	13	12
14	15	12	13	12	14	11	14	15	11	15	11	14	13	12
14	15	12	13	12	14	11	14	15	11	15	11	14	13	12

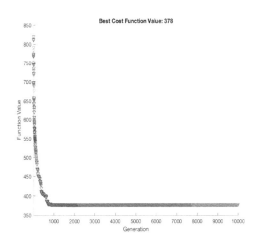

Figure 4. Simulation result of the GA method. Total processing time is 60.5 minute and best cost function value is 378.

Initial inputs (chromosome encoding) required for the population in this problem are considered as same size of the number of containers which represents our variables. Each generation has a certain amount of population size. The proper solution space searching is directly proportional with the number of population however; the computational time is inversely proportional with the number of population.

The number of repetition, a crucial option for finding the optimal solution, is defined as the number of generations based on the information in MatLab options. This repetition number is, in practice, 100*Number of variables. However, in this problem the number of generation is taken 10000 because of increasing the computational time. The result of the problem using Genetic Algorithm is as follows

As shown in Table 8, the different options were used in GA. With increasing iteration number, the computational time increases and so the best solution is also obtained.

It can be seen in Figure 4 that GA calculated the best cost function value as 378 in 60.5 minute. As shown in Table 9, the numbers in the slots represent destination port for that slot's container after GA method implementation.

3.3 *Problem solution using simulated annealing algorithm*

For this problem, the last possible experimental way of finding optimal solution is Simulated Annealing (SA). The study of SA was first carried out by Kirkpatric et al. (1983).

To begin SA process, a random trial point generates and then it calculates. In the meantime, the initial value of temperature which is crucial parameter is assigned for the problem to be optimised. Afterwards, according to a probability distribution with a scale based on the current temperature, the distance of the trial point is chosen from the current point. The trial point distance distribution is set as a function with the AnnealingFcn option. The trial point can be changed via SA, if needed, to be able to stay within bounds. Then, the algorithm compares the new point with the current point which one is better or worse. If the new point is better, the new point is accepted and used as a next point. If not, again the next point is generated using the worse point depending on an acceptance function

If the new point is better than the current point, the new point is used as a next point, otherwise the SA tries to create the next point using the worse point depending on an acceptance function (probability base). Then, the temperature is dropped systematically via the SA algorithm, recording the best point obtained so far. The algorithm used the specified the function by TemperatureFcn option to update the temperature. The annealing parameter remain stable same as the iteration number till reannealing.

After Simulannealbnd accepts ReannealInterval points, it reanneals. Reannealing assigns the annealing parameters to lower values than the iteration number, hence the temperature increases in each dimension. Also, the estimated gradients of the objective function values in each dimension describes the annealing parameters.

Once the average change in the objective function is small relative to FunctionTolerance, the SA algorithm terminates or when it satisfies any other stopping conditions (MathWorks, 2017).

The result of the problem using SA Algorithm is as follows;

As shown in Table 10, the different options were used in SA. The high iteration number enables to obtain the best solution even if it causes the increase in the computational time.

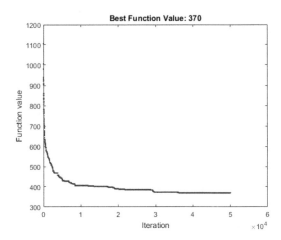

Figure 6. Simulation result of the SA method. Total processing time is around 7 minutes and best cost function value is 370.

Table 11. Outline of container distribution in a bay after using SA algorithm.

3	2	2	3	5	3	3	1	5	1	5	3	5	3	3
3	2	2	3	5	4	3	1	5	1	5	3	4	2	3
3	2	2	3	5	4	3	1	5	1	5	3	2	2	3
3	2	2	3	5	4	3	1	5	1	5	3	4	2	3
3	2	2	3	5	4	3	1	5	1	5	3	4	2	3
4	2	2	3	5	4	4	1	5	1	5	3	4	2	3
14	12	12	15	15	14	14	11	15	11	15	13	12	12	13
14	12	12	15	15	14	14	11	15	11	15	11	14	12	13
14	12	12	13	15	14	14	11	15	11	15	11	14	12	13
14	12	12	13	15	14	14	11	15	11	15	11	14	12	13

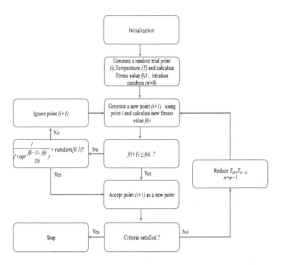

Figure 5. Simulated annealing algorithm flow chart.

Table 10. Results from the use of different options in SA.

Variable	Iteration	Reanneal interval	Best cost value	Time (Min)
150	10000	200	380	1.09
150	10000	300	380	1.08
150	20000	100	380	2.24
150	20000	300	384	2.49
150	30000	300	394	4.10
150	30000	400	398	3.89
150	40000	100	386	4.75
150	40000	500	400	5.17
150	50000	200	370	6.92
150	50000	400	376	5.66
150	100000	100	382	13.66
150	100000	500	370	14.29

It can be seen in Figure 6 that SA calculated the best cost function value as 370 in around 7 minutes. As shown in Table 11, the numbers in the slots represent destination port for that slot's container after SA method implementation.

Importantly, the best scenario to our problem data, the cost value would be 300 if whole container was travelled to the same destination. So, the results gained from three algorithms show that there is no big discrepancy among three algorithms however there is a difference among the computational time.

4 CONCLUSION

The aim of the present research was to find the optimal solution for container ship stowage plan. To solve this problem, three different algorithms were implemented Genetic, Simulated Annealing,

Hill Climbing. The results obtained from these methodologies indicated that there is no significant difference between them. However, a comparison of the three algorithm results reveals that each computational time are significantly different from each other. SA, GA and Hill Climbing covered their best solutions in almost 7, around 60 minutes and around 449 minutes respectively. Overall, these results indicate that SA algorithm is better and faster than the Hill Climbing and GA. Because their working principles are different even though the results are quite similar to each other. In future, the authors are planning to study the algorithms implemented in this work by adding more different type, size and weight of containers and vessel stability to obtain more realistic scenarios.

REFERENCES

Arriaga J, & Valenzuela-Rendón, M. (2012) Steepest ascent hill climbing for portfolio selection. In: Applications of evolutionary computation. Lecture notes in computer science, vol. 7248. Springer, Berlin, pp 145–154.

Avriel, M. & Penn, M.1993. Exact and approximate solutions of the container ship stowage problem[J]. Computers and and Industrial Engineering, 25(1–4): 271–274.

Cohen M.W. and Coelho V.N. and Dahan A. and Kaspi I. 2017. Container Vessel Stowage Planning System Using Genetic Algorithm. In: Squillero G. and Sim K. (eds) Applications of Evolutionary Computation. EvoApplications 2017. Lecture Notes in Computer Science, vol. 10199. Springer, Cham.

Kirkpatrick, S. and Gelatt, C.D. and Vecchi, M.P. Optimization by Simulated Annealing. Science, new series, vol.220, no.4598, 1983, pp. 671–680.

Liang, X. and Li, B. and Li, W. and Zhang, Y. and Yang, L. 2016. A Method Based on SNSO for Solving Slot Planning Problem of Container Vessel Bays. 9th International Conference, IDCS 2016, Wuhan, China, September 28–30, 2016, Proceedings.

Martins, P.T. and Lobo, V.J.A.S and Vairinhos, V. 2009. Container stowage problem solution for short sea shipping. Proceedings of 14th Congress of the Portuguese Association of Operational Analysis.

MathWorks, 2017. How simulated annealing works [online] Available at: http://uk.mathworks.com/help/gads/how-simulated-annealing-works.html [Accessed 22 Oct. 2017].

Nikolas, P. 2017. Optimization of Container Loading. Ph.D. thesis. Glasgow: The University of Strathclyde.

Tierney, K. 2015 Optimizing liner shipping fleet repositioning plans, operations research/Computer science interfaces series, vol. 57., p.7 Springer International Publishing, New York

UNCTAD, 2017. Review of Maritime Transport 2017. [online] Available at: http://unctad.org/en/PublicationsLibrary/rmt2017_en.pdf [Accessed 10 Oct. 2017]

Weise, T. 2009. Global Optimization Algorithms-Theory and Application. [online] Available at: http://www.it-weise.de/projects/book.pdf [Accessed 10 Dec. 2017]

A concept study for a natural gas hydrate propulsion ship with a fresh water supply function

H.J. Kang
KRISO (Korea Research Institute for Ships and Ocean Engineering), Daejeon, Korea

ABSTRACT: Natural Gas Hydrate (NGH) is noted as an alternative for lowering the cost of establishing a natural gas supply chain. However, despite many studies, the vessels carrying NGH still remain at the concept level because requirements for exclusive ships, piers and unloading facilities, including sintering, threaten the economic feasibility. From another perspective, water shortages are becoming a larger issue around the world. At the same time, many of these water shortage countries operate large ports. In this regard, this study focused on fresh water, which is 70% of NGH after the regasification process. Though NGH is disadvantageous in terms of energy storage efficiency compared to Liquefied Natural Gas (LNG), it can be used not only for fuel but also for ship ballast and fresh water supply. In this study, an NGH propulsion ship equipped with a fresh water supply function was conceptually designed via a simplified systems engineering process using a business model as a system analysis and control tool.

1 INTRODUCTION

Natural Gas Hydrate (NGH) can store 170 times the volume of methane gas in ice, and it forms at minus 20 degrees Celsius under atmospheric pressure. For this reason, NGH is noted as an alternative for lowering the cost of establishing a natural gas supply chain. However, despite many studies, the vessels carrying NGH are still not practical. The sintering effect during long-term marine transport delayed the unloading time at the quay and reduced the ship's availability. In addition, requirements for exclusive ships, piers and unloading facilities threaten the economic feasibility of the NGH supply chain. From another perspective, water shortages are becoming a larger issue around the world. In many countries, large desalination plants are built to solve water shortages. At the same time, many of these water shortage countries operate large ports. In this regard, this study focused on fresh water, which is 70% of the NGH after the regasification process. Though NGH is disadvantageous in terms of energy storage efficiency compared to LNG (liquefied natural gas), it can be used as fuel if it has a freshwater supply function for countries lacking water. In addition, NGH can be used not only for fuel but also for ballast. The suggested concept ship used NGH for propulsion, ballast and fresh water supply. For the suggested concept, diagrams for the fuel supply and regasification process have been studied. The technical possibility in terms of implementation has also been adopted on an abstract level.

1.1 NGH-carrying vessels

The exclusive NGH carrier was proposed by several research groups. A bulk natural gas hydrate pellet (NGHP) carrier has been studied and suggested to the IMO (International Maritime Organization) (IMO 2007; Oya et al. 2011; IMO 2010; IMO 2014). For massive marine transportation of NGH, the bulk NGHP carrier seems plausible. However, as the piled NGHPs huddle together into a large block of ice during the days of marine transportation by the sintering effect, delayed time for crushing and melting the NGH for unloading and regasification at the pier threaten the economic feasibility of the bulk carrier NGH supply chain. For economic feasibility compared to the bulk NGHP carrier, a tank container NGHP carrier has been proposed (Kang et al., 2016), as shown in Figure 1.

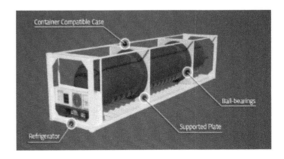

Figure 1. Concept image of the NGH tank container (Kang et al., 2016).

As the NGH tank container uses existing containers, piers, equipment and facilities, the supply chain establishment cost can be reduced. However, until now, NGH has no UN code for marine transportation. To realize the suggested concept for NGH marine transportation, further studies for convincing technical and economic solutions should be adopted including compliance with relevant regulations. Figure 2 shows an example of capital cost comparison for supply chain establishment among small-sized LNG, the NGHP carrier, the NGHP revised carrier (considering the sintering effect) and the NGH tank container in certain conditions.

1.2 Desalination plants

The global desalination market is projected to grow to $23.5 billion in 2018 (www.koreaexim.go.kr). According to the International Desalination Association, in June 2015, worldwide, the number of operating desalination plants is 18,426, producing 86.8 million cubic meters of water per day for 300 million people (Henthorne 2012). Generally, there are two types of desalination processes. One is a thermal process, and the other is a membrane process. Multistage flash (MSF) represents the thermal process that has multistep distillation methods including heating water and condensing vapor by dropping pressure. Though the thermal process has been widely used for more than 60 years, the membrane process can serve as a cost effective fresh water supply (Cotruvo 2016).

As shown in Figure 4, the cost for operating desalination plants has gradually been reduced. However, the construction costs of new desalination plants are still a huge burden.

1.3 Design motivation

As 70% of NGH can be made up of fresh water, having proper use of the fresh water after regasi-

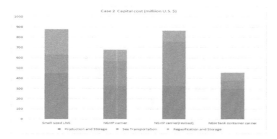

Figure 2. Example of capital cost comparison among the small-sized LNG, the NGHP carrier, the NGHP revised carrier (considering the sintering effect) and the NGH tank container under certain operating conditions. (Kang et al., 2016).

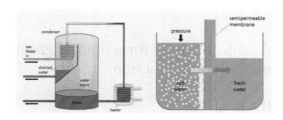

Figure 3. Examples of desalination processes (www.kictep.re.kr).

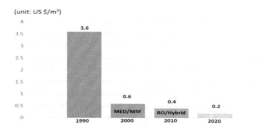

Figure 4. Cost reduction trends for desalination plant operation due to conversion of desalination technology, MED (multi effect distillation), MSF (multistage flash), RO (reverse osmosis) and hybrid (www.globalwaterintel.com).

fication at the pier is very important for establishing an NGH supply chain. In the case of the bulk NGHP carrier type, for the unloading and regasification process, a massive glycol water and purifying system is required. However, in the case of the NGH tank container, fresh water can be obtained after the regasification process without any additional purifying process (Kang et al. 2016).

If massive amounts of fresh water can be obtained through NGH transportation, the cost for desalination can be reduced for certain piers and countries. From a viewpoint of the amount of fresh water after the NGH regasification process, an NGH fueled ship can be considered. Since the storage efficiency of NGH is significantly lower than that of LNG, NGH is not considered as a source of fuel for ships. Since the volume of NGH is more than 3.5 times higher than that of LNG, it is difficult for ship designers to locate the fuel tanks in existing design specifications. In this concept study, a fresh water supply function-equipped NGH propulsion ship has been proposed.

2 DESIGN PROCESS

2.1 Design process

The systems engineering basis design process is very common these days for new system devel-

Table 1. Business model definition as a system analysis and control tool.

Task	Supply fresh water after regasification of NGH with a lower price compared to those of existing desalination plants
Value	Solution to cost effective fresh water supply
Revenue	Economic feasibility of NGH supply chain on the basis of fresh water supply
Infra	Associated laws, standards and regulations, merchant ships, other available technologies, facilities and equipment
Channel	Existing piers, ships
Stakeholder	Shipping lines, energy companies, shipbuilders, equipment manufacturers, and international organizations for cooperation and standardization

opment. In the systems engineering basis design process, not only technical issues but also other issues including economic feasibility are examined through the design process via system analysis and control tools.

In this study, to ensure the consistent identification and management of various factors for the system design, a simplified systems engineering process has been adopted. For a system analysis and control tool, the business model of Kang (2014) has been considered.

3 CONCEPT DESIGN

3.1 *Requirements analysis*

For the fresh water supply function-equipped NGH propulsion ship, first of all, the low energy storage efficiency of NGH has to be considered. In the case of the 8600-TEU container carrier, approximately 252.90 tons/day of fuel oil at NCR is required for a speed of 25.6 knots (MAN 2014, http://maritimeconnector.com).

For the same condition of operation, 885 tons/day of NGH are required to fuel the ship. To store the fuel for a 7-day voyage, at least 6,195 tons of NGH that contains 25% of the ballast water is required, since the 8600-TEU container carrier uses approximately 24,000 tons of ballast water. From the viewpoint of ship design, for this reason, NGH is not a good source for fuel. Thus, if a ship designer wanted to use NGH as a source of fuel, the fuel tank volume has to be secured without any loss of cargo capacity and major change in the ship specification (R1). For the NGH fueled ship in this paper, gasified natural gas can be gathered from the NGH fuel tanks. During marine transportation, a certain amount of NGH is gasified following the self-preservation effect (Shirota et al. 2005). Boiled off gas such as natural gas from each NGH tank has to be treated during ship operation. As a cargo, in this paper, the NGH tank container is considered since the bulk carrier-type NGHP carriers need an additional massive purifying system (Kang et al. 2014). If the NGH propulsion ship has NGH as cargo, the BOG of each NGH tank also should be treated properly (R2). When the NGH-fueled ship docks at the pier, fresh water after regasification in the fuel tanks should be gathered and unloaded for commercial use (R3). At this time, there should also be considerations for pressure in the NGH fuel tanks that may be due to over-compression during the re-gassing process. Treatment for impurities in the NGH tanks should also be considered. For realizing the fresh water supply function-equipped NGH propulsion ship, new laws, rules and regulations should be minimized since this realization takes several years (R4).

The design for a fresh water supply function-equipped NGH propulsion ship should have to test its technical and economic feasibility via system analysis and control tools. From the initial requirements for the fresh water supply function-equipped NGH propulsion ship, the total amount of fresh water production acts as a measure of effectiveness (MOE) and a key performance parameter (KPP). Then, the fresh water production cost can be a measure of performance (MOP).

As R1 is related to the generation of design alternatives, the ship building cost increases compared to that of the same kind of existing ship, and this is set to technical performance measure 1 (TPM1).

For R2, gasified natural gas from the NGH fuel tanks has to be treated with the boiled off gas (BOG) of the NGH tank containers. For this reason, TPM2 is set to the gasified natural gas treatment ability.

In the case of R3, the designed ship should serve its fresh water after the regasification of NGH at the pier with economic feasibility. Therefore, the

Table 2. Initial requirements for the alternative design.

No.	Requirement
R1	Secure fuel tank volume without major changes in ship specifications
R2	Treatment of gasified NGH during marine transportation
R3	Recovery of fresh water after NGH regasification at pier
R4	Minimization of need for new laws, rules and regulations

Table 3. TPMs for the system design.

Type		Item	Threshold	Indicator
MOE/KPP		Total fresh water production	Cost effectiveness compared to desalination plants	Production relative to the desalination plants
MOP		Fresh water production cost	0.2 US \$/m^3	CAPEX, OPEX relative to the desalination plants
TPM	1	Ship building cost increase	Same kind of existing ship	CAPEX, OPEX
	2	Gasified natural gas treatment ability	Gasified gas treatment for similar LNG engine	Design level of internal pressure for storage tank
	3	Amount of fresh water gain	Approximately 70% of NGH used as fuel	Required no. of new equipment
	4	Existing laws, rules and regulation compatibility	Fulfilled	Regulatory satisfaction

Table 4. Functions for the requirements.

No.	Functions
F1	NGH fuel supply function
F2	Boiled off gas treatment function
F3	Fresh water recovery function

amount of fresh water gain at the pier is set to TPM3. Finally, for R4, existing laws, rules and regulation compatibility are set to TPM4 to minimize the unnecessary time and cost from the new rules and regulations for realization of the design concept.

3.2 Functional analysis and allocation

To apply each requirement for the system design considering the TPMs, the business model of system analysis and the control tools, each requirement has to be functionalized one by one.

For R1, the NGH fuel supply function has to be achieved without major changes in ship specifications and building cost. The boiled off gas treatment function is required for R2, and fresh water recovery function is required for R3.

In the case of R4, existing rules and regulation compatibility is not a functional requirement but an entity of the business model of the system analysis and control tools. As shown in Table 4, each requirement of Table 2 can be set to the required functions one by one.

3.3 Synthesis

To generate a design alternative that fulfills the requirements and function list within the boundaries of the business model, the available technologies have been researched.

For F1, NGH fuel tanks should have high pressure compressors the same as other gas-fueled

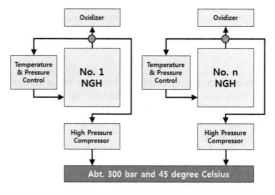

Figure 5. Concept diagram for NGH fuel tanks (derived from MAN 2014 and redesigned).

engine systems for natural gas supply to the engine as shown in Figure 5.

From the Figure 5, LNG or heavy fuel oil can be used with NGH if necessary as artificial ballasting is rarely required in full loading condition of a ship. In other words, in empty cargo condition, NGH can be used as for both of fuel source and ballasting. But in case of full loading condition which not require artificial ballasting, LNG or heavy fuel oil can be used instead of NGH for fuel. As stated in the introduction section, the required NGH fuel tank volume is 3.5 times larger than that of the LNG fuel tank, and the arrangement of the fuel tanks can be an issue for ship design. For the 8,600-TEU container carrier, the weight of NGH for 7 days of operation is assumed as 25% of the ballast water, so some part of the ballast tank can be replaced with the NGH fuel tanks because even after regasification, the weight of the NGH remains close to 70% in the form of fresh water. In the full loading condition, the 8,600-TEU container carrier can be operated with a comparatively small amount of ballast water. For this reason, the

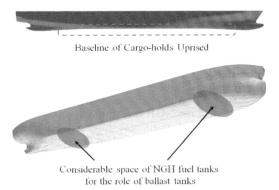

Figure 6. Example of the ballast water-free hull form utilizing the NGH fuel tank for heel and trim control of certain loading conditions (images are derived from Kang et al. 2017 and modified).

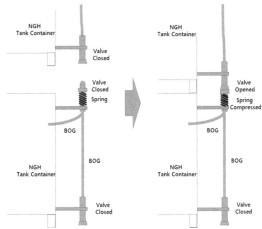

Figure 7. Valve connection concept of tank container for BOG recovery.

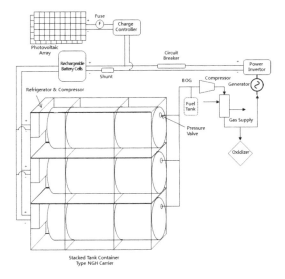

Figure 8. Concept diagram for the BOG treatment system.

NGH fuel tanks should be carefully arranged to not affect the ship's appropriate trim and heel condition. If required, the operation speed can also be adjusted as considered for the NGH fuel tanks required space. Recently, minimal ballast water ships (MIBS), non-ballast water ships (NOBS) and other ballast water-free concepts have been suggested (Shingo, 2015 and Van et al. 2014 and Parsons et al. 2011). In this study, the hull from which the baseline of the cargo holds are differentiated from the bow and stern (Kang et al. 2017) has been considered to apply the NGH fuel tanks for trim and heel control.

Figure 6 shows the design concept of the hull form for ballast water-free (or minimum use) operation. In this case, the remaining NGH can be continuously utilized as fuel or fixed ballast through adjustment of the temperature and pressure conditions. As the hull form of Figure 6 is for the Capesize bulk carrier, a new hull form for the container ship will be required at the time of design concept implementation.

For F2, the boiled off gas during marine transportation of NGH tank containers has to be gathered and treated.

As shown in Figure 7, the boiled off gas from stacked NGH tank containers can be gathered via attached pipes and valves that are automatically linkable when stacking the tank containers. Then, gathered BOG can be transported to the generator of the ship. Otherwise, to minimize the cost for the BOG pipeline, the BOG can be used for the operation compressor and freezer to maintain the self-preservation effect of the stacked NGH tank containers.

Figure 8 shows the BOG treatment system diagram to maintain the temperature and pressure on the tank container in the stack units. Utilization of the photovoltaic array is optional as it depends on the insulation performance of the NGT tank container. Without the BOG treatment system for the NGH tank containers, the thickness of the pressure vessel will not be allowable for economic feasibility (Kang et al. 2015).

In the case of F3, fresh water after the regasification process should be gathered and transferred to on-land water tanks. As shown in Figure 9, the structure of the fresh water gathering system is simple. However, to minimize the sloshing effect during ship operation, the geometry and anti-sloshing

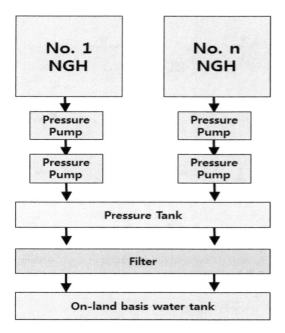

Figure 9. Concept image for the fresh water gathering system.

Table 5. Synthesis result.

No.	Chosen design alternatives
S1	Ballast functioning NGH fuel tanks
S2	Boiled off gas treatment system for stacked NGH tank containers
S3	Fresh water recovery system

equipment and facilities should be carefully considered. As the fresh water of the NGH tank container is gathered after unloading, Figure 7 does not consider fresh water gathering from the NGH tank containers. The diagram in Figure 7 considers pressure control and impurity treatment in the NGH tanks.

Table 5 shows the synthesis of design alternatives to accommodate each requirement and function. The suggested result shows only a few possible examples for the system design.

4 PROVISIONAL EVALUATION

To verify and validate the conceptual design, the technical and economic feasibility should be evaluated within the boundaries of the business model, which is the system analysis and control tool of the design process.

4.1 Technical considerations

Technically, the suggested design alternatives use existing proven technologies. However, as there are no commercialized NOBS, MIBS and other ballast water-free (or minimum use) ship concepts, immediate commercialization of the suggested design alternative S1 will not be possible. For S1, the operability of the NGH fuel supply system should be confirmed before its real ship application. As the suggested fuel supply system diagram is very similar to those of LNG fuel supply systems, the regasification process of NGH and a high compression process centered evaluation should be done. For S2, as there are no UN cargo codes for marine transportation of NGH, related standards, rules and regulations should be examined for the commercial use of the design concept. For this, related standards such as ISO 6447 (1983) should be considered in the process of prototype building and the test process. For S3, the fresh water recovery system has no risk compared to the liquefied natural gas bunkering system. However, after the regasification of NGH in the fuel tanks, the inner tank pressure and flammable gas should be carefully treated. As vapor destruction units generally adopted to gas supply systems, detonation arrest safety equipment can be considered in the detailed design process (Kang et al. 2017).

4.2 Cost benefit

From a viewpoint of the capital cost to establish a fresh water supply chain via a desalination plant, the suggested concept for the fresh water supply function equipped NGH propulsion ship will shift the cost burden from the desalination plant management companies (or countries) to the shipping company. For shipping companies, fresh water has at least 3.0 million dollars of value per 1 mega gallon per day (MGD) with the consideration of existing desalination plants (Watereuse Association 2012). Moreover, the cost for a ballast water treatment system (BWTS) will be saved by combining the suggested concept with a ballast water-free (or minimum use) vessel such as NOBS and MIBS. To evaluate the economic feasibility, the obtainable amount of fresh water from the suggested design concept considering the port traffic data should be considered. If the suggested concept is competitive compared to desalination plants in capital expenditure (CAPEX) and operational expenditure (OPEX), the fresh water supply function-equipped NGH propulsion ship will be marketed. By roughly estimating the amount of fresh water that can be obtained using NGH as fuel, an 8,600-TEU container carrier uses 6,195 tons of NGH for 7 days of voyage with 25.6 knots, and then 4,340 tons of fresh water can be recovered for a shipment. As the throughput of Dubai (Jebel Ali) container port in 2014 was 11,124,000 TEU (Elbayoumi et al. 2016), 2,806,869 tons (approximately 2.0 million gallons per day (MGD)) of fresh water can theoretically be obtained annually on the assumption that an 8,600-TEU container ship can be put down and loaded again at the pier and that all vessels sail for seven days on average. From the report of the Watereuse Association (2012), approximately 6.0 to 9.0 million US dollars can be saved to generate 2.0 MGD of fresh water with finished water quality. When enlarging the target ports to all the container ports of the Middle East, a throughput of 48,132,000 TEU can lead to approximately 12,144,934 tons of fresh water and save approximately 9.6 to 14.4 million US dollars in desalination cost.

5 CONCLUSIONS

In this study, design alternatives have been studied and suggested on an abstract level for a fresh water supply function-equipped NGH propulsion ship. First, various factors that should be considered in the design phase are reflected using the business model in the form of a system analysis and control tool. Then, requirements for the design have been identified. By functionalizing each requirement, three design alternatives for each function have been suggested in the boundaries of the business model. To defend the suggested design concept, a technical and economic evaluation process will be adopted in the next study.

ACKNOWLEDGEMENTS

This research was supported by a grant from the Endowment Project of "A basic research on next generation energy transportation technology based on safety performance improvement technology centered on modular natural gas hydrate transportation technology" funded by Korea Research Institute of Ships and Ocean engineering (PES9250).

REFERENCES

Cotruvo, J. (2016) "Desalination basics" Water technology solutions for industrial water management (https://www.watertechonline.com/desalination-basics-thermal-membrane/).
Dobryin VM, Yu P Korotavej, Plyuschev DV (1981) Gas hydrates- a possible energy resource, in R.F. Meyer and J.C. Oslon, Eds. Long term energy resources. Pitman publishing, Boston: 727–729.
Elbayoumi, C.O.F., & Dawood, C.A. (2016). Analysis of the Competition of Ports in the Middle East Container Ports Using HHI. Journal of Shipping and Ocean Engineering, 6, 339–347.
Henthorne, Lisa (June 2012). "The Current State of Desalination". International Desalination Association.
http://maritime-connector.com/ships_uploads/bremen_express-9343728-container_ship–1460.pdf.
ISO 6447 (1983) Rubber seals — Joint rings used for gas supply pipes and fittings — Specification for material
Kang, H.J., Choi, J., Ahn, H.S. and Kim, K.S., (2017) "Systems Engineering-based Feasibility Study for Development of Ballast Water-Free Vessels" In: Proceedings of IMAM 2017. Lisbon, Portugal.
Kang, H.J., Chun, T.B., Ahn, H., & Yim, G.T. (2017). Feasibility Study for Onboard Marine Debris Gathering and Recycling. Marine Technology Society Journal, 51(1), 32–39.
Kang, H.J., Yang, Y., Ki, M.S., Shin, M.S., Choi, J., Cha, J.H., & Lee, D. (2016). A concept study for cost effective NGH mid-stream supply chain establishing strategies. Ocean Engineering, 113, 162–173.
Kerr, S. 1994. Ballast water ports and shipping study. AGPS. No. 5.
MAN (2014) "ME-GI Dual Fuel MAN B&W Engines A Technical, Operational and Cost-effective Solution for Ships Fuelled by Gas", MAN Diesel & Turbo
MAN (2014) "Propulsion of 200,000–210,000 dwt Large Capesize Bulk Carrier", MAN Diesel & Turbo (http://marine.man.eu/docs/librariesprovider6/technical-papers/propulsion-of-200000–210000-dt-large-capesize-bulk-carrier.pdf?sfvrsn=14).
Parsons, M.G., & Kotinis, M. (2011) "Refinement of the Ballast-Free Ship Concept, Department of Naval Architecture and Marine Engineering", University of Michigan, 2600 Draper Road. Ann Arbor, MI, 48109–2145.
Shingo, S. (2014) "Preliminary Hull Form Planning for Non-Ballast Water Ship and Minimal Ballast Water Ship (1st Report)-Simplified Estimation Method of Propulsive performance." Journal of the Japan Society of Naval Architects and Ocean Engineers, Vol. 20.
Shirota, H., Hikida, K., Nakajima, Y., Ota, S., Iwsaki,T., (2005) "Self-preservation property of hydrate pellets in bulk in ship cargo holds during sea-borne transport of natural gas.In:Proceedingsofthe5thInternationalConference on Gas Hydrates.
Sloan ED (1998) Clathrate hydrates of natural gases. 2nd Ed, Marcel Dekker, New York: 705.
Van Kluijven, M. & Van der Valk, M. 2014. Ballastless sailing.
Watereuse association (2012) Seawater Desalination Costs (https://watereuse.org/wp-content/uploads/2015/10/WateReuse_Desal_Cost_White_Paper.pdf)
www.globalwaterintel.com.
www.kictep.re.kr.
www.koreaexim.go.kr.

Development and initial results of an autonomous sailing drone for oceanic research

U. Dhomé, C. Tretow, J. Kuttenkeuler & F. Wängelin
Centre for Naval Architecture, KTH Royal Institute of Technology, Stockholm, Sweden

J. Fraize & M. Fürth
Davidson Laboratory, Stevens Institute of Technology, New Jersey, USA

M. Razola
SSPA Sweden AB, Stockholm, Sweden

ABSTRACT: This paper describes the ongoing development of Maribot Vane, an autonomous sailing vessel at the Maritime Robotics Laboratory of KTH, the Royal Institute of Technology, Stockholm. There is an accelerating need for ocean sensing where autonomous vehicles can play a key role in assisting scientists with environmental monitoring and collecting oceanographic data. The purpose of Maribot Vane is to offer a sustainable alternative for these autonomous missions by using wind and an energy efficient self-steering mechanism. The rig is composed of a free-rotating wing fitted with a coupled control surface. A completely novel wind vane self-steering solution has been developed and is being evaluated. A key point in the development of the vessel is robustness, with a goal of being able to sail in open seas for long period of times. The paper discusses some key concepts, the development method and presents initial results of the new systems.

1 INTRODUCTION

There is a growing need for ocean-based sensing in remote areas not accessible to most researches. With the opening of arctic waters and the changing weather patterns all over the globe, reaching new destinations to carry out research observations has become vital. The realization of an energy independent autonomous surface vessel, to act as a mobile research tool, has been targeted by the Maritime Robotics Laboratory at KTH, and the Maribot Vane project is the result.

The opportunities for an autonomous robotic sailboat are extensive. Applications such as long-term position bound environmental monitoring, shallow water mapping, multi-agent missions for fish tracking (to mention a few) is where this type of vessel can complement conventional research vessels and equipment in data acquisition. For weather data collection, while it is within satellite capabilities to gather some surface weather data, being able to reach areas of shallow water where data is not possible to be gathered by satellite is crucial. This type of vessel would also be able to carry a payload of deployable sensors that could include data gathering aerial drones, or underwater instrumentation that could be deployed and regathered after any amount of time necessary for the mission objectives of the payload.

The goal of a marine research tool capable of carrying out oceanographic and environmental research is not completely new. Other sea drones have been created with similar aims, such as Saildrone (Saildrone, 2017), Sail buoy (Hole, 2016), C-Enduro (ASV Global, 2017), AutoNaut (Autonaut, 2017), or Wave Glider (Daniel, 2011). Some of these vessels have achieved their objectives well, reporting successful missions after months out at sea. The Vane project however stands apart in a few distinct respects. The Vane project approaches the problem with specific focuses that were not achieved by the previous attempts at a similar vessel, or simply out of scope.

The C-Enduro (ASV Global, 2017) developed by ASV Global for example, uses a mix of renewable energies (solar and wind) and fuel to drive its electric motors, increasing complexity somewhat but with the advantage of being able to propel even without wind. Amongst the other autonomous vessels, some get their propulsion form the environment such as the Wave Glider (Daniel, 2011) from Liquid Robotics or AutoNaut (Autonaut, 2017) from AutoNaut Ltd. With different techniques they both use waves to propel the vessel and thus have similar benefits and disadvantage as a sailing vessel: ability to propel without fuel, but with dependency on weather conditions. Their propulsion systems rely on underwater moving parts which potentially are sensitive to damage by any underwater hazard (seaweed, mammals but also plastic or trash). The Maribot Vane project aims at solving some of the described issues by sailing. With the Maribot Vane concept, the wind energy is used to drive the craft, meaning Maribot Vane too is weather dependent. However, the risk of air borne hazards is much more limited than the risk of underwater entanglement, so in comparison with the gliders, a sailing boat should be more robust.

The choice of pursuing a wind powered vessel is multifaceted. The goal of energy independency is directly achieved by this decision. Also, the removal of the drive system that relies on finite fuel, internal power transmission and mechanical propulsion is a direct simplification to avoid corresponding maintenance and limits. Compared to traditional manned research vessels with fuel and man-hours costing tenths of thousands of euros per day (NRC, 2009, MBARI, 2017), an unmanned vessel is potentially able to endure months at a time at sea at a fraction of this cost. This allows sensors and other devices deployed to stay on location longer and collect more data.

Saildrone is an established commercially viable Unmanned Sailing Vessel (USV) that has been operating regularly for the past few years. This vessel shows that there is demand for such a research tool, however Saildrone is still power reliant in a way suited for sunnier latitudes where solar power is easily generated by an onboard array. The Sail buoy project produced a very robust USV, but with somewhat limited sailing abilities compared to a fully sailing boat. Other autonomous sailing vessels have been developed for competitions such as the Microtransat but most of them are of a smaller scale and remain competition projects that will not be used for long-term missions. One ongoing project similar to Maribot Vane is the ASPire (Friebe, 2017). The principle for the boat propulsion is similar and it also contains a wind steering system but as far as the authors know the technical solutions on both the wind and the steering are very different. The Maribot Vane is intended as a gap bridging vessel that can achieve comparable performance to that of a higher energy consumption vessel with robust solutions that allow for an adaptable research tool able to expand upon what is currently possible.

In the first part of this article, an overview of the platform in its current development stage is given, describing the electronics system, the hardware systems and the rig. In the second part the novel features of the Maribot Vane are presented. The wing principle and the design methodology are explained together with the self-steering system. The third part presents the results obtained from test campaign done in the summer and autumn 2017. An evaluation of the wing's performance is presented and compared to the designed rig. The self-steering capabilities are then presented. Finally, based on the presented results, the future work planned to turn the actual prototype into a fully autonomous and energy efficient research platform is detailed.

Table 1. Sailing specific terms.

Term or sign	Definition
Apparent wind	The relative wind that the vessel experiences due to forward headway.
AWA	Apparent Wind Angle, the angle between the boat heading and the apparent wind. Measured from 0 to 180° on each side, positive when the wind comes from starboard.
TWA	True Wind Angle, the angle the wind is blowing at the sailing vessel. Measured as the AWA.
Point of sail	Boat's direction relative to the wind.
Close-hauled	A point of sail where the boat is as close as possible to the wind direction.
Downwind	A direction of the boat that is directed away from the wind (from 90° to 180° TWA).
Upwind	A direction of the boat that is directed towards the wind (from 0° to 90° TWA).
Polar Diagram	A visual plot of a sailing vessels speed at various true wind angles.
Puff	An area of increased wind sharply defined from the regular wind around it.
SOG	Speed Over Ground, the vessel's speed with respect to the earth's reference. Given by the GPS.

1.1 *Nomenclature*

Some specific sailing terms that are not common knowledge for those who do not sail are used throughout the paper. Table 1 summarizes some of these words necessary to qualify sailing performance.

2 SYSTEMS OVERVIEW

As shown in Figure 1, the Vane is based on a modified 2.4 mR hull with a custom built free rotating

Figure 1. Maribot Vane sailing upwind in Baggensfjärden, Stockholm.

wing sail arrangement. The entire boat weighs about 250 kg with a total length of 4.16 m. The philosophy behind the choice of hull and rig is based on a compromise of handling, robustness, modifiability and cost along with safety concerns. An innovative wind vane self-steering mechanism was developed (Wängelin, unpubl. 2017) and is described in more detail below. Due to the relatively small size of the vessel a rapid prototyping approach can be used to evaluate multiple technical solutions in a short amount of time. Most new parts of the boat are first 3D-printed in plastic, tested in real conditions and newer versions are iterated.

2.1 Electronics systems

An important part in an autonomous vessel such as Maribot Vane is the electronics systems. The electronics hardware are off-the-shelfs products. This enables quick and cost-effective modifications.

2.1.1 Sensors

In this first development phase of the project, the boat was only fitted with a basic set of sensors necessary for navigation such as GPS for position and velocity relative to earth and an Attitude and Heading Reference System (AHRS). A 3-Space AHRS from Yost Labs (Yost Labs, 2017), is mounted in the hull to measure the attitude of the boat: roll, pitch and yaw, along with rotational rates and linear accelerations in all three directions. A problem encountered in other sailing robots projects (Sauzé & Neal, 2006) was that the compass needed to be horizontal at all time, which is not a concern with an AHRS using on-chip Kalman filtering and tilt compensation. Weather data is collected using a rig-top mounted ultrasonic anemometer CV7-V (LCJ Capteurs, 2017) from LCJ Capteurs which provides appar-

Figure 2. The wing while sailing, with some of the sensors and actuators visible.

ent wind speed, wind direction relative to the rig along with air temperature. The top compartment on the rig also holds a 433 MHz RF antenna for real time telemetry communication at distances up to a few hundred meters with e.g. a tender boat during close vicinity tests. A second AHRS unit is placed in the mast to record the mast attitudes, rates and accelerations needed to e.g. calculate true wind direction. A similar configuration can be found in (Elkaim, 2001).

2.1.2 Actuators

The boat mainly contains two actuators, a high torque RC-servo (Hitec HS-1100 WP) for the rudder control, and a linear actuator (Actuonix L16P) for the flap control. The linear actuator was chosen for its capability to hold position even when powered off. A radio-controlled switch enables a safety override functionality where standard RC-control can override the on-board control loops. This functionality also proved to be advantageous in harbor maneuvering. Further, for ease of harbor maneuvering, an electric thruster (T200 from BlueRobotics) was mounted on the hull. In the second iteration of the platform, when the self-steering system was installed, the servo controlling the rudder was replaced by a stepper motor and its driver.

2.1.3 Microcontroller

Several microcontrollers (MCU) are used for sensing, control and communication, see Figure 5. Two Arduino DUE are used: one in the hull that serves as the main controller of the boat, and one in the mast, that is used to control the sensors and actuators in the wing. Connection between the rig and the hull is carried through a slip ring placed at the bottom of the mast to let the mast rotate freely without damaging the cabling.

2.1.4 Software

The main Arduino board holds a program responsible for all the boat functions: sensor acquisition, steering and flap control, data logging and communication with the chase boat. All sensors are updated at a frequency of 2 Hz. Transmission of data and reception of commands to/from the chaser boat are also done every half second.

Figure 3. Internal aluminum frame to support mast, electronics and batteries.

Figure 4. Wing under construction with internal tubing and other fasteners visible.

2.2 Hardware systems

2.2.1 Internal structure

With the free rotating rig comes the need for different hardware compared to the traditionally rigged sailboat. An internal aluminum frame was designed for mounting the mast and to stiffen the deck to ensure a good seal with the hatch, see Figure 3.

2.2.2 Rig construction

The rig is a sandwich structure with a Divinycell® foam core, laminated with fiberglass and epoxy resin with internal piping for electronics and cabling as shown in Figure 4. The bending moments and shear forces are transferred from the wing to the internal frame structure through a 70 mm carbon fiber tube, which acts as the wing's rotation axis.

3 DESIGN

3.1 Free rotating wing

Unlike conventional sailboats, Maribot Vane is fitted with a rigid symmetric free-rotating rig. The rationale behind this choice is robustness in combination with care-free control in the sense that once the rig is adjusted correctly to desirable lift coefficient (CL) by adjustment of the flap angle, the rig becomes self-trimming (neglecting friction and inertia) in the sense that CL is maintained regardless of wind speed and direction. Thus, the wing acts as a wind vane.

The flap is the mechanism setting the desired CL of the main wing by producing a rotational moment around the common axis of rotation, i.e. the "mast", placed at the chord-wise center of pressure of the main wing. Hence, setting the flap deflection to a non-zero value generates aerodynamic forces at a distance from the mast, creating the needed torque to rotate the whole rig, putting the main wing at the desired angle of attack. Figure 6 shows this principle on the boat with an aerial view from above.

The arrangement with main wing and flap typically causes a non-favorable mass distribution where center of mass is aft from the rotational axis. At nonzero heeling angles this leads to unwanted rig rotation due to the influence of gravity. The wing is balanced by adding mass on a rod extending in front of the rotation axis. However, this comes at the cost of higher center of gravity and larger moment of inertia, which in turn could increase the risk for roll-yaw-sway coupled dynamic instability phenomena. Such excitation has been clearly observed in the Vane experiments, as described further in this paper.

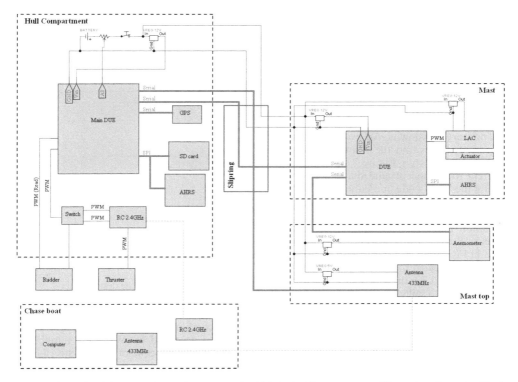

Figure 5. Schematic of the boats electronic systems. The thick dashed lines represent different locations, the thin dotted lines represent wireless transmission, the thick solid lines represent bus communication and the thin lines represent wires.

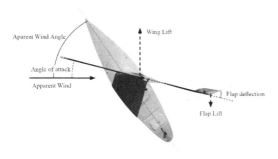

Figure 6. Aerial view from above the vessel with forces on the wing and flap drawn.

As seen in Figure 6, the flap lift, used to rotate the rig, actually acts in the opposite direction of the overall and desired rig lift. The flap also adds weight and generates resistance. However, apart from the role of generating rig yaw-rotation, the flap has the secondary role of acting as the rig tailplane, i.e. to generate rig rotational restoring moment whenever the apparent wind direction changes. In this role, the flap is performing better with increased area. Hence, a good balance between wing-to-flap areas as well as the flap location behind the wing is sought for. A too small flap at a to close distance to the rotational axis will result in a less responsive and stable configuration and a too large flap too far aft will lead to unnecessary drag, fragile construction and a heavy rig.

A key advantage of the free-rotating wing sail is that it operates using a single control, i.e. the flap, which makes for a robust system with few failure modes compared to a traditional sailboat rig. Another advantage of using a free rotating wing is that unlike on conventional sailing boats, there is no yaw moment transferred to the hull, which leads to better course stability.

3.2 Rig design

The analysis and design of the free-rotating wing is performed by combining two in-house developed analysis tools, a potential flow Vortex Lattice Method code (VLM), based on the work of Helmstad & Larsson (2013), and a Velocity Prediction Program (VPP) which enables an analysis of the sailing vessel as a complete mechanical system. The VLM model is used to compute the aerodynamic loads on the wing, while the VPP model comprises the aerodynamic model of the rig and a numerical representation of the hull to compute the hydrodynamic forces. In the VLM model, the rig lift and drag is found by solving the moment equilibrium for a given flap deflection.

Once the aerodynamics of the rig is calculated, the VPP equations (surge, sway and roll) for the entire boat can be solved with speed, heel and leeway as primary unknowns. An example of panel discretization, forces and the results of the VPP calculations are shown in Figure 7.

Different wing concepts and shapes were evaluated using the VLM code in a parametric study to identify the effects on lift, drag and stability of a few key parameters such as the rake, aspect ratio, flap position and size. The sailing performance was then evaluated using the VPP. The process is depicted in Figure 8.

It was early decided that robustness, rather than boat speed, would be the key factor for success. Hence, the ability to sail in strong winds was favored and a conservative rig configuration with a span of 3.5 meter was designed as a good compromise between sailing performance, robustness, low weight and center of gravity as well as being easy to handle.

3.3 Self-steering mechanism

3.3.1 Operating principle

A novel feature introduced in the project is a self-steering system that enables the boat to sail on a constant apparent wind angle (AWA) using only mechanical control by the wind. The main idea behind this vane steering system is to achieve a zero-electricity consumption when used. Furthermore, the system is also intended to harvest energy under

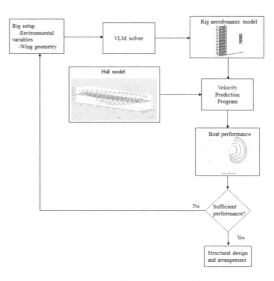

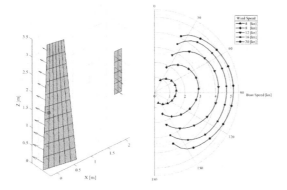

Figure 7. Left: aerodynamic model of the rig with panel discretization, panel forces and center of pressure. Right: corresponding boat polar plot for wind speeds from 4 to 20 knots.

Figure 8. Flow chart of the design method.

certain conditions. A conceptual design was developed and described in (Wängelin, unpubl. 2017). The operating principle of the self-steering system is similar to the one of a wind-rudder vane steering mechanism that is well known by long hauled offshore sailors. It is described in Figure 9a–d. According to Letcher (1976) the system can be described as a single-axis vane with the primary rudder as control.

Unlike conventional wind rudders found on the market, there is no additional wind-vane added to the boat to track the apparent wind. Instead, the main wing, given its property of self-adjustment to a given AWA, is mechanically coupled to the rudder. However, the issue of adjusting the vane control feedback to enable self-steering at varying apparent wind angles requires a somewhat intricate solution which is not described in detail here.

In Figure 9, the systems response to a wind shift is exemplified, but the same correction appears if the variation in apparent wind angle comes from a change in heading of the boat due to a wave or some other external disturbance.

Some of the key features of the self-steering mechanism are:

The system is able to transfer mast rotation to the rudder to keep the AWA and the transmission is done so that the coupling ratio (feedback gain) between the mast and the rudder angle can be modified on ground.

The system can be controlled manually, with a "clutch" system that engages or disengages the wing coupling. Manual steering is assured by a motor that

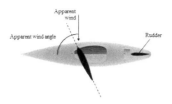

a) The boat sails on a given course, at a specific apparent wind angle.

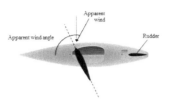

b) The wind direction changes, leading to a different apparent wind angle than previously. The previous angle is shown by the dotted line.

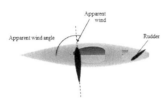

c) The change in apparent wind angle makes the wing rotate, thus turning the rudder.

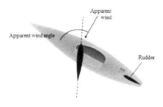

d) The boat rotates until the same apparent wind angle is reached, leading to a zero-rudder deflection, but a new course.

Figure 9. Illustration of the operating principle of the steering system. For simplicity the flap is not pictured.

doesn't prevent mast rotation when not used. As for the flap on the wing, the coupling mechanism of the clutch does not consume energy when engaged or disengaged but only when changing state.

The wing must be able to rotate freely through multiple revolutions while the rudder angle is limited on both sides, therefore a safety feature enables the wing to rotate if the rudder reaches its limits. In order to increase robustness, this safety relies on purely mechanical principle but should in the future also activate the electronic clutch to release the wing completely.

A prototype of the self-steering was built and tested in the fall of 2017 with very promising results. Based on the results, a more robust and reliable solution is under development at the time of writing this paper and should be built during 2018.

4 PERFORMANCE EVALUATION

4.1 *Test campaign*

After the final development of the prototype during the summer 2017, a test campaign was carried out in the fall with a focus on evaluation of the boat performance and the limits of usability in different conditions. The tests were performed in the relatively protected area Baggensfjärden in the Stockholm archipelago, shown in Figure 10.

4.2 *Boat experimental polar*

During the sailing experiments, the boat was kept on a constant heading using a feedback controller based on AHRS-compass, rather than at constant AWA. By steering on a constant compass course, the wind angle naturally oscillates. Therefore, sample lengths with reasonable stable apparent wind angles where limited between 2–10 minutes. An example is shown in Figure 11.

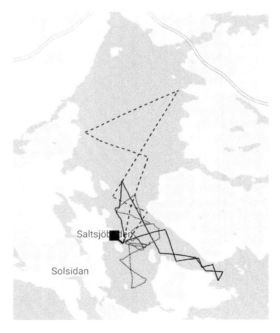

Figure 10. Map of Baggensfjärden and Ingaröfjärden where the tests took place in 2017. The three lines represent the trajectories during three testing days. Depending on the weather and wind direction different areas were chosen.

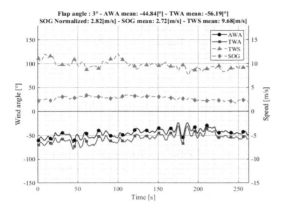

Figure 11. Typical test run used to draw the polar diagram. The apparent and true wind angles are plotted as plain lines on the left axis. The true wind speed and speed over ground are plotted as dashed lines on the right axis. The conditions for the run to be processed was that the wind angle and speed should remain fairly constant.

Figure 12 presents a polar diagram for the Maribot Vane based on experimental data and on VPP calculations. The measured boat speed is taken as the average SOG for each run and is here plotted for different average true wind angles (TWA). The wind speed varied between 8–12 knots during testing and was relatively unstable even for shorter periods of time. Therefore, the boat speed is normalized by dividing the average speed over ground by the average wind speed, and multiplying by 10 knots. This normalization also enables comparison of different wind angles.

Although the polar presented in Figure 12 is based on scattered data derived in unstable conditions it gives qualitative confirmation that the Vane behaves as planned and predicted in the design phase.

4.3 Self-steering

In order to assess the performance of the self-steering system, several tests were performed. First, the ability to sail on a straight course to the wind was visually assessed. The boat was then sailed in very unstable wind in order to evaluate how well the boat would follow a constant apparent wind angle using the self-steering system. Finally, the ability to resume a course after an external perturbation was tested by pushing the boat off its course.

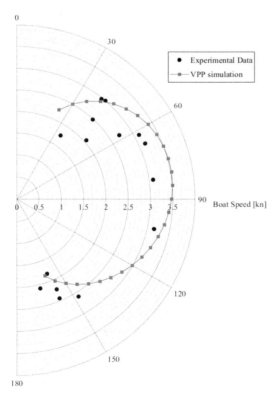

Figure 12. Polar diagram of Maribot Vane. The dots represent experimental values obtained during the test campaign in the fall of 2017. The squares are values obtained by running the Velocity Prediction Program on the actual wing design.

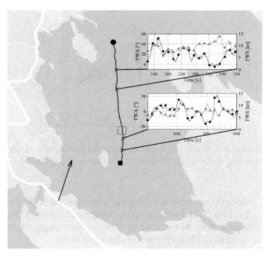

Figure 13. Trajectory of the boat in an upwind tack using mechanical vane feedback control. The rectangle shows the part of the path that is studied more in depth and shown in Figure 15. Temporal evolution of true wind angle (round marks, read on left axis) and true wind speed (triangles, read on right axis) are shown on two sections of the path.

4.3.1 Response amplitude

The ratio between the amplitude of the mast and the rudder rotations was set around 0.5, i.e. a mast rotation of 10° leads to a rudder change of 5°. This was set empirically but can be modified and the precise positioning is part of further development of the system. A long tack close-hauled was sailed in very puffy conditions in order to assess the response amplitude. Figure 13 shows the trajectory during this run, where the wind shifts can be seen in the path curvature. It starts at the round mark and ends at the square on the bottom of the picture. The arrow on the lower left corner shows the average wind direction.

The first observation that can be made from this test is the ability of the vessel to follow the wind closely. This is shown for example in Figure 14. After approximately 7 seconds, the apparent wind angle increases, showing a change in wind direction. This leads to a rotation of the mast; here the mast yaw (heading of the mast) increases which creates a change in the rudder angle and boat heading. The feedback response is slightly too large, leading first to a decrease of the apparent wind angle (between 13 and 22 seconds), thus a rotation of the rudder to the other side to compensate, but finally the boat is brought back to the same apparent wind angle as before the wind shift.

The too large response was observed several times during the run, sometimes bringing the boat completely into the wind. An example of this behavior is shown in Figure 15. Even in that case, the vessel would after some time revert to its previous course and start sailing again. A conventional

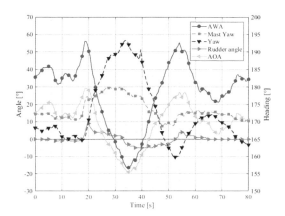

Figure 15. Example of the boat response to wind shift with too high feedback gain. The dashed lines represent the boat and mast headings and should be read on the right axis. The plain lines represent the apparent wind angle, the angle of attack on the wing and the rudder angle and should be read on the left axis. The data correspond to the rectangle in Figure 13. The response here is too large and brings the boat into the wind.

sail would first collapse when headed into the wind, eventually inflating on the wrong side, and the boat would likely not resume sailing.

Figure 15 shows a similar oscillatory behavior as Figure 14, but a significant difference is that here the angle of attack and apparent wind angle change sign, meaning that the vessel is turned into the wind, almost all the way into the other close-hauled tack. If the boat turns too far into the wind eventually the hull rotates more than the wing, which inverts the rudder angle, bringing the vessel back on the right course. This needs to be tested in stronger winds, but the steering system as it is should be able to avoid undesired tacks when going upwind.

On the upwind course described above, Maribot Vane, due to a too large rudder response amplitude, sometimes sailed at a too low AWA. The current iteration of the system therefore leads to less than optimal speed by sailing too close to the wind. An idea for further development is to have an adaptable ratio between the wing and rudder rotations in order to prevent this behavior when sailing upwind, while maintaining the good course keeping performance observed on other points of sail.

4.3.2 Off-course behavior

Another important characteristic of the vane feedback system is its ability to regain the desired course after a course disturbance, e.g. from waves etc. This can be regarded as course stability under the condition that true wind direction and speed are constant. This was tested by stressing the boat

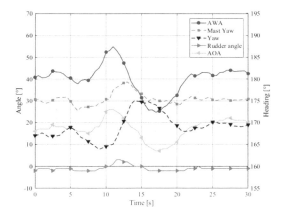

Figure 14. Example of the boat response to a change in apparent wind direction. The dashed lines represent the boat and mast headings and should be read on the right axis. The plain lines represent the apparent wind angle, the angle of attack on the wing and the rudder angle and should be read on the left axis.

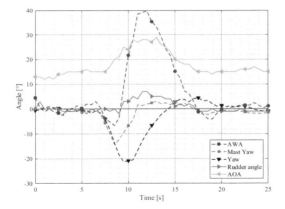

Figure 16. Perturbation "push test" downwind. The dashed lines represent variation to the average value before being physically pushed off course, and the plain lines are the angles as read by the sensors. In this case the boat was pushed away from the wind, as shown by the increase in apparent wind angle.

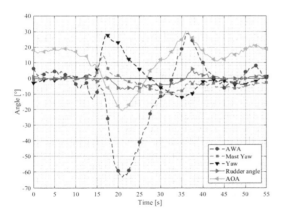

Figure 17. "Push-test" upwind. The dashed lines represent variation to the average value before being pushed off course, and the plain lines are the angles as read by the sensors. In this case, the boat is pushed towards the wind, as shown by the decrease of apparent wind angle. Note that for improved readability, the scales are different than on Figure 16.

during sailing by pushing the stern of the boat with a pole to turn it. Tests were done at different point of sail, and both pushing the boat towards and away from the wind, i.e. decreasing and increasing the wind angle respectively. It was not possible to measure the duration during which the boat was pushed and the amplitude of the course change, but the maximum course deviation reached around 20 degrees in all cases.

In this subsection, focus is put on the evolution of the boat course and apparent wind angle after a disturbance. Hence, Figures 16 and 17 show variations of the hull and mast headings and of the apparent wind angle: the average value before the disturbance are calculated and the values are shown as variation to their average.

The "push" occurred in Figure 16 between 6 s and 7 s, and a change in hull and mast yaw is seen after 7 seconds. The amplitude of the mast rotation is much smaller than the one of the boat, because as soon as the angle to the wind is too large, the wing turns. A perfectly free rotating mast should not rotate at all, but the delay in the mast rotation is due to the rig damping added by the steering system. After approximately 13 seconds, the apparent wind angle stabilizes around the same value as before the perturbation. By taking the average speed during the change, we can estimate that the initial unperturbed course was reached after approximately 5 to 6 boat lengths.

Figure 17, shows a "push test" towards the wind. As can be seen, it takes longer time to reach the initial course and true wind angle compared to the "downwind" tests. This can be explained by the fact that, as shown by the negative angle of attack, the wing points completely into the wind, therefore not generating any thrust, and it takes here approximately 8 to 9 boat length to get back on course. On the other hand, when pushed away from the wind, the wing is continuously at angles where it produces thrust, helping the boat to get back to the desired course. A similar behavior to that observed when sailing close to the wind was observed; the amplitude of the response is too large, leading to the boat turning too much away from the wind at

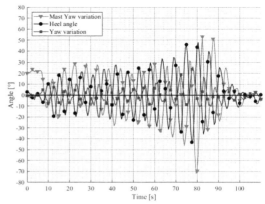

Figure 18. Mast Yaw (triangles), Vessel Heel angle (circles), and Vessel Yaw Angle (squares) plotted during one of the extreme parametric roll episodes. The yaw angles are normalized with their average value to compare the data.

Figure 19. Roadmap for the direction of the project.

first (around 32 and 42 seconds in Figure 17) but finally returning to the previous course.

4.3.3 Roll instability

During the test campaign, some instances of extreme roll behavior were observed. The behavior appeared to be a parametric wind induced coupled roll-sway motion of the hull with large yaw rotation of the rig. The motion show similarities with the unstable parametric roll behaviors common to traditional sailing boats under spinnaker at deep reach conditions.

When the motions were first recorded, the vessel was excited into roll motions by a following sea of wavelengths of roughly half the waterline length of the vessel. The wing was visibly excited, and began swinging due to the rapid changes in apparent wind. The vessel's roll motions compounded with the swinging weight aloft and the exaggerated state of roll reached angles around ±40 degrees. The roll motion was also excited in calm water with just the wing yaw being the visible cause for excitation. This coupling of mast yaw and vessel roll can be seen in the graph of the data from one of the episodes in Figure 18. Encouraging is however, that the instabilities appear to have been mitigated with the integration of the self-steering system to the platform. Further investigations of this will be conducted.

5 FUTURE WORK

A roadmap (Fig. 19) has been created to show the future directions of the project. The short-term goals include adaptations of the actual platform to reach the required robustness that are needed to sustain several days of continuous testing. This for example includes modifications to the wing to prevent the exaggerated roll motions described above and shown in Figure 20. Energy harvesting through the self-steering device and other potential onboard sources are being investigated and tested. The communications capacities of the plat-

Figure 20. Pictures of the boat at two extreme heel angles during the same instance of exaggerated roll.

form will be extended so that missions over a day length can be followed and modified from shore.

Mid-term goals include extended test missions and incorporation of other research sensors or equipment to assess the platforms ability to solve additional tasks, and determine the development need of further routing capabilities.

A long-term goal is to replace the existing platform with a new hull and rig combination. The aim is to design a robust system that utilizes the full potential of the free-rotating rig and self-steering concept. Studies on different hull concepts are already being undertaken.

Investigation is also being performed to understand the dynamics causing the exaggerated roll motions, so they can be prevented in the new hull design.

New concepts for the rig design will also be evaluated and experiments at smaller scale will be performed. The study will aim at finding solutions for improved robustness while maintaining good sailing performance.

6 CONCLUSIONS

This work shows the potential of use of a free-rotating self-adjusting wing to be used as the main propulsion for a sailing vessel. A relatively simple wing has been designed and manufactured and full-scale testing have shown good agreement with the simulated performance. A completely new steering system has been designed and tested, and its ability to keep the vessel at a desired wind angle has been experimentally proven.

The work presented in this paper and especially the performance of the two above mentioned systems open new possibilities for energy efficient autonomous oceanic research platforms. The results will be used to develop further the automation of Maribot Vane in order to reach the goal of a very robust and energy efficient oceanic sensing platform.

ACKNOWLEDGEMENT

The authors would like to acknowledge the financial support from the Swedish Maritime Administration (Sjöfartsverket) and the hospitality of the sailing club Skota Hem who let us use their crane and keep the boat at their storage.

REFERENCES

ASV Global, 2017, [online] Available at: www.asvglobal.com [Accessed November 2017].

Autonaut, 2017, [online] Available at: www.autonautusv.com [Accessed November 2017].

Daniel, T et al., 2011, The Wave Glider: enabling a new approach to persistent ocean observation and research, Ocean Dynamics 61:1509–1520.

Elkaim, G.H., 2001, System Identification for Precision Control of a Wingsailed Gps-guided Catamaran, PHD thesis, Stanford University, Stanford, California, USA.

Friebe, A, 2017, A marine research ASV utilizing wind and solar power, Oceans 2017, MTS/IEEE 19–22 June 2017 Aberdeen, UK.

Helmstad, A., & Larsson, T. "An Aeroelastic Implementation for Yacht Sails and Rigs", Master's thesis report, Centre for Naval Architecture, KTH Royal Institute of Technology, Stockholm. 2013.

Hole, L.R. et al., 2016, Directional wave measurements using an autonomous vessel, Ocean Dynamics 66:1087–1098.

LCJ Capteurs, 2017, [online] Available at: www.lcjcapteurs.com [Accessed July 2017].

Letcher, J.S., 1976, Performance and Stability of Wind-References Autopilots for Sailing Vessels, Marine Technology 13(3): 301–308.

Monterey Bay Aquarium Research Institute, 2017, [online] Available at: www.mbari.org/at-sea/mars-ship-rates/ [Accessed December 2017].

National Research Council, 2009. Science at Sea: Meeting Future Oceanographic Goals with a Robust Academic Research Fleet. Washington, DC: The National Academies Press.

Saildrone, 2017, [online] Available at: www.saildrone.com, [Accessed November 2017].

Sauzé, C. & Neal, M. 2006. An Autonomous Sailing Robot for Ocean Observation.

Tretow, C, 2017, Design of a free-rotating wing sail for an autonomous sailboat, Master's thesis report, Centre for Naval Architecture, KTH Royal Institute of Technology, Stockholm.

Wängelin, F, 2017, Energy-efficient steering mechanism for an autonomous sailboat, Bachelor's thesis report, Centre for Naval Architecture, KTH Royal Institute of Technology, Stockholm.

Yost Labs, 2017, [online] Available at: www.yostlabs.com [Accessed November 2017].

Author index

Abbott, C. 983
Abeil, B. 543
Abt, C. 203
Adamopoulos, N. 173
Ahola, M. 285, 853
Albrecht, P. 17
Alves, M. 479
Amine-Eddine, G. 983
Ando, H. 499
Andrews, D. 3, 55
Andrews, D.J. 307
Andric, J. 459
Ang, H.E. 669
Ang, J.H. 995
Arai, M. 499
Arens, E.A. 983
Arola, T. 49
Asbjørnslett, B.E. 147, 355, 367, 953
Aubert, A. 215

Banda, O.A.V. 831, 853
Bandas, J. 543
Bastide, G. 983
Bekker, A. 135
Belanger, D.G. 897
Benson, S. 439
Berg, B. 203
Berglund, R. 841
Bergström, M. 831, 841, 853, 881
Bernard, P. 1099
Bi, Y. 719
Bilde, R. 85
Billerbeck, H. 595, 701
Birmingham, R.W. 275, 439
Bø, T.I. 923
Bolbot, V. 691
Boulougouris, E. 227, 603, 617
Braidotti, L. 181
Brefort, D. 255
Brett, P.O. 31, 147, 267, 291, 355, 569
Brunswig, J. 215

Bucci, V. 181
Bucknall, R.W.G. 941
Bulten, N.W.H. 1033

Calle, M.A.G. 479
Campbell, C. 377
Chai, W. 789
Chang, H. 1007
Chen, W. 1057
Cheng, X. 1007
Choy, K.-L. 831
Christensen, C. 367
Chung, U. 491
Claus, L.R. 347
Collette, M.D. 347, 1133
Corrignan, P. 163, 215
Cui, T. 961

Daidola, C.J. 411
Danielsson, J. 491
Datla, R. 491
Day, A.H. 507
Day, S. 961
de Bossoreille, X. 163
de Jongh, M. 163, 203
de Vos, P. 1083
Demirel, Y.K. 971
Devine, T.E. 339
Dhomé, U. 633
Dimopoulos, G. 203
Dong, G. 1057
Dow, R.S. 439
Droste, K. 327
Duchateau, E.A.E. 1083
Duong, T. 737

Ebrahimi, A. 31, 267, 569
Erikstad, S.O. 147, 355, 367
Eronen, H.K. 769
Eskeland, G.S. 923

Fan, S.M. 535
Feng, B. 1007
Feng, C. 453
Feng, P.Y. 535

Ferrari, V. 543
Field, C. 471
Filip, G. 517
Florean, M. 113
Fraize, J. 633
Fürth, M. 491, 633, 737, 1121
Furth, M. 897

Gad, A. 579
Garcia, J.J. 31, 267, 569
Gaspar, H.M. 31, 123
Gernez, E. 291
Goerlandt, F. 659, 853
Goh, C. 995
Goodrum, C. 385
Grin, R. 543
Gu, X. 1019, 1043
Guegan, A. 163

Habben Jansen, A.C. 1073
Halse, K.H. 1099
Ham, S.H. 563
Han, D.-f. 747
Happonen, K. 909
Haranen, M. 917
Harries, S. 215
Hassani, V. 203
Hauge, R. 291
He, J.S. 1065
He, P. 517
Heikkilä, E. 861
Heikkilä, E.K. 193
Hekkenberg, R.G. 871
Hiekata, K. 887, 931
Hinz, T. 589
Hirdaris, S. 439
Hirdaris, S. 831, 881
Holbert, J.P. 339
Holmefjord, K.E. 163
Hopman, J.J. 3, 327, 1073
Howett, B. 961
Høyland, K.V. 789
Hu, W.-j. 747
Hu, Y. 527
Hunt, J. 85

Ichinose, Y. 1015
Igbadumhe, J. 1121
Ito, K. 887

Jalonen, R. 861
Jambrecic, L. 459
Jansen, J.E. 203
Jansen, K. 897
Jirafe, V.P. 995
Johnsen, T.A.V. 953
Johnson, M.C. 439
Jokinen, M.L. 193

Kajosaari, M. 77
Kaklis, P.D. 579
Kana, A.A. 3, 103, 327, 1073
Kang, H.J. 625
Karjalainen, M. 809
Karolius, K.B. 727
Karvonen, H. 909
Keane, A. 267
Keiramo, M.L. 193
Khan, F. 677
Kim, K.S. 613
Kinnunen, A. 777
Knight, J.T. 397, 471
Konispoliatis, D.N. 1111
Konovessis, D. 669
Kooij, C. 871
Kring, D.C. 397
Krüger, S. 595, 701
Kujala, P. 479, 659, 757, 809, 831, 841, 853, 881
Kulovesi, J. 757
Kuo, C. 377
Kuttenkeuler, J. 633
Kuuliala, L. 777, 841

Lang, J.-c. 453
Langheinrich, J. 595
Lappalainen, A. 881
Le Diagon, V. 163
Le Nena, R. 163
Lecointre, L. 215
Lee, H.W. 563
Lee, S.H. 613
Lee, S.M. 613
Leira, B.J. 789
Lensu, M. 841
Li, C. 429
Li, F. 841
Li, J. 801, 1065
Li, N. 163
Li, Y. 995
Li, Z. 831
Liao, P. 801
Lindeberg, M. 809

Lindstad, E. 923
Liu, S. 173
Loonstijn, M. 871
Lorkowski, O. 595
Lübcke, A. 701
Lundh, M. 831

Ma, N. 1019, 1043
Maasch, M. 507
Määttänen, P. 757
Maki, K.J. 517
Mallam, S. 285
Manderbacka, T. 917
Martins, J.R.R.A. 517
Martio, J. 909
Marzi, J. 215
Mauro, F. 181
Mavrakos, S.A. 1111
Mavrelos, C. 579
Mazarakos, T.P. 1111
Mazerski, G. 589
Meng, Z.-f. 453
Mitsuyuki, T. 887, 931
Murto, P. 285

Naess, A. 789
Nagel, R. 595
Ni, B.-y. 747
Nikolopoulos, L. 227
Nordby, K. 291, 831

Olsen, K.E. 203
Oshiro, R.E. 479
Ozyurt, R. 971

Papanikolaou, A. 173, 215
Pawling, R.J. 85, 307
Perälä, I. 777
Pettersen, S.S. 147, 355
Phillips, H.J. 439
Plessas, T. 173
Prebeg, P. 459
Priftis, A. 603
Prini, F. 439
Pu, Y. 421
Putu Arta Wibawa, I. 275

Qiao, D. 1019

Rafine, B. 163
Raleigh, K. 737
Rapuc, S. 543
Rautaheimo, P. 17
Razola, M. 633
Rehn, C.F. 147, 367
Reichard, L. 897
Ringsberg, J.W. 831

Roh, M.I. 563, 613
Romanoff, J. 3
Romanoff, J.M. 193
Rosén, A. 491
Rotteveel, E. 103
Routi, A.-L. 709
Rudan, S. 459
Ruponen, P. 709
Ruud, S. 819

Sagaspe, L. 163
Sandvik, E. 953
Seim, Ø. 291
Seyffert, H.C. 397
Sheppard, P.J. 439
Shi, G. 421
Shi, X. 1065
Shivachev, E. 507
Singer, D.J. 245, 255, 385
Sinsabvarodom, C. 789
Skjetne, R. 819
Smith, D. 677
Smith, M.V. 471
Soininen, M. 17
Sonninen, S. 853
Sormunen, O.-V. 809, 841, 881
Stachowski, T.-H. 983
Steen, S. 953
Stefanidis, F. 579
Stefenson, P. 831
Strickland, J.D. 339
Strøm, M.A. 147
Szlatenyi, C. 471
Szymański, P. 589

Tahara, Y. 1015
Taimuri, G.H. 757
Tan, X. 669
Tao, J. 669
Tay, Z.Y. 1121
Taylor, R. 677
Taylordean, S. 385
Theotokatos, G. 691
Thomas, G. 831
Timmerman, M. 579
Toivola, J. 809
Torben, S. 163, 203
Tretow, C. 633
Trincas, G. 181
Troesch, A.W. 397
Tsitsilonis, K.M. 579
Turan, O. 507, 603, 617, 961, 971

Ueno, M. 651
Uzun, D. 971

Valdez Banda, O.A. 881
Valtonen, V. 783
van Gelder, P.H.A.J.M. 853
van Leeuwen, S. 1083
Varas, J.M. 439
Vassalos, D. 579, 691, 719, 727
Veitch, B. 677
Visser, K. 871

Wahlström, M. 861
Wanaka, S. 931
Wang, D. 429
Wang, P.-p. 453
Wängelin, F. 633
Weltzien, C. 555
Wielgosz, C. 491
Williams, S. 685
Wincott, C.M. 1133
Windén, B. 1121
Wöckner-Kluwe, K. 595
Wu, P. 941
Wu, Y.S. 535

Xiong, X.Q. 535
Xu, S.-b. 453
Xue, Y.-z. 747

Yan, X.P. 659
Yang, C.-J. 1043
Yang, F. 1057
Yin, X. 1057
Yoshida, T. 499
Yrjänäinen, A. 113
Yuan, H. 245
Yurtseven, M.A. 617

Zanotto, D. 737
Zaraphonitis, G. 215
Zhang, D. 659
Zhang, G. 527
Zhang, M.Y. 659
Zhang, W. 1043
Zhou, R. 801
Zhou, Y. 527
Zhou, Y.H. 1065
Zymaris, A. 203